FOREST MANAGEMENT

Third Edition

Lawrence S. Davis

Department of Forestry and Resource Management
University of California, Berkeley

K. Norman Johnson

Department of Forest Management
Oregon State University

McGraw-Hill, Inc.

New York St. Louis San Francisco Auckland Bogotá
Caracas Lisbon London Madrid Mexico City Milan
Montreal New Delhi San Juan Singapore
Sydney Tokyo Toronto

This book is dedicated to Kenneth P. Davis. He served us as ranger, researcher, academic administrator, professional society leader, and, most of all, as a teacher. He was an able man who gave all that he had. For well over two decades the first and second editions of this book dominated our thinking about forest management. Those of us lucky enough to study with him learned about red pencils, second and third drafts, and what being a professional forester was all about.

This book was set in Times Roman by Bi-Comp, Incorporated.
The editor was Marian D. Provenzano;
the production supervisors were Marietta Breitwieser and Denise L. Puryear.
Project supervision was done by Cobb/Dunlop Publisher Services Incorporated.

FOREST MANAGEMENT

8 9 10 11 12 13 BRBBRB 93 9 8 7 6 5 4 3 2 1 0

Library of Congress Cataloging in Publication Data

Davis, Lawrence S.
Forest management.

Rev. ed. of: Forest management/Kenneth P. Davis.
Includes bibliographies and index.
1. Forest management. I. Johnson, K. Norman.
II. Davis, Kenneth P. (Kenneth Pickett), 1906–
Forest management. III. Title.
SD431.D36 1986 634.9'068 85-12789

ISBN 0-07-032625-8

CONTENTS

Part 2 Decision Analysis in Forest Management

Part 3 Valuation

McGraw-Hill Series in Forest Resources

Avery and Burkhart: *Forest Measurements*
Brockman and Merriam: *Recreational Use of Wild Lands*
Brown and Davis: *Forest Fire: Control and Use*
Dana and Fairfax: *Forest and Range Policy*
Daniel, Helms, and Baker: *Principles of Silviculture*
Davis: *Land Use*
Davis and Johnson: *Forest Management*
Dykstra: *Mathematical Programming for Natural Resource Management*
Harlow, Harrar, and White: *Textbook of Dendrology*
Knight and Heikkenen: *Principles of Forest Entomology*
Nyland, Larson, and Shirley: *Forestry and Its Career Opportunities*
Panshin and De Zeeuw: *Textbook of Wood Technology*
Panshin, Harrar, Bethel, and Baker: *Forest Products*
Sharpe and Hendee: *An Introduction to Forestry*
Shaw: *Introduction to Wildlife Management*
Stoddart, Smith, and Box: *Range Management*

Walter Mulford was Consulting Editor of this series from its inception in 1931 until January 1, 1952.

Henry J. Vaux was Consulting Editor of this series from January 1, 1952, until July 1, 1976.

PREFACE

Forestry in the United States began as a crusade. At the turn of the century, our forests were being rapidly cut and burned with little thought to providing a heritage for the future. The forestry profession arose as a means to protect, plant, and nurture our forests . . . to ensure a continuity of growth for future generations.

Forest management books traditionally have been instruments of that crusade. They have combined exhortations for "good" forest management with procedures and rules of thumb to make sure that such growth-maximizing management happens. Back when forestry had a well-defined litany on how to manage forests for the benefit of society, such books were a valued expression of it.

But forestry's original crusade has been largely won, at least in the United States. Our forests are protected and growing. People now wish to use their forests for all sorts of purposes, and forest management reflects this change. Rather than locking in on commercial timber production, forest management is now directed to helping people achieve whatever goals they have for forests.

Our revision of Kenneth Davis's "Forest Management: Regulation and Valuation," second edition reflects three fundamental shifts in the tenets of forest management: (1) the new dominance of economic and social goals as determinants of forest management choices, (2) the necessity of private and, especially, public managers to consider all timber management decisions in the context of a larger, socially defined, multiple use management problem, and (3) the need to justify, based on quantitative analysis, that management recommendations, decisions, and plans satisfy owner constraints and are the best of the alternative choices. Technologically, it was our good fortune that the computer came along just in time to help us with the enormous amount of quantitative analysis needed to satisfy these shifts. But computers are also now an everyday part of our professional lives.

Gone are the many bromidal prescriptions on how a forest should be managed. Rather than prescribe "good forestry" we submit that good management follows from meeting the objectives of the forest owner. Formal, systematic recognition is given in this third edition to a set of organizing principles: that decision making guided by forest owner's goals lies at the heart of forest management and that thinking about forest problems in terms of objectives and constraints provides a unifying analytical framework for decision making.

In terms of detail and example, this book primarily covers management of forests for commercial crops of timber. Management for all resources is emphasized, however, and many techniques and example problems are presented for dealing with timber while simultaneously considering one or more other forest outputs.

Lawrence S. Davis
K. Norman Johnson

USING THE BOOK

We have written the book for students of forest management—those who are out of college as well as those who are in. By including many numerical problems in the text with answers at the end of the book, we hope it will work reasonably well as a self-study text. We believe, as did Ken Davis, that management is a mix of art and science and that solving problems is nourishment for the art. Although previous courses or study in silviculture, mensuration, and forest economics is desirable, more than anything else the concepts covered here take concentration and an analytical way of thinking. Algebra is heavily used, calculus rarely crops up in the discussions, and sophisticated mathematics is avoided. Instructors could use the material in this book to support several combinations of courses depending on student background. We conceived and have taught this material for a senior course sequence in forest management.

Part 1 advances the new, central role of stand type prescriptions in forest management and then reviews the concepts and methods of timber growth and yield projection from the managerial perspective. Part 2 is substantially new material and provides a comprehensive framework for decision analysis in a multigoal setting. Chapter 6 covers linear programming basics and is essential prerequisite material for the formulation and analysis of harvest scheduling problems presented in chapter 15. The arithmetic of interest in chapter 7 is basic to financial analysis of timber stand management in chapter 13. Part 3 covers valuation in much the same manner as the second edition. The principal addition is chapter 12 which covers establishing values for nontimber forest outputs. Part 4 is the capstone of the book and presents our 1980's view of what forest management is about. Chapter 13 looks especially hard at making economically efficient decisions for the management of existing and future timber stands. Chapter 14 summarizes the classical approaches to forest management planning and harvest scheduling which were the backbone of the second edition, and chapter 15 provides extended concepts and methods of formulating forest management scheduling problems for solution by linear programming and binary search simulation. Chapter 16 closes by looking at the results of planning and scheduling analysis, draws on the evaluation framework of chapter 8, and ponders the question—so what plan and activity schedule is best?

The book is full of numbers, equations, problems, and answers. As much as we and our students have checked and proofed, there are sure to be errors. We assume full responsibility. Be confident; if it doesn't add up after a few tries and others have the same difficulty, it is probably us not you. When you find errors, please let us know.

ACKNOWLEDGMENTS

We started this project 7 years ago when we were both on the faculty of the Department of Forest Resources at Utah State University. What was a modest revision project became The Book. In the process we created an extended list of friends, colleagues, employers, secretaries, and students to thank for their help and encouragement in seeing it to completion. Thanks and thanks again!

Our wives, Claire and Betsy, and our children, Lawrence, Katharine, Alexandra, Amy, and Andy, have been our most patient and enduring supporters. Thanks are inadequate compensation for the time forgone, but thanks, love, and being done are mainly what we have to offer.

Doug Brodie, Emmet Thompson, and Doug McKinnon reviewed the first full draft and their comments greatly helped us chart major changes in the original concept. This is a much better book for their efforts.

Even though we have significantly revised this book, our debt to Kenneth Davis is incalculable. His pioneering efforts systematized the lore of forest management and framed the nature of forest management problems. His insistance on introducing valuation and inventory planning directly into forest management planning, and his fervent call for analysis and judgment form a treasured inheritance.

Lawrence S. Davis
K. Norman Johnson

CHAPTER

ONE

INTRODUCTION TO FOREST MANAGEMENT: DEVELOPMENT AND STATUS

For as long as people have thought about the future, they have managed forests. Before the start of recorded history, American Indians used fire in their woodlands to create more productive conditions for the game on which they depended, and tribes of southeast Asia burned a portion of the forest each year to free nutrients for their crops.

As civilizations developed axes and other tools for extraction, they often plundered their forests until the forests were gone or they realized the importance of saving some trees for the future. India's creation of a minister of forests many years before Christ and China's recognition long ago of the importance of forests to prevent flooding are early evidence of protective forest management.

Disputes over the rights to forests and their management dot the history of European civilization. Robin Hood's dispute with the king supposedly started over the deer that abounded in England's medieval forests. As the objectives of European forest management gradually changed from serving the king to serving the people, foresters began developing the body of knowledge that forms the basis of forest management as we know it today.

Now, as then, forest management involves the use of forests to meet the objectives of landowners and society. While the objectives may change and the means to reach them become sophisticated, forest management still is the attempt to guide forests toward a society's goals.

A forest manager is the catalyst of this effort and, as such, needs the earthy and intimate understanding of a botanist, the long-range viewpoint of a planner, the skills of an administrator, and the alertness, flexibility, and all-around resourcefulness of a successful business executive. Above all, the forest manager requires a genuine sense and feeling for the forest as an entity. Every forest offers a real and living individuality. Recognizing this uniqueness while applying the principles of management is the heart of forest management.

In this book, we cover the principles and techniques of management planning for forests. Timber production is emphasized, but the impact of timber activities on other forest outputs is also detailed.

This chapter is devoted to three subjects that should provide some perspective on what follows: the development of forest management in the United States; the current status of ownership, inventory, harvest, and growth; and the intensity of current forest practice.

DEVELOPMENT OF FOREST MANAGEMENT IN THE UNITED STATES

Organized forest management, other than as a mechanism to harvest timber and then abandon the land, has been slow to develop in the United States. The establishment of the national forests in 1905, almost 300 years after the first permanent English settlement in America, marked the first formal mandate over a large area to grow timber in addition to harvesting it. Even in 1905, no protection against fire existed on any appreciable scale, and protection against insects and diseases was practically unheard of. The first forestry schools, established at the turn of the century, had produced only a handful of technically trained foresters.

Nevertheless, inspired largely by their own zeal, early public foresters approached the staggering job of extending management to immense forest areas with prodigious enthusiasm, if not with adequate knowledge and finances. Timber was their primary concern, and silviculture received much initial emphasis. An intensity of practice was attempted on some early national forest timber sales that has scarcely been equaled since, as the influence of intensive European forestry was strongly felt through early leaders like Pinchot, Fernow, Roth, and Schenck. For about three decades, the national forests set the forestry pace of the country and employed most professional foresters.

Soon foresters recognized that detailed tending of trees, based on the European model, could not come first; taming wildlands, developing access, building fire organizations, and establishing forest administration had to take precedence. Widespread and disastrous forest fires made it abundantly clear that fire control was an indispensable prerequisite to forest management. For many years, preoccupation with the tremendous job of extending fire protection to the nation's public forests crowded out the application of other forest management.

As would be expected, lumbering considerably antedated timber growing as a business on private land. Large-scale lumbering began about 1850, made possible by the steam engine, the circular saw, and the railroad, as well as the great westward migration. Lumbering increased rapidly for the next half-century, reaching a peak lumber production in 1906. An abundance of virgin timber provided the harvest. Converting cheap standing timber into marketable forest products, mostly lumber, was the main forest management problem of industry. Plans were framed accordingly, with timber regarded as a stock to be extracted or mined much like coal or oil.

The very nature of these plans, which treated the forest as a wasting asset, made stable continuation of a forest business impossible without continued purchase of more timber. As the national supply of readily available timber declined, additional stumpage and timberland became more expensive and harder to find. A major part of the forest industry entered the twentieth century with a limited operating life ahead of it, when measured by the existing timber available for harvest.

Clawson (1979) assembled available historical evidence to track the level of forest inventory in the United States. After early land clearing for agriculture, he found the total standing inventory to stabilize in about 1900 and thereafter to fluctuate around the level of 2,000 billion board feet (bd ft) (Fig. 1-1*a*). Certainly the composition changed from 1900 to the present, with more smaller trees and a much larger hardwood component. Nevertheless, it is interesting that timber did not behave as a stock to be mined; rather, the forest persistently grew about as fast as it was extracted. Early loggers were correct in that large overmature timber was a stock to be extracted without replacement, but they were wrong in their assessment of the forest as a nonproductive asset.

Change came slowly. While desirable timber was becoming scarce, firms were still reluctant to make long-term investments in timber growing. Commercial timber growing during the first part of the century did not pay. Until it became profitable, timber growing would not be practiced on private land.

Organized management of privately owned timberlands for continued production based on growth started during the late 1920s, primarily among the larger ownerships. The depression during the 1930s halted its development, but it rapidly increased as a part of the intense industrial activity accompanying and following World War II.

Timber growing became economically viable on a large scale after World War II in the late 1940s because of a change in four major factors affecting industrial forestry: rising stumpage prices, fire control, federal

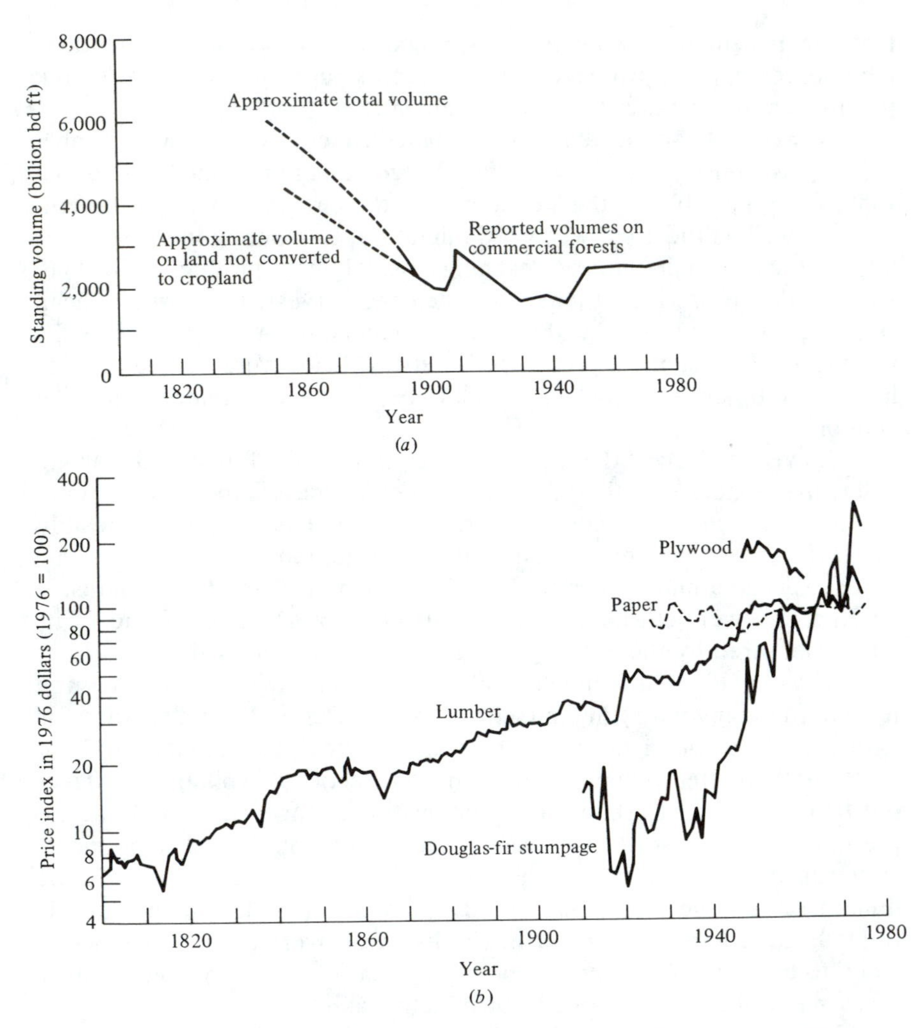

Figure 1-1 *(a)* Changes in inventory and *(b)* stumpage price trends from 1800 to 1980. (*From Clawson, 1979, © Science,* **204:**1168–1174, Figs. 2, 4, 5, 7.)

and local taxation policy, and technological innovation. The strong rise in stumpage prices since World War II (Fig. 1-1*b*) is the most obvious explanation of decisions by private individuals to buy or invest in land for the purpose of growing commercial timber. There has been a tenfold real increase in Douglas-fir stumpage prices since 1940. A similar rise is observed in the value of southern pine stumpage.

During the early part of the century, fire control was not sufficient to make investment in timber production a good business risk. Especially in the south, fires caused by humans swept millions of acres every year.

Providing adequate fire control during the 1930s and 1940s dropped the burn rates to a small percentage of the uncontrolled rates and had an almost miraculous effect on the southern forestry landscape. No other social action approaches it as a force demonstrating both the effect of lower risk on forest investment and the great recuperative capacity and productivity of southern forests.

Income derived from the sale of capital assets that have increased in value over time may be taxed as capital gain, which means the tax rate is approximately half that of ordinary individual or corporate income tax rates. This tax treatment reflects a long-standing national policy of encouraging private investment in productive capital assets.

Prior to 1944, timber stumpage was not treated as a capital asset, and stumpage growers had to pay ordinary income tax rates on stumpage income. The only exception was when noncommercial timber growers sold occasional lots of timber on a lump sum or fixed price basis. Section 117k of the Internal Revenue Code was enacted in 1944 and permitted timber cut and used by owners within their own wood manufacturing plants and timber sold to others on a per-unit volume basis to be also treated as capital gain income. In effect, the physical growth of timber was now viewed as capital appreciation. Further revisions of the code in 1954 and 1980 kept this concept intact while refining accounting, expensing, reporting, and other details. Treatment of timber growth as capital growth gave a strong incentive to own forestland and invest in timber growing, particularly for integrated forest product companies.

Local property taxes during the early part of the century were high relative to the value of annual timber growth. As taxes had to be paid every year while the timber stand could not be harvested for decades, the timber inventory had to be cut or the land sold to meet tax obligations. Change during the 1940s to reduce taxes until the land contained merchantable timber, or until the timber was harvested, greatly eased the tax burden.

Finally, investment in timber growth became attractive because of technological improvement in timber management and utilization. People learned how to plant trees and get them to grow. New manufacturing processes permitted more and more the use of what had been considered waste to make such products as particle board and kraft paper. Vertical integration in the industry allowed a firm to exploit the value of wood to the fullest extent.

Fueled by all these changes and the continuing real increase in stumpage prices, commercial timber management for sustained production became accepted business practice. By midcentury private forestry, particularly among larger owners, had become a thriving enterprise.

Until the 1950s, timber in the remote national forests was often offered for sale with buyers remaining scarce. America's fascination with the wilderness experience and outdoor recreation was just starting and the forests were largely unused, except by local residents. In the 1960s, the demand for timber increased; as the supply of western private timber decreased we witnessed a dramatic increase in the value of public timber. Concurrently the environmental movement was born, and visitors with many different interests rushed to stake their claims on the public lands.

During the 1960s and 1970s, everybody seemed to want forestland for something—timber production, recreation, wildlife habitat, reservoirs, wilderness. Pressure of this kind applied to federal lands in particular and spawned analyses by federal agencies to discover procedures and technology to accommodate these demands and to make acceptable compromises when not all demands could be met.

As the conflict increased over how the nation's forests should be managed and for what purposes, people turned more and more to the law and the courts. Federal lawmakers passed a flurry of bills increasing direction to America's forest managers: the Multiple Use–Sustained Yield Act (1960), directing the national forests to be managed under those two glorious tenets; the National Environmental Policy Act (1969), mandating the consideration of environmental impacts of major federal actions; the Clean Air Act and the Clean Water Act (1970–1977), setting air and water quality standards that potentially can affect all forestry actions; the Endangered Species Act (1973), giving maintenance of endangered species habitat priority over other uses of that land; and the National Forest Management Act (1976), redefining sustained yield and multiple use for the Forest Service. States increasingly used their regulatory power to control actions on private lands and showed their willingness to fight the federal government over actions on public land. Certainly, the state of California's (1980) successful suit against the Secretary of Agriculture over U.S. Forest Service wilderness planning provides but a beginning to this kind of maneuver. Citizens increasingly sued the federal government, state government, companies, and each other over proper use of forestland.

As this country moves through the last years of this century, deciding on the importance and manner of timber production on public and private lands, as compared to other uses, will remain a major and often bitterly fought issue.

STATUS OF FOREST MANAGEMENT

What is the management status of American forests? In a country of the length, breadth, and diversity of the United States, this question cannot be answered specifically; no one knows with precision. Forestry is on the march and changes every day. Nevertheless, some overall facts, statistics, and trends can help in understanding the general forest situation of the United States in the 1980s. Almost all of the following information was assembled to describe the U.S. forest economy as of 1976, although published in 1980 and 1982.

Forestland occupies one-third of the total land area in the 50 states (Table 1-1, Fig. 1-2). Most of this forestland, 483 million acres, is classified as commercial, that is, available and capable of growing continuous crops of wood products. The remaining forestland, 254 million acres, is classified as noncommercial, mostly because of low productivity for timber growing. However, 25 million acres of public noncommercial forests are productive timberlands reserved for recreation or other nontimber uses.

Ownership of Our Commercial Forestland

Nearly three-quarters of all commercial timberland lies in the eastern half of the United States, about equally divided between the north and the south, with private holdings predominating (Fig. 1-3*a*). Commercial timberland in the west is concentrated in the Pacific coast states and the Rocky Mountain states of Montana and Idaho, with public holdings predominating.

Millions of individuals, thousands of corporations, and hundreds of public agencies own commercial timberland in the United States. Grouping this vast assortment masks meaningful differences but is necessary to make the analysis manageable. Four ownership categories are commonly recognized and reported: (1) farm and miscellaneous private tracts, (2) forest industry holdings, (3) other public forestlands, and (4) national forests.

Forest industries, defined as companies and individuals that operate wood-using plants, own 14 percent of the nation's commercial forestland,

Table 1-1 Land area of the United States by land type and geographic region (millions of acres)

Land type	North	South	Rocky Mountain	Pacific Coast	United States	Percent
Forest						
Commercial	166	188	58	71	483	21
Noncommercial	12	19	80	143	254	11
Total	178	207	138	214	737	33
Other	450	304	417	357	1,527	67
Total, all land	628	511	555	571	2,264	100

Source: U.S. Forest Service (1982).

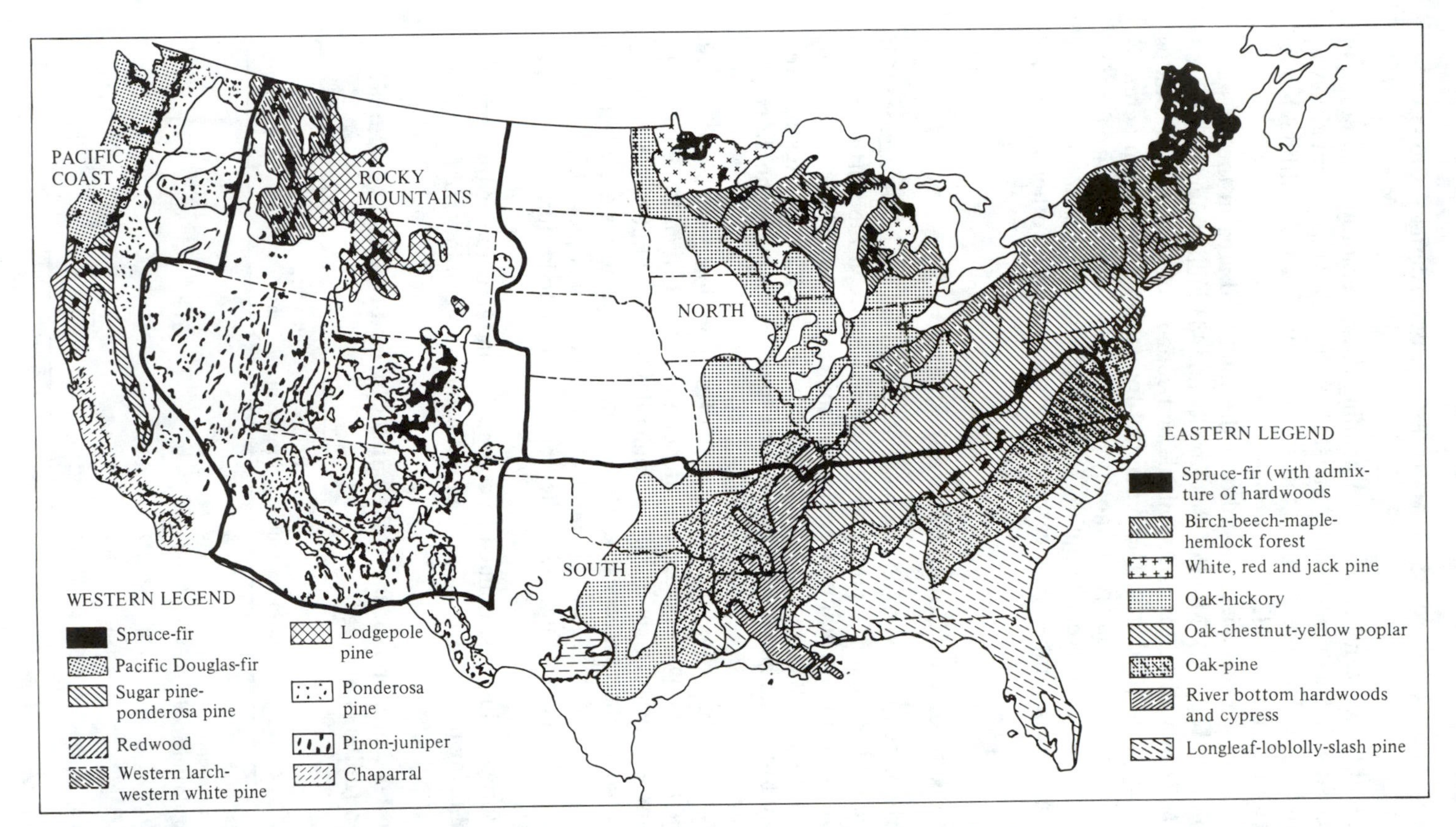

Figure 1-2 Major forest types and geographic statistical reporting regions of the 48 contiguous states. The Pacific coast region includes the spruce and fir forests of Alaska. *(From Shantz and Zon, 1924.)*

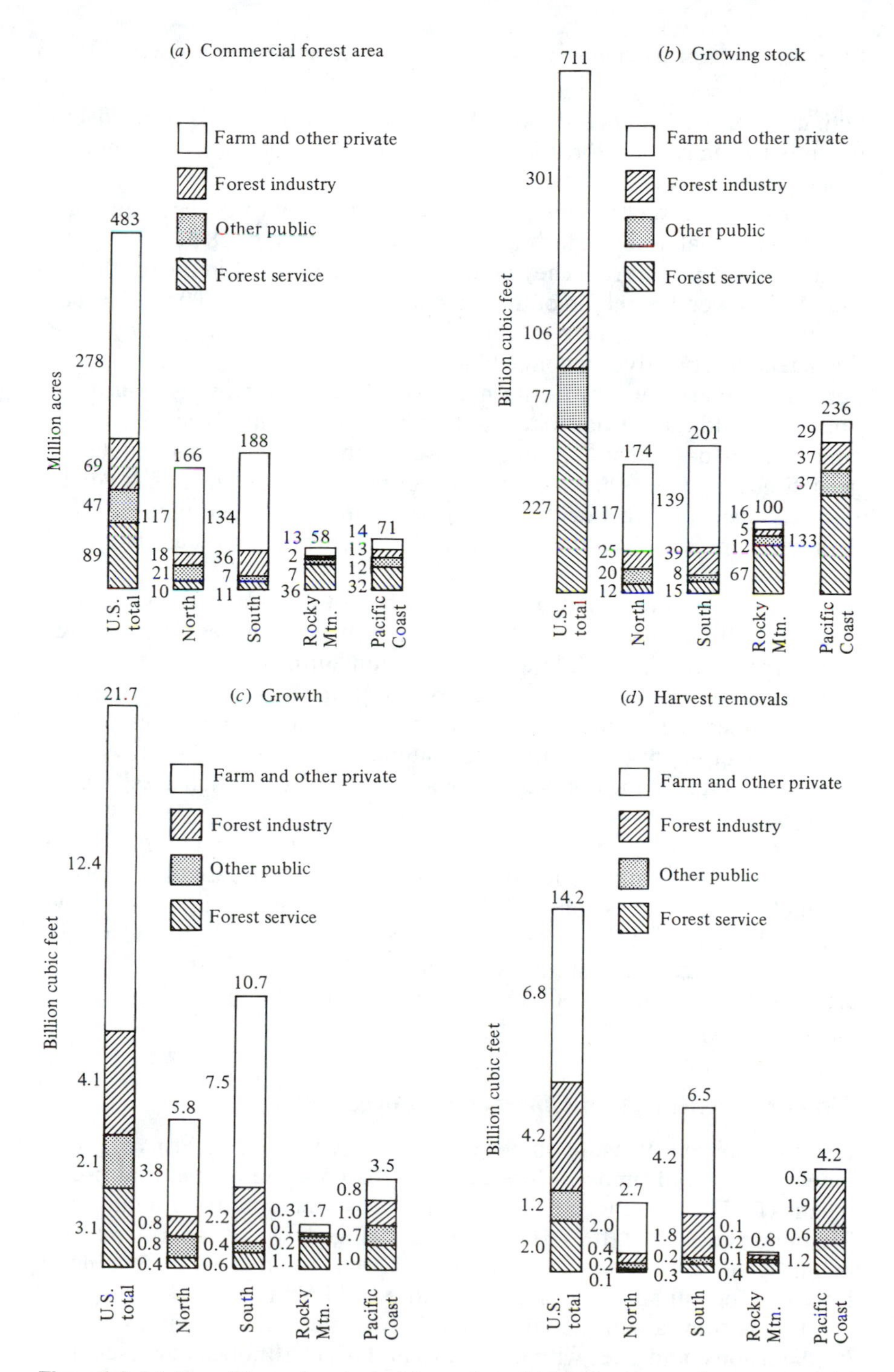

Figure 1-3 *(a)* Area, *(b)* growing stock, *(c)* growth, and *(d)* harvest removals on commercial forestland of the United States in 1977.

with most of this land in the south. In addition, the industry's stewardship extends to some other forestland through long-term lease from farm and other private owners.

Farm and miscellaneous private ownerships hold the remaining private commercial timberland—some 278 million acres, or 58 percent of the total commercial forests. They dominate land ownership in the north and the south. Over 3.8 million of these owners hold less than 500 acres each; a substantial number have less than 10 acres. At any time, many have management objectives incompatible with timber harvesting: less than 10 percent of these owners in the north say timber growing is a primary objective. Although small-sized and with management objectives which constrain management for timber production, these acres continue to grow timber. Tenures are short and objectives change as owners or conditions change. Available evidence suggests that most timber on these ownerships eventually becomes available for fuelwood or industrial wood products.

Ownership of commercial timberland by federal, state, and local government reflects a variety of forces. Much national forestland was reserved from the original federal public domain during the late 1900s, when widespread concern existed over a possible "timber famine" in the country. Considerable public domain land was deeded to the states by the federal government upon entry into statehood. Other forestland was obtained by state or local government as tax-delinquent lands, especially during the 1930s.

National forests, located mostly in the west, make up one-fifth of all commercial forestland. Much of this land is on sites of less than average productivity and accessibility. Other public lands, mostly in the west and the north, compose 9 percent of commercial forestland. In the north, other public forests are largely state- or county-owned; in the west, most are held by the Bureau of Indian Affairs or the Bureau of Land Management.

Timber Inventories on Commercial Forestlands

National surveys measure timber volume on forestlands in terms of growing stock—sound timber in trees greater than 5 in. in diameter at breast height (DBH). Commercial forestlands across the country contain over 710 billion cubic feet (ft^3) of growing stock (Fig. 1-3*b*). Almost 70 percent of this volume is in sawtimber-size trees, that is, in trees containing at least one log suitable for the manufacture of lumber.

In addition, a substantial volume is contained in unsound trees, unused portions, and trees with less than 5 in. DBH. Although at present it is not economically feasible to use much of these additional volumes, they do represent a large potential source of fiber, pulp, fuel, and future growing stock.

Softwoods predominate in the nation's timber inventory, accounting for nearly two-thirds of the growing stock. This softwood inventory is largely in the old-growth stands of the Pacific coast, in contrast to the eastern location of most commercial forestland. Softwood volumes increased slightly across the nation from 1952 to 1977. Decreases on the Pacific coast due to harvest of old growth were more than offset by increases elsewhere.

Hardwoods make up 36 percent of the nation's growing stock, almost all located in the east. The 255 billion ft^3 of hardwood growing stock in 1977 was almost half again that of 1952.

Growing-stock holdings of forest industries account for 15 percent of the total, about proportionate to their acreage. Their highest volumes per acre occur in the west, reflecting some remaining old growth there (Fig. 1-3*b*). Nearly all this timber is accessible to primary timber processing plants.

Farmers and other private owners hold more growing stock than any other class, but they also have the lowest average volumes per acre. Included in this growing stock is a major part of the nation's inventory of hardwoods—about 70 percent. While much of the volume is composed of lower-value species, it is accessible and favorably located relative to the major timber-consuming centers of the north and south.

The national forests, with one-fifth of the commercial forestland, have almost one-third of the growing stock. More importantly, they contain almost half of the softwood growing stock. Most of this timber lies in western old-growth stands.

Public agencies other than the U.S. Forest Service hold roughly 11 percent of all timber inventories. Nearly all these inventories are accessible and are important sources of timber for processing industries, especially on the Pacific Coast.

Net Annual Timber Growth

Net annual growth—the total annual gross growth less mortality on growing stock—was 22 billion ft^3 in 1976, or 45 ft^3 per acre, equal to a 3-percent annual rate. More than half of this growth occurred in the south (Fig. 1-3*c*).

Almost 60 percent of this growth occurred on farm and other private ownerships. The forest industry ranked next with one-fifth of the net growth, then came the national forests with one-sixth, and the other public lands with one-tenth.

On a per-acre basis, industry lands have the highest growth and na-

Table 1-2 Removals from growing stock by product and geographic region in 1976 (billions of cubic feet)

	Sawlogs	Veneer logs	Pulpwood	Miscellaneous	Fuel	Residue	Other	Total	Percent of total
North	0.9	—	0.7	0.1	0.1	0.3	0.5	2.7	19
South	2.2	0.6	2.4	0.2	0.2	0.6	0.6	6.6	46
Rocky Mountains	0.6	0.1	—	—	—	0.1	—	0.8	5
Pacific Coast	2.6	0.6	0.3	0.1	—	0.4	0.1	4.2	30
Total	6.3	1.3	3.4	0.4	0.3	1.4	1.2	14.3	100
Percent of total	44	9	24	3	2	10	8	100	

Source: U.S. Forest Service (1982).

tional forestlands the lowest. This relationship appears due to two reasons. First, industry lands tend to be located on higher-quality timber-growing sites than do national forestlands. Second, industry has more of its land in second-growth timber than does the national forest, which still has large acreages of old growth. In the east, for average site and stand conditions net annual growth per acre on national forests is close to or above that of other major ownerships.

The net annual growth of growing stock increased from 14 to 22 billion ft^3 between 1952 and 1976, or to 45 ft^3 per acre, a rise of 56 percent. In spite of recent increases, the net growth per acre on all ownerships is three-fifths of what can be attained in fully stocked natural stands and considerably below what could be achieved with genetic improvement, fertilization, and other intensive management measures.

Harvest Removals

Harvest removals in 1976 totaled more than 14 billion ft^3 of growing stock, including 65 billion bd ft of sawtimber (Table 1-2, Fig. 1-3*d*). This level was somewhat above the harvest of 12 billion ft^3 in the 1950s and early 1960s. Softwoods made up three-quarters of the harvest, with their harvest concentrated on the Pacific coast and in the south. Large-diameter material for sawlogs and veneer logs took over half the total harvest, pulpwood representing about a quarter. Relatively little was cut for fuel. More than five times the growing-stock fuel harvest, however, was cut for fuel from rough and rotten trees, standing dead, branches, and tops. Ten percent of the removals were left as residue.

The south is clearly the nation's primary wood basket, providing almost half the total harvest. The Pacific coast supplies 30 percent, but the large forest area of the Rocky Mountain west does not contribute much to the commercial harvest.

While the Pacific coast contributes about one-third of the total harvest volume, it provides over 50 percent of the harvest value (Table 1-3). The

Table 1-3 Value of timber products at points of delivery by geographic region in 1972 (millions of dollars)

	Sawlogs	Veneer logs	Pulpwood	Other	Total
North	346	28	186	139	699
South	980	315	708	142	2,145
Rocky Mountains	377	69	8	13	467
Pacific Coast	2,012	890	90	51	3,043
					6,354

Source: U.S. Forest Service (1982).

old-growth sawtimber cut at the Pacific coast is still our most valuable timber product.

Growth/Harvest Ratio

Growth is significantly above removals for the nation as a whole (Table 1-4) when both are measured in cubic feet. By region, growth is significantly above removals in the east and somewhat below removals in the west. These relationships reflect both the large volume of hardwoods in the east that are not being cut and the considerable amount of old growth that remains in the west.

While this aggregate growth/harvest ratio suggests that the nation does not face a timber famine measured in gross physical terms, much of this growth occurs on farms and miscellaneous private ownerships in the east. These lands are often held for recreation and other nontimber purposes, and at any given time may not be available for harvest. In addition, what is being grown is not the same as what is being removed. In the South, for example, the most recent data (Peterson, 1985) show that while the aggregate growth is above harvest, growth of the economically important softwood segment has leveled off. Furthermore, softwood removals have continued to rise, reaching softwood growth by 1985. Forecasts indicate that even with real price rises the harvest will soon exceed growth and draw down the softwood inventory that accumulated for two decades. In the South it is the softwood growth/drain ratio that is critical, not the total.

The relationship between timber growing stock, removals, and growth on the different ownerships is a continuing source of controversy. Forest industry is the only ownership cutting its growth; all other ownerships have growth surpluses. The U.S. Forest Service with 32 percent of the growing stock contributes 15 percent of the harvest, while industry with 14 percent of the growing stock contributes 29 percent. The national forests annually cut 0.9 percent of their inventory, while industry annually cuts 3.9 percent.

Such numbers mask more than they reveal. How much the national forests or private owners harvest depends on the objectives for their lands in relation to their condition rather than on some mechanical ratio.

Projected Timber Resource Use

Projecting future timber supplies, demands, and prices has been a serious undertaking since the Copeland report (U.S. Forest Service, 1933). Growing trees takes time, and deciding how many to plant and how much to invest in timber growing has long been a legitimate public policy issue in an era of perceived resource scarcity.

Table 1-4 Comparison of growing stock, growth, and harvest removals by ownership in 1976

Ownership	Commercial forestland		Growing stock		Growth		Removal		Removal as a percentage of inventory, percent	Ratio of growth to removal
	Thousand acres	Percent	Million ft^3	Percent	Million ft^3	Percent	Million ft^3	Percent		
Farm and miscellaneous private lands	278	58	301	42	12.4	57	6.8	48	2.3	1.97
Forest industry	69	14	106	15	4.1	19	4.2	29	3.9	.98
Other public lands	47	10	77	11	2.1	10	1.2	8	1.6	1.75
National forest	89	18	227	32	3.1	14	2.1	15	.9	1.48
Total/average	483	100	711	100	21.7	100	14.2	100	2.2	1.53

Source: U.S. Forest Service (1982).

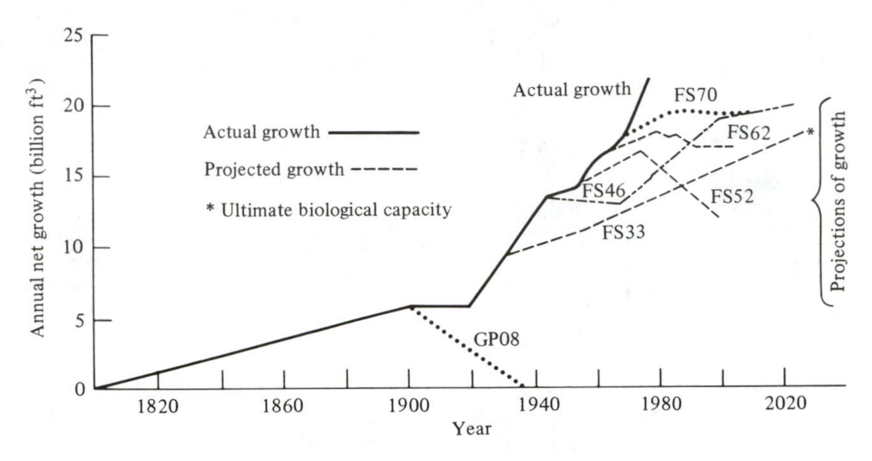

Figure 1-4 Actual growth (—) and projection of growth (---) 1800 to 2020. (*From Clawson, 1979,* © *Science,* **204:**1168–1174, Figs. 2, 4, 5, 7.)

Clawson also examined the history of the timber economy projection business and made the interesting observation that the projections of timber growth rates have consistently been pessimistic and underestimated actual growth (Fig. 1-4). The five forecasts made by the U.S. Forest Service between 1933 and 1970 all predicted a much lower rate of total growth than was actually realized. Low projections imply lower or deficit growth/harvest comparisons and tend to keep the idea of timber famine alive.

Still, periodic projections must be made as current policy and action depends on the best possible anticipation of the future. Projections published in 1982 used newer techniques, which explicitly estimate equilibrium levels of market prices and amounts purchased (Tables 1-5 and 1-6). The volume of timber products harvested is projected to rise 85 percent over the next 50 years, with softwood harvest increasing 50 percent and hardwood harvest increasing 300 percent. Thus the softwood/hardwood harvest ratio will change from 3 : 1 to 1.4 : 1 over the next 50 years. Under this projection, the cubic foot harvest will stay roughly constant on the Pacific coast, increase by 70 percent in the Rockies, and double in the south and north. In terms of board foot harvest, the Pacific coast will experience a 20-percent decline as it adjusts to smaller logs.

Forest industry harvests are projected to increase only gradually. However, within this trend lie important regional changes. Industry cubic foot harvests on the Pacific coast are projected to drop 60 percent over the next 50 years, with most of the decline by 1990. The old-growth inventory on these industry lands faces rapid depletion, and merchantable second-growth stands are insufficient to continue the recent harvest level. Steady increases in industry harvests in the east, where the old growth

Table 1-5 United States wood product harvests* by region and ownership (billions of cubic feet)

	Actual harvest 1976				Projected harvest under equilibrium prices 2000				2030			
	Softwood	Hardwood	Total	Percent of total	Softwood	Hardwood	Total	Percent of total	Softwood	Hardwood	Total	Percent of total
By region												
North	0.62	1.33	1.95	15.9	1.10	2.37	3.47	18.5	1.30	3.50	4.80	21.0
South	4.18	1.44	5.62	46.0	6.77	3.53	10.30	54.8	7.37	5.40	12.77	56.1
Rocky Mountains	0.76	—	0.78	6.4	1.17	—	1.20	6.4	1.36	—	1.36	6.1
Pacific Coast	3.81	0.09	3.88	31.7	3.76	.10	3.83	20.3	3.87	0.10	3.97	16.8
Total	9.37	2.86	12.23	100.0	12.80	6.00	18.80	100.0	13.90	9.00	22.90	100.0
By ownership												
Farm and other private lands	3.40	2.54	5.94	46.9	5.69	4.63	10.32	57.0	6.56	6.99	13.55	58.5
Forest industry	3.42	0.47	3.89	30.7	3.28	0.98	4.26	23.5	3.77	1.49	5.26	22.7
Other public lands	0.80	0.17	0.97	7.6	0.92	0.27	1.19	6.6	1.03	0.37	1.40	6.0
Forest service	1.78	0.10	1.88	14.8	2.18	0.16	2.34	12.9	2.71	0.25	2.96	12.8
Total	9.40	3.28	12.68	100.0	12.07	6.04	18.11	100.0	14.07	9.10	23.17	100.0

Source: U.S. Forest Service (1982).

* Harvest estimates by region are based on actual 1976 consumption and projected supply and demand equilibrium consumption in 2000 and 2030, while the harvest estimates by ownership are trend estimates and differ somewhat from the actual consumption estimates.

Table 1-6 United States real stumpage prices: actual 1976 prices and projections for 2000 and 2030 (1967 dollars per thousand bd ft)*

	Actual prices, 1976	Equilibrium price indices, 1979 estimates		Average percentage rate of real price increase 1976–2030	
		2000	2030	1979 est.	1983 est.†
Softwood stumpage					
North	16	35	53	2.2	2.5
South	42	104	195	2.9	1.9
Rocky Mountains	13	57	116	4.1	2.8
Pacific Northwest					
Douglas-fir subregion	68	87	163	1.7	1.5
Ponderosa pine subregion	33	65	119	2.4	1.5
Pacific southwest	42	79	163	2.2	1.7
Hardwood stumpage					
North	27	25	30	0.2	0.8
Southeast	21	20	26	0.4	0.2
Southcentral	21	25	42	1.3	0.6

Source: U.S. Forest Service (1982). Price and other forecasts were made in 1979.

* These price indices are net of inflation or deflation. Thus they indicate real price trends—stumpage is seen to be an increasingly valuable resource relative to all other commodities.

† This column is a revised set of equilibrium price forecasts made in 1983 and is the official basis for Forest Service planning in the late 1980s. (*Source:* Haynes and Adams, 1984.)

was cut long ago and the second-growth inventory is still increasing, offset much of this decline.

While harvest from public lands increases slightly, and that from industry lands stays about constant, harvest from farm and other private lands is projected to more than double over the next 50 years, with all of that increase in the east. Even with these increased harvests, stumpage prices are projected to increase over the projection period.

Four factors may cause the harvest trend to be different from that projected: three work to decrease the harvest and the fourth works to increase it, as compared to the projections. First, public timber harvests actually have declined slightly in the past 10 years. Even modest increases in harvest may not be possible if the increasing emphasis on nontimber objectives continues. Second, projections about farm and miscellaneous private ownerships reflect historical behavior combined with increasing timber inventories, rather than owner intentions. Although these owners were responsive to stumpage price increases between 1950 and 1974, uncertainty exists about reaction to future price rises, given that many of these owners have nontimber objectives which could increasingly constrain harvests and raise harvesting costs. Third, recent evidence (Peterson, 1985) suggests that the softwood harvest in the South will not increase as much as projected due to lower growth than expected. Fourth, the projection that industry harvest slightly increases is based on the assumption that utilization rates do not change much. Industry probably will move to whole-tree utilization over the next 20 years, possibly even using the stump. This changing utilization standard could provide significant volumes unaccounted for in the projections.

Throughout this period of increased harvest, projected total growing-stock inventory will also increase. Total inventory should increase 25 percent, with hardwood volume increasing slightly more than softwood volume. This inventory may be composed of less desirable species and smaller sizes than exist now, but the nation definitely is not running out of wood. Still, the question remains: how much of this inventory will be harvested and how much will be dedicated to other uses?

CURRENT MANAGEMENT PRACTICE

Most timber management aims at increasing growth and protecting against losses. Such activities include fire management, insect and disease management, stand regeneration, stocking control, and fertilization.

Reduction of Losses

The largest and most effective management effort in the United States has been in the control of forest fires. The burned area has declined drastically from the beginning of the century. While further reductions can be made, the greater challenge now lies in the use of fire as a silvicultural tool.

Nowhere nearly as much success has been achieved in insect and disease control. The recent spruce budworm epidemics in eastern Oregon and Maine, the continuing infections of white pine blister rust, and the browning of large parts of the central Rockies and the south caused by pine beetles attest to the continuing battle against insects and disease. Mechanical treatments to eradicate these pests directly, such as the great ribes eradication program to prevent white pine blister rust, have done little to control the pests. Harvesting of affected timber is now the more common practice. In the long run, reduction of losses depends on silvicultural techniques that encourage more pest-resistant stands, stand diversity by age and species, and methods of control that rely on the pests' own biological enemies and biological quirks.

Regeneration

Most forest regeneration in the United States occurs naturally or through harvest practices designed to encourage natural regeneration, but the acreages planted and direct-seeded have been increasing (Table 1-7). Nearly all this planting and seeding involves use of commercially important softwood species, chiefly southern pines and Douglas-fir. Direct-seeding now accounts for only 5 percent of the acres treated.

Trends in artificial regeneration during the late 1950s and early 1960s differed from those of the 1970s. In the earlier period, the average acreage planted and seeded increased on ownerships across all geographic regions, with the major portion of the increased regeneration on farm and other private ownerships in the south. This situation came about as a direct result of the Soil Bank Program, which made payments to farmers for cropland retirement during a period of low farm product prices. Acreages regenerated by these owners declined sharply in the early 1970s to approximately 400,000 acres per year—a level maintained to the present time.

Since the early 1960s, most increase in artificial regeneration has occurred on forest industry ownerships in the south and on the Pacific coast. Over the last decade the average acreage planted and seeded on industry lands has more than doubled. This increase reflects higher rates of harvest, concern about availability of timber from other ownerships, and the increased financial attractiveness of timber growing.

Intermediate Stand Treatments

Intermediate stand treatments include all measures taken to increase growth between the time of stand establishment and harvest. Two practices, precommercial thinning and release or weeding, account for most stand treatment. The annual area treated has remained fairly constant during the 1970s at 1.3 million acres per year, or 0.3 percent of the commercial forestland. This treatment acreage also includes forest fertilization, with 3 million acres being fertilized over the past 10 years.

Some conversion from hardwood to softwood types is being done, primarily on private lands in the south. The area of successful stand conversion is small, however, compared to unplanned reversions from softwoods to hardwoods following harvest. Between 1970 and 1976, the area in pine types dropped by 4 million acres, with nearly all of this area changing to hardwoods and hardwood and pine types.

While these numbers give some idea of management intensity, they do not represent the entire picture. The numbers give no impression of the success of artificial regeneration. In addition, they give no clue about the success of natural regeneration, the only source of future growth follow-

Table 1-7 Area planted and direct-seeded and area of intermediate stand treatments, 1950–1978 (thousands of acres)

Year	Total United States	Section				Ownership			
		North	South	Rocky Mountains	Pacific coast	National forest	Other public	Forest industry	Farm and other private
				Area planted and direct-seeded					
1950	489	137	285	15	52	45	54	153	237
1960	2,100	308	1566	14	212	134	130	521	1,315
1970	1,577	225	1025	70	257	261	131	763	422
1977	1,942	160	1301	57	424	257	120	1,138	427
				Area given intermediate treatments					
1970	1,359	209	785	86	279	304	100	592	363
1978	1,393	263	444	201	485	420	98	521	354

Source: U.S. Forest Service (1982).

ing most harvests in the United States. No mention has been made of the timber growth impact of partial cutting so prevalent in the north. Are good growers left, or do only the weak and lame remain?

In the late 1970s the overall picture remains one of *extensive* management on most lands in most places. Only 0.4 percent of the commercial forestland is planted and seeded each year and 0.3 percent is given intermediate stand treatments.

Opportunities for Achieving Further Increases in Timber Growth

Extensive, low investment management on the nation's forests continues in spite of ample physical opportunities for intensive management. A recent U.S. Forest Service study (Dutrow et al., 1983) shows that the potential exists for intensifying management on 168 million acres of commercial timberland (some 35 percent of the nation's total) using the economic screen that any investment must return at least a 4-percent real rate of return. Net annual timber growth would increase by 13 billion ft^3 with treatment of these acres—a volume equal to the total timber harvest in 1976 and to three-fifths of the total net annual growth. Achieving this growth would take time. Several decades would be required for the effects of the investments to be realized: projected softwood offering would increase 11 percent by the year 2000 and 30 percent by 2030. Furthermore, an investment of over 15 billion dollars would be required.

These treatment opportunities, on an area basis, are primarily type conversion from hardwoods to softwoods and, to a lesser extent, reforestation of existing softwood sites or culture of existing stands. Treatments of these conversion or reforestation opportunities would require an investment of 13.5 billion dollars and would eventually increase net annual timber by 11.6 billion ft^3. Most of these opportunities are on farm and other private ownerships, chiefly in the south, where a 10.7-billion-dollar investment could eventually raise net annual growth by 9 billion ft^3. The remaining opportunities occur largely on forest industry land.

While significant biological and economic opportunities exist to increase timber growth on the forestland of the United States, owner objectives and their economic situation determine forestland use and management. Many farm and other private owners of commercial timberland will not respond to these opportunities because they do not manage the land primarily for timber production or are reluctant to incur the environmental disruption that intensified timber management, especially site conversion, might bring. They like their scruffy pine and hardwoods, they cut occasional firewood, hunt a few squirrels and rabbits, and watch the colors change in the fall. Other owners do not have the needed capital, or are not interested in investments that will not pay off for 30 to 60 years, especially when their investment might be eaten by bugs or destroyed by fire along the way.

In addition, many benefits from increasing timber supplies do not directly accrue to the woodland owners who make such increases possible. Lower prices for wood products, less environmental pollution from substitutes such as steel and plastics, less dependence on foreign sources of supply, and lower use rates for nonrenewable resources are major potential benefits of increased wood production that go to society in general. With the low expectancy that private owners will undertake these investments, and because of the public benefits that come from them, publicly supported cost sharing and technical assistance programs probably will be needed for any significant action to be taken. Whether or not it is in the public interest to invest in producing more wood on private land is a central policy question.

REFERENCES

Clawson, M., 1979: Forestry in the Long Sweep of American History. *Science,* **204:** 1168–1174.

Dutrow, G. F., J. M. Vasievich, and M. F. Conklin, 1983: "Economic Opportunities for Increasing Timber Supplies in the United States," U.S. Forest Service and Forest Industries Council.

Haynes, R. W., and D. M. Adams, 1985: "Simulation of the Effects of Alternative Assumptions on Demand–Supply Determinants on the Timber Situation in the United States," U.S. Forest Service.

Peterson, R. M., 1985: Timber Supply: The Situation in the South. *Forest Farmer,* July–August, pp. 14–16.

Shantz and Zon, 1924: "Atlas of American Agriculture," U.S. Department of Agriculture.

U.S. Forest Service, 1933: "A National Plan for American Forestry," 2 volumes.

———, 1980a: "The Nations Renewable Resources, an Assessment." Forest Resource Report 22.

———, 1980b: A Recommended Renewable Resources Program—1980 Update.

———, 1982: "An Analysis of the Timber Situation in the United States 1952–2030." Forest Resource Report 23.

PART ONE

PRODUCTION OF TIMBER FROM THE FOREST

The central task of applied professional forestry is to take a parcel of forestland, decide how to treat it to grow trees and other forest outputs, implement the treatment, and predict when and how much wood and other products will result. Doing this well is the test. If we cannot decide what to do or if we cannot predict what will happen when we implement a treatment, our claim to professional status is more hype than substance.

Part I of this book is about predicting timber yields with confidence.

In Chapter 2 we look at the elements of a forest—land types, stands, management units—how they are described and classified, and how this ties in with the contemporary concept of a prescription. How the land is classified and stratified for planning turns out to be one of the most important and far-reaching decisions managers make, particularly when multiple outputs are needed from the forest.

Chapter 3 defines the terminology and concepts of timber growth and yield. The discussion of growth and yield is set in the context of even- and uneven-aged stands, and the dynamic and structural characteristics of these two stand forms are examined.

The potential ability of land to grow tree crops as indexed by site quality and the influence of existing tree vegetation on actual growth as indexed by stand density are covered in Chap. 4. Site index, vegetation typing, and environmental factor approaches to estimating site quality are reviewed. Five measures of stand density are covered and sample calculations presented.

The number, kind, scope, and sophistication of models to predict tree growth and yield have exploded in the past decade, and this is the subject of Chap. 5. When correctly entered by land type, site, and density of a subject stand, the relative accuracy of these models is what determines our confidence level in yield prediction. While the development and testing of these models are the realm of mensurationists and biometricians, it is the manager who uses them and takes the blame for bad predictions.

Our intent is to review the menu of models available and to discuss and evaluate their strengths and limitations from a managerial perspective. We begin with a classification of models within three broad categories: whole stand models, diameter class models, and individual tree models. A geographically representative sample of specific models is presented to illustrate each type. The chapter concludes with a framework for evaluating different models to guide in the selection of one for a particular application situation.

The algebra gets quite heavy in spots, but this cannot be avoided. Most of the newer models are constructed as sets of equations, and the only way to understand what is going on is to dig in. Fortunately, they all get down to basics: what portion of the trees dies and how much do the survivors grow in height, diameter, and crown?

CHAPTER

TWO

ELEMENTS OF FOREST MANAGEMENT: LAND CLASSIFICATION, PRESCRIPTIONS, AND YIELD PREDICTIONS

Developing, evaluating, and implementing prescriptions—a schedule of activities for some stand or parcel of forestland—is the central activity of applied professional forestry and is the instrumental act whereby theory and principles are translated into reality. Once the land is roaded or trees are cut, forestry theory is put to the test of empirical validation.

Prescription development has taken on a new and even more fundamental role in contemporary forest management planning. The specified prescription for a given type or parcel of land along with the quantitative estimate of timber yields and other results expected when this prescription is implemented is the building block of virtually all modern forest planning and harvest scheduling models. Prescription formulation necessarily integrates the strategies of land classification, the basic and applied biological knowledge of silviculture, the growth prediction techniques of mensuration, economic values, and the decision analysis techniques of the management economist. We see prescription formulation giving new identity and substance to professional forestry by forging stronger and much more explicit technical bonds between these traditional but often severely compartmented subjects of silviculture, mensuration, and forest management—not to mention a growing technical linkage to other resource management specializations such as wildlife management.

Getting specific, consider this prescription and prediction:

> For existing small sawtimber, mixed fir stands on erosive soils having slopes less than 30 percent, site indices over 100, and basal area greater than 200 ft^2 per acre, implement, using tractor logging, a com-

mercial thinning to reduce basal area to 140 ft^2 per acre followed in 30 years by a final harvest, site preparation, and planting to improved Douglas-fir. The expected yields per acre from this prescription are 15 thousand board feet (M bd ft) of sawtimber and 600 ft^3 of pulpwood this year through the thinning, 34 M bd ft of sawtimber and 1,200 ft^3 of pulpwood in 30 years at final harvest from the current stand, and, eventually, 65 M bd ft at final harvest from the regenerated stand.

This terse statement illustrates the three essential elements of a prescription for a land type or forest:

1. *A land-type classification,* which describes parcels or types of land by location, timber size, stocking, species, soils, slope, and other land attributes
2. *A management "activity schedule"* describing the timing, methods, and conditions by which the vegetation and other resources will be manipulated or disturbed to achieve desired outcomes, including:
   - Logging rules
   - A timber thinning and harvest schedule
   - Regeneration techniques for the next tree crop
3. *A quantitative growth and yield projection,* which numerically describes how much timber is expected for commercial harvest: specifically, volumes removed at each thinning and final harvest entry for both the existing and subsequent regenerated stands

Such prescription statements can be made with greater or lesser detail than our example, but all three elements are required to manage and plan a forest in any coherent, quantitative way. Each element has a corresponding management decision that must be made at the outset of planning:

1. How shall I organize and classify my forest by its vegetative, physical, and developmental characteristics into stand types or management units for planning and implementation purposes?
2. How many and what kind of alternative management prescriptions shall be considered for each defined stand type or parcel?
3. What methods and empirical data shall be used to make quantitative projections of growth and yield?

Guidance for answering these questions will ultimately depend on the objectives of the forest owner and the amount of time, money, and analytical detail devoted to planning and management of the forest. For example, if the owner is a land speculator with purely financial objectives and does not want to spend time and money on management and planning, then very few stand classes or prescriptions will be considered. However, if high-resolution planning and multiple-output objectives are the case, such as for the U.S. Forest Service, then we would expect a detailed classification of land, many prescription choices for each class, and quantified predictions of many outputs.

DEFINITIONS

This text will use a consistent terminology for describing forests and forest management activity. The terminology of the Society of American Foresters is the starting point. However, some modification is needed for our purpose. The most important terms we use follow:

Forest. A set of land parcels which has or could have tree vegetation and is managed as a whole to achieve tree-related owner objectives (synonyms: ownership, management unit, planning unit).

Physical characteristics. The set of attributes used to characterize the permanent, physical nature of forestlands, including topography, soils, bedrock, climate, hydrology, habitat type.

Vegetation characteristics. The set of attributes used to characterize tree and other vegetation currently growing on forestland, including height, age, basal area, volume, average diameter, diameter distribution, crown density, species, cover type, community type.

Development characteristics. The set of attributes used to characterize the organization, development, and accessibility of forestland for human use, including ownership, roads, buildings, administrative boundaries, political boundaries.

Stand type. All forestland that has the same defined combination and attribute range of the physical, vegetation, and development characteristics chosen to classify the forest into homogeneous types (synonyms: land type, site type, condition class, forest type, analysis area).

Stand. A homogeneous, geographically contiguous parcel of land, all of the same stand type and larger than some defined minimum size (synonyms: homogeneous land unit, capability unit, ecological land unit, logging unit).

Management unit. A geographically contiguous parcel of land containing one or more stand types and usually defined by watershed, ownership, or administrative boundaries for purposes of locating and implementing prescriptions. A management unit is usually larger than a stand and typically contains many stand types and individual stands (synonyms: heterogeneous planning unit, allocation and scheduling zone, administrative area).

Stand and stand-type prescriptions. A schedule of activities (cultural treatments, harvests, or other events) which, when implemented on a stand or stand type, are expected to achieve certain desired outcomes. Planting, thinning, regeneration harvesting, fertilizing, and so on, are typical activities used to achieve desired vegetation structure and timber product outcomes. Considering timing choices, hundreds of different prescriptions are technically or biologically possible for each stand type.

Management unit prescription. In addition to prescriptions for individual stand types, a management unit prescription considers the spatial integrity of the management unit. It may schedule the location and sequence of road development, the order of stand entry per harvest, and the spatial strategies and activities for protecting or enhancing water, wildlife, amenity, and other values within the management unit.

LAND CLASSIFICATION

Land classification is the first prescription element since it sets the stage and context of the activities and yield projection. It is here that we decide what we mean by homogeneous and heterogeneous, what is similar and what is not. Classification of lands into homogeneous strata allows us to generalize results from observed or studied areas to similar but unstudied areas; for example, we say that because fertilization of 10-year-old ponderosa pine on a specific type of soil, aspect, and elevation produced a 10-percent increase in average growth in some research plots, applying the same treatment to other areas with the same soil, aspect, and elevation is predicted to generate the same 10-percent response.

Historically we have not paid much attention to land classification. Virtually all timber yield tables are pegged to a single characteristic of the land: the tree site index. As interest in treatment impacts and responses other than timber yield increase, it becomes obvious that classification by site index alone is not enough.

To increase similarity or homogeneity of a land class, more physical, vegetative, and developmental characteristics are used to further subdivide and refine land classification. However, this is an exponential process, and we are rapidly swamped with too many unique classes. How many classes are enough? The answer lies in the question: for what responses and impacts do we need the most accuracy in prediction, how much homogeneity is enough to achieve this, and what will it cost?

An Example of Forest Classification

To illustrate the important relationship and distinctions between land, vegetation and development characteristics, stand types, stands, and management units, an example is helpful. Consider a small 1,000-acre forest for which a few characteristics are chosen to classify the forest into stand types:

1. *Physical characteristics*
   a. *Slopes* (2 classes)
      Moderate, 0 to 30 percent
      Steep, 31+ percent
   b. *Watershed* (2 classes)
      Elk Creek
      Fish Creek
2. *Vegetation characteristics*
   a. *Cover type* (3 classes)
      Softwoods
      Hardwoods
      Grasses
   b. *Size* (2 classes)
      Average diameter 14 in. or less
      Average diameter over 14 in.
3. *Development characteristics*
   a. *Road distance* (2 classes)
      2000 ft or less to roads
      Over 2000 ft to roads

With even this highly simplified characterization of forestland we can generate up to 48 possible stand types, each with a unique combination of characteristics.*

When our example forest is mapped for these characteristics (Fig. 2-1*a* to *c*) and the maps are overlayed to find all possible combinations of the characteristics, the stand map shown in Fig. 2-1*d* is created. Here we observe that only 11 of the 48 possible stand types actually exist in this forest. The timber is all over 14 in., so this criterion is not distinguishing. Four of the stands map out as under 20-acre "sliver" stands, which have been defined as too small to consider as operational units and need to be combined with adjacent stands. Before combining the "sliver" stands, we also see that several of the stand types occur more than once, yielding a total of 14 geographically separated stands. If we ignore the geographic location by watershed, there are only eight stand types. These observations would be typical of any stand-type classification and mapping effort. In a large forest the number of stands is much larger relative to the number of stand types.

* (2 slopes) × (2 watersheds) × (3 cover types) × (2 sizes) × (2 road distances) = 48 stand types.

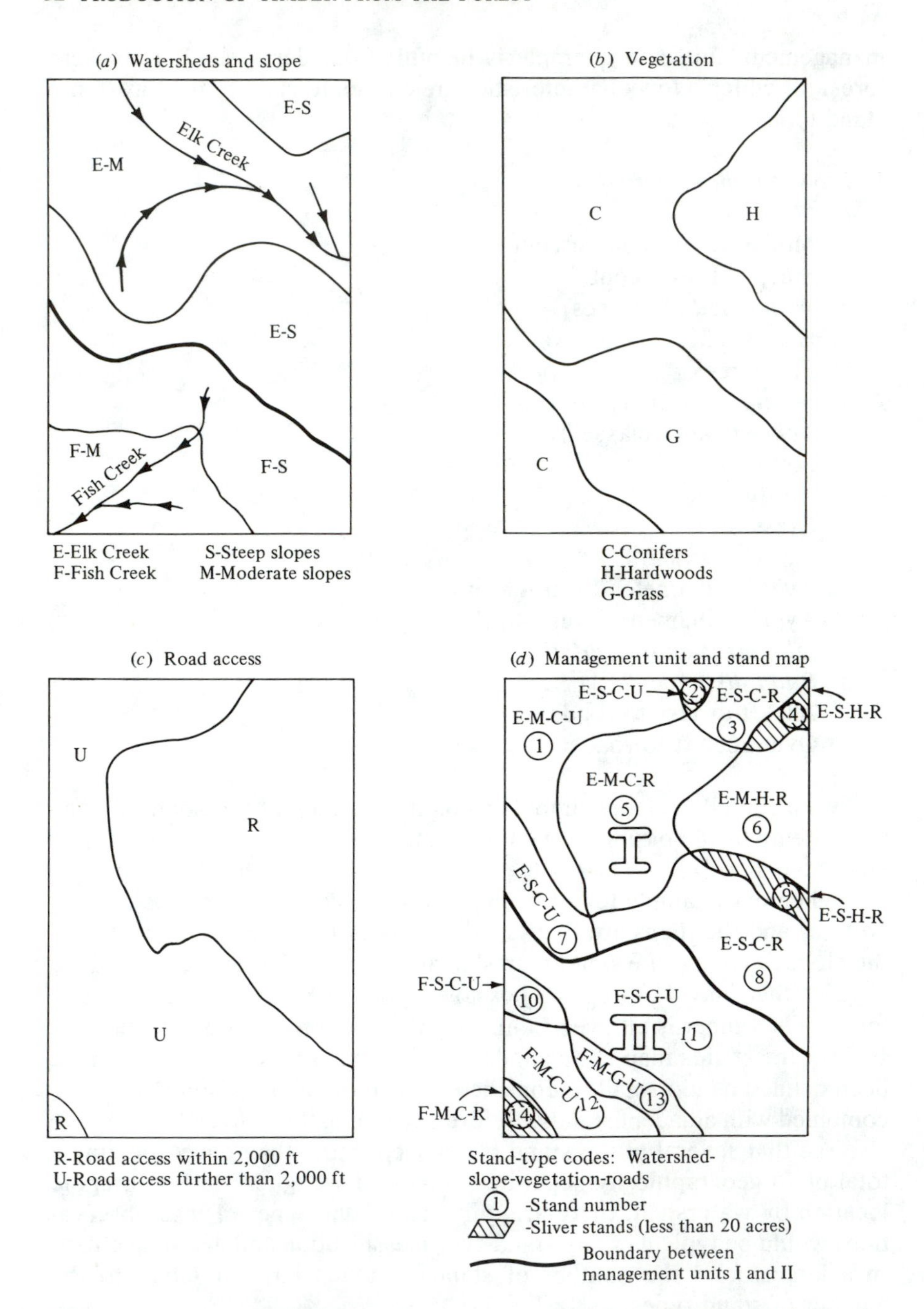

Figure 2-1 Using (*a*) physical, (*b*) vegetative, and (*c*) developmental characteristics of a forest to generate stand types, management units, and stand maps (*d*).

To illustrate management units and some management unit prescription properties, for access and logging purposes the Elk Creek drainage containing stands 1 to 9 is designated as management unit I and the Fish Creek drainage, containing stands 10 to 14, as management unit II. If the owner wished to harvest all conifer timber in Elk Creek, then additional roads must be located and constructed to access stands 1 and 7. The steep area of stand 3 is easily visible from recreational roads along Elk Creek and selection or shelterwood cutting is needed to protect visual values. The narrow pass in stand 11 between timbered stands 7 and 10 is an important big game summer-fall escape and resting area and may need special consideration by only allowing winter logging. All these considerations and activities are unique to Elk Creek and would be detailed in a prescription for management unit I. In general, the amount and the distribution of stands within a contiguous area control the unique management prescription for each management unit.

Categorical and In-Place Forest Information

The preceding example also helps illustrate the obvious but important difference between categorical and in-place or geographical information for forest management. The physical, vegetative, and developmental characteristics are used to define 48 unique stand types. The stand itself, however, can only be identified by mapping the forest and providing in-place information about the location of each contiguous parcel of land that is homogeneous by a predefined stand type. The Elk Creek and Fish Creek management units are also discovered by mapping. A stand, or management unit as we defined it, *only exists in an in-place data base*. A typical forest inventory uses a sample survey technique and puts in random or systematic sample plots. With these data we can estimate the number of different stand types present and provide categorical information showing the acres in each stand type, as shown in Table 2-1 for our example. However, a sample survey does not normally identify individual stands or, say, very much about the distribution of stands or stand types. An in-place, geographical information system, in contrast, contains both the maps of Fig. 2-1 and the categorical data of Table 2-1.

Stand Type or Management Unit Classification for Planning?

Stand types organize the land of a forest into classes that are homogeneous with regard to some basic land characteristics in order to predict timber yields and other responses of the land to treatments with confidence. Management units, in contrast, organize the land into logical spatial units for purposes of implementing a plan and to deal with concerns or impacts that are inherently spatial in character, resulting in land units that are typically nonhomogeneous.

Table 2-1 Distribution of area by stand type in example 1,000-acre forest

Stand type (not identifying watersheds)	Area in acres
M–C–U	230
S–C–R	120
M–C–R	180
M–H–R	110
S–C–U	90
S–H–R	20
S–G–U	210
M–G–U	40
Total	1,000

Historically, most forest information has been developed by sample survey inventories, and the dominant concern of management was timber production. It is therefore not surprising that most forest planning developed around stand type strata defined to be homogeneous in terms of site, age, species, and stocking. Available yield data could be keyed to these strata, and it was simply assumed that the forester could locate stands and take care of any serious spatial difficulties that emerged during implementation.

Contemporary management increasingly requires both in-place management unit information and finer distinctions in the definition of stand types. Careful economic analysis of timber management and timber sale opportunities requires information on the size and location relationship of individual stands to existing roads, streams, and soil types as well as the site quality and topographic properties of the stand in question. Accessibility of stands for adverse weather logging is an in-place question. An economic analysis using the same average unit revenues and costs for all stands of a stand type throughout the entire forest is often not precise enough in today's competitive and critical world. For managers interested in visual, water, wildlife, recreational, and other nontimber aspects of forestland, it is virtually mandatory that in-place information be used. The very concepts of viewsheds, watersheds, and wildlife habitat are location-specific. Elk habitat, for example, is defined by the spatial arrangement of different stand types in relation to topography and human activity to provide edge, winter and summer range, hiding cover, and calving cover. Visual quality depends on the character of stands at different distances from viewpoints such as roads or trails.

So we have to provide both homogeneous land classes for per-acre response prediction and contiguous but nonhomogeneous units for implementation analysis. A complete prescription has common instructions for each homogeneous stand type as well as unique site-specific instructions for each management unit. The planning models described in Chap. 15 potentially have the ability to work simultaneously with stand types and with management unit land classes. U.S. Forest Service planning has taken the lead in working out the techniques for this more sophisticated land classification and planning, but industrial and other private forest owners must also respond to many spatial concerns and are refining their planning analyses accordingly.

The growing demand for more detailed in-place information mirrors the steady change of the forestland management problem from that of utilizing a surplus of easily exploitable natural timber to one of investing in management of regenerated stands in a decision context impacted by social concern for environmental and nontimber values. The sample survey will still be used to collect accurate categorical data on current timber stands efficiently, but the timber survey will have to be coordinated with a mapping program and additional inventories to evaluate such criteria as erosion hazard or endangered species habitat. The construction and design of such multipurpose in-place information systems is just getting underway in the 1980s. The frontiers of mensuration, biometrics, and remote sensing are being extended and will likely evolve to be subsumed under a new integrated discipline of resource information systems.

Stand Prescription and Growth Projections

The empirical core of our claim to manage land scientifically and to ensure that owner objectives are met lies in our ability to predict quantitatively the future characteristics of current and regenerated stands of a given stand type managed under a specified prescription. If we cannot predict with acceptable accuracy, then it is hard to convince our clients that their goals are being met and that we foresters really know what we are talking about. Concepts are one thing, but the real world wants to know how much!

Three prescriptions for a hypothetical loblolly pine stand illustrate the potential diversity of detail in prescription development and projection. The current stand is a cutover, naturally regenerated, 60-year-old field stand growing on site 80 land at base age 25 years. It averages 12 large sawtimber-sized pine trees, 35 medium pole-sized pines, and both hardwood and pine regeneration. Overall, about 60 percent of the growing space is occupied by the pines. The future projection for each prescription is carried for 60 years to provide a common base for comparison.

Prescription 1

Owner objective. Provide pulpwood for owner's mill.

Treatment schedule. Clear-cut current stand immediately, site preparation by chopping and burning, plant 10 × 10 ft spacing with loblolly pine, no intermediate treatments, clear-cut at age 20.

Projections

Item	Year 1985	2005	2025	2045
Stand age, before/after harvest	60/0	20/0	20/0	20/0
Harvest volume, cords per acre	12	20	20	20

Prescription 2

Owner objective. Sell sawtimber stumpage, quail hunting rights, practice intensive management.

Treatment schedule. Clear-cut current stand immediately, site preparation by plowing, disking, pile and burn slash, plant 7 × 7 ft. with loblolly pine, herbicide release, precommercial thin age 10, commercial thin for pulp and sawtimber ages 20, 30, 40, 50, final sawtimber harvest at age 60. Repeat burning for fuel reduction and quail habitat improvement every 5 years after age 10.

Projections

Item	Year 1985	1995	2005	2015	2025	2035	2045
Stand age	60/0	10	20	30	40	50	60/0
Inventory (preharvest volumes per acre)							
Sawtimber, M bd ft per acre	6	0	2	10	18	23	25
Pulpwood, cords per acre	8	4	13	18	16	10	6
Average stand DBH, in.	9	4	12	16	19	22	25
Harvest (volume per acre)							
Sawtimber, M bd ft per acre	6	0	0	0	3	5	25
Pulpwood, cords per acre	8	0	3	10	10	6	6
Harvestable quail crop, 20-bird coveys per acre	0.01	0.05	0.03	0.03	0.03	0.03	0.02

Prescription 3

Owner objective. Public land. Provide the socially best mix of multiple-use outputs. Satisfying objective requires large trees and uneven-aged management to achieve wildlife species diversity.

Treatment schedule. Work with existing stand. Initial sanitation, salvage, and precommercial thinning. Mark to maintain a 60-percent pine, 40-percent hardwood average composition. Clear-cut small patches of 1 to 2 acres, regenerate naturally. Prescribe burn for fuel management, improved wildlife habitat, and hardwood control.

Projections

Item	Year						
	1985	1995	2005	2015	2025	2035	2045
Stand age	60	70	80	90	100	100	100
Inventory							
Pine sawtimber, M bd ft	6	9	11	12	14	13	14
Hardwood sawtimber, M bd ft	0	1	3	4	5	7	6
Basal area pine	40	55	75	80	95	85	90
Basal area hardwood	5	10	20	25	20	40	50
Harvest							
Pine, M bd ft	0	2	2	0	4	5	2
Hardwood, M bd ft	0	1	0	0	2	1	3
Cordwood, cords	0	5	5	0	6	4	4
Other outputs							
Fuel volume, tons per acre	30	20	18	15	15	15	15
Forage production, animal unit months (AUMs) per acre	0.1	0.15	0.2	0.2	0.2	0.2	0.2
Habitat diversity index, 1–10 scale	6	6	7	7	7	8	8
Relative deer habitat, 0–1 scale	0.6	0.5	0.5	0.4	0.3	0.3	0.3
Relative pileated woodpecker habitat, 0–1 scale	0.2	0.3	0.3	0.4	0.5	0.5	0.6
Visual quality rating, 1–10 scale	5	5	6	6	7	7	7

These three prescriptions illustrate several points:

1. The amount of detail and thus the quantity, diversity, and complexity of numbers to predict can get very large.
2. The number of items to be quantified and predicted depends largely on the owner's objectives.
3. Many different prescriptions are possible for the same stand.

The hard part is getting good numbers to put into the projections. For timber items such as volume, DBH, basal area, or stem counts, the growth and yield equations, tables, and computer simulators are the source. Many, although not all, of the nontimber items (such as quail numbers, visual quality, or erosion rates) are also related to the vegetative structure. However, a great deal of additional research is needed to establish these relationships for consistent quantitative analysis. Subjective "expert" judgment is still the primary source of nontimber numbers.

In this book we present material about predicting future tree vegetation characteristics only, recognizing that a wealth of yield and response information also is needed for nontimber concerns of different owners.

Resolution of Planning

In Chap. 15 we present and discuss in some detail the scheduling models for forest management. The size, usefulness, and resolution of these models will largely be determined by the number of stand types, individual stands or management units, prescriptions for each stand type or unit and time periods used for the analysis. The maximum number of different variables in a planning model can roughly be estimated by the equation

$$\left\{\begin{array}{l}\text{Number of}\\ \text{variables to}\\ \text{keep track of}\end{array}\right\} = \left\{\begin{array}{l}\text{number of}\\ \text{stand types}\end{array}\right\} \times \left\{\begin{array}{l}\text{average number}\\ \text{of stands per}\\ \text{stand type}\end{array}\right\} \times \left\{\begin{array}{l}\text{number of}\\ \text{prescriptions}\\ \text{per stand type}\end{array}\right\} \times \left\{\begin{array}{l}\text{number}\\ \text{of time}\\ \text{periods}\end{array}\right\}$$

Choosing 48 stand types, 30 stands per stand type, 10 prescriptions per stand type, and 10 time periods would yield a problem with at least 144,000 variables to keep track of. And this, as you will soon see, is a fairly small management problem.

Our concern here is to emphasize that initial decisions about the number of land characteristics to use in creating stand types, the degree to which individual stands will be identified, and the number of prescriptions to be used are very important forest management decisions. They are not givens or mandates from the mensurationist, silviculturists, and other specialists and must be treated as deliberated choices of managers. The decisions made here affect the entire management planning enterprise: the kinds of inventory and mapping to do, the kinds of personnel to hire, the kinds of computer hardware and software to buy, the character and effectiveness of planning models, and, ultimately, the quality of forest management decisions.

QUESTIONS

2-1 Consider the problem of land classification to establish land and stand types for the purpose of forest management planning.

(*a*) What is the difference in the role or use of information about *stand types* as contrasted to information about *stands* in forest management planning?

(*b*) What criteria would you use to decide if you had a good stand-type classification system for a specific forest and planning situation?

2-2 Suppose a forest of 170,000 acres was classified into 18 different stand types for timberlands and 10 land types for nontimberlands. The owner wanted to keep track of these stand and land types in each of six geographic ranger districts. For the timber stands the foresters wanted to look at four different silvicultural prescriptions for each stand type; the prescriptions could be initiated in any of the next five planning periods. For the nontimberlands, a decision would be made immediately as to which of six alternative land uses each land type would be assigned in each ranger district.

Estimate the size of this planning problem in terms of the number of possible choices (decision variables) that could be made over the entire forest.

2-3 Review the distinction between *categorical* and *in-place* information. For each of the following management concerns, indicate what kind of information is needed for decision analysis and one reason for your choice of information.

(*a*) Evaluating timber yields from different timber culture prescriptions applied to different stands and stand types.

(*b*) Evaluating the impact of harvesting on stream sedimentation.

(*c*) Deciding what lands can be logged in wet weather.

(*d*) Evaluating stands as habitat for squirrels.

(*e*) Evaluating stands as habitat for black bear.

(*f*) Locating log landings.

(*g*) Determining total grazing-carrying capacity for domestic livestock.

(*h*) Selecting stands for commercial thinning.

REFERENCES

Bailey, R. C., 1983: Delineation of Ecosystem Regions, *Environmental Management*, **7**(4):365–373.

———, R. D. Pfister, and J. A. Henderson, 1978: Nature of Land and Resource Classification—a Review, *J. Forestry*, **76:**650–655.

Davis, L. S., 1980: Strategy for Building a Location-Specific Multi-Purpose Information System for Wildland Management, *J. Forestry*, **78:**402–408.

Henderson, J. A., L. S. Davis, and E. H. Ryberg, 1978: "ECOSYM: A Classification and Information System for Wildland Resource Management," Department of Forest Resources, Utah State University, Logan.

Romesburg, C. H., 1984: "Cluster Analysis for Researchers," Lifetime Learning, Belmont, California.

Shute, D. A., and N. E. West, 1982: Two Basic Methodological Choices in Wildland Vegetation Inventories: Their Consequences and Implications, *J. Appl. Ecology*, **19:**249–262.

U.S. Forest Service, 1978: "Integrated Inventories of Renewable Natural Resources: Proceedings of the Workshop, Tucson," General Technical Report RM-55, Rocky Mountain Station.

CHAPTER

THREE

GROWTH, YIELD, AND STAND STRUCTURE: CONCEPTS FOR FOREST MANAGEMENT

Predicting future growth and yield of managed and unmanaged stands is absolutely essential to credible forest management planning. While few foresters aspire to the biometrician's love of statistics, it is important that they be familiar with yield estimation, for these are key numbers underpinning the management plan and the amount of cutting on the forest. Moreover, in many cases there are no adequate, documented empirical models available to estimate yield. Foresters are then frequently called on to make, or approve, best-guess, subjective estimates of yield, and they can hardly afford to be ignorant of the art of making such estimates.

In this chapter we review the basic concepts and definitions of growth and yield and then look at the meaning and measurement of growth in the context of forest stands managed under even-aged and uneven-aged silvicultural systems. The structure and dynamics of even-aged and uneven-aged stands in terms of stems and volumes by tree diameter class is reviewed to provide a foundation for the growth estimation models covered in Chap. 5.

STAND GROWTH AND YIELD CONCEPTS

Stand growth is measured as the change in a selected stand attribute over some specified time. As an example, a stand might increase in volume by 2,000 ft^3 over 10 years. Its average periodic annual volume growth equals 200 ft^3 per year. Yield has a dual meaning, namely, either the amount of some selected stand attribute that can be harvested and removed per period, or the total amount that could be removed at any time. For example, an average of 200 ft^3 could be harvested from the stand each year over the period, or 2,000 ft^3 after 10 years. If that yield equaled the growth and could be continued at the 200-ft^3 level in perpetuity, we might say the property was under sustained yield management. Also, the stand might have an inventory of 5,000 ft^3 at age 80, all of which could be harvested to provide a yield of 5,000 ft^3. Thus growth is a biological production rate concept, and yield is a harvest or removal concept measured as either a rate or an amount at a specified time. In general, the maximum that a forest can yield at any time is the growth that has accumulated up to that time, and the maximum yield that can be removed perpetually per period equals the growth per period.

Growth and yield can be measured in physical units such as volume, basal area, and weight. Also, they can be measured in value, which is often the variable of interest. We shall concentrate here on physical measures of growth and yield because they form the basis of value measures.

By their nature, trees accumulate growth over many years, a layer per growth period. Woody growth occurs on the total tree: branches, bole, and roots. Most commonly, however, only the bole to some merchantable top is removed from the forest, and bole growth is not easy to measure. Trees do not grow according to a standard form any more than people do. They grow short and thick, tall and slim, and everything in between, making measurement of the bole form a major problem.

Similarly, measuring stand growth can be difficult. By far the major source of difficulty involves the successful measurement and prediction of stand mortality. Mortality, that is, how many trees die, is, by its nature, harder to measure than how much wood is being laid down on living tree boles. Trees that die, fall down, decay, and disappear between inventories are not accurately counted without great cost. For most stand types, mortality is the great unknown in estimating periodic increases in volume or value.

Utilization of Forest Growth

Forest growth can be looked at in terms of total potential per unit area (Fig. 3-1). Level A represents total woody growth, either accumulated or currently produced, including all branches to their tips. Level B indicates a woody growth potential usable by manufacturers under present technology. Level C portrays woody growth actually removed from a stand and reflects the economics of logging.

None of these levels is fixed. The total potential can be changed by soil treatment, irrigation, or fertilizer. Technological breakthroughs can find uses for more kinds and sizes of wood. As an example, increased

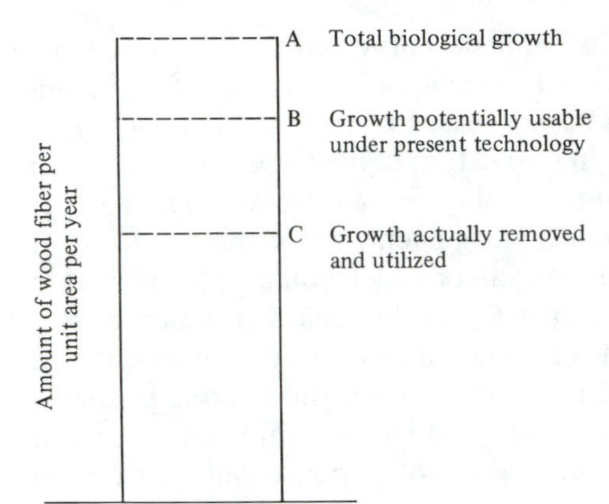

Figure 3-1 Utilization of forest growth potential.

chipping of branches and tops in the woods was made possible by portable chippers becoming available and chip product manufacturing becoming less sensitive to the variability in the raw material used. Growth actually used in an area increases as it becomes profitable to do so, and as people learn about these potential profits. On the west coast, increases in the chipping of residues and small material relate to lower costs for producing chips, to higher prices for their delivery, and to a well-organized market for these chips. Genetic manipulation that reshapes and redistributes growth within the tree also pays by making more of the total potential growth profitable to remove from the forest.

Components of Forest Growth

Increment is made by living trees, but its sum does not give stand increment because some trees die, some rot, and some are cut. For example, consider an even-aged stand measured at two successive inventories 10 years apart (Fig. 3-2). The net change in the stand between inventories shows the 8-, 10-, 12-, 14-, and 20-in. classes losing trees and the 16-, 18-, and 22-in. classes gaining trees. Overall, there are fewer total numbers of trees and the average stand diameter is larger. Dynamically, we expect most of the trees in the first inventory to have grown enough to rise into the next larger DBH class. The 8-in. trees in the second inventory must therefore come primarily from ingrowth (new trees not counted at the first inventory). Trees are lost to mortality throughout the size classes between inventories but at a higher rate in the smaller classes, which are likely to be the less vigorous, suppressed, and intermediate trees. Sorting out exactly what happens requires extremely careful tracking of every tree between inventories.

Multiplying the number of trees in each diameter class by the appropriate volume per tree permits a volume comparison of the two invento-

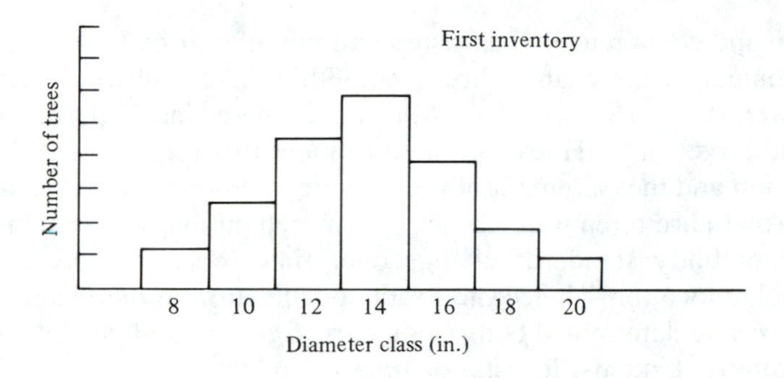

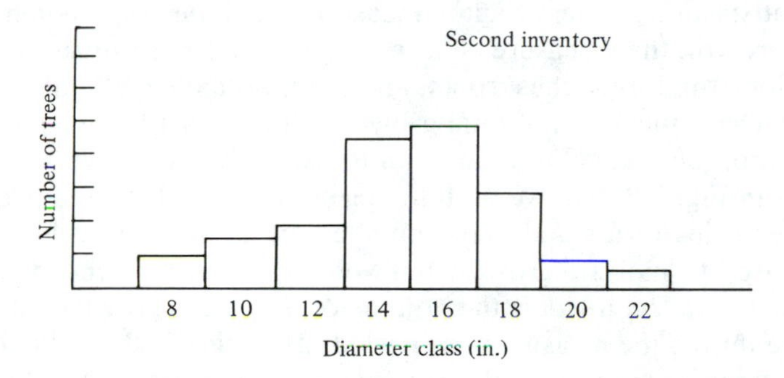

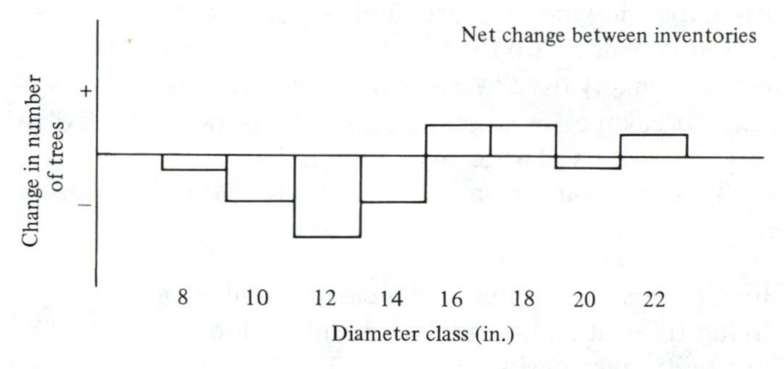

Figure 3-2 Changes in average per acre structure of an even-aged stand over a 10-year period.

ries. Through that comparison, volume growth can be measured. Ingrowth, mortality, and cut complicate growth comparisons and their accurate use requires careful definition.

Growth Definitions

The first step in defining growth is to establish the smallest tree we are going to measure and count as contributing to growth. This might be defined as the minimum size that is merchantable: a tree that can be technologically and economically harvested and manufactured into products. The statement of merchantability is something like "all trees that will provide at least one 16-ft log to an 8-in. top diameter inside bark; all trees 11 in. and larger in DBH are assumed to meet this requirement."

The definition and measurement of the volume and other characteristics of stand growth are often in reference to a merchantable size specification. Merchantability standards change over time, between species, and by geographic location. Hence the exact specification of merchantability must *always* be determined before comparing growth and yield data derived from different stands, forests, or time periods.

When scientists and foresters want to measure total biomass or potential wood fiber growth, they measure all trees larger than a very small size of 1 in. or so. Sometimes branches, roots, and other nonbole material are also measured. These measures of total growth will necessarily be larger than the amount of merchantable growth on the same trees.

Assume, as in Fig. 3-2, that we wish to measure growth between two successive inventories of a stand. Tree number, basal area, or volume could be measured to indicate growth, but volume is closely related to value and is used here. *Ingrowth* is the volume of new trees growing into measurable size during the measurement period. *Mortality* is the volume of measurable trees dying during the measurement period. *Cut* is the volume of timber felled during the measurement period. Combining these three measures of stand change with the volume of living trees at the end gives the components normally used to estimate stand growth. Deterioration, such as decay, breakage, or cracking, usually is not included in growth calculations because its change over time is hard to measure.

Symbolically, the stand components can be represented by (Beers, 1962):

V_1 = volume of living trees at beginning of measurement period
V_2 = volume of living trees at end of measurement period
M = volume of mortality over period
C = volume of cut over period
I = volume of ingrowth over period

Given these components, five different measures of increment over the growth period can be defined by equations:

1. Gross increment including ingrowth $= V_2 + M + C - V_1$
2. Gross increment of initial volume $= V_2 + M + C - I - V_1$
3. Net increment including ingrowth $= V_2 + C - V_1$
4. Net increment of initial volume $= V_2 + C - I - V_1$
5. Net change in growing stock $= V_2 - V_1$

The appropriate definition depends on the user's purpose. The pragmatic forest owner who simply wants to know how much wood is actually being produced would use definition 3, net increment, including ingrowth. A system ecologist interested in total biomass would include small trees and use definition 1. The forest researcher concerned with thinning and reduction of mortality would look to definition 2, while the national accountant monitoring the condition of the forest resource might use definition 5.

These growth definitions and concepts are illustrated in calculations on the trees measured at the beginning and end of a 10-year period on a ⅕-acre plot (Table 3-1). You should be able to trace through all five growth definition equations on a tree by tree basis.

The example records no growth for trees that die or are cut. There-

Table 3-1 Example of growth calculation using data from one permanent sample plot and a growth period of 10 years (sound volume in board feet)

Tree number	First inventory V_1	Second inventory V_2	Survivor growth	Mortality M	Cut C	Ingrowth I	Net growth
1	62.1			62.1			−62.1
2	81.3				81.3		—
3	66.8				66.8		—
4	42.4	62.3	19.9				19.9
5	63.3	122.5	59.2				59.2
6	106.0	163.8	57.8				57.8
7		34.6				34.6	34.6
8	93.3				93.3		—
9	82.0	119.8	37.8				37.8
10	147.2	246.3	99.1				99.1
Plot totals	744.4	749.3	273.8	62.1	241.4	34.6	246.3

Growth equation		V_2	M	C	I	V_1	
1. Gross increment, including ingrowth	=	749.3	+62.1	+241.4		−744.4	= 308.4
2. Gross increment of initial volume	=	749.3	+62.1	+241.4	−34.6	−744.4	= 273.8
3. Net increment, including ingrowth	=	749.3		+241.4		−744.4	= 246.3
4. Net increment of initial volume	=	749.3		+241.4	−34.6	−744.4	= 211.7
5. Net increase in growing stock	=	749.3				−744.4	= 4.9

Source: Adapted from Beers (1962).

fore the gross increment of initial volume equals survivor growth. Trees often will be measured when they die or are cut. Then some growth is recorded for these trees, and the gross increment of the initial volume exceeds survivor growth.

Growth of Stands

In the long run, we cannot cut more than we can grow. In the short run, when old-growth volumes abound, we can live on the accumulated growth of the past. Such an approach still occurs in parts of the west. But eventually yield must depend on growth. Predicting stand growth—how much a site will produce under different conditions—probably has occupied more effort by foresters than any other prediction activity. The way we predict and describe the growth of a stand depends in part on the silvicultural system used—even- or uneven-aged.

Even-aged stands are stands where all the trees are born or initiated at about the same point in time, and while tree sizes will increasingly vary as the stand ages, the calendar age of all trees is about the same when the stand is again regenerated. Even-aged stands have definite beginnings and endings and are easy to conceptualize and schedule for culture and harvest. Most even-aged stand characteristics are related to stand age, and these relationships are used to guide decisions about when to treat or harvest the stand. Control of the genetic makeup can be directly manipulated. When known seed sources are used for planting, getting desired genotypes into the regenerated stand is relatively easy to achieve, as is maintenance of stands in seral species, such as the commercially prized southern pines or Douglas-fir.

Uneven-aged stands are distinguished by lacking a definite beginning or end in time. Trees on any given acre vary by age as well as size and frequently can be of several different species. Throughout most of their lives the trees compete for light or moisture with larger overtopping or nearby trees. Management prescriptions are related to a periodic cycle of partial harvests and specify the species and size structure of the residual stand left after a harvest. Control of the genetic makeup in uneven-aged management is difficult; traditionally the new trees originate from seeds of the mature trees and grow up under the shade of the old trees or in small openings where, although light is ample, they still compete for moisture. The system favors the tolerant climax species such as true firs or sugar maple. Sometimes underplanting is used to get more desirable reproduction.

Growth and yield of even-aged stands A typical lifeline of an even-aged stand in terms of volume and age is shown in Fig. 3-3 for two complete rotations or planned birth-to-death cycles. The stand is regenerated, initi-

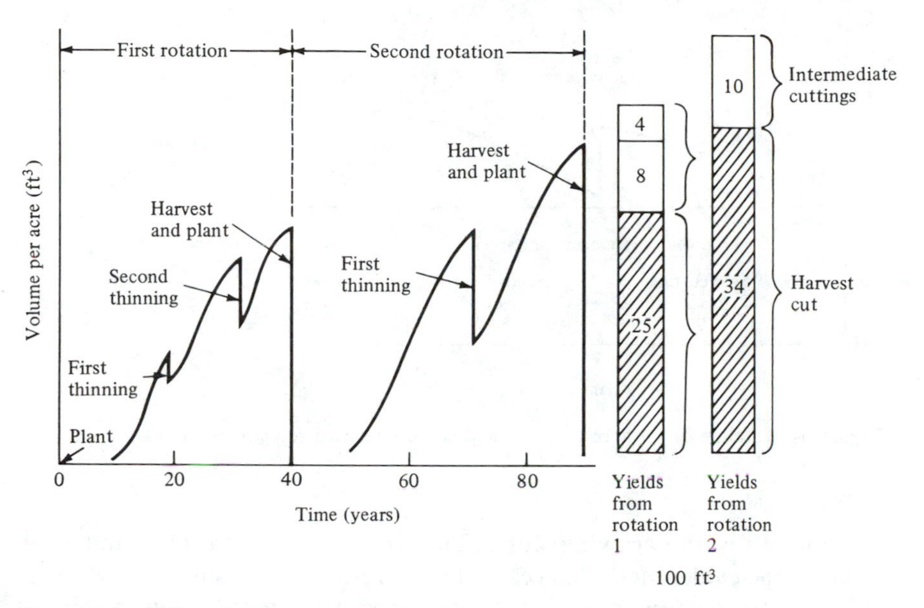

Figure 3-3 Lifeline of a 1-acre even-aged stand over two sequential rotations.

ates merchantable growth at age 10, and has a volume of 1,300 ft^3 at age 20 when 400 ft^3 are removed in a thinning. The residual 900 ft^3 grows to 2,100 ft^3 by age 30, when 800 ft^3 are removed in a second thinning. The residual again grows to 2,500 ft^3 by age 40, when it is all harvested, the land planted, and another cycle begins. The second cycle has only one intermediate harvest at stand age 30. While we typically think of successive rotation cycles as being of the same length, there is no necessity to do this. The preferred rotations in published plans have in fact changed and shortened substantially since the start of American forestry in the early 1900s. It is true that, at the time of planning, analysis will suggest some particular rotation as "best" for the owner's objectives. But assumptions and objectives keep changing and the "best" rotation can change long before a given tree matures.

The total yield, including thinnings from this even-aged stand, is 3,700 ft^3 (Fig. 3-3) or an average growth rate of 3,700/40 = 92.5 ft^3 per acre per year for the first rotation and 4,400/50 or 88.0 ft^3 per acre per year for the second rotation.

Growth and yield of uneven-aged stands The interval between harvests in an uneven-aged stand is called the cutting cycle. Just as the management framework of an even-aged forest is built around the rotation, the management framework of an uneven-aged forest is built around the cutting cycle (Fig. 3-4). A cycle starts with a harvest that leaves a certain

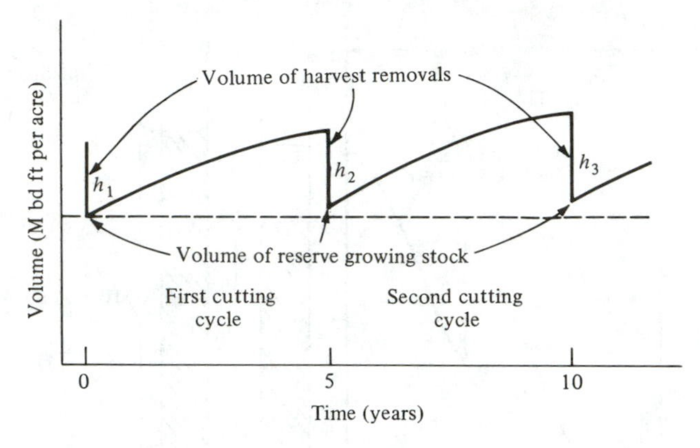

Figure 3-4 Lifeline of a 1-acre uneven-aged stand over two sequential cutting cycles.

volume of reserve growing stock. This volume grows for the number of years in the cycle, then a harvest cut removes the merchantable portion of this growth, plus or minus whatever adjustments are desired in the reserve growing stock to initiate the next cycle.

Growing stock volume can be measured at any time during the cycle: three choices are just after a cyclic cut, just before a cyclic cut, or midway through the cycle. The first gives the volume reserved for future growth. The second gives the volume available just before the next harvest. The third gives the midpoint volume of growing stock upon which growth is made. It approximates the stand's average growing stock level. Each measure has its usefulness but must be clearly differentiated from the others. Confusion results if growing stock volume figures are given without also stating what point in the cycle is involved.

Growth in a particular stand is rarely uniform from period to period because of changing growth conditions and treatments. Climatic fluctuations, disease, insects, and fire can all affect growth. In addition, the volume of reserve growing stock may be increased or decreased through harvest. Timber removed at each harvest entry combines all elements of the thinning and final harvest which are handled as separate operations at different times in even-aged management. In theory at least, the stand as a whole receives treatment during each harvest with the trees cut reflecting a mixture of reasons, including sanitation, salvage, release, final removal, and to establish regeneration.

The pattern of harvest and growth in an uneven-aged stand is best shown through an example (Fig. 3-4, Table 3-2). After an initial harvest the reserve growing stock contained 6,500 bd ft per acre at the beginning of the first 5-year cycle. A policy of adding some of the growth to the growing stock was adopted in order to increase future growth rates and

Table 3-2 Per acre growing stock, growth, and harvest in an uneven-aged stand on a 5-year cutting cycle (board feet)

Point in cycle	Volume before cut	Volume reserved after cut	Total periodic growth	Volume cut	Volume added to growing stock
Beginning of first cycle	8,770	6,500	2,690	2,270	—
Beginning of second cycle	9,190	6,820	3,770	2,370	320
Beginning of third cycle	10,590	7,290		3,300	470

harvest levels. In the first cutting cycle the stand grew from 6,500 to 9,190 for an increment of 2,690 bd ft. At the start of the second cycle 2,370 bd ft was allocated to harvest and 320 was added to growing stock, bringing it to 6,820 bd ft to initiate the second 5-year cycle. In the second cycle, total increment was 3,770 of which 3,300 was cut and 470 added to growing stock, bringing it to 7,290 bd ft. Reserve growing stock increased from 6,500 to 7,290, or 790 bd ft over the two cutting cycles, while the cut per cycle increased 1,030 from 2,270 to 3,300 bd ft. Total net increment for the 10 years equaled 2,370 + 3,300 + 790, or 6,460 bd ft for an average annual growth rate of 646 bd ft per acre per year. This example also illustrates that in uneven-aged management, each cutting cycle begins with a management decision of choosing how much of the previous cycle's inventory to harvest and how much to retain and invest to produce growth in the next cycle.

CHARACTERISTICS AND STRUCTURE OF EVEN-AGED STANDS

Normal yield tables describe how the best stands found in nature grow. The stands selected for these tables have plenty of well distributed trees, show little evidence of disease or other damage, and appear to have grown vigorously throughout their lives. While they are no longer constructed, as forestry shifts attention to the growth of managed stands, we use normal stands to begin our examination of even-aged stand structure and growth. With the exception of a few plantations, we have inadequate experience in the United States with managed stands older than 30 or 40 years.

Table 3–3 shows a normal yield table for average site Douglas-fir (McArdle et al., 1930). Net yield at any age is the volume of living trees. Gross yield at any age is the volume of living trees plus the volume of all mortality occurring up to that age. Periodic annual increment (PAI) is the average growth occurring over the time interval specified for the yield

Table 3-3 Yield per acre of pure, even-aged, fully stocked stands of Douglas-fir on site index 140 land total stand 1.5 inches DBH and larger

Stand age, years	Number of trees	Height, ft	Average tree DBH, in.	Basal area, ft^2	Net periodic annual increment, ft^3	Net yield, ft^3	Net mean annual increment, ft^3	Gross periodic annual increment, ft^3	Gross yield, ft^3	Gross mean annual increment, ft^3
					125			125		
20	1,460	37	3.4	92		1,250	63		1,250	63
					205			225		
30	865	64	5.5	140		3,300	110		3,500	117
					195			237		
40	585	84	7.4	177		5,250	131		5,870	147
					180			235		
50	430	98	9.3	204		7,050	141		8,220	164
					165			227		
60	337	109	11.1	226		8,700	145		10,490	175
					145			209		
70	274	119	12.8	244		10,150	145		12,580	180
					125			182		
80	232	127	14.3	259		11,350	142		14,400	180
					104			164		
90	205	134	15.6	272		12,390	138		16,040	178
					88			146		
100	184	140	16.9	283		13,270	133		17,500	175
					73			130		
110	166	145	18.0	292		14,000	127		18,800	171
					60			115		
120	152	149	19.1	301		14,600	122		19,950	166
					54			108		

Source: McArdle et al. (1930).

table (here it is 10 years). It can be calculated by dividing successive yield entries by the time between them. As an example, net periodic annual increment between ages 50 and 60 equals (8,700 − 7,050)/10, or 165 ft^3/yr. Mean annual increment (MAI) is the average growth of the stand up to the age in question. It is calculated by dividing yield at that age by the age itself. As an example, net mean annual increment at age 60 equals 8,700/60, or 145 ft^3/yr.

The yield table and several related figures adapted from the reference publication reveal a number of common features and characteristics of even-aged stand structure and growth. From Table 3-3 we see:

1. The number of trees decreases continuously due to mortality as the stand ages.
2. The height of dominant and codominant trees increases through the life of the stand, equaling the site index (140) at 100 years.
3. The diameter (DBH) of the average tree increases throughout the life of the stand as trees grow and the smaller trees within the stand suffer a disproportionately higher mortality rate.
4. The basal area increases throughout the life of the stand. In some species, such as ponderosa pine, the normal basal area rises to a plateau and remains fairly constant from then on.
5. Gross and net yields both rise throughout the life of the stand, with net yield gradually falling below gross yield as mortality accumulates. Eventually both gross and net yields will peak and decline as the stand begins to fall apart with the ailments of old age. Douglas-fir yields do not peak until some age past 200 years and redwood peaks even later.
6. Net yield reflects the amount of yield available for removal at any age while gross yield reflects the total amount produced on a particular site. Unless the site quality is changed through fertilization or genetically improved planting stock is used, gross yield from a normal yield table approximates the maximum potential volume obtainable from the site for the species in question. Difference between gross and net yields is the cumulative mortality and roughly estimates the maximum possible amount of additional growth that can be captured through thinning and mortality salvage. The mean annual increment and the periodic annual increment for both gross and net yields are plotted in Fig. 3-5, which shows the classic relationships between these curves.
7. Periodic growth rises, peaks, and declines as does the mean annual increment for both gross and net yields. Periodic growth (PAI) is above mean annual increment (MAI) when MAI is rising, equals it when mean annual increment peaks, and is below MAI when MAI

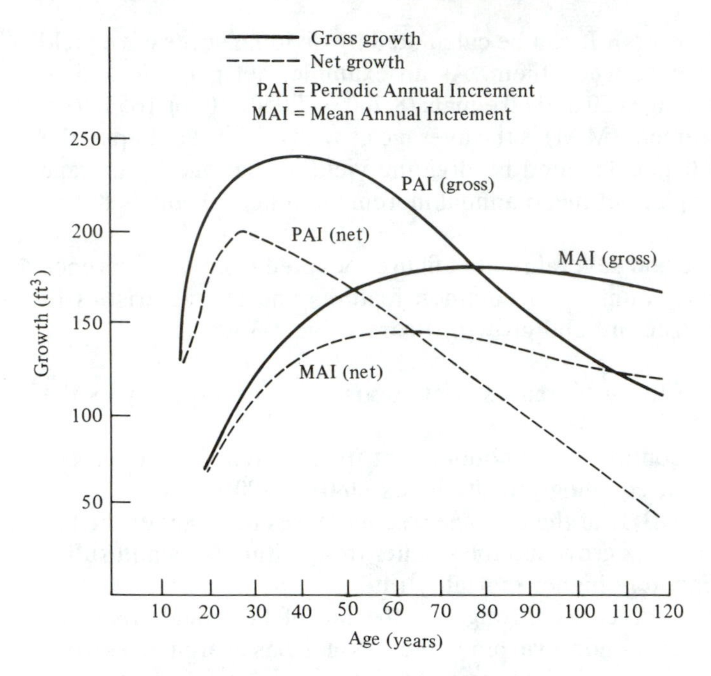

Figure 3-5 Relationship of mean annual increment to periodic annual increment for net and gross yields on fully stocked stands of Douglas-fir; site index 140.

declines. Such a relationship is required by the mathematics of the definitions.

8. Summing the area under the periodic annual increment curve to any age gives the yield to that age. In other words, yield is the integral of periodic growth.
9. Mortality is an increasing proportion of gross increment as stands age. Hence the mean annual increment culminates at a younger age when calculated on a net yield basis. Here we see that culmination is at 65 years for net yields and at 75 years for gross yields.
10. The structure of an even-aged stand in terms of the number of stems per diameter class changes markedly as the stand ages (Fig. 3-6). When young, the stand has a great many trees within a limited range of small diameters in a bell-shaped distribution. As the stand gets older, the tree count drops and the bell flattens as the trees grow at variable rates, expressing their condition, genetic makeup, and position in the stand.
11. The influence of establishing minimum tree size standards to define merchantability on our estimates of growth and yield is shown in Fig. 3-7. The minimum diameter breast height (DBH) of counted trees is

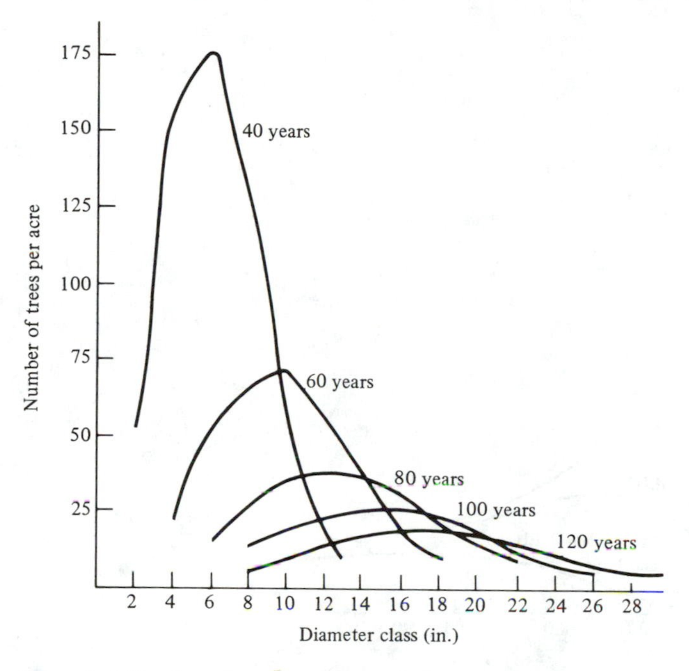

Figure 3-6 Number of trees per diameter class as a function of age for fully stocked natural stands of Douglas-fir; site index 140.

moved from 1 in. to 7 in. to 12 in. to make the comparison. The number of trees counted converges with age for the different utilization standards. We note that the 7-in. and 12-in. counts increase for some time before starting to fall off.

12. The higher the minimum size standard, the lower is the volume considered merchantable at each age, especially at the younger ages. The larger the minimum size, the older is the stand at culmination of mean annual increment.

Site and stand structure Site quality describes the inherent capability of the land to grow trees and integrates the soil, moisture, nutrient, geographic, and other important factors affecting tree growth. Because of its effect on growth, we spend some time in Chap. 4 discussing the ways site quality is measured and expressed as a numerical site index. Site quality also has a profound impact on the structure of even- and uneven-aged stands. The nature of this impact is well illustrated by the data from McArdle's Douglas-fir yield table shown in Fig. 3-8. We see two to fourfold differences in basal area, tree count, average diameter, and volume at a given age.

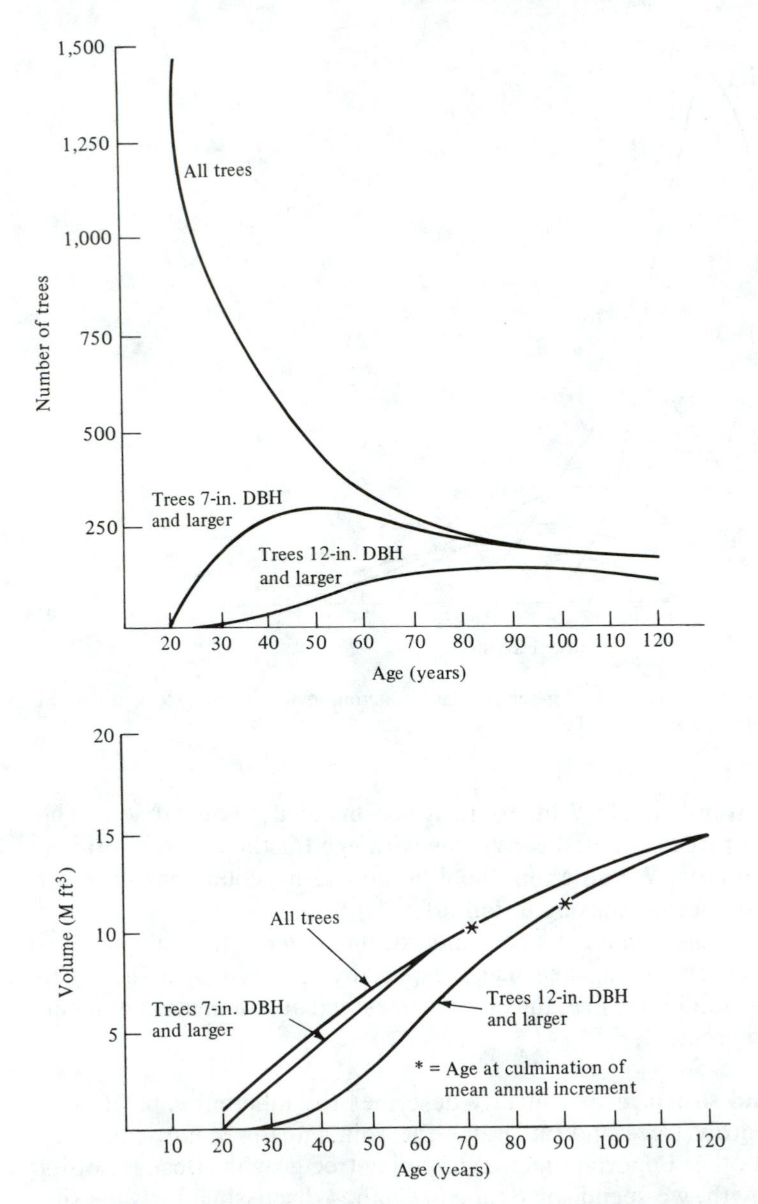

Figure 3-7 Effect of merchantability standards on number of trees and volume per acre for fully stocked natural stands of Douglas-fir; site index 140.

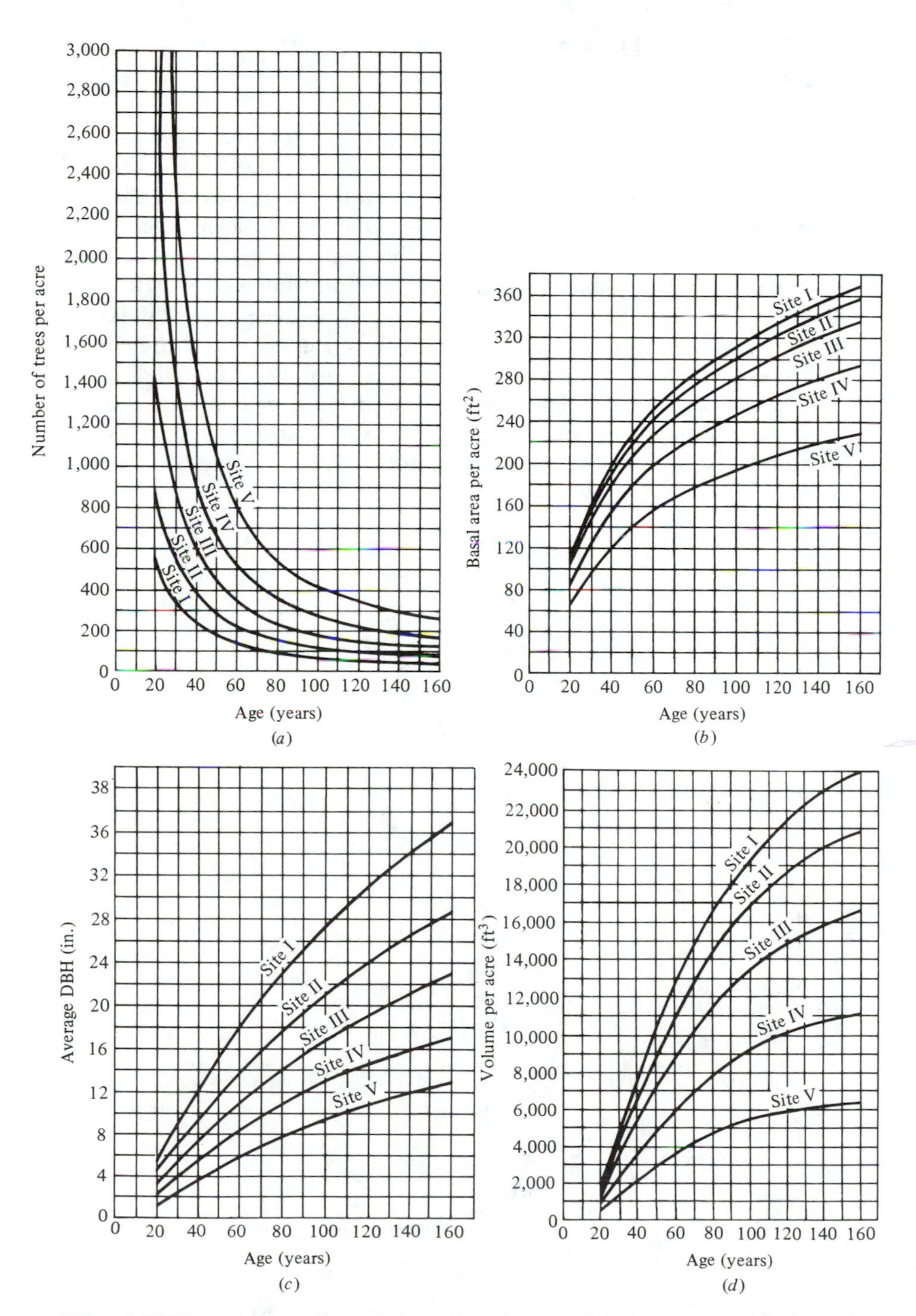

Figure 3-8 Effect of site quality on (*a*) number of trees, (*b*) basal area, (*c*) average stand diameter, and (*d*) volume for fully stocked natural stands of Douglas-fir.

CHARACTERISTICS AND STRUCTURE OF UNEVEN-AGED STANDS

The characteristic bell-shaped distribution curve for even-aged natural stands of Douglas-fir was shown in Fig. 3-6. As the stand ages, the bell shape is retained, but the total number of trees decreases and the bell flattens and shifts to the right. By contrast, the uneven-aged stand has a reverse J-shaped diameter distribution which may or may not shift with time. A generalized comparison of even- and uneven-aged structures is shown in Fig. 3-9. Actual stands, particularly uneven-aged stands, diverge considerably from these theoretical concepts. Often they have marked deficits in some diameter classes and surpluses in others.

Our operational conceptualization of uneven-aged management is dominated by the diameter size structure of a stand rather than its age structure. Management operates by periodically harvesting some of the

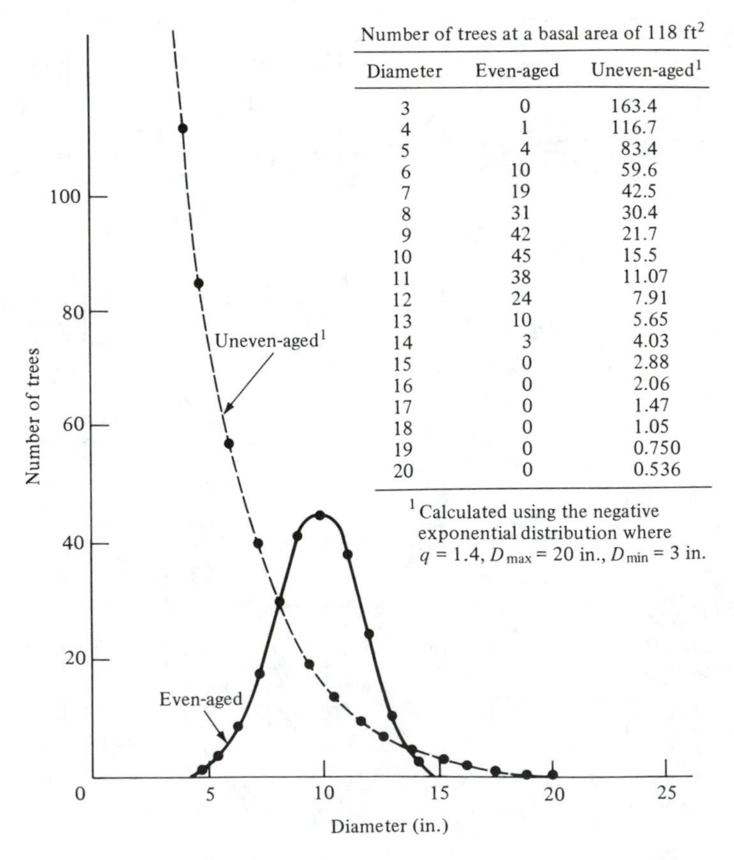

Number of trees at a basal area of 118 ft^2

Diameter	Even-aged	Uneven-aged[1]
3	0	163.4
4	1	116.7
5	4	83.4
6	10	59.6
7	19	42.5
8	31	30.4
9	42	21.7
10	45	15.5
11	38	11.07
12	24	7.91
13	10	5.65
14	3	4.03
15	0	2.88
16	0	2.06
17	0	1.47
18	0	1.05
19	0	0.750
20	0	0.536

[1] Calculated using the negative exponential distribution where $q = 1.4$, $D_{max} = 20$ in., $D_{min} = 3$ in.

Figure 3-9 Even- and uneven-aged structures having the same basal area of 118 ft^2.

trees on each acre, and the key decision variable is how many trees of particular sizes and species to remove. This requires a good knowledge of the stand structure and effective methods to predict growth by diameter class.

The larger trees tend to be older than the smaller ones, and every tree obviously has a unique age. The uneven-aged stand could be conceptualized as a summation of several different even-aged stands growing on the same parcel of land at the same time. Such an aggregation is illustrated using Schnur's (1937) data for even-aged upland oaks (Fig. 3-10). When all the bell curves for each 10-year age class are added, the characteristic uneven-aged J distribution emerges, as shown by the heavy line. An actual uneven-aged stand would have the total number of trees proportionately reduced. The age distribution by diameter classes can also be extracted, and it would show older trees in each successively larger size class.

Conceptually we assume regeneration in uneven-aged stands to be

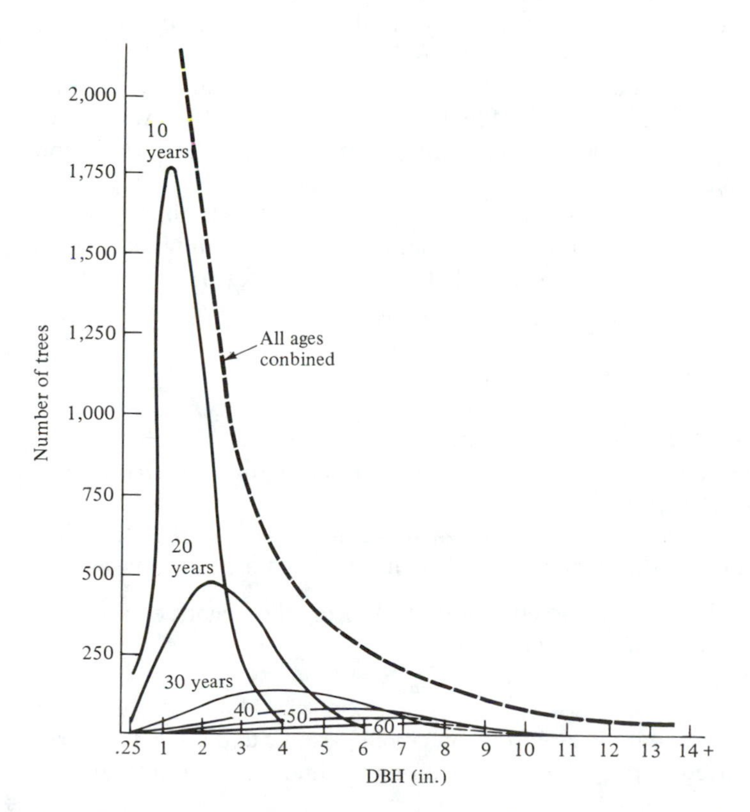

Figure 3-10 Upland oak stand structures per age class (ages 10 to 60) and for all age classes combined. *(From Schnur, 1937.)*

continuous, with new trees starting under the canopy of the residual stand at a steady rate. However, in reality regeneration is patchy and periodic, depending on seed sources, the creation of openings, and scarification of the soil during periodic harvests. As a result the uneven-aged stands often have most of the trees falling into three to five identifiable age groups.

Negative Exponential Distribution

According to Meyer (1952), deLiocourt (1898) published the first numerical studies of growing stock distribution in uneven-aged forests. Using data from selection forests in his homeland, deLiocourt observed that the ratio of trees in successive diameter classes tends to remain constant through the range of diameter classes represented in a forest. In other words, he found that the number of trees in successive diameter classes, going from largest to smallest, forms the geometric series $m, mq, mq^2, mq^3, \ldots$, where q is the diminution ratio coefficient of the series and m is the number of trees in the largest diameter considered. This idea and the diminution coefficient have been a popular conceptual device ever since for describing uneven-aged stand structures.

To illustrate, consider Meyer's data on a 21-acre beech-birch-maple-hemlock forest in Pennsylvania (Meyer and Stevenson, 1943) which are plotted in Fig. 3-11. The derivation of the parameter q for this distribution proceeds as follows.

A geometric series $m, mq, mq^2, \ldots$ plots as a straight line on semilog paper. The linear equation of this line, describing the number of trees as related to the diameter class, is of the logarithmic form

$$\log N = \log k - aD \log e \qquad (3\text{-}1)$$

where N = number of trees per diameter class
D = DBH class
e = base of natural logarithm
k = number of trees at smallest DBH recognized—an index of relative density
a = slope of line—the rate at which the number of trees logarithmically diminishes between successive diameter classes

The antilog of the logarithmic equation (3-1) forms the negative exponential equation of the reverse J shape:

$$N = ke^{-aD} \qquad (3\text{-}2)$$

When Meyer's data are plotted on semilog paper and a least-squares regression is fitted (Fig. 3-11b), the resulting log linear equation (Husch et al., 1982) is

$$\log N = 1.72242 - 0.05563D$$

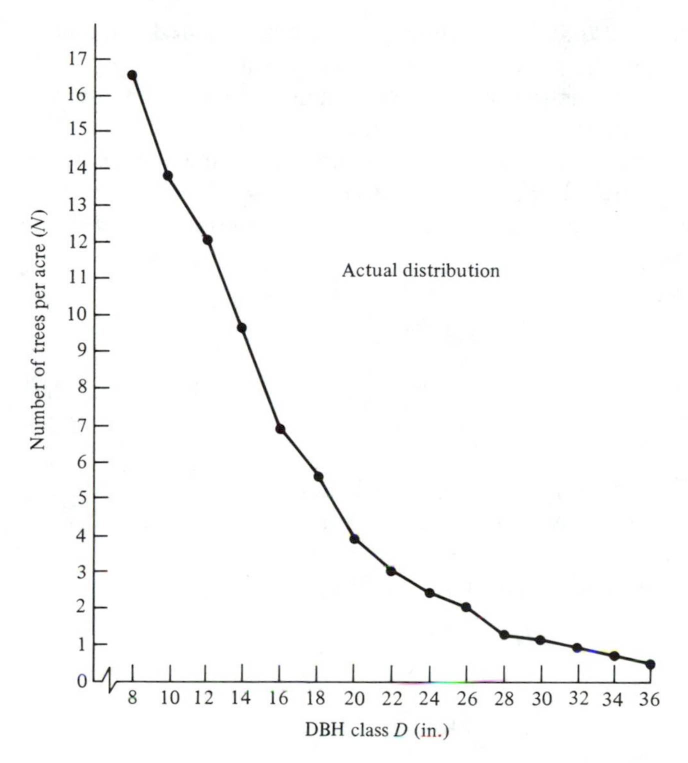

Figure 3-11 Actual diameter distributions for an uneven-aged 21.4-acre virgin stand of beech-birch-maple-hemlock in Pennsylvania. *(Adapted from Meyer and Stevenson, 1943.)*

Using the logarithmic form of Eq. (3-1) and solving for the coefficients k and a, we obtain

$$\log k = 1.72242$$

$$k = \text{antilog}(1.72242)$$

$$k = 52.77$$

$$-a \log e = -0.05563$$

$$a = \frac{0.05563}{\log e} = \frac{0.05563}{0.43429} = 0.1281$$

Substituting the values of k and a into the exponential equation (3-2), we have

$$N = 52.77e^{-0.1281D}$$

The parameter a determines the rate by which trees diminish in successive diameter classes, and k indicates the relative density of the stand.

Meyer's work showed that high values of k are associated with high values for a: the larger the number of trees in the smallest diameter class, the more rapid is the reduction in trees with each succeeding diameter.

The diminution quotient q can be estimated from the exponential equation given above. Let N_{20} and N_{22} represent the number of trees per acre in the 20- and 22-in. DBH classes. By definition, $q = N_{20}/N_{22}$. Substituting the estimate of the exponential equation for the number of trees in each diameter class,

$$
\begin{aligned}
q &= \frac{ke^{-a*20}}{ke^{-a*22}} \\
&= e^{-a*20+a*22} \\
&= e^{2a} \qquad (3\text{-}3)
\end{aligned}
$$

Notice that q is a function of class width. If 1-in. classes were used, the q value derived between, say, the 20- and 21-in. classes would be $q = e^{-a*20+a*21}$ or $q = e^{a}$.

Using Meyer's actual data in the example, we get

$$q = \frac{N_{20}}{N_{22}} = \frac{3.04}{2.43} = 1.25$$

while q, as estimated from Eq. (3-3), equals

$$e^{2a} = e^{2(0.1281)} = 1.29$$

Comparisons of q values should be made across a common diameter class width. Also, as pointed out by Husch et al. (1982), k varies with both stand area and diameter class width, while a varies with neither. Therefore, comparisons of k values should be made across common stand areas and class widths.

The q ratio and the related exponential equation are popular for two reasons. First, Meyer (1952) showed that the diameter distribution of natural undisturbed uneven-aged stands tends toward a constant ratio in terms of the number of trees between successive diameter classes. Meyer also reported q values of 1.2 to 2.0 for upland oak in Pennsylvania; q values reported for other cover types, such as ponderosa pine, northern hardwoods, Engelmann spruce–subalpine fir, and loblolly pine–shortleaf pine, tend to fall within these limits (Daniel et al., 1979). Low values of q define a flat curve for the frequency distribution of diameters: a stand with a relatively high proportion of trees in the larger diameter classes.

Second, the q ratio has long been used as a way to conceptualize and describe *desirable* diameter distributions for uneven-aged stands. Meyer (1953) defined a balanced uneven-aged forest as "one in which current growth can be removed periodically while maintaining the diameter distribution and initial volume of the forest.'' He proposed describing this ''normal'' uneven-aged stand and forest using the q ratio.

Defining a Desired Uneven-Aged Structure

The practical interpretation of Meyer's remarks is that we can construct a ''balanced'' diameter distribution for a stand given any four of the five stand parameters:

q = desired diminution coefficient
B = desired stand density, usually expressed as basal area
$D_{\min}$ = smallest diameter class counted
$D_{\max}$ = largest planned diameter class
$N_{\max}$ = number of stems in largest class

Using these parameters, the stand structure is described by the set of $N - 1$ equations for all pairs of adjacent diameter classes,

$$N_{i-1} = qN_i \qquad \text{for } i = D_{\min}, \ldots, D_{\max} \tag{3-4}$$

and by the stocking constraint,

$$\sum_{i=\min}^{\max} b_i N_i = B \tag{3-5}$$

where b_i is the contribution of volume, basal area, or other density measure per tree of size class i, and B is the stand density defined in units of volume, basal area, or other measures of stand density.

These equations can be solved for any one of the parameters if the others are given. When B, q, $D_{\min}$, and $D_{\max}$ are specified, we often want to calculate the number of trees in the largest diameter class $N_{\max}$ to set the distribution. A handy equation for this, which is valid regardless of the units used to define stand density B, is

$$N_{\max} = \frac{B}{\sum_{i=\min}^{\max} b_i q^{(D_{\max} - D_i)/w}} \tag{3-6}$$

where D_i = midpoint of diameter class i
w = diameter class width, in inches

and B, b_i, and $N_{\max}$ are as defined above.

The denominator of this expression is the density per acre of a stand with exactly one tree in the largest diameter class. Once $N_{\max}$ is calculated, the number of stems in each successive smaller class can be calculated using the diminution quotient and Eq. (3-4).

Sustainability of Structure

Distributions arrived at by these procedures show lots of trees in the smaller diameter classes. These trees provide ingrowth into the merchantable size classes that form the harvest pool. Even though we may not cut in these smaller sizes, it is important to know what is happening there to evaluate progress toward the stand structure goal. Critical is the assumption that these trees are in fact provided in adequate numbers by regeneration and that dynamically they continuously flow into the smallest size class we measure. A debatable issue is how many small trees are needed to assure this. If not enough arrive or if the species that do arrive are unacceptable, then additional regeneration will have to be provided by artificial means to attain the desired structure. On the other hand, if too many arrive and do not die naturally, their killing requires an investment in the stand. If this treatment is not done, the trees could create a hump in the diameter distribution with the result that more of the growing space may be devoted to small trees than desired. Loss in growth on the desired trees must be weighed against removal cost for the undesired. Using two q factors (or some other special distribution) has also been suggested as a compromise. Under his scheme, the larger q factor is applied to commercial size trees and the smaller q to the precommercial sizes.

In addition to the q ratio, a number of other measures of diameter distribution have been suggested: (1) a linearly increasing ratio among trees in adjacent diameter classes (Leak, 1964), (2) the Weibull distribution (Bailey and Dell, 1973), and (3) empirical forms where the relationships of numbers to size are directly built from field measurements.

Adams and Ek (1974) have moved entirely away from formula-specified distributions. They generalized the requirements for sustainable distributions, that is, structures in which current growth can be removed periodically while maintaining the diameter distribution and initial volume of the forest. As they state, "maintenance of a given distribution from one cutting cycle to the next clearly requires that just prior to cutting, the number of trees in each diameter class be at least as large as the number at the start of the cycle," or

$$N_{i,t+1} \geq N_{i,t} \qquad \text{for all initial diameter classes } i \tag{3-7}$$

Any (initial) distribution at time t satisfying this condition can be sustained indefinitely, barring catastrophes, by harvesting the excess of $N_{i,t+1}$ over $N_{i,t}$.

To illustrate this idea of sustainability, consider Table 3-4, which gives the initial distribution of trees by diameter class and three possible distributions that might appear after 20 years of growth, depending on the species, site quality, and agents of mortality.

Table 3-4 Initial and possible future distributions of trees by diameter class (number of trees per acre)

Diameter class, in.	Initial distribution	Distributions after 20 years of growth		
		A	B	C
4	70	120	10*	10*
6	50	70	30*	40*
8	40	60	50	30*
10†	30	40	60	20*
12	20	25	40	15*
14	8	10	15	15
16	4	6	10	10
18	1	2	2	2
20	0	2	1	1
22	0	1	0	1

* Fewer stems than initial distribution.
† Minimum merchantable size is 10 in.

Distribution A indicates that the initial distribution is sustainable because there is ample ingrowth in the smallest size classes and all merchantable diameter classes show an increase. It is possible that some precommercial thinning may be needed in the submerchantable classes to perpetuate this stand structure.

Distribution B would suggest that the initial distribution is sustainable if only the merchantable classes are examined. However, ingrowth to the 4- and 6-in. classes is low, and in one more growth cycle this deficiency will probably move to the merchantable size classes. Distribution B might represent the development of an even-aged stand in the understory and the change of the distribution to the bell shape in subsequent cycles.

Distribution C shows a growth pattern which indicates that the initial distribution is not at all sustainable. Mortality and slow growth rates, for whatever reason, are the problem.

QUESTIONS

3-1 The growth of a forest was estimated using the equation

$$\text{Growth} = V_2 - V_1 + C$$

where V_2 = volume at time 2
V_1 = volume at time 1
C = volume harvested over growth period

Is this an estimate of gross or net growth? Explain.

3-2 An inventory plot was measured at 10-year intervals and the following measurements taken:

Tree	Tree volume, ft³		Comment
	First inventory	Second inventory	
1	21	41	—
2	42	53	—
3	27	38	—
4	—	19	Ingrowth
5	97	—	Tree died
6	86	—	Tree cut
7	—	24	Ingrowth

Calculate the following for the plot:

(*a*) Gross increment including ingrowth
(*b*) Gross increment of initial volume
(*c*) Net increment including ingrowth
(*d*) Net increment of initial volume
(*e*) Net increase in growing stock

3-3 Fill in the missing entries in this even-aged growth analysis table (units are cubic feet).

Stand age	Periodic annual net growth	Net yield	Mean annual net growth	Periodic annual mortality	Periodic annual gross growth	Gross yield	Mean annual gross growth
20		____	30			____	35
	____			____	____		
30		1050	____			1350	____
	55			____	80		
40		____	____			____	____
	____			22.25	37.5		
50		____	____			____	____

Important: Periodic changes are on the intervals between ages.

3-4 Fill in the missing entries of this uneven-aged growth analysis table. Units are cubic feet.

Year	Growing stock after harvest	Net periodic annual growth	Periodic volume harvested	Periodic mortality	Gross periodic annual growth
0	2500				
		200		—	220
10	—		0		
		250		300	—
20	1500		—		
		300		100	—
30	—		0		
		275		100	—
40	—		2000		

3-5 Give a convincing argument why periodic annual increment *always* equals mean annual increment at the age when the mean annual increment is at its maximum value—or why it does not.

3-6 Give some reasons why there would be fewer trees per acre at a given age for better quality sites, as shown in Fig. 3-8.

3-7 A natural stand is sampled and a log linear equation is fit to the stand table data to relate the number of stems per diameter class N_i to the midpoint of the 2-in. diameter class D_i in inches. The resulting equation is

$$\log N_i = 1.9 - 0.07 D_i$$

(*a*) Find the coefficients of the negative exponential distribution of Eq. (3-2).
(*b*) Find the value of q for this distribution.

3-8 It was desired to precommercially thin a dense, young, uneven-aged hardwood stand leaving a residual stand having a negative exponential distribution with $q = 1.4$, a maximum tree size of 14 in., a minimum counted size of 2 in., and a growing stock of 200 ft^2 of basal area. Calculate this residual stand distribution of trees by diameter class and show that the basal area adds up to 200 ft^2. *Hint:* Use Eq. (3-6).

DBH	Basal area per tree
2	0.02
4	0.08
6	0.19
8	0.34
10	0.54
12	0.78
14	1.07

3-9 Review the definition of sustainability for uneven-aged structures [Eq. (3-7)]. Discuss the biological assumption about how trees regenerate and grow that are required to support a conclusion that any specified residual stand structure, including negative exponential structures, is sustainable.

REFERENCES

Adams, D. M., and A. R. Ek, 1974: Optimizing the Management of Uneven-Aged Forest Stands. *Can. J. Forest Res.*, **4**:274–287.

Avery, T. E., and H. E. Burkhart, 1983: "Forest Measurements," 3d ed., McGraw-Hill, New York.

Bailey, R. L., and T. R. Dell, 1973: Quantifying Diameter Distributions with the Weibull Function, *Forest Sci.*, **19**:97–104.

Beers, T. W., 1962: Components of Forest Growth, *J. Forestry*, **60**:245–248.

Daniel, T. W., J. A. Helms, and F. S. Baker, 1979: "Principles of Silviculture," 2d ed., McGraw-Hill, New York.

deLiocourt, F., 1898: De'lam. Reference in Meyer (1952).

Husch, B. C., C. I. Miller, and T. W. Beers, 1982: "Forest Mensuration," 3d ed., Wiley, New York.

Leak, W. B., 1964: An Expression of Diameter Distribution for Unbalanced, Uneven-Aged Stands and Forests, *Forest Sci.*, **10**:39–50.

———, and S. M. Filip, 1977: Thirty-eight Years of Group Selection in New England Northern Hardwoods, *J. Forestry*, **75**:641–643.

McArdle, R. E., W. H. Meyer, and D. Bruce, 1930: "The Yield of Douglas-Fir in the Pacific Northwest" (rev. October 1949), U.S. Department of Agriculture Technical Bulletin 201.

Meyer, H. A., 1952: Structure, Growth, and Drain in Balanced Uneven-Aged Forests, *J. Forestry*, **50**:85–92.

———, 1953: "Forest Mensuration," Penn Valley Publishers, State College, Pa.

———, and D. D. Stevenson, 1943: The Structure and Growth of Virgin Beech–Birch–Maple–Hemlock Forests in Northern Pennsylvania, *J. Agric. Res.*, **67**:465–484.

———, A. B. Recknagle, D. D. Stevenson, and R. A. Bartoo, 1961: "Forest Management," 2d ed., Ronald Press, New York.

Schnur, G. L., 1937: "Yield, Stand and Volume Tables for Even-Aged Upland Oak Forests," U.S. Department of Agriculture Technical Bulletin 560.

CHAPTER

FOUR

TIMBER PRODUCTION: SITE AND DENSITY

Site quality tells how much timber a forest potentially can produce. Stand density measures how thickly trees grow. Site and density together tell how much timber it will produce, and what the wood quality will be. Stand density is the major factor that a forester can manipulate in developing a stand. By changing how thickly trees grow—their number and spacing—a manager can influence stem quality, diameter growth, stand volume growth, and even regeneration of the next crop.

SITE QUALITY IN TIMBER MANAGEMENT

Timber productivity of forestland is defined in terms of the maximum amount of volume that the land can produce over a given time; to put it more bluntly—how good is the land, and how much timber can it grow? Site quality is an index number related to this timber productivity. To estimate site and thus timber productivity, foresters observe or measure some attributes of the land or the vegetation currently growing on it.

Measurement of Site Quality

Nothing would seem more logical than measuring site quality by the volume of the desired materials produced, thus expressing the integrated net effect of all site factors in terms of the product itself. In practice, however, the life history of a stand also affects the volume of timber on an area at any time. Density when the stand originated, character of past cutting, disease, and insects can reduce the standing volume that would otherwise occur. Therefore, site quality is rarely measured directly through standing volume.

Rather, a number of indirect methods have been developed to measure timber site productivity. These methods attempt to select a few easily measured properties of the vegetation or the land, which will represent all factors important to the growth of a particular species on a given site. Following Jones (1969), three indirect approaches for estimating the site quality are discussed here: site index, vegetation, and environment.

Site index Of all the indirect measures that have been investigated, the rate of tree height growth appears the most practical, consistent, and useful indicator of timber site quality. It is not a perfect measure by any means, but it remains the standard to which other measures, such as soil properties, are compared. The height growth of dominant free growing trees in the upper canopy is sensitive to differences in site quality, strongly correlated with volume growth, and weakly correlated with density and species composition. Diameter growth, in contrast, is strongly correlated with density.

Standard practice in the United States is to define site index in terms of the total height in feet of the largest, full-crowned trees in a stand, which are the strongest competitors for light, moisture, and nutrients. Ideally the site trees should have been in the strongly competitive growing position over their entire lives.

Numerically, the site index is defined as the total height at specified ages, usually 25, 50, or 100 years. The base age for stands maturing or primarily utilized at younger ages is 25 years. Southern pines, for example, commercially mature at around 25 to 40 years, and a 25-year base is often used.

The site is then expressed as a site index number. Site index 60 on a 50-year base means that dominant position trees will average 60 ft in total height at 50 years. Similarly, site index 140 on a 100-year base means that trees will be 140 ft tall at 100 years. Site index values on 50- or 100-year bases are commonly grouped in 5-ft, 10-ft, 20-ft, or broader classes. When expressed in broader classes, as 30 or 40 ft, they are sometimes denoted by Roman numerals. Site I is the best site, site II the next best, and so on.

Site index curves showing the average height of full crown competitive trees at various ages have been prepared for most species by region or more specific localities. Curves based on height for typical eastern and western species are given in Fig. 4-1. Note that in each case the curves pass through the index height at the base age.

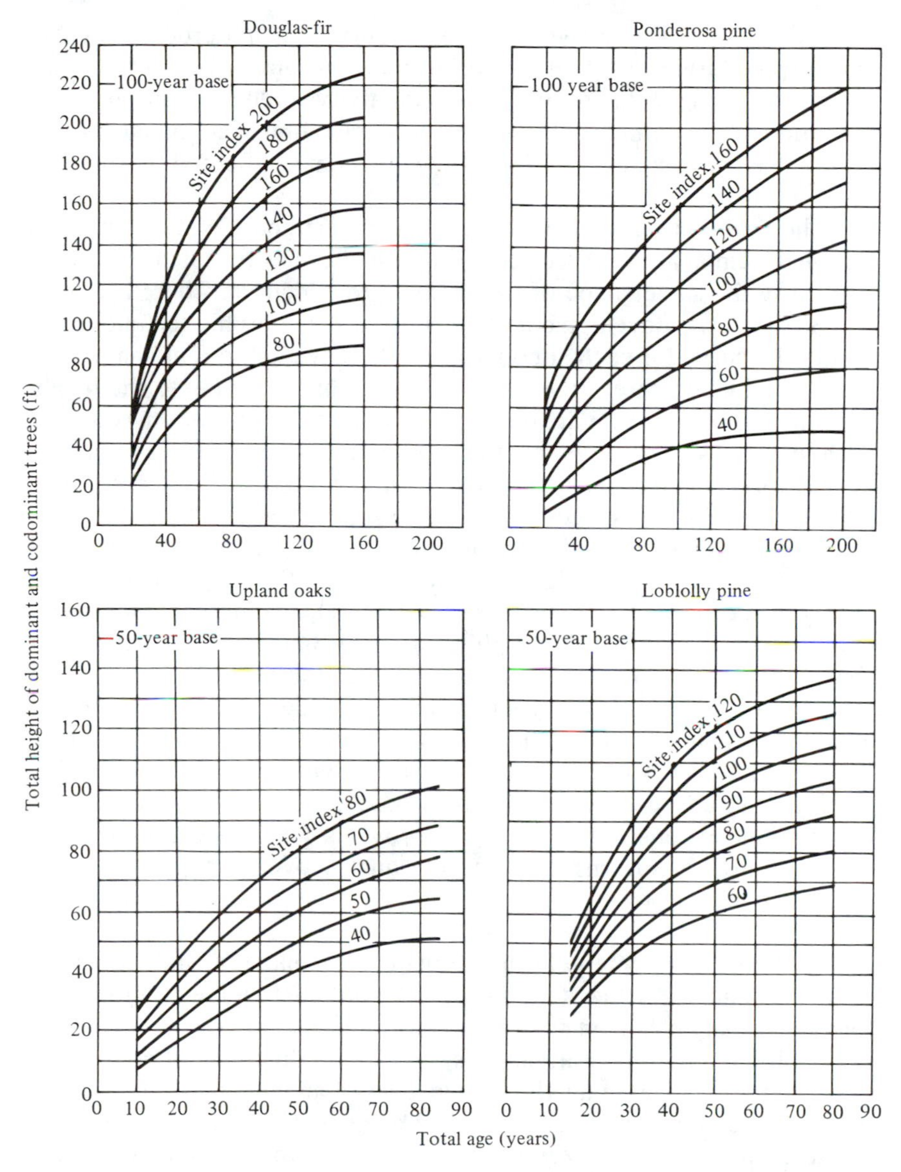

Figure 4-1 Site index curves for four important timber species.

These curves came about as a result of "normal" yield tables. Trees from stands of a particular species or species group having a full canopy of normally developed crowns were measured for age, diameter, and height. Then curves were developed by fitting an average height-over-age guide curve to these data and constructing a series of higher or lower curves with the same shape as the guide curve. Such a process is called anamorphic curving.* By relating these curves to similar volume-over-age curves, maximum net natural stand productivity could be determined as a function of the site index. Data thus determined provide managers with knowledge about stand growth and development as affected by site quality.

The traditional anamorphic guide curve method has its limitations. Detailed stem analysis of trees on different sites has revealed that not all sites have the same shape of tree height-growth curve. Therefore many recent site curve development studies use a polymorphic technique that allows the curve for each site index level to have a different shape. This technique follows three steps: (1) A curve is fit separately to the height-growth pattern of each individual tree. (2) The fitted curve is used to give each tree a site index value. (3) A characteristic, multiple-parameter curve form is fitted to each site index value of interest. See Clutter et al. (1983) for a thorough discussion of these procedures.

An interesting comparison of anamorphic and polymorphic curves for the same species is provided by superimposing Dunning and Reineke's (1933) anamorphic site curves for site indices 40, 80, and 110 at a base age of 50 years on the new polymorphic curves of Biging and Wensel (1984)† (Fig. 4-2). We see that in this instance the polymorphic curves show slower growth at both the younger and the older ages for the higher-quality site. Numerically the height differences are on the order of 5 to 10 ft at age 100. Such a difference, when translated to volume, could be significant to a forest owner. The polymorphic approach, although more complex and expensive to implement, provides a more accurate portrayal of height-growth patterns and thus site quality over a range of different sites than anamorphic curve sets.

Choosing site trees Site determinations can be made for individual species in a stand or for the stand as a whole, although in the latter instance they are usually based on one or more key species. In younger even-aged stands, it often is comparatively easy to identify dominant or codominant trees and safe to assume they have been dominant all their lives. In uneven-aged stands and older, natural even-aged stands, the process is not so simple; ages and sizes are mixed and the crown canopy is uneven.

Operationally, foresters examine trees in a stand visually and select several with full crowns in the upper canopy that appear vigorous and

* Specifically, site curves are anamorphic if there is constant relative height growth for all sites at a given age, that is, $[(dh/dt) \div h] = k$, a constant for all sites at each different age. If this relationship does not hold, the curves are polymorphic.

† Appropriate adjustments were made for the difference in base age concepts since Dunning and Reineke used total age and Biging and Wensel used breast height age.

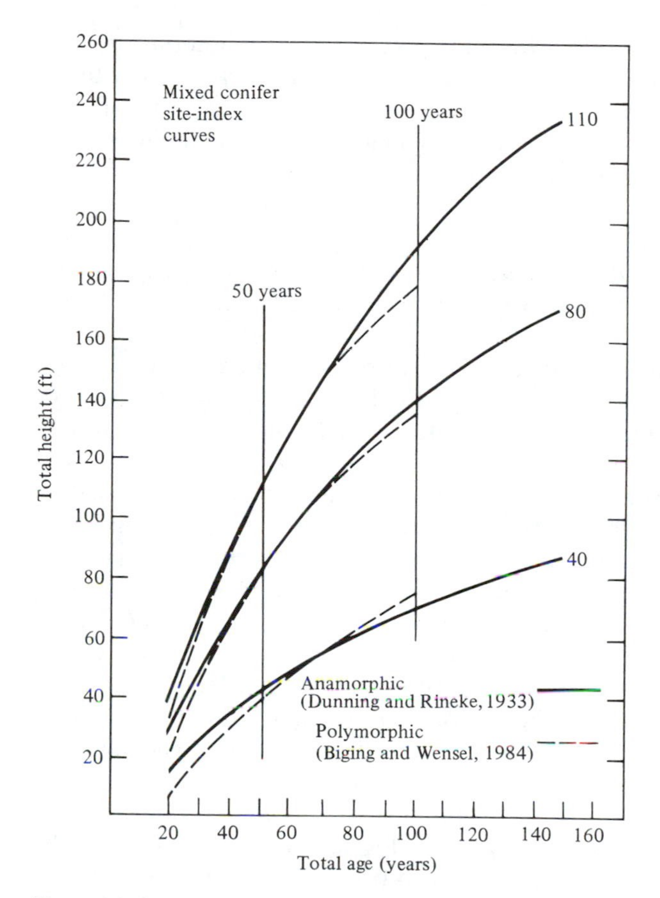

Figure 4-2 Comparison of anamorphic and polymorphic fitted site index curves for California mixed conifers. *(From Biging and Wensel, 1984.)*

strongly competitive for light, moisture, and nutrients. The height and age of each tree are measured, and a site index value is determined for each tree from appropriate site curves. The site index values of the sampled trees are then averaged to estimate the value of the plot or stand. This approach assumes that the sample trees have always been vigorous, competitive, dominant trees. If early suppression distorts the height-growth curve, the method can prove unreliable. Stage (1963) suggests classifying the site by using growth for 10 years after the dominant tree has reached a certain height, such as 55 ft. While it is probably true that ocularly estimating or measuring 10 years of growth in the upper part of a tree could drive a forester to distraction, the approach has not been tried broadly enough to rate its proficiency.

Still, the problem remains that using total height over age can underestimate the site index when trees have been suppressed when young.

This can happen when trees grow under uneven-aged conditions or when a tolerant species grows under even-aged conditions or faces early brush competition. Increment borings to assess growth near the pith can help alert you to the difficulty. Trees showing evidence of slow or suppressed growth when young should not be selected as site trees if better trees are available.

Interpreting and using the site index Some important points to remember in interpreting and using site index data are the following.

1. When making site index comparisons, be sure to identify on what age each index is based and whether or not it is DBH age or total age. For example, site index curves exist for Douglas-fir based on both 50 years and 100 years.
2. A site index based on one age cannot be converted to one based on another age through any simple numerical relationship. Site index 60 on a 50-year base does not mean the same thing as site index 120 on a 100-year base unless the trees of the species involved grow exactly twice as tall at 100 years as at 50 years. Site indices can be converted from one base age to another only by reference to the site curves themselves, which show actual tree heights at different ages. For example, dominant Douglas-fir trees of site index 140 on a 100-year base are 98 ft tall at age 50 (Fig. 4-1). Thus site index 140 at 100 years corresponds to site index 98 on a 50-year base.
3. Age is defined at breast height for many contemporary site curves. This means the tree is chronologically several years older by the number of years it took from seed to breast height. Some older site curves, such as McArdle's (1930) Douglas-fir normal yield study, used total age to define the site curves. Thus an adjustment is needed if comparisons are made with site indices at breast height age. To illustrate, consider the Douglas-fir site index 140 curve in Fig. 4-1, which was based on a total age of 100 years. If it took 7 years to reach breast height, this same height of 140 ft is achieved by a tree 93 years old at breast height. We should expect this tree to be about 145 ft high when it reaches a breast height age of 100 years (total age of 107 years). Thus a site index of 140 at breast height age of 100 years is not as good as a site index of 140 at a total age of 100 years.
4. Similar site indices for different species do not necessarily mean similar site productivity in either an absolute or a relative sense (Fig. 4-3). At 60 years, site index 140 on a 100-year base for Douglas-fir shows a fully stocked stand having 6,000 ft^3 per acre, while the same site index for ponderosa pine shows a fully stocked stand having 11,000 ft^3 per acre. Ponderosa pine is a stubbier tree than Douglas-fir; at the same age and height, a fully stocked stand of pine will show more volume

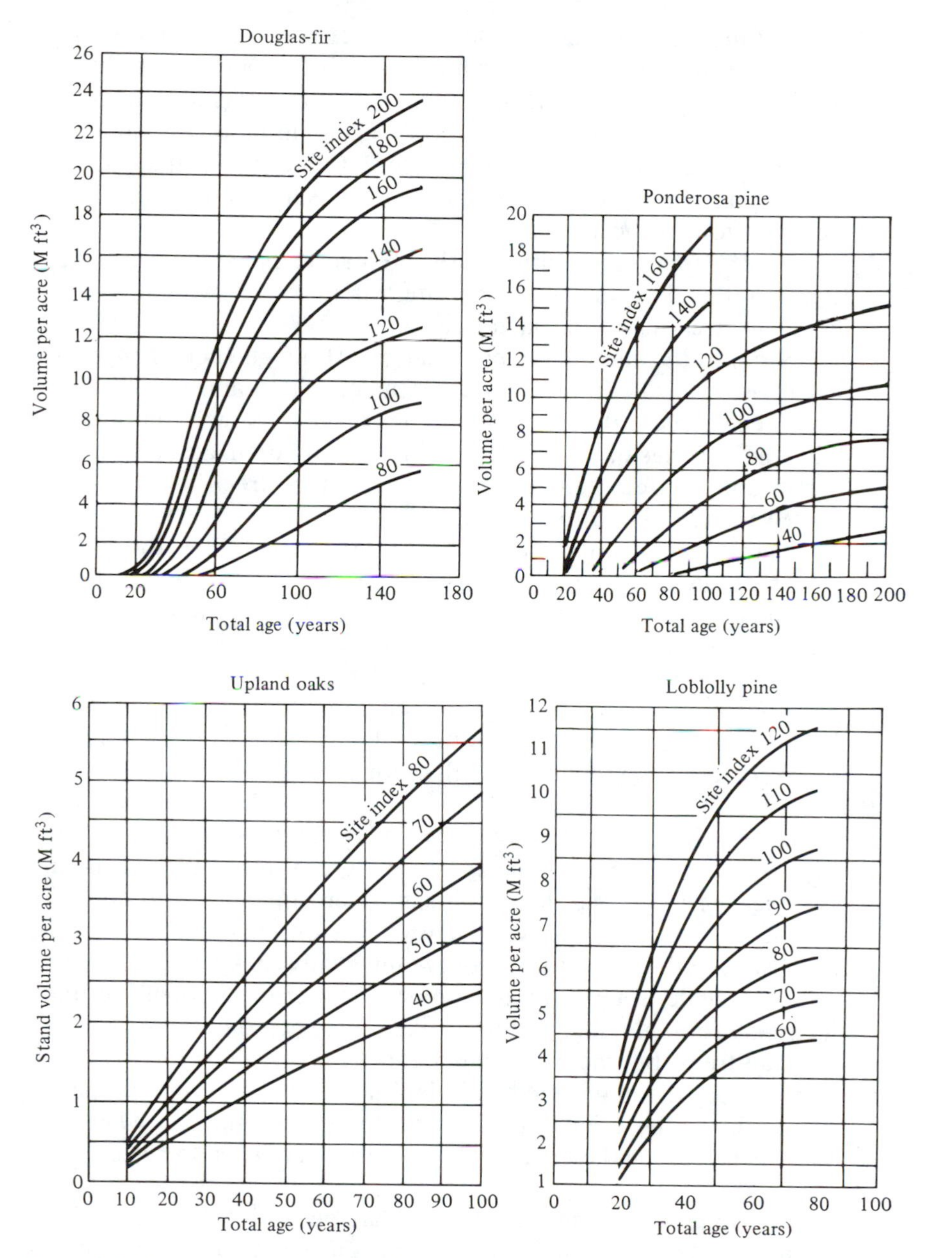

Figure 4-3 Merchantable volume yield as related to site index for four important timber species. (*a*) Douglas-fir. (*b*) Ponderosa pine. (*c*) Upland oak. (*d*) Loblolly pine.

than a fully stocked stand of Douglas-fir. However, site index 140 is very good for ponderosa pine while only fair for Douglas-fir. Pine, in general, occupies lower site indices than does Douglas-fir. Site indices give dominant tree heights at a certain age—no more and no less.

5. Site index measurements taken at different times in a stand's life can indicate different site qualities for the same stand. Four reasons can be given for this occurrence.

   First, suppression during early life can cause a stand to give a significant underestimate of the site when young, with this effect gradually wearing off through time.

   Second, the site quality can change. Although such changes are usually slow, aside from cultural treatments like drainage or fertilization, occasionally they can quickly leave their growth imprint. For example, a series of wet years may impede soil drainage and aeration significantly, affect height growth, and induce tree damage, as reported for red pine (Stone et al., 1954).

   Third, stands on some microsites may not follow the pattern shown by site curves. In Douglas-fir, Carmean (1956) found that trees on sands and gravels grow rapidly at first and more slowly later, while trees on imperfectly drained soils grow slowly initially, but do not slow down appreciably with age (Fig. 4-4).

   Fourth, the sampling basis for the construction of conventional height-over-age curves may not be well balanced. More data may be taken for certain ages and sites than for others.

6. Productivity estimates for a species given by yield tables formed on the basis of height-age relationships may be faulty, especially near the fringes of the species range. In ponderosa pine, MacLean and Bolsinger (1973) found that significant overestimates of productivity can occur from using conventional curves on lands that are on the arid fringe of the species range or are unusually rocky. In areas of low rainfall, ponderosa pine requires more room than "normal" to fulfill its moisture requirements. Soil pockets between the rocks may support tall trees, but the maximum yield per acre will be low. When in doubt about the applicability of yield table estimates to a particular site, refer to the original publication for information on where the basic data were gathered and what criteria were used to select plots.

Vegetative typing Considerable attention has been given, particularly in Scandinavia and Canada, to the use of plants as indicators of site quality and as a basis for site classification. Under undisturbed natural conditions, in northern forests in particular, certain plants or plant communities are characteristically associated with certain forest types and to a less specific degree with site quality within types. As an example, Spilsbury

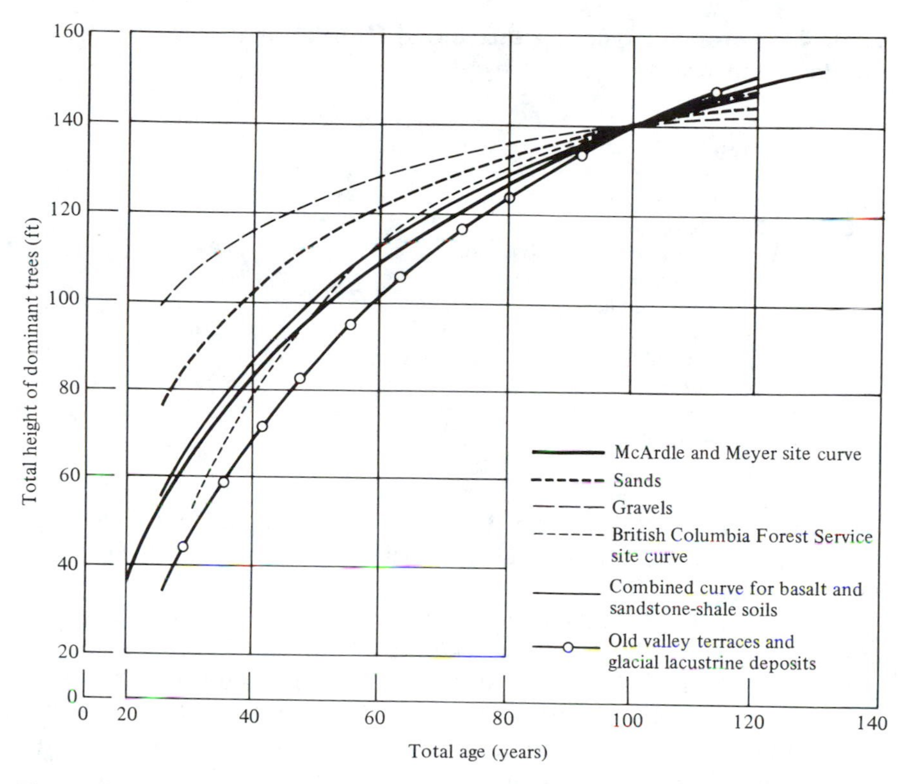

Figure 4-4 Different Douglas-fir site index curve shapes for the major soil groups of western Washington. All curves and soil types are rated site 140 at base age 100. *(From Carmean, 1956.)*

and Smith (1947) successfully identified forest site types for coastal Douglas-fir in British Columbia and Washington. They found that the dominance or codominance of a few understory species (sword fern, salal, pale-green lichen, bearded lichen) correlated well with the site index and site quality of coastal Douglas-fir stands (Table 4-1).

Habitat typing has been initiated over most of the Rocky Mountain region based on Daubenmire and Daubenmire's (1968) habitat types, which include dominant cover and ground cover. A habitat type is a class of land which is fairly homogeneous in terms of its growing environment and integrates variation in elevation, soil, moisture, and so on, in such a way that it will support the same climax overstory and understory vegetation. The name or label given to a habitat type is the binomial of characteristic overstory/understory species expected at climax. The evidence to date shows site index values for seral tree species to be only moderately associated with habitat types, and the efficacy of habitat types to predict

Table 4-1 Forest site types for the coastal Douglas-fir type in the Pacific northwest

	Forest site type	Average site index, ft	Production, ft^3
P	Sword fern (*Polystichum*)	170	19,000
PG	Sword fern salal (*Polystichum-Gaultheria*)	150	15,000
G	Salal (*Gaultheria*)	125	11,000
GPa	Salal pale-green lichen (*Gaultheria-Parmelia*)	95	8,000
GU	Salal bearded lichen (*Gaultheria-Usnea*)	70	6,000

Source: Spilsbury and Smith (1947).

site productivity is still a debatable issue. In the next chapter we shall see that habitat type is one of several site variables used by the PROGNOSIS yield projection model.

Hall (1973) used plant communities of eastern Oregon as the basis for adjusting Meyer's (1938) fully stocked levels for ponderosa pine to account for arid and rocky sites (Table 4-2). MacLean and Bolsinger (1973) tested these adjusted ratios on 30 largely undisturbed sites which appeared to have substantial limitations on stocking. One example site actually supported 96 ft^2 of basal area per acre. The normal yield table suggests that this site could support 186 ft^2, and the normal yield table after Hall's adjustment suggests that the site could support 114 ft^2. Even allowing for the sites not carrying quite their full stocking, Hall's plant community–based adjustments seem to be an innovative way of tying normal yield tables to local sites.

Table 4-2 Percentage of normal basal area that can be supported by "low productivity" plant communities in the Blue Mountain region of eastern Oregon

Plant community	Percent of normal
Ponderosa pine/wheatgrass	20
Ponderosa pine/bitterbrush/fescue	54
Ponderosa pine/bitterbrush/stipa	59
Ponderosa pine/fescue	59
Ponderosa pine/elk sedge	74
Ponderosa pine/shrub/elk sedge	79

Source: Hall (1973).

Correlating understory vegetative types with site quality takes considerable work and the technique has not been tested widely enough in the United States to tell where, and for what species, such an approach can lead to predictive plant cover–site quality equations of useful accuracy. However, the technique has intuitive appeal and could become a significant approach to site estimation. It has a time advantage in that understory indicator plants become established on a site early on, well before the trees themselves grow large enough to provide a good clue to productivity.

Environmental factors The difficulty in correlating understory vegetation with site index led to efforts to use locally significant factors in the physical environment (Coile, 1938). Linking soil to site index has been the most common approach.

Soil has a large and often controlling influence on tree growth. Site estimation from measurable soil factors offers many advantages. Soil properties change slowly over time, and soil measures can be applied in cutover, deforested, and nonforested areas in addition to areas where timber stands are present.

The correlation of site with U.S.D.A. soil survey types has not proven high, in part because these types are not classified and identified according to many of the factors that affect tree growth, such as drainage class, thickness of surface horizon, and subsurface horizon depth. As an example, Van Lear and Hosner (1967) report that little, if any, usable correlation exists between soil mapping units and site index for yellow-poplar in Virginia.

Special forest soil surveys have had better success. Two of the largest efforts were conducted by Coile and Schumacher (1953) for a group of corporations in the southeast and by Steinbrenner (1975) for Weyerhaeuser in the Pacific northwest.

Coile and Schumacher (1953) related tree growth to observable soil characteristics such as texture of certain horizons, soil depth, consistency, and drainage characteristics, plus topography, geology, and land use history. They suggested using soil site maps to aid in species selection, yield prediction, decisions on site preparation, road construction methods, and cost prediction.

The classifications used in Weyerhaeuser's soil survey of their land in the Pacific northwest were designed with predicting tree productivity as a primary objective. Also, their soil site maps are interpreted for trafficability, wind-throw hazard, thinning potential, and engineering characteristics for road construction (Steinbrenner, 1975). According to Daniel et al. (1979), Steinbrenner mapped 3 million acres in western Oregon using a site prediction method (unpublished) based primarily on soil, which accounted for most site index variations. His prediction equation included variables for depth of the A horizon, effective soil depth, texture of the B horizon, and elevation. In addition, the topography of Weyerhaeuser's mountainous terrain is strongly woven into the analysis. Steinbrenner's results have one unique feature: they provide economic site maps—maps that designate areas according to profit potential.

The early failures to correlate the site index with U.S.D.A. soil survey types discouraged many practitioners. The later studies, such as those by Coile and Steinbrenner, which paid more attention to soil characteristics relating to its moisture-holding capacity, have been more successful. Soil surveys solely for site quality, however, are expensive and take considerable imagination if they are to be done successfully. They probably will retain a minor, but intriguing, role in site estimation for the foreseeable future.

Importance of Site Quality in Management

The productivity of timberlands varies tremendously by site index. Studies of yields attainable from fully stocked natural stands at various ages give a measure of relative differentials (Fig. 4-3). The physical magnitude of the volume figures is not significant in this connection; attention should be directed to the comparative picture they give, particularly within species. For example, a Douglas-fir site index of 100 has about a third of the productivity of a site index 200 at age 90. The other species show the same general tendency.

These relative differences usually persist under management. The site quality can be changed by land treatment, such as fertilization or drainage, but only drastic treatment can make a good site out of a poor one.

DENSITY AND STOCKING

The potential of a land area to produce wood is determined by its site quality. The actual growth achieved in a given site is determined by the amount, kind, and distribution of trees currently on the site. Changing the character of this vegetative growing stock is the principal way foresters manipulate and control growth and yield. The amount of tree growing stock is evaluated quantitatively with a group of measures called stand density.

Stand density measures are functions of the stand's tree statistics such as basal area, number of trees per acre, crown competition, or various stand density indices. Stand density is thus a "fact"—a measurable attribute of the stand.

The forester has two uses for density measures, (1) to represent a stand in models used to predict the amount of future growth and yield, and (2) to decide how the subject stand is performing when compared to a criterion related to owner objectives.

Stocking, in contrast to density, is a ratio comparison of the subject stand's density to the stand density of a comparable "ideal" or reference stand. We might find that the basal area of some actual stand is 80, but it should be 120 to totally occupy the site. Hence we say it is (80/120) × 100, or 67 percent stocked relative to the reference stand.

A stand can be overstocked or understocked percentagewise relative to the reference stand but the whole idea of stocking has a meaning only if the standard or ideal stand has a meaning. When not further clarified, the term stocking in this book relates the actual density of a subject stand to the density of a reference stand growing on a site with the same productive potential. The term "fully stocked stand" has traditionally meant one that has the same density as given in the natural stand normal yield table for that site and age.

Stand Density

In Chap. 5 we survey a variety of growth and yield models, and our purpose here is to review the different stand density measures used to drive these predictive models. If you know what yield model you will use, then the stand density measure obtained from forest inventory should match the one used in the yield model.

Stand density can describe not only the degree to which a site is being utilized but also the intensity of competition between trees. At higher densities the growth rates of individual trees slow down, even though the total growth per unit area may continue to increase. When there is a price premium for larger trees—and there usually is—then we also want to know the association of the stand density measure to the size distribution and the growth rates of individual trees in the stand to better quantify value productivity from the site or stand.

In plant and animal ecology, density is commonly defined as a number of individuals per unit area. In forest stands, defining density as a number of individuals per unit area is of limited usefulness. Trees increase in size more or less indefinitely. Only in a plantation with a known initial density does the number of trees per acre serve well as a density measure. More commonly, foresters use density measures that combine size with number.

Measures of Density

Several commonly used density measures are covered here: number of trees per acre, volume, basal area, relative density, stand density index, relative stand density, and crown competition factor. In addition, the use of spacing guides to control density is discussed.

Trees per acre In homogeneous even-aged stands of known age, site, and history, the number of trees per acre is a useful stand density measure. Several plantation growth and yield models use trees per acre as the stand density input variable.

Volume Density often is measured through volume. Because so many objectives relate directly or indirectly to volume, it is a logical measure. To interpret volume, however, it is usually related to some standard, such as the volume in a stand from a yield table, and is presented as a percentage of stocking.

Basal area A measure developed by foresters, basal area is the total cross-sectional area of the trees in a stand, measured in square feet per acre, and is a common density variable in whole stand yield models. In the United States, diameter at breast height (DBH) is measured outside the bark $4\frac{1}{2}$ ft above the average ground level. When a tree's cross section is circular, its basal area equals the area of a circle with diameter equal to the DBH. Assuming the DBH is measured in inches, the formula for tree basal area in square feet is a modification of the familiar formula for the area of a circle, $A = \pi r^2$. Let b_i be the basal area in square feet of tree i and d_i the DBH of tree i in inches. Then

$$b_i = \pi\left(\frac{di}{2}\right)^2 \frac{1}{144} = 0.005454 d_i^2 \tag{4-1}$$

The basal area B for a sample of N trees is the sum of the individual trees' basal areas,

$$B = 0.005454 \sum_{i=1}^{N} d_i^2 \tag{4-2}$$

and the mean per tree basal area $\bar{b}_q$ is

$$\bar{b}_q = \frac{0.005454}{N} \sum_{i=1}^{N} d_i^2$$

The basal area per acre, BA, is the typical stand density measure and is an appropriate conversion from the sample basal area.

The quadratic mean stand diameter $\bar{d}_q$ is a useful parameter characterizing a stand and is estimated from N sample trees as

$$\bar{d}_q = \sqrt{\frac{1}{N} \sum_{i=1}^{N} d_i^2} \tag{4-3}$$

The mean per tree basal area can also be written as

$$\bar{b}_q = 0.005454\bar{d}_q^2 \tag{4-4}$$

and the basal area of N sample trees as

$$B = 0.005454N\bar{d}_q^2 \tag{4-5}$$

Thus $\bar{d}_q$ is the diameter of the tree of mean basal area and B, N, and $\bar{d}_q$ can be related in a single equation.

Several growth models and some stand density indices use quadratic mean diameter as a variable. In many publications it is not clear whether average stand diameter means the quadratic mean or the arithmetic mean diameter, and this has to be checked carefully.

Relative density Curtis et al. (1981) combine basal area and quadratic mean diameter to form a measure of density,

$$\mathrm{RD} = \frac{\mathrm{BA}}{\sqrt{\bar{d}_q}} \tag{4-6}$$

where RD = relative density
BA = basal area per acre
$\bar{d}_q$ = quadratic mean stand diameter

This density measure is widely used in the Douglas-fir region of the Pacific northwest. As a general guide, a relative density of 22 to 25 characterizes open grown stands. At a relative density of about 50, stand growth slows to where commercial thinning may be desirable, and at a relative density of 60 to 65, the ability of trees to respond to release is significantly diminished. Relative density is the density measure used by DFSIM, a popular Douglas-fir stand growth simulator.

Stand density index Relationships have been observed between number of trees, average diameter, basal area, and volume that can be used to construct indices and ratios as measures of stand density. One such measure is Reineke's stand density index.

Reineke (1933) devised a measure of density for pure even-aged stands that was independent of site index and age. He used two features of normal yield tables to construct a stocking measure: (1) Different fully stocked even-aged stands of a species having the same quadratic mean diameter were observed to have approximately the same maximum number of trees per acre. (2) A curve of the maximum number of trees per acre for different quadratic mean stand diameters, fitted with a log linear regression using normal yield table plot data, has the same slope (−1.605) for most species and yields the function,

$$\log N = -1.605 \log \bar{d}_q + k$$

where N = number of trees per acre
$\bar{d}_q$ = quadratic mean stand diameter
k = a constant for each species

Red fir and redwood with 1,000 trees per acre had the highest number of trees at a quadratic mean diameter of 10 in. Using this as a reference curve, Reineke drew a series of parallel lines relating the maximum number of trees to the quadratic mean stand diameter (Fig. 4-5). These curves permit converting the number of trees at any quadratic mean diameter into an equivalent density at a quadratic mean diameter of 10 in. The number of trees at a quadratic mean stand diameter of 10 in. is Reineke's stand density index (SDI). It is calculated as

$$\text{SDI} = N \left(\frac{\bar{d}_q}{10} \right)^{1.605} \tag{4-7}$$

As an example, a stand containing 690 trees at 3 in. has the same density as a stand with 100 trees at 10 in., and the common stand density index computed using Eq. (4-7) equals 100. The maximum stand density index for common softwoods ranges from 400 to 1,000 (Table 4-3). If the species is Douglas-fir, which has a maximum stand density index of 595, the two stands have a stocking of 100/595, or 17 percent.

Reineke's stand density index has a number of useful features. Needed data are easily obtained from inventory. Site and age need not be estimated, and densities of stands can be compared across site and age. Given the target number of trees at some diameter, the number of trees at all diameters up to the target that have the same density can be estimated. We still may ask, however, does a particular stand of trees grow according to this model?

Since the stand density index and relative density are both related to the quadratic mean diameter and the number of trees, they should them-

Table 4-3 Maximum stand density index values found in natural stands of several species

Species	Stand density index
Longleaf pine	400
Loblolly pine	450
Eucalyptus	495
Douglas-fir	595
White fir	830
Red fir	1,000
Redwood	1,000

Source: Daniel et al. (1979).

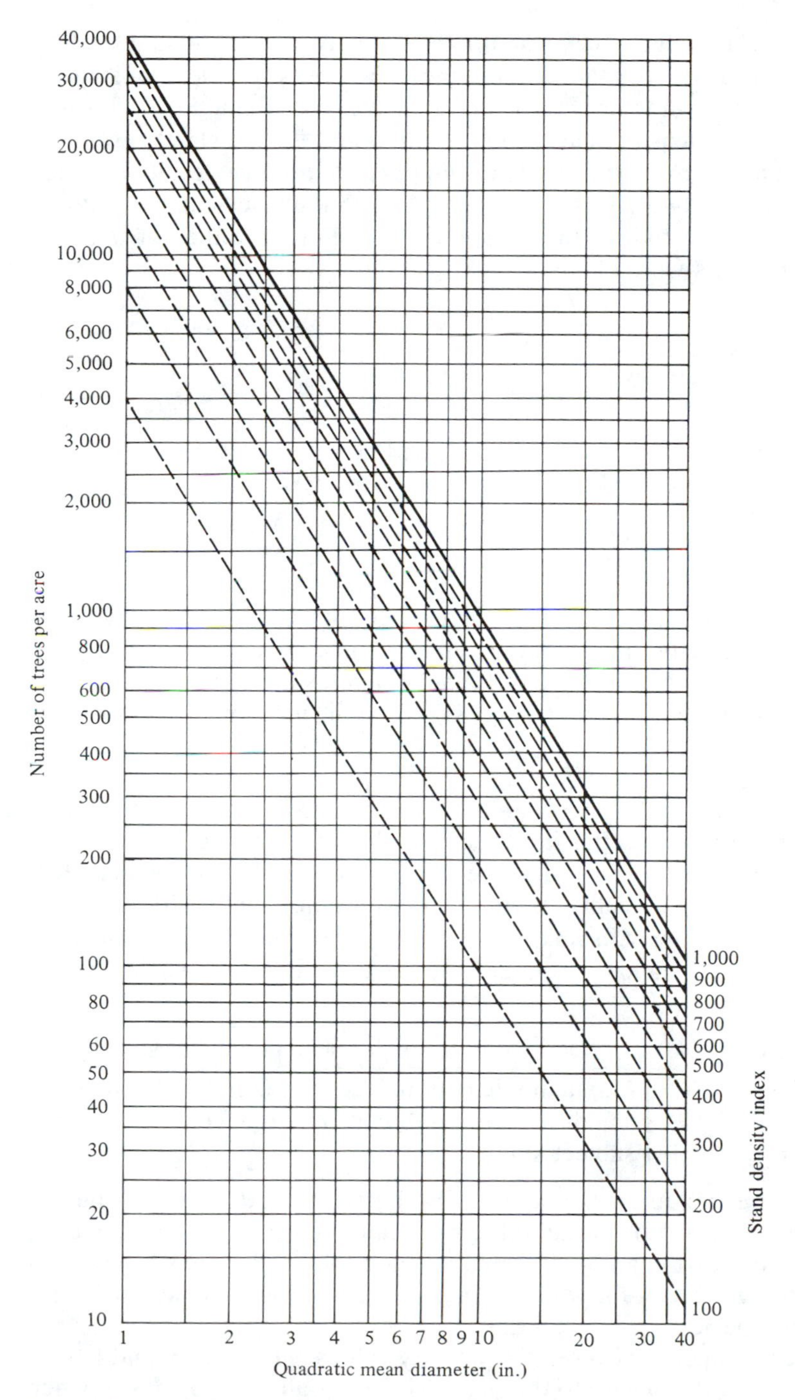

Figure 4-5 Douglas-fir stand density index as a function of number of trees per acre and quadratic mean stand diameter.

selves be related. To check, the relative density was calculated for four different values of $\bar{d}_q$ while holding the stand density index constant at 400. The number of trees was found for each $\bar{d}_q$ by solving Eq. (4-7) for N when the stand density index is 400. With N and $\bar{d}_q$, the stand basal area can be found. Then, using basal area and $\bar{d}_q$, the relative density is calculated using Eq. (4-6). Our hypothesis was that the calculated relative density values would be the same for the different values of $\bar{d}_q$. The following results were obtained:

$\bar{d}_q$	Calculated relative density with SDI = 400
2	82.3
8	70.57
10	69.00
20	64.15

This suggests that while the standard density index and relative density are not exactly the same, they appear to be similar in concept.

For a thorough discussion of Reineke's index and its calculation, see Daniel et al. (1979). Clutter et al. (1983) probe the mathematical properties of the index.

Relative stand density Drew and Flewelling (1979) developed a new stand density index similar in concept to Reineke's but recognizing that tree volume growth is related to the tree's height as well as its diameter. Independent of site quality, they observed a consistent relationship between the maximum of mean tree volume and the number of trees per acre found in natural stands. Their density measure is based on this and is defined as follows:

$$\left\{\begin{matrix}\text{Relative stand} \\ \text{density}\end{matrix}\right\} = \frac{\{\text{Density in trees per acre of subject stand}\}}{\left\{\begin{matrix}\text{Maximum density in trees per acre attainable in a} \\ \text{stand with the same mean tree volume as the} \\ \text{subject stand}\end{matrix}\right\}}$$

Since tree volume is a function of both height and diameter, through a series of transformations the linkage of stand density to stand diameter, height, and growth can be depicted in what they call a density management diagram (Fig. 4-6), which is intended to be locally calibrated for a particular species.

Three straight lines at relative density values of 0.15, 0.55, and 1.0 are shown sloping downward to the right. The 0.15 value is the density when Douglas-fir crown closure is observed to occur, 0.55 is the density where

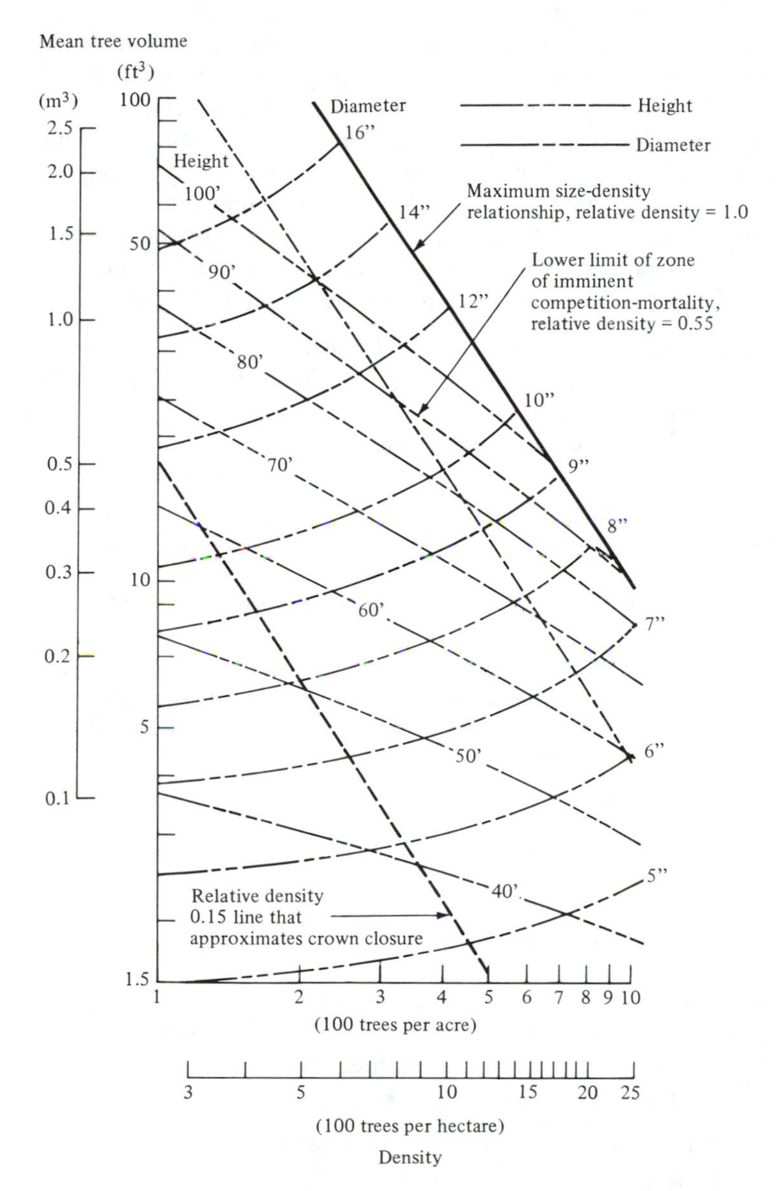

Figure 4-6 Stand density management diagram for Douglas-fir based on the relative density index. *(From Drew and Flewelling, 1979.)*

research has shown total growth per acre to level out and competition-induced mortality begins. The density of 1.0 is the empirically fitted maximum observed in nature.

The guiding concept is to keep stand densities near 0.55 when the objective is to maximize volume growth. To use such a diagram to guide management, the subject stand values for mean tree volume and trees per acre are determined from inventory, and the relative density of the stand is determined. If the stand is close to or higher than a density of 0.55, then a thinning to reduce density is indicated. The amount of thinning should reduce density to a level such that the stand will grow back to or slightly above the 0.55 density level at the next scheduled thinning. The combinations of mean stand diameters, heights, and trees per acre yielding a given density are also shown on the diagram so that thinning guides can also consider the effect of tree size (and value) in the thinning strategy.

Crown competition factor A diameter-based density measure developed by Krajicek et al. (1961) represents the percentage of an area that is covered with a vertical projection of tree crowns. It is determined in three steps. (1) Associate the crown width (CW) of open grown trees with their diameter. (2) Calculate the maximum crown area (MCA) in square feet of each tree on one acre of the subject stand as if they were all open grown. (3) Sum the maximum crown area for all trees on the representative acre, divide by 43,560, and multiply by 100 to get the percentage of the acre covered by the vertical projection of the tree crowns. This percentage is the crown competition factor (CCF).

Assuming a linear relationship between open grown trees and their diameters, we can write the equations to represent these three steps.

$$\mathrm{CW}_i = a + bd_i \tag{4-8}$$

$$\mathrm{MCA}_i = \pi(\mathrm{CW}_i/2)^2 \tag{4-9}$$

$$\mathrm{CCF} = \left(\sum_{i=1}^{N} \mathrm{MCA}_i\right) \times \left(\frac{100}{43{,}560}\right) \tag{4-10}$$

where d_i = diameter of tree i in inches.
CW_i = crown width of tree i in feet.
MCA_i = crown area of tree i in square feet.
CCF = crown competition factor expressed as a percentage.
N = number of trees representing one acre of the subject stand.

A stand can have a CCF greater than 100 percent if it has a high number of stems per acre and the crowns are overlapping. This is because the CCF assumes that all trees have a crown width equal to that of open grown trees of a given diameter. In reality, competition will reduce the crown width of all but the strongest dominants. The CCF index of stand density is used by the PROGNOSIS model reviewed in Chap. 5.

Example Calculations of Stand Density

An inventory of an overmature western pine forest might produce the stand structure shown in Table 4-4. An average acre contained 29 trees per acre that were inventoried. Starting with the stand table, columns (1) and (2), the density measures can be calculated. First the per-acre basal area (126.8 ft^2), volume (3,627 ft^3), and total MCA (25,331 ft^2) expressed as crown area in square feet are calculated. Then the volume of the average tree (125 ft^3), the basal area of the average tree (4.372 ft^2), and the quadratic mean diameter (28.3 in.) are calculated. From this set of stand attributes the density measures can be calculated as shown. The only difficulty arose in estimating the relative stand density. With a mean tree volume of 125 ft^3 and 29 trees per acre, we have run outside the range of the graph in Fig. 4-6. This particular density index chart is set up for young even-aged stands with much smaller average diameters.

Using Spacing to Control Density

Spacing rules specified as a function of diameter or height are often used to guide thinning in order to leave a desired density in developing stands. While the space actually occupied or required by the roots and crowns of individual trees is difficult to determine, average space-size relationships can be estimated. In general, trees occupy and need space in proportion to their size. Hence, spacing figures are based on some measure of tree size, whether it be diameter, height, crown spread, or crown volume.

Density as measured in basal area can be expressed readily in terms of tree diameter and number of trees per acre. Suppose that you wish to leave 90 ft^2 of basal area per acre in well distributed trees of desirable form and species. What number of trees and average between-tree spacing is necessary to obtain this density? If the trees average a quadratic mean DBH of 6 in. (basal area of 0.196 ft^2), then 90/0.196, or 459, "average-sized" trees per acre should be left. The average land area per tree is then the number of square feet per acre, 43,560, divided by 459, or 95 ft^2. Assuming that trees tend to be arranged at the corners of equal-sized squares as in a plantation, the average between-tree spacing equals $\sqrt{95}$, or 9.7 ft. For 10-in. trees, 16.2 ft works out as the average spacing between trees using similar calculations for a residual 90 ft^2 of basal area. A chart showing square and equilateral spacings for different diameters and residual basal areas is given in Fig. 4-7.

Left to their own devices, trees in the forest do not distribute themselves in any fixed geometric pattern. The square distribution is upset by inequalities in the soil or microclimate, and by anything that damages the stand. These effects are likely to become increasingly evident as the stand becomes older: even-aged mature natural stands usually exhibit a rather irregular, clumpy tree distribution pattern, and the forester has to modify idealized spacing guides constantly to local variability.

Table 4-4 Calculation of several stand density indices for a stand table representing one acre

Stand structure

Stand table		Basal area, ft^2		Volume, ft^3		Crown area		
DBH, in. (1)	Number of trees (2)	Per tree* (3)	Per class = (2) × (3) (4)	Per tree† (5)	Per class = (2) × (5) (6)	Crown width‡ (diameter), ft (7)	MCA (πr^2), ft^2 (8)	MCA/class, ft^2 = (2) × (8) (9)
40	1	8.73	8.73	211.3	211.3	45.5	1,626.0	1,626.0
38	2	7.88	15.76	197.6	395.2	43.4	1,479.34	2,958.7
36	1	7.07	7.07	184.0	184.0	41.4	1,346.14	1,346.1
34	3	6.30	18.90	170.3	510.9	39.3	1,213.04	3,639.1
32	2	5.58	11.16	156.6	313.2	37.3	1,092.72	2,185.4
30	3	4.91	14.73	143.0	429.0	35.2	973.14	2,919.4
28	4	4.28	17.12	129.3	517.2	33.2	865.70	3,462.8
26	3	3.69	11.07	115.7	347.1	31.1	759.64	2,278.9
24	4	3.14	12.56	102.0	408.0	29.0	660.52	2,642.1
22	2	2.64	5.28	88.3	176.6	27.0	572.56	1,145.1
20	0	2.18	0	74.7	0	25.0	490.87	0
18	1	1.77	1.77	61.0	61.0	22.9	411.87	411.9
16	0	1.40	0	47.4	0	20.8	339.79	0
14	1	1.07	1.07	33.7	33.7	18.8	277.59	277.6
12	2	0.79	1.58	20.0	40.0	16.7	219.04	438.1
Total per acre	29		126.80		3,627.2			25,331.2

Table 4-4 (Continued)

Computation of stand density

Volume of average tree $\bar{V}_i = 3{,}627.2/29 = 125 \text{ ft}^3$

Basal area of average tree $\bar{b}_q = 126.8/29 = 4.372 \text{ ft}^2$

Quadratic mean diameter $\bar{d}_q = \frac{1}{\sqrt{0.005454}} \sqrt{\bar{b}_q} = 13.54 \sqrt{4.372} = 28.3$ in.

	Calculation of density index	Computed value
1. Number of trees per acre	From table above	29
2. Basal area per acre, Eq. (4-2)	From table above	126.8
3. Relative density, Eq. (4-6)	$\text{BA}/\sqrt{\bar{d}_q} = 126.8/\sqrt{28.3}$	23.6
4. Stand density index, Eq. (4-7)	$N(\bar{d}_q/10)^{1.605} = 29(28.3/10)^{1.605}$	154
5. Relative stand density, Eq. (4-6)	Ave volume 125 ft³, $N = 29$ are outside range of values in Fig. 4-6	—
6. Crown competition factor	$(\text{MCA}_{\text{total}})\left(\frac{100}{43{,}560}\right) = \frac{(25{,}331)(100)}{43{,}560}$	58.2%

* Basal area per tree = 0.005454DBH^2.
† Volume per tree = $-61.9 + 6.83\text{DBH}$ (adapted from Avery and Burkhart, 1983).
‡ Crown width = $4.344 + 1.029\text{DBH}$ (Alexander, 1971).

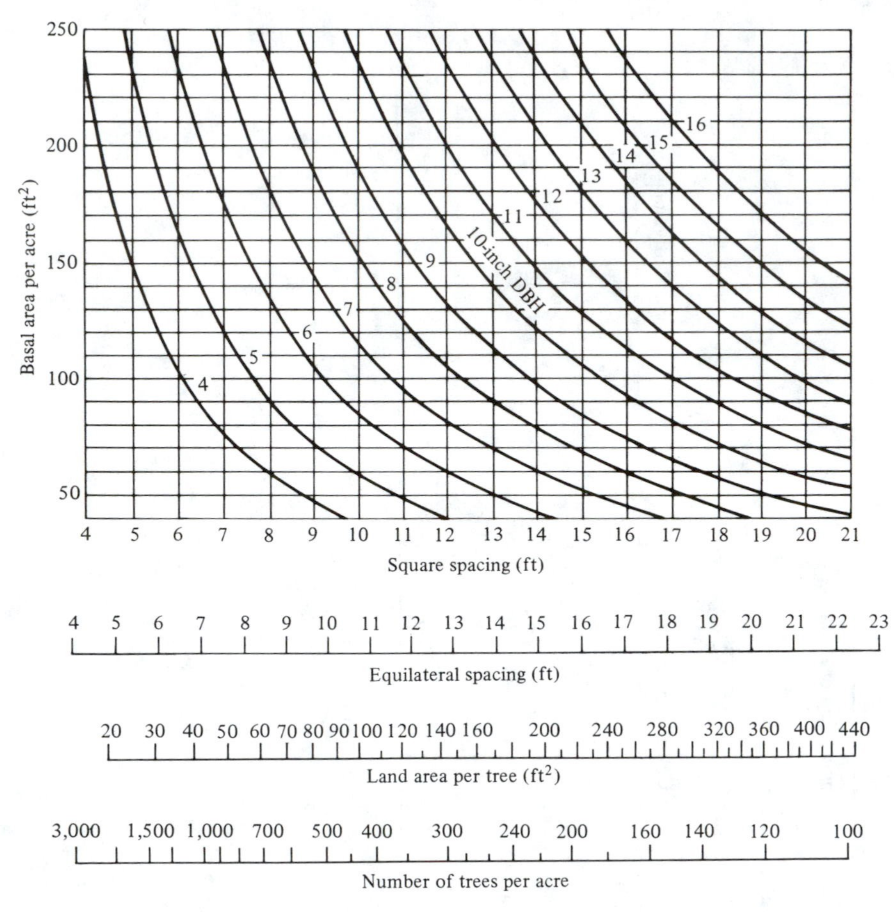

Figure 4-7 Equilateral and square spacing distances in feet and trees per acre for plantations of different basal areas and average diameters.

"Diameter times" spacing guide As a field marking guide, it is convenient to express spacing directly as a ratio of diameter. As determined previously, the average square spacing of 6-in. trees is 9.7 ft when 90 ft^2 of basal area per acre is desired. But not all trees are 6 in., and the larger trees need more space than the smaller ones. The "diameter times" ratio provides one simple adjustment for tree size. The ratio for 6-in. trees is 9.7/6, or 1.62. For every inch of diameter there should be 1.62 times as much spacing in feet ($6 \times 1.62 = 9.7$ ft). With 10-in. trees and the same basal area, the spacing should be $10 \times 1.62 = 16.2$ ft. Fortunately, for any specific basal area per acre the ratio between diameter in inches and spacing in feet is a constant.

A formula useful in computing the "diameter times" spacing figure or constant C for any given basal area per acre BA is*

$$C = \frac{15.4}{\sqrt{\text{BA}}} \tag{4-11}$$

and for 90 ft^2 of basal area

$$C = \frac{15.4}{\sqrt{90}} = 1.62$$

Given C, the spacing between trees is

$$\text{Spacing (feet)} = \text{DBH} \times C \tag{4-12}$$

The relationship of Eqs. (4-11) and (4-12) can help estimate the basal area in the field. If trees average 10 in. in diameter and are spaced approximately 14 ft apart, solving the "diameter times" formula for the basal area shows that the stand contains a little over 120 ft^2 of basal area per acre.

"Diameter plus" spacing guide Another type of spacing guide is "diameter plus." A common rule of thumb for hardwoods is "diameter plus 6"; that is, the spacing in feet should be the diameter plus 6. In this case, spacing for a 10-in. tree would be 16 ft. This spacing figure does not bear any constant relation to the basal area as a function of average diameter and cannot be used effectively in marking a stand to leave a specified amount of stocking as measured in basal area or volume.

Determining Desirable Density

Measurement of density and development of various spacing guides still leaves unanswered the fundamental question: what *is* desirable density

* The factor 15.4 and Eq. (4-11) can be derived as follows:

1. $$\text{Spacing between trees (feet)} = \sqrt{\frac{43{,}560}{\text{trees per acre}}}$$

2. $$\text{Trees per acre} = \text{BA}/0.005454D^2$$

3. $$\text{Spacing} = \sqrt{\frac{43{,}560}{\text{BA}/0.005454D^2}}$$

Simplifying,

$$\text{Spacing} = \frac{\sqrt{237.4D^2}}{\text{BA}}$$

$$= 15.4\frac{D}{\sqrt{\text{BA}}}$$

for specific species, sites, and stand conditions? Roth (1925) gave a succinct answer: "room to grow but none to waste." But how much is enough room, and what is wasted space?

It is difficult to provide easy, foolproof guides to optimum stocking. Optimum means "best" according to the owner's objectives. When these objectives are financial, the site, species, density, volume, and tree value as well as their interaction are involved in this determination.

A decision regarding density is made every time trees are cut in thinnings and other partial cuttings and every time trees are planted. Choosing a desirable density is impossible without introduction of management objectives. Calling densities that maximize growth "optimal" does not make them so if the purposes of management call for something else. Much of the rest of this book aims at developing principles and tools that will enable foresters to evaluate density in light of owner objectives.

QUESTIONS

4-1 Three trees were measured on each of three plots in order to estimate the site index of a ponderosa pine stand for growth and yield estimation purposes. Estimate the site index of this stand using the curves for ponderosa pine in Fig. 4-1.

Plot	Tree	Total age	Height
A	1	120	140
	2	160	140
	3	140	155
B	1	100	110
	2	120	140
	3	80	120
C	1	80	70
	2	90	85
	3	60	70

None of the trees showed any signs of early suppression and all appeared to be the most competitive dominant trees in the plot.

4-2 A natural, well-stocked ponderosa pine stand is growing on site quality 100 land and is now 60 years old. The silviculturist claims that by carefully fertilizing these nutrient-deficient soils, the growth rate would immediately rise to be the same as that achieved on site quality 140 land. The current stand has 3,500 ft^3, and it is believed that the yield curves of Fig. 4-3 accurately portray the growth and yield on both sites 100 and 140.

Assuming the silviculturist is correct, calculate the impact of the fertilizer program, if kept fully in effect between ages 60 and 100, on:

(*a*) The increase in expected yield at age 100 over land not fertilized.

(*b*) The MAI of the stand at age 100.

4-3 You measured the diameter of five trees on a $\frac{1}{25}$-acre plot. Their diameters are 7, 9, 13, 17, and 20 in. Calculate:

(*a*) Average and quadratic mean diameters.

(*b*) Stand basal area per acre.

(*c*) Reineke's stand density index (SDI).

(*d*) Curtis's relative density (RD).

(*e*) Krajicek's crown competition factor (CCF) when crown width = 5 + 1.3 DBH.

4-4 (*a*) What is the needed square spacing distance between trees when 700 trees are planted per acre?

(*b*) What is the square spacing between leave crop trees when a final commercial thinning reduces the plantation to 160 ft^2 of basal area in trees with a quadratic mean diameter of 12 in.?

4-5 Prescribe a "diameter times" spacing guide for your tree-marking crew when you wish them to mark a stand to leave 160 ft^2 of basal area.

REFERENCES

Alexander, R. R., 1971: "Crown Competition Factor (CCF) for Englemann Spruce in the Central Rocky Mountains," U.S. Forest Service Research Note RM-188.

Avery, T. E., and H. E. Burkhart, 1983: "Forest Measurements," 3d ed., McGraw-Hill, New York.

Biging, G. S., and L. C. Wensel, 1984: "Draft Site Index Equations for Young-Growth Mixed Conifers of Northern California," Research Note 6, Northern California Forest Yield Cooperative, Department of Forestry and Resource Management, University of California, Berkeley.

Carmean, W. H., 1956: Suggested Modifications of Standard Douglas-Fir Site Curves for Certain Soils in Southwestern Washington, *Forest Sci.,* **2**:242–250.

Clutter, J. R., J. C. Fortson, L. V. Pienaer, G. Brister, and R. L. Bailey, 1983: "Timber Management: A Quantitative Approach," Wiley, New York.

Coile, T. S., 1938: Forest Classification of Forest Sites with Special Reference to Ground Vegetation, *J. Forestry,* **36**:1062–1066.

——— and F. X. Schumacher, 1953: Relation of Soil Properties to Site Index of Loblolly and Shortleaf Pines in the Piedmont Region of the Carolinas, Georgia, and Alabama, *J. Forestry,* **51**:739–744.

Curtis, R. O., G. W. Clendenen, and D. J. DeMars, 1981: "A New Stand Simulator for Coast Douglas-Fir: DFSIM Users Guide," U.S. Forest Service General Technical Report PNW-128.

Daniel, T. W., J. A. Helms, and F. S. Baker, 1979: "Principles of Silviculture," 2d ed., McGraw-Hill, New York.

Daubenmire, R., and J. B. Daubenmire, 1968: "Forest Vegetation of Eastern Washington and Northern Idaho," Washington Agricultural Experiment Station Technical Bulletin 60.

Drew, T. J., and J. W. Flewelling, 1979: Stand Density Management: An Alternative Approach and Its Application to Douglas-Fir Plantations, *Forest Sci.,* **25**:518–532.

Dunning, D., and L. H. Reineke, 1933: "Preliminary Yield Tables for Second-Growth Stands in the California Pine Region," U.S. Department of Agriculture Technical Bulletin 354.

Hall, F. C., 1973: "Plant Communities of the Blue Mountains in Eastern Oregon and Southwestern Washington," U.S. Forest Service Pacific Northwest Area Guide 3-1, Portland, Oregon.

Jones, R., 1969: "Review and Comparison of Site Evaluation Methods," U.S. Forest Service Research Paper RM-51.

Krajicek, J. E., K. A. Brinkman, and S. F. Gingrich, 1961: Crown Competition—a Measure of Density, *Forest Sci.,* **7**:35–42.

MacLean, C. D., and C. L. Bolsinger, 1973: "Estimating Productivity on Sites with Low Stocking Capacity," U.S. Forest Service Research Paper PNW-152.

McArdle, R. E., W. H. Meyer, and D. Bruce, 1930: "The Yield of Douglas-Fir in the Pacific Northwest" (rev. October 1949), U.S. Department of Agriculture Technical Bulletin 201.

Meyer, W. H., 1938: "Yield of Even-Aged Stands of Ponderosa Pine," U.S. Department of Agriculture Technical Bulletin 630.

Reineke, L. H., 1933: Perfecting a Stand Density Index for Even-aged Forests, *J. Agr. Res.,* **46**:627–638.

Roth, F., 1925: "Forest Regulation," 2d ed., George Wahr Publishing, Ann Arbor, Mich.

Spilsbury, R. H., and D. S. Smith, 1947: "Forest Site Types of the Pacific Northwest," British Columbia Forest Technical Publication T 30, Department of Lands and Forests.

Stage, A. R., 1963: A Mathematical Approach to Polymorphic Site Curves for Grand Fir, *Forest Sci.,* **9**:167–180.

Stone, E. L., R. R. Morrow, and D. S. Welch, 1954: A Malady of Red Pine on Poorly Drained Sites, *J. Forestry,* **52**:104–114.

Steinbrenner, E. C., 1975: Mapping Forest Soils on Weyerhaeuser Lands in the Pacific Northwest, in B. Bernier and C. H. Winget (eds.), "Forest Soils and Forest Land Management," Les Presses de L'Universite Laval, Quebec.

Van Lear, D. H., and J. F. Hosner, 1967: Correlation of Site Index and Soil Mapping Units for Yellow-Poplar in Southwest Virginia, *J. Forestry,* **65**:22–24.

CHAPTER

FIVE

MODELS FOR PREDICTING GROWTH AND YIELD

The preceding chapters identified the importance to forest management of being able to make good predictions of growth and yield, defined the concept and measurement of growth, and considered the factors of site index and stand density, which strongly influence the amount of growth and yield. This chapter assembles this information to review, from a managerial perspective, the different models used to make empirical yield predictions.

A great many models are now available, presenting an often confusing array of choices. Early models from studies of healthy natural stands were packaged up and published as bulletins full of tables and graphs and were called, appropriately, yield tables. Later, as the capability to evaluate stands with different levels of stand density was added, publication length and complexity grew until many now publish sets of equations, leaving it up to the user to solve the equations and prepare hard copy tables as needed. The final evolution has seen the more complex models, particularly the individual tree models, written as computer programs. When the user wants a copy, a floppy disk or a tape arrives in the mail. This evolution of model forms from the traditional normal yield tables to variable-density stand and diameter class models to the sophisticated individual tree models mirrors the progress of our profession. Our original concern with extensive management of natural stands has given way to more intensive investment management of planted stands, and the availability of inexpensive, effective computers to accurately execute tedious computations has made possible some incredibly complex predictive models.

Forest managers do not build many of these models, but they surely are the prime users and take the rap for bad estimates. It is a dangerous mistake simply to take the results as given from some clever "black box." At least three checks can be made to determine whether a yield model is reasonably appropriate for some management situation:

1. Are the sample data used to build the model from the same geographic area and on sites with similar physical characteristics as the subject stands?
2. Is there documented evidence of the model's accuracy—validation tests where the prediction of the model was checked against known actual results?
3. Is the structure of the model in terms of its principal equations and independent variables logical and consistent with the prediction needs and inventory data available for the subject stands?

We hope this chapter will increase skill in making the third check on model logic and structure. For the other two checks, careful investigation and persistence are your primary resources. Our review and examination of growth and yield models begins with a classification of the models using the most generally accepted terminology. The main part of the chapter looks at representative models of the different types, focusing on the unique features and most important functions of each example. We then discuss how physical and economic response to intensive management is evaluated in a modern growth simulation model. We conclude the chapter by returning to the question of evaluating the suitability of growth and yield models.

We have not attempted to distinguish between models used for predicting growth of even-aged and uneven-aged stands. Most models can be adapted for both types of stands.

WRITING EQUATIONS: A BRIEF REVIEW

Growth and yield models are relationships between the amount of yield or growth and the various factors that explain or predict this growth. Such relationships are most conveniently stated as equations. These equations can be implicit or explicit in form and can be linear or nonlinear in their relationships. It is useful to review these distinctions before proceeding.

A relationship between variables is *implicit* when the variables in the equation are defined and the dependent variable(s) identified, but the relationship is not quantified. For example, the statement

$$Y = f(S, A)$$

where Y = yield per hectare of even-aged stands of loblolly pine, in cubic meters
S = site index at 25 years, in feet
A = age since regeneration, in years

is an implicit equation. It says that yield depends on or is related to site quality and age. It does not tell how to put a numerical value on the variable Y given values for variables S and A.

The equation becomes *explicit* when this relationship is specified. A yield equation, written explicitly, is

$$Y = S[A - 0.07\,(A - 20)]$$

which tells us exactly what the yield would be for any age and site index.

Linear and nonlinear equations A typical linear equation is represented by the implicit cost expression

$$\text{TC} = c_1x_1 + c_2x_2 + c_3x_3$$

where TC = total cost, in dollars—the dependent variable
x_1, x_2, x_3 = number of acres regenerated by hand, machine, and aerial methods, respectively—the independent variables
c_1, c_2, c_3 = costs in dollars per acre for the hand, machine, and aerial methods used, respectively

If a cost study for a particular forest and set of machinery, people, and technology were made, the coefficients might be determined to be c_1 = \$175/acre, c_2 = \$90/acre, and c_3 = \$30/acre. Given this information, the equation can be made explicit as

$$\text{TC} = 175x_1 + 90x_2 + 30x_3$$

This equation is called *linear* because all the independent variables are raised only to the first power and the rate of change in the dependent variable is linearly related to changes in each independent variable. If we plot the dependent variable against any independent variable, a straight line results, with the slope given by the coefficient of the independent variable. For example, consider TC and x_1 from the equation above. For every additional acre planted, TC increases by 175. Plotting this relationship, we get a straight line (Fig. 5-1).

Equations are *nonlinear* when the exponents of the independent variables are different than 1 or there are two or more independent variables

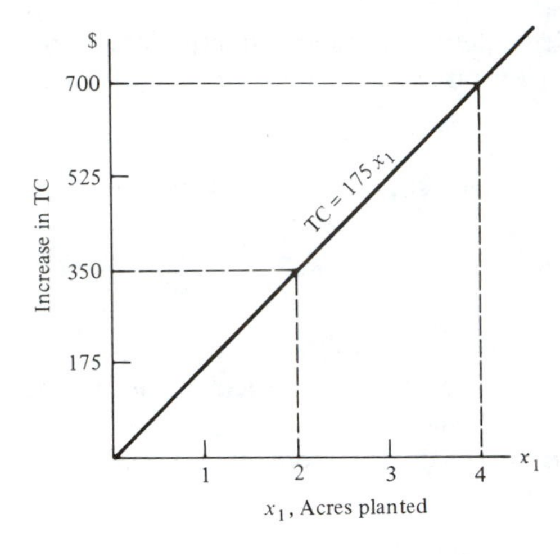

Figure 5-1 Linear relationship between acres planted and total cost.

in a term of the equation. Some examples of nonlinear equations are

$$R = 100Q - 0.25Q^2$$

$$Y = c_1x_1 + c_2x_2^4$$

$$\text{TC} = c_1x_1 + c_2(x_1x_2) + c_3\sqrt{x_2}$$

CLASSIFICATION OF GROWTH AND YIELD MODELS

Growth and yield models sort themselves out by whether they model (1) whole stands as portrayed by stand variables such as age or basal area per acre, (2) the average tree within each diameter class, or (3) each individual tree in a sample or stand. Yields per acre or hectare are provided directly by the whole stand models and are obtained in the other model types by summing the yields over diameter classes or individual trees.

Our classification (Table 5-1) is organized around these three model types and presents a broad continuum of model classes. As with all classifications, some models do not seem to fit any one class exactly, sometimes because the model itself can be used in several ways. While we give a name to each model class, a tighter definition is given by presenting the primary relationships and variables as implicit equations. For example, we describe a density-controlled model to predict current volume as $V_1 = f(A, S, D_1)$. This means the dependent variable of volume at the current time, V_1, is determined by stand age A, site index S, and current stand density D_1. Different variable-density models of this class may use different specific measures for the independent variables and a different explicit function relating the three independent variables to current volume.

We use the terminology of Clutter et al. (1983) to describe the variable-density stand models and the convention of Munro (1974) to classify individual tree models. We have chosen to distinguish diameter class models as a separate group because they have unique properties of their own. All models are assumed to be identified by species and applicable geographic region or environmental conditions, so these factors are not used as criteria in the classification.

Scanning the classification in Table 5-1, we start with whole stand models, and the first distinction is whether or not stand density D is used as an independent variable. Traditional normal yield tables do not use density since the word "normal" implies nature's maximum density. Empirical yield tables assume nature's average density. Variable-density models split by whether current or future volume is directly estimated by the growth functions or whether stand volume is aggregated from mathematically generated diameter classes. A second distinction is whether the model predicts growth directly or uses a two-stage process which first predicts future stand density and then uses this information to estimate future stand volume and subsequently growth by subtraction.

Diameter class models simulate each diameter class by calculating the characteristics, volume, and growth of the average tree in each class, and multiply this average tree by the inventoried number of stems in each class. The volumes are then aggregated over all classes to obtain stand characteristics. The two diameter class methods are distinguished by whether actual radial increment data collected from the subject stand are used to model the trees or whether generalized growth functions based on research sample data are used. Note that diameter *distribution* models are treated as whole stand models since the number of stems in each diameter class is wholly a function of the stand variables and all the growth functions are for stand variables. Diameter *class* models in contrast have the number of trees in each class empirically determined and independently model the diameter classes subject to some aggregate stand influences.

Individual tree models are the most complex and individually model each tree on a sample tree list. Most individual tree models calculate a crown competition index (CCI) for each tree and use it in determining whether the tree lives or dies and, if it lives, its growth in terms of diameter, height, and crown size. A distinction between model types is based on how the crown competition index is calculated. If the calculation is based on the measured or mapped distance from each subject tree to all trees within its zone of competition, then it is called distance-dependent. If the crown competition index is based only on subject tree characteristics and aggregate stand characteristics, then it is a distance-independent model.

Table 5-1 A classification of growth and yield models

Model stand	Implicit model equations; primary relationships and variables
I. Whole stand models	
A. Density-free models	
1. Normal yield tables	$V_A = f(A, S)$
2. Empirical yield tables for average current stands	$V_A = f(A, S)$
B. Variable-density models*	
1. Predict current volumes V_1	
a. Explicit models	$V_1 = f(A, S, D)$
b. Implicit models (diameter distribution)	$f(d_i)_1 = f(A, S, D)$ $v_i = f(d_i)$ $V_1 = \Sigma_i\, v_i(nd_i)_1$
2. Predict future growth g_{12} and volumes V_2	
a. Explicit models	
i. Direct growth prediction	$g_{12} = f(S, A, D_1)$ $V_2 = V_1 + g_{12}$
ii. Stand density prediction	$D_2 = f(S, A_1, A_2, D_1)$ $V_2 = f(S, A_2, D_2)$ $g_{12} = V_2 - V_1$
b. Implicit models (diameter distribution)	$D_2 = f(S, A_1, A_2, D_1)$ $f(d_i)_2 = f(S, A_2, D_2)$ $V_2 = \Sigma_i\, v_i(nd_i)_2$ $g_{12} = V_2 - V_1$
II. Diameter class models	
A. Empirical stand table projections	$(nd_i)_2 = f[(nd_i)_1, \text{INCR}]$ $V_2 = \Sigma_i\, v_i(nd_i)_2$ $g_{12} = V_2 - V_1$
B. Diameter class growth models	$(nd_i)_2 = f[(nd_i)_1, S, P_{12}, D_1]$ $v_i = f(d_i)$ $V_2 = \Sigma_i\, v_i(nd_i)_2$ $g_{12} = V_2 - V_1$
III. Individual tree models	
A. Distance-dependent	$\text{CCI}_k = f[\text{DIST}_k, D_1, S, (d_k, h_k, c_k)_1]$ $(d_k, h_k, c_k)_2 = f[\text{CCI}_k, D_1, S, P_{12}, (d_k, h_k, c_k)_1]$ $v_k = f(d_k, h_k)$ $V_2 = \Sigma_k(v_k)_2$ $g_{12} = V_2 - V_1$
B. Distance-independent	$\text{CCI}_k = f[D, S, (d_k, h_k, c_k)_1]$ All other operations as for distance-dependent model

* For uneven-aged stands the age variable A in these models would be replaced with P_{12}.

S = site index
A = stand age
P_{12} = length of growth period

DENSITY-FREE WHOLE STAND YIELD MODELS

A direct approach to estimating how natural stands will grow in the future is to measure how well they grew in the past. If a large number of random samples measuring age and volume per acre are taken from natural stands of a given species and site quality, a scatter diagram like that shown in Fig. 5-2 results. We would note that there is substantial variation in volume at a given age, that the average volume increases with age, and that the average volume is higher on the better-quality land. The variability is due to the fact that natural stands have openings, are clumpy, have different genetic makeups, and have different stand densities and life histories. Events in the life history of the stands such as fire, disease, insects, and cuttings are what typically causes the openings, different structures, and stand density variations.

When foresters wanted to determine the *potential* growth for natural stands on these sites, they did not use a random sample but rather selected only stands that appeared healthy, had an even distribution of trees on the ground, and were at the highest density levels observed; in short, they looked for nature's best. "Normal" stands are the ones that lie on the upper edge of the scatter envelopes in Fig. 5-2. Curves *A* and *C*, which were fitted to the data, provide an estimate of the maximum yield of natural stands at each age. Such curves are published as normal yield tables. Stand density is not an independent variable, and site index and age are needed to use the tables. Volume, basal area, average stand

---

D_1 = current stand density
D_2 = stand density at end of growth period
i = diameter class
$f(d_i)_1$ = current diameter distribution function
$f(d_i)_2$ = diameter distribution at end of growth period
V_A = stand volume at age A
V_1 = current stand volume
V_2 = stand volume at end of growth period
g_{12} = stand growth over one growth period
v_i = average volume per tree in diameter class i
d_i = size of diameter class i
nd_i = number of trees in diameter class i
$(nd_i)_1$, $(nd_i)_2$ = number of trees in diameter class i at start and end of a growth period
INCR_i = empirical periodic growth measurements for stand trees of diameter class i
k = tree k
d_k = diameter of tree k
h_k = height of tree k
c_k = crown size of tree k
CCI_k = crown competition index for tree k
DIST_k = inventoried distances of tree k to its competitive neighbors
$(d_k, h_k, c_k)_1$, $(d_k, h_k, c_k)_2$ = diameter, height, and crown size of tree k at start and end of a growth period
v_k = volume of tree k

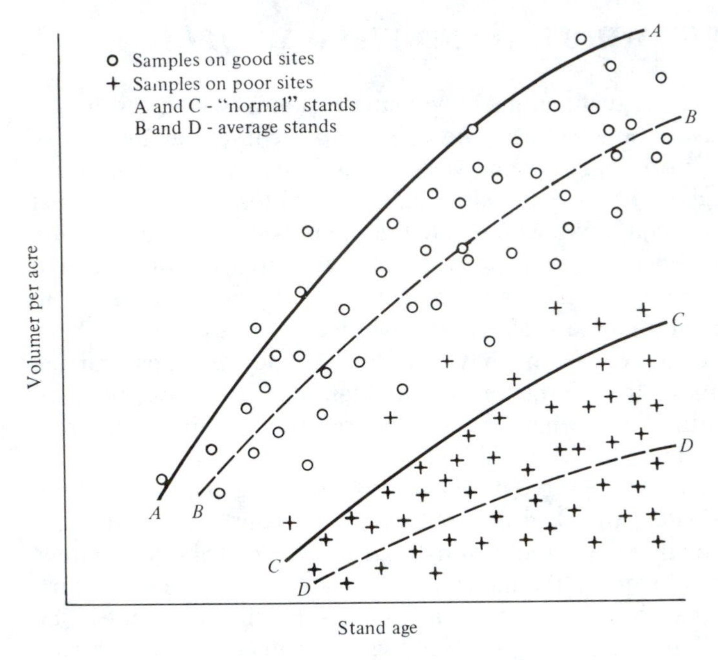

Figure 5-2 Normal and empirical yield curves fitted to volume and age observations for a random sample of natural stands on two different sites.

diameter, and a great many other stand characteristics are tabulated in a typical normal yield table, as was shown in Chapter 3 for Douglas-fir (Table 3-3).

When the foresters for a timber company or the government wanted to know how well the average natural stand grew, the logical move was to fit a curve that averaged the samples of a given site type. This resulted in curves such as *B* or *D*, which show lower volumes at a given age and site than do the normal curves. These curves are called empirical yield models, as they are empirically fit to the average of what nature provides.

Limitation of Normal Yield Tables to Describe Stand Development

Normal yield tables give a composite picture of a large number of fully stocked stands of varying age and site, each measured only once. They do not portray the actual or historical development of individual stands, except by inference. Figure 5-3 helps make this point. The solid line shows the average relation existing between stand age and volume for a particular site class as given by a yield table. The actual volume development of two individual stands *A* and *B*, which were used to help construct the yield table, is shown by the broken lines, with these curves based on

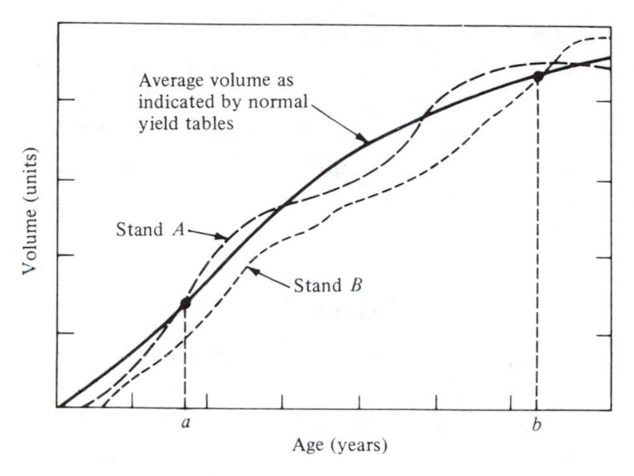

Figure 5-3 Development of two actual stands compared with the growth trend of a normal yield table.

stem analysis that re-created the stands' lives. In the yield study, stand *A* was measured at age *a* and stand *B* at age *b*. Their effect on the average yield table curve is consequently limited to these two points only. But observe the actual development history of these stands. Stand *A* has more or less followed the normal curve; stand *B* was below the normal curve nearly all of its life.

Data from stands periodically remeasured show, as would be expected, that individual stands have their ups and downs as they go through the vicissitudes of life, even if never touched by humans. Climatic cycles, insects, diseases, storms, and the like collectively affect a stand in a haphazard manner. Like people, they have good and bad periods, and observing them only at a particular time does not cast much light on the dynamics of development. Therefore, normal yield data should be used with caution in predicting what a particular stand will do; individual stands depart widely from the "normal" pattern of development.

It is unfortunate that the word "normal" has become attached as a standard of comparison to the concept of stands fully stocked at a given age; there is nothing normal about them in the sense of being "usual" or "regular." Except in a few forests, they occur in small patches only. Few stands achieve 100 percent stocking in normal yield table terms, and many natural stands have diameter structures quite different from those of carefully selected stands that form a yield table. Furthermore, researchers have consistently found that, from the standpoint of future growth, growth at near maximum levels can be attained at density levels considerably below those given by normal yield tables. The word "normal" as applied to yield tables should be thought of as a base, without any implication of normal in the sense of being usual or necessarily desirable.

Even with all these qualifications, normal yield tables remain valuable. They estimate an upper bound on the maximum growth that a site can support, and they provide a way to compare productivity among sites. They give a unified and composite picture, even though synthesized, of average development in even-aged, pure, well-stocked stands. Last, and perhaps most important, they may provide the only available base for estimating future growth.

Use of Normal Yield Tables to Predict Growth

It is rare that a subject stand measures exactly the same density as a normal yield table stand, which would allow growth and yield estimates to be read directly from the published table. Rather, the typical subject stand has a significantly lower density and thus stocking less than 100 percent relative to the yield table stand. In this case some careful thinking and assumptions are needed about how the subject stand will grow compared to the normal yield table stand. Five different assumptions span most of the possibilities. In order of increasing optimism they are:

A. Growth of the subject stand will proceed at a rate *less* than proportionate to its stocking because these are suppressed, diseased, or damaged trees.
B. Growth of the subject stand will proceed at a rate proportionate to its current stocking. With lower density, not all the growth potential of the site can be captured.
C. Growth of the subject stand will proceed at a rate proportionately *greater* than its stocking, but will still be absolutely less than the growth of the yield table stand. More than proportional growth occurs.
D. Growth of the subject stand will be exactly the same in absolute terms as the growth of the yield table stand; that is, *all* of the site's growth potential is captured by the lower density stand.
E. Absolute growth of the subject stand will be greater than that of the yield table stand. Somehow, due to better distribution, structure, site utilization, and reduced mortality, the lower-density subject stand can capture more growth potential than the yield table stand.

An example illustrates these five assumptions. Suppose a natural, even-aged stand of Douglas-fir was measured in western Oregon and was found on the average to be 60 years old, have a volume of 5,000 ft^3, and be growing on site index 140 land. The forester's prescription calls for cutting the stand in 50 years (at age 110), and wanted is an estimate of the volume at that time.

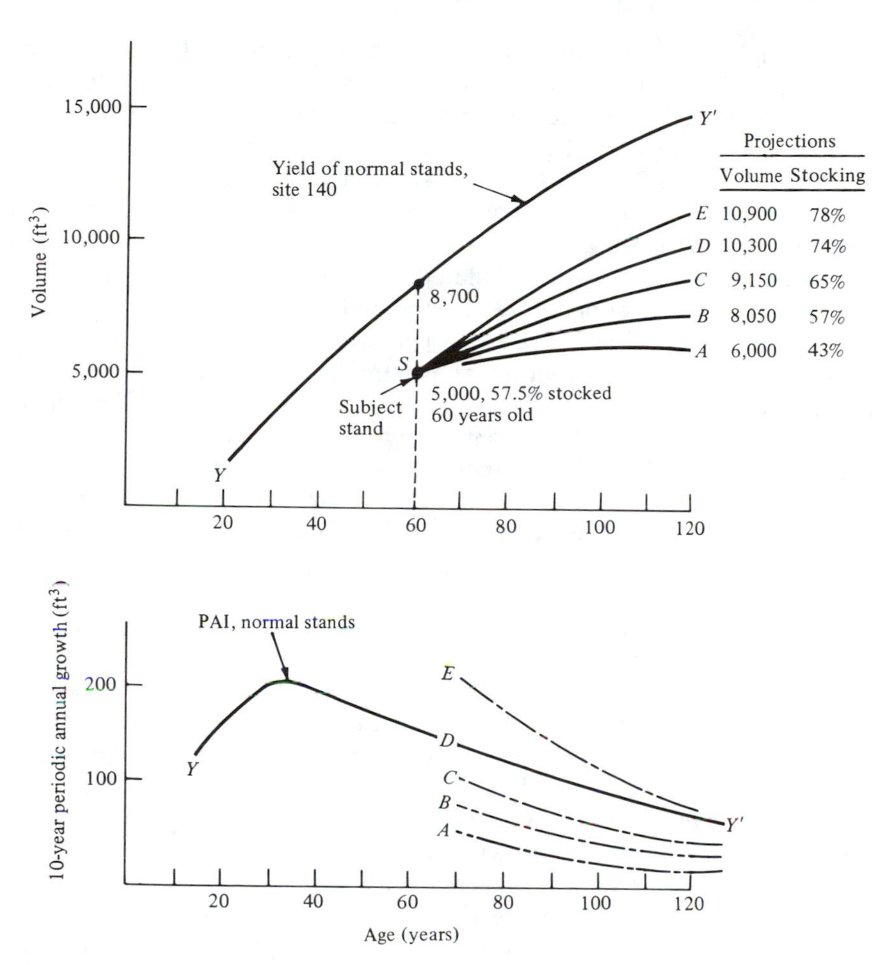

Figure 5-4 Projecting the growth of partially stocked subject stands using a normal yield table.

In Fig. 5-4*a* this subject stand is plotted on a volume versus age graph as point *S*. The graph also shows the yields of normal stands for site index 140 land (line *YY'*).* The five growth assumptions are represented by the five yield projections to age 110. The calculations are as follows. The current stocking compared to the yield table stand is 5,000/8,700, or 57.5 percent. Total growth of the normal yield table stand between ages 60 and 110 is 14,000 − 8,700, or 5,300 ft³. Projection *SA* finds the current stand in terrible shape and incapable of much growth response. From experience

* These are also the data presented in Table 3-3, which is a typical normal yield table.

it is estimated that absolute growth over the 60 years will be about 1,000 ft^3, bringing the stand to 6,000 ft^3. Compared to the yield stand, stocking actually decreases to 43 percent as many trees are expected to die. Projection *SB* says the estimated growth of the actual stand is proportionate to stocking, that is, $0.575 \times 5{,}300 = 3{,}047$. Added to the original volume of 5,000, this provides a yield estimate of 8,047 ft^3. Stocking remains at 57.5 percent. Projection *SC* assumes a relatively faster growth rate. (The specific numbers for *SC* were calculated using Gehrhardt's formula—see below—with a *K* factor of 0.9.) The estimated growth is 4,213 ft^3, bringing yield to 9,213 ft^3 and stocking to 65 percent by age 110. Projection *SD* charts the generous assumption that as much absolute growth is captured by the actual stand as by the higher-density normal stand. The yield table growth of 5,300 adds to the current stand's 5,000, providing a total yield estimate of 10,300 ft^3 and 74 percent stocking at age 110. The extremely optimistic estimate *SE* assumes an absolute growth greater than that of the yield stand, and this generates an estimated yield of 10,900 ft^3 and 78 percent stocking.

The decade-by-decade details of these growth projections over the 50-year projection are shown in the periodic annual increment curves in Fig. 5-4*b*. Projection *D* again is taken from the yield table and follows the line *YY*′.

Which is the correct projection? We cannot really say without more knowledge about the stand and its history. Given that the concept of the normal stand is that of a healthy, well-distributed stand which completely occupies the site, growth assumption *SD* seems to be a reasonable upper limit; *SE* is simply too far out without some convincing biological arguments. The most likely estimate is between *SB* and *SD*; the more the site is biologically occupied, the closer growth approaches *SD*.

Many rules of thumb and equations have been proposed to generalize how understocked stands grow and approach normality. Gehrhardt (1930) proposed an equation which caused partially stocked stands to asymptotically approach normal stocking at a rate controlled by a single factor *K*. The method consists of a formula for estimating an approach toward normality by 10-year periods. Duerr (1938) stated it as

$$g = dG(1 + K - Kd)$$

where g = 10-year growth of understocked stand
d = density, or percentage of stocking in relation to fully stocked stands, expressed as a decimal
G = 10-year growth of a yield table stand of same age and site
K = constant for species or species group for a 10-year period

The factor dG estimates the proportionate growth of the actual stand g from the yield table growth; the factor $(1 + K - Kd)$ applies a correction to accelerate growth for stands with $d < 1.0$.

Estimating rates of stocking change is difficult because every stand is unique. The best estimate will probably be made by examining all the facts from the actual stand and then making an informed judgment.

Empirical Yield Tables

Many public forests and private ownerships have developed empirical yield tables, such as the empirical yield curve for ponderosa pine on the Boise National Forest (Fig. 5-5). A regression of volume over age (and sometimes site) is fitted for all plots on the forest that meet the species definition. Yield for any acre is read directly off the curve, and growth is calculated as the difference between two values on the curve. The conceptual argument behind such an approach is that this yield function reflects "average" conditions on the forest and should therefore provide a good prediction of future yields for existing stands. A telling, practical argument is often that normal, or variable-density, tables are not available and the only data at hand are the forest inventory.

Growth measurements from inventory seldom measure mortality successfully, and only net yield is obtained when stand volumes are plotted against age. The typical empirical yield curve shows a definite leveling off of volume increase in the older ages (Fig. 5-5). Measured increment for individual plots in stands of the ages where leveling occurs, however, still often shows positive net growth. If this net growth reflects what the

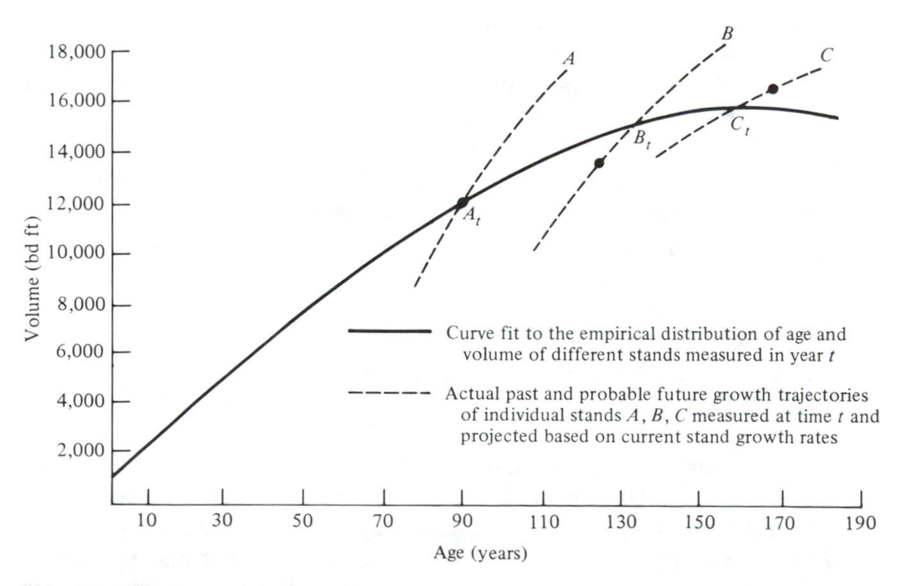

Figure 5-5 Empirical yield curve for ponderosa pine and associated species on the Boise National Forest.

stands will actually do over the next few decades, the volume-age curve should continue ever upward. What is the matter?

Most likely, mortality is either not being accurately measured or accurately predicted, or both. Often 5-year growth is obtained from increment cores, and 5-year mortality is estimated from rules of thumb, such as "count all standing dead as having died within the past 5 years." Perhaps these rules of thumb underestimate mortality systematically. Or perhaps, the flattening of the volume-growth curve after a certain age reflects past epidemics, such as insects and fire, which 5-year mortality estimates might miss. And *should* growth be reduced for these possible disasters, or has civilization gotten them under control?

Part of the problem surely lies in the methodology of using cross-sectional data to estimate growth over time. The stands sampled for empirical yield show how *different* stands grew over time, not how the *same* stand grew. If we measured all stands at time t and assumed that today's young stand will grow at the same average rate as today's old stand did, the observed empirical curve results. If most stands measured are coming out of a history of disease, overcutting, or stagnation, then their actual growth and yield trajectories may be as shown by the dashed curves in Fig. 5-5, which display much greater potential future growth than the average curve. This distortion is amplified if the sampled young stands are of better site quality than the older stands, which can happen when the better-quality and easily accessible valley bottoms and gentle slopes are logged first. (We might find this to reverse in another 50 or so years with the older stand at the bottom and the young stand in the hills.)

Whatever the reason why the plot data disagree with volume-age curves, it is not just a casual interest: harvest depends in part on projected stand growth. Empirical yield tables can provide a valuable first estimate of growth and yield for large areas of an unmanaged forest, but they are not very useful for individual stand prediction. Tying prediction of future growth entirely to the past should always be questioned—unless you believe that the efforts of foresters at improved protection and growth will come to naught.

VARIABLE-DENSITY WHOLE STAND GROWTH AND YIELD MODELS

Growth and yield models which use density as an explicit independent variable have been produced for many natural and planted species and stand conditions. Density is most often expressed as basal area, but sometimes it is also expressed as number of trees or volume per acre or the other density measures we discussed in Chap. 4. We examine representative variable-density whole stand models for four species.

Red Pine in Minnesota

Buckman (1962) published the first study in the United States that directly predicted growth from current stand variables and then integrated the growth function to obtain yield. Using permanent sample plots that reflected a wide variety of stand treatments on red pine, he derived a direct basal area growth equation for all trees 3.6 in. DBH and larger as a function of age, site index, and stand density:

$$Y = 1.6689 + 0.041066\text{BA} - 0.00016303\text{BA}^2 - 0.076958A + 0.00022741A^2 + 0.06441S \quad (5\text{-}1)$$

where Y = periodic net annual basal area increment
BA = basal area, in square feet per acre
A = age, in years
S = site index

No improvement in prediction was added by considering number of trees per acre, variability of tree diameter, thinning intensity, or thinning method (from above, from below, etc.).

Volume growth tables were developed by combining (1) the basal area growth equation, (2) a height growth table, and (3) the stand volume equation. The stand volume equation was, implicitly,

$$V = f(K, \text{BA}, H) \quad (5\text{-}2)$$

where V = volume
K = a constant representing average form for trees in stand
BA = basal area, in square feet per acre
H = average height of dominant and codominant trees, in feet

Using the basal area growth equation, Buckman then projected the basal area growth under maintenance of each of three levels of stand density (90, 120, and 150 ft^2 of basal area). With thinning cycles of 10 years, he simulated growth from a constant residual basal area at the conclusion of each cycle. Table 5-2 illustrates this computation for a residual stand density of 90 ft^2 of basal area on a site of 50 at 50 years.

Notice that the calculation of mean annual increment takes these thinnings into account. At each age, the mean annual increment is calculated by dividing the cumulative total net growth (cumulative harvests plus current residual stand) by the age. As an example, the mean annual increment at age 85 in Table 5-2 equals 7,710/85, or 91 ft^3 per acre per year.

Maximum basal area growth occurs at a residual density of 120 ft^2 throughout the life of the stand. Actually, the maximum occurs at a little above 120. Differentiating the basal area growth equation (5-1) with re-

Table 5-2 Computation of growth and yield for a red pine plantation thinned every 10 years to a residual density of 90 ft² of basal area, site index is 50 ft at 50 years

Age, yr (1)	Height of dominants and codominants, ft (2)	Residual stand* basal area (3)	Residual stand* ft³ (4)	Residual stand* cords (5)	Residual stand* bd ft (6)	Harvest removals* basal area (7)	Harvest removals* ft³ (8)	Harvest removals* cords (9)	Harvest removals* bd ft (10)	Yield: cumulative removals plus current stand* ft³ (11)	Yield: cumulative removals plus current stand* cords (12)	Yield: cumulative removals plus current stand* bd ft (13)	Periodic net annual growth,* ft³ (14)	Mean net annual* growth, ft³ (15)
25	24.5	90.0	900	8.7	—	—	—	—	—	900	8.7	—	118	36
35	35.5	90.0	1,300	12.6	—	53.7	780	7.5	—	2,080	20.1	—	125	59
45	15.5	90.0	1,670	16.2	8,500	47.4	880	8.5	—	3,330	32.2	8,500	122	74
55	53.5	90.0	1,970	19.1	10,000	41.9	920	8.9	4,700	4,550	44.0	14,700	113	83
65	60.0	90.0	2,210	21.4	11,200	36.4	890	8.6	4,600	5,680	54.9	20,500	107	87
75	66.0	90.0	2,430	23.5	12,400	31.4	850	8.2	4,300	6,750	65.2	26,000	96	90
85	71.0	90.0	2,610	25.3	13,300	27.0	780	7.6	4,000	7,710	74.6	30,900	86	91
95	75.0	90.0	2,760	26.7	14,100	23.2	710	6.9	3,600	8,570	82.9	35,300	76	90
105	78.5	90.0	2,890	28.0	14,700	19.6	630	6.1	3,200	9,330	90.3	39,100	64	89
115	81.0	90.0	2,980	28.9	15,200	16.6	550	5.3	2,800	9,970	96.5	42,400	54	87
125	83.0	90.0	3,050	29.6	15,600	13.8	470	4.5	2,400	10,510	101.7	45,200	47	84
135	84.5	90.0	3,110	30.1	15,800	11.9	410	4.0	2,100	10,980	106.2	47,500	38	81
145	85.5	90.0	3,140	30.5	16,000	10.0	350	3.4	1,800	11,360	110.0	49,500	36	78
155	86.5	90.0	3,180	30.8	16,200	9.0	320	3.1	1,600	11,720	113.4	51,300	30	76
165	87.0	—	—	—	—	98.0	3,480	33.7	17,800	12,020	116.3	52,900		73

Source: Buckman (1962).

* All data per acre.

spect to the basal area gives

$$\frac{dY}{d\text{BA}} = 0.041066 - 0.000322606\text{BA} \tag{5-3}$$

To maximize, we set this equation equal to zero and solve for X_1. This gives us the basal area that supports maximum growth, which turns out to be 0.041066/0.000322606, or 127 ft^2 of basal area. We also note that the amount of stand growth declines steadily after the initial age of 25 years (Fig. 5-6).

At young ages, maximum cubic volume (and cordwood) growth occurs at stand densities of 180 ft^2 of basal area. As the stand gets older, maximum growth shifts to 150 and then to 120 ft^2 of basal area (Table 5-3). This shifting maximum is associated with the stand's height development. At early ages, rapid height growth contributes greatly to volume. The higher the density, the more stems are available for this height growth and thus the greater the volume growth. As height growth slows down, the basal area component of volume growth becomes more important. Since basal area growth is highest at densities of 120 to 130 ft^2, maximum volume growth moves toward these densities as the stand ages and height growth declines.

Buckman's results show that cubic foot growth is positively correlated with density at young ages over the range of densities he examined. It must be remembered, however, that the higher the number of trees for a given basal area, the smaller the average tree. When he employs a utiliza-

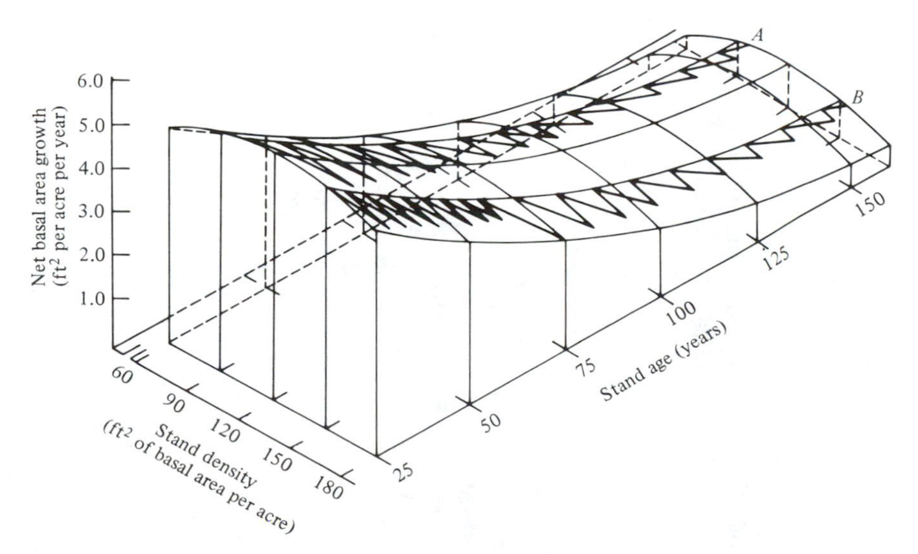

Figure 5-6 Basal area growth of red pine stands thinned to 90 ft^2 (stand A) and 150 ft^2 (stand B) after each entry. *(From Buckman, 1962.)*

Table 5-3 Periodic annual increment measured in cubic feet and board feet for red pine plantations as a function of stand age and residual stand density

Basal area per acre, ft²	Net growth per acre per year when stand age is													
	30	40	50	60	70	80	90	100	110	120	130	140	150	160
	Cubic feet													
60	87	94	92	86	80	71	61	51	41	32	25	—	—	—
90	106	115	112	106	100	92	80	71	60	52	44	37	34	30
120	122	131	127	119	113	104	92	82	70*	61*	53*	46*	42*	38*
150	135	142	134	125*	118*	107*	94*	84*	70*	60	52	44	40	36
180	143*	147*	136*	124	115	103	87	76	61	50	40	42	28	22
	Board feet													
60		480*	470	440	410	360	310	260	210	160	130	—	—	—
90			570	540	510	470	410	360	310	260	230	190	170	150
120			650*	610	580	530	470	420	360*	310*	270*	230*	210*	190*
150				640*	600*	550*	480*	430*	360*	310*	260	210	160	140

Source: Buckman (1962).

* Maximum growth at each age.

tion standard favoring large trees (board feet of all trees 7.6 in. DBH and larger), the results change slightly: growth maximizes on densities lower than the highest he examined.

Southern Pines

Compatible estimates of loblolly pine growth and yield Clutter (1963) formalized the necessary relationships between growth and yield observed in Buckman's model and laid down the conditions of compatibility between growth and yield:

> Such models are here defined as compatible when the yield model can be obtained by summation of the predicted growth through the appropriate growth periods or, more precisely, when the algebraic form of the yield model can be derived by mathematical integration of the growth model.

Clutter's procedure followed these steps: (1) Current cubic volume V_1 and basal area yield B_1 models are obtained as a function of current age A, site S, and basal area B. (2) Derivatives of the yield models (dV/dA, dB/dA) with respect to age are then taken to establish volume and basal area growth models that are compatible with the original yield models. (3) Using linear regression techniques, the coefficients of the growth equations are estimated from the research data. (4) Finally, the estimated growth equations are integrated to obtain the yield and basal area projection models.

Sullivan and Clutter (1972) refined this technique further. Given are three basic equations from the compatible set (here presented in implicit form):

$$\text{Current yield:} \quad V_1 = f(S_1, A_1, B_1) \tag{5-4}$$

$$\text{Future yield:} \quad V_2 = f(S_1, A_2, B_2) \tag{5-5}$$

$$\text{Projected basal area:} \quad B_2 = f(A_1, A_2, S_1, B_1) \tag{5-6}$$

they substituted Eq. (5-6) for B_2 into Eq. (5-5), and the result is an equation for future yield in terms of current stand variables and projected age, called a simultaneous growth and yield model:

$$V_2 = f(A_1, A_2, S_1, B_1) \tag{5-7}$$

The coefficients of Eq. (5-7) can be estimated directly. When $A_2 = A_1$, the equation resolves to the current yield equation. Clutter's procedure of defined compatibility of growth and yield and the refinement of Sullivan and Clutter provide a conceptually solid basis for developing stand growth and yield models. Eliminated is the problem of inconsistent estimates for future yield when the growth and yield equations are derived independently. Moser and Hall (1969) used this compatibility approach for natural uneven-aged stands.

An example of the simultaneous model is found in the equation published by Brender and Clutter (1970), which was fitted to 119 remeasured piedmont loblolly pine stands near Macon, Georgia. The cubic foot yield projection equation is:

$$\log CV_2 = 1.52918 + 0.002875S - 6.15851\,(1/A_2) + 2.291143(1 - A_1/A_2) + 0.93112\,(\log B_1)(A_1/A_2) \quad (5\text{-}8)$$

where S = site index
A_1 = current stand age
B_1 = current basal area
A_2 = projected stand age
CV_2 = projected volume per acre at age A_2, in cubic feet

Letting $A_2 = A_1$, this same equation predicts the current cubic foot volume:

$$\log CV_1 = 1.52918 + 0.002875S - 6.15851(1/A) + 0.93112 \log B \quad (5\text{-}9)$$

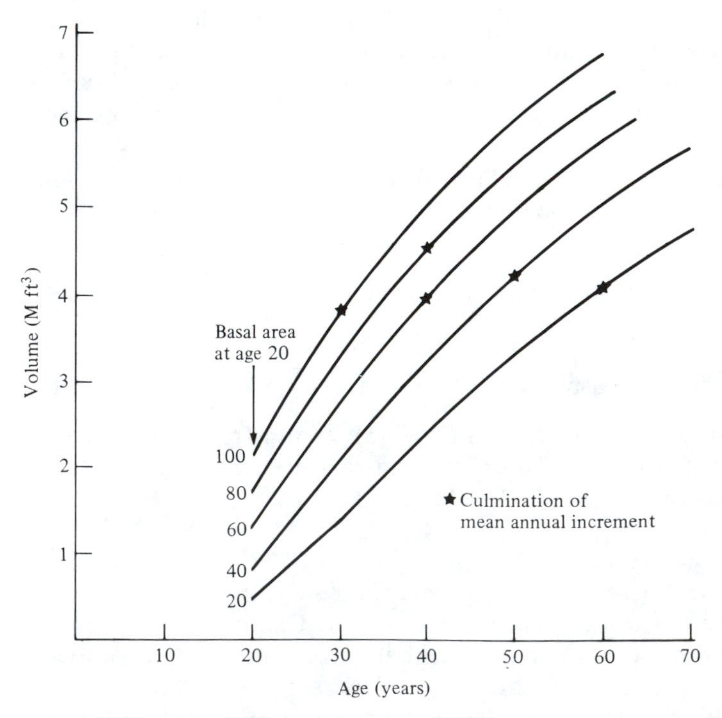

Figure 5-7 Projected yields for natural stands of loblolly pine on site index 90 and for different basal area densities at age 20.

where S = site index
A = current stand age
B = current basal area
CV_1 = current volume per acre, in cubic feet

Solving these equations for various combinations of current age, projected age, current basal area, and site index gives information on growth and yield for natural stands of loblolly pine (Fig. 5-7).

To illustrate an application of the Brender–Clutter model, assume a stand growing on site 80 is currently 25 years old with a basal area of 70 ft^2. The owner wants an estimate of current volume and the volume expected after 10 more years of growth.

Current volume is estimated using Eq. (5-9),

$$\begin{aligned}\log \mathrm{CV}_1 &= 1.52918 + 0.002875(80) - 6.15851(1/25) + 0.93112 \log 70\\ &= 1.52918 + 0.23 - 0.24634 + 1.71801\\ &= 3.23085\\ \mathrm{CV}_1 &= 10^{3.23085} = 1{,}701 \text{ ft}^3\end{aligned}$$

Volume in 10 years is estimated using Eq. (5-8),

$$\begin{aligned}\log \mathrm{CV}_2 &= 1.52918 + 0.002875(80) - 6.15851(1/25) + 2.291143(1 - 25/35)\\ &\quad + 0.93112 (\log 70) (25/35)\\ &= 1.52918 + 0.23 - 0.24634 + 0.65461 + 1.22714\\ &= 3.39459\\ \mathrm{CV}_2 &= 10^{3.39459}\\ &= 2{,}480 \text{ ft}^3\end{aligned}$$

Diameter distribution models Diameter distribution models are a widely used refinement of stand models, particularly for the southern pines. The purpose of the model is to disaggregate stand model results at each age and add additional information about diameter class structure. Height, volume, and other characteristics can then be associated with each size class. This in turn will allow a more sophisticated economic analysis, which can consider logging costs and log values by diameter class.

The method works by assuming that trees are distributed according to a specified mathematical function that provides the frequency of trees in each diameter class:

$$f(d_i) = f(T, A, H) \tag{5-10}$$

where $f(d_i)$ = frequency of trees in each diameter class i
T = number of trees per acre
A = age
H = height of dominant trees in stand

The three-parameter Weibull distribution is most commonly used in Eq. (5-10) to establish the frequencies. The estimation procedures, although lengthy, are straightforward (Bailey and Dell, 1973). Many applications for southern pine plantations have been published. The shortleaf tables by Smalley and Bailey (1974) and the slash pine tables by Dell et al. (1979) are good examples. Clutter et al. (1983) and Avery and Burkhart (1983) provide good conceptual and computational discussions with examples.

Diameter distribution models can be used for natural stands and could be used for species other than southern pines, although such excursions have so far been limited.

Comparing different loblolly projection models Burkhart et al. (1981) did an interesting comparative study to quantify the variation of volume estimates between different variable-density models for loblolly pine. Volume estimates were made for different stand ages, site indices, and stand densities. Three whole stand and four diameter distribution models were used to predict volumes of old-field plantations, and some of these results are shown in Table 5-4 for 25-year-old stands on different sites and at different stand densities.

Using the average of the seven estimates as a basis for comparison, we see that the range of the estimates is about ±20 percent from the mean. Furthermore the predictive model that gives the high or low estimate changes with site and density. All models except model 6 show a positive relationship between density and volume.

Perhaps the most important conclusion to draw is that different models give different results and that the difference in absolute or relative terms could be significant in culture, harvest, or land buying decisions. The analyst who needs to make an accurate prediction of volumes for an old-field loblolly subject stand had better study carefully the geographic source, history, and character of the sample stands underlying these models in relation to the subject stand before making a final choice of which model to use.

Growth of Managed Uneven-Aged Northern Conifer

Natural uneven-aged stands of spruce-fir-hemlock in the Penobscot Experimental Forest of northern Maine were placed under uneven-aged

Table 5-4 Comparison of the predictions from seven different current stand volume prediction models of an old-field loblolly plantation age 25

Site index, in feet (base age 25)	50			60			70		
Stand density, number of surviving trees per acre	400	600	800	400	600	800	400	600	800
Prediction model*									
1	2,030	2,373	2,609	3,305	3,944	4,426	4,916	5,937	6,737
2	1,418*	2,009	2,380	2,420	3,429	4,061	3,544	5,022	5,948
3	1,730	2,001	2,227	2,842	3,470	4,032	4,277	5,400	6,434
4	2,289	2,553	2,809	3,286	3,867	4,380	4,426	5,228	6,053
5	1,792	1,888	1,894	2,904	3,126	3,238	4,330	4,664	4,906
6	2,046	2,206	1,969	3,100	3,071	2,983	4,698	4,653	4,521
7	1,977	2,229	2,445	2,723	3,131	3,493	3,493	4,120	4,590
Average of estimates	1,897	2,154	2,333	2,940	3,434	3,801	4,240	5,003	5,598
Deviation from average estimate									
Highest	392	399	476	365	510	625	676	934	1,139
(Percent)	(20)	(18)	(20)	(12)	(15)	(16)	(16)	(19)	(20)
Lowest	479	266	439	520	363	818	747	883	1,077
(Percent)	(25)	(12)	(19)	(18)	(11)	(21)	(18)	(18)	(19)

Source: Burkhart et al. (1981).

* Maximum and minimum values for each site and density are underlined. Data are given in cubic feet per acre of 4-in. top, outside bark.

management in 1952 (Solomon and Frank, 1983). Five different residual basal areas under three different cutting cycles of 5, 10, and 20 years were set up in the experiment. All harvesting has been from the larger-diameter size classes, and 359 permanent plots in these stands have been monitored since harvesting was initiated.

Results from plots in each residual density and cutting cycle are directly averaged and presented as variable-density yield tables (Table 5-5). The cubic foot growth rates for these stands indicate total gross and net growth to be about the same for all levels of residual basal area. However, the allocation of growth between poletimber- and sawtimber-sized trees shows the sawtimber sizes increasing their share with higher residual basal areas. This is expected with large tree marking guides: with higher basal areas, more large trees are left in the residual stand. The length of the cutting cycle does not appear to affect growth rates.

Table 5-5 Gross and net annual periodic cubic foot increment of uneven-aged spruce-fir-hemlock stands at different residual densities and cutting cycles

Residual basal area, ft²/acre	Harvest interval, yr	Gross growth,	Net growth			Gross growth of residual stand			Ingrowth		Mortality	
			Pole-timber	Saw-timber	Stand total*	Pole-timber	Saw-timber	Stand total‡	Pole-timber	Saw-timber	Saw-timber	Stand total
40	5	60.35	30.64	29.00	59.64	32.20	14.67	46.87	13.48	0.70	0.00	0.70
	10	52.13	18.58	20.36	38.94	28.74	9.29	38.03	14.10	11.56	1.63	13.19
	20	62.54	22.02	33.83	55.85	33.71	14.03	47.74	14.80	6.10	.58	6.68
60	5	61.15	17.62	35.20	52.82	29.87	17.56	47.43	13.77	6.83	1.49	8.32
	10	62.02	18.63	33.29	51.92	33.36	16.18	49.54	12.53	8.02	2.08	10.10
	20	63.96	17.81	35.52	53.33	35.72	15.97	51.69	12.28	8.29	2.33	10.62
80	5	63.80	10.29	41.54	51.83	29.71	24.51	54.22	9.67	9.40	2.57	11.97
	10	64.40	12.20	38.68	50.88	32.95	21.65	54.60	9.82	9.57	3.96	13.53
	20	68.46	11.74	44.72	56.46	34.76	23.65	58.41	10.08	8.40	3.59	11.99
100	5	66.22	6.48	49.61	56.09	30.26	27.53	58.04	8.43	7.76	2.37	10.13
	10	68.70	4.63	51.94	56.57	33.39	26.94	60.32	8.34	8.00	4.13	12.13
	20	68.72	8.94	48.26	57.20	31.33	28.65	59.98	8.74	9.06	2.46	11.52
120	5	70.14	7.10	49.12	56.22	33.58	28.88	62.83	7.68	9.48	4.44	13.92
	10	61.94	.97	48.50	49.47	29.07	26.39	55.46	6.44	8.10	4.35	12.45
	20	68.81	.35	60.11	60.46	24.73	39.15	63.88	4.92	5.01	3.34	8.35

Source: Adapted from Solomon and Frank (1983).

* Net growth = gross growth − mortality.

‡ Residual growth = gross growth − ingrowth.

Growth Dynamics of Natural, Unmanaged, Uneven-Aged Lake States Hardwoods

Using detailed annual growth data from 75 long-term research plots, Moser (1972) developed and quantified a system of differential equations to characterize the growth dynamics of these stands (Table 5-6). Expressions such as dN_t/dt are differential calculus notations for the change in

Table 5-6 System of equations for the growth dynamics of natural uneven-aged Lake states hardwoods

System of equations	Purpose
1. $\frac{dN_t}{dt} = \frac{d\text{NI}_t}{dt} - \frac{d\text{NM}_t}{dt}$	Periodic change in tree numbers
2. $\frac{dB_t}{dt} = \frac{d\text{BI}_t}{dt} + dG_t - \frac{d\text{BM}_t}{dt}$	Periodic basal area growth
3. $\frac{d\text{NM}_t}{dt} = 0.00915N_t$	Mortality (number of trees)
4. $\frac{d\text{NI}_t}{dt} = 1.48106\ e^{-2.24515(B_t/N_t)}$	Ingrowth (number of trees)
5. $\frac{d\text{BM}_t}{dt} = 0.0000499N_t[-5.21295(\log x) - 7]^2$	Mortality (basal area)
6. $\frac{d\text{BI}_t}{dt} = 0.48920\ e^{-2.24525(B_t/N_t)}$	Ingrowth (basal area)
7. $\frac{dG_t}{dt} = -046569B_t^{1.0125} + 0.49818B_t$	Survivor growth (basal area)

Tree number accounting

The number of trees in the stand is defined by the initial number from inventory N_t and the change in number estimated by Eqs. (1), (3), and (4):

$$N_{t+1} = N_t + \frac{dN_t}{dt}$$

$$N_{t+1} = N_t + \frac{d\text{NI}_t}{dt} - \frac{d\text{NM}_t}{dt}$$

ending number = starting number + ingrowth − mortality

Source: Adapted from Moser (1972).

N_t = number of surviving trees greater than 7.0 in. at time t
B_t = basal area of surviving trees greater than 7.0 in. at time t
NM_t = number of trees that die over a growth interval (mortality)
NI_t = number of trees that will attain merchantable size over a growth interval (ingrowth)
BM_t = basal area of mortality trees M_t
BI_t = basal area of ingrowth trees I_t
G_t = basal area growth of survivor trees N_t
x = random variable in interval (0, 1)

the variable N_t with a specified change in the variable time t. The seven equations show the changes in stand basal area and tree number as a function of the current stand basal area and tree number. Mortality and ingrowth are explicitly modeled. The system then would be considered an explicit, whole stand growth model by our classification.

Moser's stand model has no variable to describe the distribution of trees by diameter class and stand structure is characterized by basal area and the number of trees. Caution would be needed when using such a system to predict growth of an uneven-aged stand under a periodic harvest cycle with certain diameter structure objectives, particularly if the desired stand structure was different from the stands in Moser's plots.

DIAMETER CLASS MODELS

The stand level forest growth and yield definitions developed in Chap. 3 are equally appropriate when using diameter class models. The only real difference is that we keep track of things at the class level and add up the results to get stand level statistics and information.

Consider the activity in one diameter class over a growth period t to $t + 1$. (This could also be done by species or species groups.)

Item	Definition	Example value
N_{it}	Number of trees in class i at time t	83
I_i	Number of trees growing into class i in growth interval (ingrowth)	47
U_i	Number of trees growing out of class i in growth interval (upgrowth)	61
M_i	Number of trees in class i dying in interval (mortality)	7
C_i	Number of trees in class i being cut in interval (harvest)	12
$N_{i,t+1}$	Number of trees in class i at end of growth interval	50

The solution for $N_{i,t+1}$ necessarily is the algebraic sum of all movements in and out of the class,

$$N_{i,t+1} = N_{it} + I_i - U_i - M_i - C_i \tag{5-11}$$

$$= 83 + 47 - 61 - 7 - 12$$

$$= 50 \text{ trees}$$

This accounting is exactly like that of Moser's hardwood stand model (Table 5-6) except, because we are working with diameter classes, the upgrowth term U_i is added to account for departures from each class.

Equation (5-11) is calculated and solved for each diameter class present in the stand.

Additional stand information about volume, value, and other items is calculated by developing per-tree volume or value tables by tree size and species and then multiplying and summing over all classes in the stand. In this way it is easy to compute volumes and values for different merchantability standards and product prices that vary by tree size and species.

Stand Table Projection

Stand table projection is a traditional diameter class method that estimates the future stand table of a subject stand on the basis of the present one, using actual diameter growth and other information which is collected from the subject stand for each diameter class. All trees are assumed to survive the projection period; mortality and harvests must be handled separately from the growth projection.

Two questions must be answered to use this approach: (1) What increment data should be used? 2) How should they be applied to the stand? Past increments for different tree sizes obtained from increment borings commonly provide the growth rates. But past stand growth is not a good predictor if growth conditions or stand structure change. Stand table projection gives best results with short projection periods—no more than 10 and preferably 5 years—as changes in structure and stand vigor are not pronounced over so short a time.

A decision must be made as to how growth measurements will be applied. One assumption is that a tree will maintain the radial growth it made in the past. If a tree which currently has a 12-in. DBH grew 2 in. in the past 10 years, it will also grow 2 in. in the next 10 years. Another assumption is that a current 12-in. tree will grow as did a different tree that was 12 in. 10 years ago. While both are tenable assumptions, neither may hold if growth conditions change. A 12-in. tree now may be in a better or poorer condition to grow than it was 10 years ago, perhaps 12-in. trees were mostly dominants or codominants 10 years ago, but now they are mostly codominants and intermediates. Conversely, due to a commercial thinning or an overstory harvest, the 12-in. trees that were intermediates may now be dominants. Stand table projections do not deal with stand density or site explicitly; they are assumed to be integrated by the increment data collected from the stand. Clutter et al. (1983) provide a detailed critical review of the stand projection method and found much potential for misuse. Still, when other models are not available or appropriate, a stand table projection may be the only practical way to make a growth estimate.

Differences in projection methods revolve largely around the way stems are assumed to be distributed within a diameter class and how the diameter growth information is applied to estimate class growth.

Alternative Stand Table Projection Methods

Method 1: Application of average diameter increment to the midpoint of the diameter class The simplest assumption is that all trees fall at the diameter class midpoint and grow at the average rate for the class.

Estimating future diameter classes by grouping trees on the basis of average diameter increments applied at the class midpoint is a rather crude procedure. When 1-in. classes are used, trees growing up to 0.5 in. are not credited with any growth at all but stay in the same class. Trees growing over 0.5 but less than 1.5 in. move up one full diameter class, regardless of the spread in actual increment. This difficulty is even more pronounced with 2-in. classes, where trees can grow up to 1 in. without changing diameter class at all. Depending on how the average increment is distributed in relation to the diameter structure of the stand, a considerable overestimate or underestimate of growth can result.

Application of average diameter increments to the midpoint of the diameter class ignores two important facts about tree growth and distribution: (1) actual tree diameters do not fall at the midpoint but are distributed through the class, and (2) trees within a given diameter class grow at variable rates. The next two approaches were developed to deal with these problems.

Method 2: Application of average diameter increment recognizing dispersion within classes Even though the actual distribution of tree diameters within classes is not known, it can be approximated by assuming a uniform distribution through the class. With this assumption the proportion of trees advancing into higher DBH classes can be defined as a movement ratio,

$$m = \frac{g}{i} \times 100 \tag{5-12}$$

where m = movement ratio, expressed as a percentage
g = average periodic diameter increment, in inches, of a DBH class
i = diameter class interval, in inches

Assume, for example, that $g = 1.6$ and $i = 1.0$ in. Then

$$m = \frac{1.6}{1.0} \times 100, \qquad \text{or 160 percent for the DBH class}$$

In interpreting this movement percentage ratio, the first two digits from the right indicate the percentage of the trees advancing one diameter class beyond the number of classes indicated by the third digit from the right. In the above example $m = 160$, hence 60 percent of the trees move up two classes and the rest one class. None remain in the initial diameter class.

Method 3: Application of variable diameter increment to actual diameters Trees grow at variable rates and are irregularly distributed throughout a diameter class. For example, with 2-in. classes, the 14-in. class extends from just over 13 in. to just under 15 in. If a particular tree, say 14.4 in. DBH, grew 0.5 in., it would stay in the 14-in. class, even though it was toward the upper limit of the class to begin with. Another tree, say 13.2 in. DBH, might grow 2 in. during 10 years and consequently move into the 16-in. class. Almost any such combination may occur in nature.

This third approach refines the assumption of uniform tree distribution within a diameter class used by the second method and works up the distribution empirically from the actual stand inventory and increment data. The percentage of trees moving ahead to different classes is then determined for each diameter class based on the actual distribution of growth rather than assuming the uniform distribution of trees having the same average growth. This method increases accuracy somewhat, but considerably more increment data are needed to establish the movement percentages.

Mortality, Harvest Removals, and Ingrowth

Mortality is not directly handled by stand table projections. The current stand table and radial increment data are typically collected from a one-time sample survey of the subject stand. Mortality data are not readily available from such surveys, and the implied assumption is that all trees will survive for the 5- or 10-year duration of the projection period. This, of course, is not a good assumption, and some of the trees surely will die. Mortality data normally must be brought in from some secondary source. Perhaps some information on the stand type is available from permanent plot records. Or research may provide some general mortality function that can be adapted. For example, Hamilton (1974) developed an individual tree mortality probability equation for western white pine in northern Idaho relating the mortality expectation to stand basal area and tree diameter:

$$m = \frac{1}{1 + \exp(3.25309 - 0.00072647B + 0.0166809D)} \qquad (5\text{-}13)$$

where m = mortality probability of a tree for subsequent year
B = basal area per acre, in square feet
D = tree DBH, in inches

This example also serves to illustrate the form of mortality equation used in many models:

$$m = \frac{1}{1 + \exp X} \tag{5-14}$$

where m is the probability of mortality and X is some function of tree and stand characteristics.

By iterating such an equation over the number of years in a projection period, the average cumulative periodic probability of mortality for trees of each diameter class can be estimated. This probability can be multiplied times the number of trees in each class and subtracted to obtain survivors by DBH class.

For an example of mortality calculation, assume a tree is 16 in. in diameter and is growing in a stand having 140 ft^2 of basal area. The probability that this tree will die is estimated using Eq. (5-13) as

$$\begin{aligned} m &= \frac{1}{1 + \exp(3.25309 - 0.00072647(140) + 0.0166809(16))} \\ &= \frac{1}{1 + \exp(3.25309 - 0.10169 + 0.26689)} \\ &= \frac{1}{1 + \exp(3.41829)} = \frac{1}{30.517} \\ &= 0.0328 \end{aligned}$$

The tree has a little over 3 percent chance of dying during the next year.

Permanent plot records may provide some empirical data on current mortality rates. A simple linear regression fit might yield a useful equation of the form

$$P_i = a - bD_i \tag{5-15}$$

where P_i = percent of stems in class i dying
D_i = diameter of class
a, b = coefficients

Mortality needs explicit consideration using some technique if the stand table projections are to be realistic.

Ingrowth is most easily handled by inventorying and collecting increment data for two or three size classes smaller than the merchantable size limit so that ingrowth can be estimated directly. If this is not done, then assumptions and research data on diameter distribution forms and ingrowth rates must be made. If harvests are to be considered, the number of trees to be removed by diameter class can be prescribed and subtracted before or after growth, mortality, and ingrowth calculations, leaving a residual stand for the next growth cycle.

Volume increment involves height growth in addition to diameter growth. A common practice is to assume that the height-diameter relationship does not change over time so that the volume per tree by diameter class for the present stand table applies also to the future stand table. Multiplying the volume per tree in each diameter class by the number of trees in each class gives the volume per acre. The difference between volumes in present and future stand tables gives the volume increment during the projection period under the assumption that no trees die. The average height of trees in a diameter class will typically increase with stand age. A more sophisticated projection would therefore require estimating this rate of change and adjusting the volume per tree in each DBH class accordingly.

An Example Estimating Gross Volume Growth

To illustrate the computation of a stand table projection, a stand with 135 trees per acre and initial volume of 4,369 bd ft was cruised and trees were bored for radial increment. The number and average increment of trees by diameter class are shown in columns (1) and (2) of Table 5-7. The second method of projection assuming a uniform dispersion within a diameter class is employed with the percentage and number of trees moving 0, 1, 2, or 3 diameter classes ahead, as shown in columns (4) to (11). The stand after 10 years of growth is shown in column (12) with all 135 trees accounted for. As a tree must be 8 in. to be merchantable, by tracking the growth of the 4- and 6-in. classes we have an estimate of ingrowth.

From mortality studies and prescriptions, the mortality rate by diameter class and the rate of stems harvested by diameter class are obtained and multiplied times the number of stems after growth to get the number of mortality and harvest trees in columns (13) and (14). This oversimplification implies that all harvesting and mortality occurs at the end of the growth period. Alternative assumptions might have the harvest and mortality at the beginning or in the middle of the growth period, the choice depending on the actual situation. Using a local volume table, column (16), the projection is converted to volume terms in columns (17) to (20) to complete the projection. Using our growth definition from Chap. 3,

$$
\begin{aligned}
\text{Gross growth} &= V_2 + M + C - V_1 \\
&= 8{,}518 + 260 + 2{,}104 - 4{,}384 \\
&= 6{,}498 \text{ bd ft}
\end{aligned}
$$

Table 5-7 Example stand table projection using movement percentages and

DBH class, in.	Number of trees (1)	Average 10-year DBH increment in. (2)	Movement ratio* (3)	Percentage of trees moving: 0 classes (4)	1 class (5)	2 classes (6)	3 classes (7)	Number of trees moving: 0 classes (8)	1 class (9)	2 classes (10)	3 classes (11)	DBH class, in.	Number of trees after movement† (12)
4	15.2	1.6	80	20	80	—	—	3.04	12.16	—	—	4	3.04
6	21.1	1.4	70	30	70	—	—	6.33	14.77	—	—	6	18.49
8	22.6	2.2	110	—	90	10	—	—	20.34	2.26	—	8	14.77
10	32.4	2.7	135	—	65	35	—	—	21.06	11.34	—	10	20.34
12	22.0	3.2	160	—	40	60	—	—	8.80	13.20	—	12	23.32
14	15.7	4.1	205	—	—	95	5	—	—	14.91	0.79	14	20.14
16	5.5	3.9	195	—	5	95	—	—	0.28	5.22	—	16	13.20
18	0.6	4.5	225	—	—	75	25	—	—	0.45	0.15	18	15.19
20	—	—	—	—	—	—	—	—	—	—	—	20	6.01
22	—	—	—	—	—	—	—	—	—	—	—	22	0.45
24	—	—	—	—	—	—	—	—	—	—	—	24	0.15
Total	135.1											Total	135.1

* Col (3) = [col (2)/2] × 100.

† Col (12) = sum of trees in each class in stand after growth. *Example:* Number of trees in 14-in. class = upgrowth from 10-in. class growing 2 classes [col (10)] + 12-in. class growing 1 class [col (9)] = 11.34 + 8.80 = 20.14.

This example stand is just breaking into merchantable size, and the rapid growth more than doubles the volume of the original stand with a PAI of 6,498/10, or about 650 bd ft per acre per year.

A Diameter Class Model for Long-Term Growth of Uneven-Aged Northern Hardwood Stands

Adams and Ek (1974) adapted a diameter class growth model developed by Ek (1974) as a basis for financial optimization of uneven-aged northern hardwood stands in Wisconsin. Data for the model were obtained from 5-year measurements from 132 permanent plots located on forest industry land in central and northern Wisconsin. All plots contained at least 50 percent sugar maple by basal area, along with yellow birch, white birch, northern red oak, trembling aspen, and balsam fir. Most stands had previously been harvested, but none were cut during the 5-year growth period. Plot diameter distributions showed a mixture of negative exponential, unimodal, and bimodal shapes, indicating that both even-aged and uneven-aged stands were present. Adams and Ek describe the three driving functions of stand growth for a 5-year period as follows.

1. **Ingrowth** (number of new trees growing into the smallest measured diameter class)

$$I = 7.079 \left[\frac{b_1 n_1 + b_2 n_2 + \cdots + b_k n_k}{n_1 + n_2 + \cdots + n_k}\right]^{-1.40072} \qquad (5\text{-}16)$$

considering mortality and harvest after the growth projection

Number of mortality trees‡ (13)	Number of harvest trees‖ (14)	Residual stand (15) = (12) − (13) − (14)	Volume per tree, bd ft (16)	Initial stand volume V_1 (17) = (1) × (16)	Mortality volume M (18) = (13) × (16)	Harvest volume C (19) = (14) × (16)	Residual volume V_2 (20) = (15) × (16)
0.1325	—	2.91	0	0	0	0	0
0.747	—	17.74	0	0	0	0	0
0.549	5.91	8.31	6	89	3.3	35.5	50
0.692	8.13	11.52	25	810	17.3	203.2	288
0.724	9.33	13.27	50	1,166	36.2	466.5	663
0.556	4.03	15.55	89	1,397	49.5	358.7	1,384
0.322	2.64	10.24	144	792	46.4	380	1,474
0.322	3.04	11.83	217	130	69.8	659.7	2,565
0.108	—	5.90	309	—	33.4	—	1,826
0.0003	—	0.45	421	—	3.1	—	185
0.002	—	0.15	552	—	1.1	—	83
				4,384	260	2,104	8,518

‡ Col (13) = number of stems after growth × mortality rate M = col (12) [0.05 − 0.0016 (DBH class, in inches)]

‖ Col (14) = derived from a commercial thinning prescription: cut 40 percent of all stems after growth in DBH classes 8, 10, and 12, and cut 20 percent of all stems in DBH classes 14, 16, and 18.

2. **Upgrowth** (number of trees rising from diameter class i to class $i + 1$)

$$U_i = (0.00330n_i^{0.88218})(S)(d_i^{0.48383})\exp[-0.00286(b_1n_1 + \cdots + b_kn_k)], \qquad \text{for } i = 1, \ldots, k \tag{5-17}$$

3. **Natural mortality** in diameter class i (number of trees)

$$M_i = 0.04109n_i, \qquad \text{for } i = 1, \ldots, k \tag{5-18}$$

where S = a site index measure with values of 40, 50, 60, and 70

$d_1, d_2, \ldots, d_k$ = class midpoint DBH for classes $i = 1, 2, \ldots, k$

$n_1, n_2, \ldots, n_k$ = current number of trees in diameter classes $1, 2, \ldots, k$

$b_1, b_2, \ldots, b_k$ = tree basal area, in square feet, corresponding to class midpoint DBH for diameter classes $1, 2, \ldots, k$

Ingrowth is described by Eq. (5-16) as inversely related to the average basal area per tree in the stand. This says that as the average size of the trees in the stand increases, ingrowth to the smallest class decreases.

Upgrowth is positively related in Eq. (5-17) to the number of trees in the class, site quality, and tree size, and negatively with overall stand density. The mortality equation, Eq. (5-18), simply says that the mortality rate is constant for all diameter classes.

Given these relationships. Adams and Ek then use accounting equations to characterize the number of trees at time t in each diameter class at the end of a growth period t which initiated at time $t - 1$ as follows:

For the smallest diameter class,

$$n_1(t) = n_1(t - 1) + I - M_1 - U_1 \tag{5-19}$$

for all other classes,

$$n_1(t) = n_i(t - 1) + U_{i-1} - M_i - U_i \quad \text{for } i = 2, \ldots, k \tag{5-20}$$

and for the new largest diameter class,

$$n_{k+1}(t) = U_k \tag{5-21}$$

where I = ingrowth, as calculated by Eq. (5-16)
U_i = upgrowth from a diameter class i, as calculated by Eq. (5-17)
M_i = mortality in diameter class i, as calculated by Eq. (5-18)
k = largest diameter class at time $t - 1$

Ingrowth from submerchantable trees is only added to the smallest class; addition to other classes is simply the upgrowth from the next smaller class. Note that in this model the upgrowth only moves one class.

Harvests can be introduced into the model by deducting the number of trees cut in each class at the end of each growth period. By changing the distribution of trees at the end of a growth period, harvest influences the magnitude of growth in the next period.

The model can be used to portray stand growth and development with and without harvesting. As an example, Adams and Ek (1974) portray the results of a 50-year simulation where harvesting was undertaken every 5 years to maintain cordwood volume production in the stand at 4 percent per year (Table 5-8). Harvesting was conducted by taking the largest classes first, with cordwood volume v_i for diameter class i estimated as

$$v_i = 0.000478d_i^{1.93}S^{0.35629}n_i - 0.03474n_i \tag{5-22}$$

Growth after the first period was not quite equal to 4 percent because harvest in each period was set by iterative means. Additional iterations could produce growth rates to any degree of accuracy.

The first cut was taken at year 5 and thereafter at the end of each 5-year period, with the diameter distributions shown reflecting the structure just after the 5-year cut was taken. The first cut made a dramatic change in structure by removing most of the large trees. Thereafter the change in structure over time reveals a number of interesting features:

1. All distributions have a decreasing number of trees as the diameters increase, but none conform to any particular q ratio.

Table 5-8 Simulation results using a northern hardwood growth model to manage an uneven-aged stand on a 5-year cutting cycle and a 4-percent rate of annual cordwood growth

	Diameter distribution (trees/acre) at year					
DBH, in.	0	5	10	15	25	50
6	31.0	24.6	27.9	30.7	35.1	41.7
8	19.0	20.0	19.6	20.1	22.2	28.2
10	14.0	14.3	15.0	15.3	16.2	20.2
12	10.0	10.6	11.2	11.7	12.6	15.2
14	10.0	9.5	9.5	9.6	10.2	12.1
16	8.0	8.3	8.4	8.4	8.7	8.0
18	6.0	2.4	2.4	2.2	1.8	0.0
20+	9.0	0.0	0.0	0.0	0.0	0.0
Last 5 years volume cut, cords/acre	—	13.0	2.4	2.5	2.6	3.3
Next 5 years average annual volume growth rate, percent	2.68	4.09	4.10	4.10	4.09	4.10
Average stand diameter, in.	10.9	9.9	9.8	9.7	9.5	9.2
Stand basal area, ft²/acre	83.1	54.0	55.3	56.6	59.5	65.1
Stand cord volume, cords/acre	22.2	12.3	12.7	13.1	13.8	15.0
Trees/acre, 6 in. and larger	107.0	89.8	93.9	98.1	106.8	125.4

Source: Adams and Ek (1974).

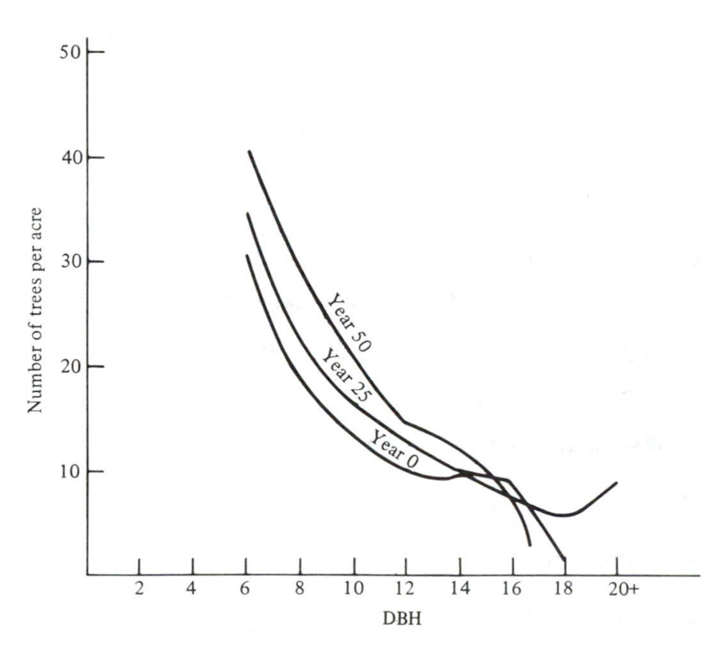

Figure 5-8 Starting, mid, and ending diameter distributions for a northern hardwood growth simulation designed to produce a 4-percent rate of annual cordwood growth. *(From Adams and Ek, 1974.)*

2. The ending distribution has more trees than the beginning distribution, but less basal area, with the distributions pivoting around the 16-in. class (Fig. 5-8), providing relatively more trees in the smaller classes over time.

To illustrate the Adams–Ek model, Eqs. (5-16) to (5-22) are used to calculate one 5-year periodic growth cycle for a stand of 120 trees distributed over diameter classes 6 to 16 in. The results of the calculations are shown in Table 5-9. First the initial stand basal area and volume were calculated. Then Eqs. (5-16) to (5-18) were used for each class to calculate ingrowth, upgrowth, and mortality, respectively. Finally the accounting equations [Eqs. (5-19) to (5-22)] were used to determine the number of trees in each class at the end of the 5-year growth period.

Given the initial basal area of 58.66 ft^2 per acre, the specific calculations for the 6-in. class are as follows:

1. *Ingrowth,* 6-in. class [Eq. (5-16)]

$$I = 7.079\left(\frac{\text{basal area per acre}}{\text{trees per acre}}\right)^{-1.40072}$$

$$= 7.079\left(\frac{58.66}{120}\right)^{-1.40072}$$

$$= 7.079(0.4888)^{-1.40072} = (7.079)\left(\frac{1}{0.3669}\right)$$

$$= 19.3 \text{ trees}$$

2. *Upgrowth,* 6-in. class [Eq. (5-17)]

$$U_i = (0.0033n_i^{0.88218})(S)(d_i^{0.48383})\exp(-0.00286\text{BA})$$

$$= [(0.0033)(40)^{0.88218}](60)(6^{0.48383})\exp[-0.00286(58.66)]$$

$$= [(0.0033)(25.9003)](60)(2.379)\exp(-0.1678)$$

$$= (0.08547)(60)(2.3795)(0.8455)$$

$$= 10.32$$

3. *Mortality,* 6-in. class [Eq. (5-18)]

$$M_i = 0.04109n_i$$

$$= 0.04109(40)$$

$$= 1.64$$

Table 5-9 Calculating 5-year growth for a northern hardwood stand using the Adams–Ek model

	First period								Second period		
Diameter class i_1, in.	Trees per class n_{i1}, per acre (1)	Basal area per tree b_i, ft² (2)	Basal area per class $n_{i1}b_i$, ft²/acre (3) = (1) × (2)	Volume per tree V_i (4)	Volume per class V_{i1}, cords/acre (5) = (1) × (4)	Ingrowth I, trees/acre (6)	Upgrowth U_i, trees/acre (7)	Mortality M_i, trees/acre (8)	Tree per class N_{i2}, per acre (9)	Basal area per class $N_{i2}b_i$, ft²/acre (10)	Volume per class V_{i2}, cords/acre (11)
6	40	0.196	7.84	0.0303*	1.22	19.3†	10.4‡	1.6‖	47.4§	9.29	1.45
8	30	0.349	10.47	0.0790	2.37		9.2	1.2	20.0	10.44	2.36
10	20	0.545	10.90	0.140	2.80		7.2	0.8	21.2	11.55	2.97
12	15	0.785	11.78	0.214	3.21		6.1	0.6	15.5	12.17	3.32
14	10	1.069	10.69	0.30	3.00		4.6	0.4	11.1	11.87	3.33
16	5	1.396	6.98	0.398	1.99		2.6	0.2	6.8	9.49	2.71
18	0	1.767	0	0.508	0		0	0	2.6	4.59	1.32
20	0	2.182	0		0		0	0	0	0	0
Total	120		58.66		14.59			4.8	124.6	69.4	17.46

* Calculated using Eq. (5-22).
† Calculated using Eq. (5-16).
‡ Calculated using Eq. (5-17).
‖ Calculated using Eq. (5-18).
§ Calculated using Eq. (5-19) for 6-in. class, Eq. (5-20) for 8- to 16-in. classes, and Eq. (5-21) for 18-in. class.

4. *Number of trees* in the 6-in. class in period 2 [Eq. (5-19)]

$$n_1(t) = n_i(t - 1) + I - M_1 - U_1$$
$$= 40 + 19.3 - 1.64 - 10.32$$
$$= 47.34$$

This model portrays two desirable attributes of a growth projection system for uneven-aged stands: (1) density-dependent growth relationships that can portray the implications of changing stand structure on growth and yield by tree size, and (2) the ability to accept objectives, such as "maintain stand cord growth at 4 percent per year," as the driving force in controlling the structure.

INDIVIDUAL TREE MODELS

Individual tree models work by simulating the growth of each individual tree in diameter, height, and crown, deciding whether it lives or dies, calculating its growth and volume, and then adding the trees together to get per-acre characteristics, volumes, and growth rates. Whether or not an individual tree lives or dies and the rate at which it grows depends on its competitive position in the stand as determined by such attributes as relative size or distance to neighboring trees. Although both stand and individual tree models can use sample survey plot data as input, only individual tree models have the ability to simulate the competitive environment of each tree. Furthermore, individual tree data are aggregated *after* the model grows each tree, while the stand model aggregates individual tree data into stand variables *before* operating the growth model. Like stand models, individual tree models are prepared for a particular species or cover type from sample data gathered over a certain range of growing environments. A model accordingly can only be applied comfortably to subject stands that fall within the range of the sample stand conditions used to build the model.

Individual tree models can be grouped into two classes by how competition among trees is handled.

Distance-independent models Each tree is modeled separately and its competitive position is determined by comparing its individual diameter, height, and condition to stand characteristics such as basal area and average diameter. These models assume that spatially all species and sizes of trees are uniformly distributed throughout the stand.

Distance-dependent models In addition to the height, diameter, and other tree variables used in distance-independent models, each individual tree is literally mapped to determine the distance to, bearing, and size of all adjacent trees that are competing with the subject tree for light, moisture, and nutrients. This allows the competitive position of each tree to be "custom" calculated with presumably greater accuracy. The disadvantages of distance-dependent models are the high cost of tree mapping and the increased computational complexity. Also we have found no clear documented evidence or case studies showing superiority of distance-dependent over distance-independent models.

This chapter can only introduce the complex and rapidly evolving subject of individual tree growth and yield models. We believe that they are an important development in contemporary forest science. As these models are developed, validated, and available for major forest types, the door is opened to improve our ability to evaluate and optimize silvicultural systems for managed stands. Proposals for complex, intensive management prescriptions can be more easily examined and forestry will take another step from art to quantitative science. Still the subtleties of how activity timing, climate, seed source, and site preparation impact stand establishment and performance will likely escape full quantification, as will the propensity of trees to grow in clumps rather than being uniformly distributed. In addition, for many forest types current individual tree models have not yet proven themselves to be significantly better at prediction than stand models.

Model Structure and Operation

While detailed in their accounting and often using elegant growth functions, individual tree models are logically straightforward. Following the flow of steps from top to bottom of Fig. 5-9, only three possibilities are considered for each tree: (1) it can die, (2) it can be cut, and (3) it can survive and grow. When this sequence of choices is worked out for all sample trees representing a stand, mortality, harvest, growth, and the residual stand structure are summarized from the individual tree results.

The flowchart suggests the functional relationships needed to build such a model as well as the input needed from the model user. Functions need to be established to determine mortality rates as well as height, diameter, and crown growth of the surviving tree. To use the model, tree list input data and prescriptions for harvesting are needed from the user.

The logic of many models is to calculate first the potential growth rates of trees which are free-growing without any competition. Then a competition index of the tree is calculated, which in turn is used to reduce the potential growth by some factor. Consider potential growth in three aspects:

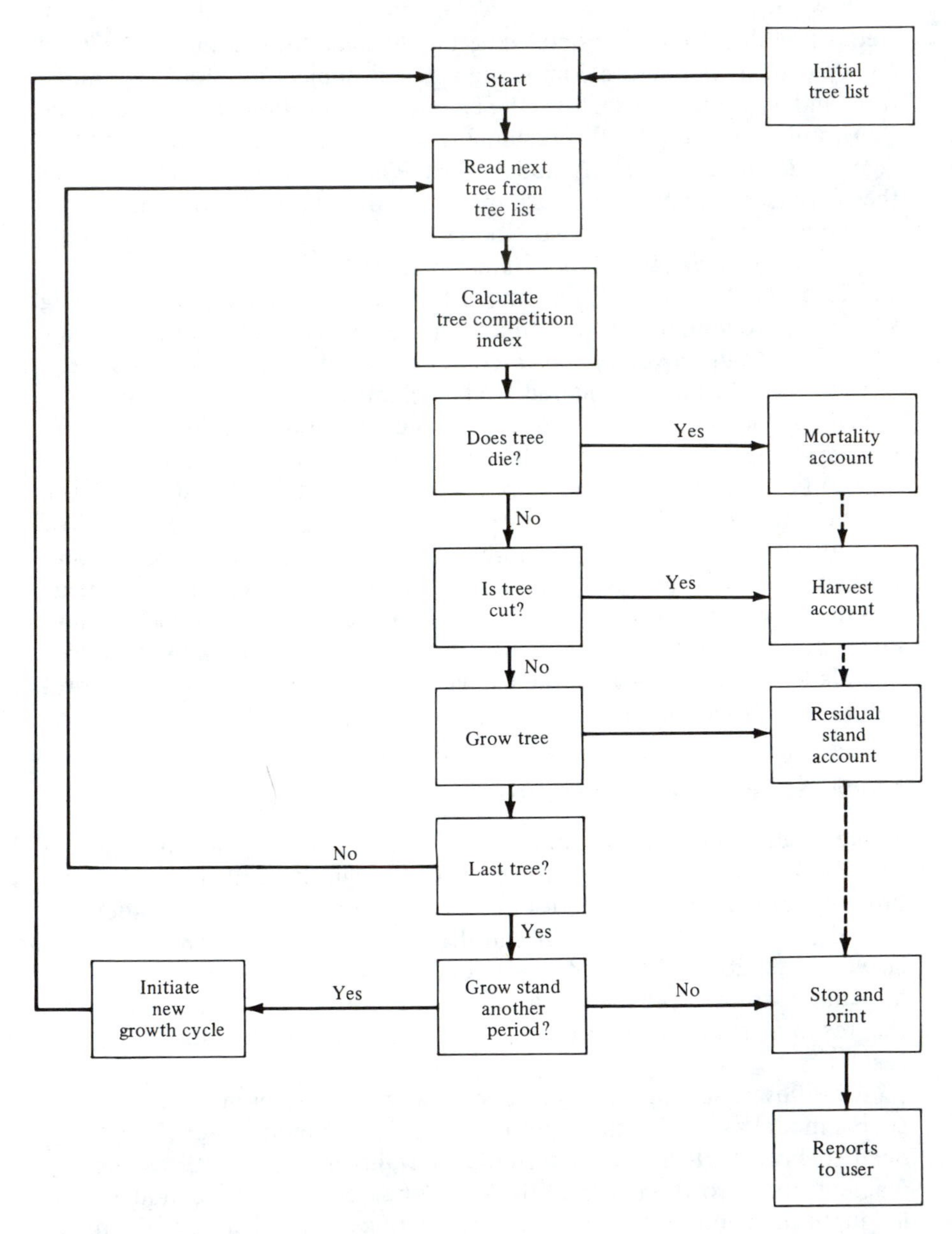

Figure 5-9 Flowchart of basic steps and decisions in an individual tree model growth simulator.

Potential Growth Estimation

Tree growth variable	Definition	Source
Δh^*	Potential height growth of free-growing trees	Site index curves, stem analysis of sample trees
Δd^*	Potential diameter growth of free-growing trees	Increment data from dominant trees under no competition
Δc^*	Potential crown growth for free-growing trees	Crown measurements of free-growing sample trees

The competitive ability of the tree can be calculated in many ways, but it typically winds up as an index scaled from 0 to 100, with 100 representing the free-growing tree. An association is then made between the competition index and the growth performance of trees, as illustrated in Fig. 5-10.

The relative growth factor k is often scaled from 0 to 1.0, as shown. As expected, diameter and crown growth are more sensitive to competition than is height growth. Given these relationships, the predicted growth of individual trees in a stand is then simply

$$\begin{aligned} &\text{Height growth:} && \Delta h = k_1 \Delta h^* \\ &\text{Diameter growth:} && \Delta d = k_2 \Delta d^* \\ &\text{Crown growth:} && \Delta c = k_3 \Delta c^* \end{aligned} \tag{5-23}$$

This two-step approach has some appeal since the potential growth is established separately from the effect of competition. It also focuses attention on how competition is calculated and how it affects growth.

Another modeling approach does not estimate potentials separately, but simply develops growth equations which directly include the tree's competition index. As an example,

$$\Delta h = f(\text{height, site, stand density, stand height, competition index, . . .})$$

Mortality in both approaches is usually handled as a probability function, where

$$\text{Prob}(m) = f(\text{competition index, tree size, stand variables})$$

As each individual tree is considered in turn, a random number is selected and evaluated according to the mortality probability function to determine whether or not that tree dies. If it dies, the tree is removed from the list.

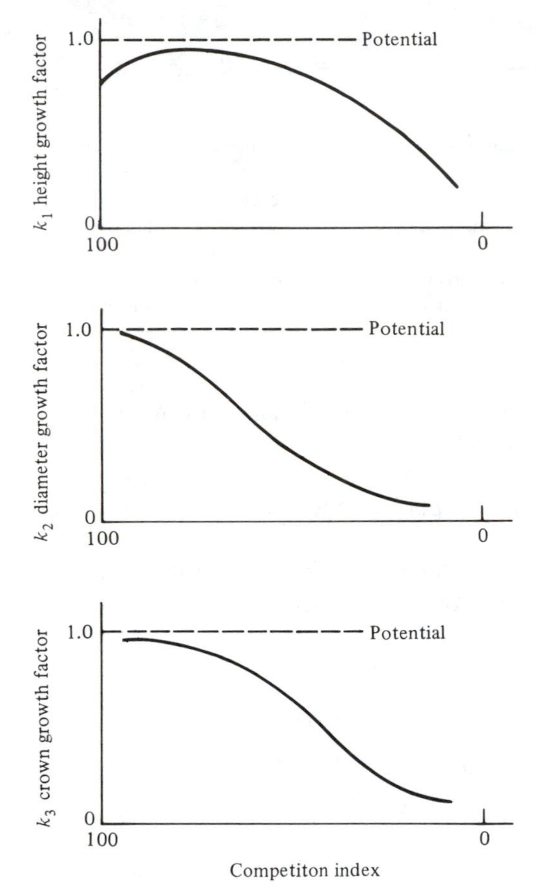

Figure 5-10 Reduction in potential height, diameter, and crown growth rates as determined by tree competition index.

Competition Indices

The unique potential of individual tree models is to individualize the competitive environment of tree growth. These indices are to the trees what stand density measures are to the stand growth models. Six representative indices are reviewed, three each for distance-dependent and distance-independent models.

Distance-independent measures When a tree is not mapped it cannot be related directly to its neighbors and the only way to handle competition is to compare the tree's size to the size distribution of all other trees in the stand. If the tree is relatively small in crown, height, or diameter compared to the stand average, then it is assumed, on the average, to lack in competitive vigor. Often this is not too bad an assumption since the smaller trees are usually the suppressed or intermediate individuals. When the stand is clumpy or patchy, however, the small trees still may be dominants or codominants.

Two competition indices based on stem size are commonly used. The first is the ratio of the subject tree basal area to the average tree basal area. The second computes the cumulative basal area of trees larger than the subject tree.

$$\text{CCI} = \frac{b_i}{\bar{b}_q} \tag{5-24}$$

and

$$\text{CCI} = \text{BAL}_i \tag{5-25}$$

where b_i = basal area of subject tree i
$\bar{b}_q$ = basal area of the average tree (tree of quadratic mean DBH)
BAL_i = total basal area per acre of trees in stand larger than subject tree i

As the ratio of the first index gets numerically larger, the tree is considered to be increasingly vigorous and will grow at rates closer to its genetic potential. In the second index as the basal area of larger trees gets numerically smaller, the tree is increasingly competitive. This second index is used by Stage (1973) in the PROGNOSIS model.

An interesting distance-independent competition factor developed by Krumland (1982) is based on the tree's crown and provides our third example. The idea is if the cumulative cross-sectional area of all tree crowns at a height two-thirds up the live crown of the subject tree is numerically small, then the subject tree must have its crown well up in the canopy and therefore be a dominant free grower. The same cumulative area of larger trees idea was used in the BAL_i index, except that here it is applied to crowns rather than stems. The precise definition of this competition index is

C_{66_i} = percentage of land area covered by live tree crowns measured at a height 66 percent of the way up the live crown of subject tree i (5-26)

Calculating C_{66}, which is used in the California redwood model CRYPTOS (Krumland and Wensel, 1982) and the California mixed conifer model CACTOS (Wensel and Daugherty, 1984), is fairly complex since the crown cross-sectional areas of all trees in the stand have to be computed at several heights.

Distance-dependent measures Most of the distance-dependent factors start with Staebler's (1951) concept of a circular zone of influence around a subject tree in which the presence of competing trees would reduce the growth rate of the subject tree. The degree of competition was reasoned to be the percentage of the influence zone that is overlapped by competitive trees. Whether it is moisture, light, or nutrient competition will depend on the species and on the particular physical characteristics of the site along with the size of the subject tree. Staebler's initial index was

$$\mathrm{CI}_j = \sum_{i=1}^{n} \left(\frac{d_{ij}\mathrm{CR}_j}{2}\right) \tag{5-27}$$

where CI_j = competition index for tree j
d_{ij} = distance of crown overlaps between competitor i and subject tree j, in feet
CR_j = crown radius of subject tree, in feet
n = number of competitors

This index will be numerically small with few competitors and little crown overlap.

Hegyi (1974) came up with a computationally simple index:

$$\mathrm{CI}_j = \sum_{i=1}^{n} \left(\frac{D_i}{D_j}\frac{1}{L_{ij}}\right) \tag{5-28}$$

where CI_j = competition index for tree
D_i, D_j = crown diameter of competitor i and subject tree j, in inches
L_{ij} = distance between competitor and subject tree, in feet
n = all trees intersected by a 10-factor basal area prism rotated at center of subject tree

Bella's (1971) competition index determines the portion of the subject tree's crown area that is overlapped by competition and weighted by diameter ratios.

$$\mathrm{CI}_j = \sum_{i=1}^{n} \left(\frac{O_{ij}}{A_j}\frac{D_i^k}{D_j}\right) \tag{5-29}$$

where CI_j = competition index for tree j
O_{ij} = area of crown overlap between competitor tree i and subject tree j, in square feet
A_j = crown area of subject tree j, in square feet
D_i, D_j = crown diameters of competitor and subject trees, in feet
k = a factor
n = all trees where O_{ij} is positive

Three trees A, B, and S are mapped as shown in Fig. 5-11 along with values for all the variables used in the Staebler, Hegyi, and Bella distance-

Subject tree: S
Competing trees: A, B

	Variable	Value		Variable	Value
Crown diameter	D_A	30	O_{SA}, O_{SB} = area crown overlap	O_{SA}	65
	D_B	43		O_{SB}	210
	D_S	34			
Crown radius	CR_A	15	d_{SA}, d_{SB} = distance crown overlap	d_{SA}	5
	CR_B	21.5		d_{SB}	11
	CR_S	17			
Distance between tree centers	L_{SA}	28	A_S = crown area of subject tree S; $A = \pi CR^2$	A_S	908
	L_{SB}	32			

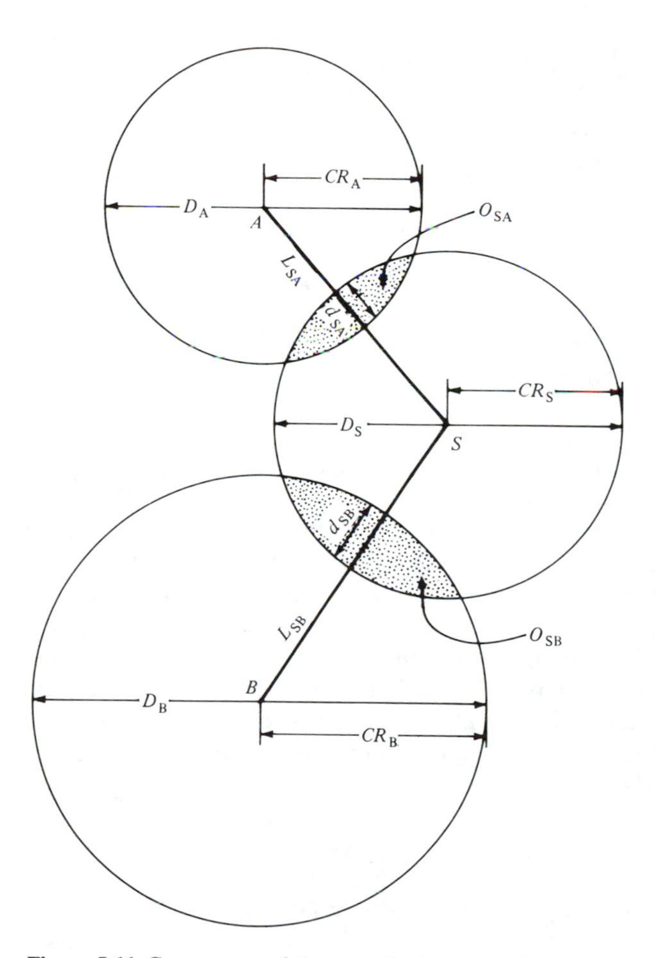

Figure 5-11 Crown map of three competing trees showing variables used to estimate distance-dependent competition indices.

dependent competition indices. Letting tree S be the subject tree, the three indices are calculated as follows:

1. Staebler's index,

$$\mathrm{CI}_j = \sum \left(\frac{d_{ij}\mathrm{CR}_j}{2}\right) = \left(\frac{d_{\mathrm{SA}}\mathrm{CR}_\mathrm{S}}{2} + \frac{d_{\mathrm{SB}}\mathrm{CR}_\mathrm{S}}{2}\right)$$

$$= \left(\frac{5(17)}{2} + \frac{11(17)}{2}\right) = 136$$

2. Hegyi's index,

$$\mathrm{CI}_j = \sum \left(\frac{D_i}{D_j}\frac{1}{L_{ij}}\right) = \left(\frac{D_\mathrm{A}}{D_\mathrm{S}}\frac{1}{L_\mathrm{SA}} + \frac{D_\mathrm{B}}{D_\mathrm{S}}\frac{1}{L_\mathrm{SB}}\right)$$

$$= \left(\frac{30}{34}\frac{1}{28} + \frac{43}{34}\frac{1}{32}\right) = 0.071$$

3. Bella's index,

$$\mathrm{CI}_j = \sum \left(\frac{O_{ij}}{A_j}\frac{D_i^k}{D_j}\right) = \left(\frac{O_\mathrm{SA}}{A_\mathrm{S}}\frac{D_\mathrm{A}^{1.2}}{D_\mathrm{S}} + \frac{O_\mathrm{SB}}{A_\mathrm{S}}\frac{D_\mathrm{B}^{1.2}}{D_\mathrm{S}}\right)$$

$$= \left(\frac{65}{908}\frac{30^{1.2}}{34} + \frac{210}{908}\frac{43^{1.2}}{34}\right) = 0.745$$

letting $k = 1.2$.

There are at least a dozen more such indices, many rather elaborately calculating the competition for growing space. Alemdag (1978) tested eight such distance-dependent indices over eight northern conifer data sets and found the Bella and Hegyi indices to be consistent performers.

Most of the popular individual tree models have, however, gone the distance-independent route, eliminating the need for costly mapping during plot inventory, data coding, input, and computation time. The forest manager planning to develop or use individual tree models needs to investigate the potential accuracy or applicability gains from their more complex and sophisticated structures against the time and costs required to develop and utilize them.

EXAMPLES OF INDIVIDUAL TREE MODELS

Two examples are briefly reviewed to get a sense of how these models work. PTAEDA (Daniels and Burkhart, 1975) is a distance-dependent model for plantation loblolly pine in the southeast. PROGNOSIS (Stage, 1973) is a distance-independent model for mixed conifers of the northern

Rocky Mountains. The models are far from completely presented since each has dozens of equations and is typically run by a computer program with at least 20 subroutines. We focus on the key growth equations and some of the more interesting or important features of the models.

PTAEDA: A Distance-Dependent Individual Tree Model for Simulation of Growth in Managed Loblolly Pine Plantations

PTAEDA was developed by Daniels and Burkhart (1975) to simulate the growth of loblolly pine plantation on different sites and with different initial tree spacings to estimate the growth impacts of site preparation, fertilization, and thinning. The model is distance-dependent since the competition index for each tree is derived from the size and distance of nearby trees. The model works only with plantations since a regular spacing of trees is assumed in generating the tree location map and between-tree distances. Most of the equations used by PTAEDA are presented here because they are fairly compact and give a good opportunity to examine the internal structure of individual tree models.

The model handles the plantation in two stages. The first is the precompetitive stage, which runs from initial planting until between-tree competition begins. In the second stage models, annual diameter, height, and crown growth of each tree are annually incremented.

A variable-size "plot" is used to represent a subject plantation and contains 100 trees in 10 rows of 10 trees each. The between-tree spacing is specified by the user when initiating a run. Also specified initially are site index and desired cultural regimes.

Juvenile stand growth The juvenile trees in stage 1 of PTAEDA are grown using a set of stand equations until the calculated crown competition factor (CCF) reaches a value of 100 percent. At this point the stand is assumed to have initiated between-tree competition. Juvenile stand mortality is then calculated and assigned at random, and individual tree dimensions (height, DBH, crown length) are generated for the residual stand using a two-parameter Weibull distribution. The equations used until CCF = 100 for the juvenile stage are as follows.

Stand parameters

HD: *Dominant stand height* (feet)

$$\log_{10} \text{HD} = \log_{10} \text{SI} - 5.87 \left(\frac{1}{A} - 0.04\right) \tag{5-30}$$

where SI = site index, in feet (base age 25 years)
A = age, in years

(TP/TS): *Ratio of planted to surviving trees*

$$\log_{10}\left(\frac{\text{TP}}{\text{TS}}\right) = A(0.013 \log_{10} \text{TP} + 0.0009\text{HD} - 0.0109\sqrt{\text{HD}}) \quad (5\text{-}31)$$

where TP = number of trees planted per acre
TS = number of trees surviving per acre

CCF: *Crown competition factor*

$$\text{CCF} = 181 - \frac{1012}{A} + 0.00347\text{HD} \times \text{TS} \quad (5\text{-}32)$$

DMIN: *Minimum stand diameter* (inches)

$$\text{DMIN} = 0.133 + 0.045\text{HD} - 0.0000187(A \times \text{TS}) + 17.3\left(\frac{\text{HD}}{\text{TS}}\right) \quad (5\text{-}33)$$

DAVE: *Average stand diameter* (inches)

$$\text{DAVE} = 2.96 + 0.054\text{HD} - 0.000052(A \times \text{TS}) + 18.5\left(\frac{\text{HD}}{\text{TS}}\right) \quad (5\text{-}34)$$

Individual tree dimensions

D: *Tree DBH* (inches), generated using the Weibull distribution from stand parameters.
H: *Total tree height* (feet)
CL: *Crown length*
CR: *Crown ratio*

$$\ln H = 1.51 + 0.71 \ln \text{HD} + \frac{0.262}{D} \ln \text{TS} - \frac{2.44}{A} - \frac{2.71}{D} + \frac{2.95}{AD} \quad (5\text{-}35)$$

CBL: *Clear bole length* (feet)

$$\ln \text{CBL} = -2.69 + 1.61 \ln H + \left(\frac{0.457}{D}\right) \ln \text{TS} - \left(\frac{8.96}{A}\right) \ln D + \left(\frac{12.7}{A}\right) - \left(\frac{1.65}{D}\right) - \left(\frac{21.7}{AD}\right) \quad (5\text{-}36)$$

Crown length is the difference between total height and clear bole length (CL = H − CBL), and crown ratio is CR = CL/H.

Here we see that the tree dimension equations are driven by the results of the stand equations. The value D, individual tree diameter, is generated from the Weibull distribution fitted to the residual stand. In natural stand models, such as CACTOS, these individual tree dimensions are usually provided by inventory data.

Competitive stand growth After the individual tree map has been generated from the juvenile stage model, all subsequent growth is calculated annually on an individual tree basis. Each year the competitive situation of each tree is calculated using the Hegyi competition index as follows:

$$CI_i = \sum_{j=1}^{n} \left(\frac{DBH_j/DBH_i}{DIST_{ij}} \right) \tag{5-37}$$

where CI_i = competition index of subject tree i
DBH_i = diameter of subject tree i, in inches
DBH_j = diameter of competitor tree j, in inches
$DIST_{ij}$ = distance between trees i and j, in feet
n = all neighboring competitive trees counted "in" by a 360° sweep using a 10-factor basal area prism centered on subject tree i

Once the competitive status of each tree is estimated, annual growth in height, diameter, bole length, crown length, and the survival probability for each tree are calculated.

Height growth Height growth is calculated first using the site index equation from the juvenile model (dominant stand height HD). By the definition of a site tree, this value of HD represents the height of a competition-free tree and is the maximum potential height obtainable.

$$HD = SI \times 10^{-5.87(1/A-0.04)} \tag{5-38}$$

Since the equation for HD expresses total height as a function of age, the annual height increment is the rate of change (first derivative) of this relationship with respect to age:

$$PHIN = \frac{dHD}{dA} \tag{5-39}$$

This potential height increment (PHIN) is then adjusted by a factor that reflects the competitive situation of the tree in question, using the tree-specific values for competition index and live crown ratio. The general form of this competition adjustment factor is

$$K_H = b_1 + b_2 CR^{b_3} e^{b_4 CI - b_5 CR} \tag{5-40}$$

where K_H = height adjustment factor
$b_1, b_2, \ldots, b_5$ = coefficients
CR = live crown ratio
CI = competition index

When fitted to the data, the resulting adjusted tree height growth equation is

$$\begin{aligned} \text{HIN} &= \text{PHIN}(K_{\text{H}}) \\ &= \text{PHIN}(0.546 + 124.9\text{CR}^{1.7}e^{-1.15\text{CI}-6.7\text{CR}}) \end{aligned} \qquad (5\text{-}41)$$

where HIN is the estimated actual height increment, in feet.

Diameter growth Potential diameter growth of the competition-free tree is developed from open grown tree data and uses a known consistent relationship between the diameter and the height and age of open grown trees,

$$D_o = -2.42 + 0.287H + 0.2095A \qquad (5\text{-}42)$$

where D_o = diameter of open-grown tree, in inches
H = total open-grown tree height, in feet
A = age, in years

As with height, the first derivative of this equation with respect to age yields the potential diameter increment relationship:

$$\text{PDIN} = 0.287\text{HIN} + 0.2095 \qquad (5\text{-}43)$$

where PDIN is the potential diameter increment, in inches.

A competitive adjustment factor for diameter growth is fitted using a relationship of the form

$$K_{\text{D}} = b_1 + b_2\text{CL}^{b_3}e^{-b_4\text{CI}} \qquad (5\text{-}44)$$

where K_{D} = competitive adjustment factor for diameter
$b_1, \ldots, b_4$ = coefficients
CL = live crown length

When fitted and the coefficients estimated, actual diameter growth adjusted for competition is calculated as:

$$\begin{aligned} \text{DIN} &= \text{PDIN}(K_{\text{D}}) \\ &= \text{PDIN}(0.086 + 0.202\text{CL}^{1.8}e^{-1.32\text{CI}}) \end{aligned} \qquad (5\text{-}45)$$

Mortality The probability that a tree remains alive in a given year was assumed to be positively related to its individual vigor, as indicated by the live crown ratio (CR), which is the ratio of live crown length to total height, and negatively related to its competitive stress as measured by the tree's competition index. The resulting probability of survival equation is

$$\text{PLIVE} = 1.086\text{CR}^{0.07}e^{-0.028\text{CI}^{1.18}} \qquad (5\text{-}46)$$

The probability calculated for each tree is then used in a random trial to determine whether the tree lives or dies.

Given incremental height, diameter, and mortality for each tree, PTAEDA then uses a series of volume equations to estimate merchantable growth and yield for each tree. The volumes for all surviving trees in the plot are then added together and converted to a per-acre basis for reporting. Many possible volume and value transformations are possible, starting from the basic tree growth model.

Model applications and results When checked against 187 old-field plantation plots, PTAEDA gave consistent and fairly accurate results, as shown in Table 5-10. These validation plots had been used to develop relationships for the juvenile growth model but were not used to estimate coefficients for any of the competitive stand equations. In a variety of trials and tests PTAEDA consistently estimated growth and yield of other old-field plantations which received no cultural treatment after planting within a ±10-percent accuracy range.

Thinning simulations are implemented in PTAEDA by specifying removal of trees between certain diameter limits or by removing whole rows of trees. Site preparation is seen as reducing the number of competing hardwoods and other vegetation. This is reflected in the model by adding a number of competing stems to the initial plantation and by setting an age when all of the competing stems are removed. This is a rather indirect way of assessing the effect of site preparation, and experience with the model is needed to do it well. Fertilization is simulated by multiplying the site index value by a factor that generates a rising and then falling positive impact of fertilizer over a finite impact period. As with site preparation, the user has to develop appropriate impact coefficients.

Table 5-10 Mean, standard deviation, and range of predicted and observed yields on 187 old-field loblolly pine sample plots*

Item	Mean	Standard Deviation	Range	
			Lowest	Highest
Number of trees per acre	729.9 (742.2)	211.5 (234.7)	228 (300)	2,028 (2,410)
Basal area, ft²/acre	143.2 (150.7)†	31.3 (32.7)	70.7 (72.0)	200.5 (217.2)
Volume, ft³/acre	2,902.7 (3,139.7)†	1,003.7 (1,123.7)	1,036 (941)	5,615 (6,275)

* Observed yields are shown in parentheses.

† Significant difference ($\alpha = 0.05$) between observed and predicted means.

PROGNOSIS: A Distance-Independent Individual Tree Model for Simulating Growth in Natural and Managed Forest Stands of the Northern Rocky Mountains

PROGNOSIS (Stage, 1973) was developed by the U.S. Forest Service for northern Rocky Mountain silviculturists. It is designed to give reliable quantitative estimates of how stands with conditions varying from decadent old growth, heavy cutover, diseased, mixed species, to newly regenerated single species would respond to cultural activities of density control, site preparation, regeneration method, and pest management. By regulation Forest Service silviculturists must conduct a stand exam, and prepare and file, as a public document, a written prescription for each subject stand in a timber sale before harvesting can commence. This, as you might expect, makes growth forecasting and prescription writing much more than an academic matter.

A more comprehensive model than PTAEDA, PROGNOSIS has the following characteristics:

1. It has the ability to initiate analysis with existing stands in almost any condition of size, stocking, species, and vigor.
2. It requires input data for height, diameter, and crown size for sampled trees in the subject stand in addition to stand and environmental parameters.
3. Although its growth equations are based on extensive sample data on tree growth within the geographic areas of application, the user can collect increment data from the subject stand and plug this new growth information into the models.
4. It is stochastic in that a random element is added to the growth of each tree based on the error distribution of the inventory data.

PROGNOSIS has become increasingly popular, and versions of the model for new species, habitat types, and environmental conditions are being developed as new research data and plot records become available to recalibrate the growth functions. A version of the model may well be available for most U.S. Forest Service regions within the decade, giving foresters throughout the country an option to use this tool. The reasons for its popularity are wide coverage of initial stand conditions, a good user manual, good system support, availability of the software and data to public and private users, and available training through forestry extension programs. On the other hand, it is a bit complicated to use, expensive to run, and currently requires a mainframe computer.

Model structure Like most individual tree models, PROGNOSIS has a set of four periodic driving functions or submodels to model each tree.

These equations have different sets of coefficients for estimating growth over period lengths from 1 to 20 or more years. Since most local and research increment data are recorded for 5- or 10-year periods, these are the recommended period lengths. The model will cycle over several periods to permit the development of the stand over longer time periods. The four driving functions, implicitly stated, are as follows:

1. Diameter growth = *f*(site characteristic, diameter, crown ratio, stand competition, tree size rank in stand) + (error term)
2. Height growth = *f*(site characteristics, diameter, height)
3. Crown ratio = *f*(site characteristics, diameter, height, basal area, tree size rank in stand, stand competition)
4. Probable mortality = *f*(tree size, stand approach to normal density, stand approach to maximum basal area)

Note that all equations use a mixture of stand and individual tree variables. The diameter growth model is the most critical as it contains the stochastic element, and after the first period, the diameter is an important independent variable for the other models. We will look at the diameter growth model in detail.

First, the model does not actually predict diameter growth directly but rather the quantity dds, which is related to the number of square inches of new wood grown over one period. This factor can readily be converted to basal area growth in square feet or to actual diameter growth. If D_1 is the initial diameter in inches and DG the amount of diameter growth in inches,

$$\text{dds} = (D_1 + \text{DG})^2 - D_1^2$$

A generalized equation to estimate dds for a given subject tree is given in Table 5-11.

The first eight terms describing the site are all constants for every tree in the sample. Note that the traditional site quality measure of tree height at a base age is not used and the site is represented by a composite of habitat type, geographic location, and topographic parameters. Competition is reflected at both the stand and the tree levels. If it is a large-diameter tree, very few trees will be larger, so the BAL variable will be small and the tree is assumed to be free-growing. Small trees by contrast will see most of the stand basal area in larger trees, BAL will be a large number, and the tree is assumed to be intermediate or suppressed.

As with PTAEDA, once a growth simulation run is initiated, each tree on the plot is modeled according to the prescription rules for the specified number of years. The driving functions cause the individual tree to grow in height, diameter, and crown or to have the misfortune of mortality. User input can include different kinds of thinning prescriptions, including

Table 5-11 Generalized equation to estimate dds in PROGNOSIS model

Dependent variable		Independent variables	Comment
			Site variables
ln (dds)	=	HAB + LOC	Constants for habitat type (HAB) and geographic location (LOC)
		$+ b_1(\text{SL} \cdot \cos \text{ASP})$ $+ b_2(\text{SL} \cdot \sin \text{ASP})$ $+ b_3\text{SL} + b_4(\text{SL})^2$ $+ b_5\text{EL} + b_6(\text{EL})^2$	Topographic position expressed by slope (SL), aspect (ASP), and elevation (EL)
			Tree variables
		$+ b_8\text{DBH} + b_{12}(\text{DBH})^2$	Subject tree diameter (DBH)
		$+ b_9\text{CR} + b_{10}(\text{CR})^2$	Subject tree vigor by live crown ratio (CR)
			Competition
		$+ b_7(\text{CCF}/100)$	Stand competition index by crown competition factor (CCF)
		$+ b_{11}(\text{BAL}/100)$	Tree competitive status as indexed by stand basal area per acre in trees of larger DBH (BAL) than subject tree
		$+ e_i$	*Random element*

thinning from above or below to basal area, to tree count limits, or to a variety of diameter limit prescriptions (Fig. 5-12).

Each period a report generated by PROGNOSIS accumulates the tree data, logs what happens to the stand character, and keeps track of the stand diameter distribution. A set of stand conditions, harvests, and scheduled activities by period is provided in the summary statistics of Table 5-12. This is the sort of information needed to evaluate stand prescriptions and to provide the yield data for the formulation of timber harvest schedules.

PROGNOSIS is blessed with a comprehensive user guide (Wykoff et al., 1982) and will likely be a permanent fixture in the forester's life, along with many other individual tree simulators using tree record inventory data as input.

INTENSIVE MANAGEMENT

Intensive management of trees is similar to intensive management of other crops. It requires prompt and adequate regeneration of desired species; protection from loss due to insects, fire, animals, and disease;

In the year 1981:

A. Remove all trees with DBH less than or equal to 3 in.

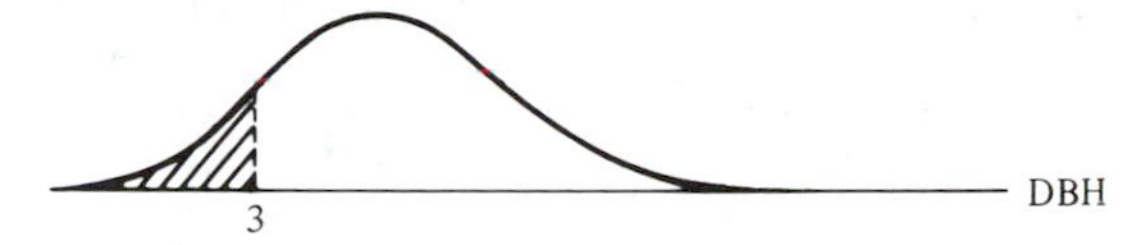

B. Remove all trees with DBH greater than or equal to 20 in.

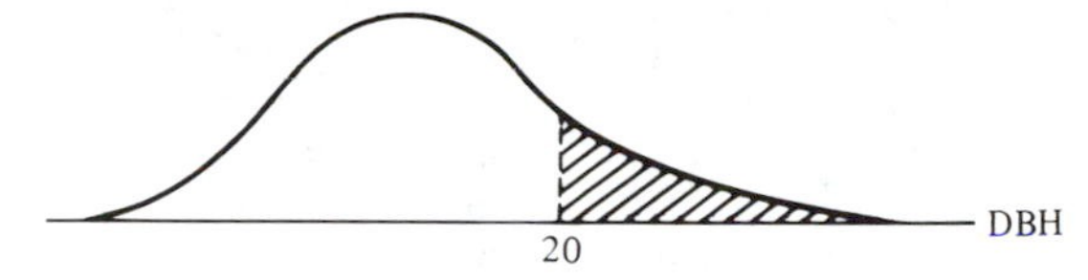

C. Remove all threes with DBH between 3 and 20 in.

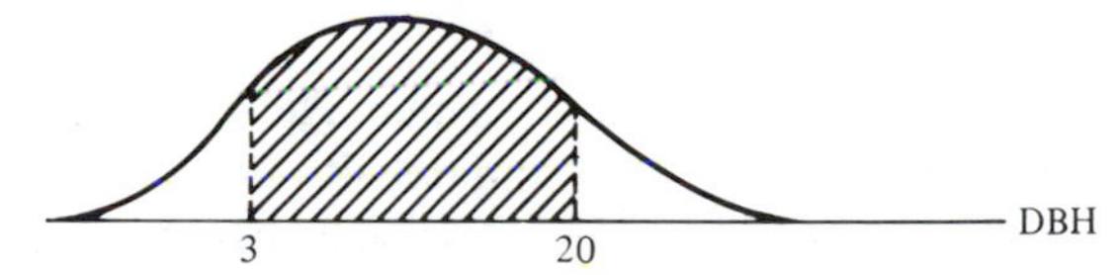

D. Leave only those trees that are between 3-and 20-in. DBH:

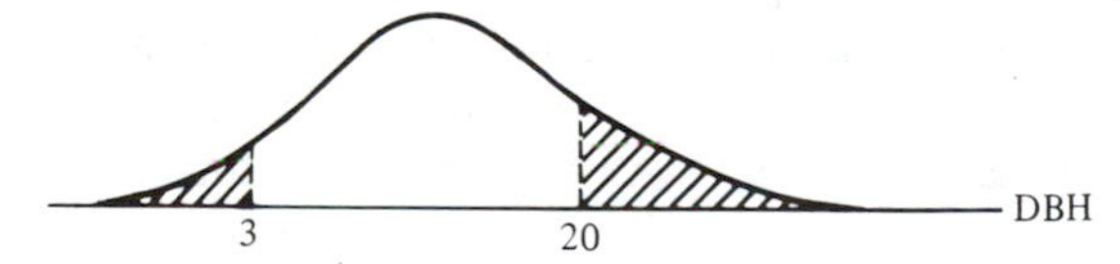

E. Leave only 50% of the trees that are between 20- and 25-in. DBH:

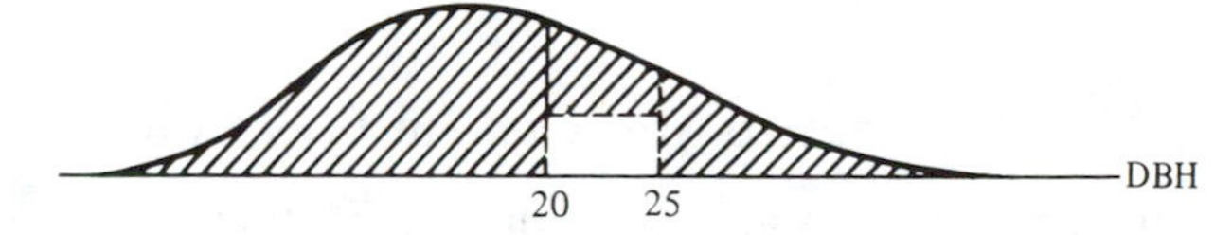

Portions of the diameter distribution removed.

Figure 5-12 Examples of PROGNOSIS thinning options. (*Adapted from Wykoff et al., 1982.*)

reduction of brush, grass, and competing tree species; and harvest of the crop at the optimum times to meet the owner's objectives. Control of between-tree spacing and competition, soil fertility improvement, pruning, and genetic improvement can affect the quantity and quality of growth and the distribution of this growth on the crop trees of the stand.

Table 5-12 Summary table from PROGNOSIS, version 4.0—Inland Empire

			Volume per acre			Removals per acre			
Year	Age	Number of trees per acre	Total, ft^3	Merch., ft^3	Merch., bd ft	Trees, per acre	Total, ft^3	Merch., ft^3	Merch., bd ft
1977	57	536	1,541	1,075	2,804	296	290	250	645
1987	67	223	1,991	1,627	4,645	0	0	0	0
1997	77	209	2,934	2,648	9,128	0	0	0	0
2007	87	196	3,887	3,606	14,518	39	945	895	3,765
2017	97	148	4,032	3,829	16,871	0	0	0	0
2027	107	139	5,126	4,914	23,122	0	0	0	0
2037	117	131	6,351	6,127	29,948	96	4,846	4,676	23,273
2047	127	33	1,760	1,703	8,034	0	0	0	0
2057	137	32	2,183	2,123	10,684	0	0	0	0

Source: Wykoff et al. (1982).

Without a doubt the single most productive management treatment for increasing growth and yield is to get the stand density up to levels that can capture most of a site's growth potential. A large portion of natural and cutover stands are simply understocked. Where competing brush is a real problem, such as on many productive west coast sites, the brush captures a healthy portion of site growth capacity, and suppressing or eliminating competing brush at the time of regeneration is critical. Uncontrolled brush can overtop young trees and can virtually take over a site for 20 to 30 years. We might say the site is understocked (with trees), but in reality it is well-stocked (with brush).

Once a site is stocked to desired density levels with growing trees, further cultural treatments, such as thinning and fertilization, are directed toward improving the quality of the wood produced for harvest.

Forecasting Yield under Intensive Management Prescriptions

While research studies provide the source data indicating how stands respond to different treatments, the response to wide-ranging packages of treatments in a comprehensive prescription is not amenable to long-term field studies because of the cost and time required. Rather we have to use the growth and yield models to simulate what is likely to happen. To illustrate, DFSIM, the Douglas-fir stand growth model (Curtis et al., 1981), is used to investigate a set of alternate rotation, thinning, and fertilization prescriptions.* A prescription is described by notation such as $(P_{15} + T_{35} + T,F_{55} + H_{70})$, which means precommercial thin at age 15,

* This material was prepared by John Helms, University of California, Berkeley, for U.S. Forest Service silvicultural short courses.

Basal area per acre, ft²	CCF	Average dominant height	Growth		
			Production cycle length in years	Accretion	Mortality
64	84	64	10	82	8
89	112	71	10	106	12
115	139	77	10	114	18
108	131	81	10	128	19
132	151	88	10	136	26
152	169	91	10	33	7
40	40	113	10	33	7
44	43	119	10	51	9
52	49	126	0	0	0

commercial thin at age 35, thin again and fertilize at age 55, and regeneration harvest at age 70. For each prescription several stand characteristics are estimated, including total yield (CVTS), cubic volume to a 4-inch top (CV4), mortality, trees per acre (TPA), mean annual increment (MAI), average stand diameter (DBH), and the economic measure of soil expectation value (SEV). We cover SEV in Chaps. 7 and 13 in more detail, but for now take it as a measure of economic payoff, and the higher the SEV, the better.

First, each prescription was evaluated over site indices (50-year base breast height) from 70 to 150, for a given rotation, and we see the predictable response in yield and an even stronger boost in SEV as the site improves (Fig. 5-13). Second, the site index and prescription were held constant while the rotation is varied, giving the results shown in Fig. 5-14. Yield and average diameter increase with the age of the stand, but we see a distinct peaking of SEV at age 70 and a falloff thereafter because of the discounting of yield at a compound interest rate of 4 percent. Since maximizing SEV is our economic objective, we might also say that age 70 is an optimum rotation.

The third analysis looks at one site and compares 15 different prescriptions at the rotation age of 70 years on site 112. In Fig. 5-15 the fifteen prescriptions have been ranked by SEV values at 70 years and presented from left to right in order of declining SEV values. Much good information results, which can help us evaluate prescription options. The prescription (T, F_{35} + T,F_{55}) has the highest SEV, and all other things being equal, would be preferable. This assumes that the projected future cost, price, and interest data used to calculate SEV are also appropriate. But it is not easy to say "this is optimum" with assurance. The big drop in SEV comes after prescription 11, when earlier thinning is no longer part of the prescription, and is reflected by a rise in the number of trees per acre.

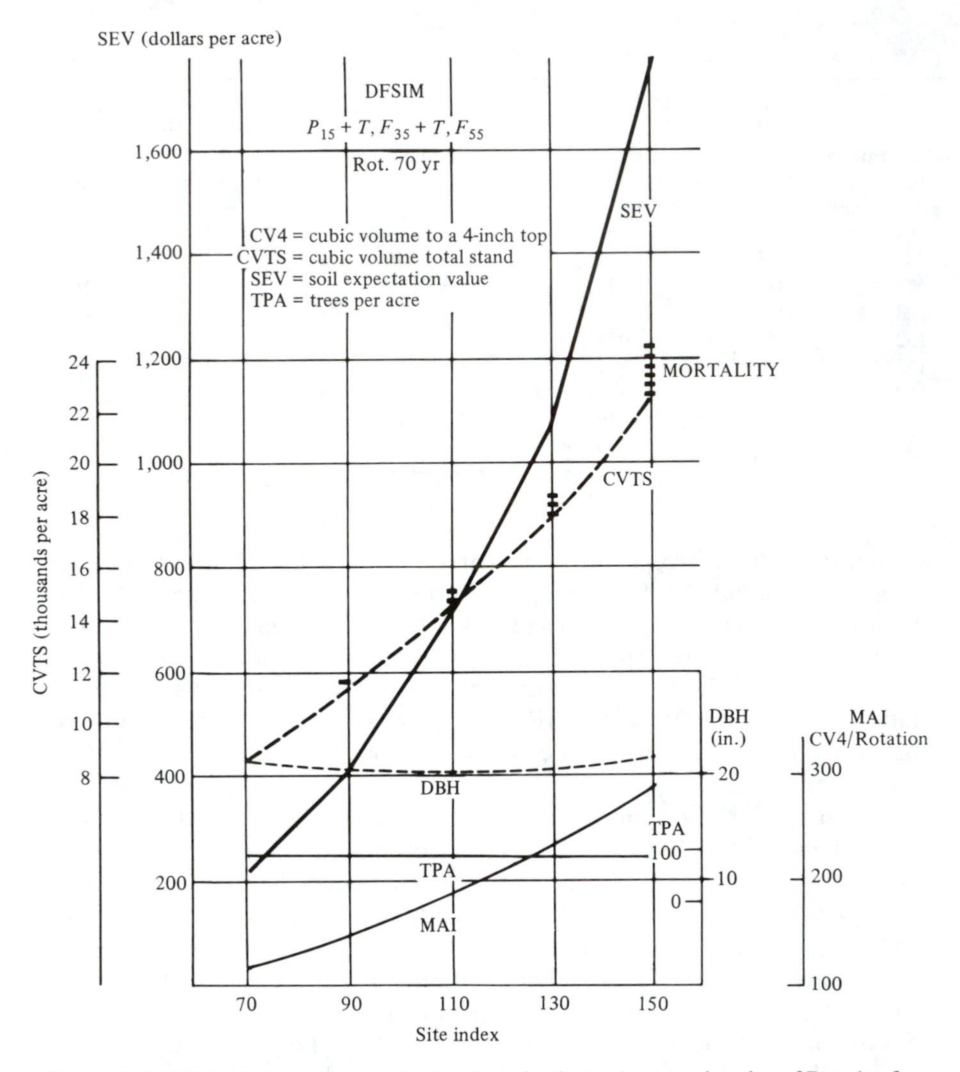

Figure 5-13 Effect of site quality on simulated production and economic value of Douglas-fir under an intensive management prescription and a 70-year rotation.

This example illustrates well the potential application of growth and yield models. It is easy to get carried away and forget that these are simulated rather than factual results. The validity depends on the accuracy and appropriateness of the yield simulator for the subject stand and the kind of prescription evaluated. Again, the manager has to understand enough about yield models to comfortably (if that is possible) accept or reject prescription evaluations as relevant to a subject stand. Mainframe and personal computers that can be used for these yield models are becoming more and more readily available, so all foresters will have ample

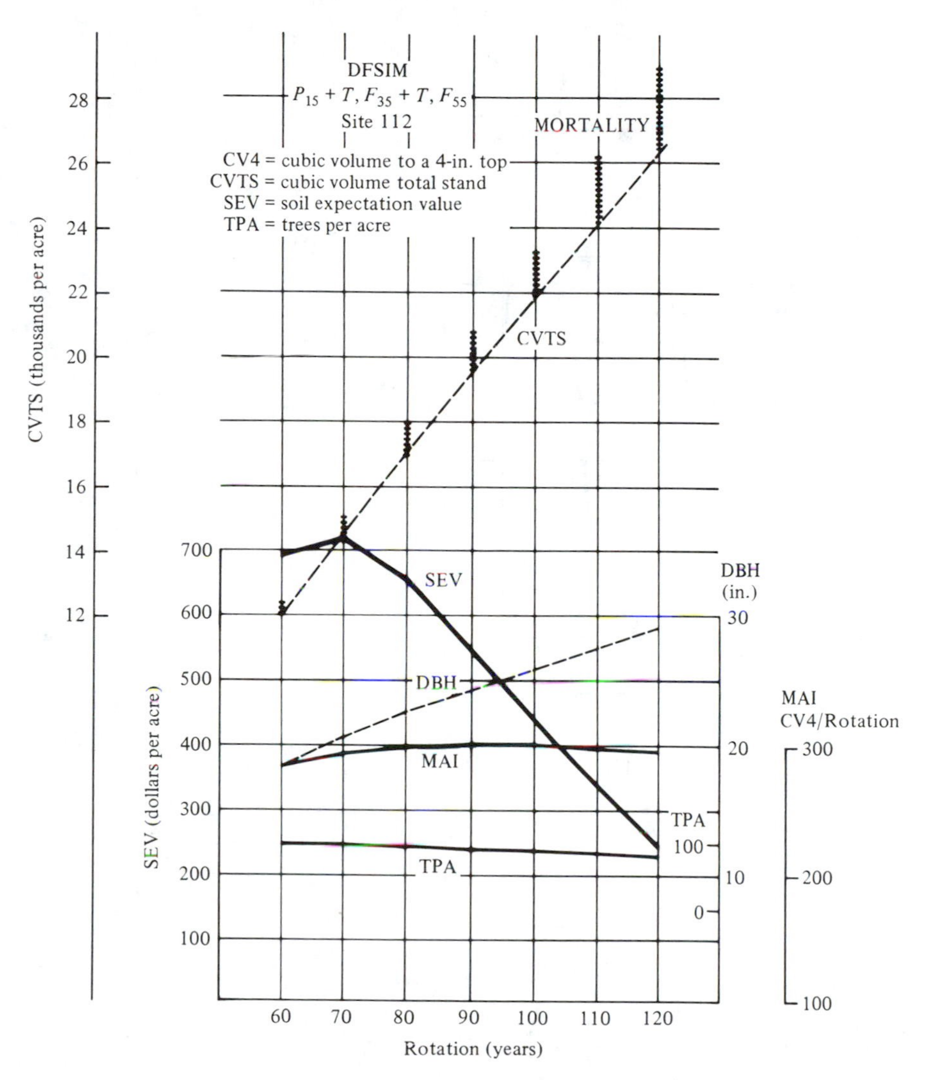

Figure 5-14 Effect of rotation length on simulated yield and economic value of Douglas-fir under an intensive management prescription on site 112 land.

opportunity to worry whether they are getting good answers and decision guidance.

EVALUATING FOREST GROWTH PROJECTION SYSTEMS

Often we have a choice of models to use in predicting the growth and yield of some species of interest. The choice may be between an empir-

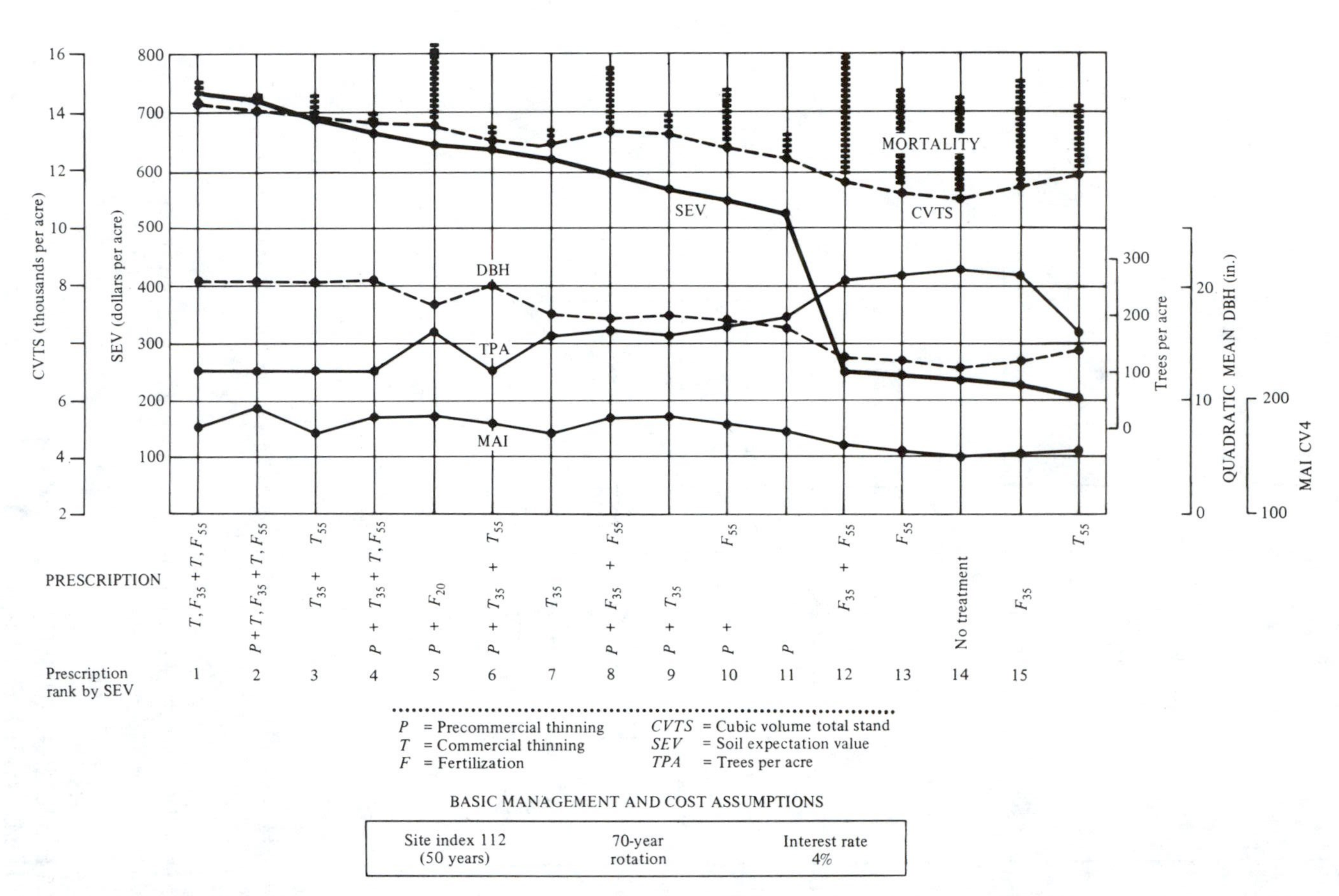

Figure 5-15 Effect of management prescription on simulated yield and economic value of Douglas-fir with a 70-year rotation on site 112 land; prescriptions are ranked by soil expectation values.

ical yield table, an adaption of a normal yield table, a new variable-density yield function, or an individual tree model published by some researchers, as often occurs these days in the North and West. Or the choice may be among a variety of research-produced variable-density yield functions, as often happens these days in the South. (See the previous discussion of loblolly pine yield tables.) How does a user decide which model to use?

To help in this decision, Buchman and Shifley (1983) reviewed the literature on the evaluation of growth and yield models. From this review, they developed a number of principles and criteria to help in this evaluation. They first point out three commonly agreed principles in the evaluation of growth and yield systems:

1. No projection system can represent perfectly the real system being modeled. Thus little is gained from proving that a projection system is an inexact copy of nature. Rather, users should be concerned about how well a projection system performs compared to available alternatives. Frequently the problem is to decide whether a new system is better than the one currently in use. (See Lundgren's analysis below where he compares the model that the U.S. Forest Service is currently using to predict growth for red pine to his proposed alternative.)
2. Projection systems cannot be evaluated in the absence of user-defined objectives and constraints. What do they want to predict and how much are they willing to spend?
3. Evaluation is partially subjective. System capabilities consistently fall short of desired performance, and for a given application each projection system has a unique set of strengths and weaknesses. Selecting the "best" system requires synthesizing a whole array of quantitative and qualitative factors.

Criteria for Evaluation

Buchman and Shifley (1983) further present three broad areas for a potential user to consider when evaluating a projection system:

1. Applications environment—those factors related to physically operating the system and organizing the outputs into usable form.
2. System performance—the accuracy and precision of projections by the model.
3. Quality of system design—the flexibility and capacity for adding or modifying components and the biological realism (biologic) of the system.

Table 5-13 Criteria for evaluation of forest growth projection systems grouped by application environment, performance, and design

Application environment	Performance	Design
User support Is the system thoroughly and clearly documented? Are user guides and sample projection runs available? Are system maintenance and revision performed by the vendor, or are they the user's responsibility? Is assistance available if the system must be modified to fit some special application? Are there other users of the system who can provide advice and assistance? Can the system help the user by providing instructions and prompts?	*Entire system* How accurately and how precisely can the system project the changes in: • Volume, basal area, number of trees, and biomass (per acre, forest, and region) • Wildlife habitat quantity and quality • Water yield • Range yield • Changes in recreation potential Do projection results show systematic bias or loss of precision with changes in stand density, forest type, site quality, tree size, tree species, geographic region, local climatic factors, or any independent variables used in prediction models? To what degree do projection precision and accuracy deteriorate with length of projection?	*Flexibility* Is the system easily modified to meet new applications? Is it programmed in modules (e.g., growth estimation module, output module, management module) that can be deleted, replaced, or modified as necessary? Can nonessential parts be easily deleted to reduce necessary computer resources? Can the system be readily modified to include projections for other resources? With what detail can management practices be simulated? Can simulated management practices be easily altered? Can models be easily recalibrated for use with different forest types?
Data What data are required? Are data requirements compatible with available data? Can the system estimate values for any missing pieces of data?	*System components* It is often possible to separate and study the	*Biologic* Are projection models formulated to incorporate basic theories of biological growth?

Does the system check for illegal data values?
Are output formats appropriate?
Are values reported in appropriate units?
Can output format and units be easily altered by the user?
If it becomes necessary to recalibrate the models, what data are required?

Computing considerations

What computing capacity is required?
What is the cost of operating the system?
How quickly does the system produce results?
What programming experience or other special skills are needed?
Is the system flexible and easy to reprogram if necessary?
Is the system portable and can it be accessed remotely by telephone?
Does the system require special computing features such as file sorting or statistical packages?
Have the program code and logic been thoroughly tested?

performance of projection system components. If possible to determine, how accurately and how precisely can the system project changes in:

- Regeneration
- Ingrowth
- Survivor growth
- Mortality (stand level)
- Tree diameter, height, crown size, and volume by species
- Other tree dimensions by species
- Mortality by species

Do these results show systemic bias or loss of precision with changes in stand density, forest type, site quality, tree size, tree species, geographic region, local climatic factors, or any independent variables used in prediction models?
Does precision or accuracy of component estimates deteriorate with length of projection?
Do models include feedback mechanisms and other controls on growth or yield?
Do system components interact logically?
For long projection periods, do trees and stands approach reasonable size limits?
Are there real or hypothetical conditions that would cause the projection system to predict obviously unreasonable results (e.g., infinite yields, negative yields, trees of excessive size)?

Source: Buchman and Shifley (1983).

The authors have assembled an evaluation checklist from the literature that covers these three areas of evaluation in detail (Table 5-13). While a potential user of a system will probably focus on a subset of the list in an evaluation, this set of criteria is impressive in terms of its comprehensiveness and completeness. Especially a system's biological realism and its performance are of importance to us here. Therefore, we will discuss these criteria in some detail.

Biological realism Our confidence in projection models increases to the degree that they conform with basic theories of biological growth. Example questions we might ask in the assessment are: Is growth positively correlated with site? Is growth positively correlated with density up to some point? Is mortality positively correlated with the number of trees? For long projection periods, do trees and stands approach reasonable size limits? Under some conditions, does the projection system give crazy results (infinite yield, negative yield)?

As an example of this evaluation, we will again look at Buckman's (1962) red pine work. Buckman's basal area growth equation is

$$\text{BAG} = f(\text{BA}, A, S) = 1.6889 + 0.041066\text{BA} - 0.00016303\text{BA}^2$$
$$-0.076958A + 0.00022741A^2 + 0.06441S$$

where BAG = basal area growth
BA = stand basal area per acre
A = age
S = site index

As we might hypothesize:

1. BAG increases with (is positively correlated with) site ($0.06441S$).
2. BAG increases with basal area up to a point and then declines.

BA	BA contribution to BAG ($0.041066\text{BA} - 0.00016303\text{BA}^2$)
60	1.877
80	2.242
100	2.476
120	2.580
140	2.553

3. BAG decreases with age.

Age	Age contribution to BAG ($-0.076958A + 0.00022741A^2$)
30	−2.104
40	−2.714
50	−3.308
60	−3.799

Therefore, we conclude that this basal area growth equation is consistent with theories on the way stands grow.

System performance A system's ultimate test is its ability to predict accurately the stand measures of interest to the user. Two major kinds of evidence can be developed to help make this assessment:

1. The system's ability to predict the growth and other information observed in the data used for model development.
2. The system's ability to predict growth and other information observed in data *not* used in model development.

Predicting the growth observed in the data used in model development Most prediction systems involve sets of regression equations of the form $y = f(x_1, \ldots, x_n)$. As an example, a typical equation is

Future basal area = *f*(future age, current age, current basal area, site)

Usually modelers specify what form of the variables they want represented in the equation (x_1^2, x_1/x_2, etc.) and then use a statistical technique on the data to obtain the equation coefficients ("ordinary least squares" is a commonly used technique) and to "test" the hypothesis that such a functional relationship will explain the variation in y found in the data. A variety of statistical measures on the goodness of fit can be produced to help assess how "good" an equation is. One common measure is the coefficient of determination r^2. It tells what proportion or percentage of the variation in y found in the data is explained by the variation in x found in the data through the hypothesized relationship between x and y. The higher the r^2, the better, with 1.00, or 100 percent, being the maximum achievable.

If we as modelers have data that represent the underlying population, set up a group of explanatory variables in a single hypothesized form that is biologically defendable, and then run a regression, a high r^2 (>0.85) should give us confidence in our equations. With the power of modern computers at our disposal, however, we often test the regressive power of hundreds or thousands of forms of our basic variables and then select those few that in combination give a high r^2. In such cases we may obtain a high r^2, but since we have abandoned the idea of a single hypothesis, we may just be modeling the idiosyncracies of our data.

Therefore we do not consider goodness-of-fit measures from regression results in our model building to be conclusive evidence that we have a model that predicts well. All they can tell us is that if the r^2 is low (<0.5, or 50 percent) after all this thrashing around, we have a model that, in an absolute sense, probably predicts poorly. But it still may be the best available.

With time series data, we may take the further step of taking a subset of the data and see how well the model predicts the major changes in these data over time. Employing data used to build a model to test it in this way is not a rigorous test, but it can still be informative, especially when the system is a set of equations to assess aggregate performance.

Predicting the growth observed in data not used in model development Here we use regression techniques to help quantify a hypothesis (build the model) with one set of data and then test this hypothesis with another set of data. This approach is the ultimate test of system performance, but usually we have so little data that we use them all up in model construction. Then we must gather more data, perhaps waiting a number of years to obtain them, before such a test can be made.

As an example of such a comprehensive test, we will again look at Buckman's red pine work. Buckman (1962) compiled yield tables for thinned stands of red pine for a range of sites and densities. His growth model made it possible to project growth and yields in stands 20 years of age or older if the initial basal area was known. Several years later, Wambach (1967) developed equations that projected the growth up to about age 35 years in Lake states plantations established at various initial spacings.

Wambach's and Buckman's models were combined by Lundgren (1981) in a computer program (REDPINE) that simulates growth and yield from the time of stand establishment to any given age. The program projects growth in unthinned as well as thinned stands, and determines mean annual increments for each stand age. It is intended for use in both plantations and natural stands. REDPINE was used for a range of site indices to develop the estimates reported here.

In assessing the quality of REDPINE, Lundgren first compared his results to those of two red pine models popular in the Lake states: (1) U.S.D.A. Forest Service charts for red pine productivity and (2) yields reported by Eyre and Zehngraff (1948). He found that his REDPINE

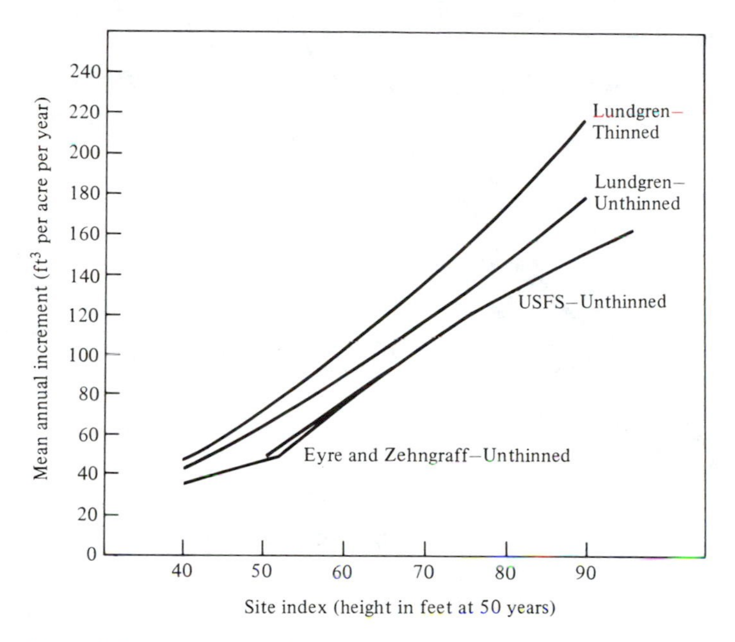

Figure 5-16 Comparison of site productivities for red pine stands in Lake states. *(From Lundgren, 1983.)*

model consistently predicts a higher site productivity than the other two models (Fig. 5-16).

Which model's predictions are closest to reality? To answer that question, Lundgren next searched the literature and found 26 observations of mean annual increments in red pine suitable for comparison to his model, with maximum mean annual increment used from the observations to the degree possible. For each observed mean annual increment, an estimate was made of the value predicted by REDPINE for the reported stand age and site index and for a stocking of 800 initially established trees per acre.

Lundgren found a high correlation between the predictions of REDPINE and the observations (Fig. 5-17). He states:

> Simple observation of the observed and predicted MAI values revealed no apparent bias (Fig. 5-17). A simple linear regression equation relating predicted (Pr) to observed (Ob) MAI resulted in the following relation:
>
> $$\text{Pr} = 6.8 + 0.905\text{Ob}$$
>
> with $r^2 = 0.73$.
>
> A confidence test (F statistic) indicated that this equation was not significantly different (0.05 level) from the simple assumption that the predicted MAI was equal to the observed MAI (that is, $\text{Pr} = 0.0 + 1.0\text{Ob}$).

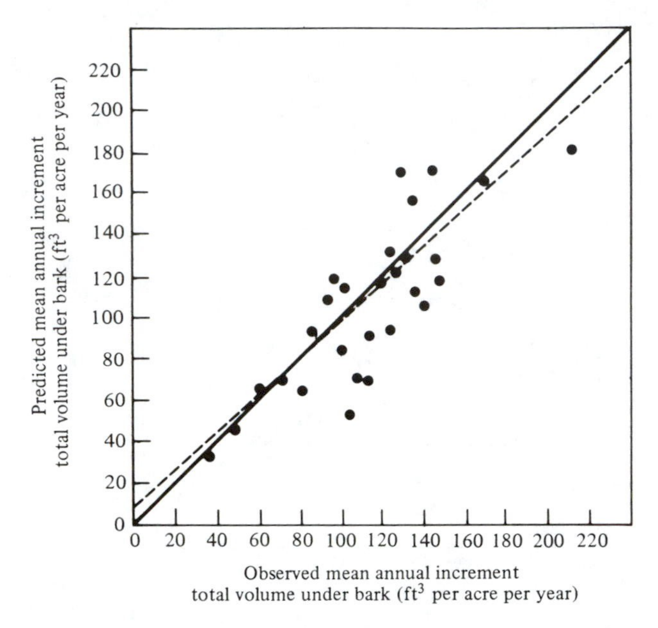

Figure 5-17 Predicted mean annual increments in red pine stands using Lundgren's REDPINE growth model. Dashed line is regression of predicted on observed. *(From Lundgren, 1983.)*

> It seems safe to conclude that the REDPINE model produces good estimates of biological potential for red pine stands in the Lake states and can be used with some confidence to estimate site productivity.

And as he further states, these results "indicate that the growth potential of red pine has been underrated in the past."

In summary, Lundgren's REDPINE model passes the tests put before it on system performance and biological realism. In addition, it appears to be a better predictor than the growth projection systems previously used. While a potential user would probably wish to assess REDPINE from the standpoint of the applications environment and system flexibility, the model shows promise as a predictor of red pine growth and yield.

QUESTIONS

5-1 A subject stand of naturally regenerated Douglas-fir is growing on site 140 land. It is currently 50 years old and has 5,000 ft^3 of growing stock. McArdle's normal yield estimates (Table 3-3) are the only models available. Make an estimate of the yield available for harvest at age 80 under the following growth assumptions:

(*a*) The growth of the subject stand is proportionate to its stocking relative to the normal stand.

(*b*) The subject stand can realize *all* of the normal growth potential of the site.

(*c*) Understocked stands grow faster—enough to increase their stocking by 5 percentage points per decade until they are fully stocked.

5-2 The equation to predict volume per acre in a stand is

$$\ln V = 4 - \frac{100}{S} - \frac{12}{A} + 0.9 \ln \text{BA}$$

The equation to predict future basal area is

$$\ln \text{BA}_2 = \frac{A_1}{A_2} \ln \text{BA}_1 + 5\left(1 - \frac{A_1}{A_2}\right)$$

where V = volume
S = site index
A_1 = current age
A_2 = future age
BA_1 = current basal area per acre
BA_2 = future basal area per acre

Derive an equation to predict future volume V_2 as a function of S, A_1, A_2, and BA.

5-3 A natural old-field loblolly pine stand is growing on site 70 land and is currently 35 years old, with a basal area of 130 ft^2 per acre. The owner has the stand scheduled for harvest in 15 years. You are using the Brender–Clutter model of Eqs. (5-8) and (5-9) and are asked to estimate:

(*a*) The expected harvest yield in 15 years.

(*b*) The periodic annual increment over the next 15 years.

5-4 A company just purchased considerable acreage of uneven-aged northern spruce–hemlock timberland. The owners want you to evaluate residual stocking levels of 60, 80, and 100 ft^2 of basal area for periodic harvests every 10 years to remove all growth. The expected value of poletimber per cubic foot is $0.20, and the value of sawtimber is $0.90. The yield data of Solomon and Frank (Table 5-5) are all that is available. The evaluation is to be in terms of periodic annual stand value growth per acre.

5-5 Given the following stand table and growth/death model, calculate the number of trees by diameter class after one growth period.

DBH class midpoint, in.	Beginning stand, number of trees	BA per tree, ft^2
6	200	0.1963
8	140	0.3491
10	100	0.5454
12		

The growth/death system is given as

$$\textit{Ingrowth} \text{ into 1st DBH class} = \frac{50}{0.01\text{BAS}}$$

$$\textit{Upgrowth} \text{ from DBH class } D \text{ to DBH class } D + 1 = \frac{10X_D}{\sqrt{\text{XAL}}}$$

$$\textit{Mortality} \text{ of DBH class } D = \frac{X_D}{\sqrt{\text{BAS}}}$$

where X_D = number of trees in diameter class D at beginning of growth period
BAS = basal area of stand at beginning of growth period
XAL = number of trees in stand at beginning of growth period

5-6 *Stand table projections.* Given the following growth rates and stand structure before growth, calculate the stand structure after growth. Use the second stand table projection method that applies average diameter increments, recognizing dispersion within classes.

DBH class midpoint (4 in. classes)	Number of trees before growth	Average class DBH growth during period, in.	Number of trees after growth
2	20	3	
6	12	4	
10	6	3	
14	2	1	
18	—	—	

5-7 A proposed per-acre stand basal area growth equation is

$$\text{BAG} = 1.5 + 0.04\text{BA} + 0.001\text{BA}^2 - 0.07A + 0.0001A^2 - 0.2S$$

where BAG = annual basal area growth
A = age, in years
BA = basal area, in square feet per acre
S = site index, in feet

Evaluate this proposed equation in terms of growth response to changes in the variables A, BA, and S to determine whether the variable is presented in a biologically defendable manner.

Evaluate over the following ranges:

$$S = 50\text{–}100 \text{ (index)}$$

$$A = 30\text{–}80 \text{ years}$$

$$\text{BA} = 50\text{–}400 \text{ ft}^2$$

Review the last section of Chap. 5 as a guide to this question.

5-8 List and discuss the important assumptions and factors you would consider before deciding that a candidate growth model is acceptable for estimating the growth and yield of a subject stand managed under a specific prescription.

REFERENCES

Adams, D. M., and A. R. Ek, 1974: Optimizing the Management of Uneven-Aged Forest Stands, *Can. J. Forest Res.*, **4**:274–287.

Alemdag, I. S., 1978: "Evaluation of Some Competition Indices for the Prediction of Diameter Increment in Planted White Spruce," Information Report FMR-X-108, Forest Management Institute, Canadian Forest Service, Ottawa.

Avery, T. E., and H. E. Burkhart, 1983: "Forest Measurements," 3d ed., McGraw-Hill, New York.

Bailey, R. L., and T. R. Dell, 1973: Quantifying Diameter Distributions with the Weibull Function, *Forest Sci.,* **19:**97–104.

Barrett, J. W., 1983: "Growth of Ponderosa Pine Thinned to Different Stocking Levels in Central Oregon," U.S. Forest Service Research Paper PNW-311.

Bella, I. E., 1971: A New Competition Model for Individual Trees, *Forest Sci.,* **17:**364–372.

Brender, E. V., and J. L. Clutter, 1970: "Yield of Even-Aged Natural Stands of Loblolly Pine," Report 23, Georgia Forest Research Council.

Buchman, R. G., and S. R. Shifley, 1983: Guide to Evaluating Forest Growth Projection Systems, *J. Forestry,* **81:**231–234.

Buckman, R. E., 1962: "Growth and Yield of Red Pine in Minnesota," U.S. Department of Agriculture Technical Bulletin 1272.

Burkhart, H. E., Q. V. Cao, and K. D. Ware, 1981: "A Comparison of Growth and Yield Prediction Models for Loblolly Pine," School of Forestry and Wildlife Resources, Virginia Polytechnic Institute and State University, Blacksburg, Publication FWS-2-81.

Clutter, J. R., 1963: Compatible Growth and Yield Models for Loblolly Pine, *Forest Sci.,* **9:**354–371.

———, J. C. Fortson, L. V. Pienaer, G. Brister, and R. L. Bailey, 1983: "Timber Management: A Quantitative Approach," Wiley, New York.

Curtis, R. O., G. W. Clendenen, and D. J. De Mars, 1981: "A New Stand Simulator for Coast Douglas-Fir: DFSIM User's Guide," U.S. Forest Service General Technical Report PNW-128.

———, ———, D. R. Reukema, and D. J. De Mars, 1982: "Yield Tables for Managed Stands of Coast Douglas-Fir," U.S. Forest Service General Technical Report PNW-135.

Daniels, A. N., and H. E. Burkhart, 1975: "Simulation of Individual Tree Growth and Development in Managed Loblolly Plantations," Virginia Polytechnic Institute and State University, Blacksburg, Publication FWS-5-75.

Dell, T. R., D. P. Feduccia, T. E. Campbell, W. F. Mann, and B. H. Polmer, 1979: "Yields of Unthinned Slash Pine on Cutover Sites in the West Gulf Region," U.S. Forest Service Paper SO-147.

Duerr, W. A., 1938: Comments on the General Application of Gehrhardt's Formula for Approach toward Normality, *J. Forestry,* **36:**600–604.

Ek, A. R., 1974: Nonlinear Models for Stand Table Projection in Northern Hardwood Stands, *Can. J. Forest Res.,* **4:**23–27.

Eyre, F. H., and P. Zehngraff, 1948: "Red Pine Management in Minnesota," U.S. Department of Agriculture Circular 778.

Gehrhardt. E., 1930: "Ertragstafeln für reine und gleichartige Hochwaldbestände von Eiche, Buche, Fichte, Kiefer, grüner Douglasie und Lärche," 2d ed., Springer, Berlin. Review by S. R. Gevorkiantz, *J. Forestry,* **32:**487–488.

Hamilton, D. A. and B. M. Edwards, 1976: "Modeling the Probability of Individual Tree Mortality," USDA Forest Service Research Paper INT-185.

Hann, D. W., 1980: "Development and Evaluation of an Even- and Uneven-Aged Ponderosa Pine/Arizona Fescue Stand Simulator," U.S. Forest Service Research Paper INT-267.

Hegyi, F., 1974: A Simulation Model for Managing Jack Pine Stands, in "Growth Models for Tree and Stand Simulation, IUFRO, Proceedings of Working Party," S.4.01-4:74–90.

Krumland, B. E., 1982: "A Tree-based Forest Yield Projection System for the North Coast Region of California," Ph.D. dissertation, Department of Forestry and Resource Management, University of California, Berkeley.

———, and L. C. Wensel, 1982: "CRYPTOS/CRYPT2 Users Guide: Cooperative Redwood Project Timber Output Simulator," Version 4.0, Research Note 20, California Co-op Redwood Yield Research Project, Department of Forestry and Resource Management, University of California, Berkeley.

Libby, W. J., and R. M. Rauter, 1984: Advantages of Clonal Forestry, *Forestry Chronicle,* **60:**145–149.

Lundgren, A. L., 1981: "The Effect of Initial Number of Trees per Acre and Thinning Densities on Timber Yields from Red Pine Plantations in the Lake States," U.S. Forest Service Research Paper NC-193.

Lundgren, A. L., 1983: New Site Productivity Estimates for Red Pine in the Lake States, *J. Forestry,* **81:**714–717.

McArdle, R. E., W. H. Meyer, and D. Bruce, 1930: "The Yield of Douglas-Fir in the Pacific Northwest" (rev., October 1949), U.S. Department of Agriculture Technical Bulletin 201.

Miller, R. E., and R. D. Fight, 1979: "Fertilizing Douglas-Fir Forests," U.S. Forest Service General Technical Report PNW-83.

Moser, J. W., 1972: Dynamics of an Uneven-Aged Forest Stand. *Forest Sci.,* **18:**184–191.

———, and F. C. Hall, 1969: Deriving Growth and Yield Functions for Uneven-Aged Forest Stands, *Forest Sci.,* **15:**183–188.

Munro, D. D., 1974: Forest Growth Models—a Prognosis, in "Growth Models for Tree and Stand Simulation," Royal College, Stockholm, Forest Research Notes 30.

Oliver, W. W., 1979: Fifteen-Year Growth Pattern after Thinning a Ponderosa Jeffrey Pine Plantation in Northern California," U.S. Forest Service Research Paper PSW-141.

Porterfield, R. L., B. J. Zobel, and F. J. Ledig, 1975: Evaluating the Efficiency of Tree Improvement Programs, *Silvae Genetica,* **24**(2–3):33–44.

Reukema, D. L., 1979: "Fifty-Year Development of Douglas-Fir Stands Planted at Various Spacings," U.S. Forest Service Research Paper PNW-253.

Smalley, G. W., and R. L. Bailey, 1974: "Yield Tables and Stand Structure for Shortleaf Pine Plantation in Tennessee, Alabama, and Georgia Highlands," U.S. Forest Service Research Paper SO-96.

Soloman, D. S., and R. M. Frank, 1983: "Growth Response of Managed Uneven-Aged Northern Conifer Stands," U.S. Forest Service Research Paper NE-517.

Staebler, G. R., 1951: Growth and Spacing in an Even-Age Stand of Douglas-Fir, M.F. Thesis, School of Natural Resources, University of Michigan, Ann Arbor.

Stage, A. R., 1973: "Prognosis Model for Stand Development," U.S. Forest Service Research Paper INT-137.

Sullivan, A. D., and J. L. Clutter, 1972: A Simultaneous Growth and Yield Model for Loblolly Pine. *Forest Sci.,* **18:**76–86.

Van Deusen, P. C., and G. S. Biging, 1985: "STAG: A Stand Generator for Mixed Species Stands," Version 2.0, Research Note 11, Northern California Forest Yield Cooperative, Department of Forestry and Resource Management, University of California, Berkeley.

Wambach, R. F., 1967: "A Silvicultural and Economic Appraisal of Initial Spacing in Red Pine," Ph.D. thesis, University of Minnesota (#67-14,665, University Microfilms, Inc., Ann Arbor, Michigan).

Wensel, L. C., and P. J. Daugherty, 1984: "CACTOS Users Guide: The California Conifer Output Simulator," Version 1.0, Research Note 10 (draft 2), Northern California Forest Yield Cooperative, Department of Forestry and Resource Management, University of California, Berkeley.

Wykoff, W. R., N. L. Crookston, and A. R. Stage, 1982: "User's Guide to the Stand Prognosis Model." U.S. Forest Service General Technical Report INT-133.

PART TWO

DECISION ANALYSIS IN FOREST MANAGEMENT

The formalization of forest management as a quantified exercise in problem solving is a distinguishing hallmark of our profession today. The fundamental nature of our problems has not changed that much with the passing decades. Now, as then, we have forest owners who want to grow trees for income, amenity, and other purposes. What is new is the number of people interested in what forest owners do. The costs and revenues from different forestland uses have risen to the point where forest management has become a high-stakes game in many areas. Now we have to explain and justify; to show, with numbers, why what we have decided to do is best.

A load of practical theory and technique from management science and economics has been transported and adapted to forestry to help with quantification, explanation, and justification. Coupled with the number-crunching power of today's computers we have more than enough tools—the analyst's problem is choosing which tool is appropriate for a particular problem situation, being aware of its assumptions and limitations, and not being overly impressed by impeccable-looking numerical results from computerized analytical packages whose inner workings are hidden or not understood.

Quantitative decision analysis has forced us to understand our forest management problems much better. Recently we have learned a lot more about trade-offs between forest outputs, where our knowledge base is strong and where it is weak.

In many respects better applied economics and management science have allowed us to say, with measurably improved confidence, that our recommended land management choices are efficient choices which best meet the objectives and satisfy the constraints of forest owners. We have concurrently found that the more we know, the more we recognize the depths of our ignorance. The data base describing the character of our commercial forestland is woefully scanty and incomplete, and we still cannot confidently predict demand and prices in the distant futures for which we plan. Experienced, sensitive managers with political acumen are still relied upon to balance the divergent goals of public and private forest owners and appraise the reliability and meaning of information.

The art and science of problem identification is initiated in Chap. 6, where the foundations are laid for all analyses in this text. Our most fundamental assumption is that you must specify goals and constraints appropriate to the problem in order to solve it and provide a useful analysis. The character and specification of decision problems are discussed using linear equations and many examples. Decision problems under uncertainty, risk, and certainty are defined and compared.

Linear programming is then introduced as a technique to solve problems that are identified by sets of linear equations. We explore the information provided by linear programming solutions and the use of multiple solutions and sensitivity analysis to evaluate problem situations when not all problem aspects are known with certainty, or when the goals and constraints of the problem are themselves contestable. Goal programming is introduced, and the chapter concludes with some philosophy on the limitations of mathematical modeling.

Financial analysis and the arithmetic of interest are two important components of required forest management expertise used to recognize that we grow trees in an economic world. Genesis of interest rates opens Chap. 7 and is followed by the formula for interest computations using the same terminology as the second edition. A decision tree is provided for matching the correct equation to a problem. Procedures to adjust for inflation and conversion to nominal and effective rates are covered.

The second half of Chap. 7 covers the quantification of cash flow schedules to portray investment projects and the various ways to apply the interest formula to summarize the present or future net worth of a project. Calculating the present net worth of perpetual timber-growing projects—called land or soil expectation value in forestry—is highlighted, and the chapter concludes with procedures to adjust for differences in project time length and reinvestment of intermediate revenues in order to make comparisons between different projects.

Useful decision analysis is articulated in Chap. 8 as an activity that produces the important, factual, and objective information on which forest owners, administrators, and politicians base their decisions. The format and the presentation of such information are a critical matter, in addition to the quality of the information itself. We take a multiple-goal viewpoint in this chapter and organize the presentation around five classes of public and private goals: economic efficiency, regional equity, distributional equity, stability, and environmental security.

Economic efficiency goals receive the greatest length of treatment. The definition of economic efficiency is elaborated and discussed in terms of identifying the problem situations in which it can give clear decision guidance and those where it cannot. Limitations of economic efficiency analysis in multiple-goal forest planning problems involving important nonmarket benefits are detailed.

At the project level, the use of the economic criteria of present net worth, internal rate of return, and benefit-cost ratio are demonstrated and compared. Their application in selecting among and ranking individual projects within sets of independent and mutually exclusive projects is detailed.

Regional impact analysis is outlined, and the important input/output modeling technique for estimating economic impacts of changes in forest production levels is presented and evaluated in some detail. Some ideas on measuring the impact of forestland management activities on benefit distribution to the members of society, on the stability of local economies, and on environmental security are covered. The chapter concludes with a presentation and critique of the benefit-cost framework for a comparison of alternative forest management plans.

CHAPTER

SIX

A FRAMEWORK FOR DECISION MAKING

A central role of the forest manager is decision making, choosing among alternative courses of action. Prerequisite to this act of choosing, however, is a perception by the manager that some situation or issue requires a decision to be made. Problem situations can be obvious—the mill is running out of wood, deer populations have declined, a new timber management plan is scheduled, or someone has filed a lawsuit to stop the use of herbicides in regeneration. Problems can be subtle as well: the rate of interest charged for forest investment capital is rising, a shortage of logs of certain sizes appears to be likely in about 10 years, or through urbanization the local population is changing to one with much more urban than traditional rural values. Becoming aware of obvious problems is not hard; it is looking for, anticipating, and responding to subtle but important problems that marks an effective manager.

Loosely defined perceptions of problem situations such as these do not by themselves provide a basis for thoughtful decision making. The details of the problem need quantification and articulation in terms of the goals of the landowner as seen by the manager and any binding limitations on the manager's activities. Even more important, alternative solutions to the problem need to be identified in order to provide reasonable choices.

Successful managers are usually people who correctly identify relevant goals, issues, and resource limitations of problems and come up with alternative solutions that are workable and beneficial. Many if not most failures at problem solving (bad decisions) can be traced to a failure to identify the problem properly in the first place.

For example, consider the student with limited funds who needs some way to travel between apartment, school, and job. The student also would like something to drive around town and to use on Friday nights. Many kinds of small and midsized cars might do, but the student got carried away and bought a fancy four-wheel-drive pickup for little down and delayed monthly payments. Within six months the fuel costs and monthly payments had the truck up for sale, and a $5,000 net loss was probable. Not the best decision; but why? The student had lost sight of the primary goal (basic transportation) and had ignored the budget constraints. Yet next year when the student graduates and has a regular and higher income, this same decision might be a good one. What changed? The budget constraint.

For another and more far-reaching example consider the slow reaction of foresters during the 1960s and 1970s to perceive the growing national evaluation of aesthetic and environmental quality as a more important goal than in the past. Fully believing they were doing a quality professional job, forest managers clung to traditional use of herbicides, pesticides, large-size clear-cutting, and other techniques which were effective for timber production, yet were upsetting to many people. Rather abruptly the National Environmental Policy Act passed and provided for public participation in the decision process. Forest practices subsequently have been changing to reflect these new social goals. Public foresters resisted these changes and developed a collective image as being insensitive, not perceptive of changing social desires, and not good managers. Without judging the virtue of decisions made during this period, it is fair to say that, collectively, the forestry profession did not identify the changing structure of public land management problem in a timely way.

The examples illustrate another and most important point. Problems change with time and managers constantly need to look for these changes. You cannot solve a problem "once and for all" and then comfortably cling to old solutions in a dynamic world.

Setting up problems for decision making by the process of problem identification is something that you can learn and get better at. Problems have a structure, and systematic procedures exist to describe this structure. The many problems and issues discussed in this book cover the most important problems in the special domain of the forest manager. A consistent language and some basic concepts are requisites to their exposition and discussion, and are the substance of this chapter.

First, some key terms and concepts are presented as a series of steps in the problem identification process, and several problems are presented to illustrate the steps. Approaches and criteria for decisions under certainty, uncertainty, and risk are then outlined. Finally, the chapter concludes with an introduction to mathematical programming as a help in finding the best solution to complex problems that must explicitly consider many constraints.

WORD PROBLEMS AND EQUATION WRITING

Word problems are sets of verbal statements describing the interrelationships of a problem. For example, consider the three verbal statements describing a reforestation problem:

1. An annual operating budget of $500,000 is available to a forestry department to plant or seed cutover acres by hand, machine, or aerial methods.
2. Average site preparation and seeding costs for hand planting is $365 per acre, for machine planting it is $310 per acre, and for aerial seeding, $245 per acre.
3. Combinations of the three planting methods will be used such that production costs use up the annual budget.

To convert these three statements to a linear equation, we define the variables and their coefficients. Let

b = annual budget in dollars = $500,000
X_1 = number of acres hand-planted
X_2 = number of acres machine-planted
X_3 = number of acres aerial-seeded
a_1 = cost of hand-planting 1 acre = $365
a_2 = cost of machine-planting 1 acre = $310
a_3 = cost of aerial-seeding 1 acre = $245

With these variables and coefficients, we can write the three sentences as the implicit equation

$$a_1X_1 + a_2X_2 + a_3X_3 = b$$

or, explicitly, as

$$365X_1 + 310X_2 + 245X_3 = \$500{,}000 \tag{6-1}$$

This linear equation represents all the combinations of X_1, X_2, and X_3 that fully use the $500,000 budget.

Suppose it was suggested that 500 acres hand-planted, 270 acres machine-planted, and 1,000 acres aerial-seeded is one possible combination of reforestation methods that satisfies this budget requirement.

To check, the suggested production mix is $X_1 = 500$, $X_2 = 270$, and $X_3 = 1{,}000$. Substituting these values into the equation, we have

$$365X_1 + 310X_2 + 245X_3 = 500{,}000$$

or

$$365(500) + 310(270) + 245(1{,}000) = 500{,}000$$

or

$$182{,}500 + 83{,}700 + 245{,}000 = 500{,}000$$
$$511{,}200 \neq 500{,}000$$

So this does not check out as a combination that satisfies the equation. Can you define a set that will work?

Inequalities

Not all mathematical formulas use the symbol =, which means "equal to." We have other relationships:

The symbol < means "less than," as

$$a < b \text{ means } a \text{ is less than } b$$

The symbol > means "greater than," as

$$a > b \text{ means } a \text{ is greater than } b$$

The symbol ≤ means "less than or equal to," as

$$a \leq b \text{ means } a \text{ is less than or equal to } b$$

The symbol ≥ means "greater than or equal to," as

$$a \geq b \text{ means } a \text{ is greater than or equal to } b$$

Equation (6-1), which described the reforestation production problem, is an *equality* in that the equal sign is used to connect the two sides of the equation. If any one of the four symbols >, <, ≥, or ≤ is used, the formula becomes an *inequality* in that the two sides of the equation no longer need to equal each other.

Inequalities are used frequently in problem identification. Suppose the third verbal statement of the reforestation problem was changed to

3. Combinations of the three planting methods will be used such that their costs do not exceed the annual budget.

This suggests an inequality of the "less than or equal to" form, that is, the total cost can be less than or even equal to the budget, but it cannot exceed the budget. If we revise the equation as an inequality to read

$$365X_1 + 310X_2 + 245X_3 \leq 500{,}000 \qquad (6\text{-}2)$$

it correctly reflects the revised third statement.

PROBLEM IDENTIFICATION—THE STEPS AND LANGUAGE

Identifying problems follows a well-ordered series of steps and the consistent use of some important terminology. First, a *decision maker* who sets the context of the problem and makes needed value judgments is determined. The problem *goals* are extracted from the decision makers and given operational *criteria*. *Activities* and *decision variables* describing actions that could be undertaken to resolve the problem are then defined to formulate *problem solutions*. One of the criteria is represented in an operational *objective function*. Finally, resource, policy, and multiple-goal *constraints* are specified to complete the *problem statement* and permit the calculation of a feasible solution. These steps are detailed below.

Decision Maker

The decision maker, who is either the landowner or a forest manager with delegated authority to act for the owner, chooses which policy or action will be undertaken to solve a given problem and accepts responsibility for that choice. The goals and constraints relevant to a problem must be those considered relevant by the decision maker. Consequently it is crucial that the decision maker be known before you, or some other analyst, attempts to identify and suggest solutions to a problem.

Often the owner of a tract of forestland or a small forest products company is also the active decision maker and is easy to locate. In larger corporations, however, the owners are the stockholders as represented by the board of directors and the executive officers. Many forest-level decisions are made well down in the corporate hierarchy by forest managers who know corporate goals and policies only indirectly through written guidelines, and who accumulate "sense" through years of meetings and interactions to develop an overall feel for the traditions of the organization.

A public forestland manager's task is perhaps the most difficult. The public manager is a hired hand who has been legislatively delegated authority for the real owners: all the citizens. More than their corporate counterparts, public managers have poorly defined guidance on the nature and relative importance of the many different goals for the many different "publics" interested in the public lands. Many citizens do not speak up or communicate with the manager while organized interest groups are informed, persuasive, and vocal and often threaten or actually use political or legal intervention to change, prevent, or reverse the forest manager's decisions.

As analysts we are trying to represent the goals and constraints of the landowners as perceived by the forest manager. When there are many owners with divergent and conflicting values and viewpoints, a single set of goals and constraints cannot represent them, and several different sets are used to cover the different viewpoints. The numerical results of such analyses are presented by the forest manager to the landowners to give an objective information base for political compromise.

Goals

These end states, aspirations, or purposes reflect what the decision maker hopes to achieve. Typically, more than one goal is relevant to a problem. Goals are often stated in general form such as:

To increase income
To provide local economic stability
To be happy or satisfied
To provide needed raw materials for the factory
To provide the greatest good for the greatest number in the long run
To build a road from (here) to (there) with minimum environmental impact

or they may be more specific:

To make an income of $50,000 a year within 10 years
To produce 1,000,000 more cords of wood per year from the land holdings managed
To increase by 100 percent the number of mallards born on the refuges managed
To maximize profit

Such specific statements of goals are often termed objectives or targets.

Goal Criteria

These numerical measures determine whether or not the activities of decision makers are moving them toward or away from their goals. For a goal to be an operational guide, we must be able to devise a criterion that measures goal achievement. For example,

Goal. To increase income
Criterion. Increase in dollars of income per year over current level

Goal. To increase the output of mallards
Criterion. Number of additional mallards produced per year

Goal. To grow wood more efficiently
Criterion. The percentage rate of return on assets and funds invested in timber growing

Goal. To have income stability
Criterion. The standard deviation σ of annual income over a decade

For some goals it is difficult to state a criterion:

Goal. To be happy
Criterion. Happiness index, number of headaches per month, number of highs (lows) per month

Goal. To maintain an aesthetic environment
Criterion. Subjective index (1–10 scale) of aesthetic quality

The existence of a suitable criterion to represent a goal, agreed upon by affected interests, is the test of whether a goal can be operationally defined. Forestland managers often invoke the goal of "multiple use" but, as yet, no criterion to measure the degree of multiple-use attainment has been agreed upon. It is hard to say if one management plan contains more multiple use than another, and by how much. Sustained yield is another long-standing forest management goal that defies a consensus operational criterion.

Often philosophical guides to forest management are hard to quantify. Gifford Pinchot's goal that forests be "managed for the greatest good for the greatest number in the long run" contains emotional and inspirational feeling and can give spirit and intent to management. But unless we can dissect such a statement and decide how to measure "good," "number," and "long run," the goal is not particularly useful as a decision guide. Private owners and corporations commonly have goals to maintain continuity of ownership, stability of income, security of wood raw material supplies, or insurance against financial disaster—notions that are hard to reduce to simple numbers or dollar measures.

Activities and Decision Variables

The idea of activities and decision variables can best be developed with a specific example. Suppose a problem involves allocating a budget to different projects for increasing the recreation opportunities on a forest. The goal is to maximize new recreation opportunities on the forest as measured by the criterion of additional annual visitor-days of recreation opportunity provided.

Activities are the type or kinds of projects or things that can be done to help achieve the goal and *decision variables* represent the amount of each type of activity undertaken. For the recreation problem, assume that three types of activities are being considered:

Activity	Decision variable
Build campsites	X_1 = number of campsites built
Build boat ramps	X_2 = number of boat ramps built
Build trails	X_3 = number of miles of trail built

It is rare, if ever, that decision variables will logically take on negative values, and for most problems the X_i are thus constrained to be nonnegative,

$$X_i \geq 0 \qquad \text{for all } i$$

This is usually assumed unless stated otherwise.

Problem Solutions

These are sets or mixtures of different amounts of activities which might be selected to achieve the decision maker's goals. Such solutions consist of specified amounts for each decision variable. Three such problem solutions are,

		Solution		
Activity	Decision variable	A	B	C
Build campsites	X_1 (number)	100	50	50
Build boat ramps	X_2 (number)	10	5	0
Build trails	X_3 (miles)	0	30	70

Solution A, for example, consists of building 100 campsites, 10 boat ramps, and 0 miles of trail.

Objective Function

An objective function is a mathematical statement of a goal combining the criterion for the goal and the decision variables specified for the problem. Normally the objective function is set up to help the decision maker to either maximize something like profit, income, or harvest volume or minimize something like cost, risk, or loss. For the recreation example we have

Goal. Maximize new recreation opportunity on the forest
Criterion. Number of additional user-day capacity
Decision variables. Number of new units of campsites, boat ramps, and trails X_i

The objective function can then be written as

Maximize Z

where
$$Z = u_1X_1 + u_2X_2 + u_3X_3 \tag{6-3}$$

with Z = total number of visitor-day capacity added per year
u_i = added visitor-days per year per unit activity
X_i = amount of activity undertaken

Here Z is the value of the objective function; in this case it gives a precise criterion for the goal and could be calculated for each possible solution. If the values of u_i are known, the objective function can be written explicitly. Letting $u_1 = 100$, $u_2 = 500$, and $u_3 = 100$, we can write

Maximize Z

where
$$Z_1 = 100X_1 + 500X_2 + 100X_3 \tag{6-4}$$

The objective function value for each of the three example problem solutions presented above is

$$Z(\text{A}) = 100(100) + 500(10) + 100(0) = 15{,}000$$

$$Z(\text{B}) = 100(50) + 500(5) + 100(30) = 10{,}500$$

$$Z(\text{C}) = 100(50) + 500(0) + 100(70) = 12{,}000$$

Since Z is the criterion for the goal, the solution with the highest Z value, other things being equal, would ordinarily be chosen as the best solution. In the example solution A gives the highest value. Thus the objective function serves as a decision guide for sorting through possible decision variable combinations to determine which one is "best" according to the goal.

Other things are not always equal, however, and it is frequently difficult to state an objective function that has one-to-one correspondence with the goal. As discussed previously, criteria for some land management goals are tough to articulate or even conceptualize. For example, specifying a criterion to represent a visual quality goal has befuddled foresters for decades. In addition, multiple goals are often relevant, suggesting two or more possible objective functions to the problem. Still, these complications can usually be handled, and one dominant or guiding goal can be selected and used to formulate the objective function.

An important point to remember when using objective functions is that you can only maximize (or minimize) one goal at a time. Suppose we defined three objective functions to represent three possible goals for the problem,

Z_1 = total new recreation opportunity (our current goal)
Z_2 = total cost of new facilities
Z_3 = total back country recreation opportunity

We may not find a problem solution that simultaneously maximizes Z_1 and Z_3 and minimizes Z_2. To minimize cost Z_2, you would pick the cheapest solution presented. To maximize backcountry opportunity, you would probably pick the solution with the largest value for trail construction. To maximize total user-days for all activities, another solution could be best.

The decision maker has to choose one goal to guide the choice among solutions. Remaining goals need to be handled as constraints.*

Constraints

Constraints are anything that limits the achievement of a goal. Constraints are invariably present in every problem and need to be identified if one is to attempt to solve the problem adequately.

Within a problem, constraints are placed on inputs and outputs. They arise from three general sources: resource limitations, decision maker goals, and externally imposed policies or regulations. Each of these three sources may require a mix of both input and output constraints.

Resource constraints These physical, human, technological, and economic restrictions limit the amount and kind of decision variables that can be selected for problem solution. Available budgets, land, work force, and time are common limits on production inputs.

Consider the recreation problem given above if the maximum amount that can be spent on construction is \$60,000. This budget constraint can be written as the linear inequality

$$c_1X_1 + c_2X_2 + c_3X_3 \leq 60{,}000 \qquad (6\text{-}5)$$

where c_1 is the construction cost per unit of each type of facility. If c_1 = \$5,000, c_2 = \$8,000, and c_3 = \$1,500, the explicit budget constraint is

$$5{,}000X_1 + 8{,}000X_2 + 1{,}500X_3 \leq 60{,}000 \qquad (6\text{-}6)$$

* Another way to handle multiple goals is goal programming, a formulation we briefly discuss at the end of this chapter.

This constraint equation limits the combinations of activities to those that do not exceed the available budget of $60,000.

Decision maker goal constraints As discussed under objective functions, decision makers typically have two or more relevant goals. Since only one can be used as the guiding objective function, the others must be formulated as constraints if they are to be formally incorporated into the problem.

For example, suppose that in addition to the goal of maximizing outputs of new recreation opportunity, the decision maker has a second goal, to provide maximum backcountry recreation opportunity. Campsites do not provide any backcountry opportunities, but canoeists are considered backcountry users and use boat ramps at a rate of 50 visitor-days per ramp per year. All trail use is considered backcountry, averaging 100 visitor-days per mile per year. We can state this second objective function as maximize Z_2, total visitor-days of new backcountry opportunity, where

$$Z_2 = 0X_1 + 50X_2 + 100X_3$$

To treat a goal as a constraint, a value for the goal must be stated. Suppose Z_1, reflecting total new recreation opportunity, was chosen as the guiding objective function and Z_2, reflecting backcountry opportunity, was specified as having to achieve at least the value of 3,000. Then a formulation of the problem is as follows:

Formulation 1

Maximize Z_1

where
$$Z_1 = 100X_1 + 500X_2 + 100X_3 \tag{6-7}$$

subject to the constraint that $Z_2 \geq 3{,}000$, or

$$0X_1 + 50X_2 + 100X_3 \geq 3{,}000 \tag{6-8}$$

Next suppose that Z_2 is chosen as the guiding objective function, with Z_1 as the goal constraint. If Z_1 is specified to be at least 5,000, a second problem is as follows:

Formulation 2

Maximize Z_2

where
$$Z_2 = 0X_1 + 50X_2 + 100X_3 \tag{6-9}$$

subject to the constraint that $Z_1 \geq 5{,}000$, or

$$100X_1 + 500X_2 + 100X_3 \geq 5{,}000 \tag{6-10}$$

The number of possible problem formulations increases rapidly as three, four, or more goals are introduced to the problem.

Setting the minimum or maximum levels for goals treated as constraints is no easy task. How do you (or the decision maker) know in advance that a minimum of 3,000 days of backcountry recreation must be provided? This specification is a value judgment and must come from the decision maker. Often the constrained goal is set at different levels to see what happens to the value of the objective function, that is, if Z_2 is set successively at levels of 2,000, 3,000, and 4,000, what is the maximum value of Z_1 obtained in each case? This *sensitivity analysis* can provide useful information to the final decision.

Choice of the goal to represent in the objective function is an important decision when there are multiple goals. The "best" solution will usually be different for each different problem formulation and typically there is not enough time and money to analyze complex real-world problems for all possible formulations and permutations of minimum (or maximum) levels on the constrained goals. Normally the most important goal is to maximize either income, profit, or aggregate output level; but that must be determined for each problem.

Policy and regulatory constraints A third group of constraints is imposed on the decision maker by laws, regulations, political necessity, and other external influences. Many such constraints set minimum required levels on certain outputs. For example, the woodlands manager has a monthly quota of wood that *must* be supplied to the mill, company policy does not allow wood sales to foreign buyers, or all available deer winter range by law must be protected on state forest lands. Constraints can also be imposed on inputs and technology, such as no more than 10 percent of a watershed can be cut in any one decade, a local politician decrees that a particular campground is to be built, or the herbicide 2-4-5-T can no longer be used.

In the forest recreation problem, higher level policy might include encouraging private campground development by restricting development of public campgrounds to no more than 40 additional units, recognizing an existing contract to build at least 1 boat ramp, and implementing a prior decision to build at least 10 new campsites at the campground. These additional policy constraints are written as

$$\text{Maximum campsites} \quad X_1 \leq 40 \tag{6-11}$$

$$\text{Minimum campsites} \quad X_1 \geq 10 \tag{6-12}$$

$$\text{Minimum boat ramp} \quad X_2 \geq 1 \tag{6-13}$$

Problem Statement

A formal problem statement defines all decision variables and specifies the goal, the objective function, and all constraint equations needed by the decision maker to describe the problem completely. To summarize our example, a formal problem statement using the formulation first presented above consists of the six linear equations plus the nonnegativity restriction on the decision variables:

Goal. Maximize Z_1, amount of new recreation opportunity, where

1. Objective function	$Z_1 = 100X_1 + 500X_2 + 100X_3$	[Eq. (6-4)]
2. Budget	$5{,}000X_1 + 8{,}000X_2 + 1{,}500X_3 \leq 60{,}000$	[Eq. (6-6)]
3. Backcountry goal	$50X_2 + 100X_3 \geq 3{,}000$	[Eq. (6-8)]
4. Maximum campsite	$X_1 \leq 40$	[Eq. (6-11)]
5. Minimum campsite	$X_1 \geq 10$	[Eq. (6-12)]
6. Minimum boat ramp	$X_2 \geq 1$	[Eq. (6-13)]

with X_1 = number of campsites constructed, $X_1 \geq 0$
X_2 = number of boat ramps constructed, $X_2 \geq 0$
X_3 = number of miles of trail constructed, $X_3 \geq 0$

Feasibility

Problem constraints determine the feasibility of alternative solutions. Simply stated, a solution is "feasible" if all problem constraints are met or satisfied. Conversely, a solution is "infeasible" if one or more of the problem constraints are not satisfied. If the constraints are real, choosing an infeasible solution is bad decision making and can lead to a variety of legal, political, and administrative difficulties, not to mention trying to put a square peg into a round role.

Problems

This word has been used in so many different ways by so many different writers that it is nearly impossible to give it a meaning on which everyone agrees. So we must do the next best thing: define it in the context of a discussion (such as this book), and give the warning that whenever you see or hear it, check to be sure how it is defined.

A *problem* is defined here as a unique set of specified decision variables, goals, constraints, and guiding objective function. As soon as a coefficient in any goal, constraint, or the objective function is changed, or a new decision variable, goal, or constraint added, a different problem is identified.

A *feasible problem solution* is defined here as a unique set of values for the decision variables which helps achieve the goal or goals, and which satisfies all stated constraints of a problem.

PROBLEM IDENTIFICATION: APPLICATIONS

Ultimately, problem identification is an art. Experience, imagination, and consistent and careful use of the procedures, concepts, and terminology can help improve skills and performance. Practice is necessary, however, to achieve professional level competence.

In this section, three problems are presented to give a better feel for some typical applications. The first problem is called *steers and trees*. It illustrates how multiple outputs from a land resource can be handled. Some variations are presented, and the steers and trees problem is used later to illustrate how linear programming can assist problem solution.

The second problem, called *timber harvesting and wildlife,* is an adaptation from Thompson et al. (1973), who investigated the relationship of timber harvesting and wildlife production in Maryland. This problem illustrates the placement of area, volume, and spatial distribution constraints on timber harvest activity over time.

The third problem, called *reforestation planning and budgeting,* is taken from Teeguarden and Von Sperber (1968) and illustrates problem formulation for allocating budgets among many project opportunities when there are several resource constraints.

For each, a verbal problem description is presented, followed by the formal problem statement as a set of linear equations.

The Steers and Trees Problem

A forestry consultant visited a small landowner and came away with these notes describing the owner's land management situation.

> Bill Jackson, a part-time southeastern farmer, has 24 acres of fallow land available and wants to use it to increase his income. He can either plant fast-gro hybrid Christmas tree transplants that mature in one year, or he can fatten steers by putting part of his acreage in pasture. The trees are planted and sold in lots of 1,000. It takes 1.5 acres to grow a lot of trees and 4.0 acres to fatten a steer. The farmer is busy and only has 200 hours per year to spend on this enterprise. Experience shows it takes 20 hours to cultivate, prune, harvest, and package one lot of trees and also 20 hours per steer. He has $1,200 of operating budget available for the year and annual expenses are $30 per lot of trees and $240 per steer. He already has contracted to his neighbors for 2 steers. At current prices, Christmas trees will return a net revenue of $0.50 each and steers will return a net revenue of $1,000 each.

The consultant decides that setting up the problem in terms of objectives and constraints might help him visualize the farmer's situation. He examined his notes and proceeded to develop a mathematical problem statement as follows:

1. *Decision maker*. Bill Jackson
2. *Goal*. Maximum income from operation
3. *Criterion*. Dollars of net revenue from property per year
4. *Activities and decision variables*

   Raise steers X_1 = number of steers fattened per year

   Raise trees X_2 = number of 1000-tree lots of fast-gro Christmas trees grown per year

5. *Objective function*

   Maximize net revenue per year, hence

   $$\text{maximize } 1{,}000X_1 + 500X_2$$

   with \$1,000 = net revenue per steer
   \$500 = net revenue per tree lot (1,000 trees at \$0.50)

6. *Constraints*

   a. *Land* 24 acres available
   4 acres per steer
   1.5 acres per tree lot

   so $$4X_1 + 1.5X_2 \leq 24$$

   b. *Budget* \$1,200 available
   \$ 240 per steer
   \$ 30 per tree lot

   so $$240X_1 + 30X_2 \leq 1{,}200$$

   c. *Labor* 200 hours available
   20 hours per steer
   20 hours per tree lot

   so $$20X_1 + 20X_2 \leq 200$$

   d. *Contract*. A minimum of 2 steers must be produced to meet a contract,

   so $$X_1 \geq 2$$

   e. *Nonnegativity*. All decision variables must be positive or zero since we cannot have negative steers or trees.

   $$X_1 \geq 0, \qquad X_2 \geq 0$$

In summary form, the problem statement is as follows:

$$\text{Maximize } 1000X_1 + 500X_2$$

where X_1 = number of steers
X_2 = number of lots of trees

subject to the constraints

a. Land	$4X_1 + 1.5X_2 \leq$	24
b. Budget	$240X_1 + 30X_2 \leq$	1,200
c. Labor	$20X_1 + 20X_2 \leq$	200
d. Contract	$X_1 \geq$	2
e. Nonnegative	$X_2 \geq$	0

The Timber Harvesting and Wildlife Problem

In this study, Thompson et al. (1973) were concerned with managing the Pocomoke State Forest in Maryland to provide a flow of timber products and huntable populations of game species. This problem illustrates how to formulate linkages of production over time and how constraints on cultural and harvesting activities for timber production can control or ensure production of a second output such as wildlife. They describe the situation as follows:

> This study's objective was to evaluate the interrelationships, or trade-offs, between timber output and intensity of wildlife management. That is, how much timber output must be foregone, over a specified time period, as increasingly intensive levels of wildlife management are practiced in an area.
>
> The Pocomoke State Forest is comprised of approximately 12,000 acres and is located on Maryland's eastern shore. The forest is composed of 10 separate tracts ranging in size from 110 to 3,870 acres. Approximately 2,400 acres are in river bottom or dedicated to parks or other recreational uses, leaving 9,500 acres available for multiple-use planning.
>
> The tracts are dispersed among land which produces both agricultural and timber crops. The principal timber species on the forest is loblolly pine (*Pinus taeda* L.). Eighty-five percent of the forest is covered with either pure loblolly or mixed loblolly-hardwood stands. The remaining 15 percent is covered with mixed hardwood stands. Whitetail deer (*Odocoileus virginianus*) is the primary wildlife species, but bobwhite (*Colinus virginianus*), cottontail rabbits (*Sylvilagus floridanus*), and gray squirrels (*Sciurus carolinensis*) are also found on the forest.
>
> The Maryland Department of Natural Resources' goal in managing the forest is to obtain a maximum, even flow of sawtimber consistent with sound management for other outputs, primarily wildlife. In addition to the even flow of timber volume, an approximately equal annual area cut is considered desirable. All harvest cutting is accomplished by clear-cutting.
>
> The forest was divided into 66 separate and essentially homogeneous stands using data from a recent inventory of the Pocomoke's timber resources. Timber growth for

existing stands was also obtained from the inventory data. The 60-year period was divided into 12 subperiods of 5 years each. The basic management activity, then, was defined as cutting a specific stand in a specified subperiod.

To represent this problem on a manageable scale and still use the same goals and constraints, assume that the following data describe a forest situation like that found on the Pocomoke:

1. *Area*

Stand type 1 (plantation)	100 acres
Stand type 2 (old field sawtimber)	200 acres
Stand type 3 (overmature sawtimber)	60 acres
Total area in forest	360 acres

2. *Volume per acre*

	Volume cut in future (M bd ft/acre)		
Stand type	First period (0–20 years)	Second period (21–40 years)	Third period (41–60 years)
1	3	10	30
2	12	17	20
3	25	20	18

Problem identification Using the data for the representative forest and the text from Thompson et al., we can identify this problem:

1. *Decision makers.* Director, Maryland Department of Natural Resources
2. *Goals*
   a. Maximum even flow harvest of sawtimber
   b. Sound management for wildlife
   c. Cut an equal area per period
3. *Criteria*
   a. Amount of timber harvested per period
   b. Distribution of timber stands by age, stand type, and spatial location meeting standards established by wildlife biologists
   c. Acres cut per period
4. *Activities and decision variables.* Each of the three stand types can be cut in each of three periods, yielding nine decision variables, X_{ij},

where i is the stand type and j the period cut:

X_{11} = acres of stand type 1 cut in period 1
X_{21} = acres of stand type 2 cut in period 1
X_{31} = acres of stand type 3 cut in period 1
X_{12} = acres of stand type 1 cut in period 2
X_{22} = acres of stand type 2 cut in period 2
X_{32} = acres of stand type 3 cut in period 2
X_{13} = acres of stand type 1 cut in period 3
X_{23} = acres of stand type 2 cut in period 3
X_{33} = acres of stand type 3 cut in period 3

5. *Objective function.* The total harvest is the sum of the acres cut in each stand type in each period multiplied by the volume per acre at the time of harvest,

maximize

$$3X_{11} + 12X_{21} + 25X_{31} + 10X_{12} + 17X_{22} + 20X_{32} + 30X_{13} + 20X_{23} + 18X_{33}$$

6. *Constraints.* Four kinds of constraints are imposed on solutions to this problem.
   a. *Total acreage cut.* Total acreage cut in a stand type over the three periods cannot exceed the acreage in the stand type. An obvious constraint, but it must be formulated. To illustrate for stand type 1,

   $$X_{11} + X_{12} + X_{13} \leq 100$$

   b. *Even flow harvest volume.* An identical requirement is imposed on the volume of timber that is harvested in each period. With repeated runs, the problem is solved to get the highest equal level of cut possible. To illustrate, the first period equation assumes that 2,000 M bd ft is the required harvest,

   $$3X_{11} + 12X_{21} + 25X_{31} = 2{,}000 \text{ M bd ft}$$

   c. *Equal acreage cut.* An equal periodic acreage cut is considered desirable for wildlife and operational purposes. Suppose it is decided that an equal periodic acreage cut of 360 ÷ 3, or 120 acres per period is wanted. This also says that the whole forest must be cut in 60 years. The first-period formulation is then

   $$X_{11} + X_{21} + X_{31} = 120 \text{ acres}$$

   d. *Limiting cut areas.* It is considered desirable by the wildlife biologists to restrict the acreage cut for each stand type in each period to provide good vegetative structure for habitat. This restriction will keep each period's cut smaller than all area in the stand type and help ensure a balanced age class distribution on the ground. A maximum area cut per stand type per period needs to be established. Suppose the decision on maximum area cut per period were for stand type 1, 40 acres; for stand type 2, 90 acres; and stand type 3, 25 acres. Nine constraint equations are needed in total, one for each stand type in each period. For period 1 these constraints are

$$X_{11} \leq 40$$

$$X_{21} \leq 90$$

$$X_{31} \leq 25$$

Putting all constraints together with the objective function and assuming nonnegativity for all decision variables, we have the following formal problem statement:

Maximize $3X_{11} + 12X_{21} + 25X_{31} + 10X_{12} + 17X_{22} + 20X_{32} + 30X_{13} + 20X_{23} + 18X_{33}$

subject to the constraints:

a. Acreage	(1)	X_{11}			$+\ X_{12}$			$+\ X_{13}$			$\leq$	100
available	(2)		X_{21}			$+\ X_{22}$			$+\ X_{23}$		$\leq$	200
	(3)			X_{31}			$+\ X_{32}$			$+\ X_{33}$	$\leq$	60
b. Required	(4)	$3X_{11}$	$+\ 12X_{21}$	$+\ 25X_{31}$							$=$	2,000
harvest	(5)				$10X_{12}$	$+\ 17X_{22}$	$+\ 20X_{32}$				$=$	2,000
per period	(6)							$30X_{13}$	$+\ 20X_{23}$	$+\ 18X_{33}$	$=$	2,000
c. Required	(7)	$X_{11}\ +$	$X_{21}\ +$	X_{31}							$=$	120
acres	(8)				$X_{12}\ +$	$X_{22}\ +$	X_{32}				$=$	120
per period	(9)							$X_{13}\ +$	$X_{23}\ +$	X_{33}	$=$	120
d. Maximum	(10)	X_{11}									$\leq$	40
cut	(11)		X_{21}								$\leq$	90
size	(12)			X_{31}							$\leq$	25
per stand	(13)				X_{12}						$\leq$	40
per period	(14)					X_{22}					$\leq$	90
	(15)						X_{32}				$\leq$	25
	(16)							X_{13}			$\leq$	40
	(17)								X_{23}		$\leq$	90
	(18)									X_{33}	$\leq$	25

The Reforestation Planning and Budgeting Problem

This problem is extracted from Teeguarden and Von Sperber's (1968) report on a study with the Bureau of Land Management in southwestern Oregon to develop economic guides for Douglas-fir reforestation. Up to a 2-year lead time is needed prior to actual planting or seeding in order to produce needed seed and seedlings from the nursery. Cutover areas needing reforestation occurred on many site types from moist, low-elevation north slopes to dry, high-elevation south slopes. On each site reforestation could be handled by planting, direct-seeding, or interplanting. Teeguarden and Von Sperber presented an example to represent their problem as follows:

Assume that the activity alternatives available to the land manager include the following four land classes:

Class 1: Site II type B bare land with a north aspect in seed zone 53
Class 2: Site III type B bare land with a south aspect in seed zone 52
Class 3: Site IV type B bare land with a north aspect in seed zone 51
Class 4: Site II substocked land with a south aspect in seed zone 52

Assume further that there are 100 acres in each land class and that any portion of the total may be scheduled for treatment. Resources available are: budget—$8,000; seedlings—30,000, 55,000, and 35,000 for zones 51, 52, and 53, respectively; and seeds—25 lb for each zone.

Assume that 0.75 lb of seed or 600 seedlings are required to treat 1 acre of bare land, and in the case of interplanting, 500 seedlings per acre. If necessary, seed and seedlings may be transferred between adjacent zones. Thus seedling supply in zone 51 is 85,000 trees; in zone 52, 120,000 trees; and in zone 53, 90,000 trees.*

The alternative activities, capital requirements, and activity values are shown in [the following table]. There are two alternative activities—seeding or planting—for bare land situations, while interplanting is the only possible treatment for substocked plantations.

Problem identification

1. *Decision maker*. Reforestation forester, Southwest Oregon district, Bureau of Land Management
2. *Goals*. Allocate resources efficiently to reforestation opportunities
3. *Criterion*. Total present net worth of lands reforested in the southwest region district†
4. *Activities and decision variables*. See the following table, which also lists resource requirements and values per acre

* In the paragraph above there are in total 30,000 + 55,000 + 35,000 = 120,000 seedlings available. Zones 51 and 52 are adjacent and hence zone 51 could have 30,000 + 55,000 or 85,000 seedlings if zone 52 used none.

† Present net worth values are the future returns discounted at a guiding rate of interest appropriate to the decision maker. See Chap. 7 for the full development of this concept.

Decision variable	Activity	Resource requirements: Capital	Seedlings	Seeds (lb)	Present net worth
X_1	Acres planted in land class 1, seed zone 53	\$31.50	600	0	\$163.20
X_2	Acres seeded in land class 1, seed zone 53	10.80	0	0.75	104.10
X_3	Acres planted in land class 2, seed zone 52	31.50	600	0	58.90
X_4	Acres seeded in land class 2, seed zone 52	10.80	0	0.75	19.30
X_5	Acres planted in land class 3, seed zone 51	31.50	600	0	6.30
X_6	Acres seeded in land class 3, seed zone 51	10.80	0	0.75	−1.35
X_7	Acres interplanted in land class 4, seed zone 52	24.00	500	0	73.80

5. *Objective function*

Maximize present net worth of acres planted, hence

Maximize

$$163.20X_1 + 104.10X_2 + 58.90X_3 + 19.30X_4 + 6.30X_5 - 1.35X_6 + 73.80X_7$$

6. *Constraints*

a. Budget constraint

$$31.50X_1 + 10.80X_2 + 31.50X_3 + 10.80X_4 + 31.50X_5 + 10.80X_6 + 24.00X_7 \leq 8{,}000 \tag{1}$$

b. Seedling constraints

$$600X_1 \leq 90{,}000 \tag{2}$$

$$600X_5 \leq 85{,}000 \tag{3}$$

$$600X_3 + 500X_7 \leq 120{,}000 \tag{4}$$

$$600X_1 + 600X_3 + 600X_5 + 500X_7 \leq 120{,}000 \tag{5}$$

c. Seed constraints

$$0.75X_2 \leq 50 \tag{6}$$

$$0.75X_4 \leq 75 \tag{7}$$

$$0.75X_6 \leq 50 \quad (8)$$

$$0.75X_2 + 0.75X_4 + 0.75X_6 \leq 75 \quad (9)$$

d. Area constraints

$$X_1 + X_2 \leq 100 \text{ (land class 1)} \quad (10)$$

$$X_3 + X_4 \leq 100 \text{ (land class 2)} \quad (11)$$

$$X_5 + X_6 \leq 100 \text{ (land class 3)} \quad (12)$$

$$X_7 \leq 100 \text{ (land class 4)} \quad (13)$$

e. A final requirement is that solution values be nonnegative

$$X_i \geq 0 \qquad i = 1, \ldots, 7$$

The example problem has an objective function and 13 constraints. Although its structure is identical, the model for the southwestern Oregon scheduling problem was far more complex than this hypothetical example. Considered were 5,954 acres, classified by treatment, site, and aspect into 271 classes. Included was the planting–seeding alternative for all bare land situations where either regeneration method is technically feasible.

TWO PROCESSES FOR GENERATING ALTERNATIVE SOLUTIONS

Given these definitions of problem and problem solution, the term *alternative,* when used as an adjective, simply means "different." Alternative is more often used as a noun, as in "the alternative," and here the term has two significantly different meanings in common forest management planning usage: (1) different solutions to the same problem, and (2) solutions to different problems.

A. Identify a single problem and represent it with one or more solutions This type A alternative generation process is what we have described as problem identification. The set of goals and constraints relevant to the decision maker is identified, and the task is to find the best of the feasible solutions. The objective function often is income or profit directed, with the "best" alternative being the one that makes the most money. Finding the most profitable combination of inputs and outputs for a production problem under constrained levels of budget and other resources is the classic economic efficiency problem in economics textbooks. Most of the budgeting, management intensification, and procurement problems in forest management are of this type, and mathematical programming is an important tool for guiding the solution of type A problem situations. For example, the steers and trees problem and the reforestation budgeting problem just presented are typical type A situations.

B. Identify several problems and represent each with one or more solutions In many situations we do not have a uniquely defined problem. The decision makers are uncertain and disagree about which set of goals and constraints represent the "true" problem. To recognize this disagreement, the type B process describes different sets of goals and constraints, each of these problems is solved, and solutions for each different problem are presented as alternatives. The critical decision is choosing which set of goals and constraints best represents the true problem.

The type B process is commonly found in public land management planning, but it is also found in private and corporate strategic planning. In public planning, problem constraints may range over high, medium, and low budget levels for management activity and, within these resource limits, may specify different sets of constraints, coefficients, and decision variables, reflecting goal emphasis ranging from amenity and preservation to intensive commercial production.

Trying to figure out an objective function that represents the overall public interest is a major difficulty for the public manager. Lack of an identifiable landowner also encourages type B solution formulations. In private management, if the decision maker is actually a committee or if the line manager does not really have authority, then alternative goal and constraint sets are defined and solved to help higher executives or boards of directors identify the true problem by reacting to the solution options presented.

When you act as a reviewer of planning documents, the type B situation is not always easy to spot. If each alternative solution is presented along with a tidy package of goals and constraints in a formal problem statement, then there is no difficulty. Often, however, only one *solution* is presented to represent each goal–constraint formulation and you have to figure out whether the solutions were generated by a type A or a type B process. They both look the same: a listing of values for the decision variables. For example, if a set of "alternative" land management plans were presented in terms of the outputs provided per year, it might appear as follows.

	Alternative		
Output type	K	L	M
Timber, M bd ft	100	80	60
Forage, animal-unit-months	400	800	1,000
Recreation, visitor-days of opportunity	5,000	3,000	4,000
Budget I, dollars/year	$10,000	$12,000	$ 9,000
Budget II, dollars/year	$12,000	$12,000	$12,000

If the data presented include budget I, then it is clear that the budget constraint has been changed between alternatives, and we are sure it was a type B process of generating alternatives. But suppose the budget II figures were provided. On the surface it appears that since the budget is the same, this is a type A process. Other stated and unstated assumptions and constraints may well have varied between alternatives, however, and in reality this also could be a type B process. For example, the solution under alternative M may have been found under an additional constraint that forage production must be greater than 1,000 animal-unit-months. Close reading of documents and reports and sometimes questioning are needed to determine whether constraints have varied, and how much.

The timber harvesting and wildlife problem, although presented above as one problem specification and thus a type A situation, was actually used by Thompson et al. to generate alternatives in a type B process. They systematically changed the cut area and cut distribution constraints to reflect increasing emphasis on wildlife habitat and wildlife production. A series of six different formulations produced the following results:

Problem formulation	Volume available for harvest (million bd ft)
1. Maximum even-flow of timber and equal area cut per subperiod	170
2. Same as 1, except hardwood cuts were limited to 25 acres and pine and pine-hardwood cut to 150 acres	167
3. Same as 2, except cuts were distributed over the forest	160
4. Same as 2, except pine and pine-hardwood cuts limited to 100 acres	166
5. Same as 4, except cuts distributed over the forest	156
6. Same as 5, except 1,900 acres carried at reduced stocking for quail habitat improvement	148

Each formulation was a different problem and produced a different timber output. Note carefully that we do not have a measure of how much wildlife or wildlife habitat is produced; thus the total output and benefit is not presented.

Making the distinction between these type A and type B problem and solution generation situations is critical. Economics, operations research, and a variety of analytical techniques can be used to create and evaluate alternative solutions in a type A process and to find which one is "best." No sure technique exists for selecting among solutions generated by a type B process, where the real issue is deciding the goals and constraints of the underlying problem. Administrators and ultimately politicians must make the judgments needed to choose among alternate conceptualizations of the problem—analysis surely cannot. And, in the end, there is no way to guarantee that the *best* problem specification has been chosen—all we can say is that people, politicians, and administrators have reached a political decision.

INTRODUCING RISK AND UNCERTAINTY INTO DECISION MODELS

Given skills at identifying problems, we still must face up to the issue of making the choice. We have seen that a particular solution to a problem consists of a set of values for the decision variables, and the payoff is the value of the objective function calculated for that solution. For the recreation example in the preceding section, the first solution package was X_1 (number of campsites) = 100, X_2 (number of boat ramps) = 10, and the payoff according to the objective function was 15,000 recreation-visitor-days of new capacity. To shorthand this description and introduce some new notation, let the solution package be denoted as action a_1 and the value of the associated outcome or payoff as return R_1. For the first action then

$$a_1 = \{X_1 = 100, X_2 = 10, X_3 = 0\}$$

$$R_1 = 15{,}000$$

We now need to consider one more important characteristic of decision problems that arise from their futuristic orientation. Seldom if ever do we know with certainty the payoff from implementing a problem solution. In the recreation problem if we are uncertain about that payoff, the simplistic calculation of the objective function leaves us uncomfortable. The future could turn out different: roads to the developments may not be built or maintained, or the physical and social requirements of tomorrow's campers and boaters may change and we could wind up producing more or less than 15,000 units of capacity.

Different decision models are distinguished by how they deal with the unknown aspects of the future, such as weather or the human values that determine the amount of payoff resulting from an action implemented today. We call this package of future factors affecting outcomes *states of nature*. We will assume that problem constraints have already been considered and satisfied and that all actions considered are feasible. With this generous assumption we can focus attention on just three elements of the decision models: the actions a_i, the states of nature s_j, and the return expected from an action under a known state, R_{ij}.

Suppose we just harvested an old pine and fir stand which had a clean, parklike understory and wish to make a good financial return from the next crop of trees. We know, however, that from a few to several millions of old but viable brush seeds may be in the soil. When germinated in the sunny, mineral soil environment following clear-cutting, the brush may seriously compete and reduce value growth of the planted timber stand. If there are few brush seeds, good seedling growth is expected, but if there are many brush seeds, tree seedling growth is very slow without brush control treatments. The states of nature in this problem are the amounts of brush seed, and the actions are the preplanting treatments that could be undertaken to kill the seed before germination. Three states of nature and three actions are defined to represent the range of possibilities for this problem. The decision problem itself is simple with a goal of maximizing the present net worth of the next tree crop and the only constraint is limiting possible problem solutions to three actions.

1. *States of nature* s_j

   s_1 = many viable brush seeds present
   s_2 = moderate amounts of viable brush seeds present
   s_3 = few viable brush seeds present

2. *Actions* a_i

   a_1 = broadcast burn, prespray; cost = \$140
   a_2 = broadcast burn; cost = \$60
   a_3 = no treatment, (plant through logging slash); no cost

Using growth simulators and assumptions about other costs and prices, the cash value expected from each action under a given state is estimated and provides a return matrix.

1. *Gross return matrix,* gross present net worth in dollars per acre

	State of nature			
Action	s_1	s_2	s_3	Cost
a_1	160	200	240	140
a_2	10	140	190	60
a_3	−400	−200	150	0

The net return is the item of interest, so \$140 must be subtracted from all entries in the first row, \$60 from those in the second, and 0 from those in the third row.

2. *Net return matrix* R_{ij}

	State of nature		
Action	s_1	s_2	s_3
a_1	20	60	100
a_2	−50	80	130
a_3	−400	−200	150

CLASSIFICATION OF DECISION MODELS

With the return matrix R_{ij} in hand, the practical next step is to make or recommend a decision: to select one of the actions. The best choice or recommendation depends on what we know about the states of nature. The literature of decision theory distinguishes between three classes: (1) decision under *certainty*, (2) decision under *uncertainty*, and (3) decision under *risk*. Here is what makes them different.

1. *Decision under certainty* assumes that we know precisely what future state of nature will occur at the time we choose the action. Certainty is like owning the proverbial crystal ball, and it makes decision making easier—particularly when the complexities of constraints and feasibility are not at issue. If state s_1 were known to occur for sure, then action a_1 is clearly the action that would maximize the net present value of the new stand. Similarly, action a_2 is best for state s_2, and action a_3 for state s_3.
2. *Decision under uncertainty* means that we do not have the slightest idea which future will occur at the time we are making the decision; all we know is that s_1, s_2, and s_3 are all possible. Ponder this, since it is hard to cope with the idea of zero information to help indicate what state will occur. It would be easy for you to be a hero or a scapegoat—particularly if you picked a_3.
3. *Decision under risk* is the case where we are able to estimate or assign occurrence probabilities to each state of nature. When we know the probabilities, then we have some information about the future and calculations can help guide the decision, even when we still do not know for sure which state will occur.

We will briefly look at some approaches to dealing with decision under uncertainty and risk. We also discuss dealing with uncertainty and risk when mathematical programming techniques are introduced later in the chapter. In forestry we tend to use certainty models even when we know the results are uncertain. For example, most of the growth models are presented in Chap. 5 as certainty models. Do they predict future yields for some subject stand with absolute certainty? Not likely; but what do we do about it? Can we assign confidence limits to the predictions? What else can we do? The problem identification examples from previous sections are also presented as certainty examples.

Decision Making under Uncertainty

The approach to decision making under uncertainty depends on the psychological makeup of the decision maker and the character of the feedback anticipated from making good or bad decisions. There are strategies for everyone: optimists, pessimists, politicians, and cool rationalists.

The rationalist reasons that since we do not know how likely any of the states of nature are, there is no rational basis for a view that presumes one is more likely than another, so it is therefore most rational to consider all states equally likely and pick the action with highest expected average net return.

The optimist and the pessimist differ on how they expect nature to treat them. The optimist believes that whatever decision is chosen, nature is benevolent and will assume the state most favorable to that action. With this rosy view, the optimist will pick the action that has the highest overall return. The pessimist in contrast takes the dark view that nature is out to get us and will always take on the worst condition for any chosen action. In this gloomy world, the thing to do is to choose actions that minimize the losses when nature does its worst.

The politician or the conservative administrator may be worried most about recriminations and "I told you so's" after the decision has been made and nature reveals herself. The difference between the return from the best possible decision for a given state of nature and the return from the action taken is a measure of potential "regret" and is something to be minimized.

The preplanting decision can illustrate all four of these views on how to make decisions under uncertainty. Table 6-1 takes the initial net outcome return matrix and adds some more calculations:

1. The maximum value of outcome for each action (row) is identified for the optimist.
2. The minimum value of each action (row) is identified for the pessimist.
3. The expected value for each outcome is calculated under the assumption of equal probabilities for each state.

To get the politician's solution, we find the best return for each state (column) and subtract it from the respective return values for each

Table 6-1 Return matrix

Action	State of nature			Row maximum	Row minimum	Equally likely expected value EV
	s_1	s_2	s_3			
a_1	20	60	100	100	*20[b]	*60[c]
a_2	−50	80	130	130	−50	53.333[e]
a_3	−400	−200	150	*150[a]	−400	−150
Equal probability	$\frac{1}{3}$	$\frac{1}{3}$	$\frac{1}{3}$			
Maximum return for state of nature	20	80	150			

Regret matrix

Action	State of nature			Row maximum
	s_1[f]	s_2	s_3	
a_1	0	20	50	*50[d]
a_2	70	0	20	70
a_3	420	280	0	420

[a] Optimist. Choose a_3, maximize overall outcome.
[b] Pessimist. Choose a_1, minimize the maximum loss.
[c] Rationalist. Choose a_1, maximize expected value for equal likelihood of states.
[d] Politician. Choose a_1, minimize the maximum regret.
[e] Sample calculation. $53.333 = 0.333(-50) + 0.333(80) + 0.333(130)$
[f] Column maximum = 20. Regrets for $a_1 = (20 - 20) = 0$; for $a_2 = (20) - (-50) = 70$; for $a_3 = (20) - (-400) = 420$.

column to develop the regret matrix. The maximum regret for each action (row) is then identified.

The best choice for each of the four viewpoints follows directly. The optimist maximizes the maximum row return—the *maximax criterion*—and chooses a_3, hoping for a net return of 150. The pessimist minimizes possible losses by finding the maximum value over the row minima—the *maximin criterion*—and chooses a_1 to hold the worst possible outcome to \$20. The rationalist uses the *maximize expected returns criterion* and also chooses a_1 for an expected yield of \$60. Our politician wants to minimize over the worst regret potential for each action—the *minimax regret criterion*—and chooses action a_1 that has the smallest potential for post-decision complaining (50 regret points).

Which is best? Which of these criteria is best to use? That depends on the problem and who you are. Any attempt to deal with uncertainty problems often seems a bit unreal—but this is part of the initial condition of knowing absolutely nothing about the future states of nature. In forestry we have considerable uncertainty when we are trying new things, such as genetic manipulation or nonchemical pest management, which depend on a close timing of activities with climatic and biological development. Forest fire control has its moments of uncertainty. Most of the time, however, we feel that instincts, past experience, weather records, and some documented research allow us to assign probabilities to the different states of nature. We should not be terribly uncertain about whether or not trees will grow—we have a lot of data that show they do. The rate at which trees in a particular stand grow in response to different prescriptions is a "probabilistic" event. But until you as an individual have experience, or have looked up the research, you can easily feel like it is an uncertain situation—much like buying your first used car.

Decision Making under Risk

Whenever probabilities can be assigned to the states of nature by subjective or objective means, it becomes a decision-making-under-risk situation. Suppose, for example, we had records on 60 other planting cases in similar clear-cuts and found that 12 of them had many seeds, 30 cases had a moderate level of brush seeds, and 18 had only a few seeds. Assuming we think the land characteristics of the sampled stands are similar to that of the subject stand, this case record provides us with some probabilities on the states of nature.

State	Number of cases	Fraction	Probability P_j
s_1, many seeds	12	12/60	0.2
s_2, moderate seeds	30	30/60	0.5
s_3, few seeds	18	18/60	0.3
Total	60	60/60	1.0

The probabilities are simply the portion of the times in repeated trials that we would expect the state or event to occur. We would expect s_1 to occur 20 times in 100 different clear-cuts, s_2 to occur 50 times, and s_3 30 times. Note that the probabilities by definition always sum to 1.0, that is, $\Sigma P_j = 1.0$.

With this new information we can reexamine the return matrix and calculate expected values for each action.

Action	State of nature			Expected value EV*
	s_1	s_2	s_3	
a_1	20	60	100	64
a_2	−50	80	130	69
a_3	−400	−200	150	−135
Probability	0.2	0.5	0.3	

* $EV(a_1) = 0.2(20) + 0.5(60) + 0.3(100) = 64$
$EV(a_2) = 0.2(-50) + 0.5(80) + 0.3(130) = 69$
$EV(a_3) = 0.2(-400) + 0.5(-200) + 0.3(150) = -135$

A satisfying choice of action to many decision makers is the one that maximizes expected values. If decisions are repeatedly made on this basis, on the average the greatest gain will be made over many such decisions. Insurance companies and large organizations take such an actuarial view when hundreds of decisions or cases are handled. For a forestry organization on which serious business impacts are seldom caused by the events on just a few acres and many such small area decisions are made over time, the maximum expected value approach is rational. Just to raise a question though, compare the range of expected outcomes for actions a_1 and a_2. Could you imagine choosing a_1 over a_2 to reduce the worst outcome potential from −50 to 20 at a cost of reducing expected value return from $69 to $64, or $5? The rationality of using average expectations to guide decisions depends on the individual and the decision context. Remember that all of this also assumes we have identified correctly the range of possible states of nature, actions, and their payoffs.

Because we often want to consider more complex, sequential decisions, a decision tree is visually a better way to present decision making under risk problems. A decision tree shows the action, states of nature, and outcomes as nodes and branches of a dendritic chart.

The tree develops in time from left to right, with the choice of action by the decision maker followed by nature's "choice" of state. Figure 6-1 depicts a decision tree for the preplanting treatment problems. The actual calculations of the tree move from right to left (backward in time) starting with the gross (not net) return values. Working down the upper branches of action a_1, the first step is to calculate the expected value of gross return and write it over the N_1 node, $204 = 0.2(160) + 0.5(200) + 0.3(240)$. Next the cost of treatment a_1 is subtracted to get the net expected value of a_1, $204 - 140 = 64$. Given that we want to maximize expected value, the optimal path through the tree is the action that maximizes expected value

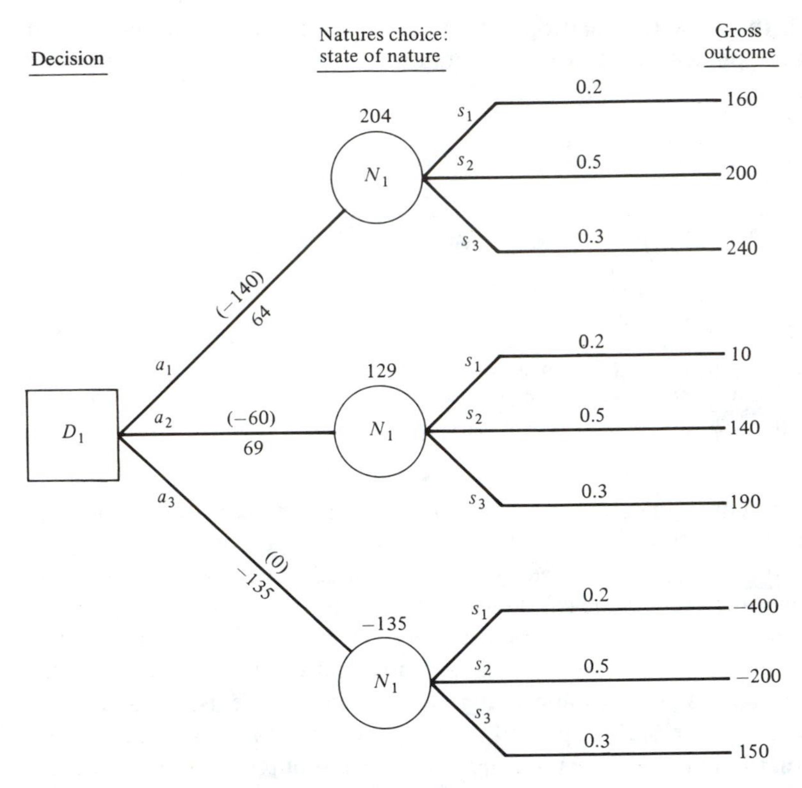

Figure 6-1 Decision tree analysis of preplanting treatment choices for brush control.

at the decision maker's node. That path is action a_2, the same result we obtained before when we used the table to maximize expected value.

Sequential Decisions

Decision tree analyses of risk problems are finding use in forestry, often for the evaluation of regeneration and prescription options that involve a sequence of treatments. The decision maker chooses, nature chooses, the decision maker chooses, nature chooses, and so on in a sequence over time. Unfortunately the tree rapidly grows branches and will no longer fit on a piece of paper—the only good storage device is a computer.

We will take our reforestation example one step further to illustrate its expansion to a sequential decision problem. The second treatment or decision is whether or not to spray the germinated brush seedlings after planting. If seedlings are present, then the second spray will normally be successful and kill most of the brush. Some of the time, however, weather conditions, timing, and application of the spray will be bad, and only part of the brush is killed and sometimes some of the planted trees are damaged. To keep track of the actions and states when there are multiple, sequential decisions we need to modify our symbols. Let a_{ij} be an action selected at decision sequence order i and j be the actual action taken at that time. Similarly let s_{ij} be the state of nature occurring after decision sequence order i and j be the actual state occurring. The states of nature of the second decision are thus s_{21} (good postplanting timing and spray success) and s_{22} (poor timing and success). The probability of these states is in part dependent on spray timing and technology and in part on the prior treatment used and the number of brush seeds actually in the soil. Table 6-2 provides the additional data for this expanded problem. The by now familiar actions, states, and probabilities of the first (preplant) decision are shown in the first three columns but using the new labels of a_{11}, a_{12}, a_{13} and s_{11}, s_{12}, s_{13}. The second (postplant) decision has two actions: a_{21} = spray and a_{22} = no spray. If a_{21} is chosen, then new outcomes are shown for both s_{21} good timing and success and s_{22} poor timing and success states. The probabilities of the states are seen as conditional on the initial decision. If a_{22} is chosen, then the outcomes shown are the same as for the first problem.

The data from Table 6-2 are used to create a decision tree (Fig. 6-2) for this two-decision problem and to calculate the combined expected values. Calculations proceed from right to left exactly as before; there are simply more of them. To illustrate, the calculations for the branches originating with preplanting broadcast burn a_{12}, and a following state of few seeds s_{13}, are discussed. If the no spray posttreatment is used, the expected outcome is 190. With zero cost, the expected value at the D_2 node is also 190. If postplanting spray is applied, the gross expected value over both states is $0.9(290) + 0.1(200) = 281$. When the spray cost is subtracted, the expected value at the D_2 node is $281 - 50 = 231$. The \$231 is higher than 190, so it is marked with an asterisk to indicate that a_{21} is the best choice if one were at the D_2 node, following an a_{12} action and s_{13} state from the earlier preplanting decision. Similar calculations are made for the s_{11} and s_{12} states of the initial burn action, and the expected value at the N_1 node is calculated as $0.2(166) + 0.5(218) + 0.3(231) = 211.5$. Subtracting the burn cost, we have a net expected value of the a_{12} burn action of \$151.5. A similar process is followed to calculate the value of the a_{11} and a_{13} branches at decision node D_1. Action a_{12} turns out to have the highest expected value.

What do these new numbers and the elaborate tree diagram tell us?

1. The addition of the postplanting treatment option increased the expected value of all three preplant treatments when compared to the

Table 6-2

	First decision: choice of preplanting treatment		*Second decision:* to postplant and spray or not; cost \$50/acre				
				a_{21}, Spray			
			a_{22}, No spray	s_{21}, Good timing		s_{22}, Poor timing	
a_{ij}, Treatment (1)	s_{ij}, State of nature (2)	Probability of state (3)	Return (4)	Return (5)	Probability (6)	Return (7)	Probability (8)
a_{11}, prespray, broadcast burn, plant; cost = \$140	s_{11}, much brush seed	0.2	\$160	\$290	0.7	\$170	0.3
	s_{12}, moderate brush seed	0.5	200	270	0.8	220	0.2
	s_{13}, little brush seed	0.3	240	260	0.9	250	0.1
a_{12}, broadcast burn, plant; cost = \$60	s_{11}, much brush seed	0.2	10	260	0.8	40	0.2
	s_{12}, moderate brush seed	0.5	140	280	0.9	160	0.1
	s_{13}, little brush seed	0.3	190	290	0.9	200	0.1
a_{13}, plant through logging slash; cost = 0	s_{11}, much brush seed	0.2	−400	100	0.8	−100	0.2
	s_{12}, moderate brush seed	0.5	−200	120	0.9	−50	0.1
	s_{13}, little brush seed	0.3	150	160	0.9	150	0.1

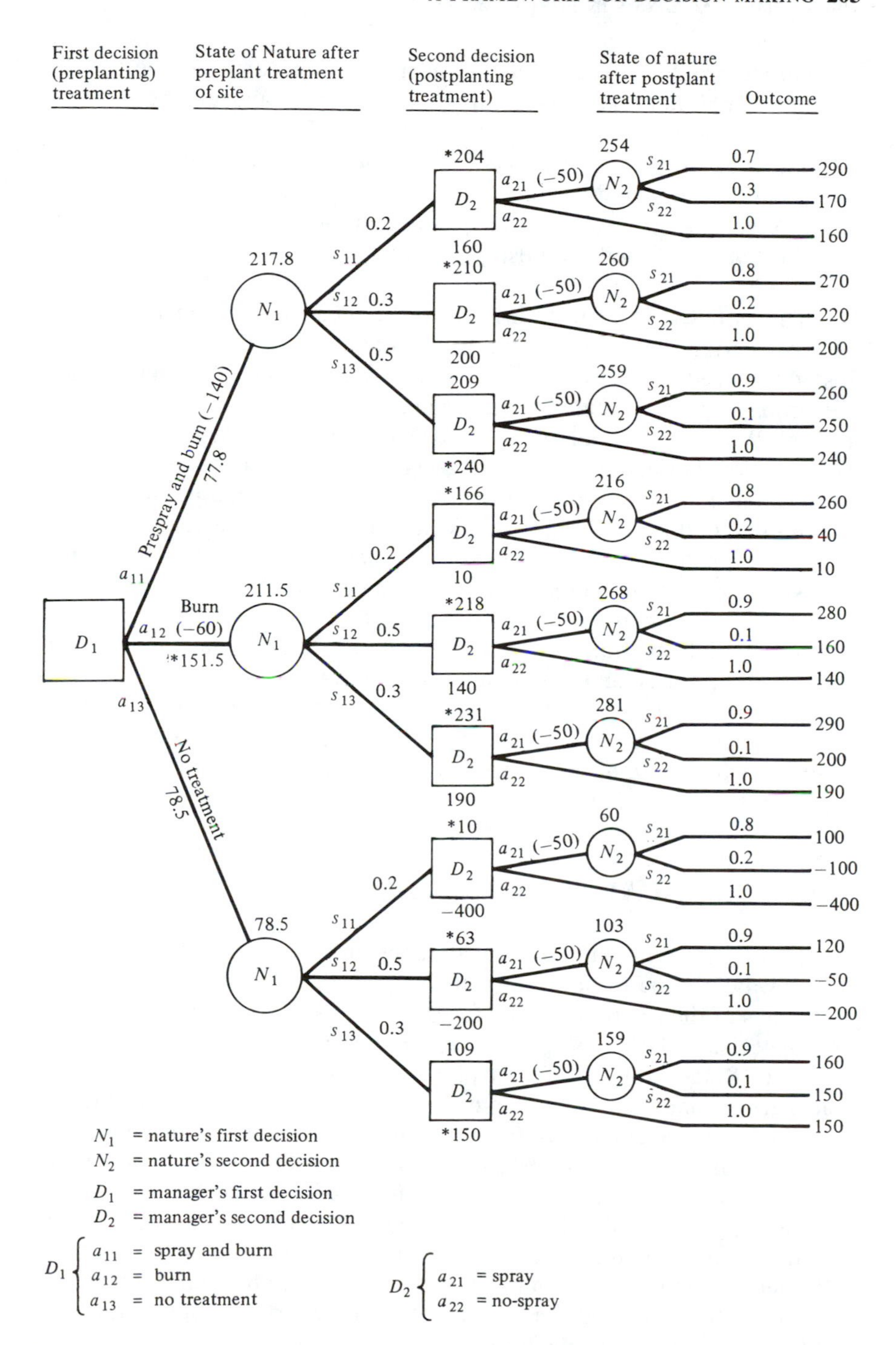

Figure 6-2 Decision tree analysis of two sequential decisions for plantation brush control.

earlier problem. The effect is most pronounced when there is no initial treatment (action a_{13}); if weeds do in fact appear, the option is there to spray them.

2. The rank of the action at the first decision can change with the addition of sequential decisions. In this case a_{13} moved from a poor third in Figure 6-1 to second in Figure 6-2, and a_{11} moved from second to third. a_{22} remained the best with these numbers we used.

This is but a scant introduction to the ideas about decision making under uncertainty and risk. Suppose we could pay $25 and have the soil tested for weed seed prior to the preplant treatment and thus better know the probability of the states of nature in advance. How would this change things? Would it be worth paying $25 for the test? Partly because of computational difficulties, but partly also because of intellectual laziness, foresters like most people are inclined to ignore uncertainty and pretend the future is certain. That is, if we take a certain action, we assume that the predicted result will in fact occur. Sometimes this is correct and sometimes it is not, and it is important to give it careful thought. Does a given growth and yield model give us average expectations for a prescription, or does it give information for exceptional states (like normal yield tables)? Are the future price and cost assumptions certain? Are we sure the goals and constraints are correctly specified? If the answers to these questions are no, then it may be prudent to start thinking in terms of defining alternate states with different price, cost, and yield estimates and assigning probabilities to them.

MATHEMATICAL PROGRAMMING

Forest management planning considers the entire forest at once and looks a long way into the future; if anyone has reason to feel uncertainty about outcomes of their actions, it is a forester. Forest planning problems have many decision variables (possible actions) and many constraints to be satisfied. Since proposed forest plan solutions themselves are complex packages of many actions to be implemented in different places at different times, it is not at all obvious or easy to determine whether a solution is feasible or not. This complexity and need to consider problem constraints explicitly makes it difficult simultaneously to integrate the uncertainty and risk approaches previously described. Instead we often turn to mathematical programming to help search out the best solution to our forest planning problems. Mathematical programming often treats problems as certainty situations, and the consideration of uncertainty and risk is handled through sensitivity analysis.

Mathematical programming refers to a group of techniques that can take problems explicitly described by an objective function and a set of constraint equations and efficiently search all possible feasible solutions to find the one that is optimum (best) according to the objective function. Mathematical programming techniques are normally grouped by whether they deal with linear or nonlinear representations of the problem. Nonlinear techniques are less well developed and, as you might expect, more expensive and difficult to use on large-scale problems. Sometimes it is useful to segment a nonlinear equation and represent it as a series of linear equations, one for each segment. By this route, computationally efficient linear programming techniques sometimes can be used on nonlinear equations.

Linear programming is a well-developed technique extensively used and applied in manufacturing, transportation, agriculture, and land management. Most mainframe computers have available powerful programs written for large-scale linear programming, and, with increasing availability of the programs and small desktop computers as well, it is a contemporary, operational tool. This brief introduction focuses, at an intuitive level, on what the linear programming solution technique does to find the best solution. The steers and trees problem developed earlier provides the example for this discussion.

Steers and Trees as a Linear Program

The steers and trees problem can be examined graphically as a linear programming problem to gain some insight into how linear programming works. The problem fits a two-dimensional graph simply because the problem contains but two decision variables. All constraints are plotted on one graph to display the combinations of the decision variables that are feasible solutions. Then the objective function is plotted as a family of lines, with each line displaying the different combinations of the two decision variables that have a given objective function value. Finally, the feasible solution region and the objective function lines are compared to identify the best solution: the feasible solution with the highest objective function value.

Constraints The steers and trees problem has five constraints, each of which is plotted on a graph. To illustrate the plotting procedure, consider the land constraints of $4X_1 + 1.5X_2 \leq 24$. First the equation is restated as an equality and solved for the intercepts. When $X_1 = 0$, we have $4(0) + 1.5X_2 = 24$, or $1.5X_2 = 24/1.5 = 16$. This result says that when no steers are produced and all the land is put into Christmas trees, enough land is available to grow 16 lots of trees. Similarly when $X_2 = 0$, then $X_1 = 6$ and the land can support 6 steers. These two intercepts (0, 16) and (6, 0) are located on Fig. 6-3 as points T and S. Since this constraint is linear, all

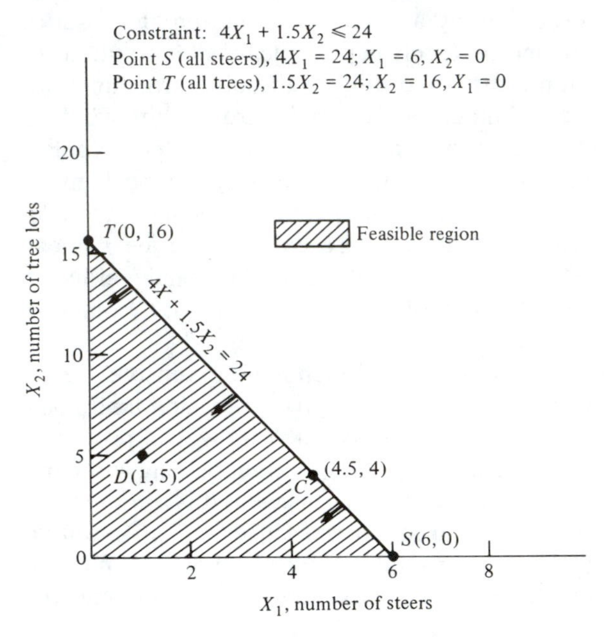

Figure 6-3 Plotting the land constraint of the steers and trees problem.

points on a straight line drawn between the two intercepts are also feasible solutions to the problem. For example, point C is on this line and has values of $X_1 = 4.5$ and $X_2 = 4$. Substituting into the equation we have $4(4.5) + 1.5(4) = 24$, or $24 = 24$. Since the constraint is actually an inequality, all points on or under the line are feasible. Consider also point D (1, 5). $4(1) + 1.5(5) = 11.5$, which is ≤ 24 and satisfies the constraint. Arrows along the line indicate the direction of the inequality. For emphasis, the *feasible region* for the land constraint (set of all points satisfying the land constraint) is shaded. Note that we have assumed X_1 and X_2 to be nonnegative. The nonnegative constraint $X_i \geq 0$ is not drawn but can be interpreted as being the axis of the graph. We will no longer need to write the nonnegative constraints because they are assumed by linear programming solution procedures.

Using the same plotting procedure, all constraints are displayed on Fig. 6-4. Arrows represent the direction of each inequality. The shaded area is the feasible region for the problem—the set of solutions to the decision variables that satisfied *all* constraints of the problem. It is quite possible to have an empty feasible region. Suppose our farmer contracts to raise 6 steer. Since 6 steer are not feasible with the budget constraint, the problem would be *infeasible*. Each point on or within the polygon *SABCD* satisfies the problem constraints and is a feasible solution. That is a lot of points. Fortunately, nearly all points are inefficient and can be

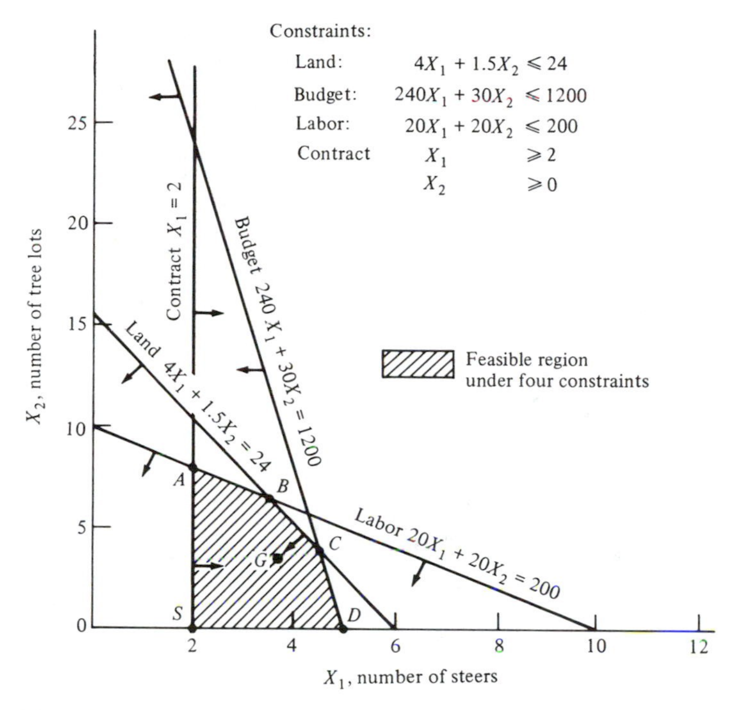

Figure 6-4 Feasible region for solution to the steers and trees problem.

discarded. Take point G, for example. Given the resources of the problem, solutions B, C, or any solutions along the line BC provide more of either X_1 and/or X_2 than point G. Since the objective function coefficients are both positive, all these points also produce more income than G and would be preferred.

The line segment $ABCD$ covers all efficient solutions in the feasible region. Because we are using linear constraint equations, this line will be segmented with distinct corner points such as A, B, C, and D.

Objective function The objective function for the steers and trees problem is the family of equations $Z = 1{,}000X_1 + 500X_2$, with Z taking on different values. The objective function is plotted on Fig. 6-5 for Z values of \$3,000, \$6,000, \$9,000, and \$12,000. When $Z = \$9{,}000$, for example (line EF), every solution for the decision variables on this line provides an objective function value of 9,000. At point F, only 9 steers are produced with a value of $9 \times 1{,}000 = \$9{,}000$. At point T, 3 steers and 12 tree lots are produced worth $3(1{,}000) + 12(500)$, or \$9,000.

Best solution The constraint and objective function graphs are combined in Fig. 6-6. Since the goal is to maximize income, the task is to find the

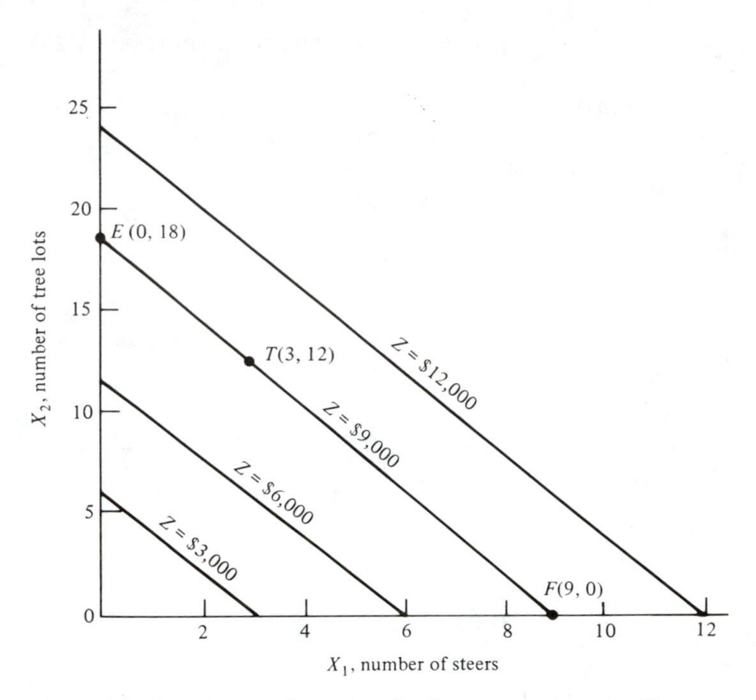

Figure 6-5 Objective function values for the steers and tree problem.

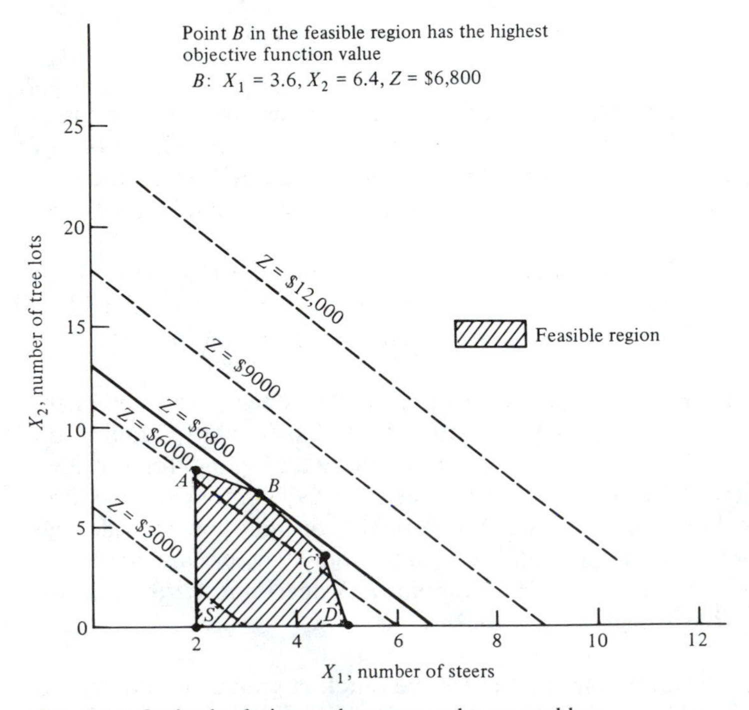

Figure 6-6 Optimal solution to the steers and trees problem.

feasible solution with the highest objective function value. Graphically this amounts to finding the point in the feasible region touching the highest (farthest from the origin) objective function line. In Fig. 6-6 the point satisfying this condition is *B*, which has income line $6,800 tangent to it. Hence, point *B*, where $X_1 = 3.6$ and $X_2 = 6.4$, is the best or optimal solution. So the consultant's recommendation to the farmer is to "raise 3.6 steers and 6.4 lots of trees." This solution is unrealistic, of course, since you cannot raise 0.6 of a steer. However, in most forestry applications the numbers are bigger and you can round off to the nearest whole cow or output unit without much loss in accuracy.

Simplex Method of Linear Programming

The simplex method of linear programming is a mathematical solution technique (an algorithm) that does what we just described. It takes the constraints, defines the feasible region, identifies the corner points on the region boundary, and iteratively searches from corner point to corner point to converge on the best solution. At each step the solution procedure selects a corner point with a higher value for the objective function. This avoids having to look at all corner points and thus has a relatively short solution time.*

Many real problems have hundreds to thousands of corner points and decision variables, but this is where the computational power of the simplex method and modern computers pays off. Graphic solutions are only possible for problems with two or three decision variables. Beyond that, algebraic procedures must be used.

For the steers and trees problem, the five corner points along with their solution values are listed in Fig. 6-7. To maximize income, it is surely important to find the objective-maximizing corner point. Even here the income values range from $2,000 to $6,800. Be honest, how long would it have taken you to produce the solution data in Fig. 6-7 if you were simply handed the consultant's notes and knew nothing about mathematical programming? More realistically, how would you have dealt with the more complex timber harvesting and wildlife or reforestation problems? Not very well, probably.

Neither did land managers 20 years ago. Typically, they described the constraints in a general way and then came up with a single plan or alternative that they judged feasible. Many of these earlier plans on problem solutions probably were much like point *G* in Fig. 6-4—feasible, but

* For details on exactly how the simplex algorithm mathematically creates and searches the feasible region see Dykstra (1984) for an introduction and any one of a number of operations research texts, including Danzig (1963), who first published the method.

Point	Decision variable X_1	X_2	Objective value Z
A	2	8	\$6,000
B	3.6	6.4	\$6,800
C	4.5	4	\$6,500
D	5	0	\$5,000
S	2	0	\$2,000

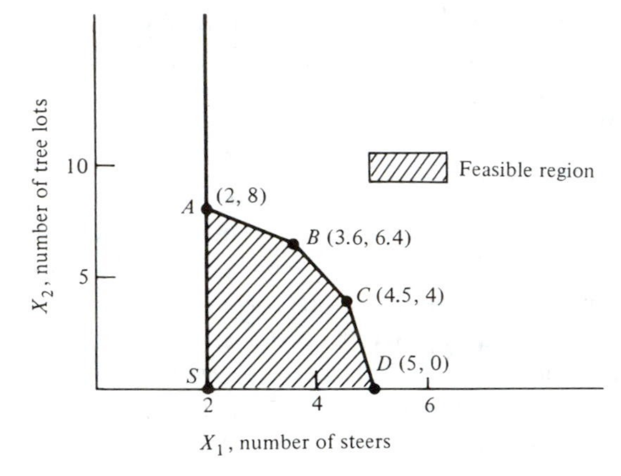

Figure 6-7 Solution values for the corner points of the feasible region.

not likely to be on the efficient boundary or optimum. However, 20 years ago computers and mathematical programming were just coming on line and no ready techniques existed to do the thousands of calculations needed to define feasible regions and find optimal solutions. Rules of thumb and rough approximations were the *only* realistic solution techniques available.

Twenty years is not a long time yet most of today's land managers were trained before this technology was available and relatively few have picked up on it through postgraduate courses and self-study. Rapid changes in technology are unsettling; both the new graduate and the experienced managers need patience. New technologists need to explain their tools and learn the pragmatic reality of politics, data availability, and the limits of mathematical applications to management. Experienced managers need to reexamine familiar procedures and replace those that the new techniques make obsolete.

Variations on steers and trees Suppose that shortly after the consultant had completed the analysis of the farmer's situation, potential for changing the technology of both steer and tree production came to his or her attention as well as knowledge that in a year or so beef prices would rise and tree prices would fall.

The new technology included buying $160 of supplemental feed per steer per year, which reduced the land requirement to fatten a steer from 4 to 2 acres. A new high-speed pruning saw was found that reduced the labor requirement per lot of trees from 20 to 10 hours per lot and increased the cost an additional $10 per lot. If the new technology proves profitable, Bill Jackson agrees to raise his operating budget limit to $2,400. These changes altered the original steers and trees constraints set as follows:

1. *New technology production constraints for steers and trees*

a.	Land	$2X_1 + 1.5X_2 \leq 24$
b.	Budget	$400X_1 + 40X_2 \leq 2{,}400$
c.	Labor	$20X_1 + 10X_2 \leq 200$
d.	Contract	$X_1 \geq 2$

Regarding future prices, a beef shortage was expected to raise the net income from selling 1 steer to $2,000. A glut of Christmas trees was expected because previous high profits had attracted many new growers, and net income from trees was expected to be only $100 per lot of 1,000.* This objective function is

2. *Future objective function*

$$2{,}000X_1 + 100X_2$$

To check all possibilities, four permutations of objective functions and constraints can be examined. Using a linear program to obtain the best solution, optimal solutions to the four problems are as follows:

Problem	Objective function	Constraints (technology)	Optimal solution values for decision variables: X_1 (steers)	Optimal solution values for decision variables: X_2 (tree lots)	Value of objective function (income)
1 (original)	Original	Original technology	3.6	6.4	$ 6,800
2	Future prices	Original technology	5.0	0	10,000
3	Original	New technology	5.0	9.23	9,692
4	Future prices	New technology	6.0	0	12,000

* A more precise analysis would estimate prices and costs separately. Here the original objective would have to change because of the higher costs of the new technology.

Whenever the future objective function is used, only steers should be produced. This result is not surprising as steers are worth relatively much more under the future price assumptions.

Suppose the farmer and the consultant are uncertain as to whether the new technology would work or that prices would change. They could view this table of four different problem specifications as alternative future states of nature. Since each problem solution represents a different problem formulation, the four problems taken together also serve to illustrate the type B process for generating possible solutions where the decision maker is considering alternative specifications of the problem before deciding which one best describes the "true" problem. Operationally, the decision choice lies between producing only steers or producing a mix of steers and trees. If the farmer produces 5 steers under current technology, assuming the steer price will rise *and it does not rise,* then the 5 steers are worth only $5,000 (5 at $1,000). He makes $1,800 less than if he had produced a 3.6-steer and 6.4-tree mix worth $6,800. On the other hand, if the original mix is produced *and prices do change,* 3.6 steers and 6.4 tree lots are worth $7,840, $2,160 less than what the 5 steers are worth at the higher future price of $2,000 per steer. If the farmer and the consultant could subjectively assign a probability that the four problem formulations would occur, then an expected value of the decision could be estimated. A decision tree approach might examine each of the four X_1, X_2 solutions as actions A_i, and for each action consider the two possibilities for future prices as states of nature S_i, yielding an 8-branch tree. Of course, many more problem formulations, price assumptions, and thus tree branches are possible.

When a type B process is used to generate solutions, these are the types of trade-offs considered. The degree of uncertainty about the future, the psychological makeup, and the financial and other personal restrictions on the decision maker will ultimately determine the choice.

Information from Problem Solutions

When a set of equations portraying a problem is solved with linear programming, the report of the optimal solution contains much information in addition to the values for the decision variables. The initial steers and trees problem, when solved using the simplex-type linear programming software LINDO (Schrage, 1981), gives the results shown in Fig. 6-8. The report can be divided into six sections, with section 1 being a report of the problem's equations. Section 2 announces that an optimal solution has been found after three steps, each step being a new cornerpoint, and that the value of the objective function at the optimum solution is 6,800. If an optimum cannot be found, section 2 shows the gloomy message "no feasible solution."

SECTION 1

```
MAX      1000X1 + 500X2
SUBJECT TO
        2)    4X1 + 1.5X2 <=   24
        3)    240X1 + 30X2 <=   1200
        4)    20X1 + 20X2 <=  200
        5)    X1 >=   2
END
```

SECTION 2

```
    LP OPTIMUM FOUND  AT STEP     3

           OBJECTIVE FUNCTION VALUE

1)          6800.00000
```

SECTION 3

VARIABLE	VALUE	REDUCED COST
X1	3.6000	0.0
X2	6.4000	0.0

SECTION 4

ROW	SLACK	DUAL PRICES
2)	0.0	200.000
3)	144.0000	0.0
4)	0.0	10.0000
5)	1.6000	0.0

NO. ITERATIONS=   3

SECTION 5

RANGES IN WHICH THE BASIS IS UNCHANGED

		COST COEFFICIENT RANGES	
VARIABLE	CURRENT COEF	ALLOWABLE INCREASE	ALLOWABLE DECREASE
X1	1000.0000	333.3333	500.0000
X2	500.0000	500.0000	125.0001

SECTION 6

		RIGHTHAND SIDE RANGES	
ROW	CURRENT RHS	ALLOWABLE INCREASE	ALLOWABLE DECREASE
2)	24.0000	1.7143	4.0000
3)	1200.0000	INFINITY	144.0000
4)	200.0000	53.3333	30.0000
5)	2.0000	1.6000	INFINITY

Figure 6-8 LINDO solution report for the steers and trees problem.

Variable values and reduced costs Section 3 gives the answer to the problem by stating the values of the variables at the optimal solution—here 3.6 for X_1 and 6.4 for X_2. The reduced cost column for the variables provides information about variables that have a zero value in the solution. In this problem both variables have positive values, so there is not much to see. Suppose a third variable, say X_3 = number of buffalo with a net return of \$250 per buffalo, had been in the problem and had a zero value, meaning no buffalo in the optimal solution. We would see a positive number, say \$385, in the reduced cost column opposite X_3. This number is the amount the current objective function coefficient for buffalo would have to increase before any buffalo would appear in solution. In this hypothetical case, the net return from raising 1 buffalo would have to increase from 250 to (250 + 385), or \$635, before a buffalo is included in the solution. If a constraint were added to force a buffalo into solution ($X_3 \geq 1$), then at current prices we would find the value of the objective function declining by a maximum of \$385. In this sense the reduced cost is the opportunity cost of forcing a unit of a nonoptimal variable into solution.

Slacks and shadow prices Section 4 of the report shows the slack and dual prices for the rows. The *slack* is the amount of the right-hand side for a constraint that is not used in the optimal solution. For example, the slack for budget is 144, indicating that this much of the budget was not used. If a slack has a zero value, then the constraint is fully utilized. A zero slack indicates that the constraint is binding or limiting and the solution can be changed by increasing or decreasing the right-hand side value to this constraint. Graphically, the zero slack or binding constraints are the ones that intersect at the corner point of the optimal solution. Look at point *B* in Fig. 6-4. The land and labor constraints intersect at *B*, and these are the constraints reported with zero slacks in section 4.

The second column of section 4 is labeled dual prices, but the more common term is *shadow prices*. Shadow price is indeed an unusual term, conjuring up images of something unreal and perhaps something not quite legitimate. Both of these instincts contain some truth. Shadow prices do not have empirical reality as they are not set in the marketplace. They are estimated as a by-product of a mathematical programming analysis and have legitimacy only with respect to the assumptions made and the data used in the particular analysis.

Shadow prices are the maximum amount that the value of an objective function would change if one more unit of a constraint were added or subtracted.

Under the four variations of the steers and trees problem, the shadow prices, in dollars of income, are as follows:

	Problem variation			
Constraint	1 Original objective, original constraint	2 Future objective, original technology	3 Original objective, new technology	4 Future objective, new technology
Land	$200	0	$307.7	0
Budget	0	$8.3	0.96	$5
Labor	10	0	0	0
Steer contract	0	0	0	0

If one more acre of land were available (25 acres instead of 24), in the original specification of the steers and trees problem, the addition of that acre would have allowed more steers and/or trees to be produced, and the objective function could be increased by a maximum of 200. In the sense that the constraints ''hold down'' the solution, shadow prices indicate the relative degree to which the different constraints restrict the solution. For example, the shadow price for budget is zero in the original problem. Increasing the budget by 1 dollar will not change the best solution or the value of the objective function—hence the conclusion that budgets are not limiting for this problem. A close look at Fig. 6-4 confirms that budgets are not limiting at the solution of 3.6 steers and 6.4 tree lots (point *B*). Since not all of the budget is used, more budget is not going to help. In fact it is the intersection of the labor and land constraints that defines point *B*, and these are therefore the limiting resources. They are also the constraints that have nonzero shadow prices.

Figure 6-9 graphs each of the four problem variations and helps to verify visually the relationship of shadow prices to binding constraints. Again these results emphasize that values obtained for shadow prices depend strictly on the problem specified. For problems 2 and 4, budget is the only constraint that restricts the solution. In problem 1 land and labor restrict the solution, while in problem 3 it is land and budget.

Shadow prices are most important as a diagnostic, indicating which constraints are limiting. In large problems, instincts and common sense often fail. Then it may be very important to know whether land, budget, or some policy is actually restricting. If budget is not limiting, it does not make sense to argue for more budget to get a higher objective function value.

If minimum constraints placed on output levels (such as requiring a minimum amount of steers) have nonzero shadow prices, these prices can be interpreted as the value foregone in the objective function to achieve the last unit of the constrained output. For example, suppose the objec-

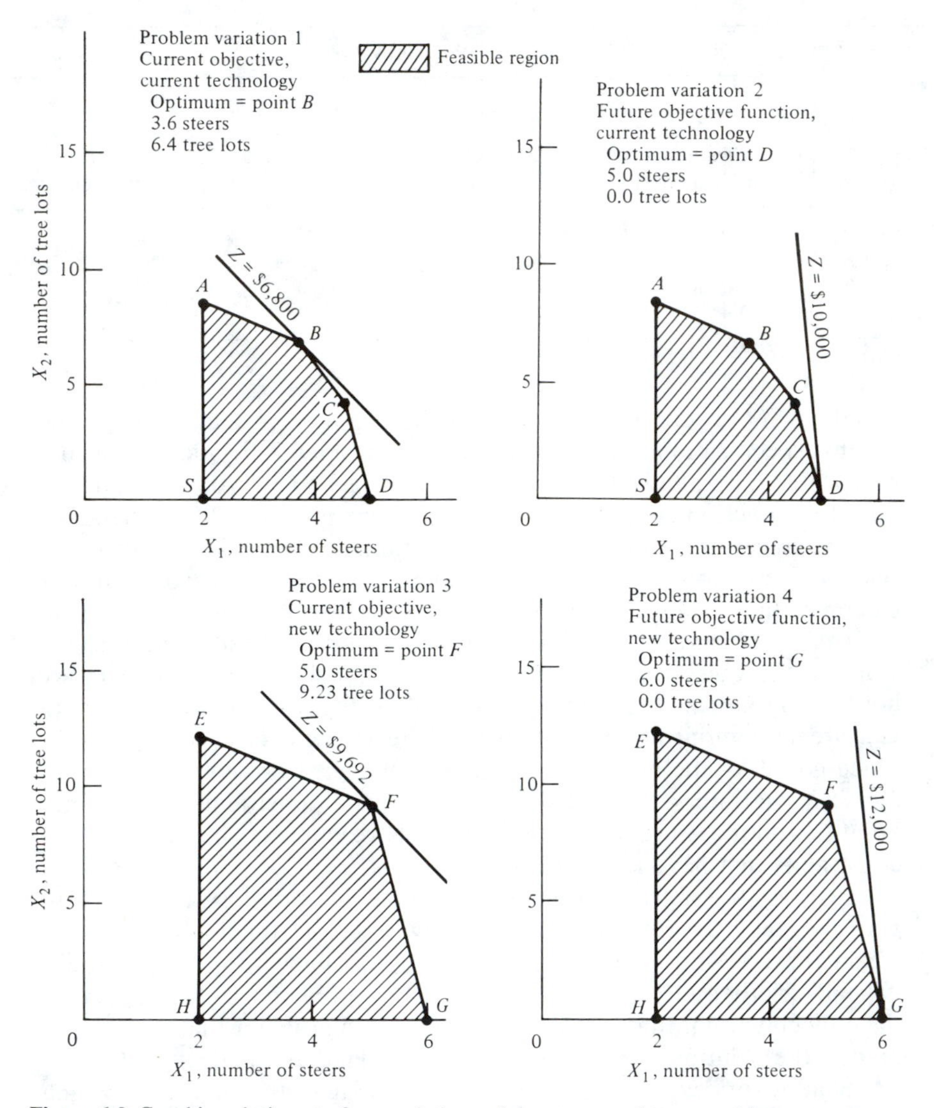

Figure 6-9 Graphic solutions to four variations of the steers and trees problem.

tive function of a problem was based on revenues from timber sales and the shadow price of $1,500 per acre was calculated for a constraint that "10,000 acres of wilderness must be provided." This says a maximum of $1,500 in timber revenue is being given up to provide the last acre of wilderness. Economists would say that "at the margin" this is the implied price or value of wilderness, because by knowingly imposing the constraint, the decision maker willingly gives up the marginal $1,500 in timber revenues. Hence shadow prices are one technique for putting marginal value or prices on nonmarket services such as wilderness.

Shadow prices must be used with caution since they are estimated at the margin and in the context of a particular problem. It is not correct to say that each constraint level in wilderness acres will yield a shadow price of $1,500 per acre. If the wilderness constraint is relaxed to say 7,500 acres, the shadow price might drop to zero. A different formulation of the same problem or the same constraint in another forest management problem might produce a very different shadow price.

Ranges Sections 5 and 6 of the report in Fig. 6-8 provide some information about the sensitivity of the solution to changes in objective function coefficients or constraint levels. (The "current right-hand side" level means the number to the right of the equality or inequality in a constraint equation.) The ranges relate to something called a *basis*. A basis is the list of decision variables and slack variables that have nonzero values in the optimal solution. Each corner point has a unique basis. The number of variables in the basis is equal to the number of constraint rows. In the steers and trees problem there are four constraints and the basis is X_1, X_2, budget slack, and steer contract slack. When a basis changes, one of the variables in the basis is dropped and a new variable, previously at a zero value, is added to the basis.

As long as any one constraint level is not varied outside its range and all other constraint levels remain unchanged, the values of the shadow prices and reduced costs will remain unchanged. This means, for example, that up to 4 acres of land could be taken away from the current land base of 24 acres before the basis and thus $200 shadow price of land would change. In the other direction, only 1.714 acres could be added before the basis and shadow prices would change. For cost or objective value coefficients, the ranges mean about the same thing—the price of steers, for example, could increase $333.33 or decrease $500 from its current level of $1,000 before the basis would change.

We cannot predict exactly what will happen when the basis changes—only that some of the decision variables not in the current basis, reduced cost, and shadow prices will change. So the ranges tell us something about the stability of the solution. If the ranges are small for important coefficients, it means small changes could change the optional solution variables.

The Detached Coefficient Matrix

Writing out problem formulations as equations is instructive, but as the problems become larger, there is a great deal of repetition in writing the same decision variable symbols over and over again. A detached coefficient matrix organizes the equation into the rows and columns of a matrix, with each column representing a different decision variable and each row representing an equation.

1. *Equations for steers and trees*

$$\text{Maximize } 1000X_1 + 500X_2$$

subject to

$$\begin{aligned} 4X_1 + 1.5X_2 &\leq 24 \\ 240X_1 + 30X_2 &\leq 1{,}200 \\ 20X_1 + 20X_2 &\leq 200 \\ X_1 + \quad &\geq 2 \end{aligned}$$

2. *Detached coefficient matrix for steers and trees*

	Column (1)	(2)		
Row	X_1	X_2	Sign	RHS
1. Objective (max)	1,000	500		
2. Land	4	1.5	≤	24
3. Budget	240	30	≤	1,200
4. Labor	20	20	≤	200
5. Steer contract	1		≥	2

You should be able to match each of the five matrix rows to the original equations and see exactly the same relationships. The column names are the decision variables, and the coefficients in the equations are "detached" and shown in the cells of the matrix. A blank in the matrix means a zero coefficient. Row 5, for example, is interpreted as $1X_1 + 0X_2 \geq 2$, or $X_1 \geq 2$. The last column appropriately is labeled RHS, which means "right-hand side." The standard convention also has no decision variables to the right of the equation sign, only a constant or zero.

A Goal Programming Formulation of the Steers and Trees Problems

Suppose after reviewing the results of our analysis and the recommendation to produce 3.6 steers and 6.4 tree lots, Bill Jackson still did not feel we had captured the essence of his problem. More discussion revealed that for tax reasons he wanted to use all of his budget allocation and that he had truly hoped for an even mix of 5 steers and 5 tree lots, one pair for each of his five children to care for.

We asked, "Do you mean achieving these three goals is more important than maximizing income?" He said, "Yes, I want to achieve the following as closely as possible":

$$X_1 = 5$$

$$X_2 = 5$$

$$240X_1 + 30X_2 = 1{,}200$$

On reflection, he stated achieving the budget goal for tax reasons is at least twice as important as either of the other two goals. Goal programming provides a problem formulation where the objective function minimizes the weighted deviation of solution values from the stated goals or target values for the decision variables. Such a formulation is useful for problems when the decision maker has two or more well-defined goals and is not comfortable with selecting one of them for the objective function.

We need to define additional decision variables to represent deviations from the goals. Let

d_1 = positive deviation (amount of overachievement) from the 5 steer goal
d_2 = negative deviation (amount of underachievement) from the 5 steer goal
d_3 = positive deviation (amount of overachievement) from the 5 tree lot goal
d_4 = negative deviation (amount of underachievement) from the 5 tree lot goal
d_5 = positive deviation (amount of overachievement) from the \$1,200 budget goal
d_6 = negative deviation (amount of underachievement) from the \$1,200 budget goal

Since any solution will either overachieve or underachieve a goal but never both over- and underachieve at the same time, for each pair of deviation variables, such as d_1 and d_2, one will have a positive value and the other will be zero, or both will be zero in the case of exact goal achievement.

The objective is to minimize the weighted sum of the deviations. A weight of 2 is used as a coefficient for the budget deviations to reflect their double importance to Bill Jackson,

$$\text{Minimize } d_1 + d_2 + d_3 + d_4 + 2d_5 + 2d_6$$

The constraints are as before, except for the budget, steer contract, and a new tree lot constraint. In each of these the deviation is treated as a slack variable to convert the constraint to an equality:

3. *Budget* $240X_1 + 3X_2 - d_5 + d_6 = 1200$

4. *Steers* $X_1 - d_1 + d_2 = 5$

5. *Tree lots* $X_2 - d_3 + d_4 = 5$

A detached coefficient matrix for the entire formulation is

Row	Column								RHS
	X_1	X_2	d_1	d_2	d_3	d_4	d_5	d_6	
1. Minimize			1	1	1	1	2	2	
2. Land	4	1.5							≤ 24
3. Labor	20	20							≤ 200
4. Budget	240	30					−1	1	= 1,200
5. Steers	1		−1	1					= −0.5
6. Trees		1			−1	1			= −1.0

When the problem was solved, the following results were obtained:

Objective value = 1.5000

Variable	Value	Row	Slack	Dual
d_1	0			
d_2	0.5	Land	0	0.875
d_3	0	Labor	30	0.0
d_4	1.0	Budget	0	−0.0
d_5	0	Steers	0	−0.5
d_6	0	Trees	0	−1.0
X_1	4.5			
X_2	4.0			

The solution minimizes the deviations at a solution point C (Fig. 6-9) where $X_1 = 4.5$ and $X_2 = 4.0$. All of the budget and land are used. Since the goal levels of the decision variables were set at high, infeasible amounts, the solution logically found a point on the boundary of the feasible region that minimized deviation. The budget goal carried greater weight, so it is not surprising that a corner point that used all the budget was found.

Goal programming has two rather demanding requirements of the decision maker: (1) that all important goals be known in advance, and (2) that weights can be assigned to the goal deviation to reflect the relative importance of goal achievement. This asks a lot since the more usual problem is to determine what the output levels should be. In some situations, however, goal programming may be appealing to decision makers exploring the implications of different goal sets or weights.

Steers and Trees as a Land Allocation Problem

Forestland management planning usually is conceptualized as a process of allocating acres of the forest to different uses and deciding what treatment prescription to apply on these acres. The decision variables that match this viewpoint are the number of acres allocated to different uses and the prescriptions. To illustrate, we now reformulate the original steers and trees problems as a land allocation problem. The new decision variables are defined as

X_1 = acres allocated to steer production
X_2 = acres allocated to tree production

Given this definition, the land constraint becomes

$$X_1 + X_2 \leq 24$$

Since we know that it takes 4 acres to produce a steer and 1.5 acres to produce a lot of trees, these factors can be used to adjust the remaining constraints and the objective function. If 1 steer is worth $1,000 and takes 4 acres of land, then 1 acre in steer production provides 1,000/4, or $250 of revenue. Similarly 1 acre of tree production provides 500/1.5, or $333.33 of revenue. Accordingly, the new objective function is

$$250X_1 + 333.33X_2$$

When all constraints are adjusted, the problem statement is as follows:*

Maximize $\quad 250X_1 + 333.33X_2$

where X_1 = acres allocated to steers
X_2 = acres allocated to trees

subject to the constraints

a. Land	$X_1 + X_2 \leq$	24
b. Budget	$60X_1 + 20X_2 \leq$	1,200
c. Labor	$5X_1 + 13.33X_2 \leq$	200
d. Contract	$X_1 \geq$	8

* You should check this to satisfy yourself that all coefficients are correct.

When solved, this problem has optimal values of

$$X_1 = 14.399$$

$$X_2 = 9.600$$

$$\text{objective} = \$6{,}800$$

Since each steer takes 4 acres, we have 14.399/4, or 3.6 steer and 9.6/1.5, or 6.4 lots of trees. Hence we get the same answer as the original formulation, but the decision variables are now more useful, since they tell the land manager explicitly how to allocate and manage the land.

A Caveat When Modeling Problems for Solution by Mathematical Programming

Every modeling effort is an attempt to simplify and describe reality, hopefully retaining the important relationships. Mathematical programming takes such a simplification of reality (the problem statement) and analyzes it to provide guidance on how to deal with or manage reality. The modeling and analysis are adequate if the guidance works when implemented.

For linear programming, some key assumptions and concerns in modeling problems as linear programs are illustrated in Fig. 6-10. The line dividing the figure separates reality (bottom half of figure) from the abstraction or model of reality (top half of figure). The abstraction exists at some time as a bunch of ideas in the analysts' and decision makers' heads or as a set of mathematical symbols written on paper or computer files.

Key questions to ask in assessing whether the abstraction is useful are:

1. Have enough goals and constraints of the real-world problem been modeled? How many are ignored, overlooked, or impossible to model? Are the omissions of sufficient importance such that solution of the modeled problem will have little meaning in reality?
2. Is the *linear* representations of *all* goals and constraints appropriate?
3. Are the numerical values for all coefficients acceptably accurate? How many are based on empirical evidence and statistically validated? How many are judgmental. If experts were used, what were their credentials?

The linear program always finds the optimal solution to the mathematical problem formulated, provided one exists. It does not care what the Xs stand for; all it sees is a set of equations. It is a strictly mechanical, neutral technique. In fact, the answer to the problem is defined the moment the problem is identified and formulated, even though it takes a bit of looking to find it.

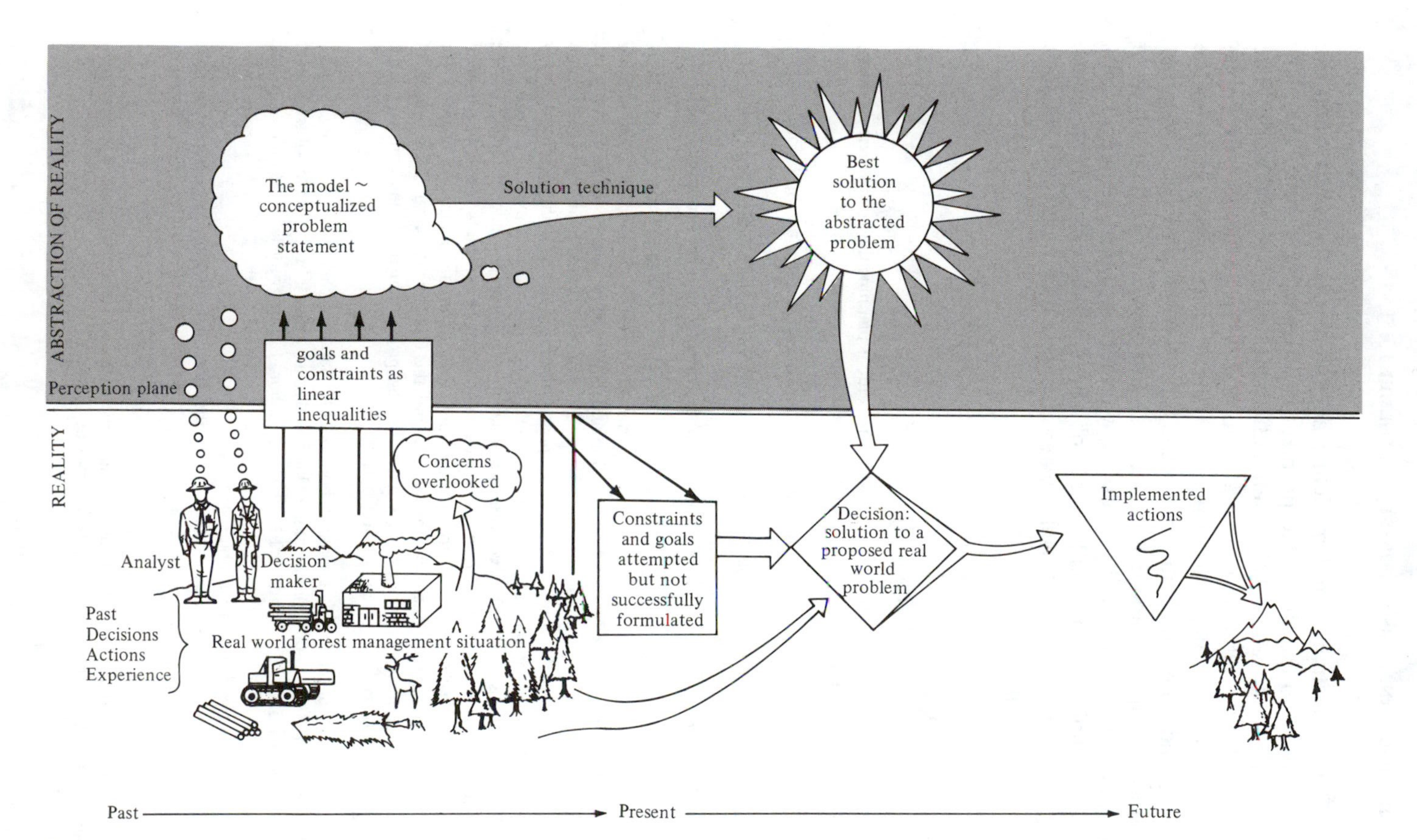

Figure 6-10 Problem identification and analysis—from reality to abstraction and back.

These questions about modeling adequacy are answered when the best solution to the abstracted problem is brought back across the plane into reality and used to help guide actual actions in the real world. The future holds the ultimate truth and reveals it when the actions are implemented. Unfortunately the analyst and the decision maker labor in the present and cannot find guidance in the future. All they can do is to adopt models that have worked for others and/or worked for themselves in other situations, continually assessing whether the approach being used is the best choice to represent their problem. Instincts born of experience and a box of salt are still useful when deciding how much of a model's solution to implement.

QUESTIONS

6-1 *Forestry club fund raising*. The school forestry club is planning the year's merrymaking, and three traditional fund raising activities are again considered.

1. Catering beer and bean feeds to fraternities and sororities at a net profit of $150 per party.
2. Selling delivered cords of oak firewood from the school forest to university faculty at a net profit of $100 per cord.
3. Selling Christmas trees donated from a lumber company at a net profit of $5 per tree.

The club wants to make as much net income for the year as possible. An inventory of its resources shows some limits to what they can do. Revenues are delayed, so the operating cost of all activities has to be paid from the club's $1,500 treasury balance left from last year. A willing-worker survey of club members showed 20 hours of skilled bean cooks, 100 hours of skilled labor, and 500 hours of general-purpose grunt labor to be available. The Forestry Department equipment manager said the club could use 500 miles of yellow cattle truck free, but additional miles must be rented from the department at $0.50 per mile. Selling Sam, the club vice-president, got enthused at these ideas and arranged contracts for two beer-bean feeds and 150 Christmas trees before club members locked him up in a growth chamber for fear that their resources might be inadequate and the club might be sued for breach of contract. The Forest Economics class analyzed the production functions and found the following inputs were required per unit output:

1. *Beer and bean feeds*. $75 capital, 5 hours cooks' labor.
2. *Firewood*. $30 capital, 5 hours skilled labor, 10 hours grunt labor, 50 miles of truck use.
3. *Christmas trees*. $2 capital, 0.2 hour skilled labor, 1 hour grunt labor, 2 miles of truck use.
   (*a*) Define the decision variables.
   (*b*) Present the problem as either a set of equations or a detached coefficient matrix, labeling each row or equation. (*Hint:* The truck rental activity takes the most thought.)

6-2 *Land use allocation*. A 1,000-acre management unit of public wildland spans a small valley and is distributed into three land types as follows:

1. North slopes 700 acres
2. South slopes 210 acres
3. Creek bottom 90 acres

Acres in each of these tracts can be devoted to timber production, forage, or set aside for recreational purposes. The forest ranger wants to decide how to best allocate his land between these three uses. The decision variables are defined as follows

Land use	Land area		
	North slope	South slope	Creek bottom
Forest	X_1	X_2	X_3
Range	X_4	X_5	X_6
Recreation	X_7	X_8	X_9

Prices for products and services produced from this tract are as follows:

1. Timber. $200 per M bd ft
2. Forage. $5 per animal-unit-month (AUM)
3. Recreation. $1 per recreation-visitor-day (RVD) on the slopes; $3 per RVD in the creek bottom

The total available budget is $5,000 per year.

The average annual yields and costs per acre for each land use and land type possibility are as follows:

	Yield (per acre per year)	Cost (per acre per year)
X_1	300 bd ft	$ 2.00
X_2	50 bd ft	0.50
X_3	800 bd ft	10.00
X_4	0.4 AUM	2.00
X_5	0.1 AUM	0.10
X_6	0.9 AUM	5.00
X_7	0.1 RVD	0.50
X_8	0.1 RVD	0.10
X_9	25 RVD	20.00

Formulate and write linear equations to reflect the following:

(*a*) A contract has been given to the local sawmill to supply a minimum of 50,000 bd ft of logs per year.

(*b*) Sam Grazemore has a historical permit to use the two slope tracts for a total of 100 AUM per year.

(*c*) The budget constraint.

(*d*) The objective function is to maximize annual *net* revenue.

(*e*) For legally mandated aesthetic and ecological balance reasons, the government requires at least 20 percent of the total land area to be in each land use.

(*f*) For every acre allocated to recreation at least 10 acres should be allocated to either forest or range to meet the government's new economic goals to produce more dollar outputs from public lands.

(*g*) Because this is cattle country, a county ordinance states that for every 1,000 bd ft sold at least 20 AUM must be offered for sale.

(*h*) Is this problem feasible if constraints (*a*) to (*g*) are all applicable? Explain.

6-3 *Log transportation*. The manager of a private company must decide how to supply his three mills with wood cut from three logging sites he will be operating next month. It costs him $1.50 per mile to run his trucks. He would like to minimize the total daily travel cost from the sites to the mills. He has gathered the following information:

	Distance in miles between logging sites and mills		
Logging site	Mill A	Mill B	Mill C
1	8	15	50
2	10	17	20
3	30	25	15

Each load requires a round trip from mill to site and back. Each mill must have a continuous daily supply of wood to keep the various operations working efficiently. The minimum number of truckloads required daily for each mill is 25. Each site can produce, at a maximum, the following number of truckloads daily: site 1, 20 loads; site 2, 30 loads; site 3, 25 loads.

Formulate the problem as a set of linear equations to determine how many loads should be shipped daily from each logging site to each mill. Let X_{ij} be the number of loads per day from site i to mill j.

6-4 *Quincy Lumber problem*. Quincy Lumber recently acquired a 640-acre section of young growth conifers and needed to allocate its acres to either even-aged or uneven-aged management systems. The allocation is to be based on maximizing the *long-term* sustained revenue potential of the forest. Close examination of the site showed that 200 acres are suitable only for even-aged management and that current forest practice rules require that a minimum of 150 acres be under uneven-aged management for the streamside zones. In a public speech, the company president (who is well-meaning but sometimes confusing) said "No more than 600 acres would be managed even-aged, and for every two even-aged acres there would be at least one acre allocated to uneven-aged management."

Under the best current silviculture, even-aged produces an average mean annual increment (MAI) of 0.7 M bd ft per acre and uneven-aged a MAI of 0.5 M bd ft per acre in managed stands. Prices are expected to be $200 per M bd ft for all stumpage produced under either system.

(*a*) Write out the problem as a set of equations.

(*b*) Plot the constraints on a sheet of graph paper to locate the feasible region.

(*c*) Plot a revenue line for an objective function value of $50,000. Locate the optimal solution. Which constraints are binding?

(*d*) Suppose future prices are expected to reflect the higher log quality of harvests from uneven-aged stands as $350 per M bd ft for uneven-aged harvests and $200 for even-aged harvests. Plot a new revenue line and locate the optimal solution. What has changed?

(*e*) If you can, solve this problem on a computer and verify that the results are the same as your graphic solution in (*c*) and (*d*).

6-5 *Shadow prices*. The objective function of a problem was to maximize the present net worth in dollars of all timber sales subject to a variety of constraints. The shadow prices for four of the constraints are given in the table below.

Interpret each of the shadow prices, stating exactly what it means in quantitative terms about the constraint.

Row	Constraint	Shadow or dual price
1	Available land, acres	+20
2	Maximum harvest, M bd ft	0.0
3	Minimum turkey habitat, acres	−300
4	Minimum wilderness output, visitor-*days*	−0.035

6-6 *Interpreting a harvest schedule solution.* A forest consisted of two timber stands and the owner wanted to plan her harvest over the next three 10-year periods using linear programming. The six decision variables were defined as acres cut by stand by period:

Stand	Period cut			Total acres
	1	2	3	
A	X_{11}	X_{12}	X_{13}	100
B	X_{21}	X_{22}	X_{23}	200

The owner wanted to maximize the total volume cut, in thousands of board feet, over all three periods, but had minimum harvest requirements in each period. With appropriate volume yields per acre as coefficients, this resulted in a problem specified as follows:

Row	Purpose	Equation
1	Objective	Maximize $10X_{11} + 12X_{12} + 15X_{13} + 7X_{21} + 10X_{22} + 18X_{23}$
2	Acres of stand A	$X_{11} + X_{12} + X_{13} \leq 100$ acres
3	Acres of stand B	$X_{21} + X_{22} + X_{23} \leq 200$ acres
4	Minimum harvest in period 1	$10X_{11} + 7X_{21} \geq 600$ M bd ft
5	Minimum harvest in period 2	$12X_{12} + 10X_{22} \geq 600$ M bd ft
6	Minimum harvest in period 3	$15X_{13} + 18X_{23} \geq 600$ M bd ft

The owner had you run this problem on your handy LINDO linear programming package and you obtained the printout in Fig. 6-11. Interpret these results by answering the following questions.

(*a*) What is the optimum harvest schedule for the owner?

(*b*) How much wood will she harvest over the three periods?

(*c*) Will there be any residual inventory and how would you estimate or calculate this?

(*d*) How much would the volume of wood per acre in stand B in period 1 have to increase before an acre of stand B would be harvested in period 1?

(*e*) How much is the minimum harvest constraint in period 1 costing the owner in terms of the overall harvest? If the constraint could be reduced to 400 M bd ft in period 1, how much would this improve the total harvest?

(*f*) How much could the acreage in stand A change before the basis changes?

```
     LP OPTIMUM FOUND AT STEP   7

       OBJECTIVE FUNCTION VALUE

1)            4584.00000

VARIABLE      VALUE            REDUCED COST
  X11           60.0000             0.0
  X12           40.0000             0.0
  X13            0.0                6.6000
  X21            0.0                2.8800
  X22           12.0000             0.0
  X23          188.0000             0.0


ROW           SLACK             DUAL PRICES
     2)          0.0              21.6000
     3)          0.0              18.0000
     4)          0.0              -1.1600
     5)          0.0              -0.8000
     6)       2784.0000             0.0

NO. ITERATIONS=    7


        RANGES IN WHICH THE BASIS IS UNCHANGED

                         COST COEFFICIENT RANGES
VARIABLE      CURRENT          ALLOWABLE         ALLOWABLE
               COEF            INCREASE          DECREASE
  X11         10.0000           11.6000            4.1143
  X12         12.0000            4.1143            6.6000
  X13         15.0000            6.6000          INFINITY
  X21          7.0000            2.8800          INFINITY
  X22         10.0000            5.5000            3.4286
  X23         18.0000          INFINITY            5.5000

                         RIGHT-HAND SIDE RANGES
ROW           CURRENT          ALLOWABLE         ALLOWABLE
                RHS            INCREASE          DECREASE
     2)       100.0000           10.0000           40.0000
     3)       200.0000         INFINITY           154.6667
     4)       600.0000          400.0000          100.0000
     5)       600.0000         1546.6672          120.0000
     6)       600.0000         2784.0000         INFINITY
```

Figure 6-11 Linear programming solution for question 6-6.

6-7 *Decisions under uncertainty (logging).* Your company needs to log a tract in northern Michigan during November and December, and you have to decide on the kind of access roads to use. Three different road systems are under consideration for which the construction and maintenance costs are significantly affected by weather. For this decision we have the actions and states of nature as follows:

1. *Action*

   a_1 = undrained dozer roads
   a_2 = drained and crowned roads
   a_3 = drained, rocked, and crowned roads

2. *States of nature*

   s_1 = dry
   s_2 = rain and snow

3. *Total costs of roads*

	s_1	s_2
a_1	20,000	170,000
a_2	60,000	130,000
a_3	100,000	100,000

Total revenue from the sale, *not including roads,* is $120,000.

(*a*) Construct a profit or net income payoff matrix for this problem.

(*b*) What is the best road system using the equally likely expected value, the optimist (maximax), and the pessimist (maximin) decision criteria?

(*c*) If a thoroughly reliable weather forecaster told you that the probability of dry weather was 70 percent and of wet weather 30 percent, would this change your decision?

6-8 *Decisions under risk—a forest nursery.* A forest nursery manager is deciding which species to plant for next year's sales program. He can only plant one species since it is a small nursery. There are three species that could be raised: regular pine, super pine, or fast oak. The nursery manager's records of past sales indicate that his annual profits varied substantially depending on whether or not it was a wet or a dry year and whether the price of lumber was rising or falling. In dry years there were many wildfires and fast oak sold well for restoration plantings. If the lumber prices were rising, customers bought the more expensive super pine seedlings, but if they were falling, it was standard brand pine. Based on past experience and from watching volcanic patterns and sunspots, the nursery manager put subjective probabilities on these events. This decision situation is summarized as follows:

Expected nursery profit per year

	Possible future conditions			
	Dry weather		Wet weather	
Action: Species planted	Prices rising	Prices falling	Prices rising	Prices falling
Pine	$ 3,000	$5,000	$ 3,000	$5,000
Super pine	10,000	−1,000	10,000	−1,000
Fast oak	6,000	6,000	1,000	1,000
Subjective probability	0.16	0.24	0.24	0.36

Draw a decision tree to represent this problem, labeling all branches and showing probabilities and outcomes. Calculate the expected value of each action.

REFERENCES

Danzig, G. B., 1963: "Linear Programming and Extension," Princeton University Press, Princeton, N.J.

Dykstra, D. P., 1984: "Mathematical Programming for Natural Resource Management," McGraw-Hill, New York.

Schrage, L., 1981: "Linear Programming Models with LINDO," Scientific Press, Palo Alto, Calif.

Teeguarden, D. E., and H. L. Von Sperber, 1968: Scheduling Douglas Fir Reforestation. Investments: A Comparison of Methods, *Forest Sci.,* **14:**354–369.

Thompson, E. F., B. Hallerman, T. J. Lyon, and R. L. Miller, 1973: Integrating Timber and Wildlife Management Planning, *Forest Chronicle,* **49:**247–250.

CHAPTER

SEVEN

FINANCIAL ANALYSIS AND THE ARITHMETIC OF INTEREST

Virtually every forestry action involves investing in land, timber, or some treatment, and then waiting several years for all the benefits to be realized. Spending $250 per acre to prepare and plant a 10-acre site and then waiting 40 years to harvest and sell the mature trees for $20,000 per acre is a good example. Here two resources, $2,500 of capital and 10 acres of land, are tied up and used for 40 years to produce the expected yield of $200,000. It costs money to use such resources over time, and the rent or price paid per unit time is measured by the interest rate. Interest costs dominate much forestry decision analysis, and facility with financial analysis techniques, including the arithmetic of compound interest, is an important part of the forester's special skills.

THE INTEREST RATE

Operation of a forest business requires a large amount of capital as defined in the sense of an aggregation of economic goods used to promote the production of other goods, instead of being valuable solely for immediate enjoyment. Accumulation of this capital, as reckoned in monetary terms, necessarily requires saving, a postponement of present enjoyment for a future benefit. In industrial countries a great deal of saving is necessary, whether voluntary or not, to accumulate the tremendous amount of capital necessary to operate a technological society.

Capital in monetary form is an extremely useful commodity; it can be used to buy things, initiate productive enterprise, and yield additional income. Its use over a period of time is accordingly worth something to its owner, and the owner likewise expects some return for it. The concept of interest stems from this basic fact; it is the return to capital. Interest is the rental price of money, the reward for waiting. Except when money appreciates in value through a decrease in price level, this reward is not received merely by holding money and waiting. Money buried in the ground will not sprout interest though it may be worth more (or less) in purchasing power when it is dug up. It is not automatically entitled to any return. Interest, as a specific payment for the use of money, arises only when it is used, i.e., borrowed. A miser gets no monetary income from his money. His return comes from the physical pleasure of ownership—and complete liquidity.

Another way to look at interest is that the rate employed gives a measure of the importance of the time element involved. The future always carries less weight than the present; a promise to pay a dollar some years hence is seldom worth a dollar, cash in hand, now. The degree to which the passage of time is discounted or valued is measured by the rate of interest; the higher the rate, the more heavily the future is discounted, and vice versa. Consequently, to the extent that time is important in financial matters, the rate of interest furnishes an expression of time preference. If an individual's time preference is nil, it means that he or she would just as soon receive a dollar at some time in the future as now. If it is high, it means that present income and gratification are valued more highly than the prospect of additional income in the future. Because of the limited span of human life, individual time preference is usually fairly high, sometimes extremely so. This fact shows up repeatedly in the selling of standing timber by an owner. The attractiveness of immediate cash often overshadows acceptance of a plan of management offering much larger total returns and a good return on the investment over a period of years.

Individual and organizational time preference varies greatly. With individuals, it varies from person to person and for the same person, depending on age, degree of education and maturity, level of income, current personal wants in relation to income, and other circumstances. A man with urgent current need for cash to meet a family emergency has a high time preference overshadowing other considerations. People living at a subsistence level on the land will often destructively utilize natural resources to meet present needs. Corporations, or other forms of continuing business organization, ordinarily have a lower time preference than individuals though the same general forces are operative. Government has a still lower time preference and in general the larger the political subdivision, the lower the time preference rate. The Federal government with its national responsibility takes the longest view and tends to have the lowest time preference. Even so, it is not realistic to assume that time preference, even for the Federal government, is zero. There can even be circumstances under which time preference may seem to be less than zero, i.e., the present is discounted in favor of the future. Where national security is involved, a government may be forced to adopt measures of current frugality reversing and overriding normal economic forces. Individuals may, for various personal reasons, and particularly during times of great stress and uncertainty, emphasize the future even more than the present in both financial as well as other matters.

The interest rate paid by the borrower is established in two ways. The first way is to go to a well-developed money market (bank or other financial institution), take out a loan, and agree to pay a certain interest rate. The rate paid to the bank is the current market price for using (renting) money and is determined by the future perceptions held by money lenders (the people who save and invest) and borrowers interacting in the market. Sometimes there are many savers and fewer borrowers, and the interest rate goes down. At other times, when people would rather consume than save and there are many borrowers, interest rates go up. The fluctuation in prime interest rates from a high of 20 percent to a low of 11 percent in 1982 indicates the volatility of money markets.

The second approach to determining the interest rate occurs when capital or other resources wanted for some specific project are provided internally by the borrower from savings or earnings generated in other economic activities. Here the borrower has choices:

1. The money can be invested (loaned) outside in the money market, say in stocks or bonds, and earn income at a known rate of interest.
2. It can be used for one or several other internal projects open to the borrower, where it could earn different estimated rates of interest.
3. It can be used for current consumption.
4. It can be invested in the project under consideration.

The interest rate charged to the project is logically the highest known rate that the money can earn elsewhere in the best alternative investment (choices 1 or 2) or consumption at the time preference rate (choice 3). We use the term *guiding rate* to describe the rate a decision maker uses to evaluate projects and make choices between them. Normally the guiding rate is the highest rate that could be earned in other investments. This guiding rate is also called the *opportunity cost of money* since there is the "opportunity" to do something else with it.

Simple and Compound Interest

After borrowing a sum of money, the lender may give the borrower a choice: (1) the borrower can pay the interest at the end of each time period for which interest is due, with the initial loan principal remaining outstanding, or (2) the borrower is not required to pay interest each period, but rather is required to pay in full the accumulated interest and principal, all being paid after two or more periods. The first choice is a *simple interest* situation, and the second is a *compound interest* situation.

To illustrate, assume the loan amount is $20,000, the interest rate is 10 percent of the outstanding principal per year, and the loan is due after 2 years. With the simple interest choice, the borrower would pay interest of 20,000 × 0.10 or $2,000 at the end of the first year and $2,000 of interest plus the $20,000 of principal at the end of the second year. Total interest payments are $4,000. For the compound interest case, the interest of $2,000 for the first year would be added to the principal, bringing it to $22,000. The interest charge for the second year is 22,000 × 0.10 or $2,200. At the end of the second year, the borrower pays 2,000 + 2,200 or $4,200 of interest plus the $20,000 principal. "Compounding" simply means that if the interest is not paid when due, interest is also charged on the unpaid interest of previous periods in addition to the charge on the original principal. Most loan and payment arrangements are set up to require compound interest calculation, and that is what all the formulas developed in this chapter are for.

Interest Rate Components

Both borrowers and lenders look at specific projects and establish the appropriate interest rate. This rate is a composite of three elements:

1. *Pure rate.** The risk-free cost of using money over time. In the market this is approximated by U.S. Government Treasury notes when there is a full employment economy with no inflation.
2. *Expected inflation rate*. With inflation, the dollar paid back in the future will not buy as much as it will today, so an inflation factor is added to the interest charge.
3. *Risk rate*. The future is always uncertain and different investments or financial ventures carry different amounts of risk. Buying federal government bonds carries a low risk of default, and speculation on common stocks of little-known, new growth companies carries a high chance of loss. A risk factor is added to the interest charge, the amount depending on the particular venture.

* We have defined the pure rate as what a defined group of savers (investors) expect to receive in a risk- and inflation-free environment; hence different groups have different pure rates. Since different borrowers do not have access to all saver groups because of legal, credit, and other reasons, the pure rate of the available lender is the relevant rate.

To illustrate these components consider the following hypothetical investment ventures, and the rationale that might be used by the borrower or lender to establish the annual guiding interest rate.

1. The federal government is investing in long-term timber production on national forests.

Pure rate	3 percent	The long-term real cost of federal borrowing through issue of bonds.
Inflation rate	4 percent	Average "official" published expectations by the government.
Risk rate	1 percent	A secure investment; little anticipated chance of failure.
Total guiding rate*	8 percent	

2. An integrated forest products company is purchasing new timberlands to support mill expansion.

Pure rate	4 percent	The stockholder's expectation, based on long-term real rates of return available from long-term internal mill expansion investments.
Inflation rate	6 percent	The company is more pessimistic than the government about inflation.
Risk rate	2 percent	While moderately risk-free, there is a concern about rising energy costs shifting demand away from forest products.
Total guiding rate	12 percent	

3. A promoter is trying to attract funds for an energy plantation of coppice tropical hardwoods in south Florida.

Pure rate	6 percent	The only source of this type of venture capital is people who demand a high return on short-term investments.

* As we shall see later, simple addition of the interest components is not exactly correct; they need to be multiplied as

$$(1.03)(1.04)(1.01) = 1.08191$$

or 8.191 percent.

Inflation rate	6 percent	As before.
Risk rate	14 percent	These species and energy plantations have not been tried or proven and there are a host of technical and legal problems. The whole project could easily be a write-off.
Total guiding rate	26 percent	

4. A private forest owner is deciding whether to reinvest some stumpage sale revenue into reforestation of the sale area.

Pure rate	5 percent	The owner does not have much other wealth and has a strong individual preference for funds now to support her growing family.
Inflation rate	6 percent	As before.
Risk rate	6 percent	The owner has health problems and considers long-term tree growing to be beyond her life expectation. She thinks land with immature timber has poor liquidity and its value is not well recognized in the market.
Total guiding rate	17 percent	

While hypothetical, these examples illustrate how the interest charge for money will vary depending on the lender and the venture in question. Forest investments are generally considered to be fairly secure, but the loan period often extends well beyond the planning horizon and lifespan of individual investors. Nearly all investment capital in forestry comes internally from the landowners, forest product corporations, or the government. Commercial lenders are not enthusiastic about long-term loans for timber production. We are, however, starting to see some pension funds, insurance companies, and limited partnerships investing in timber growing, particularly on productive sites in the south.

For most foresters, the guiding interest rate is something set by others. In federal or state agencies, the interest rate to be used for analysis of public investments and plans is a matter of policy and politics. In large corporations funds are usually obtained from retained corporate earnings, and the opportunity cost rate for their use is key policy and set at high management levels in an organization. For businesses that must actually borrow money, the current market determines the rate. Finally, for individuals or small landowners using their own time and funds, the rate is determined by each person's time preference for money, how much they discount the future, and their alternative uses of time and capital resources.

THE ARITHMETIC OF INTEREST

Compound interest formulas are used to calculate and numerically establish the relationship of costs, revenues, payments, or other benefits occurring at different points in time. A time line is a helpful way to graph and talk about events at different points in time (Fig. 7-1). The reference point, time = 0, is normally taken to mean the present, the year or moment in chronological time when you are making an analysis or when significant expenditures for a new project or plan are first incurred. Events occurring in the past or future are indexed by the subscript n, where n is the number of periods in the past or future. For example, when $n = 7$, the event occurs 7 periods into the future; when $n = -3$, the event occurs 3 periods in the past, and when $n = 0$, the event occurs exactly at the present, the reference point. Three assumptions are customarily made about analytical periods when using time in compound interest calculations.

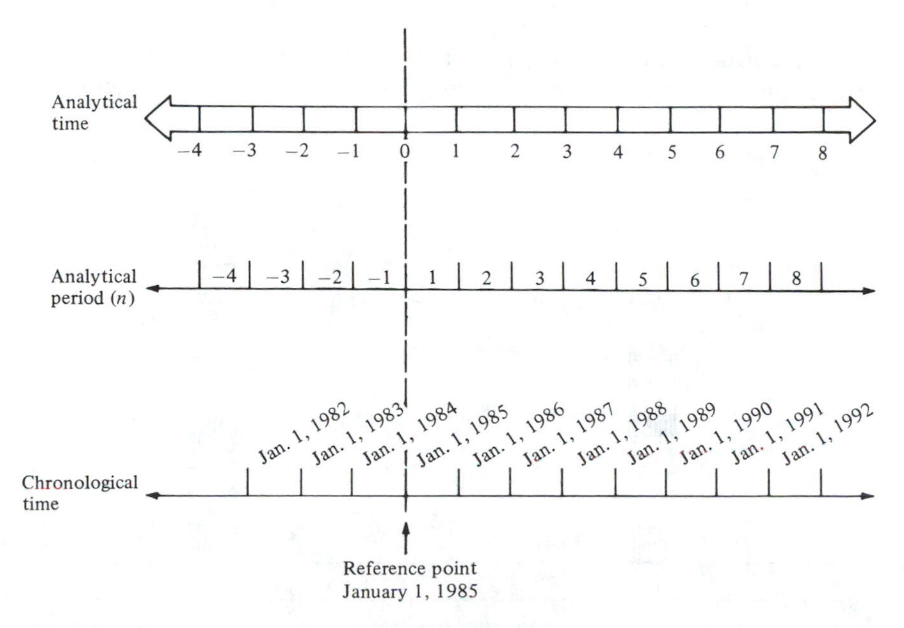

Figure 7-1 The time line concept of financial analysis.

1. The chronological date of the analytical reference point, $n = 0$, is set by the analyst.
2. Costs, returns, or events in the kth period occur at the *end* of the period. For example, a cost for the 1986 annual period, which starts January 1 and ends December 31, is treated as though it occurs on December 31, 1986.
3. Periods are 1 year long.

In most publications or studies relating to land management, these assumptions are treated as conventions and not stated. They cannot be taken for granted; bankers and realtors, for example, often calculate interest treating the period length as 1 month. As an analyst you can change any of these assumptions. If you do, you simply need to say so.

The Basic Future Value Equation

The basic formula for all compound interest calculations is

$$V_n = V_0(1 + i)^n \tag{7-1}$$

where V_n = value of an amount n periods in the future
V_0 = value of an amount at present ($n = 0$)
n = number of periods over which interest is charged and compounded
i = periodic interest rate stated as a decimal

This formula is called the *future value equation* since V_0, n, and i are known and the future value V_n is unknown and needs to be calculated.

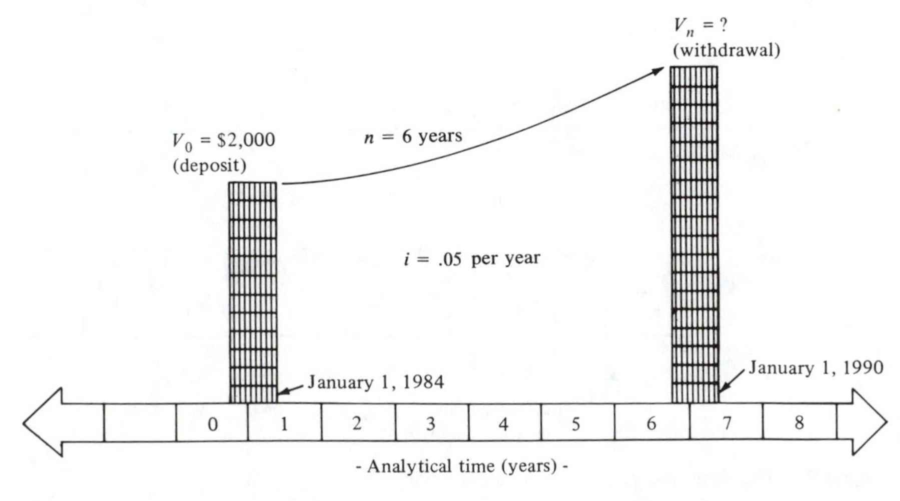

Figure 7-2 A savings account future value problem.

For example, if you put $2,000 (the principal) into a savings account at your bank on January 1, 1984, the bank promised to pay you 5 percent compound annual interest, and you planned to draw out principal and interest after 6 years, on January 1, 1990, how much would you have (Fig. 7-2)? Here $V_0 = 2{,}000$, $n = 6$, and $i = 0.05$. This can be shown on the time line as an event occurring at $n = 0$, and the problem is to calculate what it will be after 6 periods.

The values for V_0, n, and i can be directly substituted into Eq. (7-1) and a solution obtained:

$$\begin{aligned} V_n &= V_0(1 + i)^n \\ &= 2{,}000(1 + 0.05)^6 = 2{,}000(1.34010) = \$2{,}680.19 \end{aligned}$$

Most hand-held calculators handle this calculation easily.* Compound interest tables precalculate and provide values for $(1 + i)^n$ over a wide range of the variables i and n. An abbreviated version of such a table is given in Table 7-1 and can be used if a calculator is not handy. *Be aware that the answers to the problems in this chapter were calculated using a calculator and will differ slightly from answers obtained using the table.* For example, if you look in Table 7-1 to the column where i = 5 percent and read down to the row where $n = 6$, the number 1.34 is given. With this two-place accuracy we get 1.34(2,000), or $2,680.00 as an estimate of V_n in the above problem.

To verify that Eq. (7-1) actually calculates the answer we want, the problem is broken down into a separate interest calculation for each period (Table 7-2). The table shows a series of six calculations using Eq. (7-1) for each period. In the first period the principal of $2,000 earns interest of $100 (0.05 × 2,000), and the sum of principal and interest at the end of the period is $2,100. The sum of principal plus interest is calculated directly as 1.05(2,000) = $2,100. The principal V_0 at the start of the second period is the principal plus interest from the first period, that is, $2,100. Repeating the calculation in period 2 gives a principal plus interest at the end of the period of $2,205 (2,100 × 1.05). The interest earned in the second period is 2205 − 2100, or $105. Note the higher interest earned in period 2. This is the result of "compounding," where interest is earned on the original principal and on the accumulated interest earned in the preceding periods. Here the interest in period 2 is calculated as

Interest on original principal	0.05 × 2000 =	$100.00
Interest on accumulated interest from period 1	0.05 × 100 =	5.00
Total interest in period 2		$105.00

* Business calculators have special function keys or programs to do the calculation. Virtually all scientific calculators have log and y^x functions, which can do the same job.

Table 7-1

Number of periods (years) n	Interest rate per period (year)							
	1%	2%	3%	4%	5%	6%	7%	8%
1	1.01	1.02	1.03	1.04	1.05	1.06	1.07	1.08
2	1.02	1.04	1.06	1.08	1.10	1.12	1.14	1.17
3	1.03	1.06	1.09	1.12	1.16	1.19	1.23	1.26
4	1.04	1.08	1.13	1.17	1.22	1.26	1.31	1.36
5	1.05	1.10	1.16	1.22	1.28	1.34	1.40	1.47
6	1.06	1.13	1.19	1.27	1.34	1.42	1.50	1.59
7	1.07	1.15	1.23	1.32	1.41	1.50	1.61	1.71
8	1.08	1.17	1.27	1.37	1.48	1.59	1.72	1.85
9	1.09	1.20	1.30	1.42	1.55	1.69	1.84	2.00
10	1.10	1.22	1.34	1.48	1.63	1.79	1.97	2.16
11	1.12	1.24	1.38	1.54	1.71	1.90	2.10	2.33
12	1.13	1.27	1.43	1.60	1.80	2.01	2.25	2.52
13	1.14	1.29	1.47	1.67	1.89	2.13	2.41	2.72
14	1.15	1.32	1.51	1.73	1.98	2.26	2.58	2.94
15	1.16	1.35	1.56	1.80	2.08	2.40	2.76	3.17
16	1.17	1.37	1.60	1.87	2.18	2.54	2.95	3.43
17	1.18	1.40	1.65	1.95	2.29	2.69	3.16	3.70
18	1.20	1.43	1.70	2.03	2.41	2.85	3.38	4.00
19	1.21	1.46	1.75	2.11	2.53	3.03	3.62	4.32
20	1.22	1.49	1.81	2.19	2.65	3.21	3.87	4.66
21	1.23	1.52	1.86	2.28	2.79	3.40	4.14	5.03
22	1.24	1.55	1.92	2.37	2.93	3.60	4.43	5.44
23	1.26	1.58	1.97	2.46	3.07	3.82	4.74	5.87
24	1.27	1.61	2.03	2.56	3.23	4.05	5.07	6.34
25	1.28	1.64	2.09	2.67	3.39	4.29	5.43	6.85
26	1.30	1.67	2.16	2.77	3.56	4.55	5.81	7.40
27	1.31	1.71	2.22	2.88	3.73	4.82	6.21	7.99
28	1.32	1.74	2.29	3.00	3.92	5.11	6.65	8.63
29	1.33	1.78	2.36	3.12	4.12	5.42	7.11	9.32
30	1.35	1.81	2.43	3.24	4.32	5.74	7.61	10.06
31	1.36	1.85	2.50	3.37	4.54	6.09	8.15	10.87
32	1.37	1.88	2.58	3.51	4.76	6.45	8.72	11.74
33	1.39	1.92	2.65	3.65	5.00	6.84	9.33	12.68
34	1.40	1.96	2.73	3.79	5.25	7.25	9.98	13.69
35	1.42	2.00	2.81	3.95	5.52	7.69	10.68	14.79
36	1.43	2.04	2.90	4.10	5.79	8.15	11.42	15.97
37	1.45	2.08	2.99	4.27	6.08	8.64	12.22	17.25
38	1.46	2.12	3.07	4.44	6.39	9.15	13.08	18.63
39	1.47	2.16	3.18	4.61	6.70	9.70	13.99	20.12
40	1.49	2.21	3.26	4.80	7.04	10.29	14.97	21.72
50	1.64	2.69	4.38	7.11	11.50	18.40	29.50	46.90
60	1.82	3.28	5.89	10.50	18.70	33.00	57.90	101.30
70	2.01	4.00	7.92	15.60	30.40	59.10	114.00	218.60
80	2.22	4.88	10.64	23.10	49.60	105.80	224.20	471.90
90	2.45	5.94	14.30	34.10	80.70	189.50	441.10	1,019.00
100	2.70	7.24	19.22	50.50	131.50	339.30	867.00	2,199.00

Table 7-1 (Continued)

Number of periods (years)	Interest rate per period (year)							
n	9%	10%	12%	14%	16%	18%	20%	25%
1	1.09	1.10	1.12	1.14	1.16	1.18	1.20	1.25
2	1.19	1.21	1.25	1.30	1.35	1.39	1.44	1.56
3	1.30	1.33	1.40	1.48	1.56	1.64	1.78	1.93
4	1.41	1.46	1.57	1.69	1.81	1.94	2.07	2.44
5	1.54	1.61	1.76	1.93	2.10	2.29	2.49	3.05
6	1.68	1.77	1.97	2.19	2.44	2.70	2.99	3.81
7	1.83	1.95	2.21	2.50	2.83	3.19	3.58	4.77
8	1.99	2.14	2.48	2.85	3.28	3.76	4.30	5.96
9	2.17	2.36	2.77	3.25	3.80	4.44	5.16	7.45
10	2.37	2.59	3.11	3.71	4.41	5.23	6.19	9.31
11	2.58	2.85	3.48	4.23	5.12	6.18	7.48	11.64
12	2.81	3.14	3.90	4.82	5.94	7.23	8.92	14.55
13	3.07	3.45	4.36	5.49	6.89	8.60	10.70	18.19
14	3.34	3.80	4.89	6.26	7.99	10.15	12.84	22.74
15	3.64	4.18	5.47	7.14	9.27	11.97	15.41	28.42
16	3.97	4.59	6.13	8.14	10.75	14.13	18.49	35.53
17	4.33	5.05	6.87	9.28	12.47	16.67	22.19	44.41
18	4.72	5.56	7.69	10.58	14.46	19.67	26.62	55.51
19	5.14	6.12	8.61	12.06	16.77	23.21	31.95	69.39
20	5.60	6.73	9.65	13.74	19.46	27.39	38.34	86.74
21	6.11	7.40	10.80	15.67	22.57	32.32	46.01	108.4
22	6.66	8.14	12.10	17.86	26.19	38.14	55.21	135.5
23	7.26	8.95	13.55	20.36	30.38	45.01	66.25	169.4
24	7.91	9.85	15.18	23.21	35.24	53.11	79.50	211.8
25	8.62	10.83	17.00	26.46	40.87	62.67	95.40	264.7
26	9.40	11.92	19.04	30.17	47.41	73.95	114.5	330.9
27	10.25	13.11	21.32	34.39	55.00	87.26	137.4	413.6
28	11.17	14.42	23.88	39.20	63.80	103.0	164.8	517.0
29	12.17	15.86	26.75	44.69	74.01	121.3	197.8	646.2
30	13.27	17.45	29.96	50.95	85.85	143.4	237.4	807.8
31	14.46	19.19	33.56	58.08	99.30	169.2	284.9	1,010.
32	15.76	21.11	37.58	66.21	115.5	199.6	341.8	1,262.
33	17.18	23.23	42.09	75.48	134.0	235.6	410.2	1,577.
34	18.73	25.55	47.14	86.05	155.0	278.0	492.2	1,972.
35	20.41	28.10	52.89	98.10	180.0	328.0	590.1	2,465.
36	22.25	30.91	59.14	111.83	209.0	387.0	709.0	3,081.
37	24.25	34.00	66.23	122.5	242.0	456.7	850.0	3,852.
38	26.44	37.40	74.18	145.3	281.5	538.9	1,020.	4,815.
39	28.82	41.14	83.08	165.7	326.5	635.0	1,225.	6,019.
40	31.14	45.26	93.05	188.9	375.7	736.4	1,470.	7,523.
50	74.30	118.0	289.0	700.0	1,670.	3,927.	9,100.	—
60	176.0	305.0	898.0	2,565.	7,370.	—	—	—
70	417.0	790.0	2,787.	9,623.	—	—	—	—
80	986.0	2,048.	—	—	—	—	—	—
90	2,335.	5,313.	—	—	—	—	—	—
100	5,529.	—	—	—	—	—	—	—

Table 7-2 Interest calculation for each period of six consecutive periods

Period n (1)	Principal at start of period V_0, $V_0 = V_{n-1}$ (2)	Interest factor, $(1 + i)$ (3)	Principal plus interest at end of period, $V_n = V_0(1 + i)$ (4)	Interest earned over period, $(V_n - V_0)$ (5) = (4) − (2)	Accumulated interest earned, $\Sigma(V_n - V_0)_t$, $n = 1, \ldots, 6$ (6)
1	$2,000	1.05	$2,100	$100	$100
2	2,100	1.05	2,205	105	205
3	2,205	1.05	2,315.25	110.25	315.25
4	2,315.25	1.05	2,431.01	115.76	431.01
5	2,431.01	1.05	2,552.56	121.55	552.56
6	2,552.56	1.05	2,680.19	127.63	680.19

After six periods of calculation, the exact total of principal plus interest is $2,680.19, which is the same amount determined by our formula solution to the problem. So the formula works. An intuitive mathematical proof is found by substituting the symbols $V_0(1 + i)$ for the principal and interest in Table 7-2 (see Table 7-3). The principal plus interest at the end of six periods is the initial principal V_0 multiplied by the factor $(1 + i)$ six times,

$$V_n = V_0(1 + i)(1 + i)(1 + i)(1 + i)(1 + i)(1 + i)$$

which simplifies to

$$V_n = V_0(1 + i)^6$$

In general for n periods the formula is

$$V_n = V_0(1 + i)^n$$

which is the original future value equation (7-1).

The future value equation contains four variables, V_n, V_0, i, and n. Equation (7-1) is the solution of the equation for V_n in terms of the remaining three variables. By algebraic manipulation of Eq. (7-1), two additional and important equations are established and used to solve for V_0 (present value) or i (earnings rate).

Present Value Equation

Starting with Eq. (7-1) and solving to get V_0 in terms of V_n, i, and n, we proceed:

$$V_n = V_0(1 + i)^n$$

Table 7-3 Symbolic representation of future value interest calculations for six consecutive periods

Period n	Principal at start of period	Interest factor	Principal plus interest at end of period
1	V_0	$(1 + i)$	$V_0(1 + i)$
2	$V_0(1 + i)$	$(1 + i)$	$V_0(1 + i)(1 + i)$
3	$V_0(1 + i)(1 + i)$	$(1 + i)$	$V_0(1 + i)(1 + i)(1 + i)$
4	$V_0(1 + i)(1 + i)(1 + i)$	$(1 + i)$	$V_0(1 + i)(1 + i)(1 + i)(1 + i)$
5	$V_0(1 + i)(1 + i)(1 + i)(1 + i)$	$(1 + i)$	$V_0(1 + i)(1 + i)(1 + i)(1 + i)(1 + i)$
6	$V_0(1 + i)(1 + i)(1 + i)(1 + i)(1 + i)$	$(1 + i)$	$V_0(1 + i)(1 + i)(1 + i)(1 + i)(1 + i)(1 + i) = V_0(1 + i)^6$

rearranging we obtain

$$V_0(1 + i)^n = V_n$$

or

$$V_0 = \frac{V_n}{(1 + i)^n} \tag{7-2}$$

The present value equation is used to determine the value today of some future payment or event. Because the future amount V_n is divided by the factor $(1 + i)^n$, we say the future value is discounted to the present. Suppose that we expect to harvest a pine plantation 25 years from now, producing a revenue of \$750 per acre. If the guiding interest rate is 8 percent, how much is this future harvest revenue worth today (Fig. 7-3)? Our problem, as shown on the time line, is to establish the present value. The known variables are V_n = \$750, n = 25, and i = 0.08. Substituting

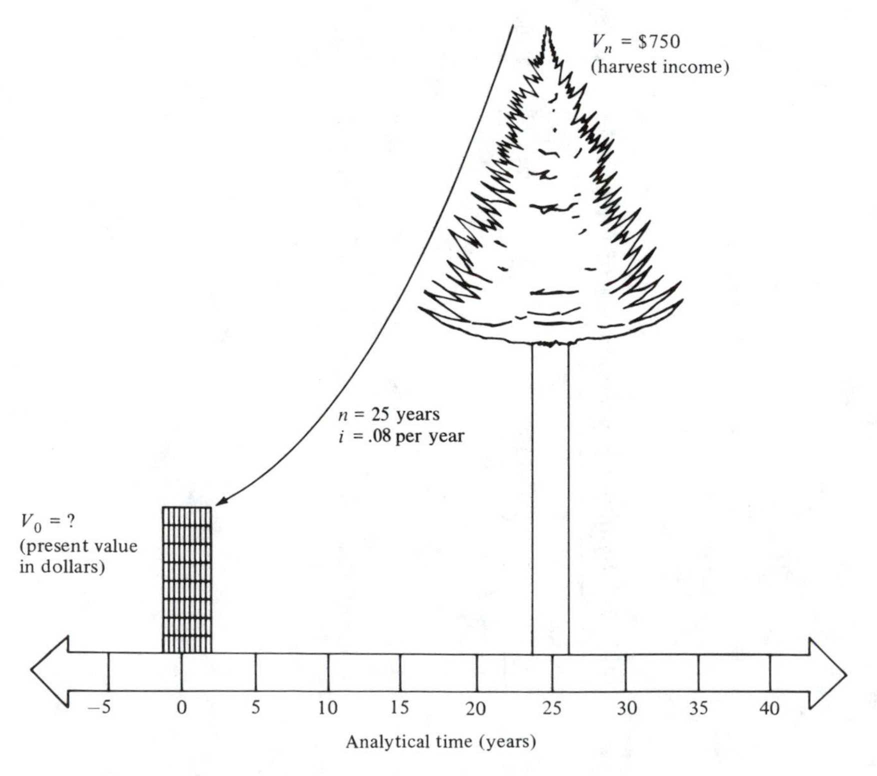

Figure 7-3 Present value of future timber harvest.

these into the present value equation, the problem is solved:

$$V_0 = \frac{V_n}{(1 + i)^n} = \frac{750}{(1.08)^{25}} = \frac{750}{6.848}$$
$$= \$109.51$$

The future \$750 return from the plantation is only worth \$109.51 today. This shrinkage of future incomes may at first seem disconcerting, incorrect, or unfair. But this is how people see the future, and they behave accordingly. If you had \$109.51 cash in hand today that could be invested at 8 percent, 25 years from now it would be worth \$750 ($V_n = 109.51 \times 6.85 = \$750$). Given the choice of \$109.51 today and \$750 in 25 years from now, presumably you would be indifferent as long as 8 percent correctly measures the rate at which you discount the future.

Earning Rate Equation

The third important variation of the basic equation is to solve for the interest rate i in terms of V_n, V_0, and n as follows:

$$V_n = V_0(1 + i)^n$$

$$V_0(1 + i)^n = V_n$$

$$(1 + i)^n = \frac{V_n}{V_0} \tag{7-3}$$

$$i = \sqrt[n]{\frac{V_n}{V_0}} - 1 \tag{7-3a}$$

By using the compound interest formula, values of i can be determined if V_n, V_0, and n are known. For example, assume you invested \$1,000 today to buy 100 shares of common stock in an oil exploration company at \$10 per share. After 10 years, the price of stock rose to \$45 per share and you sell out, collecting \$4,500. What is the earning rate of this investment (Fig. 7-4)? For this problem the known values are $V_n = \$4{,}500$, $V_0 = \$1{,}000$, and $n = 10$. Substituting into Eqs. (7-3) and (7-3a), we have

$$(1 + i)^{10} = \frac{4{,}500}{1{,}000} = 4.500$$

$$i = \sqrt[10]{4.5} - 1 = 0.162308$$

The interest rate i can be found using Eq. (7-3) by going to Table 7-1. Read across the tenth row ($n = 10$) to the interest rate column, where the value 4.5 is found. 4.41 is the value for 16 percent and 5.23 is the value for

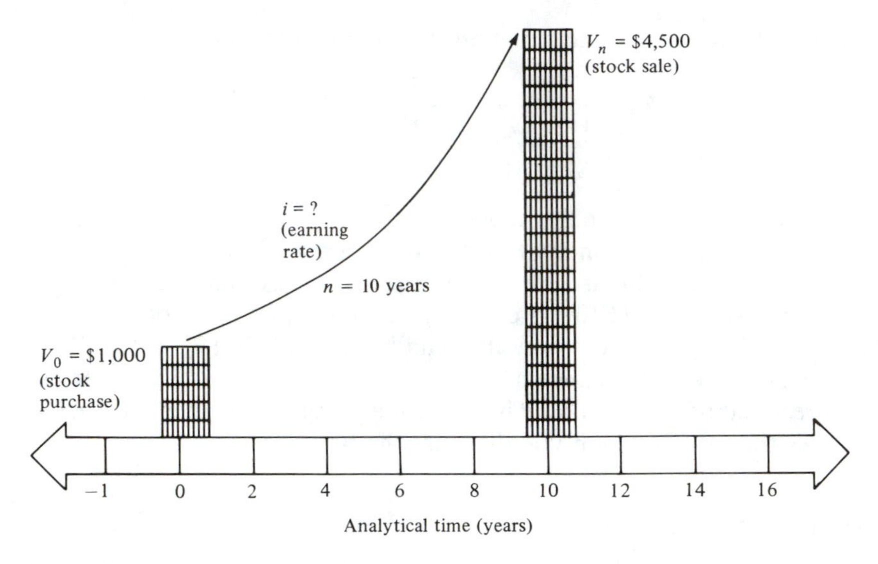

Figure 7-4 Earning rate on stock appreciation.

18 percent. Interpolation suggests about 16.1 percent. On a calculator you would evaluate $(\sqrt[10]{4.5} - 1)$ in Eq. (7-3*a*) directly.

For this problem $i = 0.162308$. Hence the interpretation is that you earned about 16 percent per year on your stock investment. Later we will calculate interest or earning rates for alternative investment opportunities.

Equation (7-3) is also used to solve for n when V_n, V_0, and i are known. For example, how long would you have to wait before an initial investment increased 10 times when the interest rate is 12 percent?

Here we can assume the initial investment is \$1 and the equation variables are V_0 = \$1, V_n = \$10 (tenfold increase over V_0), and $i = 0.12$. Substituting into Eq. (7-3), we have

$$(1 + 0.12)^n = \frac{10}{1} = 10$$

Examining Table 7-1 and reading down the 12 percent interest column until the value 10.0 is encountered, we find it takes about 20 years for an investment to increase tenfold.

Periodic Payments of Equal Amounts

The three basic equations for future value, present value, and earnings rate deal only with a single payment or event. Many times we are interested in problems where there is an equal payment or revenue at regular intervals over the time periods considered in an analysis. For example, forest managers need to calculate the present or future value of equal annual payments for taxes, protection, and basic management, a logger wants to set up a schedule of equal depreciation payments so that when his tractor-skidder wears out there will be a capital fund sufficient to purchase a replacement, or a forest owner may want to evaluate the present value of the income received from a regular, equal value forest harvest from a property every 5 years. Such a series of equal periodic payments could be evaluated by tediously applying Eq. (7-1) or (7-2) successively to each payment in the series and then adding to get the total result. Fortunately, some equations can be derived that collapse all of the individual payment calculations into one.

The timing of periodic payments can be defined by three parameters (n, j, and w) which distribute the payments of amount a by dividing the n periods of analysis into j multiperiod groups of length w. Three different payment series are shown in Table 7-4 to illustrate how these parameters are used to define the series. Series A is a payment of amount $200 made

Table 7-4 Three series of periodic payments of equal amounts

Period	Payment series A	Payment series B	Payment series C
1	—	—	500
2	—	3,000	500
3	—	—	500
4	200	3,000	500
5	—	—	500
6	—	3,000	500
7	—	—	500
8	200	3,000	500
9	—	—	500
10	—	3,000	500
11	—	—	500
12	200	3,000	500
Series parameters			
Number of periods (n)	12	12	12
Length of multiperiod groups in periods (w)	4	2	1
Number of multiperiod groups (and number of payments) (j)	3	6	12
Amount of payment (a)	$200	$ 3,000	$ 500
The guiding rate is 8 percent per period			
Future value of the series [Eq. (7-4)]	$842.28	$27,370	$9,488.56
Present value of the series [Eq. (7-7)]	$334.48	$10,869	$3,767.62

at the end of every fourth period over the 12 periods in the analysis. Here $n = 12$, $w = 4$, and $j = 3$. Note that by definition $n = jw$.

If we wanted to calculate the future value of the payments in series A we could use Eq. (7-1) to compute the future value of the three terms in the series. Assume that the length of a period is 1 year and the guiding rate is 8 percent.

Payment item (V_0)	$(1 + i)^n$ factor	Factor value	Future value at end of period 12, $V_n = V_0(1 + i)^n$
\$200, period 4	$(1 + 0.08)^8$	1.8509	\$370.186
\$200, period 8	$(1 + 0.08)^4$	1.3604	\$272.098
\$200, period 12	$(1 + 0.08)^0$	1.0000	\$200.000
Total future value			\$842.28

Future value of a terminating periodic series The above example calculation can be used to help derive a more general formula. Let a be the amount of the equal payment. Then using the parameters n, j, and w defined above and continuing to assume, as we have throughout, that payments occur at the end of a period, we can describe the items in the calculation as

$$V_n = a(1 + i)^{(j-1)w} + a(1 + i)^{(j-2)w} + \cdots + a(1 + i)^w + a$$

The series of three \$200 payments at 4 period intervals ($a = 200$, $n = 12$, $j = 3$, $w = 4$) in our example can be substituted into this general formulation to get three terms which are summed:

$$\begin{aligned} V_n &= 200(1.08)^{(3-1)4} + 200(1.08)^{(3-2)4} + 200(1.08)^{(3-3)4} \\ &= 200(1.08)^8 + 200(1.08)^4 + 200(1.08)^0 \\ &= 370.186 + 272.098 + 200 = \$842.28 \end{aligned}$$

Which gives the same \$842.28 result as the series above. The general equal perioding payment series equation above can be simplified using some algebraic manipulation. First we multiply both sides of the equation by the factor $(1 + i)^w$

$$V_n(1 + i)^w = a(1 + i)^{jw} + a(1 + i)^{(j-1)w} + a(1 + i)^{(j-2)w} + \cdots + a(1 + i)^w$$

Second, we subtract the original equation from this equation:

$$\begin{array}{lllll} V_n(1 + i)^w & = a(1 + i)^{jw} + a(1 + i)^{(j-1)w} & + a(1 + i)^{(j-2)w} & + \cdots + a(1 + i)^w & \\ \quad - V_n = & \quad - a(1 + i)^{(j-1)w} & - a(1 + i)^{(j-2)w} & - \cdots - a(1 + i)^w & - a \\ \hline V_n(1 + i)^w - V_n & = a(1 + i)^{jw} + 0 & + 0 & + \cdots + 0 & - a \end{array}$$

The remainder can be simplified and solved for V_n

$$V_n[(1 + i)^w - 1] = a[(1 + i)^{jw} - 1]$$

$$V_n = a\,\frac{(1 + i)^{jw} - 1}{(1 + i)^w - 1}$$

Since $jw = n$

$$V_n = a\,\frac{(1 + i)^n - 1}{(1 + i)^w - 1} \tag{7-4}$$

Let's use this simplified Eq. (7-4) to calculate our three \$200 payment problem

$$V_n = 200\,\frac{(1.08)^{12} - 1}{(1.08)^4 - 1}$$

$$= 200\,\frac{1.518}{0.360} = \$842.28$$

Future value of an every-period series A frequently used variation of Eq. (7-4) is established when the multiperiod group length is one period (i.e., $w = 1$) and payments occur every period

$$V_n = a\,\frac{(1 + i)^n - 1}{(1 + i)^1 - 1}$$

$$= a\,\frac{(1 + i)^n - 1}{i} \tag{7-5}$$

Suppose a 10-year contract called for a payment of \$5 per acre per year for erosion monitoring. Figure 7-5 depicts this problem when the guiding rate is 6 percent. The total of payments, with interest, is calculated using Eq. (7-5) as $\$5[(1.06)^{10} - 1]/0.06$ or \$65.90. For another example suppose the every-period payments of series C in Table 7-4 represented annual savings payments set aside by a commercial firewood cutter to pay for a hydraulic wood splitter. The savings account pays 8 percent interest per year. Assuming a period length of 1 year, we can use Eq. (7-5) to calculate the amount of money the woodcutter would have available to purchase the splitter at the end of period 12

$$V_n = \$500\,\frac{(1.08)^{12} - 1}{0.08} = \$9{,}487.50$$

The sinking fund formula The hydraulic splitter was estimated to cost \$16,500 when purchased 12 years from now and the woodcutter realized that more would have to be saved each year. The exact amount to save

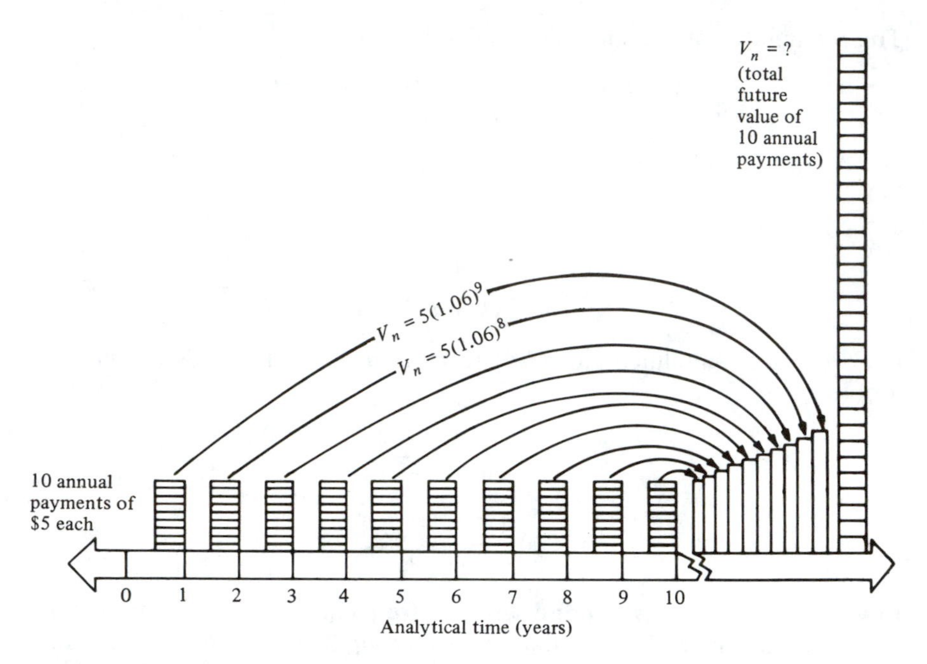

Figure 7-5 Future value of equal annual payments (guiding rate = 6 percent per year).

can be determined by revising Eq. (7-5) to solve for the payment a when the desired future value V_n is known.

$$a = V_n \frac{i}{(1 + i)^n - 1} \tag{7-6}$$

The woodcutter can directly use Eq. (7-6) to determine the needed annual payment into his savings account.

$$a = 16{,}500 \frac{0.08}{(1.08)^{12} - 1} = \$869.47$$

Present value of a terminating periodic series To obtain a set of present value equations for series payments we make use of the present value equation (7-2) and the periodic series future value equation (7-4) above. By Eq. (7-2) we know that

$$V_0 = \frac{V_n}{(1 + i)^n}$$

and by Eq. (7-4) we know that

$$V_n = a \frac{(1 + i)^n - 1}{(1 + i)^w - 1}$$

substituting the definition for V_n from (7-4) into Eq. (7-2) we obtain

$$V_0 = \frac{a \dfrac{(1+i)^n - 1}{(1+i)^w - 1}}{(1+i)^n}$$

Then simplifying,

$$V_0 = a \frac{(1+i)^n - 1}{[(1+i)^w - 1](1+i)^n} \tag{7-7}$$

Suppose the series B payments in Table 7-4 represented a $3,000 payment made every other year to a landowner by a timber company who had use of her land for the next 12 years. The landowner needed an estimate of the present value of this lease at a guiding rate of 8 percent to prepare a personal net worth statement. Equation (7-7) serves the purpose nicely:

$$V_0 = 3{,}000 \frac{(1.08)^{12} - 1}{[(1.08)^2 - 1](1.08)^{12}} = \$10{,}869$$

Present value of a terminating every period series When the payments occur every period ($w = 1$), Eq. (7-7) simplifies to

$$V_0 = a \frac{(1+i)^n - 1}{i(1+i)^n} \tag{7-8}$$

Suppose a tree farmer planned to plant an area and expected to pay an annual tax, fire protection, and management bill of $4.75 per acre for the 40 years she planned to hold the stand before harvesting the timber and starting a new crop. Her guiding rate is 10 percent per year. As part of the analysis to evaluate this plan she needed to know the present value of the annual costs. Equation (7-8) can calculate the needed information.

$$V_0 = \$4.75 \frac{(1.10)^{40} - 1}{0.10(1.10)^{40}} = \$46.45 \text{ per acre}$$

Amount of an installment payment A very common business and personal problem is installment buying, where an item is purchased for a given amount and must be paid off in a specified number of equal periodic payments at a stated interest rate. An equation for determining the amount of the periodic payment is derived directly from Eq. (7-8) by solving for a in terms of V_0, i, and n. The resulting equation is

$$a = V_0 \left[\frac{i(1+i)^n}{(1+i)^n - 1}\right] \tag{7-9}$$

For example, suppose a tree farmer considered purchasing a rubber-tired feller-buncher-skidder machine to commercially thin and selectively harvest his lands. The machine would reduce the damage to residual trees, give him work in his spare time, and by selling logs at roadside, presumably increase his income from the tree farming business. A critical part of evaluating, this option involved the amount of the annual payment for the machine, which must be financed from timber sale income.

The machine sells for $30,000 and the dealer will finance it for 5 years at an annual interest charge of 12 percent. The known variables are $V_0 =$ $30,000, $n = 5$ years, $i = 12$ percent, and $(1 + i)^5 = 1.76234$. Substituting into Eq. (7-9) and solving for the annual payment a,

$$a = V_0 \left[\frac{i(1 + i)^n}{(1 + i)^n - 1} \right]$$

$$= 30{,}000 \left[\frac{(0.12)(1.76234)}{1.76234 - 1} \right] = \frac{6{,}344.32}{0.76234} = \$8{,}322.30 \text{ per year}$$

So as part of the project evaluation, the tree farmer needs to determine if the additional revenues each year for at least 5 years will equal or exceed the payments of $8,322. Otherwise he would have a cash flow problem when the cash income in some years is not adequate to cover payments.

Present value of a perpetual periodic series If the series of payments is expected to continue indefinitely for an infinite number of periods, Eq. (7-7) can be simplified as follows:

$$V_0 = a \frac{(1 + i)^n - 1}{[(1 + i)^w - 1](1 + i)^n} = a \frac{(1 + i)^\infty - 1}{[(1 + i)^w - 1](1 + i)^\infty}$$

the ratio of the rightmost factors $[(1 + i)^n - 1]/(1 + i)^n$ approaches the value of 1.0 as the value of n approaches the value of infinity, giving the perpetual series equation

$$V_0 = a \frac{1}{(1 + i)^w - 1} \tag{7-10}$$

We use Eq. (7-10) extensively later in this chapter and in chap. 13 and call it the *soil expectation value* (SEV) where the payment a is interpreted to be a constant net periodic revenue from timber harvest received in perpetuity at intervals of w periods with a period length of 1 year. For example, if a tract of land was expected to provide a net income of $5,000 to the owner every 30 years in perpetuity starting in 30 years and a guiding rate of 6 percent per year were appropriate, the present value of this series of incomes is

$$V_0 = 5{,}000 \frac{1}{(1.06)^{30} - 1} = \$1{,}054.07$$

Present value of a perpetual every-period series Equation (7-10) also can be further simplified for the case when $w = 1$ and the payments occur every period in perpetuity.

$$V_0 = a \frac{1}{(1 + i)^1 - 1}$$

or

$$V_0 = \frac{a}{i} \tag{7-11}$$

To illustrate, a tree farm enterprise was designed to yield a stable annual income of $25,000 forever. The owner's guiding rate is 10 percent. The present value of this income stream is therefore

$$V_0 = \frac{25{,}000}{0.10} = \$250{,}000$$

Series payments starting at the beginning of each period So far we have assumed that the payments in a series occurs at the end of each multiperiod group (w) in the analysis. If instead the payments occur at the beginning of each multiperiod group, the basic series formulas (7-4) and (7-7) need to be multiplied by $(1 + i)^w$:

Future value:

$$V_n = [(1 + i)^w]a \frac{(1 + i)^n - 1}{(1 + i)^w - 1} \tag{7-4a}$$

Present value

$$V_0 = [(1 + i)^w]a \frac{(1 + i)^n - 1}{[(1 + i)^w - 1](1 + i)^n} \tag{7-7a}$$

To illustrate these modifications, if payments were at the beginning of the multiperiod groups, the series A payments in Table 7-4 would now be three payments of $200 paid at the beginning of periods 1, 5, and 9. To calculate the present value directly, these payments would be discounted 0, 4, and 8 years, respectively, using a guiding rate of 8 percent to calculate a present value of 200 + 147.00 + 108.05 = $455.05.

Using Eq. (7-7a) we have

$$V_0 = [(1.08)^4]200 \frac{(1.08)^{12} - 1}{[(1.08)^4 - 1](1.08)^{12}} = \$455.05$$

which is the same result. The future value is similarly calculated using Eq. (7-4a) as

$$V_n = [(1.08)^4]200 \frac{(1.08)^{12} - 1}{(1.08)^4 - 1} = \$1{,}145.92$$

A Decision Tree for Choosing Present or Future Value Equations

Now that the main formulas have been derived and illustrated, the user simply has to choose the correct one in a particular situation. A decision tree covering the most common situations fits this need well. By working through the tree, the appropriate interest formula can be found. To use the tree in Fig. 7-6, the user makes three sequential choices to characterize the problem to be solved.

1. Is it a single sum or series payment?
2. Is it a perpetual or terminal series?
3. Is the present or future value desired?

Adjusting for Inflation

The market interest rate was described as representing three factors: the pure earnings rate, the inflation rate, and the risk rate. If there is no inflation, the combination of the pure earnings rate and the risk rate is called the *real rate of interest*. It is used to characterize investments in

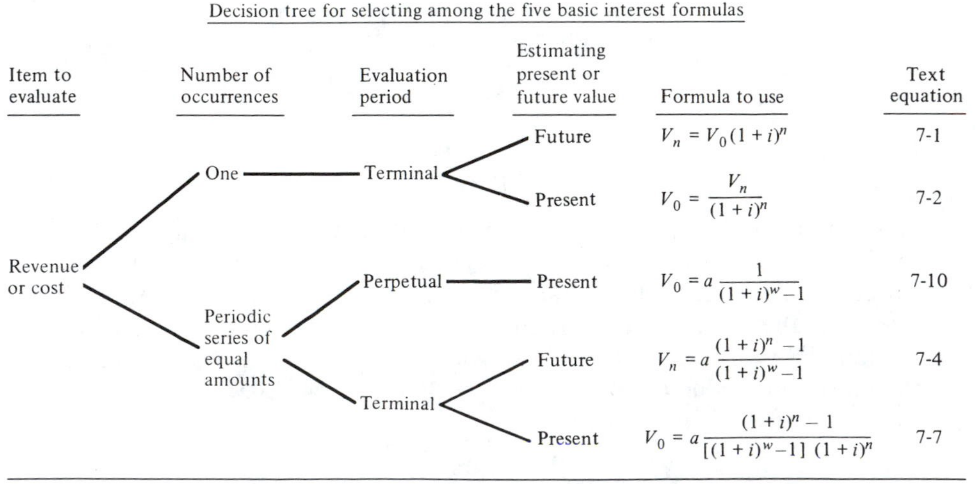

Figure 7-6 Decision tree for choosing present and future value equations.

terms of current or constant dollars. Let

r = real earnings rate expected for projects of a given risk class (guiding rate without inflation)
k = average rate of inflation
V_n = future return on a project measured in today's market prices, or deflated dollars
V_n^* = future return on a project measured at actual future market prices, or inflated dollars
$(1 + r)^n$ = discount factor for future returns measured in current dollars
$(1 + r)^n(1 + k)^n$ = discount factor for future returns measured in inflated dollars, $=(1 + r + k + rk)^n = [(1 + r)(1 + k)]$
n = number of periods

For a project whose future return V_n is measured in terms of current (noninflated) dollars the present value of the return is appropriately measured by using the real rate of return r as the guiding rate for discounting,

$$\text{Present value } V_0 = \frac{V_n}{(1 + r)^n}$$

If, on the other hand, the future return is measured in future inflated dollars, the dollar value of the inflated return is

$$V_n^* = V_n(1 + k)^n$$

But if inflation is included in the numerical amount of future returns, the discount rate must also be adjusted to account for inflation by using the expanded discount factor $(1 + r)^n(1 + k)^n$. The present value of the future return measured in inflated dollars is then

$$\text{Present value} = \frac{V_n^*}{(1 + r)^n(1 + k)^n} = \frac{V_n(1 + k)^n}{(1 + r)^n(1 + k)^n}$$

Because the inflation factor $(1 + k)^n$ occurs in both the numerator and the denominator, it cancels, leaving our initial present value estimate:

$$\text{Present value} = \frac{V_n}{(1 + r)^n}$$

The important implication is that future revenues and costs can be estimated by their current values and do not need to be inflated in an analysis as long as the discount rate does not include the inflation factor.

To illustrate we use the example presented by Klemperer (1979). Assume a project involves a future yield of 20 M bd ft of timber which sells at today's prices for \$500 per M bd ft for a total value of \$10,000. A real earnings rate of 6 percent is considered to be the guiding rate, and inflation is projected to be a 5 percent compound rate. In current dollars without inflation, the present value of the future yield is

$$\frac{\$10{,}000}{(1.06)^{10}} = \$5{,}583.95$$

If the 5 percent inflation is counted in, the 20 M bd ft will sell for $\$10{,}000(1.05)^{10}$, or \$16,288.95 in 10 years. The discount factor after adjustment for inflation is

$$(1 + r + k + rk) = 1 + 0.06 + 0.05 + 0.06(0.05) = 1.113$$

and the present value of the inflated future revenue is

$$\frac{16{,}288.05}{(1.113)^{10}} = \$5{,}583.95$$

which is the same as that calculated above. Klemperer goes on to demonstrate that if income taxes are considered, the inflation effect does not fully cancel and that after tax present values diminish with increasing inflation rates.

Real Changes in Prices or Costs

If the cost or revenue item in question is expected to persistently inflate (or deflate) in price at a rate that differs from the average inflation rate of all prices, then it should be considered in the analysis. We say that such a difference is a *real rate of price increase* (or *decrease*). To incorporate real rates of price changes, let

V_0 = present value
r = real earnings rate expected for projects of a given risk class
k = average inflation rate of all prices
p = price inflation rate of cost or revenue item under analysis
h = real rate of price increase (or decrease)
R_t = a revenue received in year t measured in current prices
C_t = a cost incurred in year t measured in current prices

When allowing for relative price inflation, the present value is calculated as

$$V_0 = \frac{R_t(1 + p)^t}{(1 + r)^t(1 + k)^t} \qquad (7\text{-}12)$$

for a revenue and as

$$V_0 = \frac{C_t(1 + p)^t}{(1 + r)^t(1 + k)^t}$$

for a cost. We often want to discuss or evaluate the real rate of price change h, which can be determined by evaluating for h

$$(1 + h)^t = \frac{(1 + p)^t}{(1 + k)^t} = \left(\frac{1 + p}{1 + k}\right)^t$$

Hence $V_0 = [V_t(1 + h)^t]/(1 + r)^t$, where $V_t = R_t$ or C_t.

Timber, for example, has had a long history of a real price inflation h at an annual rate of about 1.5 percent as compared to the average of all wholesale prices (Chap. 1). Recently energy and minerals have been showing a strong real rate of inflation. Nevertheless, the future is always a matter of forecast, speculation, or guess. Each analyst has to decide explicitly how to deal with inflation.

A big mistake is to use current prices to estimate future revenues and costs and then discount by a market interest rate that includes a healthy inflation factor to calculate present net worth. This could seriously understate the project's true value. On the other hand, using inflated prices and discounting by real rates greatly overstates the project's value. Clearly the inflation assumption embodied in prices, costs, and interest rates must be the same for correct analysis (Row et al., 1981).

Using Eq. (7-12) to evaluate forestry projects helps the analyst avoid mistakes since it forces the inflation assumptions for each cost or revenue item to be explicit.

Effective Rates and Nominal Rates

When compounding occurs monthly, quarterly, or for any period length other than 1 year, the interest rate and the payment calculations must be adjusted appropriately. Still, we need a common period length to make comparisons. Federal law now requires that all commercial interest rates be stated on an annual rate basis so that consumers can make meaningful comparisons.

The *effective rate* is defined as the annual rate equivalent to a rate charged at some other interval.

The equation for calculating the effective rate is

$$\alpha = (1 + \phi)^m - 1 \tag{7-13}$$

where α = effective annual rate
ϕ = interest rate charged for periods other than 1 year
m = number of periods interest is compounded per year

For example, if the tree farmer buying the feller-buncher-skidder machine was charged interest at a rate of 1.5 percent per month, what is the effective rate he pays? The known variables are $\phi = 0.015$ and $m = 12$ months per year. Then

$$\alpha = (1 + 0.015)^{12} - 1 = 1.1956 - 1 = 0.1956$$

So the effective annual rate is 19.56 percent. This is higher than simply multiplying 1.5 percent by 12 months (18 percent) because of the compounding effect from month to month within the year.

Nominal rate is the term used to describe a periodic rate obtained by dividing an annual rate of interest by the number of periods within the year,

$$\beta = \frac{i}{m} \tag{7-14}$$

where β = nominal rate
i = annual (stated) rate
m = number of periods within the year

The effective and nominal rates often are linked together when the nominal rate β is used to determine the periodic rate ϕ. If $\phi = \beta$, substituting the value for β from Eq. (7-14) into the equation for the effective rate (7-13) provides an equation for calculating the effective rate in terms of the nominal rate,

$$\alpha = \left(1 + \frac{i}{m}\right)^m - 1 \tag{7-15}$$

with α, i, and m defined as before.

To illustrate, the tree farmer originally was able to finance the feller-buncher-skidder at 12 percent annual rate. If the financing was set up on a monthly basis using the nominal rate, the effective annual rate using Eq. (7-15) is

$$\alpha = \left(1 + \frac{0.12}{12}\right)^{12} - 1 = 1.1268 - 1 = 0.1268$$

Hence

$$\alpha = 12.68 \text{ percent}$$

If the financing was set up on a half-month basis, the effective rate is

$$\alpha = \left(1 + \frac{0.12}{24}\right)^{24} - 1 = 0.1272$$

As shown, the effective rate starts to level off as the number of compoundings per year increases. It reaches a limit at continuous compounding defined by the equation

$$\alpha = e^i - 1$$

where e is the base of natural logarithms, i is the annual rate, and α is the effective rate of interest. Here, continuous compounding results in an effective rate of

$$\alpha = e^{(0.12)} - 1 = 0.1275$$

FINANCIAL ANALYSIS

The procedures and issues of financial analysis are now examined in some depth, always viewing it as a process of providing economic information about individual projects, plans, or policies to a decision maker. The use of such financial information in decision analysis is deferred until Chap. 8, where project acceptance or rejection, project ranking, capital budgeting, and benefit-cost analysis are developed. Financial analysis is also basic to many valuation and appraisal procedures. For example, a prospective buyer of any business, be it a tree farm or a hot dog stand, typically forecasts how the business will be handled and estimates the expected flow of revenues and costs—the cash flow schedule. Reduced to present value terms, the present value of costs and revenues along with a knowledge of the variability in the actual cash flow schedule provide guidance to the amount of money that could be paid for the business while still making a reasonable return on the investment. In Chaps. 10 and 11 financial analysis is used to establish the values of stumpage and the tree, and in Chap. 13 it provides the foundation for economic optimization of timber stand management.

Steps in Financial Analysis

Every individual project, plan, or policy considered involves a schedule of events or actions occurring over some period of time. Some of these events incur costs that will be charged against the decision maker and some produce revenues that will be received by the decision maker. Financial analysis requires four procedural steps: (1) deciding on the length of the planning period over which costs and revenues will be evaluated, (2) identifying the schedule of events associated with a project, (3) converting the events to their equivalent schedule of dollar-measured costs and revenues, and (4) adjusting the costs and revenues for time, using compound interest formulas.

A project to implement a typical timber management prescription serves to illustrate what is encompassed by financial analysis. Consider planting a tract of land, performing some cultural treatments on the immature stand, and finally harvesting the mature timber. On the time line, the first two steps of financial analysis produce the event schedule of this project as shown in Fig. 7-7.

Since only activities or events occurring up to year 50 in the future have been included, we have also answered step 1 and decided that the

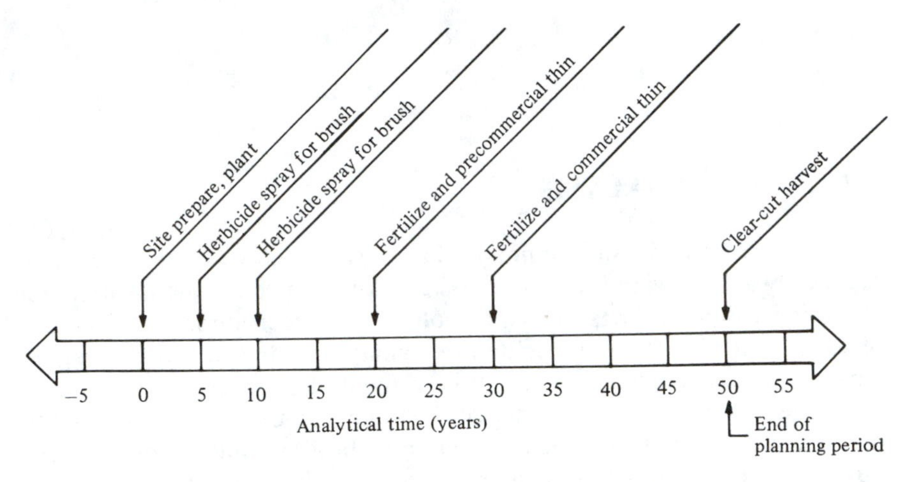

Figure 7-7 Schedule of project events for a silvicultural prescription.

planning period for this project evaluation is 50 years. More events will surely occur on the land after this, but by selecting 50 years as the planning period, they are effectively assigned a value of zero. Hence the length of the planning period is an important decision in the analysis.

By consulting appropriate data sources pricing labor, material, and yield, the third step estimates an equivalent schedule of costs and revenues for the project event schedule (Fig. 7-8). At this step the analytical decisions about price inflation rates and guiding interest rates are also made. For a private individual or corporation, the cash flow schedule can be calculated after all taxes and other, individualized, cost and revenue adjustments have been made.

In this analysis the decision maker is starting treatment after the land has been cutover but before any site preparation and planting funds have been committed. He or she might consider undertaking the proposed project, leaving the land fallow, or doing something else with it. The present value (or the future value) of the cost and revenue stream would be relevant information for this decision. To calculate the present value at time $n = 0$ for the fourth step of financial analysis, the guiding rate obtained from the decision maker is used to discount each cost or revenue item using the present value equation (7-2) (Fig. 7-9).

The four procedural steps culminating in the information presented in Fig. 7-9 are what is commonly called *financial analysis*. Financial analysis is not in itself a decision-making technique. It is really the equivalent of the mensurationist describing the physical consequences of a particular silvicultural system in terms of wood fiber production. In financial analysis we are just making a dollar-measured description of the consequences of a particular project, usually summarized at the reference time of the

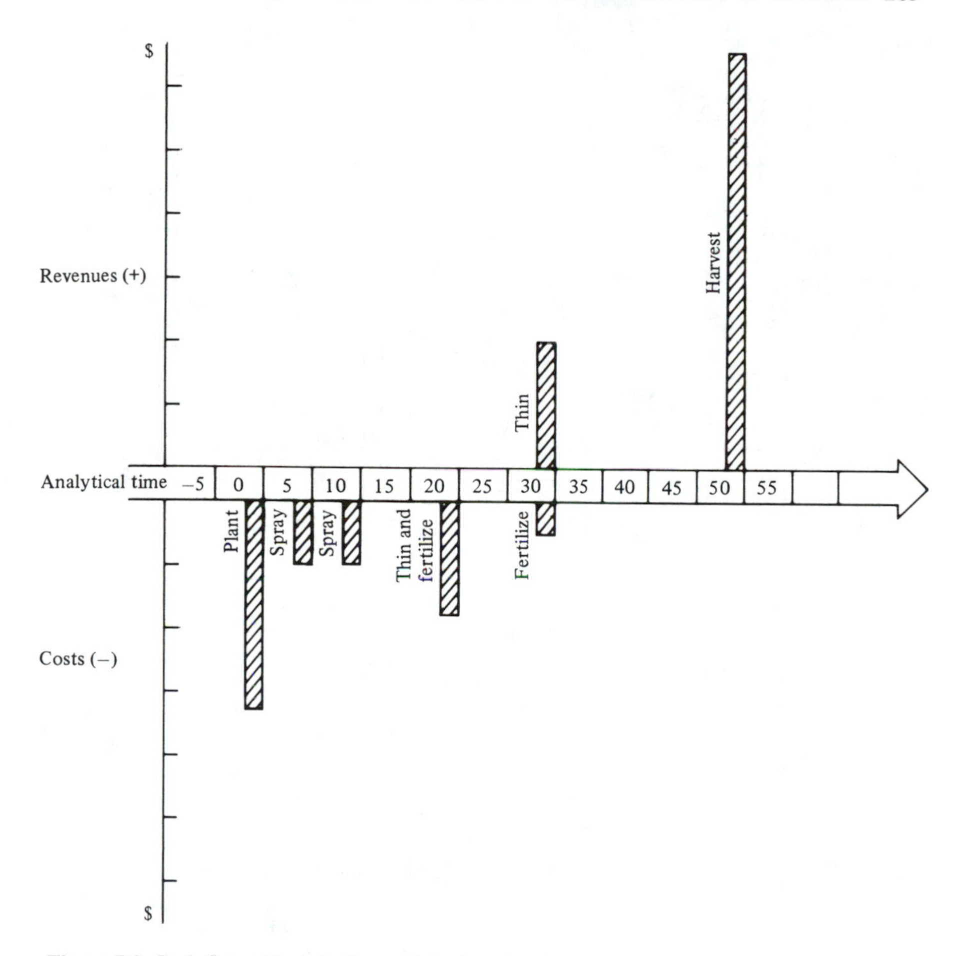

Figure 7-8 Cash flow schedule for a silvicultural prescription.

present. The decision maker still must consider the project in the context of his other goals and constraints, and must consider any other alternative projects before deciding whether this particular project should be undertaken.

Issues in Financial Analysis

The concept of financial analysis is straightforward enough. The difficulty lies in doing it correctly. Concerns about the correctness of an analysis center on the following.

1. Have the future physical events of the project been adequately forecasted with respect to magnitude and timing?

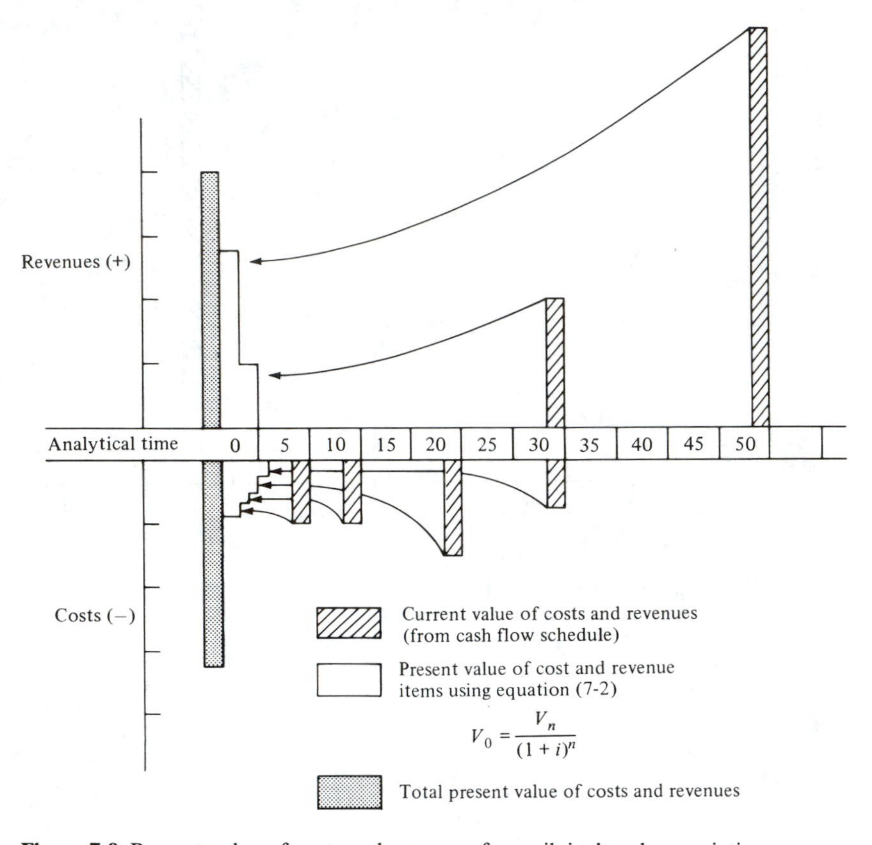

Figure 7-9 Present value of costs and revenues for a silvicultural prescription.

2. Have the prices used to establish revenues and costs considered conditions of future markets, including possible changes in supply, demand, and inflation? Is the guiding rate adjusted correctly for inflation?
3. Does the guiding interest rate correctly reflect the owner's perception of the pure rate, inflation, and the riskiness of the project?
4. Has the problem situation been properly identified in the sense that the planning period is appropriate and the project chosen for the analysis is, in fact, feasible?
5. Does the schedule of events, costs, and revenues associated with the project include all those relevant to the decision maker, and in amounts actually affected by the project in question?

The first four concerns are obviously important to accurate and relevant analysis. There is little that can be contributed here other than to encourage utilizing the best available silviculturists, engineers, economists, and other pertinent scientists, research results, and crystal balls to identify the decision problem and forecast the physical and economic consequences of projects. This is truly the hardest part of good financial analysis. Review of case study literature, administrative records, and consultation with other practicing analysts who routinely generate the numbers in cash flow schedules is where most of the time, energy, and analytical ingenuity needs to be expended. Typically, there are sharply limited time and budget constraints for doing a given analysis, and the most accurate job simply is not possible. Choices must be made, for example, between different data sources such as ordinary experts, expensive experts, published data, administrative data, and new research data. Long-run planning, such as in forestry, must also consider changing technology—for example, considering how much wood utilization standards changed in the last 50 years, how much will they change in the future, and how will this affect the amount of usable wood and its price?

The fifth listed concern is a conceptual problem and is discussed below under its better known heading of *the with and without principle*. But we shall first examine an example of financial analysis.

Growing Christmas Trees: An Example of Financial Analysis

For this study, Rudolph (1972) determined the cost and returns from growing Scotch pine, white spruce, and Douglas-fir Christmas trees for different production periods and for various selling prices. Some of the more important data used to establish the cash flow schedules are shown in Table 7-5.

To illustrate the principles of financial analysis, a schedule of project

Table 7-5

Item	Scotch pine	White spruce	Douglas-fir
Production period	7 to 9 years	7 to 12 years	7 to 12 years
Planting stock used	2-0	2-2	2-2
Selling price per tree	\$2.75	\$2.50	\$4.00

For all 3 species	
Spacing	6 × 6 ft; 1,210 trees per acre
Area in roads and lanes	10%
Net trees planted per acre	90% of 1,210 = 1,090
Survival	90% of 1,090 = 981 trees
Trees sold	90% of 981 = 883 trees
Cull trees	98 trees
Land value or cost	\$100 per acre
Management costs and taxes per year	\$30 per acre
Fertilizing and shearing labor	\$2.50 per hour
Interest rate	6%

Table 7-6

Item	Basis	Interest years until harvest	Interest factor 6%	Capitalized to end of 7 years	
				Costs	Returns
Management costs and taxes	$30 each year	7*	8.3938	$251.81	
Land value or cost	$100	7	1.5036	150.36	
Planting stock and shipping cost	$0.033/tree × 1,090 = $35.97	7	1.5036	54.08	
Planting cost	$0.015/tree × 1,090 = $16.35	7	1.5036	24.58	
Weed control at planting, 2nd year	$10	7	1.5036	15.04	
	$10	5	1.3382	13.38	
Mowing between rows, 1st year	$6	6	1.4185	8.51	
Insect control spraying	$10 each year	6*	6.9753	69.75	
Shearing, including basal pruning	3rd year, 4¢/tree × 981 = $39.24	4	1.2625	49.54	
Shearing	4th year, 2¢/tree × 981 = $19.62	3	1.1910	23.37	
Shearing	5th year, 2¢/tree × 981 = $19.62	2	1.1236	22.04	
Shearing	6th year, 3¢/tree × 981 = $29.43	1	1.0600	31.20	
Trees sold, 6th year	50% = 490 × $2.75 = $1,347.50	1	1.0600		$1,428.35
Shearing	7th year, 4¢/tree × 393 = $15.72	—	—	15.72	
Trees sold, 7th year	393 × $2.75 = $1,080.75	—	—		1,080.75
Cleanup costs, 98 cull trees	$8	—	—	8.00	
Residual land value	$100	—	—		100.00
Total accumulated returns					2,609.10
Total accumulated costs				737.38	
Future value of net income for 1 crop after 7 years					1,871.72

* Series of annual payments at the end of each year.

activities, costs, and returns is adapted from Rudolph and reprinted in Table 7-6. The first column describes the type and timing of event, the second column the amount and current value of the cost and revenue item, and the third is the number of years the item draws interest. The fourth column is the interest factor at 6 percent, using either the future value equation (7-1) or the future value of a terminating annual series equation (7-5). The future values at the end of year 7 of all costs and revenues are calculated in the last two columns. The net future revenue at the end of 7 years is $1,871.72. Note that the land cost is considered as a revenue at the end of 7 years and a cost at the beginning. This means that cost, or rent, charged for using the land over 7 years is equal to the $50.36 interest charge on using the $100 worth of land for 7 years.

The data can be converted to the project event schedule and cash flow schedule using the time line technique we developed earlier for another look at the project (Fig. 7-10). From the data in the report, this event schedule can then be described in terms of current costs and revenues as a cash flow schedule (Fig. 7-11).

The last step in this financial analysis is to calculate the future value of this cash flow schedule (Table 7-7) in order to match the results presented by Rudolph in Table 7-6. Note that in this analysis Rudolph split the first-year costs, putting the land, planting and weed control expenses at the beginning of the first period (time = 0) and the management, taxes, and mowing costs at the end of the period. This is probably how the costs were actually distributed, but it is also an example of a change in the conventional assumption that all costs occur at the end of a period.

After making a similar analysis for production periods of 7, 8, and 9 years and for four different tree selling prices, the study concluded with a table for each species, as illustrated for Scotch pine in Table 7-8.

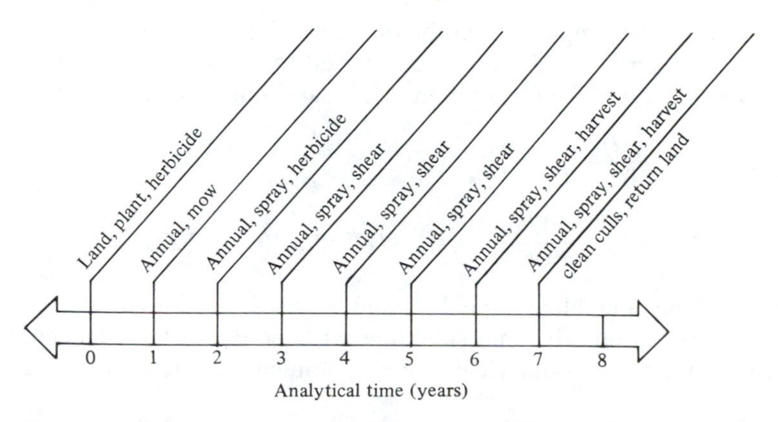

Figure 7-10 Schedule of project events for growing Scotch pine Christmas trees on a 7-year production cycle. *(After Ruldolph, 1972.)*

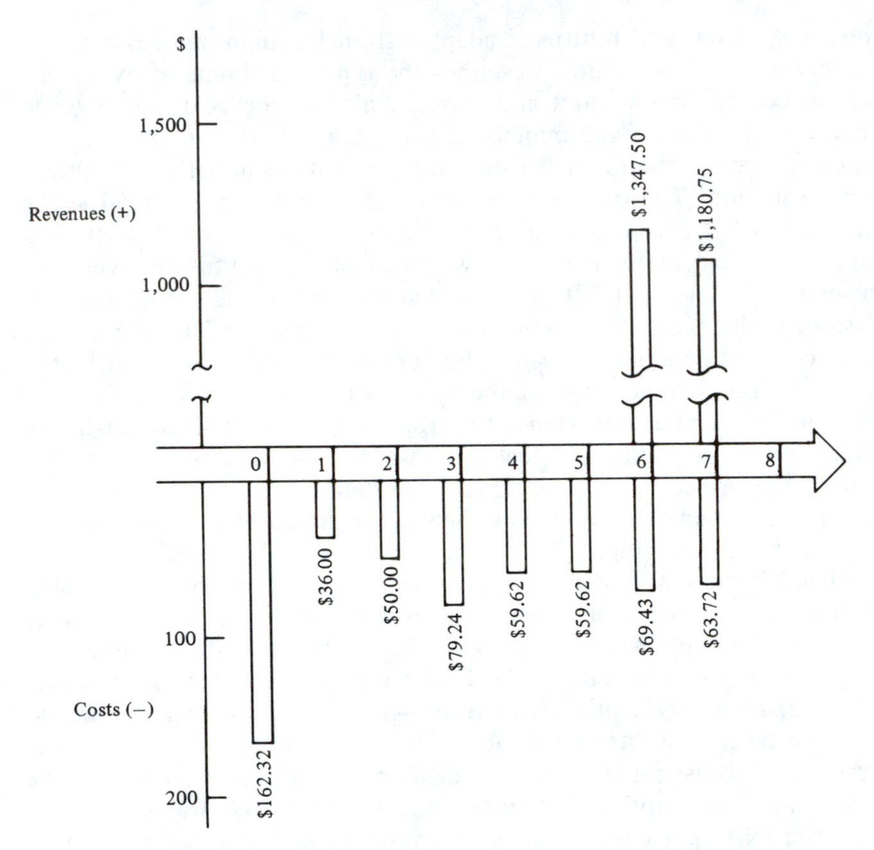

Figure 7-11 Cash flow schedule for growing Scotch pine Christmas trees on a 7-year production cycle. *(After Ruldolph, 1972.)*

The first row, showing a net income of $1,871.72, is the analysis that we used for our example. This can be converted to an annual equivalent income by solving Eq. (7-5) for a when V_n, n, and i are known,

$$a = \frac{V_n(i)}{(1 + i)^n - 1}$$

$$= 1{,}871.72 \frac{0.06}{1.5036 - 1} = 1{,}871.72(0.1191) = \$223$$

The idea of the equivalent annual income is that if the Christmas tree business with its periodic income schedule were sold and the proceeds invested at i percent, it would yield a steady annual income of $223 per acre.

The results in Table 7-8 also confirm the commonsense intuition that

Table 7-7

Period	Annual cost	Spraying	Land cost	Planting	Mowing	Weed control	Shearing	Cleanup	Total cost	Sales	$(1 + i)^n$ factor	Future value of costs	Future value of revenues
0			$100	$52.32		$10			$162.32		1.5036	$244.06	—
1	$30				$6				36.00		1.4185	51.07	—
2	30	$10				10			50.00		1.3382	66.90	—
3	30	10					$39.24		79.24		1.2625	100.04	—
4	30	10					19.62		59.62		1.1910	71.00	—
5	30	10					19.62		59.62		1.1236	66.98	—
6	30	10					29.43		69.43	$1,347.50	1.060	73.59	$1,428.35
7	30	10					15.72	$8	63.72	1,180.75	1.000	63.72	1,180.75
												$737.38	$2,609.10

Net future revenue = 2,609.10 − 737.38 = $1,871.72

Table 7-8

Production period (years)	Price per tree sold	Total net income per acre at end of period (at 6% interest)	Prorated equivalent average net income per year per acre (at 6% interest*)
7	$2.75	$1,871.72	$223.00
	2.50	1,643.62	195.81
	2.00	1,187.42	141.46
	1.50	731.22	87.11
8	2.75	1,757.69	177.59
	2.50	1,529.59	154.54
	2.00	1,073.39	108.45
	1.50	617.19	62.36
9	2.75	1,648.52	143.46
	2.50	1,420.42	123.61
	2.00	964.22	83.91
	1.50	508.02	44.21

* Calculated solving Eq. (7-6) $a = V_n(i)/(1 + i)^n - 1$.

profitability increases with higher selling prices and shorter production periods.

Project Present Net Worth

The Christmas tree example calculated the future value of the project and expressed it as an equivalent annual income. The data can also be used to illustrate the very important concept of present net worth.

Present net worth (PNW) is defined as the sum of the present value of the revenues minus the sum of the present value of the costs. As an equation,

$$\text{PNW} = \sum_{t=1}^{n} \frac{R_t}{(1 + i)^t} - \sum_{t=0}^{n} \frac{C_t}{(1 + i)^t}$$

where PNW = present net worth
R_t = revenue in period t
C_t = cost in period t
i = interest rate
n = number of planning periods

The cash flow schedule from Table 7-7 can be used to illustrate this calculation, as shown in Table 7-9.

Table 7-9

Period (1)	Costs (2)	Revenues (3)	$\frac{1}{(1+i)^n}$ (4)	Present value costs (5) = (2) × (4)	Present value revenues (6) = (3) × (4)
0	$162.32	—	1.0	$162.32	—
1	36.00	—	0.9434	33.96	—
2	50.00	—	0.8900	44.50	—
3	79.24	—	0.8396	66.53	—
4	59.62	—	0.7921	47.22	—
5	59.62	—	0.7473	44.55	
6	69.43	$1,347.50	0.7050	48.95	$ 949.95
7	63.72	1,180.75	0.6651	42.38	785.28
				$490.41	$1,735.23
				Present net worth = 1,735.23 − 490.41 =	$1,244.82

The present net worth value of $1,244.82 can be checked against the earlier calculation in Table 7-7 by discounting the present value of the net future value of $1,871.72 for 7 years,

$$\text{PNW} = \frac{1{,}871.72}{(1.06)^7} = \frac{1{,}871.72}{1.5036} = \$1{,}244.82$$

This also illustrates that the net value of a project can be calculated at the beginning, end, or any other time during the project. The net value can then be carried forward or discounted to some other point in time. The present net worth calculation is the most common since it is at time $t = 0$ that the decision to start the project is being made. If the present net worth is positive, the project is earning more than the guiding interest rate on invested funds and, other things being equal, would be an acceptable investment. If more than one project were under consideration, then the one with the highest present net worth would be the best choice.

The With and Without Principle in Project Evaluation

For many projects, an action or treatment is being imposed on an already established system and you want to know whether the treatment is worth doing. Logic suggests we must compare the costs and revenues *with* the treatment to the costs and revenues *without* the treatment. This in fact is the correct thing to do.

For example, suppose you are contemplating whether or not to intensify timber management for lodgepole pine, which has a tendency to stagnate. Currently your stands are being managed under a prescription of site preparation, plant, and final harvest. You now want to evaluate a

Table 7-10 Cash flow schedules for two silvicultural prescriptions for lodgepole pine sites

		(With) Proposed intensive management prescription		(Without) Current management prescription	
Year	Project event	Revenues	Costs	Revenues	Costs
0	Site preparation and planting	0	$100	0	$100
30	Precommercial thinning	0	100	0	0
50	Commercial thinning	$1,200	0	0	0
80	Final harvest	$8,000	0	$2,000	0
Total present value of revenues and costs at 4% interest		$515.92	$130.83	$86.77	$100
at 10% interest		14.13	105.73	1.0	100

second prescription where management intensification treatments consisting of a precommercial thinning and a commercial thinning are added to the current prescription.

A cash flow schedule representing these two silvicultural prescriptions is shown in Table 7-10.

Present net worth analysis The present values of costs and revenues from Table 7-10 (calculated just as those in Table 7-9) can be used to evaluate the intensification treatment.

1. *At 4 percent interest*
   With (intensive) prescription,

$$\text{PNW} = 515.92 - 130.83 = \$385.09$$

   Without (current) prescription,

$$\text{PNW} = 86.77 - 100 = \underline{-\$\ 13.23}$$

   Difference due to treatment (with minus without),

$$\text{PNW} = 385.09 - (-13.23) = +\$398.32$$

2. *At 10 percent interest*
   With (intensive) prescription,

$$\text{PNW} = 14.33 - 105.73 = -\$\ 91.60$$

   Without (current) prescription,

$$\text{PNW} = 1.0 - 100 = \underline{-\$\ 99.00}$$

Difference due to treatment
(with minus without),

$$\text{PNW} = -91.60 - (-99.00) = +\$\ \ 7.40$$

The present net worth analysis using the with and without principle involves subtracting the total present value of the costs and revenues *without* the treatment from the corresponding total present value *with* the treatment. For both the 4 and 10 percent interest rates the present net worth increased due to the treatment. The conclusion or recommendation therefore would be to intensify management.

A recommendation to intensify at the 10 percent rate occurs even though present net worth is negative for both the with and the without prescriptions—intensification produces a positive change in net worth [with − without = 91.60 − (−99.0) = +$7.40]. A prudent person, however, would use these negative net worths as information to question whether to grow trees under either of these silvicultural prescriptions.

The key to the positive recommendation in the 10 percent case is the assumption that you *must* pursue the without (current) prescription anyway. If you do not have to undertake either prescription, then the recommendation would be to either come up with some other prescription that has a positive PNW or let the land go out of timber production, meanwhile investing the $100 of planting cost elsewhere at the 10 percent guiding rate.

The with and without concept is very important to the correct economic analysis of programs, policies, and projects. It is critical to identify the proper (must do) without alternative and perform the comparison of with and without present net worths to see what actually happens to present net worth because of the investment.

Incremental cash flow analysis Another way of looking at the with and without idea is to establish the cash flow of the treatment itself. This involves subtracting each cost and revenue item of the without treatment from the with treatment costs and revenues using the data from Table 7-10 (as shown in Table 7-11*a*).

Table 7-11*a*

				Treatment cash flow	
Year	Activity	With − without treatment		Revenues	Costs
0	Site preparation	100 − 100	=	0	0
30	Precommercial thinning	100 − 0	=	0	$100
50	Commercial thinning	1,200 − 0	=	$1,200	0
80	Harvest	8,000 − 2,000	=	$6,000	0

Table 7-11*b*

Year	Activity	Revenues	Costs	Present value at 4%	Present value at 10%
0	Precommercial thinning	0	$100	−$ 100	−$100
20	Commercial thinning	$1,200	0	+ 547.66	+ 178.37
50	Harvest	6,000	0	+ 844.27	51.11
	Present net worth 30 years from now (PNW^{30}) =			$1,291.93	$129.50
	Present value today, = $PNW^{30}/(1 + i)^{30}$ =			$ 398.32	$ 7.40

Since no money is spent until year 30, this can be treated as the zero year for analytical time. If you were doing the analysis at the time you were spending the money on precommercial thinning 30 years from now, you would calculate the present net worth of the treatment as $1,291 and $129 for the 4 and 10 percent rates, and again conclude you should spend the money and do the precommercial thinning (Table 7-11*b*). The site preparation and planting investments are history, having already been implemented, and are no longer relevant. The only decision then is whether or not to do the precommercial thinning.

This analysis comes to the same result as our initial present net worth calculation since if we discount $1,291 and $129 thirty years to the present, the present net worth, today, of the treatment becomes $398 and $7.40, the same values as obtained earlier.

Stream Buffer Strips: An Example of With and Without Financial Analysis

In the early 1970s the Washington State Department of Natural Resources established buffer strips up to 200 ft wide on each side of streams running through their lands to protect salmon fishery values. No timber could be harvested within the buffer strips. The bottom lands within the buffer strips also can be extremely valuable for timber since they often have the largest trees and greatest potential productivity, raising the question: Just what is the best buffer strip width? Gillick and Scott (1975) made an economic analysis of timber and fishery value trade-offs with different buffer strip widths. Their study examined a 1700-ft reach of Miller Creek located on Washington's Olympic Peninsula. Their approach was to establish a cash flow schedule of timber and fishery values *with* buffer strips of different widths and compare this to the cash flow schedule *without* buffer strips.

As with most studies of this sort, establishing the cash flow schedules is the hard part. Timber values were calculated as the value of the standing timber presently on the buffer strip plus the value of future timber harvests from regenerated stands. Using a 1972 price of $33.04 per M bd ft, the value per acre of a 50-ft buffer strip was calculated as follows:

Present stand,

52 M bd ft × 33.04 =	$1,718	
Sale cost	344	
Net		$1,374

Future harvests,* per year,

1.13 M bd ft =	$37.30
Management costs	9.30
Net per year	$28.00

Capitalized net ($a \div i$)

$28 \div 0.055 =$	509
Total timber present net worth per acre	$1,883

The total value of timber harvests forgone for the 1,700-ft buffer is calculated as the number of acres in the strip times the present value of $1,883 per acre.

For the fishery values, the analysts first estimated the productivity of the test area stream as a salmon spawning area, measured in terms of the number of returning and harvestable mature salmon that were hatched on the stream segment under analysis. This estimate was based on the amount of spawning beds and a series of known conversion ratios to get the number of returning fish and potential harvest. It was estimated that the 1,700 ft of stream adjacent to the buffer area could support the fishery resource as summarized in Table 7-12.

Next Gillick and Scott estimated the value of these returning fish for both commercial and sport fishing. A synthesis of several published studies provided a commercial net value estimate per fish of $0.41 for coho and $1.47 for chinook salmon. Sport fish were estimated to provide a value of $28.00 per harvested fish. (The derivation of such values is discussed in Chap. 12.)

* For this particular calculation of realized future harvest value, they determined that the mean annual increment (MAI) for each acre of the stream bank sites was 1.3 M bd ft. They then assumed that all buffer strips, if they were managed as a unit, would provide an average annual harvest per acre equal to the MAI. This is just an approximation and may overestimate future stand values.

Table 7-12 Potential fishery resource of study area

Species	Spawning gravel (yd^2)		Maximum number of redds	Fish per 2 redd	Number of spawners	Catch-to-escapement ratio	Total run	Potential number of fish caught	
	Available sq. yards	per redd* sq. yards						Sport fishing	Commercial fishing
Coho	163	13.6	12	2	24	3 : 1	96	21.6	50.4
Fall chinook	633	24.4	26	3	78	5 : 1	468	156	234
Steelhead	163	3.5	46	2	93	1.8 : 1	260	167	—

* A redd is a spawning site or nest for one pair of fish.

Table 7-13

Fish species	Net value per fish	Expected number of fish actually harvested	Annual net benefit
Commercial coho	$ 0.41	5	$ 2.05
Commercial chinook	1.47	23	33.81
Sport coho, chinook, steelhead	28.00	25	700.00
Total annual benefit *a*			$735.86
Present net worth of fishery = a/i = 735.86/0.055 = $13,379			

Third, they recognized that high fishing pressure and low current salmon populations precluded achievement of the potential harvests of Table 7-12 and estimated the actual harvest to be under 10 percent of potential. These assumptions were used to calculate the "best estimate" of fishery values, as shown in Table 7-13. The estimated cash flow is an infinite series of annual net incomes of $735.86. When capitalized using Eq. (7-11), it gives a present value of $13,379.27 for a completely undisturbed fishery.

A fishery, however, is subject to many natural fluctuations in fish populations as well as natural disturbances from floods or windfall. With smaller and smaller buffer strips, more sediment and logging debris would reach the stream, impacting the fishery with increased frequency and severity. Gillick and Scott chose to use subjective probabilities to estimate the expected value of impact. To review from Chap. 6, if $O_1, O_2, O_3, \ldots, O_n$ are the numerical values of n possible outcomes from some action and $P_1, P_2, P_3, \ldots, P_n$ are the probabilities of these outcomes occurring, then $\Sigma P_i O_i$ is the expected value of the action when $0 < P_i < 1$ and $\Sigma P_i = 1.0$. For this problem, fisheries experts were asked to give their best guesses as to the probability that different percentage levels of fishery disturbance would occur with different buffer strips. The experts estimated that wider strips would typically have greater probabilities of low disturbance levels. The different outcomes were the net fishery values associated with each level of impact. Even with the maximum 100 percent impact, some fish would get through, spawn, and provide a fishable population. The probabilities used and the expected value calculations are shown in the summary economic analysis for the different buffer strip widths in Table 7-14.

The expected present net value of the fishery for, say, a 50-ft buffer strip is calculated as 0.9(13,379) + 0.1(12,891) = $13,330. The timber values given up by keeping the 50-ft buffer along the 1700 ft of stream are 1.95 acres worth $3,672. Treating the timber values as an opportunity cost, the present net worth of the 50-ft strip is $13,330 − $3,672, or $9,658. A similar calculation is made for each buffer strip width.

Table 7-14

Outcomes		Alternatives							
		350-ft buffer		100-ft buffer		50-ft buffer		0-ft buffer	
Fishery impact level	Present value of fishery	Probability	Value	Probability	Value	Probability	Value	Probability	Value
No harm	$13,379	0.95	$12,710	0.95	$12,710	0.9	$12,041	0	—
10%	12,891	0.05	645	0.05	645	0.1	1,289	0	—
20%	12,750	—	—	—	—	—	—	0.1	$ 1,275
40%	12,122	—	—	—	—	—	—	0.1	1,212
60%	11,490	—	—	—	—	—	—	0.2	2,298
80%	10,855	—	—	—	—	—	—	0.3	3,256
Maximum impact	7,827	—	—	—	—	—	—	0.3	2,348
Expected present value of fishery			$13,355		$13,355		$13,330		$10,381
Present value of timber harvests forgone			− 25,702		− 7,344		− 3,672		0
Present net worth of alternative			−$12,347		$ 6,011		$ 9,658		$10,381
With/without comparison to the 0-ft strip			−$22,728		−$ 4,370		−$ 723		
With/without incremental comparison to the next smaller strip			−$18,358		−$ 3,647		−$ 723		

Given this long trail of assumptions, estimates, and calculations, the analysis finally can be reduced to the with and without comparisons shown in the last two rows of Table 7-14. Treating the 0-ft strip as the without situation, its PNW of $10,381 is subtracted from the present worth of the strip in question to determine whether and how much the present net worth increases or decreases. In this particular study, the zero buffer alternative has the highest present net worth, and as the buffer strip width increases, the economic value declines. Apparently the increase in fishery values does not compensate for the decrease in timber values. This problem also illustrates the desirability of looking at several alternatives, including the do-nothing alternative. For example, if *only* the 100-ft strip had been examined, the net value of $6,011 might be considered satisfactory and a recommendation made for 100-ft strips. However, by looking at either 50-ft or 0-ft strips using with and without comparisons, we see that 100 ft is clearly not the best option.

This example has already ventured beyond simple financial analysis and has entered the subject of the next chapter, decision analysis and evaluation of alternatives. You should wonder about the accuracy of the probability and fish harvest estimates and should observe that this study used 1975 costs and prices. What about today's prices? What about wildlife and aesthetic values for buffer strips? Suppose the salmon fishery were restored. If the analysis as presented were repeated using different assumed numbers for prices and probabilities, then you would be conducting a sensitivity analysis to see how the best solution changes. If, in the unlikely event, it turned out that the best buffer strip was the 0-ft width under almost any assumption, then we might have more confidence in recommending this alternative.

Soil Expectation Value: A Special Case of Financial Analysis in Forestry

The monetary value of a productive asset to its current owner or prospective buyer is a reflection of the future stream of net income that the asset is expected to produce, be it a fast-food franchise, a hardware store, an acre of Iowa corn land, or an acre of timberland. It is this last case we foresters find interesting, and we frequently want to calculate the value of a particular parcel of bare land when used to grow timber. The result of such calculations can be helpful in deciding what prescription to use in managing the land, or to determine a reasonable price for buying or selling the parcel in question. The use of bare land values in the analysis of these forestry problems is considered in Chaps. 9, 11, and 13; our interest here is in the method used to calculate the bare land value.

Soil expectation value (SEV) is simply a forestry term used to describe the present net worth of bare forestland for timber production calculated over a perpetual series of timber crops grown on that land. The assumptions of the analysis are as follows.

1. All tree growing costs are included in the analysis, including relevant management fees, administrative costs, and taxes.
2. The guiding interest rate correctly reflects the context and outlook of the landowner.
3. The prescription for future management of the land has been decided and the same prescription will be used for each future timber production cycle.

Timber harvests typically come at regular intervals of the rotation length for even-aged management and the cutting cycle length for uneven-aged. This suggests that if we can characterize the value of one interval or period as a net income payment at the end of the period, we can use Eq. (7-10) for the present value of a perpetual periodic series to obtain the discounted value of our timberland.

Consider the even-aged management case and imagine that we are sitting on a stump in an acre of freshly clear-cut forestland with calculators at the ready. Looking around, all we see is stumps, logging slash, some beat-up submerchantable trees, and a scattering of advanced regeneration. A prescription decision is quickly made to prepare the site and plant the area with loblolly pine, to undertake brush control to release the trees from competition at age 5, to precommercial thin at age 10, and to make the final harvest at age 30. Our estimates of costs, prices, and timber yields produced the cash flow schedule for this prescription are shown in Table 7-15. The net future value of this prescription at the end of one 30-year rotation cycle is found by simply compounding all costs and revenues forward to get their future value at age 30 and then subtracting the future costs from the future revenues. This is exactly how we proceeded in the Christmas tree example earlier (Table 7-7). In general

$$a = \sum_{t=1}^{w} R_t(1 + i)^{w-t} - \sum_{t=0}^{w} C_t(1 + i)^{w-t} \qquad \text{(7-16)}$$

where period length = 1 year

a = future net income at end of one multiperiod payment interval

R_t = revenue received in year t

C_t = cost paid in year t

w = length of multiperiod payment interval in periods of 1 year

i = guiding rate of interest

Table 7-15

Year	Event (2)	Costs (3)	Revenues (4)	Interest factor $(1 + 0.05)^{30-t}$ (5)	Future values at 5% Costs (6) = (3) × (5)	Revenues (7) = (4) × (5)
0	Site preparation and planting	$250	—	4.322	$1,080.48	—
5	Release	80	—	3.386	$ 270.90	—
10	Precommercial thinning	100	—	2.653	$ 265.33	—
30	Harvest	—	$3,000	1.000	—	$3,000.00
	All annual costs	$5/acre/yr	—	—	$ 332.19	—
					$1,948.90	$3,000.00

Applying this formula to our prescription as presented in Table 7-15, we get a net periodic income every 30 years of

$$a = \$3{,}000 - \$1{,}948.90 = \$1{,}051.10$$

We can now characterize the acre of stumpy ground on which we sit as a green factory which will produce the following income stream:

Year	Income
0	0
30	$1,051.10
60	$1,051.10
90	$1,051.10
.	.
.	.
.	.

To get the present value of this perpetual periodic series, we apply Eq. (7-10),

$$V_0 = a \frac{1}{(1+i)^w - 1} = \frac{1{,}051.10}{4.3219 - 1} = \frac{1{,}051.10}{3.3219} = \$316.41$$

So finally we have the estimated bare land value of $316.41. This is the present value of net revenues from perpetually growing tree crops according to the specified prescription. In much forestry literature Eq. (7-10) is redesignated for even-aged management as

$$\text{SEV}^* = \frac{a}{(1+i)^w - 1} \tag{7-17}$$

where a = as defined in Eq. (7-16)
w = one rotation, years

For another example calculation of SEV, we can use the Christmas tree data of Table 7-5 and assume that the land is perpetually used to grow similar crops of trees at 7-year intervals. One important adjustment, however, is to add back the cost of using (or renting) land.

$$\begin{aligned} a &= (\text{maximum 7-year net income from Table 7-8}) \\ &\quad + (\text{land rental cost from Table 7-6}) \\ &= 1{,}871.72 + 50.36 = \$1{,}922.08 \end{aligned}$$

* Some writers refer to this as L_e for land expectation value, and some call it the Faustman formula in honor of the first person to formally set the equation down.

At the interest rate used of 6 percent,

$$\text{SEV} = \frac{1{,}922.08}{(1.06)^7 - 1} = \frac{1{,}922.08}{0.5036} = \$3{,}816.68$$

Of interest here is the huge discrepancy between our calculated bare land value of $3,816.68 and the $100 value of land used in the analysis. If land was really worth this much in terms of income produced and could be bought for $100, then we would expect all kinds of people to be buying land at $100 and growing Christmas trees. This would surely bid up the price of land above the $100 level. If this is such a good deal, why doesn't everyone do it? It is possible that the first people to plant trees make these returns, but as others come in and increase supply, tree prices and earnings surely will fall.

By including land as a cost in the analysis, Rudolph was saying that all profit or income was being attributed to the owner's or manager's skill rather than to the land as a capital asset. While it is always a debatable question how to split returns between entrepreneurial skill and capital assets, most theorists assume you can hire managers at a cost and thus attribute all net returns to capital (land).

An equivalent SEV for a fully regulated uneven-aged forest system with equal harvests every cutting cycle can be calculated by the SEV equation (7-17) as

$$\text{SEV(U)} = \frac{a}{(1 + i)^w - 1} \tag{7-18}$$

where a = the net return each cutting cycle
w = the length of the cutting cycle in years

Normalizing Projects

Where two or more projects are being compared for purposes of choosing which one to implement, the projects have to be comparable. To be completely comparable, projects need to have the same number of years of project life, have a similar timing of cash flows, and be of the same size in terms of investment cost. Rarely are all these conditions met. Comparisons of fertilization to precommercial thinning to site preparation to planting projects, for example, involve different project lengths and characteristic cash flow schedules.

To compare projects fairly, the projects need to be normalized to a common basis by making a series of explicit assumptions and calculations. The general approach is to adjust all projects to the lifetime of the longest project, to make explicit assumptions about how intermediate revenues are reinvested, and to make explicit assumptions about the inflation rate. We have already considered the issue of inflation, so here we consider adjustment for project life and reinvestment of intermediate revenues.

Different Project Lengths and the Reinvestment Rate

Consider two investment projects. The investor with $10,000 could either (1) buy 6-month money market certificates and earn an 18 percent rate of return, or (2) purchase a 5-year corporate bond that pays a 14 percent rate of return.

A simple comparison might suggest the money market certificate as preferable because of the higher earning rate. But in making such a comparison, we take the 18 percent of the 6-month note as its indicator of value relative to the bond. Thus we have implicitly assumed that every 6 months the investor will cash in the note and reinvest all the proceeds into another 6-month certificate that earns exactly 18 percent. In fact, a total of 10 sequential but independent investments are needed over the 5-year life of the corporate bond. Is this likely? Will the earning rate stay at 18 percent? Is there a cost in reinvesting every 6 months? These are some of the questions we should consider. The relevant comparison is the future value, after 5 years, of $10,000 invested today in each alternative. It may be that the longer term but lower yield bond will perform better over the entire period.

To adjust, the analyst must first decide on the time period for project comparison. This can be either a predetermined planning horizon or the length of the longest project. Second, an explicit earning rate for the reinvestment of project incomes into new projects is needed. Three assumptions about this reinvestment rate are possible.

1. New projects with exactly the same cash flows and thus earning rate as the initial project can be found. If the initial project were, say, precommercial thinning of red pine on site II lands, then this assumption requires that ample acreage of the same type of project opportunity be available.
2. The proceeds of the initial project will be reinvested at the guiding interest rate.
3. The proceeds will be reinvested in some other kind of specified projects with known earning rates that differ from the initial project and the guiding rate. This assumption requires that the analyst have detailed knowledge about other future projects.

To illustrate the dual adjustment for project length and reinvestment rate consider the following pair of timber management investments, one for 20 years and one for 60 years.

1. *Project A—precommercial thinning of pine*

Cost	$100 per acre
Project length	20 years
Increased value of harvest	$1,374

   Calculated project earning rate using Eq. (7-3) = 14 percent

   Guiding rate = 6 percent

2. *Project B—planting cutover land to pine*

Cost	$100 per acre
Project length	60 years
Value of harvest	$30,448

   Calculated project earning rate using Eq. (7-3) = 10 percent

   Guiding rate = 6 percent

To adjust and compare these projects, we must decide how the $1,374 received from project A at year 20 is reinvested for the next 40 years. Consider three assumptions:

1. The $1,374 is reinvested at the project earning rate of 14 percent.
2. The $1,374 is reinvested at the guiding rate of 6 percent.
3. The $1,374 is reinvested in fertilizing projects at an average return of 8 percent.

These three assumptions would normalize project A for comparison to project B as shown in Table 7-16.

Table 7-16

Project	Reinvestment rate, assumption	Initial cost	Revenue at Year 20	Revenue at Year 60	Present value of revenue at 6%	Present net worth at 6%
Project A	Project rate = 14%	$100	$1,374	$259,526	$7,867	$7,767
	Guiding = 6%	100	1,374	14,132	428	328
	Fertilizing projects = 8%	100	1,374	29,849	905	805
Project B	Earnings rate = 10%	100	—	30,448	923	823

To illustrate, the calculations for the fertilizing reinvestment at 8 percent are as follows.

1. Future revenue of the initial investment at year 20 = \$1,374.
2. Future value of the revenue reinvested for 40 additional years at 8 percent (to year 60) = $1{,}374(1.08)^{40}$ = \$29,849.
3. Present value of the revenue at year 60 = $29{,}849/(1.06)^{60}$ = \$905 at the 6 percent guiding rate.
4. Present net worth = 905 − 100 = \$805.

When project B is compared to project A on the basis of net worth, it obviously makes a difference what reinvestment rate is assumed. The assumption of reinvestment opportunities in similar projects does make project A superior. However, the other two reinvestment assumptions show project B preferable. Remember, reinvestment opportunities at the project earning rate is the implied assumption made whenever projects of different lengths are compared by the unnormalized project earning rates. Seldom is this a good assumption, and comparisons on the basis of project earnings rates tend to be in error when differences in project lengths are substantial. Since the analyst in many problems is looking 5 or more years into the future before reinvestment takes place, future opportunities are rather uncertain, and often the guiding rate is the best assumption that can be made.

These adjustments to normalize projects are rather critical in longer term forestry projects and need to be made carefully and explicitly in the specific context of the decision maker and problem in question.

The Realizable Rate of Return (RRR)

Because many people prefer to compare projects on the basis of earnings rates, an earnings rate of a project can be calculated that is the weighted average return on the project, considering the return on the initial investment and the return on any reinvestments. This earning rate is called the *realizable rate of return* (RRR) and is calculated by finding the earning rate that equates (1) the future value of earnings at the end of the normalized project life using specified reinvestment assumptions for intermediate revenues to (2) the present value of all costs, with intermediate costs discounted to the present at the same reinvestment rate as that for the intermediate revenues.

The earning rate is determined by solving for RRR,

$$(1 + \text{RRR})^n = \frac{\sum_{t=1}^{n} R_t(1 + r)^{n-t}}{\sum_{t=0}^{n} C_t/(1 + r)^t} \qquad (7\text{-}19)$$

where RRR = realizable rate of return
r = assumed reinvestment rate
n = normalized project life

R_t = revenues actually received in year t
C_t = costs actually incurred in year t

This formula will make both the adjustment for different project lengths and an adjustment for intermediate costs and returns of any project. To illustrate the calculation, the data from the earlier comparison of project A (precommercial thinning) and project B (planting) can be used. The intermediate revenue in this case is $1,374 from project A received at year 20.

Project	Reinvestment assumption	Present value of costs	Future value of revenues	RRR
Project A	14%	$100	$259,592	14%
	6%	100	14,132	8.6%
	8%	100	29,849	9.96%
Project B	10%	100	30,448	10%

To illustrate, the RRR for the 8 percent reinvestment assumption was calculated using Eq. (7-19) as

$$(1 + \text{RRR})^{60} = \frac{29{,}849}{100}$$

$$\text{RRR} = \sqrt[60]{\frac{29{,}849}{100}} - 1 = .0996$$

The comparison of the projects based on RRR is consistent with the earlier comparison based on normalized PNW. A detailed presentation of the RRR concept is presented by Schallau and Wirth (1980), followed by a spirited discussion by Klemperer (1981). We support the recommendation of these authors to use the RRR as the correct general criterion to characterize a project in terms of an earning rate. It encompasses the traditional Internal Rate of Return (IRR) criterion presented in Chap. 8 as well as the cases when the reinvestment rate is different from the earning rate of the subject project.

QUESTIONS

7-1 Site preparation and planting a spruce-fir site to Douglas-fir costs $250 per acre. When mature at 100 years, the stand is expected to have 40,000 bd ft of sawlogs per acre. At future stumpage prices of $100, $200, $300, and $400 per thousand, what is the earning rate of the planting investment at each price?

7-2 A loblolly pine stand is planted at a cost of $80 per acre and is expected to have a total stumpage value of $5,000 per acre when harvested in 60 years.

(*a*) At interest rates of 3, 6, and 9 percent, what is the present value of the harvest?
(*b*) What is the present net worth (PNW) of the investment at each guiding rate?

7-3 A ponderosa pine plantation is expected to grow according to the following yield schedule:

Age	Yield (M bd ft/acre)
30	2
50	8
70	15
90	40
110	70
130	90

Planting costs are $100 per acre. Stumpage is expected to sell for $200 per M bd ft.

(*a*) What is the rotation age that maximizes the present net worth over one rotation at 4 and 8 percent guiding rates?

(*b*) What is the rotation that maximizes the soil expectation value (SEV) at 4 and 8 percent guiding rates?

(*c*) Does the present net worth over one rotation give the same decision guidance as SEV? Explain.

7-4 A genetics research program has discovered a new strain of ponderosa pine which produces a 30 percent increase in harvest yield over the yields in Question 7-3. The increased cost for this seed runs $50 per planted acre.

(*a*) For rotations of 30, 50, and 90 years, what is the increase in harvest value of the stand at rotation age?

(*b*) For each rotation calculate the present net worth of investing in the improved seed for one 30-, 50-, and 90-year rotation at 4 percent and 8 percent. Under which condition would it pay to buy the better seed?

7-5 An investment in precommercial thinning (PCT) of lodgepole pine costs $75 per acre. The growth increase, when harvested 40 years after treatment, is an additional 5 M bd ft of sawlogs and 10 cords of firewood. Stumpage prices are $100 per M bd ft for sawtimber and $10 per cord for firewood, now and in the future.

(*a*) Calculate the earning rate on this investment.

(*b*) What is the present net worth of the investment at an interest rate of 10 percent?

(*c*) Would you recommend the treatment if the guiding interest rate were 10 percent?

(*d*) What is the maximum amount you could spend per acre for the treatment and realize a 10 percent return on the investment?

7-6 Jane Hawley inherited a wooded farm on Virginia's Eastern Shore in late December of 1985. The farm had long been used for goose and deer hunting by a club that held a valid lease signed by her father with 25 years yet to run. The lease required a payment of $15,000 to be made every 3 years. The next payment is due in December 1986 and every 3 years thereafter until the last payment in December 2010. Jane's guiding rate is 10 percent.

(*a*) Calculate the present value of the hunting club lease as of December 1985 for purposes of estimating inheritance taxes. (Use the periodic series equations.)

(*b*) If Jane put all the proceeds from the lease in a trust fund earning 8 percent interest, would there be enough to send her twin sons to Harvard graduate school in 2010 for MBA's when the tuition is then expected to be $100,000 per student?

7-7 The average annual timber yield for a 2,500-acre management area was reduced from $30 per acre to $20 per acre by a prescription to enhance wildlife and domestic forage production. If the increase in forage is valued at $1.50 per acre per year, what is the present value of increased wildlife benefits for the *whole* management area needed to *equal* the net reduction in commercial values? Assume a planning horizon of 50 years and a guiding interest rate of 7 percent. Is your answer an estimate of (1) the opportunity cost of the activities to enhance wildlife, or (2) the value of the increased wildlife benefits?

7-8 A revenue of $3,000 per acre is expected in 20 years from harvesting a timber stand and selling it at the market prices prevailing at that time. The owner expects inflation to average 5 percent over the next 20 years and she also expects a real rate of return of 7 percent on her timberland investments. What is the present value of the future harvest revenue in today's current dollars after adjusting for inflation and discounting by the real guiding rate?

7-9 A plantation that is established today at a cost of $100 is expected to provide a thinning yield of 20 cords of pulpwood at age 20 and, when regeneration harvested at age 30, a pulpwood yield of 10 cords plus a sawtimber yield of 20 M bd ft. Current stumpage prices are $15 per cord for pulpwood and $150 per M bd ft for sawtimber. The wholesale price index is forecast to inflate at a rate of 5 percent, pulpwood prices to inflate at 3 percent, and sawtimber prices to inflate at 9 percent over the next 30 years. The owner expects a real rate of return of 4 percent on timberland investments. What is the present value of all revenues for one rotation?

7-10 A minimum-level and two intensive management prescriptions for managing Virginia pine for pulpwood need to be evaluated. These prescriptions are:

1. Plant, then harvest at age 30 (minimum intensity)
2. Plant, precommercial thinning at age 10, harvest at age 30
3. Plant, precommercial thinning at age 10, fertilize at ages 10 and 20, harvest at age 30

Per acre costs: planting $100, precommercial thinning $50, and fertilizer $30 per application. The per-acre cash flows associated with these three prescriptions are as follows:

	Prescription					
	1 (minimum)		2		3	
Age	Revenue	Cost	Revenue	Cost	Revenue	Cost
0	—	$100	—	$100	—	$100
10	—	—	—	50	—	80
20	—	—	—	—	—	30
30	$400	—	$800	—	$1,200	—

Using a guiding rate of 5 percent and assuming no inflation:

(*a*) Calculate the present net worth of each prescription, treated separately.

(*b*) If the minimum prescription (1) is required by law, calculate separately the incremented present net worths of returns to the two intensification treatments of thinning only (prescription 2) and thinning plus fertilization (prescription 3) to determine which is the most worthwhile.

(*c*) Calculate the soil expectation value (SEV) for each of the three prescription options at guiding rates of 3, 5, and 7 percent.

7-11 An uneven-aged hardwood forest currently has an inventory of 17 M bd ft of growing stock. This timber sells for $250 per M bd ft now and in the future. The owner has a guiding rate of 6 percent. The forester has determined that the ideal, sustainable reserve growing stock level is 12 M bd ft and two options for initiating management need financial evaluation:

Option 1: Immediately harvest 5 M bd ft and reduce the stand to the reserve growing stock of 12 M bd ft and subsequently harvest 10 M bd ft every 20 years forever.

Option 2: Wait 10 years when the stand will have an inventory of 26 M bd ft, harvest 14 M bd ft reducing the inventory to 12 M bd ft and subsequently harvest 10 M bd ft every 20 years forever.

(*a*) Calculate the present net worth (PNW) of both options;

(*b*) Calculate the PNW of both options if a 7 percent rate of price increase for hardwood stumpage and a 4 percent rise in the wholesale price index is expected forever. The owner expects a real rate of return of 5 percent.

REFERENCES

Gillick, T., and B. D. Scott, 1975: "Buffer Strips and Protection of Fishery Resources: An Economic Analysis," DNR Report 32, Washington State Department of Natural Resources.

Klemperer, W. D., 1979: Inflation and Present Value of Timber Income after Taxes, *J. Forestry,* **77:**94–96.

———, 1981. Interpreting the Realizable Rate of Return, *J. Forestry,* **79:**616–617.

Row, C., H. F. Kaiser, and J. Sessions, 1981: Discount Rate for Long-Term Forest Service Investments, *J. Forestry,* **79:**367–369.

Rudolph, V. J., 1972: "Costs and Returns in Christmas Tree Management," Research Report 155, Agricultural Experimental Station, Michigan State University, Lansing.

Schallau, C. H., and M. E. Wirth, 1980: Reinvestment Rate and the Analysis of Forestry Enterprises, *J. Forestry,* **78:**740–742.

CHAPTER

EIGHT

EVALUATION OF PROJECT AND PLANNING ALTERNATIVES

A decision problem is framed by two or more defined alternative actions and the landowner's goals for guidance to what is best. From Chap. 6 we have the language and concepts to identify and describe the nature of decision problems, alternative problem solutions, and some ideas about how to consider solutions with certain or uncertain outcomes. From Chap. 7 we have financial analysis techniques to deal with costs and revenues over time.

This chapter centers around decision analysis—the development, evaluation, and presentation of information to forest managers and their constituencies who must make choices and implement the chosen solutions. The discussion is organized by goals which guide decisions in forest management. Difficult problem situations where two or more goals are applicable and problem situations where not all relevant goals and constraints can be acceptably quantified are examined. The goals of economic efficiency, regional development, distributional equity, stability, and environmental security are considered explicitly.

THE PURPOSE AND NATURE OF DECISION ANALYSIS

Before starting on an enthusiastic manipulation of numbers, it is well to look at the purpose of decision analysis. The manager has to make choices—and this is the mandate for decision analysis: to develop information useful in guiding choice. If we call the person preparing such information an analyst, then the analyst's role is to provide the decision maker with objective, factual information describing the consequences of alternatives. To such information the decision maker applies value judgments, intuition, political considerations, and experience to make a choice. This separation of factual analysis from subjective value judgments needed to make decisions is rather critical. When value judgments intentionally or unintentionally become embedded in what is supposed to be factual analysis, then any sure reference to truth and objectivity becomes lost or obscured.

This chapter is directed to those who are or who may become analysts and to those who need to work with analysts. From it we hope you learn something about providing useful information to decision makers. One of the first things for the analyst to remember is that decision makers are also human beings with average-sized minds. Successful managers tend to be big on self-confidence, energy, and self-discipline, are good listeners, have a basic sensitivity to other humans and their values, and have strong stomachs rather than unusually great intellectual capacity. As with other human beings, attention spans are fairly short and the ability to assimilate simultaneously a great deal of factual information is limited. Hence effective decision analysis requires keeping the presentation simple and focused on critical differences between alternatives.

For example, compare the rather elegant and comprehensive display of information given in Table 8-1 for a public land management problem to the rather terse present net worth synopsis given in Table 8-2 for the same problem.* For most of us it would be difficult to make all the necessary comparisons between the many criteria in the first presentation to decide which alternative is best. If handed only Table 8-1 and given but an hour or so to choose, what would you do? When the evaluation is simplified as in Table 8-2, it is easy enough to distinguish how the alternatives differ, but we begin to wonder how objective and factual the analysis was and whether or not all important consequences were considered. And where did all the dollar values come from? The more simplistic the analysis, the more the decision maker has to rely on the care and objectivity of the analyst.

A decision maker will want to justify and legitimize decisions he or she makes. The easiest way to do this is to use some traditional, accepted way of doing an evaluation, such as benefit-cost analysis, and to simply choose the alternative with the highest present net worth. The claim can

* The tables are extracted and modified from the originals as presented in the final environmental statement for the Smith Creek Planning Unit, Kaniksu National Forest (U.S. Forest Service, 1977).

Table 8-1 Management plan options for Smith Creek showing estimated yields, outputs, and activities

Criterion	Present condition	Estimated yields by alternatives				Estimated change in yield by alternative (as compared to present condition)				
		New alternatives			Selected plan	Current management plan	New alternatives			Selected plan
		A	*B*	*C*			*A*	*B*	*C*	
1. Water yield (acre · ft/yr)[a]	240,470	240,420	240,420	239,850	240,280	−50	−50	−50	−620	−190
2. Sediment production (tons/yr)[b]	38,740	36,620	36,620	35,830	36,450	−2,120	−2,120	−2,120	−2,910	−2,290
3. Potential accelerated erosion from timber harvest and roads (tons/yr)[c]	51,100	30,000	29,500	22,900	28,500	−21,100	−21,100	−21,600	−28,200	−22,600
4. Estimated area disturbed by timber harvest and roads (acres/yr)	1,300	780	780	600	740	−520	−520	−520	−700	−560
5. Annual timber growth potential (million bd ft/yr)	N.A.	11.1	10.6	6.4	10.2					
6. Estimated 20-year timber harvest (million bd ft/yr)	8.1	7.2	7.2	6.1	6.8	−0.9	−0.9	−0.9	−2.0	−1.3
7. Change in timber industry employment (number of jobs)						−4	−4	−4	−8	−5
8. Change in timber industry wages and salaries ($1,000 per year)						−30.46	−30.46	−30.46	−67.68	−43.99

Table 8-1 (Continued)

Criterion	Present condition	Estimated yields by alternatives: New alternatives A	B	C	Selected plan	Estimated change in yield by alternative (as compared to present condition): Current management plan	New alternatives A	B	C	Selected plan
9. Estimated 20-year recreation use (RVD/yr)										
Type I	1,520	1,470	1,490	1,560	1,520	−30	−50	−30	+40	0
Type II	2,760	2,580	2,580	2,600	2,580	−180	−180	−180	−160	−180
Type III	4,790	5,130	5,180	5,070	5,070	+390	+340	+390	+280	+280
Type IV	370	2,190	330	330	340	−40	+1,820	−40	−40	−30
Total	9,440	11,370	9,580	9,560	9,510	+140	+1,930	+140	+120	+70
10. Estimated *potential* for 20-year recreation use (RVD/yr)										
Type I	14,300	6,700	7,800	11,200	9,000	−6,500	−7,600	−6,500	−3,100	−5,300
Type II	5,800	7,400	11,900	12,800	11,400	+6,100	+1,600	+6,100	+7,000	+5,600
Type III	44,900	55,500	58,300	64,300	53,000	+13,400	+10,600	+13,400	+19,400	+8,100
Type IV	4,800	41,900	7,700	7,700	10,000	+2,900	+37,100	+2,900	+2,900	+5,200
Type V		4,000	0	0	2,100	0	+4,000	0	0	+2,100
Total	69,800	115,500	85,700	96,000	85,500	+15,900	+45,700	+15,900	+26,200	+15,700
11. Potential domestic grazing (AUM/yr)	545	640	545	475	545	0	+90	0	−70	0
12. Deer winter range quality[d]	+++	++	++	++	++	−	−	−	−	−

13. Potential number of wintering deer (annual)[e]	15	5	5	5	5	−10	−10	−10	−10	−10
14. Potential number of huntable deer (annual fall population)	20	10	10	10	10	−10	−10	−10	−10	−10
15. Fisheries habitat quality	+	++	++	++	++	+	+	+	+	+
16. Grizzly bear habitat quality	++	++	++	++	++	0	0	0	0	0
17. Mountain caribou habitat quality	++	++	++	++	++	0	0	0	0	0
18. Mineral opportunity[f]	+	++++	++++	++	+++	+++	+++	+++	+	++
19. Reconstruction and improvement of existing roads (mi in 20 yr)	N.A.	67	67	21	64	+67	+67	+67	+21	+64
20. New general-access road construction (mi in 20 yr)[g]	N.A.	12	12	12	13	+12	+12	+12	+12	+13
21. Total general-access road at end of 20 years (mi)	120 (existing)	132	132	132	133	+12	+12	+12	+12	+13
22. Construction of project development roads (mi in 20 yr)[h]	N.A.	85	89	57	78	+85	+85	+89	+57	+78
23. Construction of new trails	79 (existing)	0	9	39	39	0	0	+9	+39	+39

[a] Natural base flow = 221,310 acre · ft/yr.
[b] Base production = 30,950 tons/yr.
[c] Natural erosion for planning unit = 2,500 tons/yr.
[d] Based on vegetative succession and potential for disturbance.
[e] Based on vegetative succession and deer life equations.
[f] Based on access and other development.
[g] Includes arterial, collector, and local road functional classifications. These roads would have a long-term service life.
[h] Estimate of the amount of work roads necessary for site-specific project development. These roads could have either long-term or short-term service lives.

Table 8-2 Present index values by alternative management plan

	Current management	Alternatives *A*	*B*	*C*	Selected plan
Total discounted benefit index values*	13,253,353	14,910,000	12,980,000	12,475,000	13,300,000
Total discounted cost index values*	3,033,722	3,540,000	3,100,000	2,400,000	3,250,000
Net difference between discounted benefit and discounted cost index values (net present worth)	10,219,631	11,370,000	9,880,000	10,075,000	10,050,000
Individual benefit-cost ratio	4.37	4.21	4.19	5.20	4.09

* Discounted at 10 percent interest.

then be made that this is the best choice. We note, however, from Table 8-2 that the selected plan has neither the highest present net worth nor benefit-cost ratio. So something else must have weighed in the decision. If, however, after pondering only the information in Table 8-1, the manager chose to go with plan *B*, or any plan, he or she would be hard pressed to give a simple, uncontroversial explanation as to why this was the best choice. Few managers really enjoy going on instincts alone, saying "all things considered I think *B* is best and that is my opinion."

Analysts need to balance continually the decision maker's desire for simplicity and legitimacy against the analytical purpose of providing objective information. For the most part, as we shall see, simplification comes at a price of lowered objectivity and increased chances of error in the analysis. On the other hand, complexity extracts its price and, potentially, risks rejection of analysis as a useful exercise.

All decisions are made in the present about actions and outcomes that take place in the future. It follows that virtually all analysis must forecast future treatment responses, costs, and prices. The decision maker as well must speculate on the future and place his or her bets on uncertain outcomes. In forest management we look further into the future than anyone—even the shortest term treatment decision in forestry, such as thinning or fertilizing, has significant outcome effects lasting 5 or 10 years. The decision to plant a stand of trees for harvest 60 years in the future is more an act of faith than a deliberated forecast. What kind of an economy, technology, and society do we suppose the intrepid early foresters and their Civilian Conservation Corps (CCC) crews contemplated for the 1980s as they planted cutover lands in the depression of the 1930s?

If the future by definition is uncertain, it is hard to envision a responsible analysis that does not recognize this uncertainty in some formal way. At the least, important assumptions about the future such as the inflation rate, percentage of unemployment, or prices can be stated explicitly. Better, key assumptions can be systematically varied in a sensitivity analysis to see how the indicated best choice varies. You may even want to assign probabilities to different states or outcomes if you think it provides better information to guide decisions.

PRESENTATION OF ALTERNATIVES

In nearly every analytical report, be it a staff paper, an environmental impact statement, or a forest plan, there is a brief section with a summary table and discussion that lays out the alternatives, the criteria used to distinguish between them, and the major assumptions made in the analysis. This is what decision makers and their constituencies spend most of the time looking at. It is often called an *executive summary* for that reason. The balance of most reports is devoted to describing the problem, discussing the relevant goals, and documenting the process of estimating and quantifying the criteria shown in the summary table.

Format

The structure and layout of the summary table is important—it is the primary information transfer vehicle from the analyst to the decision maker. Tables 8-1 and 8-2 illustrate two such tables. There are three ways to format the presentation:

1. Present the total impacts and effects of the new alternatives.
2. Present total impacts of the current situation with the new alternatives.
3. Present the impacts of new alternatives as incremental or marginal changes from the current situation.

These three forms are illustrated in Table 8-3 using a simplified example, where two goals are each characterized by two criteria to evaluate two new alternatives, *B* and *C*, and compare them to the current situation, *A*.

To critique the tables, the first presentation is inferior since the current situation is not shown. Except in the rare "no activity" case (all zeros for the criteria), there is always a current level of activity and the decision maker must know whether the proposed alternatives cause improvement or deterioration. This is another situation requiring application of the with and without principle from Chap. 7. In the second presentation

Table 8-3 Three ways to present the results of a decision analysis

	Presentation 1* Alternative		Presentation 2† Alternative			Presentation 3‡ Alternative		
	B	*C*	*A* (current situation)	*B*	*C*	*A* (current situation)	*B*	*C*
Goal 1: Produce timber efficiently								
Criterion 1, M bd ft harvested	8,000	7,000	5,000	8,000	7,000	5,000	+3,000	+2,000
Criterion 2, Cost/M bd ft harvested	$40	$60	$50	$40	$60	$50	−$10	+$10
Goal 2: Provide income and employment to local economy								
Criterion 1, Number of timber jobs	10	25	15	10	25	15	−5	+10
Criterion 2, Sales in local economy	$5.2 million	$5.5 million	$5 million	$5.2 million	$5.5 million	$5 million	+$0.2 million	+$0.5 million
Summary index								
Present net worth of timber business	$750,000	$350,000	$400,000	$750,000	$350,000	$400,000	+$350,000	−$50,000

* Only the new alternatives *B* and *C* are shown.
† The current situation *A* is shown in addition to the new alternatives *B* and *C*.
‡ New alternatives are shown as changes from the current situation.

format the current situation is included, but when the table is large and the criteria are mixed, as in the first five columns of Table 8-1, the decision maker is forced to make all the numerical comparisons by sitting down with pencil and paper or by visual impression. The third presentation format makes the comparison explicitly by performing the needed arithmetic and succinctly stating the current situation as totals and the proposed alternatives as marginal changes. The algebraic signs clearly indicate what is going up or down. The right half of Table 8-1 is organized this way. To make the comparison, alternately mask the right and left sides of Table 8-1 with a strip of paper and then imagine you were trying to select between the alternatives. Which seems preferable? We like the third format best.

How Much Information?

How much information should be placed in a summary table? The number of important and substantially affected goals and criteria are the key. Table 8-1, for example, covers three or four goals, and a total of 23 criteria are used to describe the impact of each alternative on these goals. Is this too much? It seems so to us, perhaps because the goals are not really identified in the table and there is no indication of which are the most important criteria.

Criteria are quantified in units of dollars, miles, tons, jobs, and symbols such as +++. When presented as numbers, are we inclined to attach greater importance to bigger numbers? For example, do tons of sediment numbers (criterion 2 of Table 8-1) seem more important or impress us more than the number of jobs (criterion 7) just because the numbers are bigger? Does +++ symbolize three times the impact of + regarding mineral opportunity (criterion 18) and how is this compared to timber harvest (criterion 6)? No instructions for the interpretation of the criteria are included. The choice of what to put in the table is in itself a value judgment, for everything left out is by implication given a zero value. How is this decided? These questions do not have easy or general answers, but they are important. It is equally important that there be negotiation and resolution on these issues between the analyst, the decision maker, and impacted citizens before analysis is started. Remember, the information in the executive summary is the factual basis of most decisions.

Summary Indices

Table 8-2 shows two summary indices, present net worth and benefit-cost ratio, which are supposed to represent the total significance of each alternative as a single number. If the summary index is truly comprehensive, then for each and every important quantified criterion $c_1, c_2, c_3, \ldots, c_n$ a comparable value coefficient $v_1, v_2, v_3, \ldots, v_n$ needs to be defined, and the summary index is obtained as

$$\text{Summary index} = v_1c_1 + v_2c_2 + v_3c_3 + \cdots + v_nc_n$$

If the value scale is incomplete, then some of the terms in the summation are missing, and we must say the summary index only partially represents the whole alternative. For example, if v_2 and v_3 could not be defined and quantified, the second and third terms of the evaluation summary drop out and the index no longer represents goal criteria c_2 and c_3.

For the decision maker to accept the summary index as a good value representation of the alternatives, the following prerequisites must be met:

1. The criteria and their amounts $c_1, c_2, \ldots, c_n$ need to cover all relevant goals and be acceptable as objective and reasonably accurate data.
2. The value coefficients $v_1, v_2, \ldots, v_n$ are complete, comparable, and reflect the viewpoint of the decision maker.
3. The summation process appropriately categorizes the costs and revenues and uses an appropriate guiding interest rate to consider time costs.

If these conditions are not met, and they rarely are, the issue is: how useful is the summary index? More important, how do you decide when the c_i and v_i numbers are so inaccurate and/or incomplete that the summary index should be ignored by the decision maker or not even calculated? In terms of our example, does the summary index of present net worth in Table 8-2 accurately represent in value terms all the consequences and effects described in Table 8-1? Did they really find a value coefficient for +++, or for tons of sediment? A lot of checking of the full report is needed to decide whether Table 8-2 is adequate. While this is (or should be) the decision maker's burden, many times the analyst has to exercise professional integrity and raise the question about adequacy of the summary index if it is to be raised at all. All of the preceding discussion assumes that the decision maker and the analyst are two different persons. If they are the same person, that is, the decision maker does his or her own analysis, the issues remain and the individual has to try and wear two hats and separate the roles.

We return to the questions of presentation, summary value indices, and partial analysis in our discussion of benefit and cost analysis at the end of this chapter. But first we must examine some goals and their criteria in greater depths.

EVALUATION CRITERIA

Each decision problem has at least one goal, and often several goals are relevant to decision makers. Articulating these goals by identifying and quantifying goal criteria and presenting the criteria to guide choice is a demanding task in decision analysis.

Several important goals guide public and private forest management decisions:

1. Economic efficiency
2. Favorable impact on regional and local communities
3. Equity in the distribution of costs and benefits among the members of society
4. Economic and social stability
5. Security of the environment

For each goal some of the more commonly used criteria are discussed, and analytical techniques for estimating and presenting these criteria are considered.

Economic Efficiency

Economic efficiency is perhaps the most widely accepted social goal in the United States. It follows logically from generally accepted theory and our belief that self-centered, benefit-maximizing individual behavior in a competitive, capitalistic economic system gives good profits to entrepreneurs, provides products to consumers at minimum costs, and causes the economy to grow at a healthy rate. The goal itself is rooted in the materialistic desires of most humans: the more goods and services, the better. Economic efficiency means achieving this status to the fullest extent possible. At the firm, forest, or individual level, achieving economic efficiency is then equivalent to behavior that maximizes income, profits, or present net worth. For forest industry and many private forest owners, maximizing income and/or present net worth is a common objective criterion reflecting the goal of maximizing the market value of the enterprise. For federal lands, Congress in the 1970s has given legislative mandates that these lands be managed efficiently considering all goods and services.

All goals and objectives are economic in that they imply value. The modifier *economic* is defined here to mean values measured in dollars or, some other established relative value measure. If an objective is stated in physical terms, to maximize cubic foot volume for example, then all volume is assigned the same unitary value regardless of size, species, or timing. Outputs that are not counted in a problem have been assigned a zero value.

Because the definition and the operational measurement of economic efficiency have become so important in project and forest planning, a thorough understanding of this concept is needed. There are at least three distinct operational situations and hence definitions of economic efficiency, each depending on the constraints imposed on the planner.

First situation *Economic efficiency is the choice of the outputs that maximize the total net return to a fixed budget and other fixed inputs.*

This situation is similar to our initial steers and trees problem of Chap. 6. The land, labor, and budget resources are fixed or given as constraints, and the problem is to find the combination of steers and trees that maximizes total revenue. As before, we search the border of the feasible region for the product combination that maximizes total revenue. The only difference is that economists conceptualize the feasible region boundary as being continuous rather than linear segments, so instead of the now familiar polygon feasible region of Fig. 8-1*a*, in economics textbooks we see a more generalized smooth curve describing the boundary

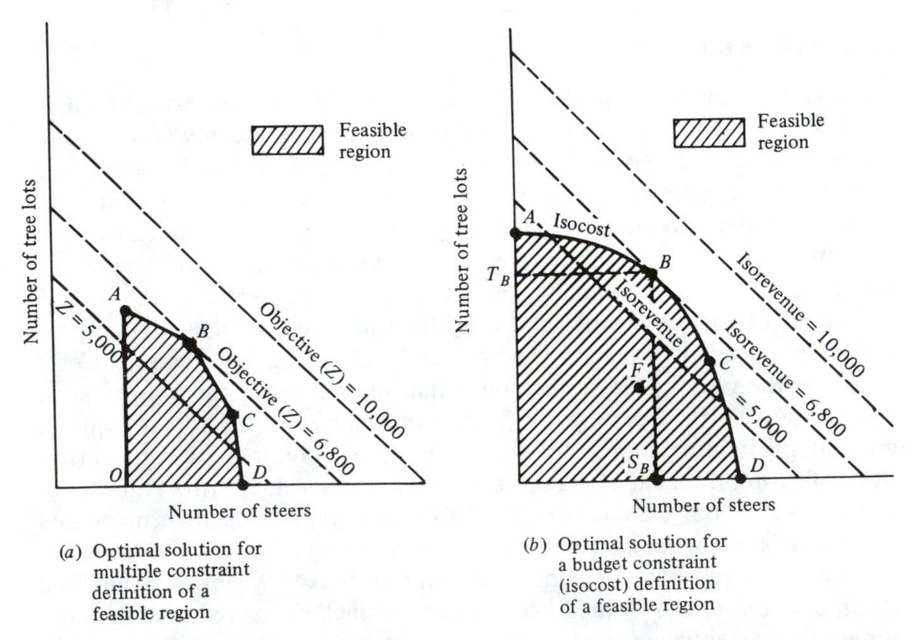

Figure 8-1 First economic efficiency problem situation: finding efficient outputs. (*a*) Optimal solution for multiple linear constraint definition of a feasible region. (*b*) Optimal solution for a budget constraint (isocost) definition of a feasible region.

of the feasible region, as in Fig. 8-1*b*. This boundary curve represents combinations of outputs that can be produced with fixed resources.

Another simplification typically used by economists is to represent all variable inputs such as labor, raw materials, or fuel by a single budget constraint. All combinations of outputs that can be produced at a given budget level are called an isocost curve, such as *ABCD* in Fig. 8-1*b*. Representing all variable inputs by a single budget constraint assumes, not unreasonably, that money can be freely exchanged for any amount of these physical inputs and that no other constraints are relevant. The combination of outputs that will produce the same amount of revenue is called an isorevenue curve. Note that the set of isorevenue lines shown in Fig. 8-1*b* are the same as the set of objective function lines of Fig. 8-1*a*.

Using the isocost representation, the first definition of economic efficiency asks that the point on the isocost curve be found that maximizes total revenue. This is obviously the same problem that we handled in Chap. 6 by finding the corner point of the feasible region that touched the highest valued objective function as point *B* in Fig. 8-1*a*. In Fig. 8-1*b* point *B* is located in the same way by finding the tangency point where the highest valued revenue line is reached.

This combination of T_B tree lots and S_B steers in Fig. 8-1*b* just touches the $6,800 isorevenue curve. Any other feasible combination on the isocost such as *A*, *C*, or *D* will produce less than $6,800 of revenue and still use up all the budget.

Hence we say that point *B* is the economically efficient production combination when resources and budget are fixed. In practice, the common application of this situation occurs when we are given a fixed annual budget and told to select the set of projects, treatments, or product mix that is economically efficient. (While we might assume that we had this all figured out when we requested the budget in the first place, this is not always the case.) In private business, a typical situation is when the firm or corporate division has received its annual operating budget and is now trying to find a specific manufacturing schedule and sales level that will maximize profit.

Consider point *F*, which is a solution within the feasible region. It is inferior to at least one point on the feasible region boundary for two possible reasons: (1) it is economically inefficient, or (2) it is technically inefficient. Economic inefficiency can be demonstrated by comparing point *F* to point *B*. At point *B*, more than twice as many tree lots can be produced with the same budget resources available at *F* without reducing the number of steers. Hence point *B* would provide greater revenue with the same set of resources and be a more economically efficient solution. Point *F* would be technically inefficient relative to point *B* if the technology (such as the cultural technique, genetic stock, machinery, and tools) which was used at *F* expended all the budget and, at the same time, better technology is actually available to the producer that would allow the higher level of production at point B with the same budget.

The economic value of research and development is the amount by which the objective function is increased as the isocost or feasible region is expanded by new inventions and knowledge. To stay technically efficient in practice means to keep abreast of research and development and to constantly modify and change production and management techniques to get the most physical output from given resources.

Second situation *Economic efficiency is the choice of the input resources that minimizes the total cost of providing prespecified outputs.*

This second situation presumes that some higher authority has, in fact, made some decisions as to what products or outputs must be achieved, and our task is to find the lowest budget (isocost curve) that will accomplish this.

As shown in Fig. 8-2, if someone has said we must provide T_X tree lots and S_X steers (point X), then the budget and input mix of isocost C_3 at \$3,000 is the cheapest and hence most economically efficient way to get the job done. C_4 and C_5 would also get the job done, but they would be more expensive. C_1 and C_2 do not provide enough budget to produce the target outputs.

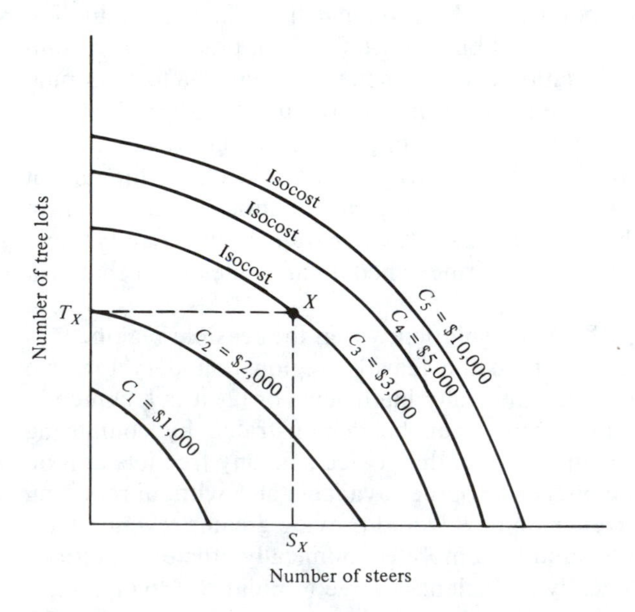

Figure 8-2 Second economic efficiency problem situation: finding cost efficient inputs.

In practice, when targets or quotas for production are handed down as rigid requirements (such as point X), then we can say that the least-cost production strategy (such as C_3) is the economically efficient strategy. It is critical, however, to be sure that someone in authority did, in fact, specify the required output mix. Otherwise this efficiency concept can be used incorrectly to rationalize any output mix and budget level.

Finding the least-cost strategy is not easy. We typically have standard or traditional ways and technologies of doing things and only want to examine these familiar options. To be reasonably sure that you have the least-cost approach, new research and outside experts should be consulted first to identify the best technology available. Then several real alternatives that are technically possible need to be defined and then evaluated for cost and overall feasibility. Of the feasible alternatives we search for the one with the lowest cost. If the production technology can be formulated as a continuous relationship (C_1 to C_5 in Fig. 8-2) or as a constrained feasible region, it can be examined mathematically for the least-cost point X.

Third situation *Economic efficiency is the simultaneous choice of both the level of inputs and outputs or products that maximize total net return.*

This third situation is simply an extension of the first situation where the budget constraint has been removed and, in fact, the key question becomes, what is the best budget level? Procedurally, we simply define a series of budget levels and for each one find the optimum (revenue-maximizing) product combination. Then the net return is calculated for each budget by subtracting the budget (cost) from the gross revenue. Finally, the budget level that produces the greatest net return or profit is called the most economically efficient budget and level of production activity.

To illustrate this third definition of economic efficiency, a series of isocost or budget curves (C_1, C_2, C_3, C_4, C_5) are shown in Fig. 8-3*a*. For each the revenue-maximizing output combination is found (points A, B, C, D, E), as with the first situation. These points are evaluated as shown on Fig. 8-3*a* to find the budget level that maximizes net return.

The budget level of \$6,000, which produce 14 steers and 8 tree lots would be recommended as being economically efficient. To make this recommendation, it is important to remember that it assumes specific output prices (steers \$500, tree lots \$700) and that the shapes (technology) of the isocost curves are known.

When this third situation is addressed using mathematical programming we do not need to systematically vary the budget constraint in a simulation to identify production possibilities. Rather we recognize that the third situation has no budget constraint and simply delete the budget constraint row from the problem formulation and solve to find the optimal

Identification of economically efficient outputs for a problem without a budget constraint

Solution	Budget	Steers	Lots of trees	Budget cost	Total revenue	Net return
A	C_1	3	6	\$3,000	\$5,700	\$2,700
B	C_2	5	10	\$4,000	\$9,500	\$5,500
C	C_3	10	8	\$5,000	\$10,600	\$5,600
D	C_4	14	8	\$6,000	\$12,600	\$6,000* (optimum)
E	C_5	20	6	\$10,000	\$14,200	\$4,200

Selling prices: steers = \$500, trees = \$700/lot

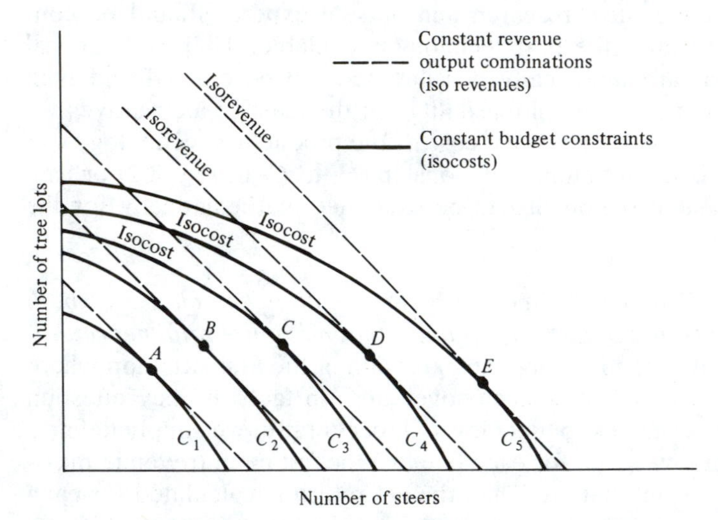

(*a*) Efficient solution identified by systematically changing the budget

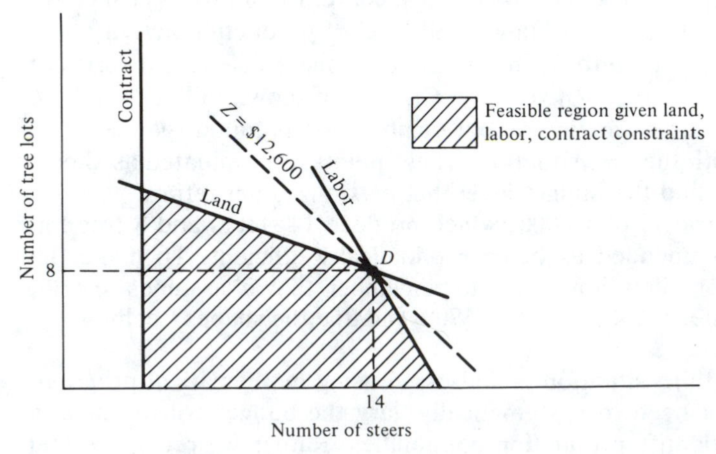

(*b*) Optimal and efficient solution identified by mathematical programming

Figure 8-3 Third economic efficiency problem situation: finding an efficient combination of inputs and outputs.

solution. A comparable figure for this mathematical programming approach is shown in Fig. 8-3*b*, where the only constraints are the contract, land, and labor and point *D*, with 8 tree lots and 14 steers, is the optimal solution. The third situation is the same as the first situation except that there is no budget constraint.

This third situation is the problem faced by budget, project, and forest planners who are trying to decide what and how much to do and then justify the choice, such as in recommending a timber management plan or a reforestation budget. When evaluating sets of timber culture, reforestation, wildlife improvement, range, or recreation projects for planning and budget preparation purposes, the essential question is, which set of projects will produce the greatest net revenue and hence define a justifiable, economically efficient budget and management plan?

Conditions for successful economic efficiency analysis Each preceding problem situation exhibited three important conditions that must be satisfied to permit a particular problem solution to be declared *the* most efficient:

1. The problem is completely and uniquely specified in terms of objective function and a set of constraints. In each of the three situations reviewed above, the efficient solution is for a uniquely and fully specified problem.
2. All outputs included in the objective or revenue function of the problem have comparable units of value. If the value of some or all outputs are indirectly expressed through constraints, then only one such constraint set can be used in identifying all candidate solutions. The efficient solution is identified, "given" the imposed constraints.
3. The full feasible region of solutions, including its boundary, is systematically searched to locate the most efficient solution. If the analytical procedure used to discover the optimal solution does not first identify the feasible region and in some systematic way evaluate several alternative solutions representing its breadth and boundary, then there is no reasonable guarantee that the most efficient solution has been found. The use of linear programming, as an example, normally ensures that this search condition is satisfied.

These conditions of successful economic efficiency analysis can now be combined with our earlier distinction between type *A* and type *B* solution generating processes to say something about which types of forest management problems and solution generating processes lend themselves to economic efficiency analysis and which do not. Recall from Chap. 6

that the type *A* process involved identifying an agreed upon set of goals and constraints and then evaluating several different solutions to this problem to select the best one. In contrast, in the type *B* process decision makers cannot agree on the problem goals and constraints and thus create different sets of goals and constraints to represent different conceptualizations of the problem. One or more solutions to each problem formulation is presented as a candidate solution to the initial problem.

The type *A* process clearly satisfies the first and second conditions for successful economic efficiency analysis: there is a unique specification of the problem, and if the problem is feasible, only one efficient solution is found. When linear programming or some similar optimization technique is used, the third condition may also be satisfied.

The type *B* process for defining alternative solutions always violates the first condition for successful economic efficiency analysis. When different input and output constraint sets are used to define alternative solutions, then the alternatives obviously do not come from the same uniquely specified problem. *Alternative solutions from a type B process therefore cannot be compared on efficiency grounds*—the "given" or "everything else equal" basis of comparison has been destroyed. When input and output constraints also change between alternative solutions, the constant and comparable value basis required by the second condition is also violated by the type *B* process.

When the alternatives from a type *B* process are presented only as solutions, we have an additional difficulty in interpretation. It is often impossible to look at a set of solutions and determine if they are (1) alternative solutions to the same underlying problem or (2) solutions to several different problems. To avoid misunderstanding, written or verbal presentations of proposed alternatives must also make the underlying goals and constraints of the problem explicit. If all the conditions of economic efficiency analysis are met for each subproblem, then solutions to the subproblems are economically efficient.

Limitations of economic efficiency analysis in forest management planning The sharply limiting conditions we have identified for successful economic efficiency analysis should raise the warning flag whenever statements are made like "this plan is more efficient than that one because it shows a higher present net worth." More often than not the alternative plans were established by a type *B* process.

Public and private planning typically is characterized by multiple goals with recognition that several different goods and services provided by the forest have value to the owners. Private owners may have a mix of financially related but competing goals relating to current income, capital productivity, market share, credit rating, and income stability. Public owners such as U.S. Forest Service are usually concerned about the production of recreation opportunity, wildlife habitat, or visual quality, goods and services that do not have well-defined market values, in addition to goods such as timber that do have a market value.

When all output-related goals cannot be explicitly valued in the objective function for either the private or the public case, a constraint package is defined to represent these goals in the analysis, or they are ignored. When used, these constraints affect the solution choices for the outputs that *are* valued in the objective function and thus, implicitly, place values on the goals and outputs represented by constraints but which are not counted in the numerical computation of the objective function. They must therefore be considered as part of the total (implicit plus explicit) benefit function of the owner.

We need to distinguish two terms in our discussions of owner goals and their criteria:

1. *Net revenue*. This is the sum of dollar (or value index)-priced outputs minus the sum of dollar-priced costs. If discounted, the comparable term is *present net worth*.
2. *Net benefit*. The term *benefit* is used to encompass all goals, output and input costs, whether or not they are dollar-priced. Included are all goals and values, even if they are quantified only by symbols or words. In the discounted sense we use the term *present net benefit*.

By definition the numerically quantifiable criterion is net revenue or present net worth. But what we are trying to estimate is present net benefit, that elusive concept of overall goodness.

If only one constraint is systematically changed, then each solution contributes to a sensitivity analysis mapping out the effect of that constraint on the objective function. The objective function value itself cannot, however, be used directly for an economic efficiency comparison between alternatives. The difference in objective values only shows the opportunity cost in terms of the outputs priced in the objective function; *not considered or counted are the benefits realized from the goals or outputs represented by the constraint or other outputs not evaluated in the objective function*.

To illustrate the fallacy of using only a partially quantified objective function to measure the efficiency of alternatives, consider two plans for a forest which provide different mixes of timber, recreation, and rare woodpeckers. Only timber is priced in the objective function. The results of the analysis are given in Table 8-4. The first four rows represent the facts that we can quantify, and plan *A* has the higher calculated net revenue. This does not tell which is better, only that the increased RVDs and woodpeck-

Table 8-4 Measuring efficiency of alternatives with partial valuation of outputs

Item		Plan *A*	Plan *B*
	Physical outputs		
1	Timber yields	1,000 M bd ft	800 M bd ft
2	Annual recreation visitor days consumed	4,000 RVD	6,000 RVD
3	Estimated average population of rare woodpeckers	1 pair	4 pairs
	Economic analysis with only timber valued		
4	Dollar value of a timber-only objective function at $200/M bd ft	$200,000	$160,000
5	Opportunity cost in timber for more woodpeckers and RVDs, row 4 (*B* − *A*)		−$40,000
6	Present value of market measured benefits	$200,000	$160,000
	Economic analysis with all outputs valued		
7	Estimated RVD value at $5/RVD	$20,000	$ 30,000
8	Subjectively estimated woodpecker population value	$30,000	80,000
9	Total nontimber value, (rows 7 + 8)	$50,000	$110,000
10	Net valuation of all benefits, (rows 4 + 9)	$250,000	$270,000

ers of plan *B* must be worth at least $40,000 to choose *B* instead of *A*. Suppose we could somehow evaluate RVDs and woodpeckers and that their dollar equivalent worth was as shown in rows 7 to 10. We see that the total benefits for both plans are actually higher and that plan *A* could be charged an opportunity cost of $60,000 in terms of RVD and woodpecker values. This more than compensates for the $40,000 timber opportunity cost of *B*. In this case, if there are no other relevant conditions, plan *B* produces the greater net benefit and hence is the more efficient plan.

More commonly, several goal and output constraints are changed between alternatives, and the analysis is a simulation of results characterizing markedly different problem perceptions. When the alternatives reflect such a procedure, opportunity cost changes between the alternative solutions can only be measured in terms of changes in bundles of constraints.

For private owners such constraint sets could reflect different emphasis on income stability, enterprise value, and mill raw-material-supply security goals. For public planning, such constraint sets often are used to produce solutions which reflect viewpoints of widely different constituencies and interest groups, such as enhance amenity values (environmental groups), enhance local economic impact (state and local governments), enhance production of commercial values (mining or timber industry), enhance output stability (professional resource managers), or, maximize net dollar contribution to the federal treasury (the Administration and Office of Management and Budget).

This sort of pluralistic decision process is inherently political and no amount of wishful thinking will reduce it to simple quantitative economic efficiency analysis where the solution with the biggest computed index value is best. The role of quantitative information from solution to different problem formulations is to illustrate the implications and impact of the different possible output levels and their distribution through society and to help guide corporate managers, public planners, and their constituencies in discussion and negotiation to politically resolve their differences in differences in viewpoint.

Although it is important to know that the solution representing each type *B* alternative is efficient, we have also found that legitimate economic efficiency comparisons cannot be made between alternatives from such type *B* processes. However, comparisons of the dollar values of timber sales, the costs of program segments, or other relevant *economic information* are useful; just remember that this is entirely different from making *economic efficiency* comparisons in which a summary criterion such as present net worth attempts to characterize the "overall goodness" of a solution. An economic efficiency comparison of net revenue in Table 8-4 would examine row 6 and incorrectly pick plan *A* as superior by $40,000. In terms of all benefits, plan *B* is superior in "overall goodness" by $20,000.

A final complication of planning problems occurs when output or goal constraints cannot be applied directly to output decision variables because the output of interest cannot be defined or given a simple unit of measure. For example, achieving the goal of biological diversity or providing certain population levels of selected wildlife species is largely done in forest planning by constraining the location, timing, and spatial mixture of roading, harvesting, planting, and other input activities that control vegetative structure and distribution. Rarely do we define an output for "diversity," nor is the wildlife species given units of measure and counted. Rather we create a complex and interacting set of *input* constraints to represent desired outputs goals indirectly. These constraints may not be well labeled or identified, making it very difficult to interpret the implicit valuation given the outputs or goals in question.

The whole conceptual argument about type *A* and *B* processes for generating alternative solutions and the error of making economic efficiency comparisons between solutions from a type *B* process holds when the analysis is conducted using a pencil rather than mathematical programming. Consider, for example, the problems of locating a main haul road through a watershed, deciding what size of dam to build, or choosing aircraft for wildfire control activities. The more traditional approach identifies a few discrete alternatives using experience and judgment. Constraints are often not formally written down, and it is not always certain that alternative solutions are actually feasible or, if there is a feasible region, that the solutions examined are close to the optimal solution. The inputs and outputs of each alternative are valued and summed to guide the choice between alternatives, the claim often being made that the selected alternative is "more efficient" than any of the others. Again, the key question is what was the mental process or model used to define alternatives? If a process of shifting goals or resource constraints was used, then it is still a type *B* process, with the result that efficiency comparisons are not valid.

Project evaluation criteria for efficiency analysis Forest management problems often involve choosing among several candidate projects, each of which satisfies all physical, biological, or economic constraints specified for the problem. These feasible candidate projects can be grouped into two classes: mutually exclusive projects and independent projects.

1. In *mutually exclusive projects* only one of the feasible candidate projects can be chosen. Choosing the best silvicultural prescription for a given acre or stand is a good example—of all the possible ways to handle the regeneration and culture of the new stand, only one can be implemented and the problem is to find the best one.
2. *Independent projects* by contrast are those where out of a group of projects two or more can be chosen. An example of independent projects is when several different geographically scattered tracts of land are being considered to find the most efficient set to harvest. Although harvesting any one tract does not preclude harvesting any other tract, the problem constraints may limit the total number of tracts harvested.

We have seen that economic efficiency involves getting the most net revenue or benefit. Choosing between or ranking projects then involves establishing measures of net revenue as criteria for the choice. Three

Table 8-5 Calculation of PNW, *B/C*, and IRR for a silvicultural prescription

				Present value costs*		Present value revenues*	
Year	Activity	Cost	Revenue	i = 5%	i = 10%	i = 5%	i = 10%
0	Plant	$100	$ 0	$100.00	$100.00	—	—
10	Herbicide and fertilizer	40	0	24.50	15.40	—	—
20	Precommercial thin	80	0	30.10	11.90	—	—
30	Commercial thin, fertilize	40	500	9.30	2.30	$115.70	$28.60
50	Clearcut and harvest	0	2,500	—	—	218.00	21.30
				$163.90	$129.60	$333.70	$49.90

Efficiency evaluation

i = 5%	i = 10%
PNW = 333.70 − 163.90 = $169.80	PNW = 49.90 − 129.60 = −$79.70
B/C = 333.70 ÷ 163.90 = 2.04	B/C = 49.90 ÷ 129.60 = 0.39

IRR = 7%

* Using the present value equation (7-2), $V_0 = V_n/(1 + i)^n$.

criteria traditionally used to measure profitability are present net worth (PNW), benefit-cost ratio (B/C), and internal rate of return (IRR). All of these criteria use compound interest to adjust for costs and revenues occurring at different points in time. To illustrate and compare these criteria, consider the silvicultural prescription project in Table 8-5.

This project involves a 50-year cash flow schedule, and the present value of each cost and revenue item is calculated for interest rates of 5 and 10 percent.

1. *Present net worth*. Present net worth (PNW) as defined in Chap. 7 is the sum of the discounted revenues less the sum of the discounted costs over a defined planning period,

$$\text{PNW} = \sum_{t=1}^{n} \frac{R_t}{(1+i)^t} - \sum_{t=0}^{n} \frac{C_t}{(1+i)^t} \qquad (8\text{-}1)$$

where R_t = revenue in year t
C_t = cost in year t
i = guiding interest rate
n = number of years in the planning period

Present net worth is dependent on and sensitive to the guiding interest rate used. In Table 8-5 the calculated present net worth changes from a positive \$169.80 at 5 percent to a negative \$79.70 at 10 percent.

2. *Benefit-cost ratio*. The benefit-cost ratio (B/C) indicates the amount of present value revenue per unit of present value cost by dividing the sum of discounted revenues by the sum of discounted costs,

$$\frac{B}{C} = \sum_{t=1}^{n} \frac{R_t}{(1+i)^t} \Big/ \sum_{t=0}^{n} \frac{C_t}{(1+i)^t} \tag{8-2}$$

The benefit-cost ratio is an index that says something about the relative productivity of each dollar spent. Like present net worth, its calculation depends on knowing the appropriate guiding interest rate. In Table 8-5 the benefit-cost ratio changed from 2.04 at 5 percent to 0.39 at 10 percent.

A variation of Eq. (8-2) is useful when the decision involves allocating a current budget or deciding whether or not to make a major capital investment in the current year. In this case the decision variable is C_0, the cost at year 0, and the investor wants to know its relationship to the discounted stream of future net revenues generated by this investment as indicated by the equation

$$\frac{NR}{C_0} = \frac{\Sigma_{t=1}^{n} NR_t/(1+i)^t}{C_0} \tag{8-2a}$$

where NR_t = net revenue in year t ($R_t - C_t$).

NR/C_0 is also the appropriate calculation for capital budgeting, a procedure which we introduce shortly. NR/C_0 is 2.698 at 5 percent and 0.203 at 10 percent for the problem of Table 8-5.

3. *Internal rate of return*. The internal rate of return (IRR) is a unique characteristic of a project and does not require a guiding interest rate for calculation. It is measured by the rate the project actually earns on money invested, and is calculated by trial and error to find the interest rate,

$$\text{IRR} = i$$

where

$$\sum_{t=1}^{n} \frac{R_t}{(1+i)^t} = \sum_{t=0}^{n} \frac{C_t}{(1+i)^t} \tag{8-3}$$

Note that this is the calculated i that makes PNW = 0. In Table 8-5, the calculated IRR is 7 percent. While the IRR has the advantage that it can be calculated without knowing a guiding rate, it has the disadvantage that multiple interest rate solutions are possible, all of which will satisfy Eq. (8-3). The number of rates is a function of the number of times the project net cash flow schedule changes sign over the planning period. Fortunately most forestry projects have costs early and revenues later, giving only one sign change, so the multiple interest rate solution is not normally a problem.

Relationship of the criteria The three criteria are mathematically related. By definition, when the guiding rate i is equal to a project's IRR, then PNW = 0 and B/C = 1.0.* For typical projects with distant future revenues and most costs close to the present, at guiding rates lower than the IRR, PNW is positive and B/C is greater than 1.0. Similarly at guiding rates higher than the IRR, PNW is negative and B/C is less than 1. Figure 8-4 illustrates these relationships and was established by calculating and plotting PNW and B/C for the project of Table 8-5 at several different guiding rates. Check out the calculations yourself.

Using the criteria to evaluate fully independent projects The fundamental goal of a firm or enterprise is to maximize its value to its shareholders or owners (Van Horne, 1980). The best estimate of this value we have is the discounted value of expected future net incomes, benefits, or dividends appropriately adjusted for risk. Since discounted net revenue is estimated by the calculated present net worth, maximizing present net worth is the correct objective for guiding investment and other management decisions to achieve the goal of economic efficiency.

When projects are fully independent, it means there are no constraints limiting the collective number of projects that can be implemented. Each project stands on its own and the only decision is whether or not to implement the project.

For the fully independent project case, all that is necessary is to calculate the PNW of each project and implement any project whose PNW is positive. Alternatively, if the calculated B/C is greater than 1.0 or the IRR is greater than the guiding interest rate, the recommendation would also be to implement the project. To illustrate consider the project of Table 8-5 as a fully independent project. At a guiding rate of 5 percent, all three criteria say to implement the project and at 10 percent all three say to not implement it. At any guiding rate at or below 7 percent, the decision would be to implement.

Special advantages and disadvantages of the criteria are as follows:

1. The project present net worth (PNW) is a direct criterion for the goal itself, and this is its greatest virtue. It is the correct criterion to make

* Using Eq. (8-3), if i = IRR, then $\Sigma_t R_t/(1 + i)^t = \Sigma_t C_t/(1 + i)^t$. Substituting this result in Eq. (8-1) for $\Sigma_t R_t/(1 + i)^t$, we have PNW $= \Sigma_t C_t/(1 + i)^t - \Sigma_t C_t/(1 + i)^t = 0$. Substituting in Eq. (8-2), we have $B/C = \Sigma_t C_t/(1 + i)^t / \Sigma_t C_t/(1 + i)^t = 1.0$.

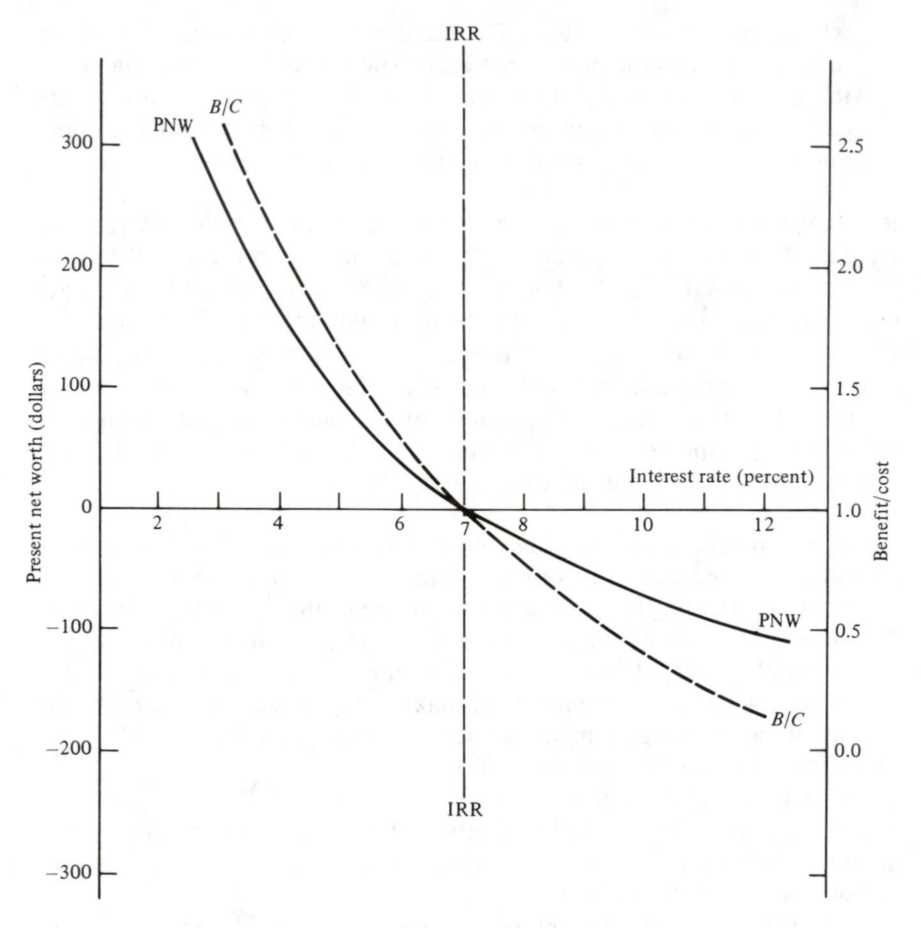

Figure 8-4 Relationship of PNW, B/C, and IRR at different guiding rates for the silvicultural prescription presented in Table 8-5.

project choices when there are no budget constraints. Because it is an absolute measure, however, it cannot rank the relative efficiencies of different projects. The concept of the present net worth criterion seems difficult for the public or nontechnical clients to understand and to accept as choice justification. Finally, to calculate present net worth the decision maker must have sure knowledge of the guiding rate.

2. The benefit-cost ratio (B/C or NR/C) is a relative measure that tells the present value of revenues provided per dollar of present costs. It is very useful to rank several different independent projects by their relative efficiencies. Its disadvantages are the same as those for present net worth.

3. The internal rate of return is similar in concept to the benefit-cost ratio, but it has theoretical and practical limits to its application. Easy comprehension of the internal rate of return is its real advantage. Virtually everyone borrows money or puts money into investment accounts where the loan cost or yield is expressed as a rate of interest. Hence everyone understands that it is not a good idea to invest in a project where its yield rate (the internal rate of return) is less than the cost of the investment money (the guiding rate), and that shopping for projects with the highest yield rate makes good business sense. Another advantage is that projects can be characterized by their internal rate of return without using a guiding rate. The disadvantages are many, and careless use of the internal rate of return can give bad decision guidance. If projects have alternating negative and positive cash flow schedules, multiple internal rates of return are calculated, and it is not easy to select the correct one to represent the project.

Mutually exclusive projects and normalization If the projects from which you are selecting are mutually exclusive, then normalization for reinvestment rate, project length, and project size is necessary to select the project for implementation that maximizes the overall PNW of the enterprise. (See p. 284 for an introduction to reinvestment rate.)

If your practical assessment suggests that the reinvestment rate will be different from either the guiding rate or the IRR of the subject project, two important implications are (1) if the reinvestment rate is different from the IRR, then the IRR is no longer an appropriate criterion for project evaluation and the realizable rate of return (RRR) as calculated by Eq. (7-19) should be used, and (2) if the reinvestment rate is judged to be different from the guiding rate, then the PNW and B/C as calculated by Eqs. (8-1) and (8-2) are no longer correct since the intermediate project revenues need to be reinvested at the specified reinvestment rate and a normalized PNW or B/C calculated and used as the decision criteria to compare projects.

Normalizing PNW and *B/C*

To illustrate the important normalization of projects prior to their comparison using the PNW or B/C criteria, we will use the data from the with and without example developed in Chap. 7. Assume the investor has two different parcels of land. Parcel A is 30 acres in size, has just been cut, and will be managed with a prescription calling for planting followed by a precommercial thinning at age 30, commercial thinning at age 50, and a regeneration harvest at age 80. Parcel B is 10 acres in size and has an existing 30-year-old stand to be given immediate precommercial thinning followed by a commercial thinning in 20 years and a final harvest in 50 years.

We will assume that these represent mutually exclusive projects to show how projects which differ in both size and length can be normalized in order to compare the two projects. Since normalization for differences in the reinvestment rate has been discussed in Chap. 7, this example will assume that the reinvestment rate is the same as the guiding rate, which is 5 percent per year, and that normalizations are needed only for size and length. The per-acre cash flow has been multiplied by the size of the parcel since the projects require all of the stand to be treated.

	Project *A*		Project *B*		Project *B* normalized for size	
	(1)	(2)	(3)	(4)	(5)	(6)
Year	Cash flow	Normalized for length	Cash flow	Normalized for length	Normalized for size	Normalized for size and length
0	−3,000	−3,000	−1,000	−1,000	−3,000	−3,000
20			12,000		12,000	
30	−3,000	−3,000				
50	36,000		60,000		60,000	
80	240,000	395,590*		483,466*	99,123†	582,588*
Decision Criteria						
PNW	4,287	4,287	8,754	8,754	8,754	8,754
B/C	2.16	2.16	9.75	9.75	3.92	3.92
IRR	6.31	—	13.80	—	9.05	—
RRR	—	6.08	—	8.03	—	6.80

Note: The reinvestment rate and the guiding rate are 5 percent per year.

* $(36{,}000)(1.05)^{30} + 240{,}000 = \$395{,}590$.

$(12{,}000)(1.05)^{60} + (60{,}000)(1.05)^{30} = \$483.466$.

$(12{,}000)(1.05)^{60} + (60{,}000)(1.05)^{30} + 99{,}123 = \$582{,}588$.

† $(2{,}000)(1.05)^{80} = \$99{,}123$.

The unadjusted cash flows for projects *A* and *B* are shown in columns 1 and 3.

Normalization for project length was accomplished by carrying all intermediate and final revenues forward to the last year of the longest project, in this case year 80 for project *A*. The normalized cash flows for the two projects are shown in columns 2 and 4.

Normalization for project size was accomplished by adding a $2,000 project that earned at the guiding rate to the $1,000 investment of project *B*, bringing the normalized size of project *B* to $3,000 with the cash flow shown in column 5. Finally, the size adjusted cash flow of column 5 is normalized for length and intermediate revenues in column 6.

The three criteria (PNW, *B/C*, and project earning rate IRR and RRR) are then calculated for the six columns and give us some insight as to the effect of normalization on economic decision criteria. First, we note that

PNW for each project is the same with or without normalization for length or size (column 1 compared to column 2 for example). This is because the reinvestment rate is the same as the guiding rate used to discount the unadjusted cash flow. Second, we note that normalization for length does not affect the calculated *B/C* (compare column 1 to 2 or column 3 to 4), but the normalization for size significantly changes the value of *B/C* (compare column 4 to column 6 for example). This means normalization for size could affect the ranking of projects by *B/C*. Third, we note that normalization for size affects the calculated IRR (compare column 3 to 5) and normalization for size and length affects the RRR (compare column 4 to 6). This is because the intermediate revenues are reinvested at a 5 percent guiding rate while the IRR for all projects is greater than 5 percent. The net effect is to lower the IRR and RRR of the projects.

Comparing mutually exclusive projects When there are several mutually exclusive projects, comparisons with an unnormalized criterion can be misleading. These problems are especially serious when *B/C* is used without normalization for size. To illustrate, consider two mutually exclusive projects.

	Project *M*	Project *P*	Incremental change project *P* − project *M*
Present value revenues	$100	$300	200
Present value costs	25	150	125
PNW	$ 75	$150	75
B/C	4.0	2.0	1.6

The higher initial investment in project *P* produces a higher PNW but a lower *B/C* than project *M*. Which project should be chosen? As suggested in the previous example, project *M* could be normalized for size by adding to project *M* an investment of $125 that returns the guiding rate. Another way to normalize for size is to assess whether the incremental investment between *M* and *P* pays off according to the criteria (PNW > 0, $B/C > 1$, RRR $>$ guiding rate). In this case, the incremental change from project *M* to project *P* increases the present net value of the costs by $125 and of the revenues by $200, giving an incremental PNW of $75 and incremental *B/C* of 1.6. Therefore, project *P* is the superior project.

With mutually exclusive projects, in summary, PNW, *B/C*, or RRR can be used to guide project selection after any needed normalization is done for size, reinvestment rate, and length. PNW must be normalized for reinvestment rate and length if the reinvestment rate differs from the guiding rate. *B/C* must be normalized for size and also must be normalized for reinvestment rate and length if the reinvestment rate differs from the guiding rate. RRR, to be used correctly, must include normalization for size, length, and reinvestment rate for purposes of comparing mutually exclusive projects.

The common forestry decision problem of choosing a rotation for a land parcel can be viewed as a mutually exclusive investment problem. The choice on a representative acre can be seen as investing the use of that acre of bare land along with some cash in one of a series of investments and harvests keyed to different rotation lengths in perpetuity. To evaluate the choices we need to calculate the PNW for each rotation in perpetuity. This is traditionally done by calculating the soil expectation value (SEV) while assuming the reinvestment rate is equal to the guiding rate so normalization is not needed. Then the rotation with the highest SEV is selected.

Independent projects with a capital constraint Capital budgeting problems arise when there are many possible independent projects in which to invest and there is not enough money to finance all of them. Hence money needs to be budgeted to the set of projects that will maximize total PNW from using the limited capital funds. As a first approximation, it is intuitively appealing to invest first in the projects that make the best use of each dollar. It turns out that when independent projects are ranked by NR/C_0 and then selected in descending order until investment funds are exhausted, PNW will be maximized. Choosing projects in the decreasing order of their PNW or B/C will not necessarily maximize overall PNW.

To illustrate capital budgeting, consider the woodland manager buying bare or brushy farmland to increase timber production capacity of company holdings. Ten different sized tracts are available for purchase, and detailed economic analyses of several prescriptions for each tract have been made to select the treatment with the highest PNW. A summary of relevant financial data is shown in Table 8-6.

Normalization of PNW for independent projects under a capital constraint is needed for reinvestment rate and length when the reinvestment rate differs from the guiding rate. We will assume here that the reinvestment rate equals the guiding rate, so this normalization is not required.

The projects are ranked for relative efficiency by NR/C_0, as indicated in the last column of Table 8-6. The projects are then reordered and presented in the order of rank, as shown in Table 8-7. In addition the cumulative budget and the PNW are obtained by keeping a running total of individual tract purchase costs and PNW. A table such as Table 8-7 is normally called a *capital budget,* since it shows which projects would be implemented at different budget levels. This same capital budget is also shown as a bar graph in Fig. 8-5. The height of each project bar is its NR/C and the width is the project cost. The area of the bar above $NR/C_0 = 1.0$ is accordingly the project PNW. For example, if the budget were

Table 8-6 Financial data for ten independent opportunities to buy tracts of cutover land for timber production (thousands of dollars)

Tract number (1)	Size of tract (acres) (2)	Tract purchase cost (3)	Present worth of net future timber incomes (4)	Present net worth (5) = (4) − (3)	NR/C (6) = (4) ÷ (3)	Rank by NR/C_0 (7)
A	40	$ 20	26	6	1.3	3
B	120	36	42	6	1.17	5
C	160	80	104	24	1.35	2
D	280	148	182	34	1.23	4
E	400	80	80	0	1.0	6
F	100	40	20	−20	0.5	9
G	250	250	87.5	−162.5	0.35	10
H	60	42	39	−3	0.93	7
I	80	40	38	−12	0.7	8
J	320	32	64	32	2.0	1

$150,000, then tracts *J*, *C*, *A*, and, if possible, $18,000 of *D* should be purchased to maximize PNW at $68,130. A capital budget shows information to guide three kinds of actions or decisions:

1. *Projects to reject,* those with $NR/C_0 < 1.0$ (projects *E*, *H*, *I*, *F*, *G*)
2. *The maximum justifiable budget request* ($316,000), cost of all projects with $NR/C_0 > 1.0$ (projects *J*, *C*, *A*, *D*, *B*)

Table 8-7 Capital budget for opportunities to purchase cutover land ranked by decreasing NR/C_0 (thousands of dollars)

Tract	Tract purchase cost*	Cumulative purchase cost	Tract PNW[2]†	Cumulative PNW	NR/C_0‡
J	32	32	32	32	2.0
C	80	112	24	56	1.35
A	20	132	6	62	1.3
D	148	280	34	96	1.23
B	36	316	6	102	1.17
E	80	396	0	102	1.0
H	42	438	−3	99	0.93
I	40	478	−12	87	0.7
F	40	518	−20	67	0.5
G	250	760	−162.5	−95.5	0.35

* From column 3, Table 8-6.
† From column 5, Table 8-6.
‡ From column 6, Table 8-6.

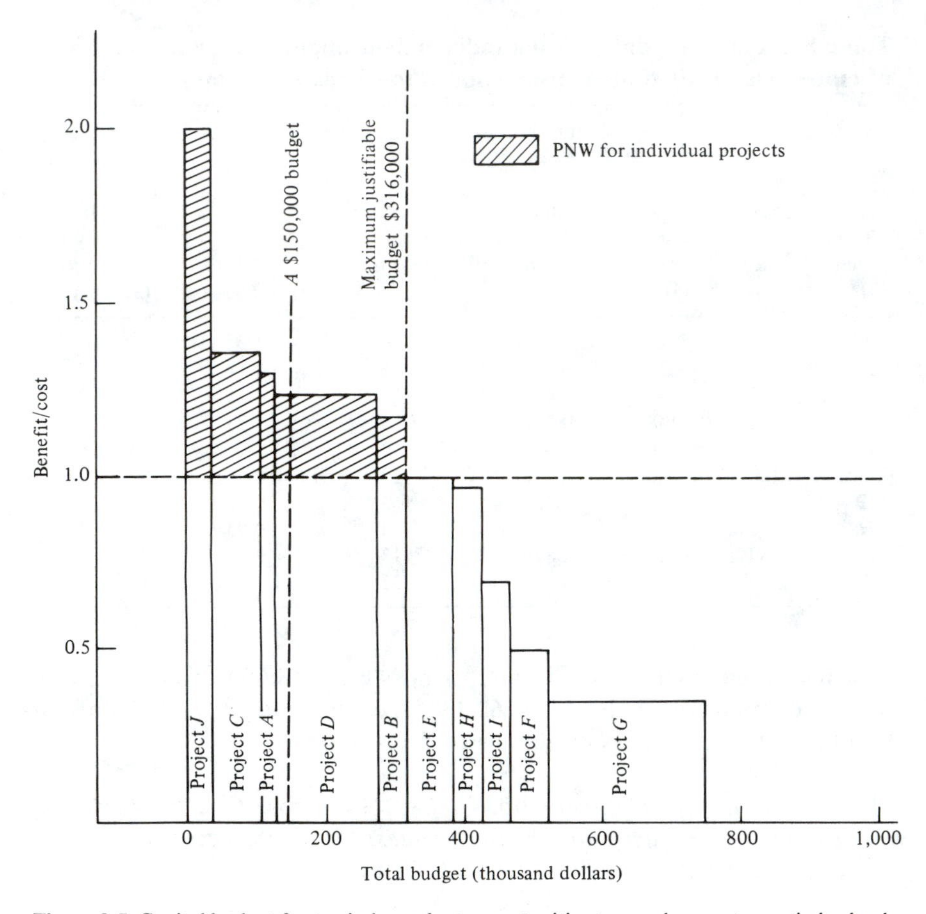

Figure 8-5 Capital budget for ten independent opportunities to purchase cutover timberland.

3. *Projects to implement with a limited budget,* implement in the order of NR/C (projects J, C, A, D, B, E)

If the projects are *discrete* and cannot be divided, then each project must be done in its entirety or not at all. This is probably the case with land buying since people do not usually sell part of a tract. Here the NR/C_0 rank may not give precisely the best answers. Discreteness amounts to an additional constraint on the budget allocation problem, and the choice of projects must now satisfy both budget and discreteness constraints. To illustrate, suppose the 10 tracts in Table 8-7 were discrete projects and the available budget were $250,000. We need to find the set of projects that (1) do not exceed $250,000 in cost, (2) include only complete projects, and (3) maximize total PNW. The relevant possibilities are as follows.

Project set	Total PNW	Total budget
Tracts *J*, *C*, *A*	62	132
Tracts *J*, *C*, *A*, *B*	68	168
Tracts *J*, *A*, *D*, *B*	78*	228 (best set)
Tracts *C*, *A*, *D*	64	248

* Highest PNW

The combination *J*, *A*, *D*, *B*, maximized PNW at $78,000 using a budget of $228,000. If strict *NR*/*C* rank order has been followed, only projects *J*, *C*, *A*, would be implemented, and projects *D* and *B* would not. The initial budget schedule was a good starting point, but some fine-tuning was needed.

Table 8-7 also illustrates the error of ranking independent projects by PNW. The order would be *D*, *J*, *C*, *A*, *B*, and when followed in the nondiscrete case, the $150,000 budget would go to *D* and part of *J* for a PNW of $36,000, as compared to the *NR*/*C* rank result of $68,130.

Many types of management treatments are divisible—such as seeding, fertilizing, and reforestation—and the straight *NR*/*C*-ranked capital budget gives good guidance. When dealing with hundreds of projects, large budgets, or when the data are rather poor, then fine-tuning is probably not worthwhile and judgment will suffice. On the other hand, if the projects are large and expensive, such as buying trucks or major mill equipment, then the fine-tuning to consider the discrete nature of the projects pays off.

A variety of other constraints can also be imposed on the general capital budgeting problem, such as saying that certain projects, or some out of a subset of projects, must be implemented for some reason. Conversely a subset of projects may all occur in the same watershed or area, and for aesthetic, wildlife, or political reasons a constraint might be imposed that only a maximum area of projects in that subset can be implemented. Many projects require cash outlays in future years and, if large, place a significant demand on future budgets. If these budgets are expected to have limits, then these future budget constraints should be defined and the simple current year capital budgeting problem expanded to a multiperiod budgeting problem. When many constraints are imposed, the budgeting problem becomes difficult to solve by inspection or two-dimensional graphics, and linear programming or integer programming can be used to effectively search out the optimal solution. Teeguarden's reforestation budget problem in Chap. 6 with its many seed source constraints is a good example of such an application.

The investment decision guides in summary To complete our discussion of decision criteria, we recommend the following steps and rules when sizing up an investment problem situation and choosing the criteria for evaluating and comparing projects:

1. Check all proposed projects against problem constraints to determine which are feasible candidates.
2. If the projects are truly independent, select all projects for which PNW > 0, $B/C > 1$, or IRR $>$ guiding rate.
3. If projects are mutually exclusive, normalize all projects for reinvestment, size, and length as needed. Choose the project with the highest normalized PNW, B/C, or RRR.
4. If the projects are independent with an initial budget constraint, normalize for reinvestment rate and length if the reinvestment rate differs from the guiding rate, and select the projects by descending NR/C_0 until the budget is exhausted.
5. If the projects are independent, capital is limited, and there are additional constraints on the projects (such as integer constraints on some projects, future period budget constraints, or various linkage and distributional constraints on the projects), then the investment problem can be formulated as a decision problem under constraints. The decision variables are dollars invested in each project, the objective function is to maximize the aggregate normalized PNW of all projects undertaken, and integer and all other relevant constraints are appropriately specified and included in the problem formulation. Mathematical programming is often useful in solving such problems.

Examples of Efficiency Analysis

Many excellent examples of capital budgeting and project evaluation are available in the published literature. Four have been partially reproduced here to suggest the diversity of application.

Reforestation investments in Oregon Our first example is the work of Teeguarden (1969) who quantified the economic returns from reforestation investments in the Douglas-fir lands managed by the Bureau of Land Management in southwest Oregon. This extensively documented study developed treatment-response relationships in a risk framework, estimated costs, used current stumpage values, and with an interest rate of 3 percent, calculated three criteria to determine the most efficient treatment prescription for each land type:

1. Present value dollar return per dollar invested (the NR/C ratio)
2. Present value of wood yield per dollar invested (this criterion assigns the same unit value to all wood produced)
3. Present net worth

These criteria were calculated for 15 regenerative prescriptions which could be applied to each of six site-aspect land classes. To apply these results, 33 tracts containing 5,954 total acres were evaluated for returns to reforestation investment. As these were independent investment opportunities, the *NR/C* criterion was used to rank the tracts, producing the classical capital budget shown in Table 8-8. This practical application of investment analysis reveals *NR/C* ratios ranging from 10.6 on the best tract to 0.4 on the worst—good information to guide decision analysis on a limited budget. Note that if you ranked the reforestation projects by their project PNW, a much different ranking with a lower aggregate PNW for a given budget would result.

Sugar pine in Oregon and California One of the earliest studies of increasing timber supply through intensive management was presented by Vaux (1954), who examined the 2.14 million acres of potential sugar pine lands in Oregon and California to estimate how much additional sugar pine could be produced at different costs. After stratifying the land into various site and condition classes, Vaux evaluated each stratum for the most efficient blister rust control and pruning treatment intensity. Then all acres in the region were presented in a capital budget that ranks opportunities by the present value of cost per thousand board feet of yield. These results are shown in Fig. 8-6.

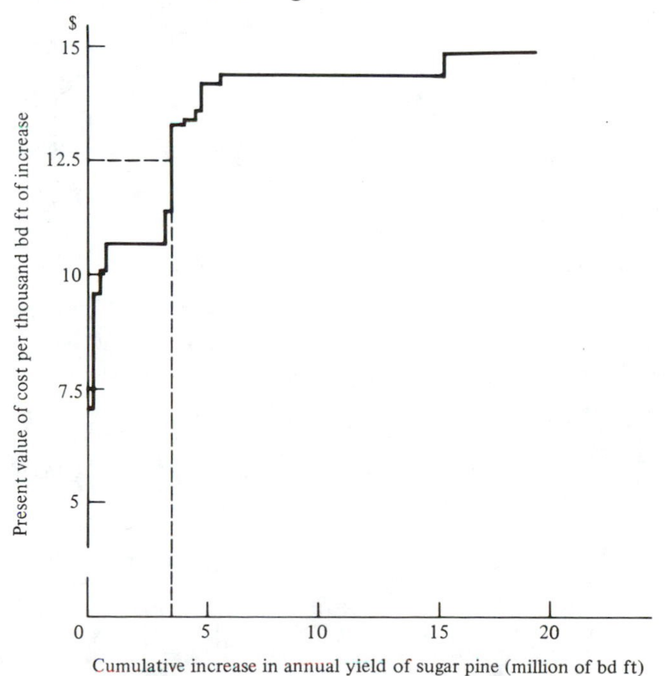

Figure 8-6 Supply schedule for sugar pine obtained from intensive management of cutover lands in Oregon and California. *(After Vaux, 1954.)*

Table 8-8 Douglas-fir reforestation investment opportunities, ranked by benefit-cost ratio

Treatment site class	Class acreage	Dollar return per dollar invested (*NR*/*C*)	Discounted wood yield per dollar invested	Project present net worth, dollars per acre
Seeding—bare land (class A and B), large tract, site II, north aspect	253	10.6	392	104.10
Planting—bare land (class B), large tract, site II, south aspect	69	6.2	228	163.20
Seeding—bare land (class A and B), small tract, site II, north aspect	109	5.9	218	53.00
Planting—bare land (class B), small tract, site II, south aspect	24	5.7	209	147.20
Interplanting—large tract, site II, north aspect	70	4.7	233	88.20
Seeding—bare land (class A and B), large tract, site III, north aspect	1,155	4.4	173	37.10
Interplanting—large tract, site II, south aspect	23	4.1	204	73.80
Planting—bare land (class B), large tract, site III, south aspect	789	2.9	112	59.00
Seeding—bare land (class A), large tract, site III, south aspect	175	2.8	109	19.30
Planting—bare land (class B), small tract, site III, north or south aspect	294	2.4	94	44.80
Interplanting—small or large tract, site III, north aspect	496	2.3	123	30.10
Interplanting—small or large tract, site III, south aspect	344	1.9	102	20.80
Planting—bare land (class A), small tract, site III, north or south aspect	34	1.9	75	36.40
Interplanting—chemical brush control, large tract, site III, north aspect	80	1.8	98	23.90
Planting—bare land (class A), posttreatment chemical brush control, small tract, site III, north aspect	11	1.7	67	31.70
Interplanting—chemical brush control, large tract, site III, south aspect	70	1.5	81	14.70
Interplanting—chemical grass control, large or small tract, site III, north aspect	256	1.4	78	16.10

Table 8-8 (Continued)

Treatment site class	Class acreage	Dollar return per dollar invested (NR/C)	Discounted wood yield per dollar invested	Project present net worth, dollars per acre
Planting—bare land (class B), large tract, site IV, north aspect	299	1.2	52	6.30
Interplanting—chemical grass control, large tract, site III, south aspect	236	1.2	65	6.90
Planting—bare land (class B), small tract, site IV, north aspect	42	1.1	46	2.30
Planting—bare land (class A), large tract, site IV, north aspect	68	1.0	42	−1.60
Planting—bare land (class B), small tract, site IV, south aspect	90	0.9	41	−1.80
Interplanting—small or large tract, site IV, north aspect	118	0.9	52	−2.20
Planting—bare land (class A), small tract, site IV, south aspect	7	0.8	33	−9.70
Planting—bare land (class B), large tract, site IV, south aspect	214	0.7	29	−10.50
Planting—bare land (class B), chemical grass control, small tract, site IV, north aspect	18	0.7	32	−11.70
Interplanting—small or large tract, site IV, south aspect	193	0.6	36	−8.90
Planting—bare land (class B), chemical grass control, small tract, site IV, south aspect	52	0.6	28	−15.80
Planting—bare land (class A), large tract, site IV, south aspect	88	0.5	23	−18.50
Planting—bare land (class A), posttreatment brush control, large tract, site IV, south aspect	62	0.5	20	−23.70
Interplanting—chemical grass control, large tract, site IV, north aspect	91	0.5	28	−19.00
Interplanting—chemical brush control, small tract, site IV, south aspect	11	0.5	28	−15.90
Interplanting—chemical grass control, small or large tract, site IV, south aspect	113	0.4	23	−22.90

Source: Teeguarden, 1969.

Vaux's application is a ranking by C/B, where the benefit B is given a unit value per M bd ft. This upward sloping curve can be interpreted as a marginal cost or supply curve. All investments with a lower present cost per thousand board feet than the present value of forecasted sugar pine stumpage prices in the period 2010–2070 would be acceptable investments. For example, if the forecasted sugar pine selling price had a present value of $12.50 per M bd ft, then investments would be justifiable in projects that provided an additional 4.5 million board feet of annual yield. Interestingly, this study was prompted by the then high prices for sugar pine stumpage relative to other species, reflecting the species' special value for millwork and pattern stock. The subsequent emergence of plastics and other substitutes eliminated this price premium, and most of the incentive for intensive sugar pine management has disappeared.

Timber management in the national forests In a classic national-level capital budgeting study Marty and Newman (1969) stratified about 97 million acres of national forest commercial timberland into 60 species site classes and examined different independent opportunities for reforestation and stand improvement for each. A cash flow schedule of costs and revenues, with and without treatment, for each land class was estimated and then the internal rate of return calculated. The treatment opportunities were then ranked by internal rate of return and are presented in Table 8-9.

A total of 14,255,000 acres of the 96,540,000 acres of commercial timberland were currently available for treatment and had positive rates of return. If all implemented, these projects would have an annual yield impact of almost a billion cubic feet. Marty and Newman showed that about 210 million dollars could be invested in the national forests in projects earning over 7 percent. This study was particularly significant as it provided an important empirical basis for much of the timber policy analysis of the 1970s.

Opportunities to increase southern pine production The last example of a capital budget is a regional analysis by Dutrow (1978), who examined the potentials of five southeastern states (Virginia, North Carolina, South Carolina, Georgia, and Florida) to increase pine timber harvests. Dutrow stratified 57 million acres of land needing some sort of treatment into nine stand conditions and three site types. Capturing a synthesis of expert judgment and available data in a series of conferences, the best prescription was determined for each stand type and its cost and timber yield estimated. Market prices in 1977 were used as a base, and a 3 percent real rate of relative price increase per year for stumpage was assumed, based on recent trend data.

This analysis was used as a part of the 1980 assessment of the nation's forest and rangelands for their current and potential productivity (U.S.

Table 8-9 Area, relative efficiency by IRR, and current timber intensification treatment opportunities by timberland class on the national forests

		Relative efficiency of management intensification			Current intensification treatment opportunities					
		Annual yield per acre			Reforestation			Stand improvement		
Timber type group and site class*	Total area† (10^3 acres)	Without intensification (ft^3)	Added by intensification‡ (ft^3)	Rate of return to intensification IRR (%)	Area (10^3 acres)	Cost (10^3 $)	Annual yield impact‖ (10^3 ft^3)	Area (10^3 acres)	Cost (10^3 $)	Annual yield impact§ (10^3 ft^3)
Sitka spruce-hemlock I	2,135	237	90	15.4	0	—	—	0	—	—
Sitka spruce-hemlock II	1,606	104	47	14.5	0	—	—	0	—	—
Longleaf-slash pine I	73	127	34	14.4	13	767	1,778	9	153	308
Longleaf-slash pine II	220	101	31	13.0	39	2,301	4,384	12	204	372
Red-E. white pine I	60	150	6	10.0	25	1,475	3,325	34	578	221
Longleaf-slash pine III	1,026	66	22	9.8	152	8,968	11,400	45	765	995
Douglas-fir I (C)	4,155	129	118	9.2	66	6,024	15,483	281	12,010	33,186
Red-E. white pine II	354	102	20	8.8	84	5,145	8,677	114	1,902	2,223
Sitka spruce-hemlock III	1,257	65	19	8.7	0	—	—	0	—	—
Oak-hickory I	107	157	25	8.2	11	649	2,152	24	438	2,088
Ponderosa pine I (C)	903	93	120	7.9	134	15,008	27,108	78	2,783	9,345
Red-E. white pine III	976	58	33	7.8	181	11,771	13,955	303	4,882	9,908
Douglas-fir II (C)	2,792	81	91	7.7	308	32,021	50,388	350	12,159	31,815
Shortleaf-lobl. pine III	2,779	69	7	7.4	892	53,349	58,159	353	5,838	2,612
Fir-spruce I (C)	2,137	163	58	7.3	93	8,229	19,530	156	7,069	9,080
Shortleaf-lobl. pine II	1,794	101	29	7.3	161	9,499	17,678	53	901	1,511
Shortleaf-lobl. pine IV	542	42	4	7.3	36	2,124	1,404	13	221	57
Douglas-fir I (M)	144	129	118	7.0	3	259	704	8	316	945
Oak-hickory II	296	110	29	7.0	27	1,656	3,759	41	745	2,624

Table 8-9 (Continued)

		Relative efficiency of management intensification			Current intensification treatment opportunities					
		Annual yield per acre			Reforestation			Stand improvement		
Timber type group and site class*	Total area† (10^3 acres)	Without intensification (ft^3)	Added by intensification‡ (ft^3)	Rate of return to intensification IRR (%)	Area (10^3 acres)	Cost (10^3 \$)	Annual yield impact‖ (10^3 ft^3)	Area (10^3 acres)	Cost (10^3 \$)	Annual yield impact§ (10^3 ft^3)
Spruce-fir II (East)	189	91	25	6.7	4	264	396	0	—	—
Spruce-fir I (East)	51	111	12	6.5	0	—	—	0		—
Maple-beech-birch II	943	98	33	6.4	34	2,244	3,546	56	1,056	1,333
Douglas-fir III (C)	3,872	54	74	6.4	129	12,423	15,789	244	9,959	18,130
Fir-spruce I (M)	151	163	58	6.2	21	1,609	4,410	3	106	174
Sitka spruce-hemlock IV	97	36	13	6.1	0	—	—	0	—	
Longleaf-slash pine IV	147	33	9	6.0	26	1,543	928	7	119	61
Red-E. white pine IV	19	40	30	5.9	2	118	118	3	51	89
Ponderosa pine I (M)	275	93	120	5.9	28	2,388	5,665	33	1,478	3,953
Shortleaf-lobl. pine I	491	136	48	5.8	36	2,124	5,641	13	221	628
Ponderosa pine II (C)	1,803	58	88	5.8	173	19,304	24,012	225	8,650	19,868
Lodgepole pine I	798	99	49	5.6	47	3,851	6,618	219	9,315	14,214
Fir-spruce II (C)	2,486	91	40	5.6	121	10,735	15,065	181	8,058	7,186
Douglas-fir II (M)	953	81	91	5.6	9	777	1,473	68	2,572	6,181
Spruce-fir III (East)	751	49	34	5.5	138	9,108	9,757	130	2,080	4,446
Maple-beech-birch I	257	139	17	5.1	0	—	—	0	—	—
Fir-spruce III (C)	3,151	64	35	5.0	108	11,151	10,227	165	7,246	5,841
Ponderosa pine III (C)	4,309	39	59	4.8	285	30,877	26,562	539	22,224	31,585
Douglas-fir IV (C)	2,329	37	56	4.8	59	5,429	5,216	110	4,657	6,204
Fir-spruce II (M)	693	91	40	4.6	80	6,155	9,961	12	342	477

Lodgepole pine II	2,279	76	50	4.5	101	8,068	12,131	581	24,314	30,126
Spruce-fir IV (East)	128	21	35	4.5	0	—	—	0	—	—
Douglas-fir III (M)	2,352	54	74	4.3	30	2,630	3,672	107	4,021	7,950
Ponderosa pine II (M)	1,002	58	88	4.2	122	9,846	16,933	123	3,550	10,861
Fir-spruce III (M)	2,472	64	35	4.2	27	2,367	2,557	64	1,880	2,266
Larch-W. white pine I	791	152	34	4.0	39	3,315	6,876	107	4,952	3,627
Maple-beech-birch III	1,281	71	40	4.0	107	7,062	9,555	441	9,236	17,684
Larch-W. white pine II	1,082	106	37	3.7	37	3,145	5,017	131	6,045	4,205
Oak-hickory III	1,671	64	40	3.6	128	8,175	10,598	810	16,860	32,157
Larch-W. white pine III	666	75	38	3.4	4	340	429	40	1,846	1,504
Ponderosa pine III (M)	3,122	39	59	3.2	195	15,743	18,174	379	11,578	22,092
Lodgepole pine III	4,617	43	52	3.1	97	8,259	8,740	680	28,189	35,360
Fir-spruce IV (C)	1,309	30	16	3.0	29	2,987	1,259	59	2,662	927
Oak-hickory IV	2,535	35	42	3.0	47	2,773	2,886	579	10,545	24,029
Douglas-fir IV (M)	2,124	37	56	2.9	21	1,869	1,856	21	588	1,185
Larch-W. white pine IV	294	40	41	2.8	2	170	154	8	366	325
Fir-spruce IV (M)	2,718	30	16	2.5	14	1,226	756	31	981	487
Lodgepole pine IV	4,253	23	53	2.4	19	1,635	1,386	324	13,243	17,269
Maple-beech-birch IV	216	36	23	2.1	0	—	—	0	—	—
Ponderosa pine IV (C)	1,750	17	28	1.5	181	19,516	7,584	218	10,100	6,017
Ponderosa pine IV (M)	4,827	17	28	1.0	151	12,883	6,327	679	22,632	11,205
Other types	7,930	55	40	—	85	5,610	3,740	0	—	—
Total	96,540	—	—	—	4,961	396,955	505,898	9,264	302,686	456,936

* Type groups and site productivity classes follow forest survey standards. Classes designated C (coastal) occur in California, Oregon, Washington, Montana, and northern Idaho, while classes designated M (mountain) occur in Arizona, New Mexico, Nevada, Utah, Colorado, South Dakota, and southern Idaho.

† Unreserved, commercial forestland as defined by forest survey.

‡ For fully occupied stands and effective stocking control.

‖ Based on varying proportions of total intensified yield (column 3 plus column 4) depending on timber type.

§ Based on yield added by intensification (column 4).

Source: Marty and Newman, 1969.

Table 8-10 Stand conditions, treatments, IRR, and marginal costs for increasing southern region softwood (after Dutrow, 1978)

Existing stand conditions	Total acres (10^3)	Recommended prescription	Annual yield response		Earnings rate* (IRR) (%)			Marginal cost ($/ft^3)
			Per acre (ft^3/acre)	Region (10^6 ft^3)	Site quality: High	Medium	Low	
Pine intensification								
1. Nonstocked timberland	4,362	Site prepare and plant	94	409	12	10	9	0.034
2. Idle cropland	3,237	Plant, with minimum site preparation	175	566	16	†	†	0.006
3. Young pine plantations being overrun by competition	573	Clean and release	39	22	15	14	11	0.032
4. Poorly stocked or damaged pine population	276	Harvest early and plant pine	121	33	11	10	9	0.026
5. Well-stocked natural pine stands over 40 years of age	5,591	Harvest and plant pine	103	573	11	10	7	0.031
6. Understocked natural pine stands with adequate seed source	3,587	Regenerate naturally	75	220	13	12	10	0.022

7. Understocked natural pine stands without adequate seed source	1,842	Harvest and plant pine	99	183	11	10	8	0.033
				2,006				
Hardwood-to-pine conversion								
8. Conversion options (to pine):								
a. Understocked oak-pine over 40 years old	6,292	Clear, site prepare, plant	111	696	10	9	8	0.035
b. Scrub oak stands	536	Clear, site prepare, plant	89	48	11	10	8	0.043
c. Chestnut oak stands	655	Clear, site prepare, plant	41	27	†	7	5	0.093
d. Oak-hickory stands	23,397	Clear, site prepare, plant	72	1,676	10	8	6	0.054
				2,447				
Hardwood intensification								
9. Bottom-land hardwood stands	6,742	Harvest old stands and regenerate	44	298	10	9	8	0.018
Regional totals‡	57,090			4,801				

* Calculated with a 3% real price increase.
† Insufficient acreage for calculations.
‡ Includes Virginia, North Carolina, South Carolina, Georgia, Florida.

Forest Service, 1980). A summary of the specific results is shown in Table 8-10. The potential gains are substantial. The first seven stand conditions are for existing pine types, and the recommended treatments involve improving stocking and growth rates. They have internal rates of return of over 10 percent and provide a total increase in annual yield of 2006 million ft^3. Opportunities for converting hardwood types to pine types provide an additional 2447 million ft^3 of annual yield, although the earning rates average a bit lower. As with Vaux's studies, the marginal costs can be compared to stumpage prices to suggest those that are good investment opportunities. A cord of wood contains about 80 ft^3, and at a selling price of only $10 per cord, this revenue has a present value of about $0.089 per ft^3,* considerably in excess of most of the marginal cost estimates. The maximum gain of 4.8 billion ft^3 per year would more than double the 1970 total growth in the region. A capital investment of over $5 billion would be needed to achieve these gains.

Regional Goals and Impact Analysis

Forests occupy lots of space, about 33 percent of the United States surface area, and as a result have a widespread political constituency. County governments depend on property taxes for incomes. When much of this taxable land is forested, maintaining extensive but little used forest roads is a major budget item for rural counties. What happens on public and private forestlands is of immediate importance to people living in the sparsely populated forested counties—they can get terribly excited about timber harvest, herbicides, grazing permits, wildlife harvest policies, access, and use regulations because it affects their livelihood and lifestyles. Not surprisingly, it is the locals who show up in force for meetings to discuss forest plans and policies. These are also the people who actively write and call their elected members of state and national Congress and bring political pressure to bear disproportionate to their numbers. As they like to say in Montana or Wyoming—one cow, one vote.

With political influence there is power to alter public and private forest management decisions, and the analyst is well advised to consider the perspective of the local constituency when evaluating the impact of proposed plans and projects on the people, economy, culture, and environment. Such an evaluation is now mandated for the U.S. Forest Service, "The physical, biological, economic and social effects of implementing each alternative shall be estimated and compared . . . including impacts on . . . receipt shares to state and local governments, and em-

* The present value was estimated using Eq. (7-12) and assuming $t = 30$ years, $r = 0.4$, $k = 0.05$, $p = 0.08$ to reflect the 3 percent real rate of price increase, $R_t = 10/80 = 0.125$.

ployment in affected areas.'' (U.S. Forest Service, 1982, Sec. 219.27.9.3). Any sensitive and successful manager knows that actions need to be acceptable to influential locals. Hence analysis of the effect of forest management decisions on local goals is needed whenever the effects are anticipated to be significant as perceived by local interests.

How do we measure in some objective way the impact of forest activity on the local area? This is the decision analyst's operational problem, for eventually criteria and numbers need to be produced for a summary evaluation table, such as Table 8-1. Procedurally, the following steps should be taken:

1. Define the geographic boundaries of the region within which impacts are to be quantified and evaluated.
2. Determine the goals held by persons within the region which are significantly affected by the proposed actions.
3. Establish a criterion for each goal to measure the impact.
4. Using appropriate analytical procedures, estimate the quantitative change over time in each criterion with and without implementation of the proposed decision alternative.
5. Summarize the impacts in some objective way so as to provide meaningful information to the decision maker.

This is easier said than done, for there is a complex array of interrelated regional goals, and numerical criteria can be defined for only a few of them. Furthermore, available quantitative procedures are limited and often expensive to apply. Even so, something can be said. The recent development of inexpensive, locally applicable input-output economic impact models now permits making estimates of impact for important regional criteria of employment and income.

Defining the region This is perhaps the easiest step. After deciding what geographic area to call ''the region,'' it is a mapping job. In regional impact analysis, the presumption is that effects on people and their environment within the region count and effects outside the region do not. This is similar to the choice of planning horizon, where we said costs and revenues occurring after the planning horizon count zero.

Most of the information needed by the analyst is recorded by political units, the smallest of which are towns, cities, and counties. Practically, regions are almost always defined as a set of contiguous counties. The size of the region depends on the decision considered. Analysis in 1980 of the proposed siting of the MX missile system in the intermountain west used a two-state region of Nevada and Utah to evaluate impacts. On the other hand the impact from closing a sawmill could probably be evaluated using a region of a few counties.

Regional goals and criteria Regional goals are not too different from national goals except that they are more personalized and take on the flavor of the local culture. Economic viability and full employment are always prime goals. In addition, the local government wants to maintain control of all actions within its jurisdiction, have a stable source of budget revenue, and particularly dislikes actions that increase social costs for schools, roads, or welfare without commensurate increases in tax and fee revenues. Forested regions are usually rural and there often is a strong, somewhat antidevelopment goal of preserving the traditional local culture—be it agriculture, ranching, or timbering.

Recent urban to rural migration has created many forest communities whose main concern is preserving the rural ambience by preventing further development and settlement. Anything that causes abrupt cultural changes, such as resort development or large-scale mineral exploitation, is resisted. Dilemmas and conflicts abound in regional goals. The desperate desire by parents for new local jobs "to keep the kids home" directly conflicts with the equally strong desire not to change the traditional culture and economic lifestyle. And businesspersons with land and other assets want to sell, rent, and develop to make more money.

Schuster (1976) provides a useful grouping of regional goals and social impact criteria for evaluating changes, as shown in Table 8-11.

Table 8-11 Regional goals and social impact criteria for use in analyzing local economic responses to forest management activities

Regional goal	Social impact criterion
Economic activity	1 Employment
	2 Payroll/wages
	3 Value added
	4 Sales
	5 Costs/gains of other outputs
Individual welfare	1 Unemployment rates
	2 Income redistribution
	3 Average wage rates
Area equilibrium	1 Economic diversity
	2 Community lifestyle
	3 Social disorganization (divorce, crime, etc.)
	4 Future development
Local government	1 In-lieu payments
	2 In-kind payments
	3 Cost to local government

Source: Adapted from Schuster (1976).

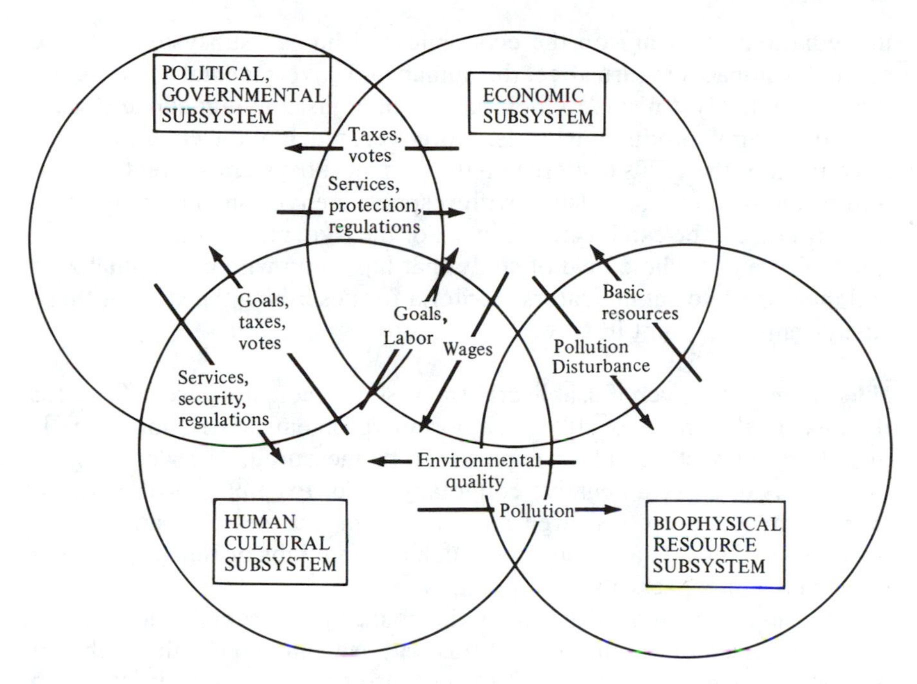

Figure 8-7 Regional subsystems for impact analysis. *(After J. J. Kennedy, Utah State University.)*

A useful conception of the internal working of a region as a set of four dynamically interrelated subsystems is illustrated in Fig. 8-7.* Some of the more important interactions between the subsystems are shown. The human and resource subsystems provide the labor and material inputs to operate the economic system, which in turn provides the employment and income to support the human subsystem. Alteration and modification of biophysical resources are continuous, and some are viewed negatively by the human subsystem as "pollution." The government subsystem is mostly a service system drawing taxes and votes from the human and economic systems and returning a variety of services and security. The government system also regulates the terms under which the human system and the economic system can utilize or exploit the biophysical resource system (such as fishing regulations, residential zoning, pollution control).

For the forest management decision analyst, the actions under consideration are almost wholly located within the resource system, and the

* This conceptualization is adapted from teaching material developed by J. J. Kennedy at Utah State University.

immediate impacts are on the economic and human subsystems. Since economic impacts in turn affect the human and governmental subsystems, the careful analyst needs to look further than just the immediate impact on, say, sawmill production levels. Moreover, much of the environmental movement of the 1970s centered on the human subsystem's direct evaluation of the resource subsystem with respect to environmental quality.

Criteria can be established to index or measure many of these interactions. There is a whole field of study that has to do with conceptualizing and measuring social indicators—criteria for describing the state of these subsystems at a point in time.

What is impact? Even if usable criteria or social indicators are defined for regional goals, there is still an important problem of defining what is meant by "impact" and how impact is to be measured. The word *impact* itself tends to carry a negative connotation. For example, most environmental impacts are considered bad, hence impact means "bad." Obviously, some impacts are also good, such as more jobs or reduced erosion through a soil conservation program.

Impact is best defined as simply the change in the social indicator due to the proposed decision. Defined this way we can employ the with and without principle from Chap. 7 to measure the amount of impact. To illustrate, suppose the goal in question is to provide full employment, and one important social indicator is the number of forest product manufacturing jobs within the region. Proposed is a policy to provide federal reforestation and timber culture subsidies and technical assistance to landowners in order to increase stumpage production and marketing on small nonindustrial private lands. Two alternatives for this program are considered: (1) a perpetual program with a budget of $1,000,000 per year, and (2) an intensive program with a budget of $3,000,000 per year for just 5 years. The level of timber growth can be converted to employment with appropriate assumptions and analytics, providing the information shown in Fig. 8-8.

What, quantitatively, is the impact? Both programs are expected to reverse a historical downtrend in employment, and program effects are cumulative and persistent over time. What numbers shall we use to measure impact? If the date of program implementation is 1985 and the effects last for some time, then the analyst must first establish a planning horizon, the length depending on the decision maker's viewpoint and the time preference rate at which the future is discounted. Suppose, 15 years is chosen. At least three possible measures of employment impact are possible.

1. Measure the difference between the forecasted level of annual employment at the planning horizon with the program and the *cur-*

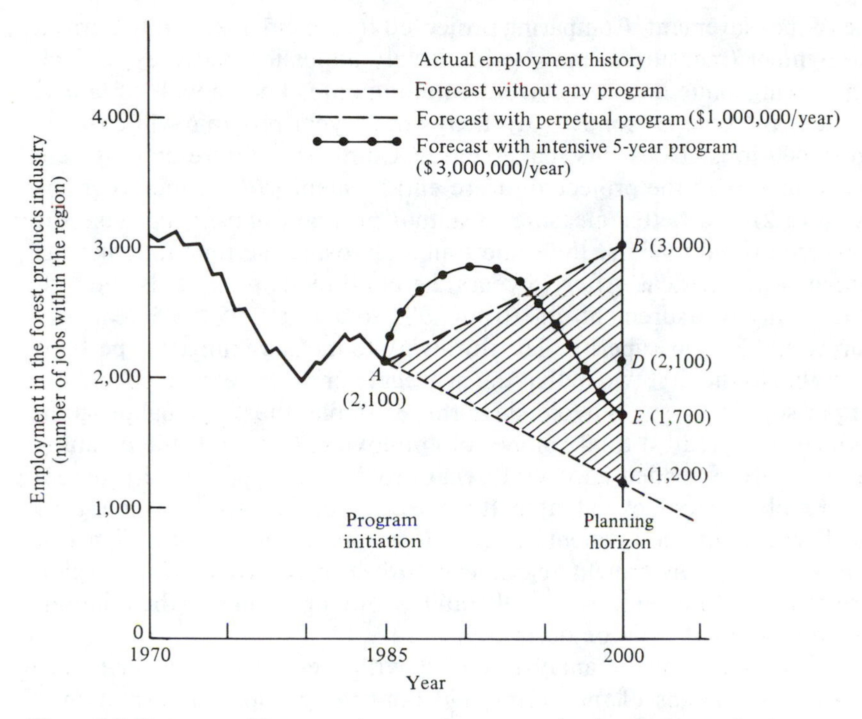

Figure 8-8 Forecasts of forest product employment under three reforestation programs.

rent level of employment. Impacts are:

$$\text{Perpetual program, } B - A = 3{,}000 - 2{,}100 = 900 \text{ jobs}$$

$$\text{5-year program, } E - A = 1{,}700 - 2{,}100 = -400 \text{ jobs}$$

2. Measure the difference between forecasted annual employment with and without the program at the planning horizon. Impacts are:

$$\text{Perpetual program, } B - C = 3{,}000 - 1{,}200 = 1{,}800 \text{ jobs}$$

$$\text{5-year program, } E - C = 1{,}700 - 1{,}200 = 500 \text{ jobs}$$

3. Measure the cumulative impact of the program over the planning period. Impacts are:

$$\text{Perpetual program} = \text{area } ABC = 13{,}500 \text{ person-years}$$

$$\text{5-year program} = \text{area } AEC = 13{,}000 \text{ person-years}$$

The first two numbers are point estimates at some reference year and are a "sample" of the impact. They are strongly influenced by the chosen time of measurement. Comparing projected future employment to current employment (measure 1) may be intuitively appealing because the decision is being made now and current employment is known with certainty. Consider the comparison $E - A$; it says the 5-year program will cost the region 400 jobs. Does this make sense? Comparing future employment *with* programs to the projected future employment *without* the programs (measure 2) is a better measure of actual program impact, because the measures are all made at the same time. Choosing the time to make the comparison is critical. After 15 years the continuous program is superior while at any measurement time from 1985 to about 1995 the 5-year program would appear superior. The third way to measure impact is perhaps preferable to the first two in that the aggregate impact is evaluated and the comparison time issue is avoided. In this example, the perpetual program produces the greatest total number of employment years. If the planning horizon were extended another 10 years to 2010, the perpetual program would probably look even better. If we used a guiding rate to discount the social benefits of employment, the 5-year program would look even better because most of its benefits came early. Still, if the goals of the region were strongly to even growth and stability of employment—the continuous program might still be desired.

It is crucial that the analyst consult with the decision maker before choosing the means of measuring and conveying impact information—lots of different summary or index numbers can be produced from the same set of information. The most objective information is simply to provide the entire schedule of effects as depicted in Fig. 8-8. But this may be more information than the decision maker wants and/or budgets and analytical techniques can make available. This conceptual problem of how to describe impacts applies to all goals and social indicators, whether numbers, symbols, or words are used to measure change. No absolute answers can be offered other than that it is a choice the decision maker(s) and the decision analyst must jointly make.

Biologists, historians, sociologists, cultural anthropologists, landscape architects, and economists all ply their trade in trying to estimate impact. The methods are descriptive and judgmental for many of the indicators, particularly in the human, cultural subsystem. One important method used to estimate regional economic impact is the input-output model. If the conceptual assumption of the model and the empirical data used do not seriously violate reality in a particular application, then input-output analysis can be quite helpful to compare alternatives on the criteria of employment, wages, total economic activity (gross regional product), and selected resource utilization rates, such as water consumption.

Input-output models The input-output (I-O) model classifies all regional industries into a set of sectors, such as agriculture, sawmilling, or law enforcement and then statistically describes the economic subsystem by the dollar sales and purchases of all sectors to and from each other. Such a transactions table for a five-sector economy is illustrated in Table 8-12*a*. In this example the first row of the table shows the sales (output) of the forest products sector to all other sectors for total sales of \$600. \$350 of sales are exports (buyers outside of the region). The first column shows the purchases (input) of the forest products sector from all other sectors, which also total to \$600. Household (labor) input at \$300 is half the total production cost of \$600. All profits and wages are accounted for in the household sector and all input shortages are provided by imports from suppliers outside the region. This example economy is heavily dependent on timber as its primary industry. A local resource (stumpage) is processed and sold outside the region (exports) to bring dollars back into the local economy. It is also the largest employer in the region because of the \$300 of purchases from the household sector (for labor). The total economic activity of the region is \$3,170.

Given the transactions table, we can use fairly simple matrix algebra to derive a predictive model that linearly relates the internal level of regional economic activity to the level of exports outside the region.

The derivation of this impact multiplier matrix from the transactions matrix uses the three submatrices X, Y, and Z which are shown partitioned in the transactions matrix (Table 8-12*a*) and presented in variable form in Table 8-12*b*.

Let x_{ij} = amount of sales from sector i to sector j. For example, x_{24} represents \$50 of sales from agriculture to households. It also is the amount of purchases by sector j from sector i.

y_i = sales of producing sector i outside the region through exports. y_3, for example, is \$150 sales of the other manufacturing sector to buyers outside the region. Sales outside the region are also called final demand in input-output models.

z_i, z_j = total sales of sector i or total purchases of sector j. By definition sales and purchases of a given sector are equal, that is, $z_i = z_j$.

Using these we can define the technical coefficient a_{ij} as

$$a_{ij} = \frac{x_{ij}}{z_j} \tag{8-4}$$

where a_{ij} = portion of total purchases by sector j from sector i
x_{ij} = amount of purchase by sector j from sector i
z_j = total purchases by sector j

For example, $a_{31} = 145/600 = 0.24$, or 24 percent of all purchases by the forest products sector are obtained from the other manufacturing and

Table 8-12 Input-output model

(*a*) Transactions matrix

Producing sector (i)	Purchasing sector (j): Forest products (1)	Agriculture (2)		Other manufacturing and services (3)	Households (4)		Exports (5)		Total sales
(1) Forest products	30	70		100	50		350		$ 600
(2) Agriculture	25	100	*X*	25	50	*Y*	200	*Z*	400
(3) Other manufacturing and services	145	50		155	100		150		600
(4) Households	300	130		100	0		170		700
(5) Imports	100	50		220	500		0		870
Total purchases	600	400	*Z*	600	700		870		3,170

(*b*) Coefficient matrix

Producing sector (i)	Purchasing sector (j)					Total sales
	(1)	(2)	(3)	(4)	(5)	
(1) Forest products	x_{11}	x_{12}	x_{13}	x_{14}	y_1	z_1
(2) Agriculture	x_{21}	x_{22} X	x_{23}	x_{24}	Y y_2	Z z_2
(3) Other manufacturing and services	x_{31}	x_{32}	x_{33}	x_{34}	y_3	z_3
(4) Households	x_{41}	x_{42}	x_{43}	x_{44}	y_4	z_4
(5) Imports						
Total purchases	z_1	z_2 Z	z_3	z_4		

(*c*) Multiplier matrix

Internal producing sector (i)	Purchasing sector (j)			
	(1)	(2)	(3)	(4)
(1) Forest products	1.25	0.42	0.34	0.17
(2) Agriculture	0.18	1.46	0.15	0.14
(3) Other manufacturing and services	0.58	0.53	1.58	0.31
(4) Households	0.78	0.78	0.48	1.18
Total impact on internal sectors—final demand (export) multipliers*	2.79	3.19	2.55	1.80

* The multipliers include direct, indirect, and induced impacts.

services sectors. Rearranging Eq. (8-4) and substituting z_i for z_j, we have

$$x_{ij} = a_{ij} z_i \tag{8-5}$$

Using the definitions of a_{ij} and the X, Y, and Z submatrices designated in the transaction matrix, we have four matrices needed for the derivation. The capital letters A, X, Y, Z stand for the whole matrix,

$$A = \{a_{ij}\}, \qquad X = \{x_{ij}\}, \qquad Y = \{y_i\}, \qquad Z = \{z_i\}$$

From Eq. (8-5) we know that

$$X = AZ \tag{8-6}$$

and from the transaction matrix we know that

$$X + Y = Z \tag{8-7}$$

Substituting relationship (8-6) for X into Eq. (8-7), we obtain

$$AZ + Y = Z$$

Then rearranging terms,

$$Y = Z - AZ$$

$$Y = (I - A)Z$$

$$(I - A)Z = Y \tag{8-8}$$

Multiplying both sides of Eq. (8-8) by $(I - A)^{-1}$, we obtain

$$Z = (I - A)^{-1}Y \tag{8-9}$$

Equation (8-9) is the important final result. It gives us a way to link the level of internal economic activity Z to final demand (exports) Y by the matrix $[I - A]^{-1}$. This matrix, which is derived from the transactions data, is the multiplier matrix shown in Table 8-12*c*. Equation (8-9) also can be written as

$$\Delta Z = [I - A]^{-1}\Delta Y \tag{8-10}$$

Equation (8-10) is the application form of the input-output model that is used to estimate impact. The change in local economic activity ΔZ is estimated from a change in final demand ΔY. Planning alternatives are evaluated to estimate ΔY.

The assumption of the input-output models are as follows:

1. The transaction matrix adequately represents the current economic structure and dollar flows of the subject region.
2. The production relationships depicted by the technical coefficients a_{ij} are constant over the range of changes in production levels consid-

ered and over the time required to adjust production to meet new final demands.

To interpret the multiplier matrix (Table 8-12*c*), read down the first column. The numbers shown are the dollar amounts by which purchases of inputs from each sector would increase due to a $1 increase in export sales of the forest products sector. For example, the first element of column 1 is 1.25. This says a dollar of increased exports by the forest products sector, in addition to the definitional $1 increase of direct sales, generates an indirect increase of $0.25. The indirect increase arises because payments to other sectors for inputs such as labor from the household sector in turn generate some additional demand and more sales of forest products to these sectors. Households pick up $0.78 of income from the dollar increase of forest product sales. The total impact on the economy is $2.79 from the $1 increase of exports: $1 of initial or direct impact and $1.79 of all indirect and induced effects.

Darr and Fight (1974) present an interesting, readable, and fully developed case study describing the use of an input-output model to estimate the impact of reduced harvest levels on Douglas County, Oregon. It is also a good introductory reading on input-output models.

The input-output model has been expensive to use because of the high cost of obtaining a reasonably accurate transactions matrix from primary survey data. To construct a table for a region empirically requires detailed interviewing of firms within the region. For tax and competitive reasons, many of them do not want to identify their customers or detail how they spend their money. A table for a region of a few counties might, for example, cost over $100,000 to build, take a year or longer, and still not be very accurate. This kind of cost is too high for virtually any individual decision analysis to finance.

A detailed 425-sector national transactions table is constructed every 8 or 10 years by the U.S. government. While the national model is not directly a good representation of a small region, a technique has been developed (Alward and Palmer, 1983) to use current county business data on employment and total sales by industrial sectors to transform the national table to the local level. The model, called IMPLAN, effectively constructs a regional transactions table for any analyst-specified set of counties and industrial sectors.

This development is significant for it will surely encourage private industry, county and state governments, interest groups, and lobbyists to use the system for the evaluation of forest management and other decision alternatives. This in turn means solid familiarity with the mathematics, data requirements, and demanding assumptions of the technique is needed for responsible interpretation of the results. An example illustrating IMPLAN's use to evaluate a grazing enhancement program is shown in Table 8-13. This is a 14-sector model of the subject region, and we see the increased grazing on the national forest increasing the livestock sector's (sector 2) final demand (exports) by $234,000. This in turn works through the regional economy, stimulating a total increase of $435,765 (last entry, column 2), distributed as shown in the total gross output column. The grazing program also stimulates creation of 8.1 new jobs, and $121,000 of new personal income is created with the distribution by sector shown in columns 4 and 5. The impact multiplier for grazing is (change in total gross output)/(change in total final demand) = 435/234 = 1.86.

These are easily understood results, and if clipped into a project report, local interests are likely to take them at face value. Are they good numbers? As with many easy to use computerized techniques, very impressive outputs are produced and it is easy to give results too much credibility. The two assumptions of an input-output model listed above are restrictive and must be thoughtfully considered. People will use the model, and as managers and analysts, we must be prepared for close review of impact estimates.

Equity Goals

Equity refers to how people perceive fairness or acceptability in the distribution of benefits and costs among the individuals of society. We can observe and measure the distribution of income or employment, for example, but whether a distribution is equitable or not is a value judgment. What is equitable to one person may be social injustice to another.

When considering a decision alternative, there are three sets of distributional information which can help decision makers and their constituencies consider equity issues:

1. The current distribution of benefits and costs among individuals in the affected population before any decision alternative is implemented.
2. Any formalized standards or goals describing the distribution of benefits and costs which different groups or interests consider ideal.
3. The distribution of changes or increments of benefits and costs to individuals in the population expected due to the decision alternative—who gains and who loses.

The first information item describes the status quo and the second provides a standard by which decision makers can judge whether or not the changes described by the third information set are desirable. Normally we would judge a decision to have desirable distributional effects if it

Table 8-13 Impact of increased grazing on a regional economy

Sector	Change in final demand (ΔY) ($1000)	Changes: Total gross output (ΔZ) ($1000)	Consumptive water use (acre · ft)*	Labor (person-years)	Income ($1000)	Value added ($1000)
1 Dairy and poultry	0.000	86.801	0.000	3.239	33.575	30.068
2 Livestock and products	234.000	332.324	0.000	4.538	79.459	87.468
3 Metal and clay mining	0.000	0.146	0.000	0.003	0.064	0.088
4 Construction	0.000	3.991	0.000	0.044	1.364	1.335
5 Stone clay manufacturing	0.000	0.159	0.000	0.003	0.065	.080
6 Communications	0.000	1.339	0.000	0.021	0.594	0.837
7 Wholesale trade	0.000	0.389	0.000	0.013	0.212	0.263
8 Retail trade	0.000	0.341	0.000	0.024	0.216	0.269
9 Eating and drinking	0.000	0.401	0.000	0.029	0.255	0.316
10 Auto repair and services	0.000	0.479	0.000	0.034	0.304	0.377
11 Finance/insurance/real estate	0.000	2.537	0.000	0.000	1.693	1.707
12 Hotels and motels	0.000	5.364	0.000	0.104	2.762	2.698
13 State and federal government	0.000	0.265	0.000	0.012	0.202	0.254
14 Miscellaneous and dummy industries	0.000	0.127	0.000	0.000	0.000	0.076
Private sectors total	234.000	434.664	0.000	8.062	120.764	125.834
Government and miscellaneous		1.102	0.000	0.066	0.782	1.173
Total	234.000	435.765	0.000	8.128	121.546	127.007

* Water consumption factors not available.

moved the actual distribution closer to the desired, detrimental if moving away.

The most common attributes characterizing the welfare status of a population are the amount and distribution of income per capita or per household. In a market economy, income measures the command over goods and services and is the measure most people use to describe their economic well-being. The amount of assets or wealth per capita might also be of interest. Federal, state, and local governments publish distribution data on a variety of characteristics, including income, age, and employment. If the population impacted by a decision can be roughly defined as some set of counties or states, then the current income distribution of the region's population can be constructed from published data.

The desired or ideal distribution is a value judgment, which varies considerably with the person or group making the judgment. Industry, management, labor, religious institutions, philosophers, politicians, bureaucrats, and each of us have some concept of what is a good distribution of income for society with a given sized economic pie. We have choices ranging from a few very rich and many poor; few poor, no rich, and nearly everyone the same; to a distribution which tries to set a subsidized minimum income and still provide a stimulus for economic growth through the wealth reward incentive for risk-taking entrepreneurs. Operationally, there is no absolute standard or ideal distribution, and the relevant standard becomes the perception of individual decision makers as constrained by laws and the particular objectives of the organization they represent.

Estimating the distributional impact of a proposed decision by income class is not easy and rarely done with great precision. The impacted population has to be identified, and after the aggregate future impacts of the decision in question are forecasted, some sort of sampling, survey, interview, or analytical technique is needed to distribute the aggregate impacts over the income, age, employment, or other classes of the impacted population.

To illustrate the three types of information needed, the work of YoungDay and Fight (1979) is presented in Table 8-14. They analyzed the distribution of cost impact from management policies that increase the price of stumpage. To do this, YoungDay and Fight obtained the current income distribution (column 1) from census data. Then they used an input-output model to estimate the impact of a stumpage price increase on industrial activity and used consumer expenditure data to calculate and distribute the increased wood cost by household income class. These results are shown in columns 4 and 5. They conclude that this price increase impacts the lower income classes proportionately more, as indicated by the cost per household as a percentage of income (column 5). We extend their work in columns 2, 3, 6, and 7 to develop a few more ideas.

Table 8-14 Evaluation of equity considerations: impact by income class of forest management decisions

				Changes due to decision*			
		Hypothetical desired income distributions		Decision 1: A forest policy that raises stumpage price $55/M bd ft		Decision 2: Creating more wilderness	Decision 3: City recreation development
Household income class (thousands of 1960 dollars)	1970 actual U.S. income distribution (% of all households) (1)	Pure equality (%) (2)	Welfare capitalism (%) (3)	$ cost per household (1960 dollars) (4)	Percent of household income (5)	Percent of participants (6)	Percent of participants (7)
0–2	18.3	0	0	8.6	77	5	40
2–4	25.7	0	35	15.4	52	10	35
4–6	25.7	100	50	21.3	39	25	15
6–10	24.5	0	12	26.9	36	25	10
10–15	4.5	0	2	37.2	32	20	0
15–25	0.8	0	0.9	48.3	27	10	0
25+	0.2	0	0.1	60.3	19	5	0

* Calculated present net benefit: 1970 actual distribution, 100; Decision 2, 110; Decision 3, 102.

Source: Adapted from YoungDay and Fight (1979).

Two hypothetical income distributions with the same $5,000 mean income of YoungDay and Fight's study are introduced in columns 2 and 3 to suggest two divergent viewpoints of what constitutes an ideal distribution. The first (column 2) represents total equality and the second (column 3) eliminates the lowest income class of the actual distribution for welfare reasons, increases the middle class, and reduces the number of the wealthy to achieve this. By either of these standards the stumpage price increase is also an undesirable move as it further skews the present distribution away from the desired distributions.

A second forest management decision alternative is advanced that considers adding more acres of wilderness or backcountry areas (Decision 2). It is postulated that most of the participants in the new areas will come from the upper income classes, as shown in column 6. Assuming that voluntary participation indicates some net benefit is being obtained, then the decision has undesirable distributional characteristics relative to the income standards of columns 2 and 3 in that benefits increase with income level, and the disparity in income distribution is worsened. In contrast, a third decision allocates recreation development funds to acquire and develop free use of beaches and parks in or near metropolitan centers (Decision 3). It is assumed to increase greatly participation by the lower income classes (column 7) and hence has a desirable equity impact. Suppose the present net benefit of each decision is also calculated to estimate the aggregate social benefit for the wilderness and city park options (see the footnote of Table 8-14). It shows that while the wilderness decision, with an increase of +10 units over the current distribution, is more "efficient" than the city beach decision with +2 units, it is not as good on equity grounds. This is the essence of the characteristic trade-off between equity and efficiency goals.

Stability Goals

Stability seems to be a universal goal of humanity. Stability for most individuals means a secure job with reasonable advancement potential, a community of friends and family, and a physical, biological, and cultural environment that changes only gradually. The desire for stability is another manifestation of the human need to reduce uncertainty in their lives. A region or community is a collection of individuals. When some public or private action adversely affects stability—particularly the employment and income status of more than a few individuals—one or more interest groups will likely initiate political action to remedy the situation. Forest management decisions can significantly affect economic stability (a mill closure or opening a resort, for example), and a change in timber production and harvest levels can directly affect the local economy. Because of this, sensitivity to and evaluation of possible instability from proposed decisions are often important.

If the economic subsystem of a region is the focus of stability issues, then employment, income, tax, production, prices, and sales are the appropriate social indicators to describe the status of that system at some point in time. Since stability is concerned with the rate of change, we really must be concerned with the rate of change in these variables over time. Schuster (1976), Schallau (1974), and Waggoner (1969) all agree that "orderly change" or a "viable dynamic equilibrium" is the appropriate conceptualization of a system that is changing at acceptable rates.

These words—stable, dynamic, orderly, viable—give feeling but little operational guidance. What do they mean analytically? How do we measure it? We have few tested benchmarks to go on. Figure 8-9 illustrates one way of conceptualizing and quantifying stability. Suppose three alternative forest management plans involved different changes in harvest levels over time, and when converted through an input-output analysis to

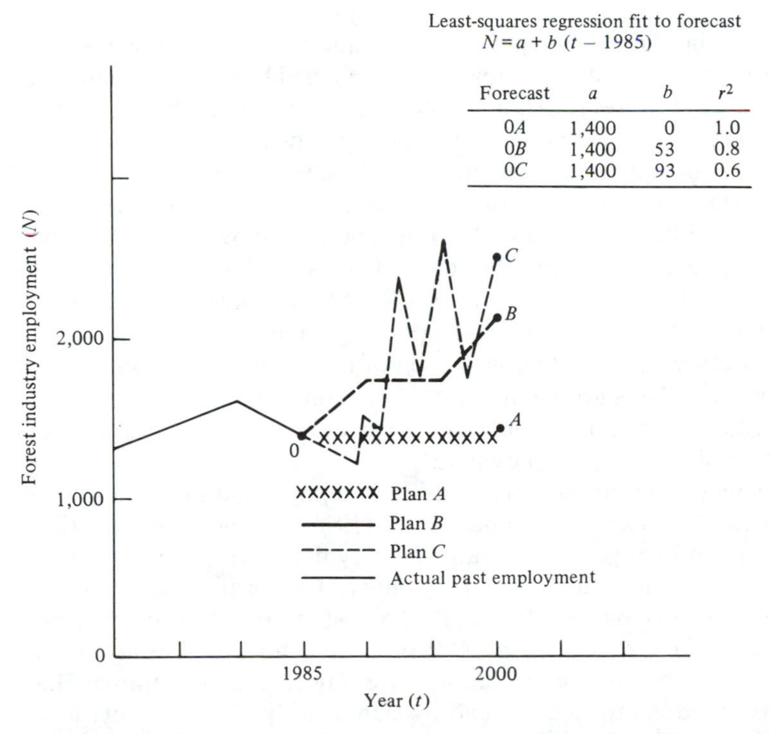

Figure 8-9 Stability analysis of future employment resulting from different forest management plans.

changes in employment, produced the three different forecasted employment trajectories shown. Plan *A* has no change or variation in harvest level and does not change employment at all. Plan *B* gradually increases the harvest with limited year-to-year variation and provides a relatively steady, even rate of employment growth. Plan *C* has an erratic year-to-year timber harvest, but on the average, it provides the greatest growth in employment. If a regression were fitted to these trajectories, the slope coefficient would provide a measure of the average rate of growth and the coefficient of variation r^2 a measure of stability. The steeper the slope, the greater the rate of economic growth. The lower the r^2 value, the greater is the instability and the worse off the region with regard to its goal of economic stability. In the example we have some goal trade-offs as efficiency increases from *A* to *B* to *C* while stability decreases in this order. Plan *A* represents complete status quo. The "orderly change" idea might best be represented by plan *B*. Plan *C* might be too unstable. While employment is shown, any other social indicator can be used as the dependent variable and the stability analogy would be the same. Politicians and decision makers still have to come to grips with the issue, how much instability is too much? This is a value judgment.

Many studies have given particular attention to the relationship of the forester's devout faith in even flow and the alleged benefits to community stability. The literature, legislation, and culture of forestry have endorsed the idea that regular, even timber harvests are socially and economically desirable because they promote stability. While benefits such as regular work for wood crews may accrue to the management of the forest business from even flow, Schallau (1974), Waggoner (1969), and others find little actual evidence or logic that even flow per se promotes economic stability, except in a very few heavily timber-dependent communities. Even flow is somewhat analogous to plan *A* in Fig. 8-9, and if anything, retards flexibility, diversity, reallocation of resources, and acceptance of change that stimulates growth in a viable economy. In fact, as technology improves, the labor requirement per thousand board feet decreases, and even flow implies falling employment.

Changing timber production levels can be important in timber-dependent communities. Dickerman and Butzer (1975) estimated that to stimulate a 3 percent change in all county employment requires a 20 percent increase in harvest level in Trinity County, California, and almost 40 percent in Lincoln County, Montana. They estimate a 25 percent harvest increase would increase jobs by 600 in Trinity and by 420 in Lincoln County. They conclude that in the heavily timber-dependent areas, harvest levels can affect the local economy significantly. At the county level, 400 jobs in a basic industry would be viewed as rather important and would surely stimulate a strong political response, even if it is only a modest change in total employment.

Environmental Security

A final goal of considerable importance to forest management is the rather ill-defined concept of environmental security. Like economic stability, we all have instincts for a healthy, secure environment, but have a hard time measuring or articulating the idea. How do we know one environment is more secure than another, and by how much? Environmental security has at least three important aspects: (1) maintaining the productive capacity of the renewable resource base, (2) providing the genetic capability of individual plant and animal species to evolve, adapt, and survive in a changing physical environment, and (3) keeping a broad diversity of healthy, dynamically stable community ecosystems.

Maintaining a productive, renewable resource has long been a dominant goal of the forestry profession. Gifford Pinchot rode to fame on the conservation movement to protect, preserve, and use the forest for all future generations. The forest reserves were originally set aside on the rationale of timber and watershed protection. Federal and state forest practice laws and programs specify prompt regeneration as a basic land management constraint. Sustained yield became a banner of the conservation movement as it represented a reversal of the timber exploitations that had occurred at the turn of the century. Today, as illustrated in Table 8-1, we attempt to quantify the trend of long-run resource productivity by indexing sediment production, soil loss rates in inches per year, miles of road and acres disturbed each year, and acres unstocked with timber. Nonreversible pollution impacts on land or water are monitored. These are the indicators of long-term productivity.

Providing viable evolutionary capability tends to focus on maintaining a diversity of genotypes for each plant and animal species at sufficient population levels to reproduce readily. Concern with endangered plant and animal species manifests this goal. Providing a broad "gene pool" is the way geneticists or ecologists might refer to the issue. There is not much agreement on how many genotypes a species requires and what minimum population levels are necessary or adequate to keep the species perpetually reproducing. To keep track of what is happening, however, we keep species lists and estimate the population level of each species. These at least indicate trends and, if forecasted, can suggest improvement or deterioration.

Maintaining healthy ecosystems is a third goal which also contributes to the goal of species diversity. If we describe ecosystems as contiguous plant and animal communities of sufficient size to show community rather than individual properties, then the goal might be achieved by a large number of different ecosystems, geographically replicated and well distributed. This arrangement provides the edge effects, shelter, food, reproductive habitat and living environments for a large number of plant and animal species. Indicators of this goal would be the distribution of acreage by tree, shrub, and forb plant communities, the age and size of plants, and the size distribution of these stand or range site types. Extensive monocultures are not usually viewed as contributing to this goal. When wilderness candidate areas were selected in California from the Rare II study, meeting minimum acreage for different plant community and land types was a dominant goal constraint (Teeguarden, 1981).

BENEFIT-COST ANALYSIS

Benefit-cost analysis is a general technique for comprehensively evaluating solution alternatives by two or more goals and for reducing these evaluations to a summary index. Tables 8-1 and 8-2 are good examples of what would be called benefit-cost analysis. The intended result is useful objective information on which to base a decision.

Problems addressed in a benefit-cost analysis typically have the following characteristics:

1. Only a few discrete alternatives for problem solution are being considered.
2. For the problem and each alternative there are many relevant goals, and many of the important consequences of each alternative do not have a market-determined value.
3. For public projects, the scope of benefits and costs is total—theoretically *all* positive and negative effects on everyone, no matter when or where they occur, are evaluated.
4. For policy, legal, administrative, and pragmatic reasons there is strong pressure to reduce all consequences, benefits and costs, to dollar terms and compare the alternatives in terms of a summary value index such as present net worth or net social benefit.

Note that benefit-cost *analysis* is much different than the benefit-cost *criterion* discussed earlier, since it refers to a comprehensive analytical process rather than one of several criteria for comparing alternatives.

The U.S. Army Corps of Engineers pioneered the use of benefit-cost analysis to evaluate their flood control and water resource development projects in the 1940s and 1950s. In the 1960s program budgeting became federal doctrine and gave the mandate for comprehensive dollar-based evaluations of the benefits and costs of nearly all federal activities, including defense, research and development, welfare, and investments in forest management. The National Forest Management Act of 1976 has firmly fixed the requirement for comprehensive benefit-cost analysis by the U.S. Forest Service.

Discrete alternatives are typical when dealing with large, complex package deals like a new tax policy, a river basin development, locating a new industrial complex, or preparing a comprehensive forest management plan. Such problems seldom are handled by alternatives generated in a simple type *A* process. Alternatives are more typically generated by a type *B* analytical process with many goals or constraints changing between alternatives. Moreover, despite our best efforts, many goals and impacts defy quantification and dollar valuation.

The effort to reduce a problem evaluation to a single summary value index is a hallmark of benefit-cost analysis. This has given incentive to place values on goods and services not sold in the market. It also has a de facto effect of uprating the economic efficiency goal and downrating all goals or impacts that cannot be quantified and explicitly entered in the calculated project present net worth. Since the best project is conceptualized as the one providing the greatest net benefits, there is a strong tendency by decision makers to choose the one with the highest calculated present net worth, even if the calculated present net worth does not include many important nonquantified benefits or costs.

The Mechanics of Benefit-Cost Analysis

Procedurally, the mechanics follow seven steps in a process of carefully reducing a complex problem to a single summary index of present net worth or benefit-cost ratio:

1. First all the goals and their social indicators considered relevant to the analysis are identified and listed.
2. Next a suitable numerical unit of measure is specified for each indicator (such as board feet, jobs, or tons of sediment).
3. The forecasters then use experts, crystal balls, and a variety of empirical and mathematical techniques to predict the physical amount of each indicator in units that will occur each year or planning period for each alternative.
4. A set of dollar prices is forecasted for each indicator.
5. The dollar prices of item 4 are multiplied by the forecasts of item 3 to create dollar impact predictions.
6. Finally, a guiding interest rate is specified, present values of the dollar impact predictions are calculated, and the benefits and costs summed.
7. The last step is to calculate the overall present net worth by subtracting all costs from the benefits. This number is often called present net

benefit (PNB) because it includes many nonmarket benefits. Still the number rarely includes all nonmarket benefits and is thus partway between market present net worth and full capture of present net benefit.

By implication, any goal or consequence that cannot be quantified at each step of this procedure and is left out receives an effective value of zero with regard to the final calculation of present net worth. While many analyses include these unquantifiables in footnotes, the discussion section, or in other places in the report presented to the decision maker, it is still important to recognize that they *do not* appear in the formal present net worth calculation.

It is important also to recognize that benefit-cost analysis does not say anything directly about the distribution of costs and benefits among the people and geographic regions of the country; it ignores equity goals. Mishan (1978) in the introduction to his primer on cost-benefit analysis gives this point particular emphasis:

> Any adopted criterion of a cost-benefit analysis, that is, requires *inter alia* that all benefits exceed costs, and therefore can be vindicated by a social judgment that an economic rearrangement which can make everyone better off is an economic improvement. The reader's attention is drawn to the fact that such a judgment does not require that everyone is *actually* made better off, or even that nobody is actually worse off. The likelihood—which, in practice, is a virtual certainty—that some people, occasionally most people, will be worse off by introducing the investment project in question is tacitly acknowledged. A project that is adjudged feasible by reference to a cost-benefit analysis is, therefore, quite consistent with an economic arrangement that makes the rich richer and the poor poorer. It is also consistent with manifest inequity, for an enterprise that is an attractive proposition by the lights of a cost-benefit calculation may be one that offers opportunities for greater profits and pleasure to one group, in the pursuit of which substantial damages and suffering may be endured by other groups.
>
> In order, then, for a mooted enterprise to be socially approved, it is not enough that the outcome of an ideal cost-benefit analysis is positive. It must also be shown that the resulting distributional changes are not regressive, and no gross inequities are perpetrated.

Implications of Using a Benefit-Cost Analysis When Some Goals Cannot Be Quantified or Valued

To illustrate how a benefit-cost analysis can give misleading decision information and guidance, we consider a forest planning problem where four outputs are of value: timber, forage, wildlife, and visual quality. Three different plans are under consideration, each providing different amounts of the four outputs while incurring the same costs. This example also can serve to illustrate a type *B* process for defining solution alternatives where problem output constraints are systematically altered or deleted to define the different solutions.

Table 8-15 True present values and cost of all alternatives

	Alternative plan		
Item	*A*	*B*	*C*
Timber output	100	75	50
Forage output	10	20	50
Wildlife output	5	50	80
Aesthetics output	30	25	−50*
Total cost	100	100	100
Total benefits	145	170	130
Present net worth	45	70 (best)	30
Benefit-cost ratio	1.45	1.70	1.3

* Requires much cutting, burning, mess making, high percentage in grass and brush; ugly.

To examine this problem, we first assume we know how to value everything in dollars and, in fact, know the truth. We later retreat from this idealistic state and accept the reality that we cannot put meaningful numbers or values on everything. You might imagine that the analytical angel appeared and gave us the true values. When the true values are summed, we get a calculated present net worth and benefit-cost ratio for each alternative plan (Table 8-15).

All three alternatives are seen to provide positive net social benefits as indicated by present net worth values. By these calculations, plan *B* is optimal, providing $70 of present net worth.

Now suppose we face reality and admit we cannot really quantify visual quality, much less put a dollar value on it. What we do is, we try to verbally articulate the visual qualities of the plans in a footnote. The dollar numbers for visual quality are removed from the present net worth calculations, giving a revised analysis (Table 8-16).

The present net worth calculation now shows plan *C* to be the best. If we had not sneaked a look at the earlier "true table" and Table 8-16 was all we had to go on, the essential question is, would the decision maker have chosen plan *C* over plan *B*? Probably so, unless the subjective interpretation of the footnote outweighed the calculated present net worth difference between plans *B* and *C* of $80 − $45, or $35.

We now take another step and further admit that while we can roughly estimate wildlife population levels, we do not know how to attach a meaningful dollar value to these outputs. Population levels are indicated in the analysis, but wildlife is no longer explicitly valued in the present net worth calculation. The third table (Table 8-17) now is obtained.

Table 8-16 Evaluation of alternatives omitting aesthetics

Item	Alternative plan *A*	*B*	*C*
Timber output	$100	$ 75	$ 50
Forage output	10	20	50
Wildlife output	5	50	80
Aesthetics*	N.A.	N.A.	N.A.
Total cost	100	100	100
Total benefits	115	145	180
Present net worth	15	45	80 (best)
Benefit-cost ratio	1.15	1.45	1.80

* Visual quality standards are easily met for alternatives *A* and *B*, but compliance is somewhat below standard for alternative *C*, where it is ugly.

Plan *A* now has the highest PNW. In addition, this is now the only plan that shows a positive present net worth. Plan *B* appears to be the worst of the lot. Would the combined subjective evaluation of the visual and wildlife information have led to choosing plan *B*?

Our point is that the objective information base for decisions is the data, calculations, and summaries presented to the decision makers. Since the analytical angel rarely appears and we do not get shown any

Table 8-17 Evaluation of alternative plans omitting aesthetics and wildlife

Item	Alternative plan *A*	*B*	*C*
Timber output	$100	$ 75	$ 50
Forage output	10	20	50
Wildlife output	Low population	Medium population	High population and density
Aesthetics*	N.A.	N.A.	N.A.
Total cost	100	100	100
Total benefits	110	95	100
Present net worth	10 (best)	−5	0
Benefit-cost ratio	1.10	0.95	1.0

* Visual quality standards are easily met for alternatives *A* and *B*, but compliance is somewhat below standard for alternative *C*, where it is ugly.

true tables, it is hard to know when the analysis is, in fact, misleading. Tables 8-15 to 8-17 were rigged to show how misleading total reliance on calculations can be.

If there is intuitive or political indication that nonvalued outputs comprise a significant portion of the total plan values, then the decision makers might be served better by simply not calculating the present net worth or the benefit-cost ratio summary indices and printing them as the "bottom line" in summary tables. Then the temptation is not there to use them incorrectly. When you as an analyst write any criteria into a report, you are implicitly telling the decision maker it is useful and valid information to guide a decision. When a bottom line is provided, you are suggesting it is a useful summary value index.

Returning with our discussion to Tables 8-1 and 8-2, which opened this chapter, we might now conclude that since the present net worth number of Table 8-2 only covered part of the outputs, then Table 8-2 probably should be deleted. But than what? The decision still has to be made. What do we show the decision makers?

Social Accounting

Davis and Bentley (1967), on reviewing this sort of situation, concluded that society was better served by not mixing facts and values in analysis. The burden of value judgment is still on the decision makers to deal with complex and conflicting goals, to make decisions as best they can, and to be accountable for their choices. A social account format for analysis was presented as a means of organizing a mixture of valued and nonvalued information.

The objective of a social account presentation is simply to organize and convey the consequences of alternative actions as objective, factual information to decision makers. Value judgments used in the analysis are explicit and clearly labeled. The benefits and costs are not summarized to a single net value index. The result is a rather detailed table that forces the decision maker to take responsibility for value judgments.

To illustrate the social account approach, we take the opening table, Table 8-1, and reorganize the rows as shown in Table 8-18. The result is a presentation grouped by goals and criteria which are identified by whether they are quantified market values, quantified nonmarket values, or only quantitative and qualitative indices.

Three major goals for the Smith Creek analysis were economic efficiency, regional development, and environmental security. The criteria for economic efficiency split into timber and forage yields, which had empirical market price valuations, and recreation, water, and hunting, which had assigned or imputed dollar prices. Note that approximately 75 percent of the total benefits claimed in Table 8-2 are revealed in Table 8-18

Table 8-18 Social account for Smith Creek planning. Plan *A*: Economic output emphasis; Plan *B*: Regional benefit emphasis; Plan *C*: Amenity benefit emphasis

Goals and indicators	Present condition	Estimated yields by alternatives				Estimated change in yield by alternative (as compared by present condition)				
		Alternatives			Selected plan	Current management plan	Alternatives			Selected plan
		A	*B*	*C*			*A*	*B*	*C*	
1. *Economic efficiency goal*										
A. Market-valued outputs										
Numerically quantified										
Estimated 20-year timber harvest (million bd ft/yr)	8.1	7.2	7.2	6.1	6.8	−0.9	−0.9	−0.9	−2.0	−1.3
Potential domestic grazing (AUM/yr)	545	640	545	475	545	0	+90	0	−70	0
Present value of market-value outputs at a guiding rate of 10% (thousands)										
Timber @ $44 M bd ft		3,071	3,071	2,844	2,979					
Grazing @ $2 AUM		11	9	8	9					
B. Nonmarket-valued outputs										
Numerically quantified										
Estimated 20-year recreation use (RVD/yr)										
Type I	1,520	1,470	1,490	1,560	1,520	−30	−50	−30	+40	0
Type II	2,760	2,580	2,580	2,600	2,580	−180	−180	−180	−160	−180
Type III	4,790	5,130	5,180	5,070	5,070	+390	+340	+390	+280	+280
Type IV	370	2,190	330	330	340	−40	+1,820	−40	−40	−30
Total	9,440	11,370	9,580	9,560	9,510	+140	+1,930	+140	+120	+70

Water yield (acre · ft/yr)[a]	240,470	240,420	240,420	239,850	240,280	−50	−50	−50	−620	−190
Potential number of huntable deer (annual fall population)	20	10	10	10	10	−10	−10	−10	−10	−10
Present value of nonmarket outputs with assigned or imputed dollar prices at a guiding rate of 10% (thousands)										
Water @ $4.60		9,251	9,251	9,241	9,251					
Recreation @ $20 RVD		566	566	567	565					
Hunting @ $430/animal		7	7	7	7					
C. Potential supplies										
Quantitative										
Estimated *potential* for 20-year recreation use (RVD/yr)										
Type I	14,300	6,700	7,800	11,200	9,000	−6,500	−7,600	−6,500	−3,100	−5,300
Type II	5,800	7,400	11,900	12,800	11,400	+6,100	+1,600	+6,100	+7,000	+5,600
Type III	44,900	55,500	58,300	64,300	53,000	+13,400	+10,600	+13,400	+19,400	+8,100
Type IV	4,800	41,900	7,700	7,700	10,000	+2,900	+37,100	+2,900	+2,900	+5,200
Type V		4,000	0	0	2,100	0	+4,000	0	0	+2,100
Total	69,800	115,500	85,700	96,000	85,500	+15,900	+45,700	+15,900	26,200	+15,700
Potential number of wintering deer (annual)[f]	15	5	5	5	5	−10	−10	−10	−10	−10
Annual timber growth potential (million bd ft/yr)	N.A.	11.1	10.6	6.4	10.2					
Qualitative										
Mineral opportunity[g]	+	++++	++++	++	+++	+++	+++	+++	+	++
D. Costs										
Present value of all costs (thousands) at 10%		$3,540	$3,100	$2,400	$3,250					

Table 8-18 (Continued)

Goals and indicators	Present condition	Estimated yields by alternatives				Estimated change in yield by alternative (as compared by present condition)				
		Alternatives			Selected plan	Current management plan	Alternatives			Selected plan
		A	*B*	*C*			*A*	*B*	*C*	
2. *Regional development goals*										
Quantitative										
Change in timber industry employment (number of jobs)[d]						−4	−4	−4	−8	−5
Change in timber industry wages and salaries ($1,000)[d]						−30.46	−30.46	−30.46	−67.68	−43.99
3. *Environmental security goals*										
Quantitative										
Sediment production (tons/yr)[b]	38,740	36,620	35,830	36,450	−210	−2,120	−2,120	−2,910	−2,290	
Potential accelerated erosions from timber harvest and roads (tons/yr)[c]	51,100	30,000	29,500	22,900	28,500	−21,100	−21,100	−21,600	−28,200	−22,600
Estimated area disturbed by timber harvest and roads (acres/yr)	1,300	780	780	600	740	−520	−520	−520	−700	−560
Roads and development										
Reconstruction and improvement of existing roads (mi in 20 yr)	N.A.	67	67	21	64	+67	+67	+67	+21	+64
New general-access road construction (mi in 20 yr)[h]	N.A.	12	12	12	13	+12	+12	+12	+12	+13
Total general-access roads at end of 20 years (mi)	120 (existing)	132	132	132	133	+12	+12	+12	+12	+13

Construction of project development roads (mi in 20 yr)[i]	N.A.	85	89	57	78	+85	+85	+89	+57	+78
Construction of new trails	79 (existing)	0	9	39	39	0	0	+9	+39	+39
Qualitative										
Deer winter range quality	+++	++	++	++	++	–	–	–	–	–
Fisheries habitat quality	+	++	++	++	++	+	+	+	+	+
Grizzly bear habitat quality	++	++	++	++	++	0	0	0	0	0
Mountain caribou habitat quality[e]	++	++	++	++	++	0	0	0	0	0

[a] Natural base flow = 221,310 acre · ft/yr.
[b] Base production = 30,950 tons/yr.
[c] Natural erosion for planning unit = 2,500 tons/yr.
[d] Factors used for Schuster, Godfrey, and Koss, 1975.
[e] Based on vegetative succession and potential for disturbance.
[f] Based on vegetative succession and deer life equations.
[g] Based on access and other development.
[h] Includes arterial, collector, and local road functional classifications. These roads would have a long-term service life.
[i] Estimate of the amount of work roads necessary for site-specific project development. These roads could have either long-term or short-term service lives.

to be associated with nonmarket outputs. Moreover almost all of this is the value of water. Many will argue that water has a zero value at the forest since it cannot be stored or allocated and only takes on value in use equal to the cost of delivery (dams, pipes, etc.). Note also that the water, recreation, and wildlife values are virtually constant across all alternatives and that most of the differences in calculated present net worth are due to differences in the value of timber harvested. We deliberately do not add the two kinds of dollar benefits in Table 8-18 because addition implies they are equivalent numbers. The high guiding rate of 10 percent likely includes an inflation assumption, and we should wonder if output prices were also inflated. The descriptive data on potential supplies are not converted to dollar values and are difficult to classify. We show them under the economic efficiency goal since they seem to relate best to future output levels.

The information presented in Table 8-18 does not give a decision maker an easy way out—in fact, it is hard to "add up" and compare the alternatives. For this particular problem we also observe that the alternative plans are not strongly differentiated, and all provide close to the same mix of benefits. The amenity plan *C* shows the greatest differences. This lack of differentiation makes us wonder whether the feasible region of different solutions was really so small.

Although better organized, we still may be overloaded with information in Table 8-18. To refine and simplify further we need to review the criteria to find the ones most sensitive to the alternatives and possibly eliminate unimportant or unchanging criteria.

QUESTIONS

8-1 Four independent forestry investment projects are being considered for implementation using this year's capital budget.

1. (50 acres) Precommercial thin overstocked pine stands at a cost of $85 per acre and net an increased yield value of $1,100 when the stands are regeneration harvested in 30 years.
2. (100 acres) Precommercial thin fir stands at a cost of $70 per acre and net an increased yield of $1,000 per acre when the stands are regeneration harvested in 20 years.
3. (100 acres) Underplant understocked pole stands at a cost of $150 per acre and net an increased yield of $5,000 in 40 years.
4. (200 acres) Release premium hardwood crop trees by selective girdling and then fertilize at a cost of $60 per acre to net a harvest value increase of $400 per acre in 10 years.

Assume a guiding rate of 10 percent, a reinvestment rate of 8 percent, and a planning horizon of 40 years.

(*a*) Calculate the per acre internal rate of return (IRR) for each project.

(*b*) Calculate the per acre present net worth of each project after normalizing the projects for length and reinvestment.

(*c*) Calculate the realizable rate of return (RRR) for each project on a 40-year normalized life for all projects.

(*d*) How many dollars could be used for projects whose RRR is greater than or equal to the guiding rate?

8-2 Given the economy depicted in Table 8-12:

(*a*) What are the technical coefficients a_{ij} that describe the input proportions for the agricultural sector?

(*b*) Considering the proportion of total sales in exports and the multiplier effect of increased export sales, rank the importance of the four producing sectors [(1) to (4)] in terms of the impact of a $100 increase in *total sales* from that sector.

(*c*) Trace the distribution of economic impacts from a $100 increase in *exports* from the agricultural sector.

8-3 *Discussion Questions*. For each of the following problem situations, identify whether it is likely to be a type *A* or a type *B* process to generate alternative solutions, and which of the three economic efficiency problem situations is addressed. (Several could go either way depending on what you assume.)

1. The range specialist for a private ranch is trying to increase forage production by 100 annual-unit-months (AUM) on an allotment.
2. The recreation staff of a state park system is deciding which campgrounds to renovate with this year's budget of $30,000.
3. The timber staff of an industrial corporation is asked to estimate the justifiable increase in budget for intensified management above current levels for the major site species timber stand groups of its 3 million acre land holdings.
4. The regional timber office wants the local forester to provide an economically justifiable schedule of timber sales from a forest, which will meet next year's timber production targets.
5. The national forest planning team is deciding which overall forest management plan to recommend, considering long-term harvest schedules and management intensity for the 1.2 million acre forest.
6. A silviculturist is recommending a prescription for a particular timber sale and stand on a 100-acre nonindustrial private forest ownership.
7. The timber company wants to locate and build an efficient tree seedling nursery for a region.
8. A fish hatchery manager is using a linear program to select an efficient feed mix.
9. A private logging engineer is trying to decide whether a cable, high-lead, or helicopter logging system is the most efficient for a particular timber sale.
10. A farmer is trying to decide whether or not to plant some of her fields in pine next year.
11. A hamburger stand operator is trying to decide whether to buy an automatic french fry cutter and cut his own potatoes, or to continue buying frozen, precut fries from a distributor.
12. A public recreation specialist is working with the timber specialist to determine an efficient balance of timber production and visual values for different planning units.

REFERENCES

Alward, G. S., and C. J. Palmer, 1983: "IMPLAN: An Input-Output Analysis System for Forest Service Planning," Draft User's Guide, U.S. Forest Service Rocky Mountain Forest and Range Experiment Station, Fort Collins, Colo.

Bonnen, J. T., 1969: The Absence of Knowledge of Distributional Impacts: An Obstacle to Effective Public Program Analysis and Decisions, in "The Analysis and Evaluation of Public Expenditure: The PPBS System," U.S. Government Printing Office, Washington, D.C.

Darr, D. R., and R. D. Fight, 1974: "Douglas County Oregon: Potential Economic Impacts of a Changing Timber Resource Base," U.S. Forest Service Research Paper PNW-179.

Davis, L. S., and W. R. Bentley, 1967: The Separation of Facts and Values in Resource Policy Analysis, *J. Forestry,* **65**:612–620.

Dickerman, A. R., and S. Butzer, 1975: The Potential of Timber Management to Affect Regional Growth and Stability, *J. Forestry,* **73**:268–270.

Dutrow, G. F., 1978: "A Study of Economic Management Opportunities to Increase Timber Supplies in the Southeast United States: Some Preliminary and Tentative Results," Forum, Spring 1978, School of Forestry and Environmental Studies, Duke University, Durham, N.C., pp. 5–10.

Marty, R., and W. Newman, 1969: Opportunities for Timber Management Intensification on the National Forests, *J. Forestry,* **67**:482–485.

Mishan, E. J., 1978: "Elements of Cost Benefit Analysis," 2d ed., Unwin Brothers, London.

Palmer, C. J., and G. D. Keaton, 1978: Economic Input-Output Analysis: An Approach to Identifying Economic and Social Impact of Land Use Planning, in "Proceedings of Land Use Planning Workshop," University of Kentucky, Lexington.

Richardson, H. W., 1972: "Input-Output and Regional Economics," Weidenfield and Nicholson, London.

Schallau, C. H., 1974: Forest Regulation—Can Regulation Contribute to Economic Stability?, *J. Forestry,* **72**:214–217.

Schuster, E. G., 1976: "Local Economic Impact, a Decision Variable in Forest Resource Management Study Report," Montana Forest and Conservation Experiment Station, School of Forestry, University of Montana, Missoula.

Teeguarden, D. E., 1969: "Economic Guides for Douglas-Fir Reforestation in Southwestern Oregon," Bureau of Land Management, U.S. Department of the Interior.

———, 1981: A Method for Designing Cost-Effective Wilderness Allocation Alternatives, *Forest Sci.,* **27**:551–566.

U.S. Forest Service, 1977: "Smith Creek Planning Unit: Final Environmental Statement and Land Management Plan, Kaniksu National Forest, Idaho," USDA-FS-R1(04)-FES-Adm-76-2.

———, 1980: "The Nations Renewable Resources, an Assessment." Forest Service Report 22.

———, 1982: "National Forest System Lands and Resource Management Planning" (the regulations), *Federal Register,* September 30, 1982, **7**(190):43026–43052, 36 CFR Part 219.

Van Horne, J. C., 1980: "Financial Management and Policy," 5th ed., Prentice-Hall, Englewood Cliffs, N.J.

Vaux, H. J., 1954: "Economics of the Young Growth Sugar Pine Resource," California Agricultural Experiment Station, Bulletin 738.

Waggoner, T. R., 1969: "Some Economic Implications of Sustained Yield as a Forest Regulation Model," Institute of Forest Products, College of Forest Resources, University of Washington, Contemporary Forestry Paper 6.

Weisbrod, B. A., 1977: Collective Action and the Distribution of Income: A Conceptual Approach, in R. H. Haverman and J. Margalis (Eds.), "Public Expenditures and Policy Analysis," 2d ed., Rand McNally, Chicago.

YoungDay, D. J., and R. D. Fight, 1979: Natural Resource Policy: The Distributional Impact on Consumers of Changing Output Prices, *Land Economics,* **55**:11–27.

PART THREE

VALUATION

To do forest planning, evaluate proposed projects, write timber culture prescriptions, or engage in buying and selling of forestland, knowing the dollar prices for land, timber, and other forest outputs is essential. It has always been difficult enough to estimate the value of timber stumpage and land. Now, by social mandate, we also include recreation, visual amenities, water, wildlife, and all other forest outputs in our public forest management calculations, and even in some private ones. The valuation methods for these nontimber outputs are varied and debatable, but it is certain that when numerical values for nontimber outputs are entered in decision models with the dollar as common denominator, their magnitudes affect the plan or action that appears superior.

The theoretical underpinning of valuation strategies and procedures used to make appraisals of specific forest properties and outputs are introduced in Chap. 9. Three concepts of value are considered: value in use, market value, and social value. The types of forestry problems that require prices, from planning to taxation, are organized and reviewed. A presentation of six general methods of appraisal concludes the chapter.

Chapters 10 and 11 highlight timber valuation. Chapter 10 presents the procedures for appraising the value of an individual tree in terms of the lumber or wood product value of the logs contained in the standing tree. This classical residual-value conversion return derivation remains virtually unchanged from the second edition; we feel that little improvement is needed on the detailed sugar maple example; prices may change but the procedure is timeless.

Chapter 11 moves from the individual tree to the valuation of stumpage, a collection of trees on a parcel or parcels of land that is or could be offered for sale. A U.S. Forest Service example sale is updated, and some recent transaction evidence approaches to stumpage valuation are added. Also considered are the value of land, the immature stand, and the value of the managed forest when used for commercial timber production. Chapter 11 does not consider the costs and receipts of nontimber outputs from the land. To extent that these additional cost and revenue items can be quantified, as is considered in Chap. 13, they could be added to the timber cash flows to develop a comprehensive appraisal of forestland value.

Chapter 12 is new material for this management text and explains the seemingly impossible art of generating dollar market prices and values for goods and services that are not sold in the market. The assumptions, procedures, and limitations of the different methods are covered, with particular emphasis on the travel cost model for estimating recreation demand. We support willingness to pay as the correct concept of value and explore the conditions for the appropriate use of consumer surplus measures in decision analysis.

CHAPTER

NINE

VALUATION AND APPRAISAL

Valuation is determining what things are worth to people. The worth of things in turn directs the decision making behavior of individuals and organizations. In the preceding chapters we saw how present net worth and income objective functions are instrumental in the search and selection of the best solution to problems where the goal is economic efficiency. Such objective functions are the quantified evaluation by humans of the relative worth of things—Bill Jackson said that to him the steer was worth $1,000 of net income and a lot of trees was worth $500—a 2 : 1 ratio of relative value.

If values were readily available for everything foresters and forest owners analyze, plan, and decide about, valuation as a subject would not be particularly important. Like a supermarket manager or a stockbroker, the forester could consult the daily price listings and rather easily determine the absolute and relative values of forest products and services. Unfortunately such price listings just are not available, and we resort to a variety of creative and often debatable techniques to estimate value indirectly.

Valuation, as a subject area, is concerned with developing appropriate concepts and methodology for estimating the value of goods and services. *Appraisal* is the application of these concepts and methods to make a specific estimate of the value of a particular item to a particular individual at a point in time. Valuation is essentially an academic topic. In contrast, appraisal is a legally guided, professional craft of actually coming up with acceptable numbers which are used to decide about the transfer of wealth and income, often in substantial amounts, between individuals or individuals and the government.

SOME SPECIAL CONSIDERATIONS IN FORESTRY

Timber and forage have fairly well established markets, particularly when processed into lumber or hamburger. However, in the forest there is not always an active market with lots of buyers and sellers to indicate clearly through their bidding what the standing tree or growing grass is worth in dollar exchange. Even when there is sales activity, the buyer or seller may have a local monopoly or some other special advantage and we often feel that the selling price may not be a "fair" or competitive value appropriate to guide decisions.

Forestland and timber have another special valuation problem shared with orchards, home mortgages, flood control, power dams, and producers of certain distilled spirits: it takes a long time to grow a tree to merchantable size. Excluding coppice, most timber crops take at least 20, usually 40 to 50, and often over 100 years to mature. By contrast, few individuals seriously plan further ahead than 2 or 3 years, and most business enterprises expect to amortize capital expenditures and turn an acceptable profit within 5 or, at the most, 10 years. Foresters with their long planning and projection periods obviously need two things: (1) a great deal of faith in a reasonably stable, or improving wood-using future, and (2) acceptance of interest charges for using physical, biological, and monetary resources over time as the dominant cost of doing business.

Foresters and land managers must increasingly deal with and make decisions about goods and services that have no market at all. Many wildlife, visual, environmental, and recreational services are not traded in markets and are provided free or made available for very low user fees. Hunting rights and some types of recreation have commercial markets and established prices, but these typically are experiences not available on public or open private lands. No reliable market information is provided to put values on rare and endangered species, nongame wildlife in general, or the view across the Smokey Mountains on a clear day. Establishing values for this diverse spectrum of goods and services is a necessity for decision analysis in forestland management, for choices must be made and values assigned, whether or not such values are supported by objective evidence. Assigning proxy market prices to those things not sold in markets in order to make comparisons to market-priced goods and services is the current trend in public planning. However, this assignment frequently exceeds the imagination and analytical ability of even the most skillful and optimistic practitioner of the valuation arts.

The forester, more than most, often makes decisions about *public goods*. Public goods are things such as air and water quality, migratory wildlife, visual amenities, and television broadcasts, where the owner of the goods or service does not control distribution or use of the goods and hence receives little or no direct revenues. Distribution control may be possible, but its cost may be much higher than any possible revenue and hence is not implemented. Air quality and visual quality are things that the forester on public lands must legally protect or provide, but the forestry enterprise cannot collect for the value produced. Only at great cost can people be kept from looking at a view or breathing the air. The increased costs of providing such public goods are passed on to the taxpayer or the purchaser of lumber from the forest.

Television broadcasts are an interesting case of public goods in partial transition to private goods. Initially all broadcasts were air transmitted and were free public goods since there was no technology to control who tuned in. With the development of cable technology, signal scramblers, and with legal blessing, the transfer cost for controlling distribution has been lowered to the point where much broadcast material is now a pay-as-you-use private good. Control of the use of wilderness and other remote areas is a related problem, since the costs of full and equitable control may be high. Who knows, perhaps future electronics will allow even the obscure backcountry hiker to be identified and sent an itemized bill.

That public goods have value is undeniable, but what is the value, who benefits, and who pays? On public lands, the net value of public goods is generally viewed as a benefit or revenue item when the government is trying to increase overall net public benefit. On private lands, the provision of public goods often is seen as a net cost when their values cannot be captured as dollar revenue, but there are real cost outlays to provide them.

These important special considerations in forestry: the market imperfections, the large element of interest costs for using resources over time, and the abundance of nonmarket and public goods, require that the contemporary forester take the art and science of valuation seriously. Techniques need mastery, and a perspective must be developed by the forest manager to use wisely the many indirect, subjective, and other methods suggested for estimating the values of nonmarket and public goods.

CONCEPTS OF VALUE

Value is a human perception—it is the worth of something to a particular individual, at a given place and moment in time. Utility, satisfaction, and pleasure are other words that connote worth or value received. The measure of worth is determined by the time, goods, or money an individual is willing to give up to obtain, possess, or use the good or service in question. Valuation, the process of quantifying values, must accordingly operate from the perspective of some individual (or group) of humans; in a specific problem situation, it is the decision maker who establishes the context or point of view for valuation.

We use three different but related viewpoints to establish values. The first is market value, where the dollar price is established by trading activity in established markets. The second is the value in use of something to a given individual. The third perspective is that of society as a whole, which goes beyond and is different from the combined views of its members. Social values or benefits are established subjectively by legislators, public administrators and, sometimes, by citizens voting on bond issues and other special elections. Society emphasizes the goals of collective security, growth, and distributional equity more than do the individuals of society.

Market Value

The standard of value most frequently invoked is the price at which fully informed, willing, and numerous buyers and sellers exchange goods and services—the price in a competitive market. Economic theory provides strong and extensive support for this notion of value since a reasonably competitive market ensures that the value received for a unit of product or labor input at the margin will equal or exceed the cost or resources given up. Hence competitive market prices measure what decision makers are "willing to pay" for goods and services. The notion of competitive markets argues not only that individuals can exchange their own time and other resources to receive maximum personal benefit, but also that when all individuals selfishly pursue their own interests, the aggregate use of resources in society will be socially efficient in the sense of providing the greatest total net benefit to society as a whole.* Hence we have strong arguments that prices revealed in competitive markets are socially desirable standards of value. If we use these prices in our evaluations and decision making, socially desirable outcomes presumably result.

Few markets are truly competitive and many do not reflect human values accurately. For example, markets with only one or a few buyers or sellers (such as, a monopsony or monopoly) can be manipulated and the price offered or paid will not equal the marginal value of resources exchanged. When either buyers or sellers are being exploited in some way, the prices established in such markets are not considered to be good

* This is the economic efficiency goal reviewed in Chap. 8. Remember, however, that achieving maximum net social benefit does not say anything about how well equity goals, the desired distribution of goods among the members of society, are achieved.

measures of value. The federal government is the dominant supplier of many amenity, wildlife, and recreational services and thus, by default, is a monopolist. Public land grazing allotments have often been provided on an historical basis to select ranchers at prices considerably below the market price of grazing rights. In many local timber sale situations there is only one log or stumpage buyer, and small-volume sellers may consistently have to sell at prices less than the value to the buyer. On the other hand, the U.S. Forest Service is sometimes the only seller of sawtimber and has difficulty transferring the stumpage at "competitive" prices.

Markets have one great virtue, they establish prices empirically and objectively. The market does not ask why a person wants to buy or sell; it only wants to know for how much. The prices of completed transactions then show how much, in fact, humans were willing to pay for goods and services. For this reason market prices, if available, are the preferred source of information for appraisal.

Market prices are clearly the pragmatic and only way to establish values for an individual who plans to buy or sell something in that market. Subjective values just do not convert to dollars received in the market. The sale value of a "priceless" heirloom, say granddad's gold watch, is only what someone is willing to pay. If the seller is in a hurry, it is what the local pawnbroker will pay.

Value in Use

Forest valuation questions typically center on the value in use of forestland and the tree vegetation growing on the land. Land and trees are evaluated by both the potential buyers and the seller (owner) of the land. Each potential buyer of forestland calculates the value of a particular parcel for the specific uses he or she is contemplating. Consider the views of four buyers: a tree farmer, a speculator, a recreationist, and a timber buyer. The tree farmer estimates the value of the land by how much future income it will provide from stumpage sales. The speculator evaluates the property by the expected future increase in the market price of the land and timber together. The recreationist subjectively evaluates the satisfaction or utility received from owning and using the property for camping, cabin building, hiking, or other purposes. The timber buyer looks only at the trees and determines stumpage value by figuring the sale value of lumber that can be made from the trees and subtracting what it costs to log, transport, manufacture the logs into lumber, and, adding back in, the resale value of the land after harvest.

Forest owners as land and/or timber sellers may calculate the value three ways: (1) the value to themselves in current or probable future uses, (2) the price it would sell for on the current market, and (3) the value in use to each possible buyer. The first calculation establishes a base for the decision to sell or retain the property—any offered price should exceed the value in current use before the owner will sell. The second calculation establishes an approximate market value, and the third provides information about each buyer or bidder in order to facilitate individualized price negotiation, which may result in higher than "market" prices if the candidate buyer is not fully informed or has some unique attraction to the property. For example, the land might have certain tree sizes and species of critical importance to a sawmill owner, or the aesthetic setting may strike a particularly responsive cord with one possible buyer.

Social Value

The total size of the national economic pie, roughly measured by the gross national product, although important, is only one of many social concerns. Society is also concerned about the distribution of goods and services—who benefits and who pays. Does the tax, pricing, and market system lead to the rich getting richer and the poor getting poorer? Are we collectively destroying the land, water, and cultural commons? Is the economic system encouraging stable growth and increased opportunities for all segments of society. Does it enhance international strength and national security? Are desirable public goods being provided? All of these social goals and concerns suggest that the values subjectively assigned by legislators and public administrators to evaluate actions, programs, plans, and policies may be considerably different than the values that would be assigned by individuals or by the market. Although it is very difficult to put quantified values on these social goals and ideas, the legislative and administrative bureaucracies of any government, through law making, rule making, regulations, and budget appropriations, do in fact make decisions that articulate social values.

DECISION MAKING AND VALUES

As long as managers or politicians make choices between mixes of market and nonmarket goods, values are inescapably assigned to the nonmarket items. The only options are (1) whether the value of the nonmarket items will be made explicit through specific value judgments *before* the decision and used to guide the choice, or (2) whether the choices among alternatives will be made directly on subjective grounds and the decision maker's valuation of goods and service implied by this choice deduced *after* the decision. In short, do we make choices based on established values or do we establish values by our choices?

To illustrate the important difference between these two options, suppose a manager is making a choice between two alternative plans for using a parcel of land. Both choices have the same costs and satisfy other constraints. The first plan yields 100 M bd ft of timber per year and 10 coveys of huntable quail. The second plan involves modified timber production practices to encourage quail and is expected to yield 80 M bd ft of timber and 20 coveys of huntable quail per year.

To base the choice on established values, the manager would make the valuation of quail *before* the decision, and might say "timber sells for $300 per M bd ft and I believe quail have a value of $100 per covey." The first plan has a total value of $31,000 per year (100 M bd ft × $300 + 10 coveys × $100). The second plan will annually produce $26,000 (80 M bd ft × $300 + 20 coveys × $100). The first plan presumably would be chosen as it has the highest total value.

When values are established by choice, the manager makes the decision directly and might say something like "we don't have unit values for quail, but all things considered I think the second solution is the best one." From the choice we can infer something about the worth of quail to this manager. The second solution involved the opportunity cost of giving up 20 M bd ft of timber per year worth $6,000 in order to get 10 more coveys of quail. Assuming the manager was rational and informed of this trade-off (and this assumption is critical to the argument), we could deduce that the marginal 10 covey unit of quail must be worth at least $600 per covey for this choice to have been made.

Most real choices are not this simple, so it is usually difficult or impossible to infer the values implied by decisions. Still, values are being implicitly assigned.

We could further reflect that all values are in fact revealed by actions and decisions. Today's prices are no more than yesterday's revealed values of buyers and sellers acting in a market where the terms of trade are measured in dollars. Is there any way values can be articulated other than by individual human decisions? Hypothetical questions yield promises of future actions (if the price is $100 next year, I will . . .) and can suggest values, but it is what we actually *do* that counts. "Watch what they do, not what they say" is a well-worn axiom of psychologists, political scientists, and other students of human behavior.

Since the trend is to provide a more open, quantified, and analytical decision process, particularly in the public sector, establishing values before the decision is increasingly preferred. We find ourselves more and more attempting to place dollar values on goods and services which are not traded in any market. By using the values in the same analysis with market-determined values, we are also asserting that they are acceptably comparable to the market values. The dollar prices used in national forest planning value everything from timber to water to dispersed recreation—see, for example, Table 12-2.

APPRAISAL

Valuation is the general study of methodology and concepts to determine the worth of things. Appraisal is the specific act of establishing the value of a particular item to a particular individual in a particular situation at a point in time. Be forewarned, however, that people still use the terms valuation and appraisal interchangeably, this distinction notwithstanding. People who do appraisals are, not surprisingly, called appraisers, and it is a recognized, certified, and licensed profession in its own right. Virtually all appraisals are concerned with buying or selling assets, with establishing a basis for taxation, or with establishing asset values for settling a variety of legal actions. Market value is the most common and the legally preferred viewpoint and conceptual framework used to guide the appraiser's efforts.

Purpose of Appraisal in Forestry

Foresters are frequently involved with valuation issues and asked to make specific appraisals. Determining values for individual logs and trees and for groups or stands of trees offered as stumpage sales are two routine problems with which every forester needs familiarity. Some of the most common forestry situations requiring an appraisal are useful to review.

Buying or selling land and/or timber Both the buyer and the seller of stumpage or timberland need an estimate of its value before entering into negotiation. If the market or bid value is higher than the value of the timber in use to the seller, the bid will probably be accepted. Similarly if the value in use to the buyer is higher than the market or asking price, an offer will likely be made. Determining the value in use is the key, and it is not a simple process. Many factors, such as lumber selling prices, manufacturing cost, logging cost, hauling cost, contracts, transportation, labor problems, market interest rates, and equipment available, have to be considered.

Virtually every timber sale of consequence is inventoried and appraised by both the buyer and the seller. It is an open procedure in public timber sales. The nonindustrial private landowner with small acreage and infrequent transactions is usually the least knowledgeable about timber appraisal procedures and is the most susceptible to exploitation by knowledgeable buyers. Assisting such landowners in appraisal and sales is consequently a major activity of consultants and extension foresters.

Planning Whenever management advances to the level of formalized planning for future production and harvest activity, then appraisals have to be made to provide the timber and other product values needed to quantify objectives and constraints or to rank discrete investment opportunities. These appraisals differ from those of the current buyer and seller in that they are forward looking forecasts of relative levels and trends rather than searching for today's exact market value. Also, the concern is more with average per-acre or per-unit prices for different stand types and product types rather than with the value of a particular set of trees to be sold or purchased.

Formalized planning is most evident on public lands, industry, large estates, and other large nonindustrial holdings. The U.S. Forest Service is raising the planning art to its most sophisticated level of quantification under the mandates of the 1976 National Forest Management Act and places dollar values on almost everything.

Damage and other legal claims Many forestry-related civil lawsuits involve a claim by one party against another to obtain financial compensation for damages. Fire, mechanical, or chemical damage to timberlands caused by negligence, condemnation of land by a public agency, and timber theft are some examples. Inheritance of an estate that includes forested land requires an appraisal of land and timber values to set the basis for estate taxes and for the distribution of assets to the heirs. Appraisal for such legal purposes has to follow customary and acceptable procedures, be done by a recognized and qualified appraiser, and be as accurate as possible since faulty work will promptly be challenged by at least one of the contesting parties. Each contesting party usually has an appraisal done, and the court may also have an independent third party do an appraisal to get a more objective, unbiased opinion. Fair market value at the time of damage action or death is the dominating concept of value used by the courts.

Timber damage appraisal gets particularly tricky when the damaged trees are only partly destroyed, or if the appraisal is made several years after the fact. Critical issues become estimating the stand condition and structure before damage, and how the undamaged stand *would* have grown compared to the observed growth in volume and value of the partly damaged stand.

Taxation From the beginning of civilization governments have sent out tax collectors to appraise and tax the assets and income of the citizens. Today most city and county governments depend on property taxes to finance themselves. In some areas of the West and South, timberland contributes much of the tax base for rural counties. Implementing the property or *ad valorum* tax process requires an assessment of land and timber value, and this is where the appraisal is needed. The guiding value concept is again the market value, but the tax base itself is usually 20 to 60 percent of the appraised market value, and the appraiser is somewhat more concerned with an equitable appraisal of similar and adjacent properties than with absolute market values.

In some states the taxable value of timberland is determined by its *value in use* to grow trees rather than market value which reflects the highest valued use. Called productivity tax assessments, this appraisal assigns a consistent assessment of, say, $200 per acre to good pine sites, even if they are in agricultural areas where the land has a market value of $1,000 per acre for agricultural purposes. The property tax is then levied using the $200 assessment whenever the land is used to grow pines. This assessment is designed to encourage tree growing by (1) taxing fallow lands at the $200 assessment rate, which penalizes lack of productive use, and (2) protecting tree growing from other encroachment such as urban development by taxing at the lower $200 rate (rather than $1,000 or higher urban value levels) as long as the land is actually used to grow timber.

Loans Banks and other financial institutions often provide the working capital needed to operate a forest business. Land and timber is the usual collateral for such loans, and the banks may require a market value appraisal as a condition of the loan. The bank will prefer an independent third-party appraisal, and usually it will loan only a percentage of the appraised market value.

Methods of Appraisal

Several methods and associated techniques are available to actually produce the numbers required for an appraisal. These can be grouped into six strategies: (1) market evidence, (2) calculated present net worth of value in use, (3) derived residual value, (4) market quantification, (5) cost of replacement, and (6) subjective judgments.

Market evidence Obtaining actual sales records (transactions evidence) from recent sales of similar and nearby land, timber, or other assets is the favored approach to setting market value. It directly measures what actual properties have sold for in the market. The only real issue is the comparability of the recently sold properties to the subject property. The appraiser obtains from the courthouse or real estate books the descriptions and locations of completed sales, evaluates them for comparability to the subject property, and makes informed estimates of value.

To illustrate, suppose the subject property was a 30-acre farm woodlot in a midwestern state. The appraiser might find data on three other properties which are similar and had been sold within the past 6 months.

Table 9-1 Blackwood farm woodlot appraisal (3/6/83)

Item	Blackwood Farm	Property *A*	Property *B*	Property *C*
Size of property (acres)	30	21	60	36
Sawtimber volume (M bd ft/acre)	6	12	8	4
Distance from subject property (miles)	—	30	3	10
Distance to town (miles)	10	16	7	3
Date of sale	—	10/1/82	2/15/83	1/5/83
Selling price per acre	—	$430	$500	$650

These properties are then compared by several criteria with the subject property (Table 9-1).

None of the properties are an exact match; property *A* is in a more rural area, property *B* is larger, and property *C* is sufficiently close to town to introduce some development value. Still this may be all the market evidence there is. The average value, weighted by acres, is $533.58, and the appraiser might conclude that this is a bit high due to the influence of property *C* and render the opinion that "a fair value for the Blackwood woodlot would be $480 per acre, all things considered."

Present net worth of value in use To determine what land, timber, or other assets are worth in use to an individual, one or more specific plans of use need to be generated by the prospective user. This plan will necessarily be a forecasted schedule of cost and revenue activities associated with the plan. Given assumptions about future prices, costs, technology, the interest rate, and other details, a financial analysis of each proposed use plan can be summarized by an estimate of its present net worth. The details and techniques of financial analysis, present net worth, and soil expectation value calculation have already been presented in Chaps. 7 and 8. Here the important concern is its interpretation: *present net worth is the economic value of the asset to the user in the specific plan of use under consideration.* It is what the buyer could pay for the asset and still make the guiding rate of return on the invested capital, given all the other assumptions about prices, costs, and yield.

The detailed Michigan Christmas tree growing problem in Chap. 7 is a good illustration. If we simply view the analysis as being an appraisal for a buyer of land planning to grow Christmas trees, then *if* the cultural regime of Table 7-5 were followed, the estimated costs and returns realized, and a 6 percent guiding rate of return required, the calculated present net worth of the land asset is $1,278.29* for one 7-year cycle. The SEV is $3,816.68 calculated over perpetual 7-year crop cycles, which is the maximum amount the buyer could pay and still earn a 6 percent return. If the buyer required some additional reward for entrepreneurship and risk, then the maximum value of the land to this buyer would be somewhat less.

When actual market evidence is not available or is highly questionable, then the present net worth–value in use approach is often used to make the next best estimate. For example, in one situation the construction of a dam by a federal agency caused the flooding of an industrial southern pine seed orchard that was just about to start commercial seed production. The problem was to find an acceptable cash settlement price for the flooded orchard. Since there were no recorded sales of tree seed orchards, the estimated value was made by an involved present value analysis requiring many judgmental assumptions about seed production rates and higher timber yields expected in plantations from the orchard seed. The agency and industry seed orchard people both made the same calculations using different experts to make these assumptions and substantiate their judgments. This resulted in orchard present net worth estimates of approximately $500,000 and $1,000,000, respectively. The out-of-court settlement rather neatly split the difference.

It is critical to remember that the estimated present net worth from one individual's perspective does not give an estimate of the market value, except by happenstance. Market value is the result of buyers and sellers interacting, each with their own present net worth estimates. Among other things, it also depends on how many buyers and sellers there are. A present net worth estimate gives some information and guidance as to how each individual will behave in a market—what he or she will bid or ask for the item in question. When several buyers are bidding for a single asset, such as a stumpage sale, then the buyer with the highest calculated present net worth for the use of that timber will likely be the highest and successful bidder.

Derived residual value The value of stumpage is often calculated using what is called the derived residual-value approach for the valuation of raw materials or other production inputs. The approach first establishes the selling value of the manufactured product and from this subtracts all of the manufacturing and raw material costs, leaving a residual, which becomes the maximum that could be paid for the raw material in question.

* This is the present value of the net future value of $1,871.2 plus the land rental cost of 50.36 (Tables 7-7, 7-9, p. 282).

For timber stumpage this calculation in outline form is as follows:

	Dollars per M bd ft
Lumber selling value	+$300
Sawmill costs	− 120
Normal mill profit	− 10
Log hauling costs	− 45
Logging costs	− 60
Maximum net value of this stumpage to this mill	+$ 65

The specific numbers in the example represent the appraised value of a particular stumpage sale to a specific stumpage buyer (the sawmill) to be used in a specific way (sawed into lumber in this mill) after contracting a particular logger and his equipment. If there is a few years' delay between the purchase and payment for the stumpage and its conversion and sale as lumber, then discounting future revenues and costs is appropriate, and the evaluation becomes a typical financial analysis and present net worth calculation. If everything occurs in the same year, then discounting is not necessary.

Different buyers at different distances from the sale area and with different mills and products would come up with different numbers. For some, the calculation would produce a negative value (do not bid) and for others a high positive value (bid aggressively). Chapter 11 is devoted largely to this problem of stumpage valuation and appraisal, making considerable but not exclusive use of the derived-value approach. Chapter 10 uses a similar approach to look at the value of logs and individual trees. Appraisals of water and forage often use this derived-value approach to establishing values.

Market quantification Another valuation approach involves actually numerically quantifying the past and present supply and/or demand curves for a specific market and product and then using the curves to estimate prices and values. This approach is more frequently used to predict the condition, sales, and prices in future markets than to determine current value. Market quantification is done two ways, distinguished by the source and type of data used to estimate supply and demand. Econometrics is the more refined and uses historical records of market prices and quantities sold to estimate statistically the supply and demand functions implied by (or which logically explain) the observed prices on quantities sold. The resulting equations can be used to forecast, if we assume people will behave in the future as they did in the past and no major events such as war or depression occur. Econometric techniques are obviously limited to products that are sold in the market and for which good historical records have been kept on prices, quantities sold, and other market factors. In forestry this pretty much restricts the methodology to lumber and other wood products. The projected equilibrium quantities and prices for wood in the United States, which we saw in Table 1-5, were such econometric forecasts.

For other forest activities, such as camping, hiking, hunting, and fishing, a second market quantification approach is frequently used. Since there is no actual market with recorded prices, the approach involves indirectly estimating the demand for the recreational service by observing behavior (frequency of visit, choice of site, distance traveled) or by interviewing, using structured hypothetical questions such as "how much would you pay for a hunt at swampy marsh?" or "would you visit Yellowstone park if the gate fee were $50?"

These indirect demand estimation techniques, each with a strict and difficult to satisfy set of assumptions, can produce a quantified demand function that can be treated just like one produced by econometrics and used to estimate prices or values. Chapter 12 gives an introduction to the indirect demand and other techniques currently used to value nonmarket forest outputs.

Replacement cost In cases of damage, theft, or loss of an asset, good, or service, an approach to valuation is to determine the current cost of replacement with a comparable item. The damaged item is likely older, worn, and worth less in the market than the cost of a new item to replace it. Still the new item is an even up functional compensation when a serviceable used item cannot be found. Replacement cost is a difficult concept to use when the replacement item is an antique.

The replacement cost approach is not extensively used in forestry because vegetation or soil cannot be replaced instantly on the same land where it was lost. This concept works best with equipment, structures, and other property that is movable, easily defined, and widely sold in the market.

Expert judgment The final class of valuation techniques uses human judgment as the method and conceptually is not much more than asking an individual "what, in dollar or relative terms, is some item worth?" The issues are to decide (1) who should make the judgments, (2) what do the resulting numbers mean, and (3) how should they be used to make decisions.

Citizens, "experts," administrators, public servants, professionals, and legislators all can and do offer opinions on the worth of things. It is reasonable to assume that elected administrators and legislators were elected because they shared or reflected the values of the majority of the voters. In addition they are accountable to the citizens and are subject to recall in subsequent elections. These administrators, legislators, and their political appointees make value judgments through their decisions, rules, and legislation. If the political system is assumed acceptable, then the argument can be made that *the collective value judgments of the elected politicians are the best quantification of social and other nonmarket values available*.

The value judgments of such elected and appointed politicians are often made directly and explicitly. The longer such prices are used, adjusted, and stand the repeated testing of political, administrative, and public review, the more likely they are to become the implied consensus values for the goods and services in question.

There is a lot of debate about the accuracy and wisdom of the political process in picking people and on the objectivity and competitiveness of the legislative and administrative organizations. For example, the administrators or politicians making value judgments about endangered species or the importance of timber to employment also make judgments on tax rates, defense, education, foreign policy, and social security, and their political accountability is probably dominated by their behavior on these larger issues. Politicians and high-level administrators get much of their information and personal contact from skillful, articulate representatives of interest groups like the Sierra Club, the National Forest Products Association, or the Western Wool Growers. There is always partial, conflicting, and incorrect information in this pluralistic environment, and we have to wonder what set of facts a politician is using when the final value judgments are made. But what alternatives are there to put a social price on an endangered species, clear air, or a visitor-day of wilderness recreation in a pluralistic, political democracy?

CHAPTER

TEN

VALUATION OF THE TREE

This chapter applies the principles and methods of valuation to appraise an individual tree in terms of the wood products it contains. When marking a stand to remove some of the trees, choosing wisely which trees to cut and which to leave to meet economic and biological objectives requires a knowledge of individual tree values. Effective and profitable marketing of timber marked for harvest similarly requires a sharp knowledge of tree values.

What is the wood product value of a tree in the woods? To prospective buyers, a tree represents a storehouse of raw material. They see in a tree, and collectively in a tract of timber, a certain quantity of logs, pulpwood, bolts, piling, and other items that can be made into salable products. They measure tree and timber values by what they can get for these products less the necessary costs of cutting, processing, and placing them on the market.

To the seller, landowner, and forest manager, a tree represents not only a marketable commodity, but also capital growing stock. They must appraise the value of the products presently derivable from the tree and balance this return against future growing-stock values if the tree is left to grow. Their problem involves finances, biology, and time. They must forecast growth in volume and value, making decisions as to what to cut accordingly. As always, the forest manager seeks the best overall result in accordance with the purposes of management.

Estimating individual tree values, consequently, is a problem of wide significance. Its importance in uneven-aged management has often been emphasized; effective application of an uneven-aged system directly depends upon the knowledge of individual tree performance measured in volume and value terms. Even-aged management requires this same knowledge of how trees grow and what they are worth. Several intermediate cuttings are often made, with the objective of obtaining the greatest possible return from the growing stock. Every time a tree is cut, an appraisal of tree and stand value, both now and in the future, is implied—even if the tree marker or feller is unaware of it at the time.

Tree valuation essentially is a matter of estimating the conversion return of individual logs or other units contained in the tree using the derived-value approach introduced in Chap. 9. Difficulties arise in the mechanics of identifying logging costs, milling costs, and selling prices in terms of individual tree or log units. Trees, logs, and processing practices are not standard; in fact, they vary endlessly. A multitude of sizes, shapes, and grades of finished or semifinished products are involved, often complexly interrelated.

Numerous logging and milling studies have measured individual tree and log values; a general body of techniques has developed with a more or less common pattern. Given the basic procedure, particular applications are largely variations on the same theme. Here we shall follow a logging and milling study through all its essential details, pointing out possible variations and alternatives.

Logging and milling studies usually take a sample of trees from some cutting operation or group of operations. Individual trees are described, with logs or other products obtained from them carefully measured and identified. The logs may be estimated in the tree, but in any event, they are checked and measured after the tree is felled. These logs are then individually followed through the various steps of logging and subsequent manufacture to finished products in whatever form sold by the primary producer. Selling prices and costs incurred in each step are collected and brought together to estimate the value of each log handled and eventually of the tree from which it came.

Two general procedural choices are possible in estimating individual tree values. The first uses the whole tree for the basic computational unit. Standing trees are classified or graded as a whole according to some qualitative scheme, based on observable differences in form, average log grade content, branchiness, and other external factors affecting value. Results are presented as value per tree by size, species, and whole-tree grade classes. The second approach separately identifies and values each log in a tree, reporting the results by size and grade.

WHOLE-TREE METHODS

When one follows a whole-tree approach, individual logs are identified as they pass through the logging and milling process, but the processing costs and selling values are summed and reported as totals for the tree unit. These data form the basis of tree value classes, which may be keyed by grade of the butt log only, by grade of the butt and second log, or by grade recovery from the tree as a whole.

The grade of the butt log gives a good measure of the tree quality for many species; the butt log often contains much of the tree's value. Also, the grade of the second and higher logs tends to correlate with the grade of the butt log. Log values by diameter, length, and grade are transformed into tree grades or classes by developing a representative log grade structure for trees of different butt characteristics. For example, Campbell (1951, 1955) bases his tree classification for Appalachian hardwoods entirely on the classification of the butt log. The field use of such tree classes depends on the ability to recognize log grades of butt logs in standing trees, and the general utility depends on how well the system distinguishes between trees of significantly different values.

Economic tree classes are useful in appraising the utilization value of the tree as a whole. It is easier to recognize tree classes in the woods than to estimate individual log grades, diameters, and lengths in a standing tree, particularly for the upper logs. Marking rules can then be based on tree classes in conjunction with silvicultural specifications. Building tree classes out of log grades, however, requires that some average sequence of log grades be assumed. Where the variation in log grade sequence within a tree is large, the sample small, and the values are high, it may be desirable to estimate the log grade structure of individual trees as shown in the next section. What to do depends on the values involved and the circumstances.

LOG-BASED METHODS

A second and more basic procedure concentrates attention on values of individual logs described by diameter, length, and grade. With continued development and use of log grades to estimate product mix recovery, their recognition in valuation becomes important. If an appraiser knows the value of individual logs by size and grade as they stand in the tree, he or she can estimate the value of any particular size, kind, or class of tree on the basis of its log grade structure. Measuring log values by size and grade provides a more comprehensive, flexible, and widely usable procedure for establishing tree values than does use of tree value classes. We follow the individual log approach in our major example below.

The usefulness of log and tree value studies is proportional to the value and variability of the timber involved. For this reason, they have been largely confined to sawtimber and other products of comparatively high value. Lumber yield studies of sawtimber species are by far the most typical, and one such study is considered here.

Four Steps in Log-Based Tree Appraisal

Following the normal procedure in stumpage appraisal, the lumber price as sold on the market provides the starting point. These prices give the value of lumber sawed from logs. The net conversion value of a log as it stands in the tree is calculated by successively deducting the costs of milling and logging. Four major steps are involved, each with several subsidiary operations:

1. Determine lumber value per thousand board feet by log grade and diameter. Estimating these values requires information on:
   a. Lumber prices by grade as received by the lumber manufacturer
   b. Thicknesses, widths, and lengths of lumber sawed
   c. Lumber grade recovery from logs of different diameters, lengths, and log grades.
2. Determine value of logs per thousand board feet of log scale by log diameter, log grade, and length as they reach the mill. Estimating this value requires deduction of milling costs obtained by:
   a. Cost analysis of the various steps in milling
   b. Time studies to determine the varying times required to saw logs of different diameters and lengths
   c. Analysis of overrun or underrun to convert values from mill or lumber scale to the log scale used to measure logs in the woods
3. Determine stumpage value of logs per thousand board feet of log scale and per log by diameter, log grade, and length as they stand in the tree. Estimating this value requires deduction of logging costs and, consequently, information on:
   a. Cost analysis of the various steps in logging
   b. Time studies to establish the varying times required to handle logs of different diameters and lengths
   c. Conversion of log values per thousand board feet to values per log on the basis of log volume
4. Apply stumpage value by log grade to standing tree values. This step requires information on the log structures of individual trees.

Some special terms are used in our tree and stumpage appraisal discussions in this and the next chapters and need definition at this time.

Variable costs (direct costs). These costs change with the amount and quality of logs, lumber, and other wood products processed (e.g., the amount of labor or gasoline used).

Fixed costs (indirect costs). These costs do not change with the amount or quality of log or wood product processed over a year or some other defined period (e.g., the annual salaries of supervisors or professional foresters, an annual building lease payment, or property taxes).

Conversion return at a processing stage. Conversion return is the end-product selling value per unit of wood input minus all fixed and variable costs that must be incurred after the processing stage being evaluated to produce the end product.

Conversion surplus at a processing stage. Conversion surplus is the end-product selling value per unit of wood input minus all variable costs (but not the fixed costs) that must be incurred after the processing stage being evaluated to produce the end product.

Conversion surplus and conversion return can be calculated and used for analysis of the wood or some other input at any stage in the production process. When wood is the input we might evaluate conversion surplus at the stage of green lumber, logs at mill deck, logs at roadside, or as standing trees (stumpage) in the forest. In forest management-economics literature the terms have been most commonly applied at the stumpage stage.

Estimating the Value of a Sugar Maple Tree: An Example of Log-Based Tree Appraisal

The mechanics of estimating individual log and tree values can best be illustrated by an integrated case. Sugar maple, a species of high value widely cut into standard lumber grades, will be highlighted. Its individual tree values tend to be highly variable, and log grades are extensively used in the buying and selling of maple timber. The case will be carried through to the tree in terms of logs measured by grade and size. The same general techniques apply equally to other species; however, only sugar maple is considered in the interest of simplicity in presentation.

A small logging and milling operation in the southeastern United States handling sugar maple along with a number of other hardwood species provides the basic information (James, 1946). The sawmill cuts about 10 M bd ft per day. Lumber is piled in the yard immediately after sawing and sold on a rough green basis.

Lumber values by log grade and diameter The market price received by grades for the lumber sold is the first step. If the study is being made for the immediate benefit of a particular operating company, current prices actually received by that company would be appropriate. If the study is made for wide application and longer range use, average prices received over a period by one or several companies would be logical. Average prices gathered by trade associations can also be used. In this case study, average regional prices of rough green lumber were used and adjusted to prices received at the sawmill.

Because sugar maple lumber prices vary by board thickness, but do not vary significantly by board width and length within the range of standard log lengths cut, they were weighted by the average percentage distribution of thickness cut. After weighting by the thickness actually cut at the study mill, sugar maple selling prices were:

Lumber grade	Selling price per M bd ft
First and seconds	$234.73
Select	159.96
No. 1 common	145.45
No. 2 common	80.30
No. 3 common, A	56.57
No. 3 common, B	38.58

Next these lumber prices are applied to lumber grade recovery from logs of different diameters and log grades. These log grades are readily identified by surface characteristics to segregate logs suitable for lumber into high-, medium-, and low-quality log grade classes (Forest Products Laboratory, 1953). Grade recovery information was obtained by sawing logs of various log diameters and grades classes and measuring the actual volume and grade of lumber contained in each log class (Table 10-1). Then the grade recovery information was combined with lumber price data to give lumber values per thousand board feet by log grade and diameter (Table 10-2). Table 10-3 illustrates the calculation of lumber values for an 18-in grade 1 log, one of the entries in Table 10-2.

Obtaining value per thousand board feet by log grade and diameter completes step 1 of our four-step appraisal process and constitutes the base from which milling and logging costs are subsequently deducted to derive the log value of the standing tree. While the specific mechanics of estimating log values vary with the situation, the basic pattern remains the same: estimate lumber selling prices, weight them by lumber thicknesses, lengths, and widths to the extent these factors influence price, and apply the result to lumber grade recovery data to get lumber values for logs of different sizes and grades.

Values of logs at the mill In step 2, milling costs are estimated and deducted from the log values.

Table 10-1 Percent distribution of lumber grade recovery by log size and grade

Log diameter (in.)	Lumber grade						
	F and S	Select	No. 1	No. 2	No. 3CA	No. 3CB	Total
			Grade 1 logs				
10							
12							
14	26	11	35	16	12		100
16	27	12	35	15	10	1	100
18	29	13	34	14	8	2	100
20	30	14	34	13	7	2	100
22	31	14	34	12	7	2	100
24	32	14	34	11	7	2	100
26	33	15	33	10	8	1	100
28	34	15	33	8	10		100
			Grade 2 logs				
10							
12		2	38	29	22	9	100
14	3	7	39	27	17	7	100
16	6	9	39	27	14	5	100
18	9	9	40	27	12	3	100
20	10	10	40	27	11	2	100
22	11	10	41	27	9	2	100
24	12	10	41	28	8	1	100
26	13	10	41	29	6	1	100
28	14	10	42	30	3	1	100
			Grade 3 logs				
10			18	45	37		100
12			26	43	29	2	100
14	1	2	28	42	23	4	100
16	3	3	28	41	19	6	100
18	4	3	30	40	16	7	100
20	4	3	31	40	15	7	100
22	4	4	31	39	15	7	100
24	4	4	32	38	15	7	100
26	5	3	33	38	15	6	100
28	6	3	33	39	15	4	100

Table 10-2 Lumber value per M bd ft by log size and grade

	Log grade		
Log diameter (in.)	1	2	3*
10			$ 83.26
12		$ 98.01	89.50
14	$149.18	108.96	95.10
16	151.56	116.75	98.53
18	154.84	123.32	101.67
20	157.42	126.32	102.57
22	158.97	128.85	
24	160.53	131.19	
26	162.39	133.22	
28	163.88	136.12	

* Grade 3 logs larger than 20 in. are seldom cut of this species and so are not included.

Cost analysis The sawmill in this case has a circular head saw, edger, trimmer, cutoff saw, and a small skidder to haul logs from the yard to the deck. Seven people operate the mill; usually the mill superintendent assists in addition to his management duties. Workers receive hourly rates. Administrative and general overhead costs are low.

The mill operates rather steadily throughout the year, obtaining logs from various sources. The head saw sets the sawmill's pace. Consequently in this case sawing time controls sawmill costs per thousand board feet cut.

The cost per hour of operation was obtained by dividing the average total yearly milling cost by the number of hours operated per year. The

Table 10-3 Calculating the lumber value of an 18-in., grade 1 log

Lumber grade	Grade recovery percentage from Table 10-1	Lumber selling price per M bd ft	Weighted log value per M bd ft, lumber scale
Firsts and seconds	29	$234.73	$ 68.07
Select	13	159.96	20.79
No. 1 common	34	145.45	49.45
No. 2 common	14	80.30	11.24
No. 3 common-A	8	56.57	4.52
No. 3 common-B	2	38.58	0.77
Total	100		$154.84

Table 10-4 Sawmill operating costs and lumber yard costs

Cost item	Sawmill costs per hour of operation	Lumberyard costs per M bd ft
Wages, including social security, workman's compensation, and unemployment insurance	$15.34	$4.70
Other direct costs, including operating supplies, maintenance, and depreciation	3.06	1.05
General costs, including supervision, administrative overhead, land rent, insurance, and taxes	3.69	2.39
Total	$22.09	$8.14
Total per minute	$ 0.3682	

average operating time was figured on a 250-day, 8-hour per day work year. Direct labor and yearly costs such as fire insurance, taxes, and depreciation are added together in this instance as they are all a part of the average total mill operating cost per hour (Table 10-4).

After sawing, the lumber is piled in a yard for initial drying and sale. Two persons do this work, using small carts to move the lumber from the mill platform to the yard. While there is some variation in yard cost occasioned by thicknesses, lengths, and widths of boards handled, the differences are small. Because of this, log size does not correlate significantly with yard costs. These costs are logically figured on the basis of the volume of lumber handled and hence are fixed per thousand board feet. The cost per thousand was determined by summing all costs chargeable to the yard for a year and dividing by the volume handled during this same period (Table 10-4).

Time studies From the above cost analysis two kinds of cost figures were obtained, costs per thousand board feet and costs per unit of time. The lumberyard costs do not vary by log size and can be applied directly to the log values per thousand board feet. Sawmill costs are expressed per hour of mill operation; they vary per thousand in accordance with the time it takes to saw a thousand board feet of lumber from logs of varying diameters and lengths. To convert hourly sawmill costs to per thousand costs, a time study of the sawing operation—the pacer of the rest of the mill—was needed.

After a preliminary test of sawing times, the various hardwood species going through the mill were divided into two significant groups: the "hard" hardwoods, of which sugar maple is one, and the "soft" hardwoods. For each group, stop-watch studies estimated the time required to saw logs of various diameters and lengths. The volume of lumber cut was also recorded so that the time required to cut a thousand board feet could be calculated. Log grade was assumed not to influence the sawing time, though it undoubtedly does to some extent.

For the diameter range studied in the "hard" hardwoods, the sawing time per thousand is a maximum for 10-in. logs and decreases rapidly with increasing diameter up to 16 in. (Table 10-5). The sawing time above 16 in. continues to decrease, though at a much slower rate, until logs of about 28 in. are cut, after which time increases somewhat. The log length affects the saw time dramatically at the smaller diameter classes.

Typical of most sawmills, the smallest logs always take the most time to saw in relation to their volumes. As larger logs are sawed, the time required diminishes up to a point and then increases as a size is reached that is larger than what the equipment can handle efficiently. This is particularly apparent in small mills when logs are handled that are oversize for the head rig equipment.

Sawmill costs The next stage of the analysis combines the cost analysis with the time study data to estimate the costs per thousand board feet in the sawmill. Two kinds of information are needed: the cost per unit of

Table 10-5 Sawmill crew person-minutes required per M bd ft to convert hard hardwood logs by diameter and length of log

	Log length (ft)			
Log diameter (in.)	10	12	14	16
10	146	130	115	99
12	93	86	81	76
14	73	70	66	62
16	63	59	57	54
18	57	54	52	49
20	52	50	49	46
22	48	46	46	44
24	46	45	44	43
26	45	44	43	41
28	45	45	45	44
30	46	46	46	46
32	48	48	48	49
34	52	52	53	54
36	55	58	59	61
38	59	62	64	67
40	64	68	73	73

time and the time required for a constant unit of volume. In this case the cost analysis (Table 10-4) shows that the cost per hour of sawmill operation was $22.09, or $0.3682 per minute. The sawing time study gives the variable sawing time in minutes (Table 10-5). Multiplying the appropriate numbers from these two tables gives the cost per thousand board feet for a log of a specified diameter and length.

To give a specific example, the cost per thousand board feet for an 18-in. 16-ft log equals the sawing time of 49 minutes times $0.3682, or $18.04. These calculations were made for the log diameter range of sugar maple occurring in this study plus a little more to show trends (Table 10-6).

Value of logs at the mill log deck Sawmill and lumberyard costs can now be combined and subtracted from log lumber values (Table 10-2) to give the log value as logs reach the mill. Table 10-7 shows the calculations in detail for grade 1 logs, 12 ft long. Looking over all grades and lengths, the effect of log grade and diameter on the value per thousand board feet is rather large, but the effect of log length is rather small (Table 10-8).

Table 10-6 Total sawmill manufacturing cost per M bd ft for hard hardwoods by diameter and length of log

	Log length (ft)			
Log diameter (in.)	10	12	14	16
10	$53.76*	$47.87	$42.34	$36.45
12	34.24	31.67	29.82	27.98
14	26.88	25.74	24.30	22.83
16	23.20	21.72	20.99	19.88
18	20.99	19.88	19.15	18.04
20	19.15	18.41	18.04	16.94
22	17.67	16.94	16.94	16.20
24	16.94	16.57	16.20	15.83
26	16.57	16.20	15.83	15.10
28	16.57	16.57	16.57	16.20
30	16.94	16.94	16.94	16.94
32	17.67	17.67	17.67	18.04
34	19.15	19.15	19.51	19.88
36	20.25	21.36	21.72	22.46
38	21.72	22.83	23.56	24.67
40	23.56	25.04	26.88	26.88

* Entries are respective entries of Table 10-5 multiplied by $0.3682.

Table 10-7 Calculating the log value for 12-ft, grade 1 sugar maple logs of different diameters

Log diameter (in.)	Lumber value of log per M bd ft (Table 10.2) (1)	Sawmill cost per M bd ft (Table 10.6) (2)	Yard cost per M bd ft (3)	Total cost per M bd ft (4) = (2) + (3)	Log value per M bd ft mill scale at mill (5) = (1) − (4)
14	$149.18	$25.74	$8.14	$33.88	$115.30
16	151.56	21.72	8.14	29.86	121.70
18	154.84	19.88	8.14	28.02	126.82
20	157.42	18.41	8.14	26.55	130.87
22	158.97	16.94	8.14	25.08	133.89
24	160.53	16.57	8.14	24.71	135.82
26	162.39	16.20	8.14	24.34	138.05
28	163.88	16.57	8.14	24.71	139.17

Log values in Table 10-8 are in mill scale, or lumber scale, as it is usually termed. This is the amount of lumber actually recovered from the logs at the mill. Sawlogs, however, are normally bought, and logging costs are reckoned in log scale according to some accepted log rule. Measurement by log scale indicates only approximately the actual board foot lumber recovery in the mill. This lack of agreement between log scale and lumber scale necessitates the calculation of overrun (or underrun), one of the more perplexing problems of logging and milling analyses. Overrun or underrun measures the difference between log scale and lumber scale. When expressed as a percentage, as is commonly done, log scale provides the base:

$$\text{Percent overrun (underrun)} = \frac{\text{lumber scale} - \text{log scale}}{\text{log scale}} \times 100$$

The difference between log and lumber scale results from the inherent nature of the log rule and scaling practice applied, the size and character of logs sawed, the kind of lumber cut, grading practices, and sawmill efficiency. The amount of overrun obtained can and often does vary substantially from mill to mill, even when the same log rule is used. Since the overrun (or underrun) obtained has a direct effect on log values, it should be checked for each particular sawmill, even though regional or local area averages by species and log size may be available. Overrun is particularly important in hardwoods because of the variable nature of the logs usually sawed and the peculiarities of lumber grading.

Table 10-8 Value per M bd ft lumber scale of sugar maple logs delivered to the mill

Log diameter (in.)	Log length (ft) 10	12	14	16
	Grade 1 logs			
14	$114.16	$115.30	$116.74	$118.21
16	120.22	121.70	122.43	123.54
18	125.71	126.82	127.55	128.66
20	130.13	130.87	131.24	132.34
22	133.16	133.89	133.89	134.63
24	135.45	135.82	136.19	136.56
26	137.66	138.05	138.42	139.15
28	139.17	139.17	139.17	139.54
	Grade 2 logs			
12	55.27	57.84	59.69	61.53
14	73.94	75.08	76.52	77.99
16	85.41	86.89	87.62	88.73
18	94.19	95.30	96.03	97.14
20	99.03	99.77	100.14	101.24
22	103.04	103.77	103.77	104.51
24	106.11	106.48	106.85	107.22
26	108.53	108.90	109.27	110.00
28	111.41	111.41	111.41	111.78
	Grade 3 logs			
10	21.36	27.25	32.78	38.67
12	47.19	49.69	51.54	53.38
14	60.08	61.22	62.66	64.13
16	67.19	68.67	69.40	70.51
18	72.54	73.65	74.38	75.49
20	75.28	76.02	76.39	77.49

In the study being followed here, overrun and underrun were estimated at the sawmill, using gross scales by the log rule as the base. The results, expressed as percentages, show high variability in both diameter and log grade (Table 10-9). The size and frequency of the negative percentages, indicating underrun, result partly because a gross log scale was used and partly because sawing hardwood lumber from lower grade logs loses significant footage in edging and trimming.

Applying these overrun percentages to the lumber scale values of Table 10-8 converts them to log scale (Table 10-10). Because percentage

Table 10-9 Percentage lumber scale overrun or underrun from log scale by log diameter and grade

Log diameter (in.)	Log grade		
	1	2	3
10			26.5*
12	24.5	13.5	13.5
14	9.5	4.0	1.5
16	0.0	− 1.0	− 9.0
18	− 5.0	− 4.0	−16.5
20	− 8.0	− 5.5	−20.0
22	−10.0	− 6.5	
24	−11.5	− 8.5	
26	−12.5	− 9.5	
28	−14.0	−11.0	

$$\text{* Overrun (underrun)} = \frac{\text{lumber scale} - \text{log scale}}{\text{log scale}} \times 100$$

overrun conversions are tricky, several formulas are available to assist movement between log and lumber scale (Table 10-11).*

Percentage overruns, which range from substantial positive values for the smaller diameters to strongly negative values for the larger diameters in this example, give a peculiar twist to the log values per thousand board feet. In terms of lumber scale (Table 10-8), unit values increase with log diameters for all log grades. In log scale (Table 10-10) this trend is reversed for grade 1 logs. For the other two grades, some increase in value with diameter remains, but the overrun flattens the trend. An overrun that is high for small diameters and thereafter decreases with increasing log diameters increases the value of small logs relative to the value of large

* As Herrick (1946, 1948) points out, overrun and underrun can usefully be expressed directly in board foot units, in which form it is initially determined, rather than as a percentage. For example, if a 16-in. 16-ft log, which contains 144 bd ft log scale Doyle rule cuts out 190 bd ft of lumber, overrun can be expressed as 46 bd ft instead of 32 percent. Overrun expressed in board feet is much less variable by log size than are percentages, Doyle scale in particular, which gives a sharply increasing percentage as the log size is reduced. The curve of overrun in board feet plotted over log diameter is much flatter. When overrun is determined by log diameters (and length), as in the case presented here, it makes little difference whether percentages or actual board feet are used. In making conversions based on the average log diameter, however, the use of an average board foot instead of a percentage figure is likely to give a more accurate answer, since it is much less sensitive to errors in estimating the average log size.

Table 10-10 Value per M bd ft log scale of sugar maple delivered to the mill after adjustment for overrun

Log diameter (in.)	Log length (ft)			
	10	12	14	16
	Grade 1 logs			
14	$125.00	$126.25	$127.83	$129.44
16	120.22	121.70	122.43	123.54
18	119.42	120.48	121.17	122.23
20	119.72	120.40	120.74	121.75
22	119.84	120.50	120.50	121.17
24	119.87	120.20	120.53	120.86
26	120.45	120.79	121.12	121.76
28	119.69	119.69	119.69	120.00
	Grade 2 logs			
12	62.73	65.65	67.75	69.84
14	76.90	78.08	79.58	81.11
16	84.56	86.02	86.74	87.84
18	90.42	91.49	92.19	93.25
20	93.58	94.28	94.63	95.67
22	96.34	97.02	97.02	97.72
24	97.09	97.43	97.77	98.11
26	98.22	98.55	98.89	99.55
28	99.15	99.15	99.15	99.48
	Grade 3 logs			
10	27.02	34.47	41.47	48.92
12	53.56	56.40	58.49	61.15
14	60.98	62.14	63.60	65.09
16	61.14	62.49	63.15	64.16
18	60.57	61.50	62.11	63.03
20	60.22	60.82	61.11	61.99

logs. This particular overrun pattern consequently reduces value differences associated with the diameter when measured in log scale.

Overruns should be recognized as a phenomenon of inconsistent measurement. Unit values per thousand board feet lumber scale do increase markedly with increasing diameter. The change to log scale, which here considerably underestimates the lumber yield of small logs and overestimates that of large logs, partially offsets this value trend.

Estimating log values per thousand board feet in log scale (Table 10-10) completes step 2 of the appraisal process. If the study is made by a mill owner who buys logs or by a producer selling logs at roadside, the

Table 10-11 Conversion formula for overrun and underrun

Conversion desired	Conversion formula*	Example†
Overrun:		
Log scale to lumber scale	$M_s = L_s(1 + p)$	$M_s = 1{,}000 \times 1.20 = 1{,}200$ bd ft
Lumber scale to log scale	$L_s = M_s \div (1 + p)$	$L_s = 1{,}000 \div 1.20 = 833$ bd ft
Log value to lumber value	$M_v = L_v \div (1 + p)$	$M_v = \$100 \div 1.20 = \$83.33$
Lumber value to log value	$L_v = M_v(1 + p)$	$L_v = \$100 \times 1.20 = \$120$
Underrun:		
Log scale to lumber scale	$M_s = L_s(1 - p)$	$M_s = 1{,}000 \times 0.80 = 800$ bd ft
Lumber scale to log scale	$L_s = M_s \div (1 - p)$	$L_s = 1{,}000 \div 0.80 = 1{,}250$ bd ft
Log value to lumber value	$M_v = L_v \div (1 - p)$	$M_v = \$100 \div 0.80 = \$125$
Lumber value to log value	$L_v = M_v(1 - p)$	$L_v = \$100 \times 0.80 = \$80$

* L_s = log scale; L_v = log value per M bd ft; M_s = lumber scale; M_v = lumber value per M bd ft; p = percentage overrun or underrun expressed as a decimal.

† Assume 20 percent overrun or underrun and units of 1,000 bd ft and $100 per M bd ft, respectively.

analysis stops here. Table 10-10 gives the conversion return of logs, in log scale, as they reach the mill, and consequently the basis for an appraisal of log values. The seller of logs at roadside, however, would have to subtract the additional costs of loading the log truck and hauling the logs to the mill in order to obtain roadside log values.

Value of logs in the standing tree In step 3, logging costs must be estimated and deducted from conversion returns of logs as they reach the mill to obtain conversion returns of logs as they stand in the tree. The time per thousand board feet, and hence the cost of felling and bucking,* skidding, and log loading, vary by log size. These costs must be calculated per unit of time and distributed per thousand board feet in accord with the variable time required to handle logs of different sizes. A check on log hauling and

* Strictly, felling is a joint cost applicable to the tree as a whole but not to individual logs. It is different from bucking, which directly applies to individual logs. The felling cost was not segregated from bucking in this study since the two operations were handled together in the woods. Although it would be technically desirable to deduct the felling cost directly from tree values, it was not done here, and in this case it makes little practical difference.

unloading showed that these costs were not significantly affected by log size and consequently could be considered as fixed per thousand board feet. Board and camp costs of the logging crew were handled by direct charges to the people employed and need not be considered in estimating operation costs.

Felling and bucking were handled by two-person crews using a power saw and paid on an hourly basis. Skidding was done by horses with the driver receiving some assistance from a swamper. A machine loader with an *A*-pole frame loaded the logs. Hauling was done by $1\frac{1}{2}$-ton trucks, and a small amount of road construction and maintenance was charged to hauling. Average costs in this instance are:

Felling and bucking	$5.25 per crew-hour
Skidding (average slope and distance)	2.52 per person-hour
Log loading at landing	4.20 per crew-hour
Log hauling, landing to mill	8.40 per M bd ft, log scale
Log unloading at mill	0.43 per M bd ft, log scale

Felling and bucking, skidding, and log loading hourly costs were distributed on a per thousand board foot basis by log diameter and length, through application of the variable times required for these operations (Tables 10-12 to 10-14). These costs were then combined with hauling and

Table 10-12 Felling and bucking: crew-minutes and costs by log diameter and length

	Crew-minutes per M bd ft log scale for log length of:				Cost per M bd ft log scale for log length of:*			
Log diameter (in.)	10 ft	12 ft	14 ft	16 ft	10 ft	12 ft	14 ft	16 ft
10	118	116	96	68	$10.32	$10.15	$8.40	$5.95
12	103	93	79	62	9.01	8.14	6.91	5.42
14	92	78	67	57	8.05	6.82	5.86	4.99
16	83	68	59	52	7.26	5.95	5.16	4.55
18	77	61	52	48	6.74	5.33	4.55	4.20
20	72	56	49	44	6.30	4.90	4.29	3.85
22	68	53	45	42	5.95	4.64	3.94	3.67
24	65	50	44	40	5.69	4.37	3.85	3.50
26	62	48	42	39	5.42	4.20	3.67	3.41
28	60	47	41	38	5.25	4.11	3.59	3.32

* The cost per crew-hour is $5.25, or $0.0875 per crew-minute. The cost per M bd ft is obtained by multiplying crew-minutes by this figure.

Table 10-13 Skidding: person-minutes and costs by log diameter and length

	Person-minutes per M bd ft log scale for log length of:				Cost per M bd ft* log scale for log length of:			
Log diameter (in.)	10 ft	12 ft	14 ft	16 ft	10 ft	12 ft	14 ft	16 ft
10	427	364	314	271	$17.93	$15.29	$13.19	$11.38
12	315	269	239	215	13.23	11.30	10.04	9.03
14	247	213	192	180	10.37	8.95	8.06	7.56
16	208	179	163	154	8.74	7.52	6.85	6.47
18	178	157	144	138	7.48	6.59	6.05	5.80
20	157	141	132	127	6.59	5.92	5.54	5.33
22	141	129	121	117	5.92	5.42	5.08	4.91
24	130	119	113	109	5.46	5.00	4.75	4.58
26	123	113	107	103	5.17	4.75	4.49	4.33
28	116	107	102	98	4.87	4.49	4.28	4.12

* The cost per person-hour is $2.52, or $0.042 per person-minute. The cost per M bd ft is obtained by multiplying person-minutes by this figure.

Table 10-14 Loading: crew-minutes and costs by log diameter and length

	Crew-minutes per M bd ft log scale for log length of:				Cost per M bd ft* log scale for log length of:			
Log diameter (in.)	10 ft	12 ft	14 ft	16 ft	10 ft	12 ft	14 ft	16 ft
10	149	125	101	77	$10.43	$8.76	$7.07	$5.39
12	89	77	65	54	6.23	5.39	4.55	3.78
14	61	53	46	40	4.27	3.71	3.22	2.80
16	46	40	35	30	3.22	2.80	2.45	2.10
18	37	32	27	24	2.59	2.24	1.89	1.68
20	30	26	22	19	2.10	1.82	1.54	1.33
22	24	21	18	14	1.68	1.47	1.26	0.98
24	21	17	14	12	1.47	1.19	0.98	0.84
26	18	14	13	11	1.26	0.98	0.91	0.77
28	16	13	11	10	1.12	0.91	0.77	0.70

* Average cost per crew-hour is $4.20, and the cost per crew-minute is consequently $0.07. The cost per M bd ft is obtained by multiplying crew-minutes by this figure.

Table 10-15 Total logging cost per M bd ft log scale by log diameter and length

	Log length (ft)			
Log diameter (in.)	10	12	14	16
10	$47.51	$43.03	$37.49	$31.55
12	37.30	33.66	30.33	27.06
14	31.52	28.31	25.97	24.18
16	28.05	25.10	23.29	21.95
18	25.64	22.99	21.32	20.51
20	23.82	21.47	20.20	19.34
22	22.38	20.36	19.11	18.39
24	21.45	19.39	18.41	17.75
26	20.68	18.76	17.90	17.34
28	20.07	18.34	17.47	16.97

unloading costs per thousand board feet, which were not affected by log size, to give total logging costs per thousand board feet by log diameter and length (Table 10-15).

To complete step 3 of the appraisal process it is necessary only to deduct these costs, which apply to log grades equally, from log values per thousand board feet at the mill (Table 10-10). Completing this subtraction gives the conversion return of logs per thousand board feet, gross log scale, as they stand in the tree (Table 10-16). Also shown is the value per log, which is obtained by multiplying the value per thousand board feet by the board foot volume per log.

Table 10-16 gives the end result of the interrelated computations needed to derive the conversion value of logs in the standing tree. The table applies specifically to a particular species, sugar maple in this case, and to specified logging and milling conditions. However, the procedures used are typical for studies whose objective is to obtain conversion return information applicable to logs in standing trees.

Directly applying conversion return information obtained under one set of circumstances to another can be hazardous. However, information underlying the conversion return calculation can often be shared. The computation of lumber values per log can be reduced through use of an average log quality index (Herrick, 1946, 1956), eliminating much of the detail required in lumber grade recovery data. Sawmill costs or sawing times may be fairly consistent between mills of the same general class, and milling cost differences between species are usually small. Logging cost components, such as loading, hauling, felling, and bucking, often are fairly standardized in contract rates. Time study data can be applied to these average contract rates to obtain the logging cost variation by log

Table 10-16 Conversion return per log and per M bd ft log scale in standing trees by log diameter and grade

Log diameter (in.)	Value per M bd ft for log length of:				Value per log for log length of:			
	10 ft	12 ft	14 ft	16 ft	10 ft	12 ft	14 ft	16 ft
				Grade 1 logs				
14	$93.48	$ 97.94	$101.86	$105.26	$ 6.54	$ 8.61	$10.19	$11.58
16	92.17	96.60	99.14	101.59	9.22	11.59	13.88	16.25
18	93.79	97.49	99.85	101.72	12.19	15.60	18.97	21.36
20	95.90	98.93	100.54	102.41	16.30	20.78	24.13	28.67
22	97.46	100.14	101.39	102.78	20.47	25.03	29.40	33.92
24	98.42	100.81	102.12	103.11	24.60	30.24	35.74	41.24
26	99.77	102.03	103.22	104.42	30.93	37.75	45.42	52.11
28	99.62	101.35	102.22	103.03	35.86	44.59	52.13	59.76
				Grade 2 logs				
12	25.43	31.99	37.42	42.78	1.27	1.92	2.62	3.42
14	45.38	49.77	53.61	56.93	3.18	4.48	5.36	6.26
16	56.51	60.92	63.45	65.89	5.65	7.31	8.83	10.54
18	64.78	68.50	70.87	72.74	8.42	10.96	13.47	15.28
20	69.76	72.81	74.43	76.33	11.86	15.29	17.86	21.37
22	73.99	76.66	77.91	79.33	15.54	19.16	22.59	26.18
24	75.64	78.04	79.36	80.36	18.91	23.41	27.78	32.14
26	77.54	79.79	80.99	82.21	24.04	29.52	35.64	41.10
28	79.08	80.81	81.68	82.51	28.47	35.56	41.66	47.86
				Grade 3 logs				
10	−20.49	−8.56	3.98	17.37	−0.61	−0.26	0.16	1.04
12	16.26	22.74	28.16	34.09	0.81	1.36	1.97	2.73
14	29.46	33.83	37.63	40.91	2.06	3.04	3.76	4.50
16	33.09	37.39	39.86	42.21	3.31	4.49	5.58	6.75
18	34.93	38.51	40.79	42.52	4.30	6.16	7.75	8.93
20	36.40	39.35	40.91	42.65	6.19	8.26	9.82	11.94

size. Once the basic pattern and procedure are understood, logging and milling data from one operation often can be adjusted to another.

Application of Log Values to Tree Values

The development of conversion return values per log as it stands in the tree (Table 10-16) completes the logging and milling phase of the job. The next step, and the reason for making the analysis, applies this information to estimating the value of standing trees by size and grade.

Let us assume the timber inventory estimated the diameter, length, and grade of each log contained in a tree. The necessity for and utility of log grades based on surface characteristics, such as are employed here, become immediately apparent. Data on 30 sugar maple trees illustrate the kind of information needed and its use (Table 10-17). Values per log are taken from Table 10-16, interpolating for odd-inch diameters not given in that table.

While log lengths, diameters, and grades obviously cannot be determined as readily for the standing tree as for logs on the ground, they can be estimated with surprising speed and accuracy. Most hardwoods contain no more than two or three merchantable logs; the butt log usually contains about half the total tree value itself. Of the 26 sugar maples having three or more logs, 51 percent of the total value resides in the butt logs, 36 percent in the second logs, and only 13 percent in the third and fourth logs (Table 10-17). The practical importance of errors in estimating log sizes and grades diminishes rapidly with increasing height above the ground and distance from the observer.

The worthwhileness of making a detailed, quality cruise depends primarily on the magnitude of the values involved and their variability among trees. For sugar maple, the species studied here, conversion returns per thousand board feet are high, and individual trees warrant careful study. For example, compare trees 2 and 16:

Tree no.	Tree diameter (in.)	Utilized length (ft)	Utilized volume (bd ft)	Log grades	Conversion return
2	24	32	370	2-3-3	$17.44
16	24	30	390	1-1	40.35

They share the same diameter, but tree 16 contains 2.3 times the value of tree 2, almost entirely because of the high-grade logs it contains.

Knowledge of log values by grade also helps in cutting a tree into log segments that yield the greatest total value. Here a 12-ft 20-in. grade 1 log is worth $20.78 in conversion return, whereas a 14-ft grade 2 log of the same diameter is worth $17.86 (Table 10-16). In cutting a tree into logs, total conversion return can be increased substantially by taking advantage of bole quality in selecting log lengths.

Using Log Grades in Buying and Selling Timber

Making lump-sum tree volume estimates of a timber tract and applying average per thousand board foot values can result in large errors in total value when considering a particular sample of trees with highly variable

Table 10-17 Conversion return value of 30 sugar maple trees

Tree no.	DBH (in.)	First log Dib (in.)	Length (ft)	Grade	Value	Second log Dib (in.)	Length (ft)	Grade	Value	Third log Dib (in.)	Length (ft)	Grade	Value	Fourth log Dib (in.)	Length (ft)	Grade	Value	Total tree value	Total* tree vol	Conversion value per M bd ft
1	24	19	16	2	$18.32	18	10	2	$ 8.42	17	10	3	$ 3.80					$ 30.54	490	$ 62.33
2	24	18	12	2	10.96	17	10	3	3.80	15	10	3	2.68					17.44	370	47.14
3	21	17	12	1	11.59	15	14	1	12.03	13	12	3	2.20					25.82	330	78.24
4	22	16	16	1	16.25	14	16	1	11.58	13	10	2	2.22	11	10	3	$0.10	30.15	370	81.49
5	18	13	12	2	3.20	11	12	3	0.55									3.75	110	34.09
6	23	18	12	2	10.96	16	12	2	7.31	15	16	3	5.62	14	10	3	2.06	25.95	490	52.96
7	26	20	16	1	28.67	18	12	2	10.96	16	14	3	5.58	13	10	3	1.43	46.64	640	72.87
8	23	19	14	1	21.55	18	16	1	21.36	15	14	3	4.67					47.58	540	88.11
9	19	15	16	2	8.40	13	16	3	3.61	12	12	3	1.36					13.37	300	44.57
10	20	16	16	1	16.25	14	14	2	5.36	12	12	3	1.36					22.97	320	71.78
11	28	22	12	1	25.03	20	16	2	21.37	18	12	2	10.96					57.36	690	83.13
12	18	14	12	2	4.48	13	12	2	3.20	11	12	3	0.55					8.23	200	41.15
13	24	18	12	1	15.60	17	16	2	12.91	13	12	3	2.20					30.71	410	74.90
14	30	22	16	1	33.92	18	16	1	21.36									55.28	540	102.37
15	21	17	14	1	16.42	15	14	1	12.03	14	12	2	4.48	11	12	2	0.70	33.63	410	82.02
16	24	19	14	1	21.55	17	16	1	18.80									40.35	390	103.46
17	18	15	14	1	12.03	13	12	2	3.20									15.23	190	80.16
18	20	16	16	1	16.25	14	14	1	10.19	13	10	2	2.22	11	12	3	0.55	29.21	360	81.14
19	21	16	14	1	13.88	15	12	1	10.10	14	12	3	3.04	11	10	3	0.10	27.12	380	71.37
20	17	15	12	1	10.10	14	14	1	10.19	11	16	3	1.88					22.17	280	79.18
21	26	24	6	Cull	0.00	17	16	1	18.80	15	14	2	7.09	12	12	3	1.36	27.25	360	75.69
22	25	20	16	1	28.67	18	14	1	18.97	17	12	2	9.13					56.77	610	93.07
23	19	16	16	1	16.25	15	16	2	8.40	14	14	3	3.76					28.41	400	71.02
24	23	18	16	1	21.36	17	16	1	18.80	15	10	2	4.41	14	10	3	2.06	46.63	550	84.78
25	22	16	14	1	13.88	16	10	2	5.62	13	10	3	1.43	11	16	3	1.88	22.81	370	61.65
26	26	21	14	1	26.76	19	12	1	18.19	18	10	2	8.42	16	8	3	2.10	55.47	660	84.05
27	21	17	12	1	13.59	15	12	2	5.89	14	12	3	3.04					22.52	340	66.24
28	18	15	16	2	8.40	13	10	3	1.43	11	10	3	0.10					9.93	240	41.37
29	26	18	12	1	20.78	16	16	1	16.25	15	10	3	2.68					39.71	390	101.82
30	26	21	12	1	22.90	20	12	2	15.29	19	10	2	10.14					48.33	590	81.92
																		$941.33	12,320	76.41

* Scribner decimal C log rule, gross scale.

individual values. Owners selling timber on a lump-sum basis frequently get less than the market value for their timber, even when the total volume is estimated accurately. Selling by log grade often can raise the price received substantially. Knowledge of the individual tree values equally helps a purchaser recognize a good buy and avoid overbidding.

Conversion return information by tree can be used directly in stumpage appraisal. The 12,320 bd ft included in Table 10-17 have an average conversion return value of $76.41 per thousand. Suppose these 30 trees were a representative stand table based on many sample plots in a large tract of maple timber. The $76.41 would then be a starting estimate of the stumpage value per thousand board feet in that tract. This approach is used for the stumpage appraisal example given in Chap. 11 which does not explicitly deal with log grades and assumes that grade considerations are imbedded in average lumber values per thousand board feet lumber scale. When you get to this example think about how you could expand the analysis to consider log grade explicitly.

All trees in this sample show a positive conversion return. Sugar maple has a high lumber value overall, and the large overrun obtained from small trees partially offsets the lower quality of their lumber. If smaller trees or trees of general lower quality had been included in the sample, some logs and trees might show a negative conversion return.

With a negative conversion return, no margin appears to exist, and apparently the tree can be cut or the log handled only at a loss. This may not be true. Costs forming the basis of the conversion return can and usually do include some fixed costs for the particular operation. As an example, the sawmill has fixed general administrative costs as indicated in Table 10.4. A log or tree is profitable to take when it returns some positive margin over variable costs and contributes something toward the fixed costs. Only items having a negative conversion surplus (the difference between selling value and variable costs only) are unprofitable to handle. Fixed costs, therefore, must be excluded in deciding whether a tree or log should be handled.

In the particular situation considered here, practically all the costs involved are variable costs. The logging and milling operation had little fixed overhead, and our first estimate of the conversion surplus is conversion return. In this case, consequently, a negative conversion return of any size indicates that the log or tree involved is nearly always unprofitable to handle. If the timber was purchased at some flat rate per thousand board feet, subtraction of the purchase price from the conversion return immediately shows the margin for profit and risk to the buyer and whether a particular tree is profitable to cut. If stumpage was purchased at $45 per thousand board feet, there are four trees (numbers 5, 9, 12, and 28) in the group of 30 given in Table 10-17 (and a considerably larger number of logs) that would return no margin for profit and risk when using conversion return to approximate conversion surplus. From Table 10-4, the fixed cost in the sawmill averages about $5 per M bd ft. Thus the last column of Table 10-17 should be increased by $5 per M bd ft. When this is done, all the trees would be included except tree 5.

USE OF TREE INFORMATION AS A GUIDE TO FUTURE VALUE

The same information that permits the determination of the current value of a tree of specified value and log grade structure also gives a basis for estimating future value. For a simple example, assume that tree 20 can grow 2 in. in 8 years with bole taper, log lengths, and log grades remaining the same except that the third log increases to grade 2. On the basis of present price levels, the tree's value will increase from $22.17 to $32.31 over the 8 years, an annual compound rate of 4.8 percent.

Log		Log values today			Log values in 8 years		
Position	Length (ft)	Diameter (in.)	Grade	Value	Diameter (in.)	Grade	Value
Butt	12	15	1	$10.10	17	1	$13.59
Second	14	14	1	10.19	16	1	13.88
Third	16	11	3	1.88	13	2	4.84
Total value of tree 20				$22.17			$32.31

If a decision to cut or hold this tree also needs to consider the cost of using the land, more information is needed. We discuss this in Chap. 13.

REFERENCES

Campbell, R. A., 1951: "Tree Grades, Yields, and Values for Some Appalachian Hardwoods," U.S. Forest Service, Southern Forest Experimental Station Paper 8.

———, 1955: "Tree Grades and Economic Maturity for Some Appalachian Hardwoods," U.S. Forest Service, Southeastern Forest Experimental Station Paper 53.

Forest Products Laboratory, 1953: "Hardwood Lumber Grades for Standard Lumber," U.S. Forest Service, Forest Products Laboratory Report D1737.

Herrick, A. M., 1946: "Grade Yields and Overrun from Indiana Hardwood Sawlogs," Purdue University, Agricultural Experimental Station Bulletin 516.

———, 1948: Accuracy in Estimating Overrun, *South. Lumberman,* **177:**57–58.

———, 1956: "The Quality Index in Hardwood Sawtimber Management," Purdue University, Agricultural Experimental Station Bulletin 632.

James, L. M., 1946: "Logging and Milling Studies in the Southern Appalachian Region," U.S. Forest Service, Southern Forest Experimental Station Technical Notes 62, 63, 64, and 65.

CHAPTER
ELEVEN

VALUATION OF THE TIMBER STAND, THE LAND, AND THE FOREST

Land and the timber stand on it are the basic capital of any forest enterprise. Through their sale, they are also the major source of ready cash for any forest enterprise. In addition, the ability to sell timber separately from the land means that forest owners face the continual dilemma, once the stand is merchantable, of whether to sell part of the timber-growing factory (the timber) or keep it to produce still more wood. Thus, valuation of forest land and the associated timber stand can take two general forms: market value—the price for which the stand or stand and land could be immediately sold—and value in use—the discounted value of the stream of products that could be produced from the land and stand over time.

Sometimes, market value and value in use coincide. This occurs especially when the major portion of the forest area is in stands of mature timber. Often, however, market value and value in use give numerically different appraisals, especially when a stand has been recently cut or contains immature timber. It is important to recognize both value concepts and valuation strategies with the stipulation that the wealth-maximizing individual should not sell the timber or timberland unless the market value exceeds the value in use.

We will discuss both types of value in this chapter, starting with the valuation of stumpage—the market value of a merchantable stand of timber. Historically, timberland values were based almost entirely on the existing value of timber, so this is a good point of departure. Next, we will value the land and immature timber stands in terms of what they can produce over time—their value in use. With the development of commercial timber-growing enterprises entailing significant investments, the valuation of forests as a basis for continuing production has emerged as an important analysis. Finally, we will value the regulated forest as a going concern. Many forest enterprises already have large tracts under management, and valuation of a forest producing a regular income has some interesting similarities to and differences from valuing an acre of forest land.

VALUATION OF STUMPAGE

Stumpage is timber in unprocessed form as found in the woods. Normally it means the physical content of standing trees, within a contiguous area, whether live or dead. The term also can apply to timber that is windthrown or cut in connection with right-of-way clearing, as long as it remains in place and is not cut into logs.* The term *stumpage* is sometimes used to mean the *value* of such timber.

Valuation of stumpage is an important and recurring activity in the commercial timber business. Stumpage is the raw material from which all wood products are derived. Not only is it frequently bought and sold in the daily conduct of a forest business, but it also provides the landowner with timber income. Small differences in contracted stumpage prices can greatly impact the profitability of timber harvesting and conversion. Because stumpage values affect the business so vitally, the practicing forester should know the methods employed in their estimation whether he or she grows, sells, or buys timber.

As stumpage is only a collection of trees in a contiguous area that is or could be offered for sale, much of what we learned in Chap. 10 about the value of individual trees applies here. The value of an aggregation of trees is substantially the sum of the individual trees making up the sale. The important additional considerations are roading and access costs, the effects of sale size on logging cost, and the institutional, legal, and market

* Timber can also be cut by the owner and taken to a market point accessible to buyers, such as roadside, streamside, railroad, or other delivery points, and is sold there as logs or bolts. This common European practice is not widespread in the United States. When done, the cost of cutting and market point delivery must be added to the stumpage value in appraisal. Such a procedure has won the endorsement from many economists as potentially increasing timber values by easing selective log purchase. In the United States, however, timber sells preponderantly as stumpage. The appraisal of logs and individual trees is described in Chap. 10.

factors affecting the price and conditions under which timber stumpage is sold and harvested. Stumpage appraisals ideally start with a good knowledge of lumber values and the cost of converting trees of different sizes, species, and grades. This information is then combined to estimate the conversion return.

Stumpage appraisal is done using both value in use and market value. Stumpage buyers who process the stumpage into other products take the value-in-use point of view to estimate the return from using the stumpage in their own conversion business. Stumpage sellers, either private tree farms or public lands, take the market-value perspective and try to estimate the likely selling value in the current market. The courts and the tax assessor also take the market view since they are usually mandated to determine "fair market value."

The U.S. Forest Service, the nation's biggest stumpage seller, has been mandated since its inception to offer timber sales with a minimum appraised price. A synopsis of policy directives (U.S. Forest Service, 1984a) as of 1984 is:

> For the purpose of achieving the policies set forth in the Multiple-Use Sustained Yield Act of 1960 and the Forest and Rangeland Renewable Resources Planning Act of 1974, the Secretary of Agriculture, under such rules and regulations as he may prescribe, may sell, at not less than appraised value, trees, portions of trees, or forest products located on National Forest System lands. [16 U.S.C. 472a].

Thus, national forest timber must be appraised before it is sold. For nearly 80 years the U.S. Forest Service has been appraising national forest timber in the west with the residual-value method. The method is outlined in the Secretary's regulation [36 CFR 223.4(a)]:

> The basic procedure will be analytical appraisal under which stumpage value is a residual value determined by subtracting from the selling value of the products normally manufactured from the timber the sum of estimated operating costs, including costs to the purchaser for construction of roads or other developments needed by the purchaser for removal of the timber and margins for profit and risk. Costs and product values under the residual value method shall be those of an operator of average efficiency and related to the operating difficulties and to size and quality of timber. . . .

In addition to the description of the "basic procedure" quoted above under "Residual Value Appraisals," the regulation states:

> Other valid appraisal methods, including transaction evidence procedures, or independent estimates based on average investments, may be used subject to approval of the chief, Forest Service. Under the approved procedures, pertinent factors affecting market value shall be considered, including but not limited to, prices paid and valuations established for other purposes for comparable timber. Consideration of such prices and valuation may recognize and adjust for factors which are not normal market influences.

In addition to the above, the *Forest Service Manual* [2420.2] states:

> Fair market value or appraised value as used by the Forest Service is based on the operator of average efficiency and is aimed at a market value which will interest sufficient purchasers to harvest the allowable cut under multiple use and sustained yield principles. In accomplishing this objective consideration must be given to providing an adequate margin for profit and risk which will be sufficient to maintain operations over the long run and thus provide a stable market for National Forest timber.

These regulations and directions have resulted in appraisal methods and timber sale procedures that are constantly debated and continuously evolving. Estimating the milling costs of an operator of "average efficiency" is contestable, as well as specifying an appropriate "allowance for profit and risk." Both forest industry buyers and efficiency-in-government interest groups periodically form task forces to review the appraisal and bidding procedures used.

Private timber sellers have sometimes been characterized as being ignorant of their timber and woodlot values and thus easy prey to slick stumpage buyers. While perhaps overstated, the timber owner is not likely to have a good estimate of volume or value unless he or she is actively in the timber sale business. Here is where an appraisal by a hired consultant or an extension forester provides solid information to guide the owner in negotiations with the timber buyer. Often private owners negotiate with a single buyer and arrive at a price without open competition, using the appraisal as a base.

Even when competition sets the stumpage price, an appraisal guides the judgment of buyers and sellers and influences their market behavior. Timely appraisals can signal present or prospective change in cost-price relationships; an alert seller or buyer will employ this information to his or her advantage. The market evidence or transactions analysis method of appraisal to estimate market value is presented first, followed by the derived residual value method to estimate value in use.

MARKET EVIDENCE METHODS OF STUMPAGE APPRAISAL

Market evidence methods set the stumpage value of a subject stand through comparison with the prices of stumpage received for stumpage recently sold from stands with similar characteristics as the subject stand. If you want to appraise pulpwood, you check the current pulpwood market; if you want to appraise old-growth sawtimber, you locate its market and get a price. Unfortunately it is usually not quite this simple. Reported stumpage prices for pulpwood or sawtimber typically reflect the average of many different sales, each with unique conditions of sale and extraction. Unless the subject stand situation is like the average of recent transactions, the average price may be too high or too low. Perhaps the subject timber is far from a road and most timber traded recently was near a road. Perhaps the subject sale is a partial cut and most recent sales were larger volume clearcuts. Or perhaps the subject timber is all spruce, and the average sale contains spruce and fir. Unlike corn, where the farmer brings so many pounds of standard kernels to market, no two offerings of timber are alike. Therefore market evidence methods for stumpage appraisal usually start with an average stumpage price and then make an adjustment for the many ways in which the sale seems unusual. Such adjustments are best seen through case studies, and we cover two here. The first is a trial equation used to estimate high bid prices in the northern Rocky Mountains. The second is a set of regionalized look-up tables used by the state of California as a basis for yield taxes.

Regression Analysis in Northern Idaho and Montana

The equation approach to estimating the appraised value examines hundreds of actual sales and fits a regression equation to these sale data which relates the many factors (variables) that cause differences in the sale price. If such an equation has a large data base, is shown to consistently explain most of the variations in sale prices in successive years, and has a logical structure, then it can be a useful and acceptably accurate way to set an appraised value on the stumpage in a specific timber sale.

The equation approach has worked reasonably well in the South where land, logging, and market conditions are the most uniform and there are a large number of individual sales. The equation approach has thus far had the poorest results on the west coast, where timber and logging conditions are extremely variable, and not enough sales and sale conditions are documented to fit equations with acceptable accuracy.

Because of criticism of the derived or residual-value approach, fresh efforts to base U.S. Forest Service appraisals on market evidence are being made in the 1980s. Jackson and McQuillan (1979) examined 52 sales on the Lolo National Forest from 1968 to 1977 and explained 80 percent of the variations in sale prices with their equation. Johnson (1979) examined 379 sales in the northern Rockies and also explained 80 percent of the variations. Preliminary results from a comprehensive study of over 1,000 sales from October 1977 to December 1982 (U.S. Forest Service, 1983) provided an equation that explains around 70 percent of the variations in sale price. One such equation from this study is presented here to show the kinds of variables that are used.

$$
\begin{aligned}
Y = {} & -106.07 + 0.493(\text{SPLS}) + 0.385(\text{CL}) - 5.83(\text{SALV}) \\
& -0.080(\text{PVJA}) - 0.494(\text{PVSK}) - 0.387(\text{ALPM}) + 1.74(\text{ADBH}) \\
& +17.4(\text{HSM}) + 7.74(\text{LNVPA}) - 6.68(\text{LNDEF}) + 3.01(\text{LNVOL}) \\
& -0.135(\text{TH3}) - 1.00(\text{OTCSTS})
\end{aligned}
$$

where Y is the expected high bid price per thousand board feet. The variables are defined in the following list:

Sale characteristics

ADBH = average tree diameter, breast high
ALPM = average number of logs per thousand board feet
LNVPA = natural logarithm of volume per acre harvested in M bd ft
LNDEF = natural logarithm of percentage of volume defective
SALV = salvage versus nonsalvage timber; salvage = 1, nonsalvage = 0
PVJA = percent of volume requiring ground-lead logging*
PVSK = percent of volume requiring skyline logging*
TH3 = haul, = miles of paved road + 3(miles of unpaved road)
OTCSTS = estimated cost of brush disposal, erosion control, road maintenance, and temporary developments in dollars per thousand board feet adjusted for inflation

Administrative variables

LNVOL = natural logarithm of total volume in million board feet
CL = contract length in months

Market variables Estimated average end-product price weighted by species volumes and adjusted for inflation:

SPLS = price per M bd ft on a log scale basis
HSM = seasonally adjusted monthly housing starts in millions, as reported two months prior to bid date

Suppose a subject timber sale was examined and the following values characterized the situation:

Variable	Value	Variable	Value
1. SPLS	$375	8. HSM	1.75
2. CL	48 months	9. LNVPA	2.562
3. SALV	0	10. TH3	65.0
4. PVJA	10	11. LNDEF	2.303
5. PVSK	40 percent	12. LNVOL	1.723
6. ALPM	15 logs/M bd ft	13. OTCSTS	$18.75/M bd ft
7. ADBH	15 in.		

* Remainder of logging is by tractor.

Table 11-1 Equation predicted and actual high bids on Forest Service timber sales in the northern region in 1982

Forest name	Number of sales (1)	Average high bid (2)	Average predicted high bid from regression model (3)	Advertised price based on residual value appraisal (4)
All data	213	$51.29	$56.16	$26.25
Bitterroot	13	57.60	58.15	46.82
Idaho panhandle	54	52.05	64.62	20.28
Clearwater	24	52.66	84.96	29.44
Flathead	38	49.22	46.36	26.33
Kootenai	43	57.94	54.25	23.76
Lolo	8	52.70	39.99	32.14
Nezperce	18	51.15	62.83	35.69
All east side*	15	26.54	26.22	17.24

* Based on 60-month base period instead of 90-month base period used for other national forests.

Source: U.S. Forest Service (1983), p. 11.

When substituted into the above equation, the estimated high bid turns out to be $109.30.

A test of the prediction equation was made using 213 sales in 1982. The results are as shown in Table 11-1 and show the predictions (column 3) to track the actual high bids (column 2) reasonably well. Still, some forests (such as Clearwater) are at considerable variance. On the other hand, the actual high bid prices are always substantially higher than the advertised minimum price (column 4), which is determined by the traditional residual-value method. Further testing and development of a set of more area-specific equations may improve the predictive power of the models.

Regionalized Harvest Value Tables for California Yield Tax

Since 1977 California county property taxes on timber have been paid by the timber owner as a percentage of the market value of the timber when it is harvested. This system replaced the earlier and more common approach of having the land owner pay annual taxes as a percentage of assessed land and timber values, whether or not the timber was actually being harvested.

A key to acceptable implementation of yield taxes is the ability to establish reasonably accurate stumpage prices for the range of species, logging conditioning, and timber marketing environments in the diverse state of California. To do this, the state was divided into nine timber value areas. Within each area all public and private timber sales are reported, and empirical tabulations of market stumpage prices by species, sale size, and logging costs are made. Because of their extraordinary value, coastal redwoods and Douglas-fir are further stratified by log size as an indicator of grade.

To illustrate, from October 1982 to September 1984, 1,417 timber sales containing 2.7 billion board feet of public timber and 1.1 billion board feet of private timber were studied. The transactions data were smoothed and processed into a standard set of six tables showing the average current market price of stumpage for each timber value area. The old-growth and young-growth tables for the north coast area are shown in Table 11-2. These tables are updated every 6 months by the state board of equalization.

To use the table, the taxpayer determines the appropriate logging cost code and enters the table by species and grade. If the sale size is smaller than 300 M bd ft or if less than 10 M bd ft per acre is harvested, further adjustments can be made as indicated.

DERIVED RESIDUAL-VALUE APPRAISAL METHODS

Residual-value appraisal is used extensively in national forest stumpage appraisals and is also widely used by states and private industry. As we did with tree valuation, the value of standing timber is calculated as the difference between the selling value of the products made from it and the stump-to-market processing costs. Stumpage value is considered a residual item—a difference.

Why is this so? Why should stumpage value, the return to timber production, be calculated as a residual? Stumpage derives its value from the same source as any other raw material: its demand originates in products that can be made from it. Buyers cannot afford (for long) to pay more for stumpage then they receive from sales of the products made from timber minus their processing costs.

An Overview Example

Suppose that you operate a small sawmill that manufactures two-by-fours, which sell for $200 per thousand board feet. A farmer down the road wants to sell you some long-leaf pine thinnings. How much can you bid? For this type of timber, variable milling costs run $50 and logging costs $75 per thousand board feet. This indicates a conversion surplus of $200 − $50 − $75, or $75. In addition, you include a cost of $5 per thousand board feet to cover some worries you have that the bottom may drop out of the two-by-four market before you process this purchase. That leaves [$75 − $5] $70 as the maximum amount you would bid.*

* See p. 388 for definitions of conversion surplus and conversion return.

Table 11-2 Harvest value tables for yield tax computation based on transactions evidence from October 1982 to September 1984

(a) Size-quality rating for old growth (average net volume per log, bd ft)

	Size-quality code		
Species	1	2	3
Redwood	Over 850	850–530	Under 530
Redwood, residual	Over 535	535–360	Under 360
Douglas-fir	Over 430	430–310	Under 310

(b) Old-growth harvest values*'† (per thousand board feet)

			Logging cost code				
Species	Species code	Size-quality code	1	2	3	4	5
Redwood	RO	1	$335	$325	$310	$290	$170
		2	310	300	285	265	170
		3	270	260	245	225	170
Redwood, residual	RRO	1	270	260	245	225	105
		2	245	235	220	200	105
		3	205	195	180	160	105
Douglas-fir	DFO	1	150	140	125	105	1‡
		2	125	115	100	80	1
		3	80	70	55	35	1
Ponderosa pine	PPO		145	135	120	100	5
Sugar pine	SPO		160	150	135	115	20
Fir, inland	FO		55	45	30	10	1
Incense cedar	ICO		55	45	30	10	1
Whitewoods, coastal	WWO		50	40	25	5	1
Port Orford cedar	PCO		650	640	625	605	510
Conifers, miscellaneous	CMO		30	20	5	1	1

Actually, you would rather offer less than $70 so that you could make some return on your investment in equipment which averages $12.50 per M bd ft. So you offer the farmer $55, a little less than your $57.50 conversion return. He shrieks in disbelief. He relates that carefully tending his crop—planting it, protecting it from fire, bugs, and disease, weeding it—has cost him at least $100 dollars per thousand board feet. You sympa-

Table 11-2 (Continued)

(c) Young-growth harvest values (per thousand board feet)

Species	Species code	Logging code 1	2	3	4	5
Redwood	RY	$180	$170	$155	$135	$ 40
Douglas-fir	DFY	80	70	55	35	1
Ponderosa pine	PPY	60	50	35	15	1
Sugar pine	SPY	60	50	35	15	1
Fir, inland	FY	55	45	30	10	1
Incense cedar	ICY	55	45	30	10	1
Whitewoods, coastal	WWY	50	40	25	5	1
Port Orford cedar	PCY	375	365	350	330	235
Conifers, miscellaneous	CMY	30	20	5	1	1

* Deduction if total volume on harvest operation is under 300 M bd ft, $15; under 100 M bd ft, $25; under 25 M bd ft, $40. Subtract deduction from table value.

† Deduction if average volume per acre on harvest operation is under 10 M bd ft/acre, $15; under 5 M bd ft/acre, $25. Subtract deduction from table value.

‡ $1 M bd ft is minimum harvest value after adjustment.

thize and tell him that you would like to cover his costs of production but cannot afford to pay more than $70 per thousand board feet. After complaining about the cosmic injustice of it all, he will either (1) sell the timber to you at $70, (2) try to sell it to someone else, or (3) hold the timber for future sale.

In assessing a tract of timber for a buyer, appraisers care not one whit whether the timber grew by itself (which is the case for most timber cut until now in the United States) or had a platoon of expensive foresters taking its pulse every day. Appraisal determines what buyers will pay for something, not what it costs the seller to produce. The buyer's appraisal process and immediate market demand forces are exactly the same.

In the long run, however, timber production costs do affect market prices. If our farmer recovers from his despair and thinks about it for a while, he may reduce his timber production efforts—rely more on natural regeneration or spend his idle Saturdays fishing instead of girdling hardwoods. As producers grow less wood, its price will rise over what it otherwise would be. The rising price and falling costs of production (due to less effort) eventually should enable the price the farmer is offered to cover all production costs.

The difference between product selling price and variable milling plus logging cost plus fixed equipment costs is termed the *conversion return*—the key figure in any indirect stumpage appraisal. A portion of this return goes to the tree buyer for entrepreneurial profit, including return on the investment, and for noninsurable risks associated with converting the logs and selling wood products. The remainder goes to the stumpage seller. Dividing this margin between buyer and seller is a central problem in appraisal, and many procedures and formulas have been devised to estimate the "appropriate" division.

In the end, however, its division is the bargaining point between stumpage buyers and stumpage sellers. In our example, the conversion surplus equaled $75, over which the farmer and the log buyer could then argue. While our log buyer wanted to pay only $55 for stumpage, he eventually might agree to $70, essentially wiping out his profit, including return on his investment. Such are the results of negotiation.

Procedures to Estimate Conversion Return: A Forest Service Timber Sale

With the general nature of residual-value stumpage appraisal in mind, it helps to consider a specific example. The following is an appraisal made by the U.S. Forest Service in the Westside Sierra mixed conifer type of California (U.S. Forest Service, 1984b). It illustrates an analytical derivation of stumpage values from lumber selling values and production costs, logging cost, roads cost, and division of the conversion return between profit and risk and stumpage. This appraisal is made by the Forest Service to determine the minimum bid it will accept for the timber and this is the amount the Forest Service thinks a bidder of average efficiency would pay for the stumpage; a similar appraisal could be made by each buyer to guide in setting his offering price.

Area and purpose of cutting The cutting area includes 498 acres in mixed conifers that will be clearcut, saving some patches of natural reproduction. Cutting will occur over four full operating seasons. These seasons normally run from May 1 to October 31.

Inventory The area was cruised estimating log grades in the standing tree. The standard error of the mean for total volume is estimated at ±5.8 percent. Estimated volumes to be cut (in millions of board feet) are

Douglas-fir	7.080
Sugar pine	7.000
Ponderosa/Jeffrey pine	6.000
White/red fir	3.600
	23.680

The average volume per acre, all species, is 47.55 M bd ft.

Table 11-3 Lumber value and manufacturing costs for California mixed conifer timber per M bd ft

Item	Douglas-fir	Sugar pine	Ponderosa/ Jeffrey pine	White/ red fir	Average, all species (weighted)
Selling value					
Lumber and veneer	197.42	397.78	362.00	175.97	295.11
Chips and bark value	+23.66	+23.66	+23.66	+23.66	+23.66
Totals	221.08	421.44	385.66	199.63	318.77
Manufacturing cost	135.85	179.35	167.17	157.08	159.88

Logging and processing The topography is moderate with sensitive soils, requiring care to prevent erosion. All acres can be tractor logged under dry soil conditions. Following a careful analysis of alternative destinations, the tract is appraised for timber processing by mills in Redding, California. To log the timber, the operator must construct 11.6 miles of permanent roads built to U.S. Forest Service utilization road standards and 14.0 miles of temporary roads.

Selling values and manufacturing costs Average selling values and manufacturing costs were estimated from data collected by the U.S. Forest Service (Table 11-3) in a series of studies of regional mills. Veneer logs and sawlogs were estimated separately by log grades and weighted by volume.

Logging costs Logging costs were estimated for all species together by adjusting average data for the area to suit this particular sale (Table 11-4).

Road construction Road construction costs often are a substantial part of the total cost necessary to get out the timber. In this appraisal, the operator must build 11.6 miles of permanent road for $323,323 and 14.0 miles of temporary road for $39,440. Total estimated road cost is $362,763, which, divided by the total volume to be cut of 23.68 million board feet, gives an average road cost per thousand board feet of $15.32. These road costs were apportioned to species groups based on the volume of stumpage in each group. Roads are joint costs because they serve all species. Allocation of such costs by species was done here because the agency wants appraised stumpage values by species or species groups to reflect road costs.

Table 11-4 Average logging costs per M bd ft

Stump to truck		
Falling and bucking (and snag disposal)	$10.43	
Skidding and loading	+16.12	$26.55
Log transportation		
Hauling	$22.82	
Road maintenance	+9.10	31.92
Administrative		
General logging overhead	$28.71	
Logging depreciation	+(0.00)	28.71
Other		
Slash disposal	$2.34	
Erosion control	1.13	
Required slash deposit BD 031	+2.75	6.22
Total logging costs (excluding roads)		$93.40

Calculating conversion return Bringing together the different parts of the appraisal permits the calculation of the conversion return (Table 11-5). For example, the conversion return of sugar pine (row 7, column 2 of Table 11-5) works out to be $133.37. Starting with a lumber content selling value of $421.44/M bd ft (row 2), the total of manufacturing, logging, and road construction costs are subtracted ($179.35 + $93.40 + $15.32, or $288.07) to obtain the conversion return. Note that the two fir species groups have negative conversion return values, mainly because of much lower lumber selling values than the pines. When all four species groups are combined, the weighted average conversion return of $50.17 lets the higher priced material carry the lower. Because the conversion return contains some fixed cost such as the logging depreciation in Table 11-4, the buyer still may want to take the fir to help pay fixed costs.

Margin allowance for profit and risk The timber buyer needs to get some return for actual time and effort, some profit for his entrepreneurial spirit, and something to cover the expected losses on an occasional sale that is not profitable. This total reward is called the *margin,* and the appropriate amount is a matter of some debate. The U.S. Forest Service currently allows from 11 to 13 percent of total stumpage, logging, and milling costs for this margin. This sale uses a profit ratio* of 0.11 for Douglas-fir, white fir, and red fir and 0.12 for sugar pine, ponderosa pine, and Jeffrey pine.

* The profit ratio is the percentage rate applied to stumpage and manufacturing cost to calculate a margin for profit and risk. It is popular among appraisers because a variation permits direct calculation of the margin or amount of profit and risk from an assumed profit

Table 11-5 Calculation of conversion return and indicated stumpage value

Item	Value per M bd ft					Sale total				
	Douglas-fir	Sugar pine	Ponderosa/ Jeffrey pine	White/ red fir	Total, all species (weighted)	Douglas-fir	Sugar pine	Ponderosa/ Jeffrey pine	White red fir	Total, all species
1. Volume to cut (M bd ft)	7,080	7,000	6,000	3,600	23,680	7,080	7,000	6,000	3,600	23,680
2. Selling value (lumber)	$221.08	$421.44	$385.66	$199.63	$318.77	1,565,246.4	2,950,080	2,313,960	718,668	7,547,974.4
3. Manufacturing costs	$135.85	$179.35	$167.17	$157.08	$159.88	961,818	1,255,450	1,003,020	565,488	3,785,776
4. Logging costs	93.40	93.40	93.40	93.40	93.40	661,272	653,800	560,400	336,240	2,211,712
5. Road construction costs	15.32	15.32	15.32	15.32	15.32	108,465.6	107,240	91,920	55,152	362,797.6
6. Total production costs	$244.57	$288.07	$275.89	$265.80	$268.60	1,731,555.6	2,016,490	1,655,340	956,880	6,360,285.6
7. Conversion return	−23.49	133.37	109.77	−66.17	50.17	−166,309.2	933,590	658,620	−238,212	1,187,688.8
8. Profit ratio	0.11	0.12	0.12	0.11	0.115					
9. Margin for profit and risk	21.91	45.15	41.32	19.78	33.377	155,122.8	316,120	247,920	71,208	790,370.8
10. Indicated stumpage value	−45.40	88.22	68.45	−85.95	16.79	−321,432	617,470	410,700	−309,420	397,318

Computations for table

lines 1, 2, 3, and 4: from Tables 11-3, 11-4, inventory
line 6: (line 3) + (line 4) + (line 5)
line 7: (line 2) − (line 6)
line 8: policy
line 9: [(line 8) × (line 2)]/(1 + line 8)
line 10: (line 7) − (line 9)

The margin for profit and risk and the stumpage value are calculated for each species. The margin for sugar pine equals:

$$M = \frac{PR}{1 + P} = \frac{0.12 \times 421.44}{1 + 0.12} = \$45.15$$

where M = margin
P = profit ratio
R = lumber selling value

The sugar pine stumpage value equals the conversion return minus the margin for profit and risk, \$133.37 − \$45.15 = \$88.22 (row 10, Table 11-5). The average indicated stumpage value for all species is \$16.79.

After the appraisal the sale is advertised and bids solicited. Both sealed and oral bidding are used, depending on the forest, and the sale is normally awarded to the high bidder. The Forest Service monitors the sale to be sure the contract specifications are met.

Special treatment of road costs In making an appraisal, costs are normally expressed in unit terms, such as so much per thousand board feet. Many costs, however, must largely be incurred if the sale is undertaken at all, rather than being incurred per unit cut. Road construction in the sample appraisal is a good case in point. A total of \$362,763 is needed for the construction of roads to make the timber accessible, and most of these roads must be built before any timber can be removed. The average cost of \$15.32 per thousand board feet allowed in the appraisal was calculated by dividing this estimated total road cost by the estimated total volume of 23,680 M bd ft.

---

ratio and a given selling value, without first defining stumpage:

$$M = P(C + S) \text{ and } S = R - C - M$$

where R = product selling value, C = production costs, M = margin for profit and risk, S = stumpage value, and P = profit ratio. Substituting for S,

$$M = P(C + R - C - M)$$

simplifying and rearranging

$$M = PC + PR - PC - PM$$

$$M = PR - PM$$

$$M + PM = PR$$

Consequently,

$$M(1 + P) = PR \qquad \text{or} \qquad M = \frac{PR}{1 + P}$$

Therefore, the margin M for profit and risk is the product of the profit ratio and the selling value divided by the profit ratio plus 1.

If this volume is cut and the road cost estimate is accurate, there is no problem. As is well known, however, estimates of timber volumes and road construction costs may be too high or too low. Assume that the road cost estimate is correct, but that the buyer harvests 15 percent more timber during the sale period than planned. The harvest volume consequently climbs to 23,680 × 1.15, or 27,232 M bd ft. At $15.32 per thousand board feet, $417,194 is allowed the buyer for building the road, whereas the actual cost of the road is estimated at $362,763. The difference, $54,431, goes to the buyer as a potential addition to profit. The situation reverses if the sale undercut the estimate or the road cost was higher than expected.

To handle this problem, the U.S. Forest Service now uses a system of purchaser credits to cover road costs. The cost of the road is estimated by the Forest Service, and this amount is stated in the offering. The bidders bid as though the road is already built (zero road costs). The successful bidder is then given credit in the amount of the estimated road cost as a deduction from the amount paid for the stumpage. The buyer must build the road to contract specifications. If the actual road cost is less than the estimate, additional profit is made.

GENERAL PROBLEM OF STUMPAGE APPRAISAL

Appraising any particular stand of timber involves as much art as science. No two stands are quite the same, and the possible uses of their wood are endless. Some problems of stumpage appraisal seem to recur, however, and we shall discuss here seven aspects of stumpage appraisal that continually give appraisers fits.

The Seller's Interest in the Buyer

In stumpage appraisal, as with any valuation problem, the approach taken depends on the purpose of the appraisal and how it might be used. Many industrial timberland owners cut their timber through contract loggers or sell stumpage to an independent logger with the expectation of buying back the timber as cut logs or pulpwood to supply their processing plants. These companies deal with a network of established timber buyers and loggers who cut timber and sell logs. These independent operators in turn are heavily dependent on company timber.

The company can dominate stumpage price negotiations but must temper its desire to make money by the necessity to enable these small operators to stay in business and reliably handle company logging needs. Both short-run and long-run objectives are brought to bear when the company appraises the value of its timber.

The U.S. Forest Service looks at stumpage appraisal in somewhat the same way. It sets a minimum appraised price using the manufacturing and logging costs of operators of average efficiency. Practically, the U.S. Forest Service wants to interest sufficient operators to purchase the planned quantity of sale offerings.

In any distribution of manufacturers, those with *above* average efficiency will have lower operating costs and thus higher conversion return values for stumpage than the U.S. Forest Service appraisal which is for operators of average efficiency. The below average efficiency operators will have lower conversion return estimates. Competitive bidding would therefore be expected to reveal two things: (1) the high bids for federal stumpage will be higher than the appraised price, and (2) the below average operators will lose money (if they are successful bidders) and tend to go out of business.

The first expectation comes true as seen in the data for the primary western timber regions from 1979 to 1984 (Table 11-6). The bid premium

Table 11-6 Weighted average data for all species sold (dollars per thousand board feet)

	1979	1980	1981	1982	1983	1984 (1st qtr)
Montana, Northern Idaho						
Advertised rate	49.41	16.66	19.44	13.45	21.32	25.00
Bid premium*	57.24	65.94	71.88	36.33	50.23	40.84
Product value†	375.48	311.33	348.12	342.05	365.80	396.15
Production cost‡	284.51	309.31	341.36	353.59	343.66	343.54
California						
Advertised rate	79.83	35.61	32.53	23.64	31.67	31.75
Bid premium	172.86	282.22	149.80	41.43	61.09	68.36
Product value	356.01	314.94	319.94	302.84	325.33	335.05
Production cost	237.42	258.44	280.54	293.37	298.68	302.32
Westside Oregon, Washington						
Advertised rate	140.62	109.16	104.83	35.82	49.88	72.00
Bid premium	220.05	290.56	208.30	74.28	101.06	72.76
Product value	465.88	443.06	472.21	433.84	465.22	513.93
Production cost	262.92	281.98	324.22	372.13	383.04	400.31
Eastside Oregon, Washington						
Advertised rate	101.26	46.56	48.36	30.90	50.59	65.05
Bid premium	82.34	111.49	113.89	30.15	39.98	34.23
Product value	395.73	341.02	371.60	358.00	399.88	410.64
Production cost	244.24	271.23	299.83	317.59	329.61	327.63

* Bid premium = high bid − advertised rate.
† Equivalent to line 2 of Table 11-5.
‡ Equivalent to line 6 of Table 11-5.

(the difference between high bid and appraised price) is always positive and at least as much as the appraised price. In 1980 and 1981 the timber sold for almost 300 percent of appraised value (and consequently many successful bidders lost their shirts trying to actually harvest and sell the lumber in the depressed markets and higher production cost environment of 1983–1984).

The less efficient operators are hard to discourage, and new operators emerge to take the place of the fallen. An expectation of a continually rising stumpage rate has kept many marginal timber operators in the hunt. Also, appraisal procedures are crude enough to preclude precise definition of the "average operator."

The U.S. Forest Service prohibition on offering timber with a negative appraisal is an important reason why its timber sometimes does not sell. The timber value in some parts of the country cannot cover sizable removal costs and needed reforestation. Road building costs to get the timber and the higher cost of environmentally protective logging often greatly exceed the stand value, especially in the Rockies. In this case the timber is sometimes offered at an arbitrary low minimum bid price of a few dollars. If no bids are received, then none of the buyers can make a positive value in use appraisal. Often, however, some bids are received, partly because the bidders with better than average conversion efficiency can calculate a positive appraisal. The U.S. Forest Service generally will not pay people to remove commercial timber. Even though the purchaser-built road system is projected to pay off from a present net worth standpoint when considering future stands and sales made accessible by the roads, today's sale must pay for the road built today. Setting minimum stumpage at a low but positive value means sales will not be made unless an operator chooses to make a positive bid and pay the deficit.

Conditions of Sale

In selling stumpage, a contract is made and a buyer enters upon the property of an owner to cut timber. This is a disruptive operation and can severely impact the land, vegetation, and amenities. In addition to wanting a high price for the timber to be cut, the seller often wants to control the cutting to meet other objectives. Normally the seller designates the area and the particular trees to be cut, specifying when and how they shall be removed and the condition in which the cutting area shall be left. A number of restrictions may be imposed to protect the trees to be left from damage, specify utilization standards, guard against erosion, minimize damage to other forestland values such as recreation or wildlife, and specify the location and construction of roads and their improvements, which may be built in connection with timber cutting. Sale conditions imposed by the seller can be long and varied.

To the buyer, these conditions of sale are part of the "deal," and the costs of meeting them is considered part of the costs of cutting timber, even though many often are not included in the appraisal. The differences in the conditions of sale are often important (differences between the sale of private, state, and federal timber, for example) and can cause controversy and frequent misunderstanding in relation to stumpage appraisal.

Estimating Product Selling Value

Because stumpage value derives from the value of products made from the timber, wood product selling price is the beginning point for an appraisal. In general, the price should be that received by buyers for the most valuable products obtainable at the earliest processing stage for which a competitive market exists. In the case of a stumpage buyer who sells logs, the log price received at the mill, or wherever the buyer's handling of the timber stops, provides the beginning point. In the case of a sawmill operator who sells rough lumber on the market, the prices he receives for his lumber are the beginning of the appraisal. If the lumber is dried and planed, market prices of finished lumber provide the starting point when weighted by the dimensions and grades that can be produced from the stumpage under consideration.

Complications occur when a log buyer produces several kinds of primary products such as pulp chips, veneer logs, and transmission poles in addition to sawed products, or is an integrated producer who manufactures and sells a variety of products. As an example, it is impracticable, if not impossible, to appraise pulpwood based on all the types of paper and cardboard made from it. However, people market pulpwood and markets establish prices for it. A mill will set a price based on its judgment of how much it must pay to get the pulpwood it wants. Prices so established adjust to the local pulpwood supply and demand situation and can be used as the beginning point in an appraisal. Market prices provide better guides than derived values in this case.

Determination of Production Costs

All costs directly required to produce the commodity under appraisal should be included. Other costs such as income taxes or cost of the stumpage itself are not to be included when estimating conversion return. They are considered in setting an allowance for profit and risk.

Accurate cost estimation is hard. No two tracts of timber are exactly alike. The size, condition, and quality of the timber and the difficulty of logging and accessibility can vary widely from place to place, and good estimates require detailed inventory. Similar problems are encountered in calculating manufacturing costs. Variation in machinery and managerial skill and organizational efficiency of the operator can result in large cost differences. Should appraisal be for a particular operator, for the most efficient operator, or for the average operator in a position to buy the stumpage? These are difficult questions, and the answers depend on the appraisal purposes. The timber appraiser can expect to spend a major portion of his or her time estimating costs.

Problems of Measurement

The difficulty in measuring trees for the products made from them, and the variations in the measurement units used complicate stumpage appraisal. Standing trees are sold and stumpage prices are expressed in forest units, of which board foot, cubic measure, and cord measure are the most common. Qualitative tree and log grades are often applied. Cut timber also sells by the piece, lineal foot, pole unit (by size classes), weight, and as bolts of various sizes and descriptions.

Felled trees are handled in the woods, sometimes in tree lengths but mostly in pieces, such as logs or pulpwood bolts. Logging costs are figured by the tree, piece, lineal foot, board foot, cubic measure, or cost, according to circumstances and local practice.

Processing and selling units are still different. Lumber manufacture in the United States employs the board foot lumber measure. Consequently, sawmill and processing costs and selling prices are expressed in lumber measure. The number of board feet measured in log scale often differs widely from that measured by lumber scale owing to defects, sawing and manufacturing methods, scaling practice, and peculiarities of the log rule employed. Converting between log scale and lumber tally causes continual problems in appraisal. See Chap. 10 for a discussion of this problem. Similar problems in measurement and conversion are encountered in plywood manufacture when the end product is measured in square feet, and in pulp manufacture when the end product is measured in tons.

The Time Element

Because timber is uncut when it is appraised, stumpage valuation should be based on an expectation of what the timber will be worth when cut and processed. If economic conditions are not likely to change, as when the sale period is short or business conditions are stable, current costs and prices for logging or lumber are adequate for appraisals. If, however, the timber will be cut and processed over a period when prices are expected to rise or fall significantly, forecasting these changes becomes important to buyer and seller. The prudent buyer, if two or more years' delay between purchase and conversion is expected, would be well advised to set up a full cash flow schedule and do a financial analysis to appraise the sale.

Residual value appraisal is essentially a backward-looking procedure that uses yesterday's prices and inflation rates. Bidding is forward-looking, anticipating tomorrow's prices and inflation rates. In periods of rising market prices we would consequently expect bid prices well in excess of appraisals and the reverse in falling markets. This principle is borne out by the huge bid premiums in 1980 and 1981 (Table 11-6) when the bidders saw price inflation rates of over 15 percent and real price increases of 2 percent.

Dividing the risk in value changes affects stumpage values. It is one thing to buy 10 years' supply of timber at one time and assume the risks that may be involved. It is another thing to buy approximately the same amount of timber on a pay-as-cut basis, with the seller assuming the risks on uncut timber. Contracts can include escalation clauses to hook stumpage payments to conversion prices when cut, removing some of this uncertainty.

Differences in Species Values

In many situations, as in our earlier residual-value example summarized in Table 11-5, several species of quite different values are offered together. Handling these differences presents a recurring problem. When different species are cut into different products, such as some for pulpwood and some for sawtimber, and these products are cut and handled separately, logging costs can be segregated. If such is not the case, logging costs must be lumped, as in the sample appraisal, even though differences exist. Manufacturing costs also are difficult to segregate by species. Selling values usually can be segregated fairly easily.

A difficult valuation problem arises when the appraisal indicates that some species have very low or negative stumpage values but the seller wants these species removed. Such a situation can occur when an owner wants hardwoods in a pine/hardwood stand removed so that they will not interfere with the next crop. If a species can carry all variable costs of logging and manufacture and give a positive return in comparison to the selling value, it will contribute something to fixed costs, and the buyer gains from it. If not, the species can be cut only at a loss that someone must meet. Normally the seller pays for the removal of these worthless species by treating the offering as a whole and not selling only the high-valued species. This situation was also shown in Table 11-5.

In an appraisal, fixed costs such as roads are added to variable costs in obtaining the conversion return. To clearly guide decisions, the fixed costs must be highlighted. Consider, for example, data from an appraisal made in the Lake states that included aspen, a species often of low stumpage value. Essential data per cord are as follows:

Aspen selling price at mill	$13.00
Fixed road and other harvesting costs	3.75
Variable logging costs	8.72
Total logging cost	12.47
Conversion return	$ 0.53
Margin for profit and risk under a profit ratio of 0.10	1.18
Indicated stumpage value	−$ 0.65

In this case, aspen is sold as pulpwood to the mill, and further manufacturing costs need not be considered; only cutting and delivering the pulpwood to the mill are involved. Fixed costs have been prorated on the basis of volume. Even though aspen stumpage is −$0.65 per cord, the owner wishes to get rid of the aspen so that it will not compete with his next pine crop.

According to the appraisal, the seller should pay the buyer $0.65 for every cord of aspen the buyer removes. As a practical matter, however, stumpage rarely sells at a negative value. Commonly the stumpage value of other higher value species is reduced sufficiently to give the aspen some arbitrary positive value—the seller partially subsidizes the cutting operation. Or the aspen is offered at some nominal figure like $0.50 per cord with other stumpage values left unchanged—in this case the buyer partially subsidizes the operation. Even in this latter case, it may still be better in the short run for the operator to take the aspen than to let his logging operation be idle and receive no contribution toward fixed costs: a $0.50 per cord stumpage price leaves $13.00 − $8.72 − $1.18 − $0.50, or $2.60 to help pay the fixed costs that have to be paid whether or not the aspen is cut. If the operator persists in operating in this aspen, he will eventually go bankrupt as there is still a loss of $3.75 − $2.60, or $1.15 overall on the fixed costs.

VALUATION OF FORESTLAND

"Land is the one enduring asset of a permanent forest business" (Matthews, 1935). The great importance of site quality upon forest productivity has been stressed in Chap. 4 and elsewhere. It is the inherent vigor of the site that determines the potential productivity of the forest. In acquisition for forest production purposes, the basic quality of the land is consequently of the utmost importance. Consider the purchase of bare land, perhaps a parcel withdrawn from agriculture. There is no timber stand to appraise, and it is purely the productive quality of the land for forest growth that is being purchased. Similarly, an area may support a timber stand of little current value. Perhaps the present stand is more of a liability than an asset and should be so regarded in valuation. But if the site is good, the land will respond to treatment.

As with any other kind of land use, the value of land for forest uses necessarily stems from the future forest crops or other services it can produce. Forestland may and often does have value for other uses of one kind or another and must be appraised accordingly for each relevant use.

A separate bare land value for timber-growing purposes cannot be identified from market transactions evidence except from land sales where there is no, or only a negligible amount of, growing stock present and timber growing is obviously the highest valued land use. Where there is a substantial growing stock present, land and timber values are usually considered together. Forestland has a value and the fact that it cannot easily be identified from examination of market data does not invalidate either its existence or minimize its fundamental importance.

Factors Determining Forestland Value

Since forestland takes its value from the crops it produces, estimation of this value necessarily requires forecasting and appraising these crops over a considerable span of time. The problem, in basic outline, is to determine the future income-producing potential of the land itself. Four controlling factors are involved: (1) site quality, (2) the kind, intensity, and cost of future management prescriptions, (3) the market value of the timber products, and (4) the time interval involved as measured by the rate of interest employed.

While potential productivity is determined by site, the volume and quality of timber actually produced are influenced tremendously by the management prescription. The degree and promptness of regeneration, species composition, stocking, and growth can be largely controlled by the silvicultural practices applied. Depending on how it is handled—or mishandled—a given site may produce crops ranging from practically none at all to those approaching its biological potential.

The value of the timber product produced entails not only an estimate of what the products will bring in the market, but also the cost of getting the products to the market. Accessibility factors are often of controlling importance in determining the value of the timber crop as it stands in the forest.

Finally, the growing of timber takes time and usually a lot of time in relation to the length of most productive processes within the range of business experience. A guiding interest rate must be specified by the buyer or seller to measure the time cost for using invested land, timber, and money capital.

Estimation of Forestland Value for an Even-Aged Stand

We have seen that the present net worth of a continuing set of rotations is given by the soil expectation formula from Chap. 7 [equation (7-17)]:

$$\text{SEV} = \frac{a}{(1 + i)^w - 1}$$

where a = net income received at rotation age every w years starting at the end of year w
w = rotation age in years
i = interest rate expressed as a decimal

To apply this formula, receipts and expenditures pertinent to a particular schedule of management must all be brought to a common point in time (rotation age) so that net income can be determined and capitalized. Since interest measures the cost of waiting, this means that each cost and revenue item must be carried forward with interest to rotation age.

For example, consider the cash flow schedule of receipts and expenditures from an acre of an even-aged loblolly pine managed for pulpwood production on a rotation of 30 years given in Table 11-7. The net value of the periodic income a at rotation age is estimated at \$80.69 per acre assuming a guiding interest rate of 5 percent. Capitalization of this value received once every 30 years represents the value of the land itself as a productive asset. Consequently,

$$\text{SEV} = \frac{\$80.69}{(1.05)^{30} - 1} = \frac{\$80.69}{3.3219} = \$24.29 \text{ per acre}$$

Significance of Forestland Values

When the mechanics of calculating SEVs are understood, as illustrated above, the question remains: what do they mean? It must be clearly recognized that a SEV is a calculated figure entirely controlled by the data used in its determination. Its estimation includes all the four factors affecting forestland values. Anything affecting receipts and expenditures should be included in the calculation. Site quality and the intensity and kind of management are integrated in the physical level of productivity estimated. The costs of management are embodied in the costs assumed. Market and accessibility factors enter into the net stumpage prices used. The importance of the time element is represented by the guiding rate of interest used. Because all these factors can vary widely SEVs should not be thought of as anything fixed or immutable, but always in relation to the particular assumptions upon which they are determined. That these assumptions may seem difficult to make and of a long-range nature only serves to point up the practical difficulties of making a direct estimation of a land value for a crop of such long duration as timber.

SEVs are extremely sensitive to the rate of interest employed. The higher the rate, the lower the land value for a given production schedule. For the schedule given in Table 11-7, the land value at 5 percent was \$24.29. At 3 percent it would be \$108.79, and at 6 percent only \$5.46 for the same receipts and expenditures. With fairly high guiding rates, and especially for long rotations and/or poor sites, SEVs often are negative. A negative SEV is not necessarily conclusive evidence that "forestry does not pay"; it merely indicates, for the assumptions made, that a forest could not economically be built up on bare land, even if the land were free. Perhaps a more intensive level of management would result in a

Table 11-7 Costs and returns per acre from an even-aged southern pine stand managed for pulpwood production, production on a 30-year rotation (interest at 5 percent)

Receipt or expenditure item and assumptions made	Value per acre	Formula for accumulation with interest	Value of interest factor	Amount of receipt or expenditure item at rotation age
Receipts				
Major harvest at age 30, 37 cords at \$6.50 per cord	\$240.50	—	—	\$240.50
Thinning at age 15, 2 cords at \$3.00 per cord	6.00	$(1.05)^{15}$	2.0789	12.47
Thinning at age 20, 3.5 cords at \$4.00 per cord	14.00	$(1.05)^{10}$	1.6289	22.80
Thinning at age 25, 7 cords at \$5.00 per cord	35.00	$(1.05)^{5}$	1.2763	44.67
Total receipts at rotation age				\$320.44
Expenditures				
Site preparation and planting in year 1	30.00	$(1.05)^{29}$	4.1161	123.48
Annual costs for taxes, protection, and administration	1.75	$\frac{(1.05)^{30}-1}{0.05}$	66.4388	116.27
Total expenditures at rotation age				\$239.75
Net value at rotation age				\$ 80.69

positive SEV or perhaps a longer or shorter rotation would cause such a result. For a particular buyer or investor, the SEV of import is the maximum SEV that is possible from feasible management prescriptions for the land in question. This is the most the buyer could afford to pay for the land and still earn the assumed guiding rate on the investment. Procedures to help make a systematic search for this maximum SEV are discussed in Chap. 13.

When timber is the highest valued land use and the land market is competitive with many buyers and sellers, the market value for bare forestland will tend to approach a SEV calculated at a "going" or market rate of interest. Timberland markets are often not competitive and forestland has often been purchased at far less than its economic value in timber production, at least as revealed by hindsight. The fact that this is so accounts in part for the very large profits that have been made on some forestland purchases. The SEV of established but immature timber stands is often undervalued by the market, providing opportunities for the shrewd land buyer. As land and growing timber values have become better understood and demand for land and timber has increased, the market value of timberland has increased accordingly. SEV calculations are consequently a useful guide to forestland value, indicating ceiling prices that could be paid on an investment basis.

VALUATION OF IMMATURE STANDS

The value of the timber stand upon the land increases as the stand grows. Land does not grow; it remains. With even-aged management, the initial value is that of the land itself. Following establishment of a new stand, the combined value of land plus timber increases to a maximum defined by economic maturity—the rotation age or time of major harvest. Mature stands with merchantable inventory can be appraised using the stumpage appraisal methods outlined above. But what about the 10–20 year old stand with little or no merchantable timber? Intuitively we feel it is worth more than bare land—but how much more?

As in any appraisal problem of this nature, there are two commonly used valuation approaches: current market value based on comparable timberlands, and capitalized income value (value in use).

In some areas, the market recognizes the potential of immature stands, and transactions evidence from recent sales gives a reasonable approximation of a subject stand's value. In many areas, however, the market does not recognize the value of immature timber and prices the submerchantable forest at close to bare land values. When this is the case, foresters turn to value in use appraisals to get a better estimate of potential timber values.

Estimation of Value for an Immature Even-Aged Stand Using the Value in Use Method

Methods of estimating values for even-aged stands follow logically from those used in land value determination except that the focus is on the timber stand. The mechanics are essentially the same.

The problem is to estimate, for a stand of given age, the present net worth at some assumed rate of interest and under an assumed future management prescription of all future receipts and expenditures. To illustrate, the value at age 15 of the same stand for which a soil expectation value was determined (Table 11-7) is determined as shown in Table 11-8. The assumptions are again the same as in Table 11-7, except that the stand is now 15 years old. According to our calculations, the stand has a value of $141.66 at age 15. If the land with this 15-year-old immature stand could be purchased for less than $141.66, it would be an investment that returned more than the guiding rate of 5 percent.

Note that the value of this immature stand has two parts—the discounted net value of the current stand of trees that will be harvested in 15 years plus the discounted value of the bare land SEV, which reflects the delay in using this land for subsequent tree crops. This two-part appraisal is represented by the equation

$$\mathrm{PNW}_t = \frac{NR_w + \mathrm{SEV}}{(1+i)^{w-t}} \qquad (11\text{-}1)$$

where PNW_t = present value of an immature stand currently at age t and the land

Table 11-8 Present net worth per acre of a 15-year-old even-aged southern pine stand with a rotation of 30 years (interest at 5 percent)

Receipt or expenditure item	Value per acre	Formula for compound	Value of interest factor	Value at age 30
Major harvest cut at rotation age	$240.50			$240.50
Thinning at age 20	14.00	$(1.05)^{10}$	1.6289	22.80
Thinning at age 25	35.00	$(1.05)^{5}$	1.2763	44.67
Total receipts				$307.97
Annual costs	1.75	$\frac{(1.05)^{15}-1}{0.05}$	21.58	37.96
Net return at age 30 (NR_w)				$270.21
Land value (SEV_w)				24.29

$\mathrm{PNW}_{15} = (NR_w + \mathrm{SEV}_w)/(1+i)^{w-t} = (270.21 + 24.29)/(1.05)^{15} = \$141.66$

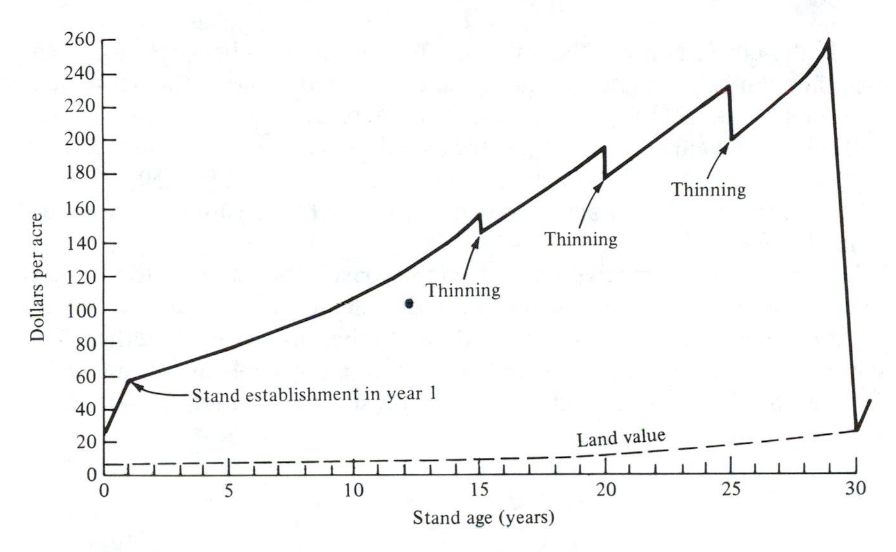

Figure 11-1 The value in use of land and timber over a 30-year rotation.

NR_w = net return of the immature stand at rotation age w
SEV_w = soil expectation value with a rotation of w years
$w - t$ = number of years before the immature stand reaches rotation harvest age

Using the data from Table 11-7, equation (11-1) was solved for each possible year t from ages 0 to w and the results are shown in Fig. 11-1. As you would expect, the present value of the current stand rises to its maximum for a rotation of length w as t increases from 0 to w. At the same time the land value rises to the SEV for rotation age w just after the stand is harvested and the land is again bare.

Value of Uneven-Aged Stands

The nature of uneven-aged stands necessitates some modification in the application of value in use methods. A land value here never exists separately from the timber value unless the stand is all cut, in which case there is no longer any uneven-aged stand. Land and timber necessarily go together. Therefore, in an uneven-aged stand, there is a combination of site quality and growing stock that produces yields and receipts either annually or periodically, depending on the length of the cutting cycle. Certain expenditures are also necessary to produce these returns. There are two elements of productive capital involved: the land and the reserve growing stock.

If the cut is annual, the value of the land and timber-growing stock together can be estimated as being the capitalized value of the net annual income [equation (7-6)]. For example, an acre that on the average yields 450 bd ft per year worth $30 per thousand board feet and costs $2.50 to protect and administer has a capital value at 5 percent of [(0.450 × $30) − $2.50]/0.05, or $220 per acre. Growing stock and land values cannot realistically be separated.

If incomes are received on a periodic basis, the value of the tract is the capital value of a permanent periodic income in the same way as a SEV is determined, except that land and timber are valued together. The capital value (or present net worth) of a continuing series of cutting cycles is given by the uneven-aged version of the soil expectation formula from Chap. 7 [equation (7-18)]:

$$\text{SEV(U)} = \frac{a}{(1 + i)^w - 1}$$

where a = net income received at the end of each cutting cycle every w years

i = guiding rate expressed as a decimal

w = length in years of the cutting cycle

To illustrate, assume that $125 worth of forest products can be cut per acre every 10 years starting 10 years from now and that annual costs of production, administration, etc., are $2.50 per acre per year. The guiding rate is assumed to be 5 percent. Using (7-18), the present net worth of the land and timber producing this income is

$$\text{SEV(U)} = \frac{\$125 - \left[\$2.50 \times \dfrac{(1.05)^{10} - 1}{0.05}\right]}{(1.05)^{10} - 1} = \frac{\$125 - \$31.44}{1.6289 - 1}$$

$$= \frac{\$93.56}{0.6289} = \$148.77 \text{ per acre}$$

This $148.77 is the capital value of a net income of $93.56 received every 10 years, the first payment coming 10 years hence. It represents the productive value of land and timber together just after a cyclic cut is made.

If the stand has a net capital value of $148.77 just after a cyclic cut, it obviously increases in value during the cycle as the stand approaches its next cyclic cut. The value for each year during the cycle is calculated in Table 11-9.

The table needs explanation. It shows the capital value of regulated uneven-aged stands at different years since entry. Each number in the last

Table 11-9 Present net worth of an uneven-aged stand by the number of years since last regular entry for a 10-year cutting cycle*

Years since last entry (k)	Years to discount ($w - k$)	Present net worth at year k†
0	10	$ 148.77
1	9	158.70
2	8	169.13
3	7	180.09
4	6	191.59
5	5	203.68
6	4	216.36
7	3	229.68
8	2	243.65
9	1	258.36
		$2,000.00

* Per-acre harvest revenue = $125, annual costs = 2.50, and guiding rate = 5 percent

† $NR_w = 125 - 2.50 \dfrac{(1.05)^{10-k} - 1}{0.05}$

SEV(U) = 148.77

$$\text{PNW}_k = \frac{125 + 148.77 - 2.50[(1.05)^{10-k} - 1]/0.05}{(1.05)^{10-k}}$$

column of Table 11-9 is based on the sum of two end-of-cycle receipt items. The first is the value of the periodic cut; the second is the capital value of land and timber together, $148.77, which represents the net value of all future harvests. This sum is discounted by the number of years to the end of the cutting cycle using the equation

$$\text{PNW}_k = \frac{NR_w + \text{SEV(U)}}{(1 + i)^{w-k}} \tag{11-2}$$

where NR_w = the net revenue at a regular cutting cycle entry
w = length of a cutting cycle in years
k = the number of years since the last regular harvest entry
$w - k$ = the number of years before the next regular entry
SEV(U) = as defined in equation (7-18)

For our example forest, NR_w = $125 less costs, SEV(U) = $148.77, w = 10, and k ranges from 0 to 9. Consider the stand that has grown 7 years since the last entry in Table 11-9. It has 3 years before the next entry and will incur three more annual cost payments of \$2.50.

$$NR_w = 125 - 2.50\,\frac{(1.05)^3 - 1}{0.05} = 117.12$$

$$\text{SEV(U)} = 148.77$$

$$\text{PNW}_k = \frac{117.12 + 148.77}{(1.05)^3} = 229.68$$

A point to be emphasized is that the above valuations for even-aged and uneven-aged stands are all based on the assumption of continued management of a stand according to a specific plan and well illustrate the fact that values depend on uses contemplated. Values would be different if some other plan of use was intended. Whether or not the values estimated are a good guide to current market value depends on the accuracy of the estimates and how closely they approach intended use. If, for example, the aim is to liquidate the stand at a certain time during the rotation or the cutting cycle, the value would be merely the current stumpage value of the timber at that time plus the land value for whatever use to which it might be put.

If a subject stand does not have a structure that is expected to provide regular, equal payments of amount NR_w, then an individualized cash flow schedule will have to be estimated and the PNW calculated using equation (8-1) without using shortcut formulas such as equation (11-2).

VALUATION OF THE REGULATED FOREST—THE GOING CONCERN

All the land and timber valuations so far have dealt with a specific stand producing an annual or periodic income. What about the value of a continuing business based on a forest organized for sustained yield? This is a common business situation and often a major aim in timber management. Receipts in more or less equal amounts are received each year and expenses are currently paid. The difference represents the yearly operating net income or loss, as the case may be.

Even-Aged Management

Consider, for example, a regulated 1,200-acre even-aged forest with management prescription, costs, and incomes per acre as were assumed in Table 11-7 for calculation of SEV.

Since the rotation is 30 years, a 30-acre unit with one acre of each age class will represent such a forest. On this representative forest, there would be the following annual incomes and costs:

Receipts:	
Major harvest cut at age 30 on 1 acre	$240.50
Thinning at age 15, 1 acre	6.00
Thinning at age 20, 1 acre	14.00
Thinning at age 25, 1 acre	35.00
Total receipts	$295.50
Expenditures:	
Site preparation and planting, 1 acre	$ 30.00
Net annual costs, $1.75 per acre on 30 acres	52.50
Total expenditures	82.50
Net annual income	$213.00

This net annual income of $213.00 is received on 30 acres. The capitalized value of the 30-acre unit, again using 5 percent, is accordingly $213.00/0.05, or $4,260, and the average per acre is $142. The whole forest is thus appraised at $142 × 1,200 acres or $170,400. This is the value of land and growing stock together as a producing entity, as estimated by capitalization at this rate of interest. It measures what this forestry business, operating on the continuing basis assumed, is worth as an investment proposition on the basis of its current earning capacity.

But is this the best combination of our land and timber resources in timber production? The highest present net worth of a regulated forest for timber production is associated with a rotation and cultural prescription that maximizes soil expectation value (SEV). The cost of achieving any regulated forest, however, must also be considered.

On the face of it, this simple capitalization process seems to ignore the fact that different acres in the regulated forest have different income values as shown for age 15 in Table 11-8 and Fig. 11-1. The income value of each unit is the sum of the present net worth of the standing timber and the present net worth of the series of rotations that begins after the stand is harvested. If, in the above illustration, the income value of each of the 30 units ranging from age 1 to 30 were separately calculated, as done for age 15 in Table 11-8, the sum of these would be exactly $4,260.

This point is important and merits explicit demonstration. Because calculation of 30 individual units for a 30-year rotation is tedious, an illustration will be given for a model forest of just 5 units of area on a 5-year rotation.

The receipts and expenditures assumed, their accumulated value at rotation age, and annual receipts and expenditures from a regulated "for-

Table 11-10 Costs and returns from a regulated forest model managed on a 5-year rotation (interest at 8 percent)

	Net returns per management unit			Regulated forest:
Receipt or expenditure item (1)	Value per unit (2)	Years to accumulate (3)	Value at rotation age (4)	annual receipts and expenditures (5)
Receipts				
Final harvest at age 5	$100	0	$100.00	100
Intermediate harvest, age 3	10	2	11.66	10
Total receipts			$111.66	110
Expenditures:				
Establishment cost, year 1*	30	4	40.81	30
Improvement cost, year 2	10	3	12.60	10
Annual costs	5	5	29.33	25
Total expenditures			$ 82.74	65
Net value at rotation age			$ 28.92	45

* Assumed establishment will be completed before the growing season so 5 years of growth will be made.

est'' of 5 one-acre management units are given in Table 11-10. As shown (column 4), the net value at rotation age is $28.92. The SEV value when a = $28.92 is consequently

$$\frac{\$28.92}{(1 + i)^5 - 1} = \frac{28.92}{0.4693} = \$61.62 \text{ per unit of area}$$

The ''going concern'' cash flow for 5 regulated units is shown in column 5 and $45 of net income is produced each year. Capitalized, the 5-unit forest is worth

$$\frac{\$45}{0.08} = \$562.50$$

Table 11-11 gives the complete structure of income and cost value calculations for each of the 5 units, one in each age class 0, 1, 2, 3, 4. As shown in column 20 the total is $562.50, which checks exactly with the ''going concern'' or capitalized net annual income from the 5 units. As an item of technique, note all receipts and expenditures are carried forward with interest to rotation age and the net income for each year is discounted back to its present age (columns 18–20).

Table 11-11 Values by years for a forest on a 5-year rotation with a guiding rate of 8 percent

			Intermediate cut, yr. 3				Annual costs			Establishment cost, yr. 1			Improvement cost, yr. 2						
Management unit (age) (1)	Land value (2)	Final harvest cut (3)	Amount (4)	Years to accumulate (5)	Value at age 5 (6)	Total income at age 5 (7)	Amount (8)	Years to accumulate (9)	Value at age 5 (10)	Amount (11)	Years to accumulate (12)	Value at age 5 (13)	Amount (14)	Years to accumulate (15)	Value at age 5 (16)	Total costs at age 5 (17)	Net income at age 5 (18)	Years to discount (19)	Present net worth (20)
0	$61.62	$100.00	$10.00	2	$11.66	$173.28	$5.00	5	$29.33	$30.00	4	$40.81	$10.00	3	$12.60	$82.74	$ 90.54	5	$ 61.62
1	61.62	100.00	10.00	2	11.66	173.28	5.00	4	22.53	—	—	—	10.00	3	12.60	35.13	138.15	4	101.55
2	61.62	100.00	10.00	2	11.66	173.28	5.00	3	16.23	—	—	—	—	—	—	16.23	157.05	3	124.67
3	61.62	100.00	—	—	—	161.62	5.00	2	10.40	—	—	—	—	—	—	10.40	151.22	2	129.64
4	61.62	100.00	—	—	—	161.62	5.00	1	5.00	—	—	—	—	—	—	5.00	156.62	1	145.02
Total																			$562.50

Computations:

(7) = (2) + (3) + (6)

(17) = (10) + (13) + (16)

(18) = (7) − (17)

(20) = (18)/(1.08)$^{(19)}$

Uneven-Aged Management

The capital value of a regulated forest managed on an uneven-aged basis is determined in the same general way as for even-aged management, only the process is simpler. Some cultural work like pruning may be done currently on an annual basis. It cannot be identified by stand age, since stands are of mixed ages, and therefore applies to the area as a whole as an annual cost. Harvest cutting and thinning operations are made simultaneously when a cyclic cut is made. Assuming a 10-year cutting cycle and the same receipt and expenditure data used in Table 11-9, the capital value per acre based on a sample area of 10 acres (1 acre for each year of the cutting cycle) is estimated as follows:

Receipts	
Net stumpage value of cyclic cut made on 1 acre	\$125.00
Expenditures:	
Annual costs of taxes, administration, protection, and an annual budget for stand improvement work not done as a part of the cyclic cut, \$2.50 per acre	25.00
Annual net income	\$100.00

Again using 5 percent, the capital value of this net annual income is \$100/0.05, or \$2,000 on 10 acres, or \$200 per acre for land and timber together as a continuously productive entity. Note that in Table 11-9 giving values for individual years in the cutting cycle, the sum of present net values for each year since entry shown in the third column is exactly \$2,000, providing a check on the arithmetic and explicitly relating values for individual years with a "going concern" for the same operating period.

In conclusion, these sections on methods for estimating forestland and timberland values on the basis of their continued capacity to produce wood have demonstrated the interweaving of the four major components of timber-growing value: (1) site quality, (2) the kind, intensity, and cost of management, (3) market value of the product, and (4) time interval required as measured by the interest rate. As shown, time is a major factor, perhaps the major factor, in all of the calculations. This is inescapably true as timber can be grown only over time. This fact necessitates the use of interest as a means of estimating the importance of the passage of time. Interest is always more or less difficult to understand fully and to rationalize. But it cannot be escaped in timberland valuation. It is perfectly true that, in practice, no one attempts to keep historical cost and return records on individual stands duly recording interest accumulation. Forestland and timber stand valuations are nonetheless an indispensable managerial tool in appraising the relative desirability of silvicultural and other management alternatives (as we will see in Chap. 13) and in giving a basis for market value determination. The mechanics are worth mastering so that they can be meaningfully applied.

REFERENCES

Jackson, D. H., and A. G. McQuillan, 1979: A Technique for Estimating Timber Value Based on Tree Size, Management Variables, and Market Conditions, *Forest Sci.,* **25:**620–626.

Johnson, R. N., 1979: Oral Auction versus Sealed Bids: An Empirical Investigation, *Natural Resources J.,* **19:**315–375.

Matthews, D. M., 1935: "Management of American Forests," McGraw-Hill, New York.

U.S. Forest Service, 1983: "Transactions Evidence Appraisal Process," Northern Region Timber Management.

———, 1984a: "USDA Forest Service Timber Pricing in the West," Timber Management, Washington, D.C. Interim Report of the National Appraisals Working Group (mimeographed).

———, 1984b: "Timber Appraisal Handbook," FSH 2409.22 Region 5 Amendment 54, August 1984.

Weiner, A. A., 1982: "The Forest Service Timber Appraisal System: A Historical Perspective 1891–1981," U.S. Forest Service FS-381.

CHAPTER
TWELVE

VALUATION OF NONTIMBER FOREST OUTPUTS

INTRODUCTION

Foresters, particularly public land foresters, now actively manage forest lands for the production and consumption of many products besides timber. Some of these are sold and others are not. Some, such as forage or recreation, are consumed on the forest and others, such as water or blueberries, are transported off the forest for consumption. Today the production of nontimber goods and services is an issue because the demands for these goods and services have increased to compete strongly with timber for the use of the land and management resources. But timber sells for dollars, while most of these other goods bring in goodwill and often ample, but unquantified human satisfaction.

It is legitimate to question the trade-off: if timber and other monetary revenues are reduced, what value is received in return? Ultimately any such comparison must use a common measure of value, whether it is utiles, value index, or dollars. Assigning dollar prices to the nonmarketed goods is appealing as a direct approach to trade-off analysis. Once prices are assigned, they can be used in a variety of mathematical programming or decision models to jointly optimize the production of multiple forest outputs.

Forest outputs can be classified by how they are priced and where they are consumed (Table 12-1). Timber, minerals, and some recreation

Table 12-1 Classification of selected forest outputs by method of pricing and where they are consumed

Location where consumed	Method of pricing on public lands: Market prices	Administrated nominal prices	Not priced
Consumed off forest	Timber, minerals	Fuelwood	Water
Consumed on forest	Hunting and recreation leases	Forage, developed recreation	Dispersed recreation, visual amenities, nongame wildlife, endangered species

are market-priced, while water, wildlife, and much dispersed recreation including hunting and fishing are provided free. Priced below market value and often lower than their supply cost are fuelwood, forage, and much developed recreation. The exact placement of an output within this table obviously depends on the particular forest and owner in question. On private lands more of the outputs will be market priced. It is the nonpriced and nominally priced activities that decision makers desire to place on par with timber.

The outputs associated with endangered species and elusive ideas such as visual quality are particularly difficult valuation problems. The difficulty occurs largely because these outputs are not countable entities and cannot be associated directly with a unit of individual human satisfaction. Rather they represent ideas or concepts that are collective in nature—a public good in most respects. Keeping a species of plant or animal viable has to be a collective judgment. Opportunity costs can be estimated and informative, but this still begs the question of what the spotted owl or view is worth. As a case in point, national forest planning calculations made in 1981–1982 have revealed a total timber opportunity cost for spotted owls to be somewhere in the neighborhood of 3.6 billion dollars, or about $500,000 per breeding pair, depending on the national forest. This high cost prompts vigorous debate (Heinrichs, 1983).

> . . . That sentiment was reflected in a 1981 speech given by John B. Crowell, Jr., the Agriculture assistant secretary. Crowell, who oversees the Forest Service, asked, "How many pairs of owls can we afford to protect?" He claimed that the spotted owl management plan would cost about $3.6 billion in lost timber sales. When that remark was quoted to one wildlife biologist, she responded, "How many owls can we afford to lose?"

Nonmarket pricing is no longer an academic issue. The necessity of coming up with these prices is demonstrated by the price table used by the U.S. Forest Service for the 1980 assessment (Table 12-2). With some local

Table 12-2 Average per-unit values for forest outputs used for national and forestland planning (dollars per unit)

Element and output description*	Unit of measure*	National forest regions								
		Western U.S.						Eastern U.S.		Alaska
		1	2	3	4	5	6	8	9	10
Timber										
Timber products sold, HW ST	M bd ft	1.13	1.64	6.32	2.80	17.15	22.54	37.51	49.18	1.01
Timber products sold, SW ST	M bd ft	78.51	34.25	76.46	48.69	127.66	173.23	118.15	38.17	32.72
Timber products sold, HW RW	M ft^3	—	45.13	49.09	10.73	52.56	101.30	21.82	32.51	5.56
Timber products sold, SW RW	M ft^3	31.57	13.83	33.90	58.02	40.10	208.78	115.69	87.62	3.67
Minerals										
Common variety (sand and gravel, building stone	Tons	0.30	0.30	0.30	0.30	0.50	0.30	0.30	0.30	0.30
Leasable phosphate rock, barite, and lead	Tons	1.60	1.60	1.60	0.32	1.60	1.60	1.37	36.90	1.60
Energy related (coal, petroleum, natural gas, geothermal)	Billion Btu	239.00	82.00	220.00	89.00	230.00	8.60	199.00	167.00	248.00
Water										
Water yield	acre · feet	7.50	5.00	5.00	5.00	2.60	1.50	1.50	1.00	0.50
Net sediment reduction (average)	tons	2.00	2.00	2.00	2.00	2.00	2.00	2.00	2.00	2.00

Recreation										
Developed recreation use, public	RVD	3.00	3.00	3.00	3.00	3.00	3.00	3.00	3.00	3.00
Developed recreation use, private	RVD	3.00	3.75	3.00	3.50	3.50	3.50	2.50	2.00	3.00
Dispersed recreation use	RVD	3.00	3.00	3.00	3.00	5.50	3.00	5.50	5.50	3.00
Visitor info. service use	RVD	3.00	3.00	3.00	3.00	3.00	3.00	3.00	3.00	3.00
Wilderness										
Wilderness use	RVD	8.00	8.00	8.00	8.00	10.00	8.00	12.00	12.00	8.00
Wildlife and fish										
Big game use	RVD	10.50	10.50	10.50	10.50	10.50	10.50	10.50	10.50	10.50
Nongame use	RVD	7.25	7.25	7.25	7.25	7.25	7.25	7.25	7.25	7.25
Inland sport fish use	RVD	5.25	5.25	5.25	5.25	5.25	5.25	5.25	5.25	5.25
Sport fish, anadromous	RVD	19.50	—	—	19.50	19.50	—	19.50	19.50	19.50
Commercial anadromous fish	Pounds	1.61	—	—	1.61	1.57	1.61	—	0.62	0.62
Waterfowl use	RVD	8.00	8.00	8.00	8.00	8.00	8.00	8.00	8.00	8.00
Range										
Grazing use (livestock)	AUM	5.57	5.88	5.51	4.96	6.00	5.90	3.76	4.41	3.00

* AUM = animal-unit-month; HW = hardwood; RVD = recreation-visitor-day; RW = roundwood; ST = sawtimber; SW = softwood.
Source: U.S. Forest Service (1980).

modification, these same numbers are being used for the planning of all 125 individual national forests, often directly entering the objective functions of the planning models. Somebody had to come up with the numbers—how was it done and are the numbers any good?

This chapter will direct itself almost exclusively to nontimber outputs such as dispersed recreation which are not sold in markets but which are countable entities. We are not going to discuss commodities such as minerals which have a well-defined commercial market, and industrial, agricultural, and culinary water and forage which are factors of production that have a well-defined end use or final product values. The value of forage and water can be derived in a budgeting approach in a manner similar to the procedures described for timber stumpage in Chap. 11.

RELATIVE VALUES AND DOLLAR VALUES

A review of some basic characteristics of relative value systems is helpful before we embark on a discussion about estimating and comparing nonmarket-determined values to market values.

Consider three forest products, which, through barter and trade in a remote village, have developed the following exchange terms:

Terms of trade for three forest outputs (units for units)

Forest output item	Units	Fuelwood	Fish	Ducks
Fuelwood	Bundles	1	0.5	2
Fish	Pounds	2	1	4
Ducks	Number	0.5	0.25	1

Each row describes the exchange rate for one output to all others. Consider fish (row 2). It takes two pounds of fish per bundle of fuelwood, four pounds of fish per duck: each row sets the relative terms of trade in units of a different output. Measured in ducks (row 3), the bundle of fuelwood is worth 0.5 of a duck. Look closely—each row has the identical ratio between any pair of items. For example, the ratio of ducks to fish for all three rows is 2/0.5 = 4/1 = 1/0.25, or one duck is worth four pounds of

Relative value of three forest outputs

Item	Value in pounds of fish
Ducks	4
Fuelwood	2
Fish	1

fish. Fish is the least valued of the items in terms of exchange, so let us give it unit value (1.0) and use it to rank all three items by relative value.

Suppose a nearby village used paper money for more convenient exchange and called the unit value of the money "one dollar." They start trading with our village for fuelwood and after a period of time the price of fuelwood is established as $25 per bundle. Trade continues within the village as before, but now dollars of paper currency are also circulated and used. Because the other village takes them in exchange for products, they have a believable value to the villagers. Soon, ducks and fish also take on dollar values at the same ratio of exchange established by barter before.

Relative to market value conversion for three forest outputs

Item	Units	Relative value per unit of fish	Market value
Ducks	Number	4	$50.00
Fuelwood	Bundles	2	$25.00
Fish	Pounds	1	$12.50

Suppose now that water, which has always been plentiful and free, becomes scarce and requires someone to go and fetch it. By law, however, the village decides that water cannot be sold for dollars in the market. Nevertheless people start bartering fish, ducks, and fuelwood for water at the rates shown.

Physical and currency exchange rates for four forest outputs (units for units)

Forest output item	Units	Fuelwood	Fish	Ducks	Water	Dollars
Fuelwood	Bundles	1.0	0.5	2.0	0.2	0.04
Fish	Pounds	2.0	1.0	4.0	0.4	0.08
Ducks	Number	0.5	0.25	1.0	0.1	0.02
Water	Buckets	5.0	2.5	10.0	1.0	0.2
Paper money	Dollars	25.0	12.50	50.0	(5)	1.0

Here water fits into the established exchange rate as being worth 0.1 of a duck, 0.4 of a pound of fish, or 0.2 of a bundle of fuelwood. These exchange rates are the real worth of water to this society. Because money is used to facilitate trade, the value of water in trade, if it were permitted, would necessarily stabilize at $5 per bucket. In a free trade system the ban on water sales is not easy to enforce—one could easily imagine secret transactions at $5 per bucket. The point of this example is that the new commodity, by virtue of being exchanged with already marketed items, has to take on an implied market value whether society likes it or not.

If someone set up a linear programming problem to plan this economy, it would make no difference whether the units of the objective function coefficients were bundles, pounds, ducks, buckets, or dollars. With the same relative value structure the same optimal solution would result. The important values are the relative exchange rates established by trade.

To summarize the important aspects of this discussion:

1. The relative values of forest outputs determine the solution to management problems within the constraints. Money prices, whether dollars, pesos, or pounds, only restate the relative values.
2. If only a single item in an array of items with established relative value exchange rates is assigned a market price, every item in the set is given a market price by implication.
3. If a new nonmarket-priced item is entered into exchange with market-priced items, it is necessarily assigned a market value. Hence any analytical comparison of a nonmarket good to a market good implies the imposition of a market price to the nonmarket good.

When administrators, politicians or their hired analysts start assigning dollar values directly to nonpriced goods and services, potential conflicts and problems arise. Trade and exchange is the means by which society's relative values are assigned. Outputs such as forest recreation do in fact have a relative value to the members of society; for example, the camping trip is exchanged for time and money which could have been used doing something else. But this exchange and thus the relative value is not revealed in the market where we can look at it. When dollar values are administratively but subjectively assigned to nonmarket goods, these administrated prices can easily imply an incorrect relative value to the output and thus misdirect the decision analysis. In our example, water became scarce and, through exchange, established a relative value of 0.2 of a bundle of firewood, which translated into the implied dollar value of $5. Suppose a politician arbitrarily declared "the value of water is $10 per bucket for purposes of public planning." By implementing this decision the analyst would effectively assign a new relative value structure such as 0.4 for a fuelwood bundle per bucket. The planning solution would then provide more water and less of other goods than society's true relative valuation warrants.

4. Given that the relative values developed through free exchange are the true valuation of goods and services of society, then every assignment of dollar values to outputs whose value is not established by

such an exchange must make every effort to emulate what these prices would be if the outputs were traded.

WILLINGNESS TO PAY, CONSUMER SURPLUS, AND VALUE

The preceding argument asks that nonmarketed outputs be evaluated using the same definitions, concepts, and criteria as marketed outputs. This follows because the essential purpose of the prices is guiding decisions about multiple-output trade-offs and the exact same value concepts for each output are requisite for valid analysis and comparison.

To elaborate on the introduction in Chap. 9, the most widely accepted standard of value is *the amount consumers are willing to pay for each increment of output used.* Payment is normally measured in dollars. To illustrate, a schedule of willingness to pay for successive increments of a good by an individual or groups of individuals is shown with a linear, negative-slope form in Fig. 12-1. The direct interpretation of this schedule is, for example, that the consumers are willing to pay $15.62 for the 60th unit of the good, but the willingness to pay falls with successive increments to only $6.25 for the 120th unit.

A market demand curve is defined as the amount of a good or service consumers will collectively buy in a given market over a time period at different prices. After it has been adjusted for any income effect, the

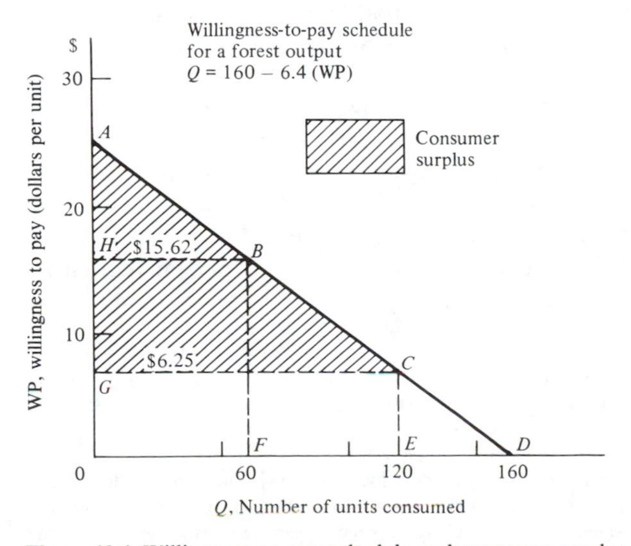

Figure 12-1 Willingness-to-pay schedule and consumer surplus.

willingness-to-pay curve we have been discussing is the same as this market demand curve. The income effect is simply that if the price of a subject good falls, the amount a consumer has to spend to get a given amount of the good also falls, releasing some income to buy other things. This extra income can be used to buy more of the subject good or other goods, the amount of increased purchase of the subject good being called the income effect. If there is no income effect, then the willingness-to-pay curve and the market demand curve are the same. In practical application, analysts nearly always estimate and use market demand curves with the assumption that there is no appreciable income effect.

If this output were actually sold at a market price of $6.25 and we had the willingness-to-pay schedule in hand, we would expect that consumers would purchase a total of 120 units.* The total dollars paid for this amount is 120 × $6.25 = $750. This is represented by area *OGCE* in Fig. 12-1.

But it can be successfully argued that the consumers would have paid a lot more for these 120 units. For example, the users would have paid $15.62 for the 60th unit; this is $15.62 − $6.25 = $9.37 more that would have been paid for the 60th unit than was actually paid. This excess of willingness to pay over actual price is called *consumer surplus*. If this excess of willingness to pay over actual price is added up for all the units used, the total gives us the consumer surplus created in this market for this good. In Fig. 12-1 this total is the shaded triangular area *ACG*, which has a dollar value of [($25 − 6.25) × 120)] ÷ 2 = $1,125.†

Willingness to pay is thus only partially captured by the market with the sales revenue of $750. An additional $1,125 accrues to the consumers as additional satisfaction for which they did not pay.‡ The total willingness to pay at a consumption level of 120 units is simply the sum of actual payments plus consumer surplus, $1,125 + $750, or $1,875 (area *ACEO*). The average willingness to pay is $1,875 ÷ 120, or $15.62. This last number, $15.62, is the unit price that can be "assigned" to the output to reflect its total value to the consumers when the market price is $6.25.

Suppose the unit of output under discussion is dispersed recreation, which is free to the users. If the estimated market demand after income adjustment is the willingness-to-pay schedule represented in Fig. 12-1, we would expect to see 160 recreation-visitor-days of dispersed recreation consumed at the zero price. None of the willingness to pay is captured by

* The 120th unit with the willingness to pay of $6.250 just equals the cost of purchase of $6.25. The willingness to pay for the 121st unit is only $6.094, which is less than the cost, so it would not be purchased.

† The area of a triangle can be calculated as area = (base × height)/2

‡ This apparent "free lunch" holds for every good and service that has a downward sloping willingness-to-pay schedule. The relative amount of the surplus depends on the steepness (elasticity) of the schedule and the level of price. The steeper (more inelastic) the schedule, the greater is the consumer surplus; while the flatter (more elastic) the schedule, the greater is the portion valued in the market.

market payments, and the entire valuation of area *ADO* is consumer surplus in the amount of [(25 × 160) ÷ 2], or $2,000. The average willingness to pay with a zero price is $2,000 ÷ 160, or $12.50. It is a reasonable argument that the consumers of dispersed recreation actually received and would have paid an average of $12.50 for the satisfaction of each visit. Therefore if the demand schedule (willingness to pay schedule) can be reliably estimated and adjusted for income effects, the total willingness to pay is a reasonable approximation to the value of goods consumed. Also, if the demand schedule is individually estimated for each forest or area, then the average willingness to pay is a good estimate of unit satisfaction value for that forest and recreation activity at the level of consumption projected.

This argument underlies most current techniques used to assign dollar prices to zero-priced or underpriced outputs, such as recreation. While the total willingness-to-pay procedure and the consumer surplus idea are conceptually acceptable, some real difficulties can occur in application, including obtaining a reliable empirical estimate of the demand schedule and making comparisons between different outputs.

We also have to be careful to distinguish between the "value" of an output which we use for purposes of analysis and the "price" of the output established in the market. In the previous discussion we have seen that value can be derived as part price and part consumer surplus, or it can be entirely consumer surplus when the market price is zero.

There is a rich literature on consumer surplus and willingness to pay as measures of value. The subject gets heavily into consumer behavior and welfare economics theory and can become rather technical. The consensus view is that total or average willingness to pay derived from an appropriate income-adjusted demand schedule is the proper measure of value to use. Most of the research, empirical work, and real debate have been on the questions of (1) how, methodologically, can an appropriate demand schedule be estimated, and (2) whether the empirical estimates currently used are in fact accurate, conceptually consistent, and comparable across different outputs and different geographic areas.

Estimating Demand

Estimating a demand schedule requires that we define (1) the geographic market area to which the schedule applies, and (2) the units for measuring output. For many marketed outputs transportation costs are a very small percentage of the output value, the market area is defined nationally, and selling prices are about the same wherever the output is sold. Consumer electronic goods, jewelry, or clothing are good examples. For products and services with higher relative transportation costs, such as building products, food, energy, and tourism, the markets are usually defined regionally. The relevant market area becomes much smaller and localized for products and services with high transportation costs or which must be consumed on site, such as camping, minerals, standing timber, forage, water, or a commuter college education. Many outputs from the forest are in this last category and require localized demand estimates. Remember, the demand schedule represents the amount of output for which people would actually show up at the forest gate, cash in hand and ready to buy at different prices. At an offering price of $50 per M bd ft you could sell a lot more timber stumpage from a forest 30 miles from Portland, Oregon, or Savannah, Georgia, than you could from forests in central Idaho, Utah, or West Virginia, which are located 200 miles from the nearest major town or large sawmill.

Estimating Demand by the Travel Cost Method

Searching for a plausible alternative to techniques that rely on market data, Clawson (1959) capitalized on the idea that for recreation-type outputs where the consumers had to travel to the recreation site in order to consume the recreation service, *the consumers would react in the same way to a change in admission fees at the gate as they would to a change in travel costs to the site*. Since a demand schedule relating the total number of site visits to the admission fee charged is the desired result, Clawson reasoned that by observing the frequency of visits by people with different travel costs, such a schedule could be estimated.

The possibilities of using travel costs to estimate site demand for recreation services has subsequently received a great deal of attention. Dwyer et al. (1977) presents a summary review of 28 studies from 1961 to 1976, which developed and explored the technique with application to a wide range of recreation situations. These demand schedules estimated an average willingness to pay that ranged from 50 cents to $5.00 per unit-day of participation.

The basic model underlying these demand estimates is straightforward. Envisioned is an individual located in his or her household contemplating participation in recreation activities. The individual is assumed to have a demand for recreation which declines as the total cost of participation rises (Fig. 12-2*a*). For this individual the assumed at-home demand is $Q = 30 - 2 \times$ (total cost), and at a zero cost, 30 visits per year would be made (point I). Only one recreation site (site XYZ) is available to satisfy all individual demand for this activity.

Individual A lives some distance from site XYZ, and it costs $10 to travel to the site. Given the basic assumption that the behavioral response is the same to an equal increment of admission fee or travel cost, we would expect individual A to treat the $10 travel cost the same as a $10 gate fee and a zero travel cost, solve their demand equation for this price [$Q = 30 - 2(\$10) = 10$], and take 10 visits per year (point F). Further if an

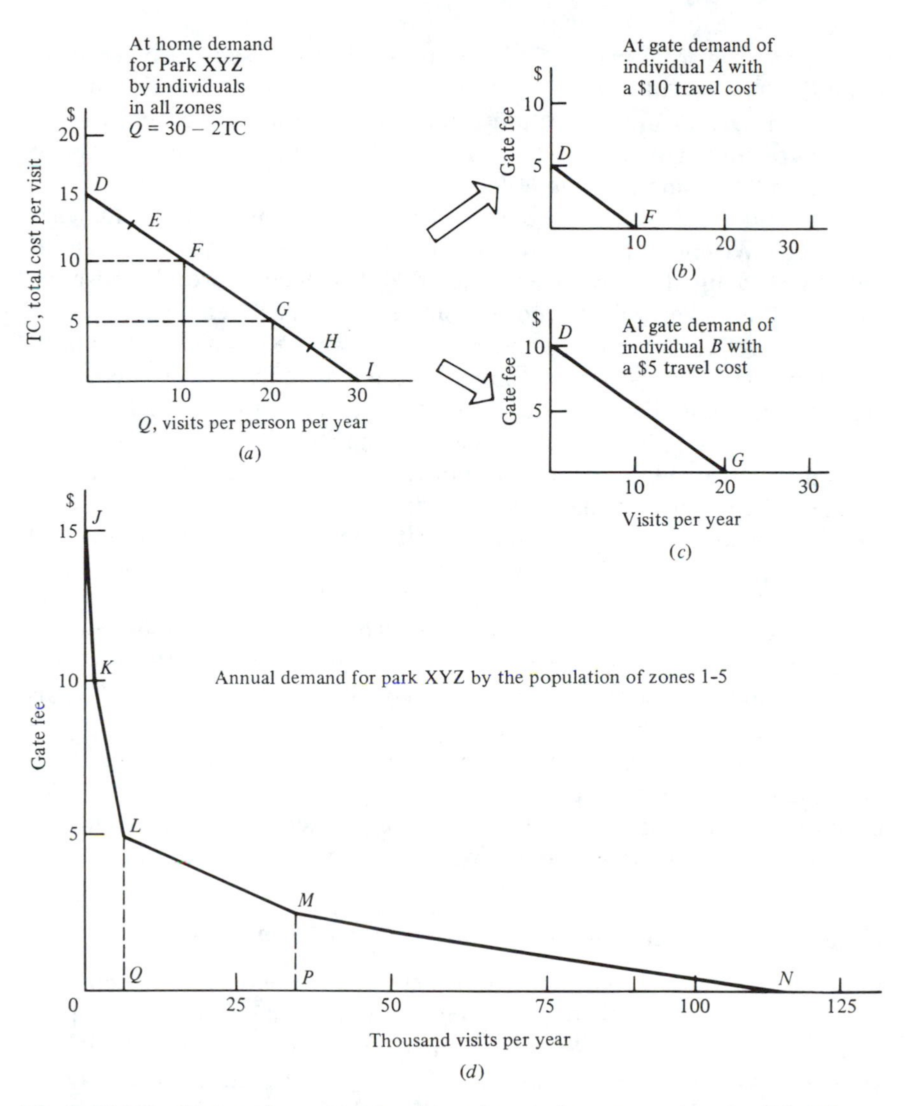

Figure 12-2 Derivation of an aggregate demand curve for a recreation site. (*a*) At-home demand for park XYZ by individuals of all zones. (*b*) At-gate demand of individual *A* with a $10 travel cost. (*c*) At-gate demand of individual *B* with a $5 travel cost. (*d*) Annual demand for park XYZ by the population of zones 1 to 5.

admission price of $5 at the gate of XYZ were suddenly imposed, the total price ($10 travel + $5 admission) would rise to $15, and we would expect the individual to stay home with zero visits (point *D*).

Individual *B* lives closer to site XYZ, so the travel cost will be lower, say only $5. If this second person had the same underlying at home demand for recreation, we would expect him or her to participate 20 times per year with a total price of $5 (point *G*). For this second person the admission price would have to rise to $10 per visit before visits dropped to zero (point *D*). Figure 12-2*b* and *c* illustrates the assumed behavior of these two individuals at the gate of site XYZ as the price of admission rises from 0 to some positive level.

The at-home demand curve of our park users is unknown and the gate fee is zero. We can, however, record how many times individuals *A* and *B* actually show up at site XYZ and what their travel costs are. Suppose we observe *A* to make 10 visits and *B* to make 20 visits. When questioned, *A* said his travel costs were $10 and *B* said hers were $5. Since individual *B* is assumed to have the same at-home demand as individual *A*, we would expect individual *B*'s visits to drop to 10 if a $5 admission fee were imposed because individual *B* now has the same total trip cost ($5 travel + $5 admission) as individual *A* before the fee was imposed ($10 travel + $0 admission). Because we have assumed that all individuals have the same at-home demand curve, the different visitation rates of individuals *A* and *B* (and others like them) can be treated as observations on the common individual at-home demand schedule with travel cost differences serving as a proxy for admission fee differences. This is not unlike the logic we used in Chap. 5 to accept the results of cross-section yield table methods, where we assume today's 20-year-old tree will look like today's 30-year-old tree after 10 years of growth.

Once the at-home demand curve is established, an at-the-gate total demand for the site can be constructed using the procedures below. The only new information used is the population of each distance zone around the park and the average travel cost from each zone to the park.

Empirical Estimation of the Total Demand for a Park

Suppose all visitors to park XYZ were believed to come from five concentric zones at different distances from the park. For a full year the park managers asked every visitor who came to the park two questions: (1) what zone do you live in, and (2) what are your round-trip travel costs between the park and your home? At the end of the year this information was summarized as shown in columns 3 and 4 of Table 12-3*a*. A little mapwork and use of published census data provided the population of each zone shown in column 2.

Assuming all visitors reacted the same to gate fees and travel costs, when plotted, the data in columns 3 and 4 are points *H* (zone 1) to *E* (zone 4) in Fig. 12-2*a* and provide an estimate of the at-home demand curve. This at-home demand curve is then used to work out the expected visitation rates from each zone if gate fees of $2.50, $5.00, and $10 are imposed, as shown in Table 12-3*b*. The travel cost from each zone is added to the gate fee to determine visitation rates.

Table 12-3 Empirical estimation of total demand for a park

(*a*) Survey data from visitors to park XYZ

Distance zone from park XYZ (1)	Zone population (2)	Average travel cost from home to park (3)	Current visits per capita per year (4)	Gate fee (5)
1	100	$ 2.50	25	0
2	500	5.00	20	0
3	5,000	10.00	10	0
4	10,000	12.50	5	0
5	1,000,000	18	0	0

(*b*) Expected visitation per capita per year to park XYZ by distance zone for different gate fees*

Distance zone	Gate fee			
	$0.0	$2.50	$5.00	$10.00
1	25	20	15	5
2	20	15	10	0
3	10	5	0	0
4	5	0	0	0
5	0	0	0	0

* To illustrate the calculations, consider the row for zone 2, which has a travel cost of $5. Using the at-home demand curve of Fig. 12-2*a* and the data of Table 12-3*a*, a gate fee of $0 would have a total visit cost of $0 + $5 and expect a participation of 20 visits. At a gate fee of $2.50 the total cost is $2.50 + $5.00, or $7.50, and the participation is 15 visits.

(*c*) Total consumption at different gate fees*

Distance zone	Gate fee				
	$0	$2.50	$5	$10	$15
1	2,500	2,000	1,500	500	0
2	10,000	7,500	5,000	0	0
3	50,000	25,000	0	0	0
4	50,000	0	0	0	0
5	0	0	0	0	0
Total, all zones	112,500	34,500	6,500	500	0

* Calculated by multiplying per capita participation rates (Table 12-3*b*) for each zone by the total population of the zone (Table 12-3*a*).

The visitation rates per capita from Table 12-3*b* are then multiplied by zone populations to produce the total visits from each zone at the different gate fees shown in Table 12-3*c*. The sum of the visits from all zones at each gate fee is our estimate of total annual demand for park XYZ, and is shown graphed in Fig. 12-2*d*.

Suppose the park managers were planning to charge a fee and wanted to know how the consumer valuation of the experience would be distributed between gate receipts and consumer surplus at different gate fees. These calculations are shown in Table 12-4. For this particular park the gate fees are maximized at the fairly low fee of $2.50. Total visits, willingness to pay, and consumer surplus each show a large additional increase as the fee is reduced from $2.50 to zero. The choice of fee would require trading off higher visitor benefits with practical needs for gate receipts to operate the park or, for commercial operations, to show a profit.

Application of the travel cost model is difficult enough and, if we are to believe that the results represent an at-the-gate demand curve for a recreation site, a rigorous set of assumptions must be satisfied. According to Scott (1965) and Dwyer et al. (1977), virtually all travel cost models make the following important assumptions:

1. The visitor reacts identically to an increase in entry fees and to an increase in travel costs.
2. Travel brings no pleasure or satisfaction in itself (allowing all travel costs to be assigned to the trip's destination).
3. Alternative recreation sites provide essentially the same quality of recreation experience.
4. Multiple-site, multiple-visit visitors are detected and adjusted for.
5. The sites in question have not reached capacity, with people being turned away.
6. Tastes and preferences, income, and thus the underlying at-home demand schedule are the same for persons in the geographically different travel or distance zones.

These are difficult assumptions to satisfy and different investigators have used a variety of empirical and conceptual techniques to control for them. Still the procedure has not gained universal consensus or support, and a lot of debate continues.

Issues in Application of the Travel Cost Demand Estimation Method

Demand curves often must be estimated for large regions or states and cannot be directly estimated for a specific forest. If average regional values for willingness to pay are assigned to individual forests such as in

Table 12-4 Demand for park XYZ, consumer surplus, and average willingness to pay at different gate fees

Gate fee (1)	Number of visits (2)	Incremental visits (3)	Gate receipts (4) = (1) × (2)	Incremental willingness to pay* (5)	Total willingness to pay† (6)	Average willingness to pay (7) = (6) ÷ (2)	Consumer surplus (8) = (6) − (4)
$ 0	112,500	78,000	0	$ 97,500	$253,750	$ 2.26	$253,750
$ 2.5	34,500	28,000	$86,250	105,000	156,250	4.54	70,000
$ 5.00	6,500	6,000	32,500	45,000	51,250	7.88	18,750
$10.00	500	500	5,000	6,250	6,250	12.50	1,250
$15.00	0		0		0	0	0

* Incremental willingness to pay can be calculated as the change in the number of visits between two different gate fees multiplied by the average of the two fees. For example, between fees of $5 and $2.50 the visits increased 34,500 − 6,500 = 28,000 (column 3). The average fee is ($5.00 + $2.50)/2 = $3.75. Incremental willingness to pay is then 28,000 × 3.75 = 105,000. In Fig. 12-2*d* this is the area of polygon *QLMP*.

† Total willingness to pay is the cumulative total of incremental willingness to pay as the gate fee is reduced from its highest level. For example, at a price of $2.50, the total willingness to pay is the sum of three increments; ($6,250 + $45,000 + $105,000) = $156,250. In Fig. 12-2*d* this is the area of polygon *OJKLMP*.

Table 12-2, they can seriously overestimate the value of outputs for a remote forest and underestimate the value of outputs for a forest close to urban areas. The remote forest has high travel costs, hence a much lower at-the-gate demand (like individual *A* in Fig. 12-2*b*), and then a lower-than-average willingness to pay gate fees.

A second issue in application occurs if a demand curve has been estimated and the nonmarketed output is to be explicitly priced in the objective of a linear programming planning model. Care must be taken not to value opportunity that is provided but not consumed. Assume a demand curve for recreation at the gate of Big Tree forest had been estimated using travel cost methods as shown in Fig. 12-3*a*. Each year 25,000 visitors have been coming to the forest at the low gate charge of $2.00. The calculated average willingness to pay is $5.33. If all sites are fully utilized for the season, the usable recreation capacity of the forest is in the neighborhood of 100,000, hence current use does not exceed capacity. In their analysis the planners envision more capacity due to road and facility development. Without a change in prices or a drastic change in promotion activities, however, they feel it is unlikely that actual visitor use growth will deviate from its historical rate of a 6 percent increase per decade. At this rate approximately 29,800 visitors are expected after 30 years, an increase of 4,800 over today's level.

The appropriate expression of recreation demand in a linear programming model which evaluates forest outputs 30 years hence is to say that $5.33 per visit (the average willingness to pay) will be assigned for up to 29,800 visits, but a price of zero will be assigned to visit opportunities provided in excess of 29,800 (Fig. 12-3*b*).* This approach recognizes that the value of recreation has to be related to the amount actually consumed at the given gate price. It is incorrect to estimate increased levels of recreation supply and multiply by the average willingness to pay. For this park there is already an overcapacity of about 70,000 visitor-days, so additional capacity is unlikely to produce new benefits.

A third issue arises when an output such as recreation visits is assigned an average willingness-to-pay value that contains a large consumer surplus component and this value is used to compare incremental changes in recreation output to incremental changes in a market-valued output such as timber. The market price of timber does not have a consumer surplus component. For small increments of timber output, the increment typically is multiplied by the current market price to estimate the value increase (or decrease). When compared to changes in outputs whose price

* If the increase in recreation demand were expected to be substantial, it would be more precise to reestimate the demand curve and recalculate the average willingness to pay each decade.

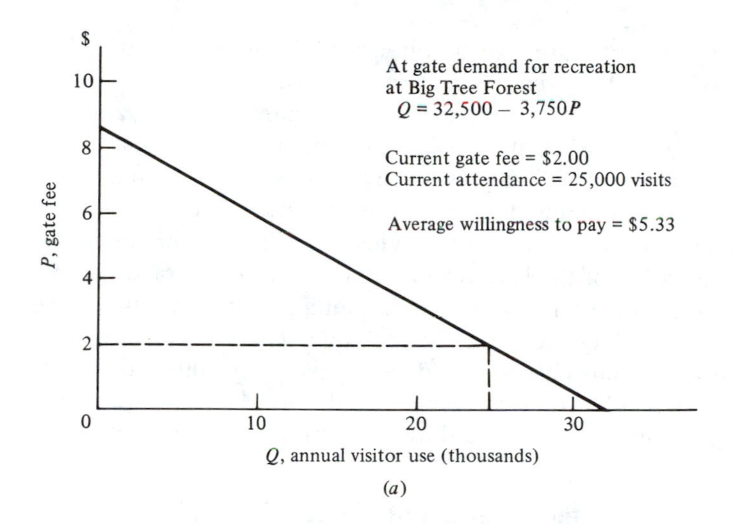

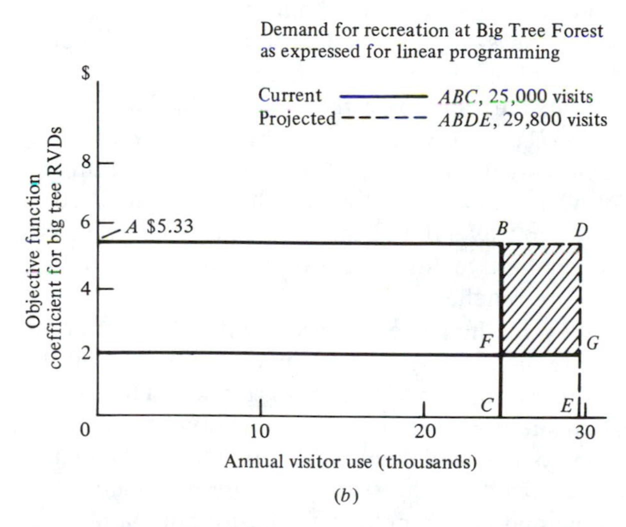

Figure 12-3 Application of willingness-to-pay analysis in linear programming models. (*a*) At-gate demand for recreation at Big Tree forest. (*b*) Demand for recreation at Big Tree forest as expressed for linear programming.

includes consumer surplus, the market-valued changes are unfairly disadvantaged, the degree being a function of demand elasticity. In Fig. 12-3*b*, for example, the increased 4,800 units of recreation are priced at $5.33, with a total value of $25,584. Of this, $15,984 is the consumer surplus (*BDGF*) and $9,600 is market valued (*CFGE*) ($2.00 × 4,800). Timber, by contrast, if it were represented by the same demand curve as Fig. 12-3*a* would only have a price of $2, the market price, to be multiplied by the increased supply of 4,800 units of output, resulting in the estimated increased value of $9,600 in area *FGEC* (Fig. 12-3*b*).

The point is obvious, if you are going to compare and analyze trade offs between two or more outputs using explicit prices, then all prices should either be average willingness-to-pay values which include consumer surplus, or they should all be estimates of market prices that do not include consumer surplus. From another viewpoint, market prices represent marginal valuation of the last unit purchased, while average willingness to pay is an average valuation over all units purchased—again two completely different value concepts. Unfortunately, it is an imperfect world and we do mix values in order to do analysis; if you must do so, just remember that the results will tend to be biased toward outputs valued by average willingness-to-pay prices and away from market-priced outputs.

Estimating Demand by the Contingent Value Method

Like the travel cost method, the objective of the contingent value (survey) method is to estimate a willingness-to-pay schedule. It does so by directly asking visitors what they would pay for their experiences. The method is inherently subjective, since its data are answers to hypothetical questions such as "what *would* you pay" or "how many times *would* you visit *if* the price were . . . ?" In contrast, the travel cost is more objective and is based on observations of what people actually did. The survey method is more flexible and is used when travel data cannot be collected, when the analysis needs to evaluate qualitative differences in recreation activity, or when capacity constraints are reached.

Procedurally, the survey method uses two sequential but different surveys. In order to estimate the at-the-gate demand curve by the contingent value method, two surveys or questionnaires are used. The first is conducted at the recreation site to determine the association of value and frequency of participation to the socioeconomic strata of participants such as age, income, and education. The second survey is usually conducted by mail or telephone and is a sample of the entire population. Its objective is to determine the percentage of persons in each socioeconomic stratum who participate at all in the activity. The size of the population weighted by the propensity to participate and multiplied by the value-participation rate information gained from the on-site survey of known participants can estimate a demand curve for the site.

Many studies only sample on-site users without data on total participation by strata. In this case we learn much about the active participants but may not be able to estimate an aggregate annual demand function. The survey method rests on three important assumptions:

1. Appropriate, representative, and comparable samples of the user population are implemented for both the first and the second surveys.

2. For the first on-site survey, visitors have a sufficiently well developed sense of relative values such that with subjective estimates they can accurately assign a dollar price to the nonmarketed recreation experience.
3. Interviews and questionnaire methods used in the on-site survey can objectively elicit values from the visitors without bias or distortion.

These are also difficult assumptions to satisfy, and research has yet to establish the conditions or situations under which they reasonably hold. The third assumption is particularly difficult since it is very easy to ask questions in such a way as to lead the answer of the interviewee.

Hammack and Brown (1974) conducted a well-conceived mail questionnaire study trying to determine the average value of a season of waterfowl hunting. The two critical questions asked of the 2,455 respondents in the questionnaire were:

1. What is the smallest amount you would take to give up your right to hunt waterfowl for a season?
2. About how much greater do you think your costs would have been before you would have decided not to have gone hunting at all during that [the previous] season?

The first question really asks how much the user would require to "sell" his or her rights to hunt—and thus not hunt at all for a season. The second asks how high the price (cost) would rise before hunting would be discontinued—the willingness-to-pay question. These are reasonably neutral questions. If you hunt ducks, how would you respond? Could you improve the phrasing of the question?

The results of this study produced an average willingness to pay of $247 per season and a willingness-to-sell estimate of $1,044 per season. (Why would a duck hunter place a much higher value on "willingness to sell" than on "willingness to pay?")

A recent contingent value study (Bishop et al., 1984) compared actual cash bids to hypothetical bids for a permit to hunt deer in special area of Wisconsin. The objective of this well-designed and carefully controlled experiment was to see how the actual and contingent values compared and to test different bidding or auction methods for eliciting the contingent value estimates. They found that the different bidding methods all gave about the same numerical results but that the highest hypothetical bids were consistently about 60–80 percent higher than the highest cash bids. These results suggest a possible high side bias in the contingent value method but the consistency of the hypothetical to cash bid ratio over several sample groups and methods offers hope that the contingent value method could help us to objectively define reliable market proxy prices for some recreation services not currently marketed.

VALUATION METHODS NOT BASED ON WILLINGNESS TO PAY

There are several methods used to assign values that are not based on the demand or willingness-to-pay concept. Still, despite the lack of conceptual underpinning, they have and will continue to be used or recommended and thus warrant our attention.

Expenditures

This is the most widely used of the questionable techniques and amounts to adding up the total expenditures of the recreationist on equipment, vehicles, travel, fees, etc., which are associated with participation in one or more recreation-type activities, such as hunting, hiking, boating, or camping. These expenditures are variously classified as occurring within the recreation area, en route, or at home. Many expenditure items, such as a four-wheel drive vehicle, physically last several years and are used for both recreation and nonrecreation purposes. Such expenditures have to be depreciated over several years and jointly apportioned to business, family, and recreation activities.

The main assumption of the expenditure method is that all of these expenditures are costs that bring no satisfaction in themselves and all are required and necessary to participate in the activity. If this were true, then the expenditures could be viewed as opportunity costs paid to participate, and hence are a minimum measure of value received. The assumption just does not hold up.

Ownership of recreational equipment does bring a great deal of user satisfaction apart from its use in recreation participation. It is hard to believe that the often cleaned, admired, but little-used boat, fly rod, rifle, or jeep carries no intrinsic value or satisfactions to the owner. Or that time in restaurants, hotels, or in traveling brings no pleasure and is only a cost paid to see the park. For a more familiar example, the cost argument would say that the value of a movie includes the cost of the dinner beforehand and the cost of gas to go driving afterward. Most of us would say the value of the movie is the price of the ticket.

The expenditure method does provide good information for evaluating the local or regional economic impact of recreation and tourism activity. Maximizing local recreation expenditures might be called the "chamber of commerce objective," and from the perspective of recreation-based economics it is certainly legitimate. One could use expenditure data to qualify input-output models or to set coefficients for an objective function to design programs by the chamber of commerce objective. Remember, however, that this is a measure of recreation value to the local economy, not to the recreation participant.

Cost of Supply

Sometimes the supply cost of providing recreation opportunities is suggested as a minimum value of the experience to the user. For example, we argue that because it takes an average $5.95 per visitor-day to build, maintain, and operate a campsite supporting 200 visitor-days per year, this is a value to assign to visitor-days of camping.

This argument has no credence at all. If only 100 visitors showed, would we say the value per visit is $11.90 when all the costs are assigned to actual consumption? Second, since the recreationist did not decide to build or finance the campground, how can the supply cost be considered an opportunity cost to the recreationist? Third, if the supply cost method is used, the recreation producers can tautologically rationalize and justify whatever they wish to spend by simply assigning a benefit to the experience equal to the cost of supply. This could easily lead to big budgets for supply organization and agencies.

Market Value of Commodities

For some recreation activities, notably hunting, fishing, and firewood gathering, the product itself is tangible and has a market value, for example, the deer, the catfish, or the firewood. Some suggest that the market value of these commodities be used to assign a value to the recreation experience. For example, hatchery trout can be purchased for $2.49 per pound at the local fishmarket. Since the average fisherman catches 1.65 trout of this size per visit, his visits are worth at least 1.65 × $2.49, or $4.11. This argument would have some merit if the only objective of the participant were to obtain and eat the trout in question. Numerous studies show, however, that usually it is much cheaper to buy the commodity in the local butcher shop, fish store, or firewood depot than it is to gear up and procure it yourself. So some other objective and satisfaction must be involved in recreation. Fuelwood cutting is closer to a commercial objective than the catch-and-release fly fisherman, but even so, firewood gathering is recreation for many.

The issue here is that the commodity is not always the primary or even secondary reason for participation. Therefore using commodity values to estimate the value of a visitor-day almost always underestimates the value of the whole experience.

Unit-Day Value

Starting with U.S. Senate Document 97 (1962), the federal legislature has decided that even though acceptable willingness-to-pay values were not available from empirical research, they still wanted to use dollar numbers for analysis. Moreover, they were willing to assign these values. In its principles and standards the U.S. Water Resource Council (1973) states:

> In the interim, while recreation methodology is being further developed, the following schedule of monetary unit values may be used for planning:

	Range of unit-day values
General A recreation day involving primarily those activities attractive to the majority of outdoor recreationists and which generally require the development and maintenance of convenient access and adequate facilities. This includes the great majority of all recreation activities associated with water projects such as swimming, picnicking, boating, and most warm water fishing.	\$0.75–\$2.25
Specialized A recreation day involving primarily those activities for which opportunities in general are limited, intensity of use is low, and which may also involve a large personal expense by the user. Included are activities less often associated with water projects, such as big game hunting and salmon fishing.	\$3.00–\$9.00

Where these numbers came from or what they represent is unclear—they are probably derived from the prices charged for commercially available recreation opportunity. The important point is that these price ranges are *legislated* and represent the value judgment of the elected representatives of the people. Because of this they may be the best subjective judgments available assuming (1) legislators are responsive and accountable to the electorate, and (2) the prices assigned are of sufficient importance that review and periodic adjustment are made to keep them current.

We strongly prefer methods with a solid conceptual and analytical foundation. However, for a variety of reasons, such methods are often not possible to implement, and judgment is called for. The issue becomes, whose value judgment is appropriate? We feel it has to be the landowners' or their elected representatives'.

Valuation of Forestland Revisited

In Chap. 11, we presented methods for calculating the value of forestland that will be used for timber production. In the discussion, we valued only the timber that would be produced over the rotation or cutting cycle. We did not value the other outputs such as water and wildlife that often would also be produced. To the extent an owner is willing to accept and use the methods for valuing these outputs from this chapter, he can expand the valuation of forestland to include the multiple outputs that can be produced.

Calish et al. (1978) did such a multiple-output valuation of forestland. Basically, they broadened the soil expectation formula [equation (7-17)] to include revenues accruing from the production of nontimber outputs as the stand ages.

The details of these calculations will be covered in Chap. 13 when we discuss finding the income-maximizing rotation age considering all forest outputs. It is interesting to note, however, that their calculations using "reasonable values" for the nontimber outputs produced over a rotation of coastal Douglas-fir showed 75 percent of the total SEV coming from nontimber sources.

QUESTIONS

12-1 A study established the terms of trade in a barter forest economy as follows:

Forest output	Units	Number of units exchanged for 1 M bd ft of sawlogs
Sawlogs	M bd ft	1
Water	Acre · ft	100
Forage	AUM	275
Recreation use	RVD	175
Sport fishing	RVD	120
Commercial fishing	Pounds	50

(*a*) Construct a full terms-of-trade table showing the exchange rates in terms of each output similar to the table on p. 449.

(*b*) If sawlogs sold for $100 per M bd ft, what is the imputed dollar value per unit of water, recreation use, and sport fishing, which are not sold in markets?

12-2 The following survey data were collected describing all of the visitors coming to the Big Stoney Creek recreation area for 1 year.

Distance zone from Big Stoney Creek park	Zone population	Average travel cost from home to park	Visits per capita per year	Gate fee
1	200	5	9	$5
2	20,000	10	7	5
3	10,000	20	5	5
4	100,000	35	3	5
5	500,000	40	2	5

(*a*) Derive the at-home demand curve for Big Stoney Creek visits by individuals in all zones (assuming it is the same for all zones). To do this, plot the survey data on graph paper to fit a nonlinear curve of total cost and visit rates. Then interpolate the needed data for the at-home demand curve. Check part (*a*) of the answers before proceeding. You may have slightly different numbers. This is OK but use the data in the answer to part (*a*) for the rest of the questions.

(*b*) Estimate the at-gate demand curve for individuals from zone 3 for different gate fees.

(*c*) Calculate and plot the aggregate at-gate demand curve at different gate fees.

(*d*) What are the estimated total willingness to pay, average willingness to pay, and consumer surplus for a gate fee of $15?

REFERENCES

Bishop, R. C., T. A. Heberlein, M. P. Welsh, and R. M. Baumgartner, 1984: "Does Contingent Valuation Work?", University of Wisconsin, Madison, mimeographed.

Burt, D. R., and D. Brewer, 1971: Estimation of Net Social Benefits from Outdoor Recreation, *Econometrica,* **39:**813–827.

Calish, S., R. D. Fight, and D. Teeguarden, 1978: How Do Nontimber Values Affect Douglas-Fir Rotations? *J. Forestry,* **76:**217–221.

Clawson, M., 1959: "Methods of Measuring the Demand for and Value of Outdoor Recreation," Resources for the Future Inc., Washington, D.C., Reprint 10.

Dwyer, J. F., J. R. Kelly, and M. D. Bowes, 1977: "Improved Procedures for Valuation of the Contribution of Recreation to National Economic Development," Water Resources Center, University of Illinois, Urbana-Champaign, Research Report 128.

Hammack, J., and G. M. Brown, 1974: "Waterfowl and Wetlands: Toward Bioeconomic Analysis," The Johns Hopkins University Press, Baltimore, Maryland.

Heinrichs, J., 1983: The Winged Snail Darter, *J. Forestry,* **81:**212–215.

Scott, A., 1965: "The Valuation of Game Resources: Some Theoretical Aspects," Canadian Fisheries Report 4, May 1965.

U.S. Forest Service, 1980: "The Nations Renewable Resources: An Assessment," Forest Resources Report 22.

U.S. Senate, 1962: Supplement 1, Evaluation Standards for Primary Benefits, in "Policies, Standards, and Procedures in the Formation, Evaluation, and Review of Plans for Use and Development of Water and Related Land Resources," S. Doc. 97, 87th Congress, 2d Session, May 29, 1962.

U.S. Water Resources Council, 1973: "Water Related Land Resources, Establishment of Principles and Standards for Planning," *Federal Register,* September 12, 1973, **38,** part III, no. 174.

Willig, R. D., 1976: Consumer Surplus without Apology, *Amer. Econ. Rev.,* **66:**345–356.

PART
FOUR

FOREST MANAGEMENT PLANNING

In the preceding chapters we developed the essentials of forest classification, growth and yield, decision analysis, and valuation of forest outputs. Now we are ready to assemble these components to aid in identifying and solving forest management planning problems.

To understand forest management problems, it is helpful first to envision the forest as a vegetation management enterprise—a green factory, if you will (Fig. A). This factory of land, vegetation, and animals is continually converting the inputs of labor, money, sunshine, and moisture into a variety of tree- and land-related outputs, including commercial timber, water, and recreational opportunities. Because our primary concern is how the forest relates to human needs, we define the production period of a forest as 1 year to match the most common time frame of human planning. Trees to be sure take a lot longer than 1 year to mature—and this is one of the things that makes forest management problems both difficult and interesting—but since the problem should be defined in terms that are meaningful to the forest owner, annual accounting is used.

This linkage of today's forest structure and management activity to the level and kind of forest outputs 20, 40, or more years into the future

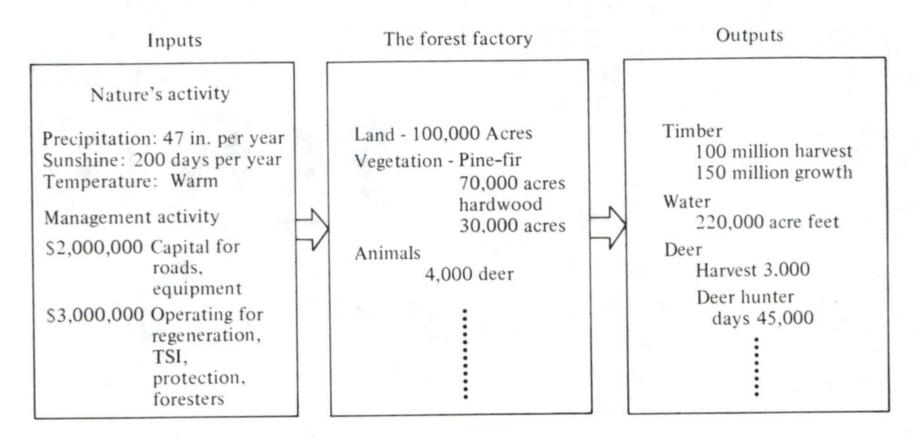

Figure A The forest production system.

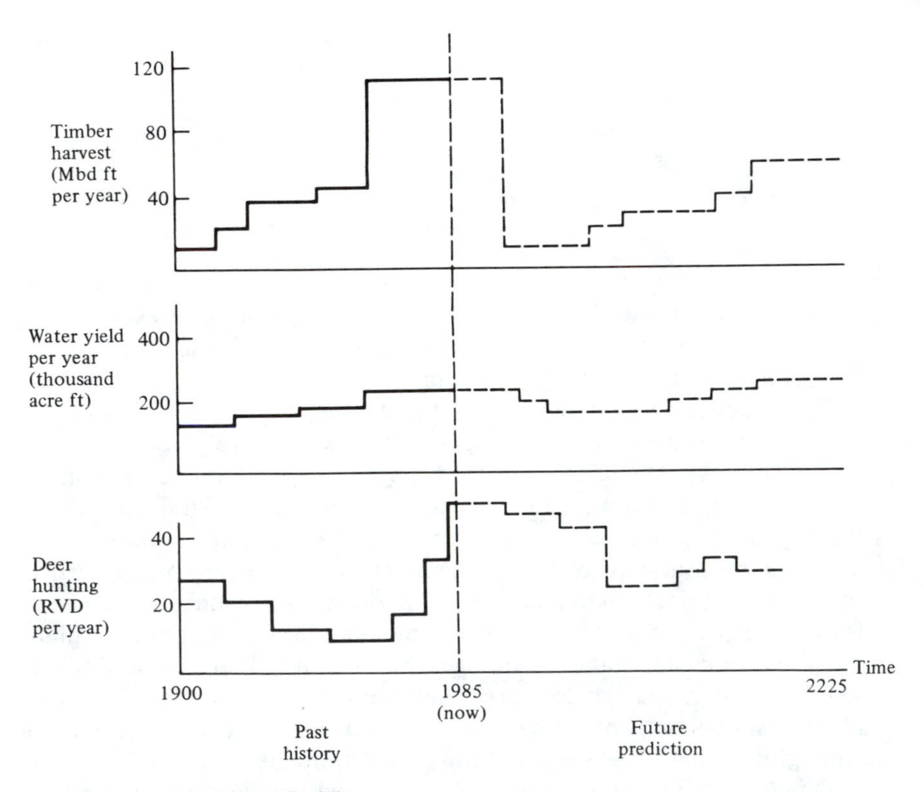

Figure B Forest output schedules.

gives special impetus to foresters to gaze in their green crystal ball for guidance on today's decisions. Such forecasts are shown in Fig. B, which displays historical outputs and forecasted future outputs from our forest factory under continuation of current management. History is a fact, but it does establish a perspective by which to forecast future outputs and their associated schedule of activities.

A forest management problem arises when a decision is needed about some aspect of forest management. Often the impetus for this decision is stimulated by the landowner's discontent with the flow of outputs forecasted under continuation of current management. The owner of fallow farmland who wants pine stands for quail habitat, the owner of pine sawtimber stands that have not been successfully regenerated, or the owner of a hardwood swamp all may wish to consider whether to take the actions needed to produce softwood timber. Their problem is whether to change the harvesting, treatment, and developmental input activities such that the future forest factory will have a plant structure supporting an annual output mix that includes softwood timber. The decisions needed vary by owner: the first of our three owners might consider initiating a continuing planting program; the second might consider burning, spraying, and seeding; while the third might consider draining, harvesting, plowing, and planting. The results could be about the same, with each owner's future forest eventually composed of a balanced age distribution of pine stands. The problem gets even more complex when the owners are concerned with a combination of wildlife, recreation, water, and other forest outputs in addition to timber—more outputs to forecast and more activities, costs, and values to consider.

SOLVING FOREST MANAGEMENT PROBLEMS

A forest management problem can be approached at the tree, stand, or forest level. A forest owner may simply wish to find out which trees to cut to reach his or her objectives over time. Or an owner may want to think of forest management in terms of the timber stands on the property and plan a series of actions for each one, based on its condition and the owner's objectives. Or the owner may wish to understand and control actions across the entire forest, coordinating the actions on different stands to meet a variety of objectives relating to issues such as financial return, harvest flow, and the forest structure that will be achieved when the planning horizon is reached. All of these views of forest management represent legitimate formulations of forest management problems, and each is covered in Part 4 of this text.

In Chap. 13 we address forest management problems at the tree and stand level, when the landowner's situation permits such decisions to be made independently for each tree or stand as it is encountered. Given the basic objective that the owner wishes to make money from the use of his or her forest, we present decision guides for the multitude of actions that can be taken in both uneven-aged and even-aged stands.

In Chaps. 14 and 15 we address forest management problems in terms of the overall forest. Chapter 14 covers the time-honored approaches to "forest regulation" that have been used by foresters since the 1500s. Foresters traditionally defined a target forest that would give a stable output once achieved. Foresters then used a variety of formulas and procedures to "convert" the current forest into a desired "regulated" forest, producing harvest schedules like that shown at the top of Fig. C. Some actions would convert the forest over one rotation while others that achieved a more stable harvest would take longer.

By and large, traditional regulation techniques employed biological measures of the forest, such as inventory volume and growth, in the decision guides, leaving economic considerations almost wholly aside. In this process foresters tended to assume that their personal and professional objectives for the forest (allocate most land to timber production, maximize growth, have an even flow of harvest, and achieve a regulated forest) in fact also represented the objectives of the forest's owners.

As wood became more and more valuable and also people began to demand that timber production be coordinated with other uses, the traditional objectives of foresters began to diverge from those of the forest owners. When that happened (in the 1960s and 1970s), the simple classical formulas of Chap. 14 no longer proved adequate. Simultaneously with this development came procedures such as linear programming, and the high-speed computers to process them, that made more detailed, more "objective" analyses possible. In Chap. 15 we examine forest management planning in light of all these developments. Through a series of examples, we introduce modern forest management planning with its requirements for explicit objectives and constraints. Compared to the classical approach, we describe in Chap. 15 a process called harvest scheduling under constraints that pays a lot more attention to the near-term schedule (the conversion period) than to the far-term schedule (the regulated forest).

Such a harvest schedule is shown in the bottom graph of Fig. C. Here, in contrast to the classically regulated forest, the short-term harvest does not fall; in this case it rises.

Schedules and plans are typically reevaluated within 10 years to consider new goals and changing constraints, and the longer-term forecast is only tentative. In general a forest will never be fully regulated in the classical sense and the harvest will rise and fall over time. When the acres are stocked and growing, we would call this a productive forest.

We use three different expressions to talk about scheduling in Part 4. *Harvest scheduling* is the traditional term for scheduling a timber harvest

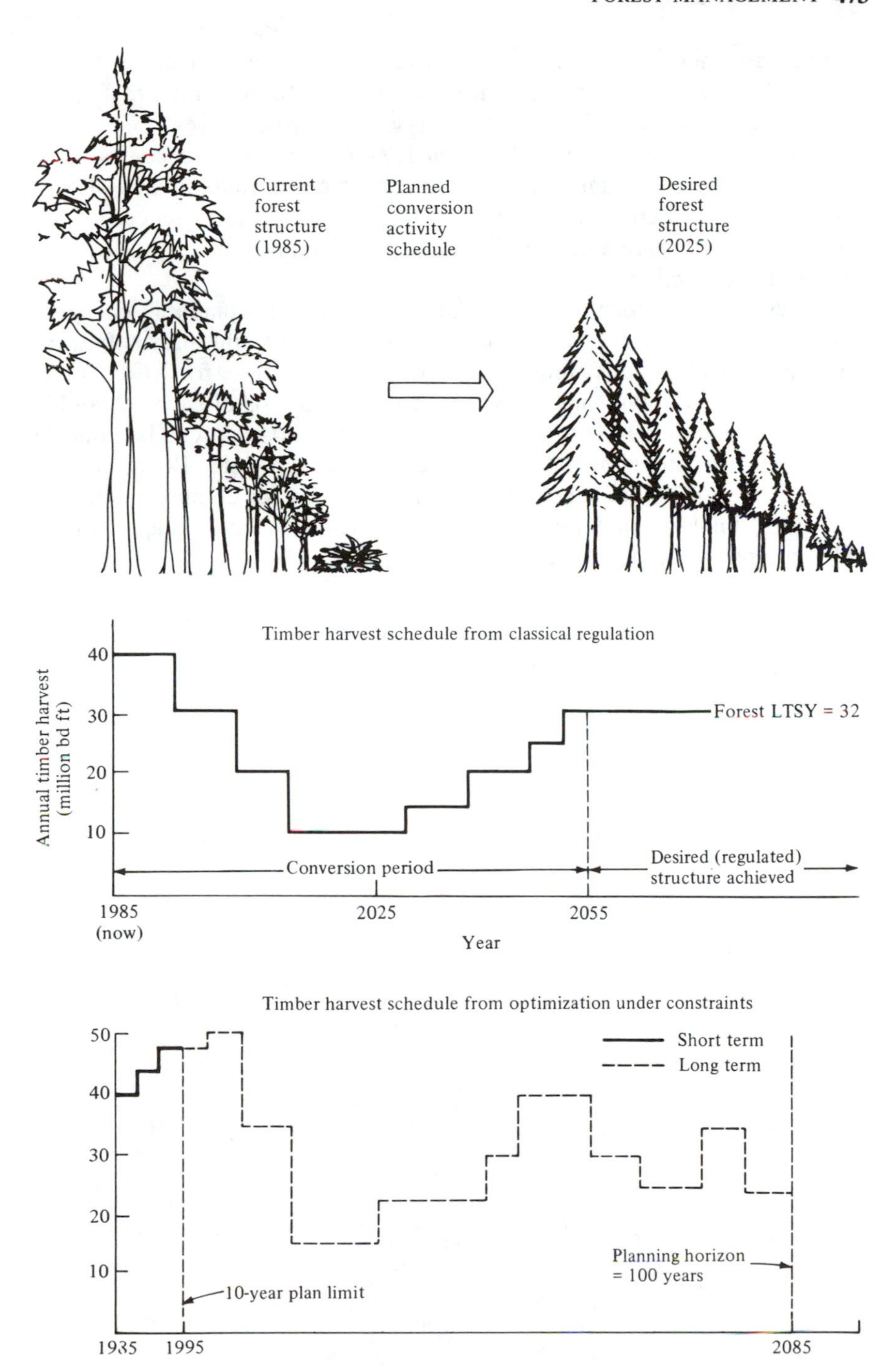
Current
forest
structure
(1985)
Planned
conversion
activity
schedule
Desired
forest
structure
(2025)
Timber harvest schedule from classical regulation
Annual timber harvest
(million bd ft)
40
30
20
10
Forest LTSY = 32
Conversion period
Desired (regulated)
structure achieved
1985
(now)
2025
2055
Year
Timber harvest schedule from optimization under constraints
50
40
30
20
10
Short term
Long term
Planning horizon
= 100 years
10-year plan limit
1935
1995
2085

Figure C

by periods when there were few, if any, prescription choices for future stands. *Timber management scheduling* covers both timber harvest activities and cultural and other activities associated with present and future stand production. *Forest management scheduling* describes comprehensive analysis that considers timber and nontimber outputs in land allocation and activity scheduling decisions. Output and cultural schedules for nontimber outputs (for example, forage seeding and forage production) can be made explicit.

These modern techniques for forest management planning can produce a vast, almost overwhelming amount of results and information on the schedule of actions, inputs and outputs, over time for a forest. In Chap. 16 we set down some principles for the analysis of such schedules to assist in deciding how to interpret the results and what can be done to ameliorate the effect of the constraints on the schedule. Being able to interpret a timber management schedule is a critical component in the successful analysis of forest management problems, and Chap. 16 provides an introduction to this skill.

CHAPTER
THIRTEEN

TREE AND STAND DECISIONS UNDER FINANCIAL OBJECTIVES

Suppose that you have just gotten you first job in professional forestry, working for a tree farm in the upper midwest that sells stumpage from its woodlands. On the first day on the job your boss takes you on a grand tour of the company's holdings. Stopping at a typical stand, he sits down next to a three-log, 18-in. DBH sugar maple and says, ''Our main objective is to make money from the harvest and growth of our timber. We will be cutting timber from this stand during the next few months and don't plan to be back for 8 full years after that. Should we cut this tree or let it grow?''

Remembering some vague ideas on financial planning from your forest economics class, you ask him a few questions and find out that the company will not invest in anything that does not yield at least a 4 percent rate of return. Armed with this knowledge, you return to the company office and dig into its little-used library of forest economics and management. After much searching, you stumble upon a publication entitled ''Financial Maturity: A Guide to Profitable Timber Growing'' by William Duerr, John Fedkiw, and Sam Guttenberg (Duerr et al., 1956). Reading this booklet, you find:

> One of the most common decisions foresters make is which tree to cut and which to leave. Periodically, the forester visits each acre of his holdings and marks for cutting those trees whose quality or growth rate is below par. Bearing in mind how the typical

> tree develops—initially low in value but increasing rapidly, then gradually slackening in rate of increase until the stage is fully reached where the tree ceases to pay its way—the forester appraises the trees with the object of putting the ax to those that are reaching this crucial point.
>
> This is the point of financial maturity. It is the stage beyond which the expected value increase no longer equals or exceeds the net return possible—the so-called alternative rate of return—from liquidating the tree, using its cash value elsewhere, and turning the growing space over to other trees. If the owner expects a rate of return of 4 percent, the forester will cut any tree that is not producing or will not produce at this rate. The net effect of cutting financially mature timber is to maximize the net return to the forest business in line with the owner's business policy.
>
> The financial maturity idea can be applied to any species and, if properly adapted, under any silvicultural system and for any product or combination of products. The chief problem is to determine the rate of value increase that can be expected of trees over a given period of time. Once these rates are determined, simplified marking guides can be prepared to define the diameters at which trees of given species, vigor, grade, and height are economically ripe for harvest for desired alternative rates of return.

Sweeter words were never written about the forester's perennial problem of how much growing stock to leave and how much to cut. You have come upon the pioneering work by William Duerr and his colleagues that helped begin the view of the tree as an economic asset and whose "financial maturity" technique can be used to decide whether to cut a tree, a clump of trees, or a stand.

We use the financial maturity approach to forest management, and related approaches, to help make the harvest and growth decisions faced by every forest landowner. We look first at uneven-aged management. Starting with the cut/leave decision for a tree isolated from its neighbors, we move to a group of trees where the subject tree is competing with its neighbors and finally to an uneven-aged stand.

Next we study harvest and growth decisions for an even-aged stand. First we look at the rotation decision when we plant an acre of bare ground, considering first timber values only and then timber and nontimber values. Then we consider investments such as precommercial thinning and fertilizing, introduce the possibility of commercial thinning, and search for the objective-maximizing level of growing stock at each age. Once we have found the "best" timber management regime for an acre of bare ground, we move to the case where we already have timber existing on the acre and search for the set of decisions that best meet our objectives considering the stand there now and the generations of stands that come after it.

Finally, we introduce two major kinds of taxes that private owners must pay (income and property taxes) and discuss their effect on stand decisions, and also examine stand decisions when the possibility of multiple outcomes is recognized in stand management.

Basic Assumptions

Throughout this chapter we make five basic assumptions about forest owners:

1. They are interested in maximizing the present net worth of their woodlands.
2. Growing timber is the most profitable use of forestland.
3. They have no constraints that require coordination of their actions on different acres over time and space, such as a forestwide even-flow constraint on timber harvests or a budget constraint.
4. A single guiding interest rate can be used to evaluate their initial investments and any reinvestments.
5. They wish to do the analysis in real terms so the effect of inflation has been removed from all prices, costs, and the guiding interest rate.

These assumptions have been traditionally made in much economic analysis of timber production. While in many cases they unrealistically simplify the problem, they are reasonable for many owners with commercial timber growing objectives. However, many owners also have nontimber objectives such as maintaining wildlife habitat or are concerned with the long-term stability of activities and outputs. For these owners there often are very real temporal and spatial constraints on their management of the forest. How to formulate and analyze the joint stand and forest management problem when such objectives and constraints are important is the subject of Chaps. 14–16.

UNEVEN-AGED MANAGEMENT

Uneven-aged stands have no beginning and no end over time. Trees of many different sizes and ages occur in close proximity, and each may need a uniquely determined time of harvest to contribute the most to the owner's objectives.

Making the Cut/Leave Decision for an Isolated Tree

Let us assume that your sugar maple tree is growing distant enough from other trees in the stand that it feels little competitive stress from them. For this tree, the financial maturity guide presented by Duerr et al. (1956) can be stated in four steps:

1. Determine the guiding rate of return.
2. Calculate the current stumpage value of a tree.

3. Estimate the stumpage value of the tree at the various ages when the tree might be cut.
4. Compare the highest prospective value increase of these future years, in terms of a compound annual rate of value increase, with the guiding rate of return to judge whether the tree is financially mature and should be cut. The rule: the tree is financially mature if the highest projected rate of value increase is smaller than the guiding rate of return.

For simplicity, we will assume that there is only one future period of interest (8 years from now). Your boss has already told you your guiding rate of return (4 percent), so the next step is to calculate the value of your three-log, 18 in.-DBH tree today and 8 years hence. Using a site table to estimate height growth, increment borings to estimate diameter growth, your boss's estimate of log grade now and in 8 years, and a table giving the stumpage values by log and grade size, you estimate the tree stumpage values today and in 8 years (with all costs but interest already netted out of stumpage price):

Log		Today			8 years hence		
Position	Length (ft.)	Diameter (in.)	Grade	Value	Diameter (in.)	Grade	Value
First	12	15	1	$10.10	17	1	$13.59
Second	14	14	1	10.19	16	1	13.88
Third	16	11	3	1.88	13	2	4.84
Stumpage value				$22.17			$32.31

Next we need to find out the compound rate of value increase achieved by $22.17 growing to $32.31 over 8 years. Given

SV_0 = stumpage value now
SV_n = stumpage value after 8 years
i = compound rate of value increase expressed as a decimal,

we want i such that $SV_0(1 + i)^n = SV_n$. Using Eq. (7-3)

$$(1 + i)^n = \frac{SV_n}{SV_0} \qquad \text{or} \qquad 1 + i = \sqrt[n]{\frac{SV_n}{SV_0}} \tag{13-1}$$

In other words, we need to find the earning rate of investing the capital value of the tree in another 8 years of growing. In your case this rate is

$$(1 + i)^8 = \frac{32.31}{22.17} \qquad \text{or} \qquad 1 + i = \sqrt[8]{1.46} = 1.048$$

Hence, $i = 0.048$

According to the financial maturity rule,

> If value growth percent (here 4.8%) is greater than guiding rate of return, leave the tree.
> If value growth percent is less than guiding rate of return, cut the tree.

With your company's guiding rate of return of 4 percent, you would recommend leaving the tree.

In checking whether the compound rate of value increase is greater than the guiding rate of return, we found out that the capital represented by the tree would grow to a larger value over 8 years in the tree than if we placed the money in the best alternative investment, as expressed by your guiding rate of return. We did this analysis in terms of a comparison of two compound rates, but we can also do it in terms of a comparison of the absolute amounts of value increase.

1. *Predicted increase in tree value,*
   Value of tree now = $22.17
   Value of tree in 8 years = $32.31
   Value increase = $32.31 − $22.17 = $10.14
2. *Predicted return in alternative investment.* Cut tree and invest return at 4 percent for 8 years,
   Value of tree now = $22.17
   Value of investment in 8 years = $22.17 × $(1.04)^8$ = $30.34
   Value increase = $30.34 − $22.17 = $8.17

The value increase from the best alternative investment ($8.17) is known as the *cost of capital.* As we found before in comparing the rate of value increase with the guiding rate of return, leaving the capital in the tree will produce a bigger return ($10.14) than placing it elsewhere ($8.17).

The financial maturity guide used here to determine the time of harvest stipulates that a tree should be cut if its projected rate of value growth is less than the guiding rate of return. We assessed the choice of investing the capital value of the tree in more wood growth or converting it to cash and investing it elsewhere by calculating the future value of both choices (value of tree in 8 years = $32.31, value of tree capital in alternative investment in 8 years = $30.34) and then calculating the net gain for both ($10.14 from leaving the tree, $8.17 from cutting it).

We could also make this assessment in terms of present values. Now we would state the steps in our financial maturity analysis as:

1. Find the discounted return (present net worth) from harvesting the sugar maple tree in each possible year.
2. Harvest the tree in the year that gives the highest present net worth.

Mathematically, this rule can be expressed as

$$\text{PNW} = \max_n \left\{ \frac{\text{SV}_n}{(1 + i)^n} \right\} \tag{13-2}$$

where

SV_n = stumpage value of sugar maple tree in year n
i = interest rate
n = years in which sugar maple tree could be harvested

and the notation $\max_n$ means: find the value of n (year) that maximizes the term in brackets. In this example, the components of the decision rule can be calculated for two decision choices: (1) *cut* immediately or (2) *leave* the tree to grow 8 additional years.

Item	Decision: Harvest year:	Cut $n = 0$	Leave $n = 8$
1. SV_n		\$22.17	\$32.31
2. $\dfrac{\text{SV}_n}{(1 + i)^n}$		22.17	23.61

Since the present net worth from keeping the tree for 8 more years (\$23.61) is greater than the present net worth from cutting it now (\$22.17), we should keep the tree for 8 more years—the same result as was found when future values were used.

You show your boss this analysis. He thinks you are on the right track, but points out that by holding the tree for 8 years, you will be delaying the start of a new tree or trees by 8 years. He wants you to consider this "cost" in your calculations. In other words, he wants you to include the opportunity cost of delaying the start of future generations by holding the tree for 8 years. Now you must consider the cost of delaying future generations on the site—the cost of the land—in addition to the cost of capital.

Considering Future Generations of Trees in the Cut/Leave Decision

Assume that we are again looking at your sugar maple tree in isolation from surrounding trees, but now you are concerned with maximizing the present net worth from the use of both the capital in the tree and the land on which it sits. Your best alternative use of the capital is to place it in an investment that will return 4 percent. Your best alternative use of the land is to grow another tree (trees) on it.

Suppose that when you cut the sugar maple, a sugar maple seedling will appear in its place, which your boss will consider cutting at age 30, 40, 50, or 60. We wish to choose the harvest age (rotation for the individual tree) that will produce the most income over time. It would be difficult to compare the discounted income produced for a single rotation since we are comparing choices involving different lengths of time. To be consistent in your analysis, you would need to compare the return from two 30-year rotations to that of one 60-year rotation. Therefore you will compare your choices in terms of a continuing series of rotations of the designated length under the assumption that whenever you cut the sugar maple occupying the space, another sugar maple seedling will appear to take its place.

As we learned in Chap. 7, the present value of this continuing series of rotations can be calculated as [Eq. (7-17)]:

$$\mathrm{SEV} = \frac{a}{(1 + i)^w - 1}$$

where a = the net periodic return that occurs every w years starting in year w

w = period length

i = interest rate

Here the period length is one rotation and it will be denoted by the symbol R. To show the effect of the interest rate, we have done this calculation at two rates (4 and 6 percent):

Tree harvest age R (rotation age)	Stumpage value per tree at harvest	$i = 6\%$		$i = 4\%$	
		$(1 + 0.06)^R - 1$	SEV	$(1 + 0.04)^R - 1$	SEV
30	$30	4.74	$6.33*	2.24	$13.37
40	55	9.29	5.92	3.80	14.47*
50	70	17.42	4.02	6.11	11.46
60	80	31.99	2.50	9.52	8.40

* Highest SEV.

Not surprisingly, the higher the interest rate, the shorter the harvest age with the highest SEV. The highest SEV occurs at 40 years for i = 4 percent and at 30 years for i = 6 percent.

These SEV values give the present value of infinite series of individual tree rotations of the designated length. We wish to discover the cost of delaying the start of this series for 8 years.* To do this, we find the value

* Adapted from Teeguarden (1968).

increase from investing the SEV for 8 years at the guiding rate of return:

$$\text{Value increase} = V_0(1 + i)^n - V_0$$

where V_0 = value now (SEV)
i = interest rate

At $i = 0.06$, delaying future rotations by 8 years costs

$$6.33(1.06)^8 - 6.33 = \$3.76$$

At $i = 0.04$, delaying future rotations by 8 years costs

$$14.47(1.04)^8 - 14.47 = \$5.33$$

To decide whether to cut the tree and turn its growing space over to a seedling, we must compare the benefit from leaving it (value growth of \$32.31 − \$22.17 = \$10.14) with the total opportunity cost of leaving it, namely, (1) the stock holding cost (lost income from not investing tree capital in its best alternative investment), and (2) the land holding cost (lost income from not turning land over to its best alternative investment):

Return from keeping tree 8 years	i	Stock holding cost	Land holding cost	Total cost	Net gain from holding
\$10.14	0.06	\$13.17	\$3.76	\$16.93	\$−6.79
10.14	0.04	8.17	5.33	13.50	−3.36

Because under both interest rates the value growth is less than the sum of the stock holding cost plus the land holding cost from keeping the tree for 8 years, we should cut the tree, releasing the capital to its best alternative investment (at 4 or 6 percent) and releasing the land to its best alternative investment (the next generation). Including the land holding cost in our calculation causes us to reverse the decision on whether to keep the tree.

Our revised financial maturity guide, based on future values, for deciding whether to hold the tree for some time into the future or cut it can be summarized as follows:

1. Keep the tree if the value growth is greater than the stock holding cost plus the land holding cost.
2. Cut the tree if the value growth is less than the stock holding cost plus the land holding cost.

Again, this decision rule can be portrayed in another form that expresses it in terms of present values:

1. Find the stumpage value of the subject tree in each candidate year of harvest.
2. Find the highest SEV (MSEV) from future rotations of trees.
3. Add the discounted return from harvesting the subject tree in each possible year to the SEV discounted from that same year.
4. Harvest the tree in the year that gives the highest total present net worth.

Mathematically this rule can be expressed as

$$\text{PNW} = \max_n \left[\frac{\text{SV}_n}{(1+i)^n} + \frac{\text{MSEV}}{(1+i)^n} \right] \tag{13-3}$$

where SV_n = stumpage value of the subject tree in year n
MSEV = maximum soil expectation value across all possible rotation lengths
i = interest rate
n = year in which to harvest the sugar maple tree

and the notation $\max_n$ means: find the value of n (year) that maximizes the term in brackets.

In this formula and throughout this chapter we assume that the return from the infinite series of rotations after the existing stand is cut (SEV) is not sensitive to the period in which that initial harvest takes place. That assumption does not always hold. Including a constant rate of price increase, as an example, will make SEV a function of the period in which it begins. In such a case, MSEV must be calculated for each period (n) in which the series of future rotations might begin and those values used in the formula above (see the next section for more discussion).

In our example, the components of the decision rule expressed in (13-3) can be calculated as follows:

	Guiding rate:	$i = 0.04$		$i = 0.06$	
	Decision:	Cut	Leave	Cut	Leave
Item	Harvest year:	$n = 0$	$n = 8$	$n = 0$	$n = 8$
1. SV_n		\$22.17	\$32.31	\$22.17	\$32.31
2. $\frac{\text{SV}_n}{(1+i)^n}$		22.17	23.61	22.17	20.27
3. MSEV		14.47	14.47	6.33	6.33
4. $\frac{\text{MSEV}}{(1+i)^n}$		14.47	10.57	6.33	3.97
PNW (rows 2 + 4)		36.64	34.18	28.50	24.24

Under both interest rates, the highest present net worth is associated with cutting the sugar maple now. As we discovered earlier, the value growth just exceeds the stock holding cost at 4 percent. This result shows up here in that the discounted value at 4 percent of cutting the tree in the future is slightly greater than the value of cutting it now (\$23.61 > \$22.17). Delaying the future series of rotations by 8 years at 4 percent costs \$14.47 − \$10.57 = \$3.90. This is the discounted value of the land holding cost. The land holding cost was given previously as a future value of \$5.33. Discounting that amount by 4 percent gives $5.33/(1.04)^8$ = \$3.90—the same discounted cost as calculated above. Therefore the two approaches to the decision are equivalent and give the same result.

Until now we have assumed that the next generation will start when your sugar maple is cut. Therefore leaving the tree for 8 years involves a cost of delaying future generations by a like period of time. Sometimes young trees can start growing whether or not a mature tree occupies the site, and at other times the mature tree is needed as a seed source or as a shelter to protect the seedling during its early years.

The land holding cost changes as a function of these three possible effects of the existing tree on the next generation (delays the next generations, neutral toward the next generation, assists the next generation) and may change the harvest decision (Table 13-1). When leaving the existing tree delays the next generation, which is the case we have been discussing with a delay of 8 years, the tree should be replaced under both 4 and 6 percent interest rates. When leaving the existing tree is neutral toward the next generation, such as when advanced regeneration is easily established, only the stock holding cost is relevant, and the tree should be left under a 4 percent interest rate and cut under a 6 percent interest rate. When leaving the existing tree assists the next generation, the land holding "cost" becomes a benefit. Since we already determined that the tree should be left at 4 percent when the tree is neutral toward future generations, it should definitely be left when it assists them. At 6 percent, we would have to do further analysis to see whether the gain from speeding up future generations would offset the loss caused by the tree's value growth being less than the capital growth achieved by cutting the tree and taking the capital elsewhere.

Including a Real Price Rise

The value of timber in real terms (after deducting inflation) has risen at a compound rate of somewhat more than 1 percent per year for many decades. We can consider this price rise in our decision of when to cut the sugar maple tree, but our formulas must be somewhat altered.

Assuming that this price rise applies to net stumpage value, we can alter the value of the tree in year *n* by multiplying the stumpage value in

Table 13-1 Future value analysis of the harvest decision for a sugar maple tree under two interest rates and three possible types of impact of the existing tree on the next generation[a]

Return from leaving tree 8 more years[b]	i	Stock holding cost[c]	Land holding cost[d]			Total cost[e]			Net gain from keeping tree[f]			Decision[g]		
			D	N	A	D	N	A	D	N	A	D	N	A
$10.14	0.06	$13.17	$3.76	0	—?	$16.93	$13.17	<$13.17	−$6.79	−$3.03	>−$3.03	C	C	?
10.14	0.04	8.17	5.33	0	—?	13.50	8.17	<8.17	−$3.36	1.97	>1.97	C	L	L

[a] D—delays; N—neutral; A—assists.

[b] Value in 8 years − value now = 32.31 − 22.17 = $10.14.

[c] Stock holding cost = value now × $(1 + i)^8$ − value now. When $i = 0.06$, it equals $22.17 \times (1 + 0.06)^8 - 22.17 = \$13.17$.

[d] Land holding cost = $SEV(1 + i)^n - SEV$. When $i = 0.06$, it equals $6.33(1.06)^8 - 6.33 = \$3.76$.

[e] Total cost = stockholding cost + land holding cost.

[f] Net gain = return from leaving tree for 8 years − total cost.

[g] Decision rule: cut the tree if value growth < stockholding cost + land holding cost. C—cut; L—leave.

year n by $(1 + m)^n$, where m is the rate of real price rise:

$$SV'_n = SV_n(1 + m)^n$$

where SV_n = stumpage value of tree in year n with no real price rise
SV'_n = stumpage value of tree in year n with a real price rise

The present net worth of cutting the tree in year n when r = the guiding rate expressed in real terms is now:

$$PNW_n = \frac{SV_n(1 + m)^n}{(1 + r)^n} = \frac{SV_n}{\left(\frac{1 + r}{1 + m}\right)^n}$$

Assuming that the real price rise applies to the net return a in the soil expectation value calculation [Eq. (7-17)]:

$$SEV = \frac{a}{\left(\frac{1 + r}{1 + m}\right)^w - 1}$$

Adjusting decision rule (13-3) for the real price rise,

$$PNW = \max_n \left(\frac{SV_n}{\left(\frac{1 + r}{1 + m}\right)^n} + \frac{MSEV}{\left(\frac{1 + r}{1 + m}\right)^n} \right) \tag{13-3a}$$

If the net return a is composed of costs a_2 and revenues a_1 that appreciate at different real rates, then the equation must be broken into its components for calculation of SEV. If $a = a_1 - a_2$, and a_1 appreciates at a compound rate of m_1 and a_2 appreciates at a compound rate of m_2, then

$$SEV = \frac{a_1}{\left(\frac{1 + r}{1 + m_1}\right)^w - 1} - \frac{a_2}{\left(\frac{1 + r}{1 + m_2}\right)^w - 1}$$

MSEV in (13-3a) has now become $MSEV_n$, as a different MSEV must be calculated for each year n. Each component (a_1, a_2) must be compounded for n years at its real rate of price increase (m_1, m_2) in forming SEV. The second component of (13-3a) becomes $MSEV_n/(1 + r)^n$ because we have now included the real rate of price increase in $MSEV_n$.

Taking Account of Adjacent Trees

Usually trees grow close enough together to compete for sunlight, water, and nutrients. In such a case our decision rules must recognize that we do not necessarily face an all or nothing decision, but that a strategy of removing some trees and holding others may be the choice that maximizes present net worth. This decision arises in any kind of thinning or selection harvest.

Suppose that two sugar maple trees grow in a group: tree 1 (our original candidate for removal) and tree 2. Further suppose that we plan for one future tree to occupy the space now used by the group, and that tree will not grow until those two trees are removed. We now have four choices: (1) leave both trees for 8 years, (2) remove tree 1 and leave tree 2, (3) remove tree 2 and leave tree 1, and (4) remove both trees. The amount of stumpage value increases under the different possibilities that retain at least one tree, and the associated stock holding costs and land holding costs (for a 4 percent interest rate) are as follows:

Item	Leave both: Tree 1	Leave both: Tree 2	Leave tree 2	Leave tree 1
1. Value now	$22.17	$34.00	$34.00	$22.17
2. Value in 8 years	32.31	41.50	51.00	44.50
3. Value growth	10.14	7.50	17.00	22.33
	17.64			
4. Stock holding cost, = value now × $(1.04)^8$ − value now	8.17	12.53	12.53	8.17
	20.70			
5. Land holding cost	5.33		5.33	5.33
6. Total cost (4 + 5)	$26.03		$17.86	$13.50
Net gain (3 − 6)	−$8.39		−$0.86	$8.83

Leaving both trees would return a negative net gain, so one or both trees should be removed. Leaving tree 2 would return −$0.86 and leaving tree 1 would return $8.83, so tree 1 should be left and tree 2 removed to maximize present net worth. If all three choices forecast a negative net gain, both trees should be cut.

We can now summarize our financial maturity guide based on future values as follows:

> Leave the combination of trees (growing stock) in the group that gives the largest excess of value growth over stock holding cost and land holding cost as long as this net return is positive. If no combination of trees gives a positive net return, cut them all.

Again, this decision rule can be expressed in terms of present values:

> Leave the combination of trees that give the highest present net worth considering both the return from existing trees and the return from the future generations that come after them.

Mathematically, this rule can be expressed as

$$\text{PNW} = \max_{mn}\left[\frac{\text{SV}_{mn}}{(1+i)^n} + \frac{\text{MSEV}}{(1+i)^n}\right] \tag{13-4}$$

where SV_{mn} is the stumpage value of tree combination m in year n and the other symbols are as before. The notation $\max_{mn}$ means: find the combination of trees m and year n that maximizes the term in brackets.

In our case we have four combinations and two periods to search over to locate the maximum present net worth.

	Leave both		Leave tree 2		Leave tree 1		Leave none	
Item	Tree 1 $n = 8$	Tree 2 $n = 8$	Tree 1 $n = 0$	Tree 2 $n = 8$	Tree 1 $n = 8$	Tree 2 $n = 0$	Tree 1 $n = 0$	Tree 2 $n = 0$
1. SV_n	\$32.31	\$41.50	\$22.17	\$51.00	\$44.50	\$34.00	\$22.17	\$34.00
2. $\dfrac{\text{SV}_n}{(1+i)^n}$	23.61	30.32	22.17	37.27	32.52	34.00	22.17	34.00
	53.93		59.44		66.52		56.17	
3. MSEV	14.47		14.47		14.47		14.47	
4. $\dfrac{\text{MSEV}}{(1+i)^n}$	10.57		10.57		10.57		14.47	
PNW (rows 2 + 4)	64.50		70.01		77.09		70.64	

As before, the best approach is to leave tree 1 and cut tree 2. In general, then, the decision of what trees to keep and what trees to remove in uneven-aged management requires the financial evaluation of each tree or group as it is encountered to assess whether it should be kept or removed. Doing such a detailed calculation as shown here on each tree or group could slow timber marking to a crawl. Luckily, the world is somewhat simpler than that. As you move through the forest, many trees should obviously be kept or removed. They are young and fast growing, or they are decrepit, diseased, or lightning-struck. In addition, many of the groups encountered with two, three, or four trees growing together can be anticipated, classified into categories, and analyzed in the office. Still, there is no getting around the fact that the management of trees and land as financial assets in uneven-aged management requires lots of skill and judgment. But this is why we need foresters.

Considering the Entire Stand

Up until now we have considered individual trees and groups of trees in the uneven-aged stand without worrying about the stand itself. We have not considered what our objective-maximizing cutting policies will mean for the stand in the long run—what type of stand will be produced and what revenue and products will be provided.

Foresters traditionally have thought in terms of target stands and target forests, so it is not surprising that attention has been paid to finding the target uneven-aged stand that maximizes financial return—to finding the level and composition of growing stock that maximizes present net worth over an infinite series of cutting cycles. Growth on the stock will be cut at the end of each cutting cycle, with the stand being in a balanced (equilibrium) condition so that this growth will be the same in each cutting cycle.

Given that we have already netted all production costs but interest out of the stumpage price, Duerr (1960) used future values to find the growing stock level that maximizes net revenue for a given cutting cycle length in one of two ways:

1. Find the level of growing stock that maximizes the quantity,

$$\begin{pmatrix}\text{value growth over}\\ \text{cutting cycle}\end{pmatrix} - \begin{pmatrix}\text{interest charges}\\ \text{on growing-stock}\\ \text{investment to achieve}\\ \text{that value growth}\end{pmatrix}$$

2. Find the level of growing stock that equates the marginal value growth on the last increment of growing stock with the guiding rate of return.

Duerr used both approaches to find the best level of growing stock in a loblolly pine and shortleaf pine selection forest in southern Arkansas (Table 13-2) with a 1-year cutting cycle and a 4 percent interest rate. He looked at levels of growing stock from 5,000 to 11,000 bd ft per acre in 1,000-ft steps. For each level he estimated the annual growth that could be expected. To simplify the problem, he assumed that this growth gives the highest value of output procurable from each level of growing stock. Thus he ignored questions about whether the growing-stock level has the best species mix and diameter distribution to meet his objectives—a point to which we will come back in the next example.

Table 13-2*a* finds the level of growing stock that maximizes the difference between value growth over the cutting cycle and the interest charges on the growing stock investment to achieve that growth (the cost of capital). After valuing the growing stock at the beginning of the cutting cycle (column 3), valuing the growth (column 4), and estimating the interest charges on the growing stock for the 1-year cutting cycle (column 5), we can find the level of growing stock that gives the maximum difference between the value growth and the interest charges (column 6). Here that difference is maximized at 9,000 bd ft.

Table 13-2 Best stocking calculation of a loblolly pine and shortleaf pine selection forest on good site in southeastern Arkansas based on annual per-acre data, using a 4 percent interest rate

(a) Based on maximizing the difference between value growth and interest charges

Initial growing-stock level (bd ft) (1)	Annual growth (bd ft) (2)	Value of initial growing stock ($) (3)	Value of growth ($) (4)	Interest charges on growing stock ($) (5)	Net revenue ($) (6) = (4) − (5)
5,000	468	210.00	21.46	8.40	13.06
6,000	534	257.30	25.00	10.29	14.71
7,000	592	306.50	28.20	12.26	15.94
8,000	638	357.40	31.06	14.30	16.76
9,000	676	410.10	33.54	16.40	17.14*
10,000	704	464.70	35.62	18.59	17.03
11,000	722	521.00	37.26	20.84	16.42

Source: Duerr and Bond (1952), Duerr (1960).

* Highest value.

(b) Based on equating the marginal value growth percentage with the alternative rate of return of 4 percent

Initial growing-stock level (bd ft) (1)	Annual growth (bd ft) (2)	Value of initial growing stock ($) (3)	Marginal value of growing stock ($) (4)	Value of growth ($) (5)	Marginal value of growth ($) (6)	Marginal value growth (%) (7)*
5,000	468	210.00		21.46		
			47.30		3.54	7.5
6,000	534	257.30		25.00		
			49.20		3.20	6.5
7,000	592	306.50		28.20		
			50.90		2.86	5.6
8,000	638	357.40		31.06		
			52.70		2.48	4.7
9,000	676	410.10		33.54		
			54.60		2.08	3.8†
10,000	704	464.70		35.62		
			56.30		1.64	2.9
11,000	722	521.00		37.26		

* Calculated using Eq. (7-3*a*) and expressed as a percentage as

$$\left(\sqrt[c]{\frac{(4)+(6)}{(4)}}-1\right)\times 100$$

where c is the length of the cutting cycle in years. Here the cutting cycle is 1 year and this calculation is

$$\left(\sqrt[1]{\frac{(4)+(6)}{(4)}}-1\right)\times 100=\left(\frac{(4)+(6)}{(4)}-1\right)\times 100$$

† Highest value.

Table 13-2*b* finds the level of growing stock that equates the marginal value growth on the last increment of growing stock with the alternative rate of return. Given the value of the growing stock at each level at the start of the cutting cycle (column 3), we can calculate the marginal value of each growing-stock increment (column 4). As an example, the marginal value of the 1,000 bd ft added in increasing the growing-stock level from 5,000 to 6,000 bd ft is $257.30 − $210.00 = $47.30. Given the value of the growth at each level of growing stock (column 5), we can similarly calculate the marginal value growth on each growing-stock increment (column 6). Combining the marginal value growth with the marginal value of the associated growing-stock increment, we can find the marginal value growth percentage (column 7) and compare it to the guiding rate of return. The investment of $47.30 to increase the growing stock from 5,000 to 6,000 bd ft increased the value of growth from $21.46 to $25.00, a net increase of $3.54. Over one year, $3.54 is a 7.5 percent return on the $47.30 investment. Here the marginal value growth percentage comes closest to the alternative rate of return at a growing stock level of 9,000 bd ft.

As Duerr (1960) showed, the decision rules can also be presented in graphical form (Fig. 13-1). These graphs suggest that the optimum growing-stock level actually is 9,200 bd ft.

Adams and Ek (1974) generalized the approach developed by Duerr and Bond to permit the determination of the objective-maximizing diameter distribution simultaneously with the objective-maximizing level of growing stock. Using Ek's growth model for northern hardwoods (see Chap. 5), they determined the optimum sustainable diameter distribution for a given level of stocking, and by varying the stocking level, developed the information needed to make the optimum stocking decision.

Basically they found a starting distribution for each level of stocking (basal area) which, when grown for 5 years, would provide the maximum value of trees that could be harvested while returning the stand to this starting distribution. Mathematically, their problem can be written as

$$\max_{\{X_D(t)\}} \sum_{D=1}^{N+1} V_D \Delta_D \qquad (13\text{-}5)$$

subject to

$$\Delta_D \geq 0, \qquad D = 1, \ldots, N+1 \qquad (13\text{-}6)$$

$$\sum_{D=1}^{N} Y_D X_D(t) = L \qquad (13\text{-}7)$$

$$X_D(t) \geq 0, \qquad D = 1, \ldots, N \qquad (13\text{-}8)$$

where $X_D(t)$ = number of trees in diameter class D at time t (beginning of cutting cycle)

N = number of diameter classes at time t

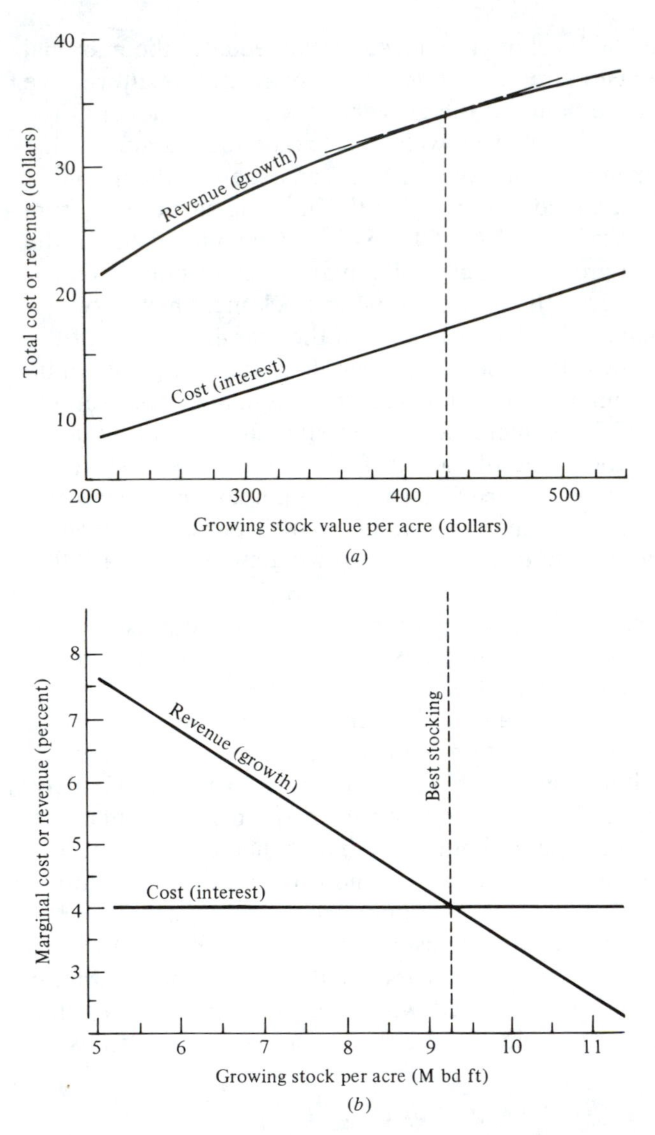

Figure 13-1 Best stocking of a loblolly pine and shortleaf pine selection forest. *(From Duerr and Bond, 1952, Duerr, 1960.)*

V_D = stumpage value per tree in diameter class D
Δ_D = change in number of trees in diameter class D over cutting cycle
Y_D = average basal area per tree in diameter class D
L = growing-stock level (basal area) whose optimum diameter distribution is being determined

The objective function Eq. (13-5), is written with all measured diameter classes considered merchantable. The first set of $N + 1$ constraints (13-6) is a restatement of the sustainability conditions for a selection harvest treatment over a cutting cycle:

$$\Delta_D = X_D(t + 1) - X_D(t) \geq 0, \qquad \text{for all } D$$

Constraint (13-7) requires that the sum of the basal areas in all classes at the start of the cutting cycle must equal some fixed level L. Finally, constraints (13-8) disallow negative numbers of trees in the initial distribution.

Expressions (13-5) and (13-6) can be written as functions of the initial numbers of trees in each class $X_D(t)$ if appropriate substitutions are made in the equations from Ek's growth model discussed in Chap. 5. The resulting expressions are nonlinear in the $X_D(t)$. The problem posed then is to maximize an objective function composed of the sum of $N + 1$ nonlinear terms subject to a set of linear and nonlinear constraints. Optimization results for site $S = 60$ are shown in Table 13-3 for basal areas between 60 and 120 ft^2.

The marginal value growth percentage associated with each increment of growing stock is shown at the bottom of Table 13-3. These rather low percentages increase to a peak at 90 ft^2 of basal area and then decline. With these low marginal value growth rates it is doubtful that many owners would choose a density above 60 ft^2 of basal area. We might actually wonder if the stand should be clearcut.

Both Duerr and Bond (1952) and Adams and Ek (1974) developed optimum levels of growing stock for an uneven-aged stand. In addition, Adams and Ek simultaneously developed the optimum diameter distribution. Usually, though, existing stands will differ in structure from the desired equilibrium conditions of stocking and diameter distribution. Assuming that the manager wants to attain these equilibrium conditions over some period of time (the transition period), a methodology is needed to determine the appropriate cutting schedule during this transition.

Using an expansion of the nonlinear programming technique that they employed to find the optimum equilibrium conditions, Adams and Ek found the cutting schedule that maximizes the present value of the yields during the conversion period, under a 5 percent interest rate, subject to achieving a specified equilibrium level of stocking and diameter distribution 10 years hence. Results of the optimization are given in Table 13-4 for the case when the terminal distribution corresponds to the 80-ft^2 basal area from Table 13-3.

The optimum cutting schedule by diameter class is given in columns 2, 5, and 8. The present worth of material cut during the 10-year conversion period totaled \$99.23/acre. Adding the present value of subsequent equilibrium cuttings at 5-year intervals, the total present net worth of the conversion and equilibrium that follows would be \$188.98 per acre. This

Table 13-3 Optimum diameter distributions, growth, and growing-stock data from value growth maximization subject to basal area constraint*

	Number of trees per acre subject to stocking level constraint per acre							
Diameter (in.)	60 ft²	70 ft²	80 ft²	90 ft²	100 ft²	110 ft²	120 ft²	Value per tree
6	159.1	119.4	103.6	96.7	91.8	89.3	76.3	$0.04
8	25.3	36.6	43.1	50.0	55.2	59.8	54.5	0.55
10	12.3	21.0	27.6	33.1	38.7	44.6	41.2	1.15
12	10.0	16.8	22.1	26.4	30.6	35.2	32.4	2.06
14	2.9	5.0	6.7	8.1	9.6	10.7	26.2	3.37
16	1.0	1.7	2.3	2.8	3.3	3.7	6.2	4.89
18	0.5	0.8	1.1	1.3	1.6	1.7	0.9	6.60
20+								8.52
1. Value of initial growing stock ($)	72.98	114.10	146.20	173.38	200.62	225.63	271.70	
2. Marginal value of growing stock ($)	41.12	32.10	27.18	27.24	25.01	46.07		
3. 5-year value growth ($)	35.00	37.78	40.39	42.76	44.88	46.76	47.91	
4. Marginal value of growth ($)	2.78	2.61	2.37	2.12	1.88	1.15		
5. Marginal value growth (%)†	1.3	1.6	1.7	1.5	1.5	0.5		

Source: Adams and Ek (1974).

* Adams later showed (1976) that these results could be slightly improved by maximizing value growth subject to achieving a specified economic amount of growing stock ($) rather than maximizing value growth subject to achieving a specified physical amount of growing stock (basal area) as done here.

† Calculated as $[\sqrt[5]{(\text{row 2} + \text{row 4})/\text{row 2}} - 1] \times 100$. As an example, the marginal value growth percentage between a basal area of 60 and a basal area of 70 is $[\sqrt[5]{(41.12 + 2.78)/41.12} - 1] \times 100 = (\sqrt[5]{1.0676} - 1) \times 100 = 1.3$ percent.

number can be calculated using a modification of Eq. (7-18):

$$\text{SEV(U)} = \frac{a}{(1 + i)^w - 1}$$

where w = cutting cycle length in years

a = equal net return every w years, the first return w years from now

Here we have a series of entries during the conversion period, and the PNW equation is

$$\text{PNW} = \text{PNW}_c + \frac{\text{SEV(U)}}{(1 + i)^c} = \text{PNW}_c + \left[\frac{a}{(1 + i)^w - 1}\frac{1}{(1 + i)^c}\right]$$

Table 13-4 Illustrative example of optimal conversion cutting schedule with an objective of maximizing present worth of conversion period returns. Conversion period is 10 years in length with cutting allowed at years 0, 5, and 10 to reach desired terminal distribution in column (9).

Diameter (in.)	(1) Initial (starting) distribution (trees/acre)	(2) Cut at year 0 (trees/acre)	(3) Residual at year 0 (trees/acre)	(4) Distribution at year 5 before harvest (trees/acre)	(5) Cut at year 5 (trees/acre)	(6) Residual at year 5 (trees/acre)	(7) Distribution at year 10 before harvest (trees/acre)	(8) Cut at year 10 (trees/acre)	(9) Terminal (desired) distribution (trees/acre)	(10) Total PNW of conversion cuts ($/acre)
6	140.00	0.00	140.00	138.46	0.00	138.46	135.41	31.80	103.61	
8	50.00	0.00	50.00	63.69	37.87	25.82	46.49	3.40	43.09	
10	28.00	0.00	28.00	31.37	0.00	31.37	27.60	0.00	27.60	
12	23.00	3.81	19.19	20.37	0.00	20.37	22.11	0.00	22.11	
14	7.00	3.09	3.91	8.97	5.87	3.10	9.01	2.34	6.67	
16	6.00	6.00	0.00	1.90	0.01	1.89	2.29	0.00	2.29	
18	2.00	2.00	0.00	0.00	0.00	0.00	1.07	0.00	1.07	
Present value of conversion cutting ($/acre)		60.81			31.76			6.66		99.23
Present value from immediate conversion ($/acre)		32.94			31.64			24.79		89.37

Source: Adams and Ek (1974).
PNW of cutting during conversion, $99.23/acre.
SEV of all subsequent cuts, $\$40.39/(1.05^5 - 1)(1.05)^{10} = \$89.75$/acre.
Total PNW of management program, $188.98/acre.

where PNW_c = present net worth of the harvest during the conversion period
c = length of the conversion period

Other symbols as before.

For a target basal area of 80 ft^2, the present net worth here is

$$PNW = 99.23 + \left[\frac{40.39}{(1 + 0.05)^5 - 1}\frac{1}{(1.05)^{10}}\right] = 188.98$$

As Adams and Ek note, it is possible but not desirable to convert the stand immediately to the desired terminal condition. As indicated at the bottom of Table 13-4, immediate conversion increases the present net worth of the last cutting in the conversion period but reduces the value of the initial cut. The result is a net reduction of $9.86/acre in the present net worth of conversion period revenues.

Haight et al. (1985) recently generalized the work of Adams and Ek such that they could find the optimal diameter distribution of the equilibrium level simultaneously with the optimal diameter distribution of the transition period. As expected, the more general formulation found a higher present net worth and a somewhat different equilibrium diameter distribution.

As elegant and intellectually satisfying as these uneven-aged stand models of Duerr and Bond, Adams and Ek, and Haight et al. are, only limited application has been made of them as of yet. Few people have the underlying growth models for their stands that the analysis requires. In addition, the Adams and Ek and Haight et al. approaches require access to nonlinear programming techniques that are not readily available. The techniques of the previous section, which involve inspection of each tree and group as they are encountered and a few calculations, perhaps fit the needs of managers who face an irregular and diverse forested area, have chiefly financial objectives, and do not have access to the models and techniques needed for the more sophisticated approaches.

EVEN-AGED MANAGEMENT

Unlike uneven-aged stands, which go on forever, even-aged stands have a definite beginning and end in time. The trees in even-aged stands regenerate naturally or are planted at about the same time. The stand then grows for a number of years, perhaps undergoes stocking control, fertilization, and one or more commercial thinnings along the way, and finally has all stems removed in one or more regeneration cuts.

The Rotation Decision

Our investment decisions for an even-aged stand involve what cultural treatments to undertake, when and at what intensity to undertake them, and when to remove the growing stock that is created naturally or through such activities. Perhaps it is best to start our analysis with a single acre of bare ground in the Douglas-fir region that we estimate as having a 50-year site index of 105. Given that we plant the acre tomorrow, we wish to find out how long to grow the stand before clearcutting it.

Using DFSIM, a coastal Douglas-fir whole stand growth and yield simulator (Curtis et al., 1981) for site index 105 we find that the stand achieves maximum average growth at age 80, with a mean annual increment of 146 ft^3/yr (Table 13-5).

Prices and costs are adapted from a recent publication on Douglas-fir by Tarrant et al. (1983). The costs are as follows:

Site preparation	$120
Planting	100
Annual cost of management	3

We make two different stumpage assumptions to illustrate the effect of a price premium for larger tree sizes on the rotation decision:

1. An average net stumpage price over all tree sizes of $0.75/$ft^3$
2. A net stumpage price as a function of tree size:

Volume per tree (ft^3)	Price ($/$ft^3$)
5–10	0.20
10–20	0.50
20–30	0.75
30–40	0.95
40–50	1.15
50–60	1.35
60+	1.50

We use a guiding rate of return of 4 percent.

To calculate the present net worth from one rotation PNW_1, we take either of two approaches. First we could develop a cash flow schedule for all costs and revenues over one rotation. Then we could discount the revenues and costs to year 0 using the formulas of Chap. 7:

$$PNW_1 = \frac{V_R P_R}{(1+i)^R} - \frac{T[(1+i)^R - 1]}{i(1+i)^R} - G \tag{13-9}$$

Table 13-5 Yield/acre and financial return/acre from a stand of Douglas-fir for site index 105 (50-year breast height) using a utilization standard of 7-in. DBH to a 4-in. top

						Financial return with constant stumpage price				Financial return with stumpage price a function of tree size			
Age (years)	Number of trees	Net yield (ft^3)	Volume per tree (ft^3)	Quadratic mean DBH (in.)	Mean annual increment (ft^3)	Value ($\$/ft^3$)	Net revenue at harvest (volume × value) ($)	PNW_1 ($)	SEV ($)	Value ($\$/ft^3$)	Net revenue at harvest (volume × value) ($)	PNW_1 ($)	SEV ($)
30	130	1,140	8.8	9.0	38	0.75	855	−8	−12	0.20	228	−202	−292
40	240	3,718	15.1	10.2	93	0.75	2,789	302	381	0.50	1,859	108	136
50	258	6,093	23.6	11.6	122	0.75	4,570	359*	417*	0.75	4,570	359	417
60	244	8,212	33.6	13.1	137	0.75	6,159	298	329	0.95	7,801	454*	501*
70	230	10,080	43.8	14.3	144	0.75	7,560	195	209	1.15	11,592	454*	485
80	203	11,665	57.5	15.8	146*	0.75	8,749	88	92	1.35	15,748	391	409
90	181	13,011	71.9	17.2	145	0.75	9,758	−7	−7	1.50	19,517	279	288
100	163	14,060	86.2	18.5	141	0.75	10,545	−85	−86	1.50	21,090	124	127

* Highest value.

where V_R = cubic foot volume per acre harvested at age R
P_R = price per cubic foot for timber harvested at age R
T = annual management cost per acre
G = regeneration cost
i = guiding interest rate per year
R = rotation age in years

As an example, consider age 60 under the constant price assumption,

$$\text{PNW}_1 = \frac{8{,}212 \times 0.75}{(1 + 0.04)^{60}} - \frac{3[(1 + 0.04)^{60} - 1]}{0.04(1 + 0.04)^{60}} - 220$$
$$= 585 - 67 - 220 = \$298$$

Second, we could compound all costs (and revenues) to the end of the rotation, form a net return N_R, and discount that to year 0,

$$N_R = V_R P_R - \frac{T[(1 + i)^R - 1]}{i} - G(1 + i)^R \tag{13-10}$$

$$\text{PNW}_1 = \frac{N_R}{(1 + i)^R} \tag{13-11}$$

Again, considering age 60 under the constant price assumption,

$$N_R = 8{,}212 \times 0.75 - \frac{3[(1 + 0.04)^{60} - 1]}{0.04} - 220(1 + 0.04)^{60}$$
$$= 6{,}159 - 714 - 2{,}314 = \$3{,}131$$
$$\text{PNW}_1 = \frac{3{,}131}{(1 + 0.04)^{60}} = \$298$$

In our example the present net worth for a single rotation PNW_1 peaks at age 50 under the constant price assumption and at age 60 or 70 under the price as a function of tree size assumption. But comparing present net worths over one rotation for different rotation ages ignores the problem that investments are being compared that cover different lengths of time. Difficulties with finding a common year to use as a base have led foresters to estimate the discounted return from each rotation in terms of an infinite number of rotations of the designated length.

As shown in Chap. 7 and used in the single tree selection example earlier in this chapter, the present net worth (soil expectation value) of an infinite cycle of repeating rotations of length R on our acre of bare ground can be estimated as

$$\text{SEV} = \frac{N_R}{(1 + i)^R - 1} = \text{PNW}_\infty \tag{13-12}$$

For age 60, this return is

$$\text{SEV} = \frac{3{,}131}{(1 + 0.04)^{60} - 1} = \$329$$

It is the sum of the discounted return from a series of 60-year rotations, each with a net return of $3,131, starting in year 60,

$$\text{SEV} = \frac{3{,}131}{(1 + 0.04)^{60}} + \frac{3{,}131}{(1 + 0.04)^{120}} + \frac{3{,}131}{(1 + 0.04)^{180}} + \cdots$$

$$= 298 + 28 + 3 + \cdots = \$329$$

Because SEV includes the first rotation plus the infinite series of rotations that come after it, it will always be larger in magnitude and of the same sign as the PNW_1 (except that it equals PNW_1, when $\text{PNW}_1 = 0$). By including the return from these later rotations, and thus implicitly measuring the cost of delaying future rotations, SEV tends to reach a maximum at shorter rotations when PNW_1 is positive. In our cases (Table 13-5), the maximum SEV occurs at the same age (50 years) as the maximum PNW_1 when stumpage price is not a function of tree size and at the earlier of the two possible maximum PNW_1 rotation ages (60 years) when stumpage price is a function of tree size.

To summarize our results, the rotation ages that give highest financial returns under a 4 percent interest rate are as follows:

	Constant price		Price as a function of diameter	
Item	One rotation	Infinite cycle of rotations	One rotation	Infinite cycle of rotations
Optimum rotation age	50	50	60 or 70	60
PNW/acre	359 (PNW_1)	417 (PNW_∞)	454 (PNW_1)	501 (PNW_∞)

In contrast, the physical growth-maximizing rotation occurs at age 80 with a MAI of 146 ft^3 per acre per year.

Including Nontimber Outputs and Values

Calish et al. (1978) expanded the analysis of economic rotations to include nontimber yields, such as wildlife, water flows, and aesthetics, in addition to timber yield. Basically they broadened the soil expectation formula to include the revenues accruing from the production of nontimber outputs as the stand ages.

To calculate the economic rotation for timber only, in which the only costs incurred were from stand establishment, Calish et al. defined the soil expectation formula (in our symbology) as

$$\mathrm{SEV}_T = \frac{V_R P_R - G(1+i)^R}{(1+i)^R - 1} \tag{13-13}$$

Because nontimber outputs typically occur annually as continuous flows, as opposed to timber yields that occur periodically, they chose a different form of the SEV equation to measure the value of nontimber outputs. With V_t^{NT} being the amount of a nontimber output in year t ($t = 1, 2, \ldots, R$) and P_t^{NT} being its dollar value per unit in year t, they calculated the soil expectation value attributable to a nontimber output (SEV_{NT}) over a timber rotation of R years as

$$\mathrm{SEV}_{NT} = \frac{\sum_{t=1}^{R} V_t^{NT} P_t^{NT} (1+i)^{R-t}}{(1+i)^R - 1} \tag{13-14}$$

To calculate the best joint rotation for timber and a nontimber yield, Eqs. (13-13) and (13-14) were combined, giving

$$\begin{aligned} \mathrm{SEV} &= \mathrm{SEV}_T + \mathrm{SEV}_{NT} \\ &= \frac{V_R P_R - G(1+i)^R + \sum_{t=1}^{R} V_t^{NT} P_t^{NT} (1+i)^{R-t}}{(1+i)^R - 1} \end{aligned} \tag{13-15}$$

Now, the best rotation is the year R that maximizes this joint soil expectation value. Since both time-dependent revenue functions $V_R P_R$ and $V_t^{NT} P_t^{NT}$ influence SEV, the best joint rotation may differ from the best timber-only rotation. To calculate SEV, yields and values for all outputs must be estimated.

Timber-only rotation McArdle and Meyer's (1930) normal yield table for Douglas-fir, site index 170 (site class II), was used as the basis for calculating a timber-only rotation, with yields expressed in thousand cubic feet per acre, reduced by a factor of 0.94 to account for deer browsing on early growth. Stumpage was valued from \$700 per thousand cubic feet for 10-in. trees to about \$875 for 21-in. trees. Regeneration cost was set at \$75 per acre. Under a 5 percent rate of interest, the best economic rotation for timber production is 36 years, resulting in an SEV_T of \$643 per acre. In contrast, if culmination of mean annual increment (MAI) is used, rotation length is 64 years (Fig. 13-2), with an SEV_T of \$329 per acre.

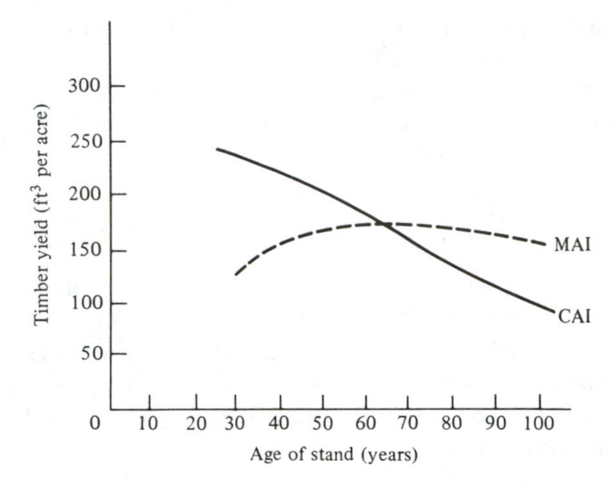

Figure 13-2 Timber yield in relation to rotation. CAI—current annual increment; MAI—mean annual increment. *(From McArdle and Meyer, 1930.)*

Nontimber yield functions Figure 13-3 displays the seven nontimber rotation-dependent yield functions developed by Calish et al. from published data. Just as with timber, the current annual increment (CAI) functions measure the marginal contribution of a stand to the total output as the stand ages. The authors use the current annual increments in Eqs. (13-14) and (13-15). Again, as with timber, the mean annual increment functions measure the average productivity over the life of the stand. The highest point on this function corresponds to the best rotation if the objective were to maximize production of the goods or service involved.

The mean annual increment functions have different shapes; maximum productive capacity is achieved at different rotation ages. For example, the deer function peaks at 25 years, while elk peaks at 35 to 40 years. The other functions are either continuously increasing or continuously decreasing throughout the range of ages, so the rotations that maximize the mean annual increments are either 0 or longer than 100 years. Faced with such functions, the manager cannot simultaneously maximize all outputs—some compromise is necessary.

As Calish et al. point out the wildlife functions are meaningful only in the context of a regulated forest with a mix of age classes in close proximity to each other. Consider elk. The carrying capacity in terms of forage production is greatest in 5-year-old stands, but elk require the protective cover provided by older stands. Therefore the productivity of any particular age class depends on the whole mix of age classes. In Fig. 13-3 the rotation where the current annual increment first exceeds zero is the

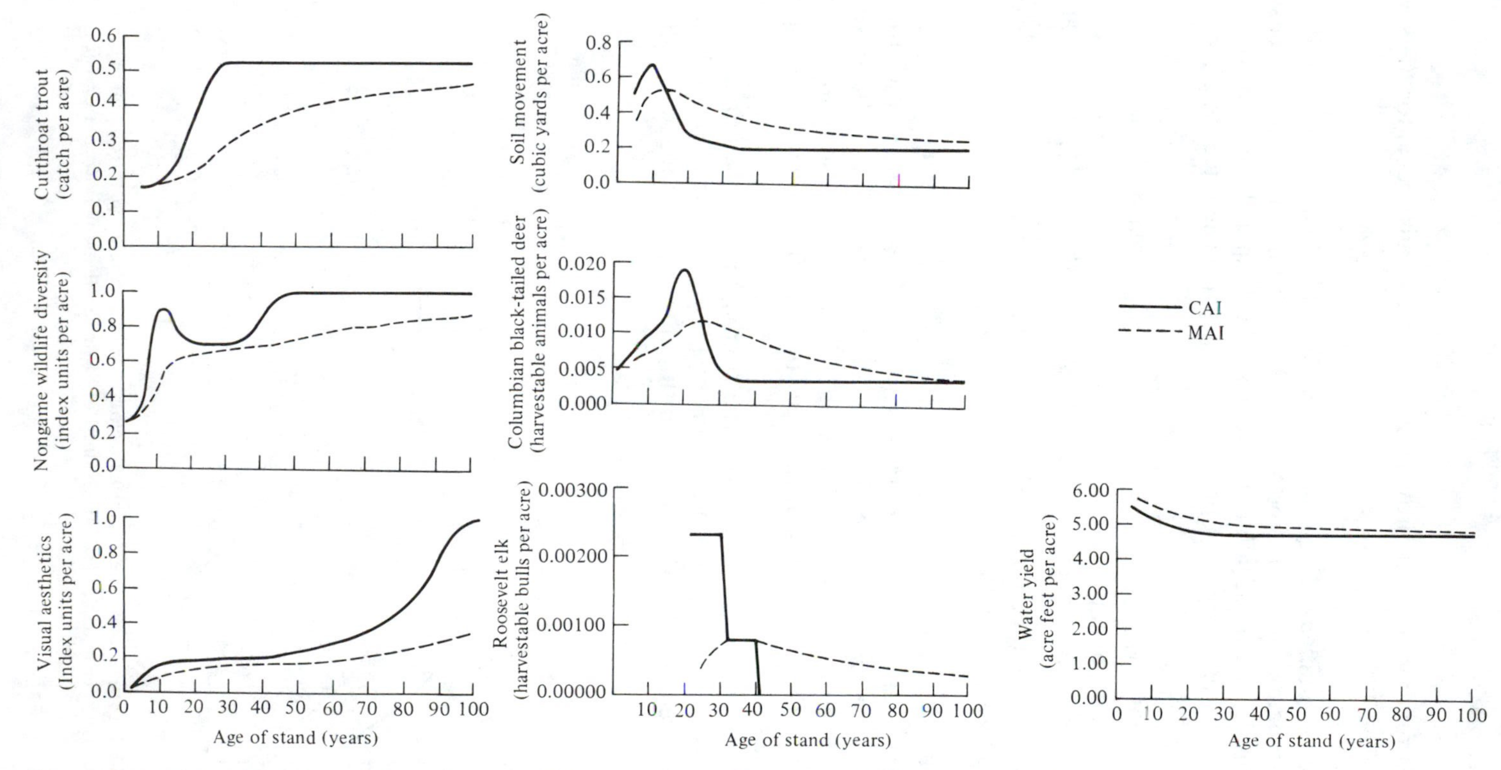

Figure 13-3 Nontimber yield in relation to rotation. CAI—current annual increment; MAI—mean annual increment. *(From Calish et al., 1978.)*

shortest rotation that will provide the ecologic requirements for the species in question.

Nontimber rotations Given the data in Fig. 13-3, Calish et al. computed SEV_{NT} functions for each type of yield. They assumed that no extra costs are incurred to provide nontimber values. They imputed prices to nontimber yields to test a range of prices rather than a single price. In price per unit:

All yields except mass soil movement	\$5, \$10, \$50, \$100, \$500
Mass soil movement	−\$0.05, −\$0.10, −\$0.50, −\$1.00, −\$5.00

Calish et al. found that the SEV_{NT} functions for the nontimber yields fell into two general patterns:

1. The SEV_{NT} functions of the type *A* group reach a peak level relatively soon after stand regeneration, then decline slowly as the forest matures. An example is the SEV_{NT} function for deer (Fig. 13-4). Elk and water flow display a similar shape. So does mass soil movement, except that the function involves a cost (negative value) rather than a benefit.
2. The SEV_{NT} functions for the type *B* group increase continuously as the stand ages, but at a decreasing rate beyond a certain point. Examples are aesthetics (Fig. 13-4), cutthroat trout, and nongame wildlife.

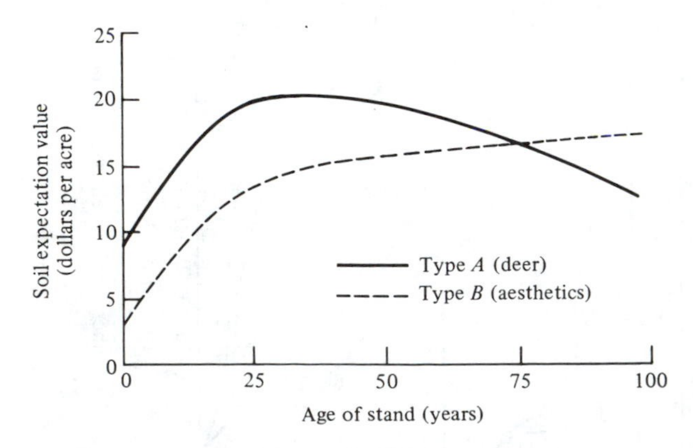

Figure 13-4 Type *A* and type *B* soil expectation values in relation to rotation age. Deer valued at \$100 per animal; aesthetics valued at \$5 per aesthetic unit. *(From Calish et al., 1978.)*

If land is managed only for the nontimber output and there are no costs, the SEV_{NT}-maximizing rotations are as follows:

	Rotation (years)
Deer	25
Elk	40
Water flow	0
Cutthroat trout	Indefinitely long
Nongame wildlife	Indefinitely long
Visual aesthetics	Indefinitely long
Mass soil movement	Indefinitely long

With no costs, any of the prices mentioned above will give the same SEV_{NT}-maximizing rotation age for an output. The price chosen will affect the value of SEV_{NT}, but not the rotation that gives the highest SEV_{NT}.

Joint rotations Optimum joint rotations for seven pairs of timber and nontimber yields were computed by Calish et al. using Eq. (13-15). Table 13-6 summarizes results for the three pairs where rotations are significantly affected by nontimber values.

Their calculations show that the economic joint rotation may be different from a timber-only rotation. As they say,

> Type *A* functions, except for mass soil movement, tended to shorten rotation; type *B* benefit functions tended to lengthen it. Yet the joint rotations differ little from timber-only rotations for the range of unit values examined. In the case of deer, elk, mass soil movement, and visual aesthetics, rotations change by no more than 3 years, even if yield is valued at $500 per unit. Cutthroat trout and nongame wildlife must be assigned unit values greater than $100 to significantly change rotation; water affects rotation only in the range of $50 to $500 per unit. These general results were also found to hold for Douglas-fir site III and for rates of interest from 3 to 7 percent.

These results do not, however, mean that nontimber values are trivial. As values for nontimber forest outputs were added to timber, the SEV of land used for forestry increased. Indeed, much of the value accruing to the land derived from nontimber sources. As an example, when Calish et al. assigned prices to nontimber outputs of:

Elk	$500 each
Deer	$200 each
Wildlife index	$10/acre
Water	$15/acre-ft

Table 13-6 Changes in the best timber rotation due to the inclusion of nontimber outputs

		Length of rotation period (years)	
Type of nontimber output (output unit)	Price per unit of output ($)	Best joint rotation	Change in rotation*
Water flow (acre-feet)	5	36	0
	10	35	−1
	50	33	−3
	100	2	−34
	500	1	−35
Cutthroat trout (harvestable fish)	5	36	0
	10	36	0
	50	37	+1
	100	39	+3
	500	47	+11
Nongame wildlife (diversity index)	5	36	0
	10	36	0
	50	37	+1
	100	38	+2
	500	53	+17

Source: Calish et al. (1978).
* Best rotation for timber alone is 36 years.

Trout	$10
Soil	$10/yd^3
Visual aesthetics	$10/acre

and combined all these outputs with timber in a joint SEV calculation, the SEV of the site was $2,520 per acre. Of this total, 75 percent was from nontimber sources. While we may question the prices they assigned to the nontimber outputs in coming up with this number, there can be little question that these outputs often contribute significantly to the site value in Douglas-fir forests of the Pacific Northwest.

The practical issue is, can we actually get together enough good information on the relationship of nontimber outputs to timber age (Fig. 13-3) and acceptable prices for these outputs? In most situations this will be difficult. Production relationships are often but subjective estimates, and the output can be scaled as an index or percent. In Chap. 12 we reviewed the problems of reliably pricing nonmarketed outputs. When all the numerical data are subjective, combining them with the more objectively defined timber data opens the door for results that are difficult to defend. Beyond the problems of quantification we need to point out that many of these nontimber outputs are difficult to evaluate on an average per-acre basis. Visual quality relates to the viewpoint, the location of roads, and the topography. Elk and deer migrate and need mixes of different kinds of acres at different times of the year. Many of these outputs do interact with timber management, and we may find that the interaction can be better handled by formulating constraints in a forest planning model as we introduce in Chap. 15. Still, the effort by Calish et al. is an innovative attempt to expand SEV analysis to nontimber outputs.

Evaluating Intensive Management Investments

Precommercial thinning, herbicide application, and fertilization often are suggested as cultural treatments for forest stands. In addition, commercial thinnings often are considered. By adding the costs or revenues of these treatments to the return calculation in the year that they occur, their financial attractiveness for any particular rotation and their effect on rotation age that gives maximum SEV can be measured. Our net return equation now will be (timber values only)

$$N_R = V_R P_R + \sum_{t=1}^{R-1} V_t P_t (1+i)^{R-t} - \frac{T[(1+i)^R - 1]}{i} - G(1+i)^R - \sum_{j=1}^{U} C_{jt}(1+i)^{R-t} \tag{13-16}$$

where

V_t = cubic foot volume per acre for commercial thinnings harvested at age t
P_t = price per cubic foot for commercial thinnings harvested at age t
C_{jt} = cost of cultural treatment j that occurs at age t

The other symbols are as before, and $(1 + i)^{R-t}$ compounds the treatment cost or revenue for the years that remain in the rotation after its occurrence.

As an example, DFSIM (Curtis et al., 1981) predicts that precommercial thinning at age 10 in site index 105 at a cost of \$100 increases yield at age 60 to 8,968 ft^3 from the 8,212 ft^3 shown at age 60 in Table 13-5. Our net return equation under the constant price assumption is

$$N_R = 8{,}968 \times 0.75 - \frac{3[(1+0.04)^{60} - 1]}{0.04} - 220(1+0.04)^{60} - 100(1+0.04)^{60-10}$$

$$= 6{,}726 - 714 - 2{,}314 - 711 = \$2{,}987$$

and our soil expectation value equals

$$\text{SEV} = \frac{2{,}987}{(1 + 0.04)^{60} - 1} = \$314$$

Since precommercial thinning causes a reduction in SEV ($314 instead of $329), it is not a good investment for a 60-year rotation that involves planting and final harvest.

As another example, DFSIM predicts that precommercial thinning at age 10 plus a commercial thinning at age 40 that removes 20 percent of the standing volume yields a thinning volume of 837 ft^3 and a final harvest volume of 8,368 ft^3 at age 60. Assuming that thinnings have a stumpage value of $0.30 per cubic foot, our net return equation is

$$N_R = 8{,}368 \times 0.75 + (837 \times 0.30)(1.04)^{60-40} - \frac{3[(1 + 0.04)^{60} - 1]}{0.04}$$

$$- 220(1 + 0.04)^{60} - 100(1 + 0.04)^{60-10}$$

$$= 6{,}276 + 550 - 714 - 2{,}314 - 711 = \$3{,}087$$

and our soil expectation value is

$$\text{SEV} = \frac{3{,}087}{(1.04)^{60} - 1} = \$324$$

Since precommercial thinning plus commercial thinning causes a reduction in SEV ($324 instead of $329), the combination of treatments is not a good investment for a 60-year rotation that involves planting and final harvest. Perhaps, though, a heavier commercial thinning or a commercial thinning without precommercial thinning is a good strategy. Further analysis would be needed to find that out.

Finding the Best Combination of Commercial Thinnings: A Dynamic Programming Approach

While it is easy to account for the value of commercial thinnings in the soil expectation formula (13-16), it is not always easy to calculate when to thin and how much volume to remove in each commercial thinning to achieve a specific objective. Often we are faced with many potential ages at which intermediate harvest can take place and many potential residual growing-stock levels at each of these ages. Suppose that we wish to examine the possibility of thinning at four ages with five levels of residual growing stock considered at each age. Potentially we would need to do $5^4 = 625$ simulations to examine all combinations of residual growing-stock and associated removals. While the actual number of simulations would probably be less than 625 because some combinations of residual growing stock at the different ages cannot occur together, we still would need to do an extensive analysis.

Given the possible levels of intermediate harvest at each age, we need to examine the benefit from different sequences of intermediate harvest (and residual growing stock) over time to find the sequence of intermediate harvests that best meets our objective for each given rotation. By locating the objective-maximizing amount of thinning volume to remove at each age for each rotation length and then comparing the results for the different rotation lengths, we can determine the combination of thinnings and final harvest that maximizes some objective.

Recently people have begun turning to a technique called *dynamic programming* to do such analysis. To understand the use of dynamic programming in finding the objective-maximizing level of growing stock at each age, we will look at it under two objectives: (1) maximize the volume (mean annual increment) over the rotation, and (2) maximize the soil expectation value over the rotation. In this analysis we make the following assumptions:

Planting cost	\$220
Volume at age 15	500 ft^3/acre
After age 15, $\log \text{Vol}_{t+5}$	$= 1.5 + \left[\frac{t+5}{t} \log \sqrt{\text{Vol}_t}\right]$
Net stumpage price	\$0.75/ft^3/acre
Interest rate	0.04
Rotation age	35 years

where Vol_{t+5} = volume at age $t + 5$
Vol_t = volume at age t

Between any two ages (t and $t + 5$), grow the stand from t to $t + 5$ and then consider harvesting some part of it.

To simplify our example further, we make two kinds of restrictions: (1) once we start thinning, we must keep thinning every 5 years until final harvest, and (2) in addition to growing-stock levels that naturally occur, we will consider thinning to a residual density of 1,000 ft^3/acre and 1,500 ft^3/acre at each age. While these particular constraints are applied here to cut down the amount of computation required, other constraints often are placed on the permissible thinning regimes in "real-life" applications, such as constraints that any thinning must remove at least some minimum volume or that the residual growing-stock level must not drop below some minimum amount.

Using our growth function, we estimate the thinning volume associated with each possible pattern of growth and harvest (Table 13-7). As an example, the stand at age 15 will produce a volume of 500 ft^3/acre, which will grow into a volume of 1,992 ft^3 at age 20. At age 20 the stand can be

Table 13-7 Stand volume and harvest at different ages and growing stocks (cubic feet per acre)

				Residual volume at age + 5			
				All	1,500	1,000	0
Age (t)	Node	Volume	Volume at age (t + 5)	Harvest at age (t + 5) to achieve residual volume			
15	*A*	500	1,992	0	492	992	
20	*B*	1,992	3,648	0	2,148	2,648	
	C	1,500	3,055		1,555	2,055	
	D	1,000	2,371		871	1,371	
25	*E*	3,648	4,337	0	2,837	3,337	
	F	1,500	2,545		1,045	1,545	
	G	1,000	1,995		495	995	
30	*H*	4,337	4,185				4,185
	I	1,500	2,253				2,253
	J	1,000	1,778				1,778

cut back to 1,000 ft^3 providing a harvest of 992 ft^3, cut back to 1,500 ft^3, producing a volume of 492 ft^3, or left alone at the volume of 1,992 ft^3. Each of these three residual volumes (1,000, 1,500, and 1,992 ft^3) will then grow for 5 years and have a number of harvest levels considered at age 25.

We can diagram the permitted levels of residual stock, and the associated harvests, in terms of a network containing *nodes, arcs,* and *rewards*. Each possible growing-stock level at each age is a node. With 11 possible growing stock-age combinations, we have 11 nodes from *A* to *K*. Arcs enter and leave each node, connecting the permitted levels of growing stock through time. A number is shown on each arc, representing the reward associated with going between the nodes that the arc connects.

The network for our maximum volume objective is shown in Fig. 13-5, with each reward indicating the associated harvest volume. Beginning with a volume of 500 ft^3 at age 15 (node *A*) as an example, we recognize three choices between ages 15 and 20: (1) let the stand grow to 1,992 ft^3 at age 20 with no harvest (node *A* to node *B*), (2) let the stand grow to 1,992 ft^3 at age 20 and then cut it back to 1,500 ft^3, producing a harvest of 492 ft^3 (node *A* to node *C*), or (3) let the stand grow to 1,992 ft^3 at age 20 and then cut it back to 1,000 ft^3, producing 992 ft^3 (node *A* to node *D*).

Finding the growing-stock levels that maximize volume In searching for the growing-stock levels at ages 20, 25, and 30 that maximize volume over a 35-year rotation, we use the following terminology (with examples from

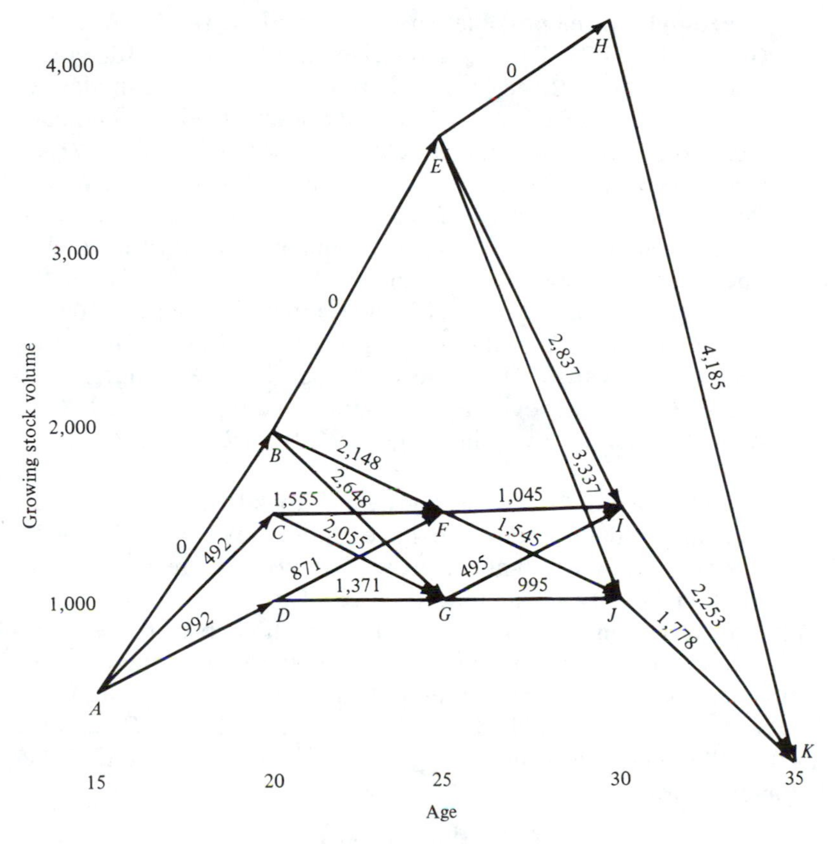

Figure 13-5 Network of possible growing-stock levels with harvest volume from moving between any two levels shown on the arc connecting the levels.

the network in Fig. 13-5, which shows the harvest volume as the reward for moving between any two nodes):

r_{XY} = reward for going on an arc from growing-stock level X at age t to growing-stock level Y at age $t + 5$. As an example, the reward (harvest) from going from a growing-stock level of 1,500 ft^3 at age 20 (node C) to a growing-stock level of 1,000 ft^3 at age 25 (node G) is the associated harvest volume of 2,055 ft^3, so r_{CG} = 2,055.

R_Y = maximum reward achievable from one of the permitted paths to growing-stock level Y. Usually numerous paths exist through the growing-stock network to a node. We seek to discover the path that gives the highest cumulative reward to that node. As an example, three possible paths go to node G (A–B–G, A–C–G, and A–D–G). Path A–B–G gives a reward of 0 for going from A to B and 2,648 for going from B to G for a cumulative reward of 2,648. Path A–C–G gives a reward of 492 for A–C and 2,055 for C–G for a cumulative reward of 2,547. Path A–D–G gives a reward of 992 for A–D and a reward of 1,371 for D–G, for a cumulative reward of 2,363. Among the three paths to node G, A–B–G has the highest cumulative reward of 2,648. Therefore we say that $R_G = 2{,}648$.

P_Y = previous growing-stock level X (node) that gives R_Y at node Y. In our example, X is the growing-stock level at age t on the maximum reward path to growing-stock level Y at age $t + 5$ years. As shown above, the maximum reward path to node G is A–B–G. Therefore we should get to G from B, or $P_G = B$.

We wish to find R_Y and P_Y for each growing-stock level (node) recognized at each age. Suppose that we have three growing-stock levels (nodes) at age t (X, W, and Z) that can grow naturally to node Y at age $t + 5$ or grow and then be cut back to node Y at age $t + 5$. We can describe the cumulative reward from following the path through each of these three nodes to node Y in two parts: (1) the maximum reward for getting to each of the three nodes (R_X, R_W, or R_Z), and (2) the reward for getting from each of the three nodes to node Y (r_{XY}, r_{WY}, r_{ZY}). Therefore R_Y can be described as the maximum from among $R_X + r_{XY}$, $R_W + r_{WY}$, and $R_Z + r_{ZY}$. Put another way,

$$R_Y = \max \begin{Bmatrix} R_X + r_{XY} \\ R_W + r_{WY} \\ R_Z + r_{ZY} \end{Bmatrix}$$

Continuing our example, R_G can be described as

$$R_G = \max \begin{Bmatrix} R_B + r_{BG} \\ R_C + r_{CG} \\ R_D + r_{DG} \end{Bmatrix} = \max \begin{Bmatrix} 0 + 2{,}648 \\ 492 + 2{,}055 \\ 992 + 1{,}371 \end{Bmatrix} = \max \begin{Bmatrix} 2{,}648 \\ 2{,}547 \\ 2{,}363 \end{Bmatrix} = 2{,}648$$

To maximize the total volume produced over the rotation, we calculate R_Y for each growing-stock level (node), starting with A and moving across to K.

Age 15

$$R_A = 0$$

No choices on how to get to A and no rewards or penalties.

Age 20

$$R_B = R_A + r_{AB} = 0 + 0 = 0, \qquad P_B = A$$

$$R_C = R_A + r_{AC} = 0 + 492 = 492, \qquad P_C = A$$

$$R_D = R_A + r_{AD} = 0 + 992 = 992, \qquad P_D = A$$

B, C, D, each have only one path to them with an associated reward.

Age 25

$$R_E = R_B + r_{BE} = 0 + 0 = 0, \qquad P_E = B$$

$$R_F = \max \begin{Bmatrix} R_B + r_{BF} \\ R_C + r_{CF} \\ R_D + r_{DF} \end{Bmatrix} = \max \begin{Bmatrix} 0 + 2{,}148 \\ 492 + 1{,}555 \\ 992 + 871 \end{Bmatrix} = \max \begin{Bmatrix} 2{,}148 \\ 2{,}047 \\ 1{,}863 \end{Bmatrix}$$

$$= 2{,}148, \qquad P_F = B$$

$$R_G = \max \begin{Bmatrix} R_B + r_{BG} \\ R_C + r_{CG} \\ R_D + r_{DG} \end{Bmatrix} = \max \begin{Bmatrix} 0 + 2{,}648 \\ 492 + 2{,}055 \\ 992 + 1{,}371 \end{Bmatrix} = \max \begin{Bmatrix} 2{,}648 \\ 2{,}547 \\ 2{,}363 \end{Bmatrix}$$

$$= 2{,}648, \qquad P_G = B$$

Age 30

$$R_H = R_E + r_{EH} = 0 + 0 = 0, \qquad P_H = E$$

$$R_I = \max \begin{Bmatrix} R_E + r_{EI} \\ R_F + r_{FI} \\ R_G + r_{GI} \end{Bmatrix} = \max \begin{Bmatrix} 0 + 2{,}837 \\ 2{,}148 + 1{,}045 \\ 2{,}648 + 495 \end{Bmatrix} = \max \begin{Bmatrix} 2{,}837 \\ 3{,}193 \\ 3{,}143 \end{Bmatrix}$$

$$= 3{,}193, \qquad P_I = F$$

$$R_J = \max \begin{Bmatrix} R_E + r_{EJ} \\ R_F + r_{FJ} \\ R_G + r_{GJ} \end{Bmatrix} = \max \begin{Bmatrix} 0 + 3{,}337 \\ 2{,}148 + 1{,}545 \\ 2{,}648 + 995 \end{Bmatrix} = \max \begin{Bmatrix} 3{,}337 \\ 3{,}693 \\ 3{,}643 \end{Bmatrix}$$

$$= 3{,}693, \qquad P_J = F$$

Age 35

$$R_K = \max \begin{Bmatrix} R_H + r_{HK} \\ R_I + r_{IK} \\ R_J + r_{JK} \end{Bmatrix} = \max \begin{Bmatrix} 0 + 4{,}185 \\ 3{,}193 + 2{,}253 \\ 3{,}693 + 1{,}778 \end{Bmatrix} = \max \begin{Bmatrix} 4{,}185 \\ 5{,}446 \\ 5{,}471 \end{Bmatrix}$$

$$= 5{,}471, \qquad P_K = J$$

Tracing the volume-maximizing path back to A gives $P_K = J$, $P_J = F$, $P_F = B$, and $P_B = A$, so the volume-maximizing path is $A \rightarrow B \rightarrow F \rightarrow J \rightarrow K$, with a total reward of 5,471. Therefore 5,471 ft^3 is the maximum volume for a 35-year rotation given our assumptions and constraints, which means a maximum mean annual increment for this rotation of 5,471/35 = 156 ft^3/acre.

In this example we sought the maximum reward path to each node with the reward measured in volume harvested. Using the terminology of dynamic programming, we can define this maximum reward as

$$f_t(S) = \text{value of volume-maximizing path to stand age } t \text{ and stocking level } S,$$
$$= \max_{(s)}\{f_{t-5}(s) + H_t\} \qquad (13\text{-}17)$$

where H_t is the volume harvested per acre at age t and (s) is the set of all feasible stocking levels at age $t - 5$ that can grow naturally to stocking level S at age t or grow and be cut back to stocking level S at age t. Matching $f_{t-5}(s)$ to our R_Y and H_t to our r_{XY}, we have the same value function as used in our example.

Finding the growing-stock levels that maximize soil expectation value To find the path through our growing-stock network that maximizes the soil expectation value for a 35-year rotation, we must first convert the harvest reward between each set of nodes to a dollar reward (harvest volume × harvest value per unit). As an example, the harvest reward between node *C* and node *G* is 2,055 ft^3. With a stumpage price of \$0.75 per cubic foot, the dollar reward between node *C* and node *G* is 2,055 × 0.75 = \$1,541.

Next we must convert this dollar reward to a soil expectation value. We can do this conversion in one of two ways:

1. Compound the dollar reward associated with each arc to the end of the rotation and then discount that result to reflect receiving the reward at the same age in rotation after rotation forever. Putting these two calculations together, we would multiply each dollar reward by $(1 + i)^{R-t}/[(1 + i)^R - 1]$, where i is the interest rate, R the rotation age, and t the age at which the harvest occurs. In our case we receive a harvest value of 1,541 at age 25 for going from node *C* to node *G*. Therefore the soil expectation contribution (reward) of this harvest is

$$1{,}541 \, \frac{(1 + 0.04)^{35-25}}{(1 + 0.04)^{35} - 1} = \$774$$

2. Discount the dollar reward associated with each arc to age 0, solve the problem obtaining the highest present net value over the 35-year rotation, and then find the soil expectation value implied by this present net worth by compounding the present net worth to the end of the rotation and discounting the result to reflect receiving this present net value at the same age in rotation after rotation forever. Symbolically, we should multiply the highest cumulative present net worth

over the 35-year rotation by $(1 + i)^R/[(1 + i)^R - 1]$, or in our case, $(1 + 0.04)^{35}/[(1 + 0.04)^{35} - 1]$.

We will take the second approach. It has the advantage that the rewards shown between the nodes are not a function of rotation length since the rotation length adjustment needed for the soil expectation calculation is made after the highest cumulative reward for the rotation has been calculated. Therefore we can use the same reward between any two nodes in the soil expectation analysis for different rotations.

Multiplying all the harvest volume rewards from the previous network (Fig. 13-5) by the stumpage price of \$0.75 and discounting the dollar return back to age 0 gives the dollar rewards shown in Fig. 13-6. Notice that a reward of −\$220 is associated with node *A* to reflect the reforestation cost at age 0. This reward is the present net worth of all actions up to

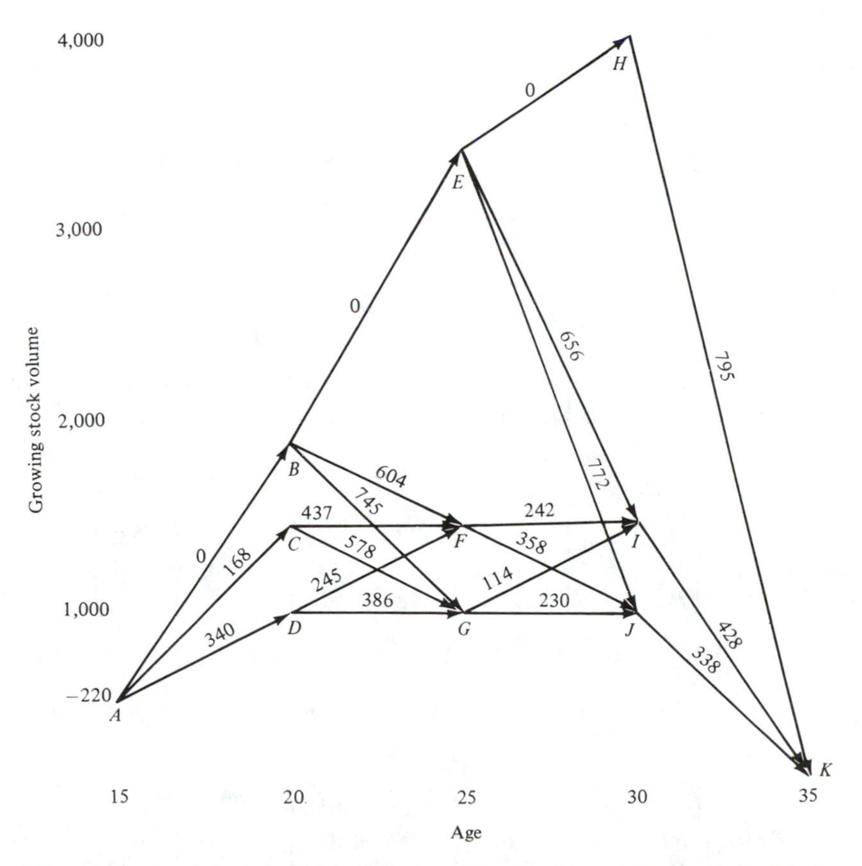

Figure 13-6 Network of possible growing-stock levels with present net worth of the harvest volume from moving between any two levels shown on the arc connecting the levels.

age 15. If precommercial thinning had occurred at age 10, with a cost of \$100, this cost would be added to the reforestation cost, giving a total reward at age 15 of

$$-220 - \frac{100}{(1 + 0.04)^{10}} = -\$288$$

To maximize the present net worth over a 35-year rotation, we calculate R_Y for all growing-stock levels starting with A and moving across to K:

Age 15

$$R_A = -220$$

No choices in how to get to A, but R_A is given a penalty equal to the planting cost.

Age 20

$$R_B = R_A + r_{AB} = -220 + 0 = -220, \qquad P_B = A$$

$$R_C = R_A + r_{AC} = -220 + 168 = -52, \qquad P_C = A$$

$$R_D = R_A + r_{AD} = -220 + 340 = 120, \qquad P_D = A$$

Age 25

$$R_E = R_B + r_{BE} = -220 + 0 = -220, \qquad R_E = B$$

$$R_F = \max \begin{Bmatrix} R_B + r_{BF} \\ R_C + r_{CF} \\ R_D + r_{DF} \end{Bmatrix} = \max \begin{Bmatrix} -220 + 604 \\ -52 + 437 \\ 120 + 245 \end{Bmatrix} = \max \begin{Bmatrix} 384 \\ 385 \\ 365 \end{Bmatrix}$$

$$= 385, \qquad P_F = C$$

$$R_G = \max \begin{Bmatrix} R_B + R_{BG} \\ R_C + R_{CG} \\ R_D + R_{DG} \end{Bmatrix} = \max \begin{Bmatrix} -220 + 745 \\ -52 + 578 \\ 120 + 386 \end{Bmatrix} = \max \begin{Bmatrix} 525 \\ 526 \\ 506 \end{Bmatrix}$$

$$= 526, \qquad P_G = C$$

Age 30

$$R_H = R_E + r_{EH} = -220 + 0 = -220, \qquad P_H = E$$

$$R_I = \max \begin{Bmatrix} R_E + r_{EI} \\ R_F + r_{FI} \\ R_G + r_{GI} \end{Bmatrix} = \max \begin{Bmatrix} -220 + 656 \\ 385 + 242 \\ 526 + 114 \end{Bmatrix} = \max \begin{Bmatrix} 436 \\ 627 \\ 640 \end{Bmatrix}$$

$$= 640, \qquad P_I = G$$

$$R_J = \max \begin{Bmatrix} R_E + r_{EJ} \\ R_F + r_{FJ} \\ R_G + r_{GJ} \end{Bmatrix} = \max \begin{Bmatrix} -220 + 772 \\ 385 + 358 \\ 526 + 230 \end{Bmatrix} = \max \begin{Bmatrix} 552 \\ 743 \\ 756 \end{Bmatrix}$$

$$= 756, \qquad P_J = G$$

Age 35

$$R_K = \max \begin{Bmatrix} R_H + r_{HK} \\ R_I + r_{IK} \\ R_J + r_{JK} \end{Bmatrix} = \max \begin{Bmatrix} -220 + 795 \\ 640 + 428 \\ 756 + 338 \end{Bmatrix} = \max \begin{Bmatrix} 575 \\ 1{,}068 \\ 1{,}094 \end{Bmatrix}$$

$$= 1{,}094, \qquad P_K = J$$

Tracing the present net worth–maximizing path back to A gives $P_K = J$, $P_J = G$, $P_G = C$, and $P_C = A$, so the present net worth–maximizing path is $A \rightarrow C \rightarrow G \rightarrow J \rightarrow K$ with a total reward of \$1,094. Therefore the maximum present net worth on a 35-year rotation (given our constraints) is \$1,094. To convert the maximum present worth over the 35-year rotation to a soil expectation value, we simply multiply the present net worth (\$1,094) by $(1 + i)^R/[(1 + i)^R - 1]$. Here this factor is $(1 + 0.04)^{35}/[(1 + 0.04)^{35} - 1]$, giving a soil expectation value of \$1,465 for a 35-year rotation.

In the example we again sought the maximum reward path to each node, with this reward measured in present net worth. Using the terminology of dynamic programming, we can define this maximum reward to each node as:

$f_t(S)$ = value of maximum present net worth path to stand age t and stock level S (this reward here consists of harvest revenues and regeneration costs),

$$= \max_{(s)} \left\{ f_{t-5}(s) + \frac{P_t H_t}{(1 + i)^t} \right\} \tag{13-18}$$

where P_t = price per unit of harvest at age t
H_t = volume harvested per acre at age t
i = interest rate

and (s) is the set of all feasible stocking levels at age $t - 5$ that can grow naturally to stocking level S at age t or grow and be cut back to stocking level S at age t. Matching $f_{t-5}(s)$ to our R_Y and $P_t H_t/(1 + i)^t$ to our r_{XY}, we have the same value function as that used in our example.

Two important points about the dynamic programming approach to finding the yield regime that meets a specified objective and set of constraints are illustrated by our soil expectation solution.

1. The "optimum" path does not necessarily select the highest cumulative reward at each age. As an example, the highest cumulative re-

ward at age 20 is \$120 gained by going through node D, but our optimum path goes through node C. We cannot simply retain information on the node at each age that gives the highest cumulative reward, but we must retain the highest cumulative reward for each node at each age. We do not know what node at each age will be on the optimum path until we reach the end of our problem and trace this optimum path backward through the network.

2. Not all nodes at an age achieve their highest cumulative reward by passing through the same node at the previous age. At least one node (the "natural growth" node) at ages 20, 25, and 30 comes from a different node at the previous age than do the other nodes for these ages. Therefore we cannot necessarily find a single node at the previous age that gives the highest reward for all nodes at the age we are examining. Rather we must calculate and save what node we should come from at the previous age for each node at the age we are examining. Incidentally the natural growth option could easily become the optimum path. Suppose that our final harvest stumpage price is three times that of the thinning price (\$2.25 per cubic foot for final harvest and 0.75 per cubic foot for thinnings). Now R_K is calculated as

$$R_K = \max \begin{Bmatrix} R_H + r_{HK} \\ R_I + r_{IK} \\ R_J + r_{JK} \end{Bmatrix} = \max \begin{Bmatrix} -220 + 2{,}385 \\ 640 + 1{,}284 \\ 756 + 1{,}014 \end{Bmatrix} = \max \begin{Bmatrix} 2{,}165 \\ 1{,}924 \\ 1{,}770 \end{Bmatrix}$$

$$= 2{,}165, \qquad P_K = H$$

Tracing backward from K to A gives $P_K = H$, $P_H = E$, $P_E = B$, and $P_B = A$ for a present net worth path of $A \to B \to E \to H \to K$ with a total reward of \$2,165.

Plotting the volume-maximizing path and present net worth-maximizing path for a 35-year rotation against the natural growth path (Fig. 13-7) shows that thinning is prescribed under each of the two maximization regimes. As is often the case, the present net worth-maximizing path generally carries less growing stock than the volume-maximizing path.

Calculating the total volume harvested in MAI, PNW, and SEV for each path shows the two maximization regimes fairly close to one another by these four criteria and superior to the natural growth regime in all of them:

Path	Volume harvested	MAI	PNW	SEV
Maximize volume (MAI)	5,471	156	1,080	1,447
Maximize PNW (SEV)	5,320	152	1,094	1,465
Natural growth	4,185	120	575	770

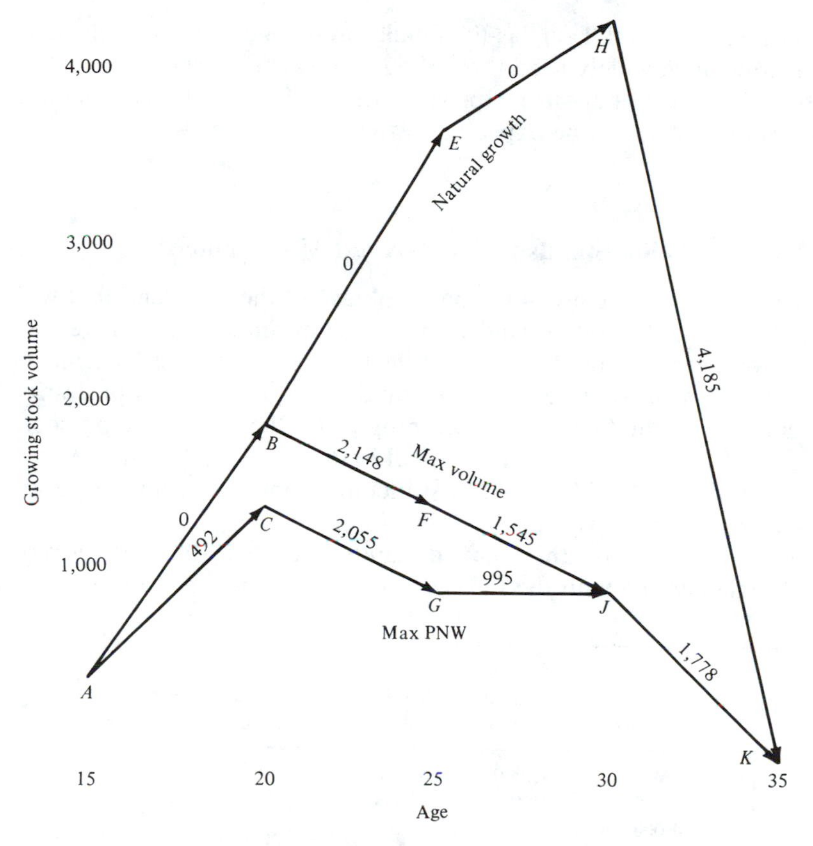

Figure 13-7 Three possible paths through the growing-stock network with the associated harvest volume shown on the arc connecting any two growing-stock levels.

By allowing nodes that represent a zero growing-stock level at ages other than 35, like node *K* at age 35, the objective-maximizing paths for other rotation lengths can be determined. Then the MAIs or SEVs for all different rotations can be examined to find the objective-maximizing combination of thinnings and final harvest over all rotations under each objective.

With the increasing availability of growth projection models and the increasing emphasis on objective-oriented yield regimes, the use of dynamic programming to find the best combination of thinnings and final harvest under specified objectives and constraints is probably here to stay. Brodie and his colleagues at Oregon State University have built numerous dynamic programming models, including those for coastal Douglas-fir (Brodie and Kao, 1979) and southwestern ponderosa pine (Ritters et al., 1982), and have done much to further the application of dynamic programming to stand management. Also, a dynamic program for the DFSIM coastal Douglas-fir simulator by Sleavin and Johnson (1983) is now being widely used throughout the Pacific Northwest by the U.S. Forest Service and various private firms. More of these dynamic programming models can be expected in the next few years.

Considering Existing Stands in Even-Aged Management

Often the investment we are studying involves not the bare land that we have studied so far but rather land with a stand of timber. Should we cut the stand or keep it? Just like our single tree selection case, we must compare the benefit from keeping the stand—its value growth plus any partial cuts—with the two costs of keeping it: (1) the stock holding cost (lost income from not investing the capital in the stand in its best alternative), and (2) the land holding cost (lost income from not turning the land over to its best alternative: the next rotation).

Suppose that we have three mixed stands of Douglas-fir–alder growing on Douglas-fir site with the following characteristics (per acre):

Stand	Age	Timber value now	Timber value in 5 years	Timber value growth	Compound value growth percent
A	40	$1,000	$2,000	$1,000	14.8
B	60	2,500	3,100	600	4.4
C	120	4,000	4,500	500	2.4

Stand *C* has a value growth percent (2.4) below the 4 percent guiding rate of return. This result means that the value growth ($500) is less than the stock holding cost [stock holding cost $= 4{,}000 \times (1 + 0.04)^5 - 4{,}000 =$ $867]. Therefore we can say that stand *C* should be replaced without even looking at the land holding cost.

Stands *A* and *B*, on the other hand, have timber value growth rates greater than the guiding rate of return, which means that their value growth is greater than the stock holding cost. If we keep these stands for 5 more years, we will delay future rotations by that time.

Our maximum SEV for a Douglas-fir site index 105, equals $417 under the constant price assumption and $501 under the assumption that the stumpage price is a function of tree size (Table 13-5). To discover the cost of delaying the start of the series of rotations underlying each SEV, we find the value increase from investing SEV for five years at the guiding rate:

$$\text{Value increase} = \text{SEV}(1 + i)^5 - \text{SEV}$$

For the constant price assumption, delaying future rotations for 5 years costs

$$417(1.04)^5 - 417 = \$90.34$$

For the assumption that price is a function of tree size, delaying future rotations for 5 years costs

$$501(1.04)^5 - 501 = \$108.54$$

In addition, we must include the future value of an annual management cost of \$3 per year for existing stands that is already included in the SEV calculation for future stands:

$$\frac{3[(1 + 0.04)^5 - 1]}{0.04} = \$16.25$$

Comparing the value growth for 5 years with the total cost for keeping the stand for 5 years (Table 13-8), we see that we should keep stand *A* and replace stands *B* and *C* under both sets of price assumptions for future stands.

Our guide for deciding whether to hold each Douglas-fir–alder stand for some time into the future or to replace it can be summarized in terms of future values as follows:

1. Leave the stand if the value growth is greater than the stock holding cost plus the land holding cost plus the mangement cost.
2. Cut the stand if the value growth is less than the stock holding cost plus the land holding cost plus the management cost.

This decision rule can also be formulated in terms of present values:

1. Find the stumpage value of the stand in each possible year for harvesting each stand.
2. Find the highest SEV (MSEV) value for the land.
3. Add the discounted return from harvesting each stand in each possible year to the MSEV discounted from that same year.
4. Subtract the annual management cost for existing stands.
5. Harvest the stand in the year that gives the highest total present net worth.

Mathematically this rule can be expressed as

$$\text{PNW} = \max_n \left[\frac{\text{SV}_n}{(1 + i)^n} + \frac{\text{MSEV}}{(1 + i)^n} - \frac{\text{AMC}[(1 + i)^n - 1]}{i(1 + i)^n}\right] \quad (13\text{-}19)$$

where SV_n = net stumpage value of a Douglas-fir–alder stand in year n
MSEV = maximum soil expectation value across all possible rotations
AMC = annual management cost for existing stands

Table 13-8 Analysis of harvest decisions for three Douglas-fir–alder stands

Stand	Value now	Value in 5 years	Value growth	Stock holding cost	Land holding cost	Management cost	Total cost	Net gain from holding	Decision
			Under assumption of constant stumpage price in future stands						
A	$1,000	$2,000	$1,000	$217	$90	$16	$323	$677	Leave
B	2,500	3,100	600	542	90	16	648	−48	Cut
C	4,000	4,500	500	867	90	16	973	−473	Cut
			Under assumption of stumpage price being a function of tree size in future stands						
A	$1,000	$2,000	$1,000	$217	$109	$16	$342	$658	Leave
B	2,500	3,100	600	542	109	16	667	−67	Cut
C	4,000	4,500	500	867	109	16	992	−492	Cut

and the notation $\max_n$ means: find the value of n (year) that maximizes the term in brackets. In our example the components of this decision rule can be calculated as follows (constant price assumption):

	Stand A		Stand B		Stand C	
	Cut $n = 0$	Leave $n = 5$	Cut $n = 0$	Leave $n = 5$	Cut $n = 0$	Leave $n = 5$
1. SV_n	1,000	2,000	2,500	3,100	4,000	4,500
2. $\frac{SV_n}{(1+i)^n}$	1,000	1,644	2,500	2,548	4,000	3,699
3. MSEV	417	417	417	417	417	417
4. $\frac{MSEV}{(1+i)^n}$	417	343	417	343	417	343
5. $\frac{AMC[(1+i)^n - 1]}{i(1+i)^n}$	0	13	0	13	0	13
PNW (rows 2 + 4 − 5)	1,417	1,974	2,917	2,878	4,417	4,029

As before, we should keep stand A and replace stands B and C.

INCLUDING TAXES IN EVEN-AGED MANAGEMENT

Until now, all of our analysis has been done without the consideration of taxes of any kind, be they income, property, estate, or other. Using the basic calculation of soil expectation value for an acre of bare ground, we now illustrate how two of the major kinds of taxes, income and property taxes, enter into our analysis to determine the economically optimum rotation and present value for the acre.

Income Taxes

Almost everyone has some experience with federal and state income taxes. As usual with such calculations, we first calculate the taxable income and then apply the appropriate tax rate to find the amount of these taxes that must be paid on timber production.

For federal tax purposes, income is classified into one of two categories: capital gains or ordinary income. Capital gains occur through profit from the sale of capital assets, where capital assets are generally defined as assets not bought or sold in the ordinary course of a person's or firm's trade or business. Ordinary income is any income that will not qualify for capital gains treatment.

Capital gains generally are taxed at a lower rate than ordinary income, so it is in the taxpayer's interest to have as much of his or her income classified as capital gains as possible. Because Congress has long wished to encourage investment in timber growing, U.S. tax regulations include special provisions (as of 1985) that make it possible for a taxpayer who owns and disposes of timber to receive capital gains treatment of timber on the income received, even though the taxpayer is in the business of growing and selling timber. While not all timber sales qualify for such treatment, we will assume that all timber income in our analysis qualifies [see Clutter et al. (1983) and Leuschner (1984) for more discussion].

Before applying the applicable income tax rate, the taxable income must be calculated. This income is equal to the gross income minus the permitted deductions. For tax purposes, all legitimate business expenditures may be deducted. A deduction may be taken in one of two ways: (1) *expensing* it—entirely deducting the expense in the tax year in which it occurs, or (2) *capitalizing* it—not entirely deducting the expense in the year in which it occurs, but rather prorating its deduction over some year or years in the future.

Costs that contribute directly to future productivity generally must be capitalized, while the rest can be expensed. In timber growing, costs that must be capitalized include site preparation, reforestation, and permanent road construction. Costs that are generally expensed include timber sale preparation, protection, maintenance. When a person or firm has taxable income, expensing an item is generally more valuable than capitalizing it because doing so lowers the immediate tax bill.

When an item can be deducted in the year in which it occurs (expensed), the after-tax cost of the item can be less than the actual cash outlay. Suppose that we have an annual management cost per acre of \$3.00. Deducting this expense will reduce the taxable income by \$3.00. Assuming that this expense is deductible from ordinary income and that the ordinary income tax rate is 40 percent, this annual management cost actually will reduce profits by only $3(1 - 0.40) = \$1.80$. Therefore, the management cost to the person or firm is only \$1.80 per acre instead of \$3.00, since the tax savings are \$1.20. In this form, the after-tax cost of an item that can be expensed is equal to (before-tax cost) $\times$ (1 $-$ tax rate).

As long as the person or firm has taxable income, this relationship applies. Otherwise the after-tax cost of these expenditures equals their before-tax costs. In our examples we will assume that the person or firm being studied has enough income taxable at the ordinary income rates for the reduction in the cost of items that can be expensed to apply as shown above.

Many forms of capitalization exist for tax purposes, but a particular type called *cost depletion* is used for timber-growing costs. Basically, cost depletion permits deducting the original investment cost against the value of that timber when sold. As an example, reforestation costs to produce a stand of timber are deducted from the value of the stand for tax purposes when the stand is sold. If a number of thinnings occur before the final harvest, the reforestation cost can be proportioned among these sales and the final harvest.

In addition to federal income tax, most people also pay state income tax. Capital gains rates are not available, in general, at the state level, and timber sales are taxed at the ordinary state tax rate for the person or firm in question. Because as of 1985 we do not have to pay federal tax on the income that goes for state tax, the effective state tax rate is equal to (applicable state tax rate) × (1 − applicable federal tax rate). To obtain the total income tax rate in each instance, we must add the applicable federal rate to the effective state rate.

Adjustment of the soil expectation formula Previously we calculated the value of an acre of land for timber production using the soil expectation formula [Eqs. (13.10) and (13.12)],

$$\text{SEV} = \frac{N_R}{(1+i)^R - 1} = \frac{V_R P_R - G(1+i)^R - T[(1+i)^R - 1]/i}{(1+i)^R - 1}$$

where N_R = net return at rotation age R
V_R = cubic foot volume per acre at age R
P_R = price per cubic foot for timber at age R
T = annual management cost per acre
G = regeneration cost per acre
i = guiding interest rate per year
R = rotation age in years

Because the annual management cost is independent of the rotation length, an equivalent formulation that we will use here is

$$\text{SEV} = \frac{V_R P_R - G(1+i)^R}{(1+i)^R - 1} - \frac{T}{i}$$

Considering income taxes, we would expand that equation to

$$\text{SEV} = \frac{V_R P_R - \pi(V_R P_R - G) - G(1+i)^R}{(1+i)^R - 1} - \frac{(1-\phi)T}{i} \quad (13\text{-}20)$$

where π = effective tax rate on capital gains, (federal capital gains tax rate) + [(state income tax rate) × (1 − federal ordinary income tax rate)]. If the capital gains rate does not apply, substitute ϕ for π.

ϕ = effective tax rate on ordinary income, (federal ordinary income tax rate) + [(state income tax rate) × (1 − federal ordinary income tax rate)]

$\pi(V_R P_R - G)$ = amount of income tax when G is capitalized
$(1 - \phi)T$ = effective annual management cost
Other symbols are as before.

Notice that the reforestation cost G is deducted as a depletion cost from the timber revenue at rotation before the application of π, and that the annual management costs T are expensed and deducted from ordinary income in the year of occurrence by multiplying them by $(1 - \phi)$.

Using the prices and costs from our previous example and some typical tax rates, we would have the following at a rotation age of 60:

Cubic feet at rotation age	= 8,212
Price per cubic foot	= \$0.75
Site preparation	= \$120
Planting	= \$100
Annual cost of management	= \$3
Capital gains tax rate	= 0.28
Federal ordinary income tax rate	= 0.46
State ordinary income tax rate	= 0.06

Therefore,

$$V_R = 8{,}212$$
$$P_R = 0.75$$
$$G = 220$$
$$T = 3$$
$$\pi = 0.28 + 0.06(1 - 0.46) = 0.28 + 0.032 = 0.312$$
$$\phi = 0.46 + 0.06(1 - 0.46) = 0.46 + 0.032 = 0.492$$

and

$$\text{SEV} = \frac{8{,}212 \times 0.75 - 0.312(8{,}212 \times 0.75 - 220) - 220(1 + 0.04)^{60}}{(1 + 0.04)^{60} - 1} - \frac{(1 - 0.492)(3)}{0.04}$$

$$= \frac{6{,}159 - 0.312(5{,}939) - 2{,}314}{9.5} - \frac{0.508(3)}{0.04}$$

$$= \frac{1{,}992}{9.5} - 38.1 = \$172$$

Therefore the inclusion of income taxes has lowered the soil expectation value for a 60-year rotation from \$329 (Table 13-5) to \$172 per acre.

Property Taxes

Nothing can make forest landowners' blood boil like a discussion of property taxes. From the early days of forestry in this country, when landowners would give up their cutover land rather than pay their property taxes, to the recent years, when numerous states have passed laws changing their property taxes, there has been a continuing controversy over what kinds and levels of property tax are fair to the landowner and society alike. We will look at three forms of the property tax for forestland: (1) the ad valorem tax (unmodified property tax), (2) the productivity tax, and (3) the yield tax.

The ad valorem tax (unmodified property tax) Property taxes ordinarily are based on the value of a property. Once this value has been appraised, the tax is established by multiplying the appraised value by the tax rate. Suppose that the tax rate is 0.5 percent of the assessed valuation, and the property is assessed at $20,000. Then the annual property tax would be $0.005 \times 20{,}000 = \$100$.

In applying this approach to an acre of bare ground that we are about to regenerate, we are faced with an unusual problem. At year 0, the only property to tax is the land itself, but over time as the land regenerates and develops merchantable timber, the property value will increase until rotation age, when it will again fall to the land value. If the assessed value rises with the property value, the property tax will rise as the stand ages. The landowner is then faced with the prospect of paying an ever-increasing tax bill year after year until the rotation age is reached. By increasing the cost of waiting for the timber to mature, such a tax tends to shorten rotations and thus reduces the investment in timber production, since much of the tax can be avoided by simply cutting the timber. In fact, as Fairchild (1935) first pointed out, the 0.5 percent property tax here acts much like a 0.5 percent increase in the guiding rate of return.

In terms of our soil expectation equation, inclusion of this form of the property tax is as follows:*

$$\text{SEV} = \frac{V_R P_R - G(1 + i + m)^R}{(1 + i + m)^R - 1} - \frac{T}{i + m} \tag{13-21}$$

where m is the tax rate as a proportion and the other symbols are as before.

Because of the impact of taxing full forest value annually, the ad valorem tax is almost always modified to exempt a portion of timber value. In many cases it is replaced by other types of taxes, as discussed below.

* Conditions needed for the property tax rate to be treated strictly as a change in interest rate are: (1) the forest market value in any year is the present value of the future revenue minus the present value of the future costs, (2) the assessed value is equal to the previous year's forest market value, and (3) taxes are fully capitalized into lower land values. See Klemperer (1976, 1981) for more discussion of these points.

The productivity tax To overcome the difficulties with the unmodified property tax, an alternative is to put a constant tax on the land, perhaps based on productivity. Now our soil expectation equation is modified to

$$\text{SEV} = \frac{V_R P_R - G(1 + i)^R}{(1 + i)^R - 1} - \frac{M + T}{i} \tag{13-22}$$

where M is the amount of land tax per acre per year and the other symbols are as before.

Here the rotation decision is independent of the tax (since the tax is the same for all rotation lengths, and this decision should not be influenced by it. Given that the tax does not become confiscatory (cause the SEV calculation to drop below 0), it should not affect the level of investment in timberland. Still, the owner has to pay the tax for a number of years before the next rotation can be harvested, and the need for such cash may cause some premature harvest.

The yield tax To make more of the property tax fall at the time timber revenues occur, another alternative is to obtain part or all of the property tax from forestland through a tax directly on timber revenue. Now our soil expectation equation is modified to

$$\text{SEV} = \frac{V_R P_R - Y(V_R P_R) - G(1 + i)^R}{(1 + i)^R - 1} - \frac{T}{i} \tag{13-23}$$

where Y is the yield tax as a proportion and the other symbols are as before.

Such a tax largely overcomes the problems with the unmodified property tax and the productivity tax. However, when the taxable value exceeds actual stumpage receipts, the yield tax can reduce the number of stands and the number of trees within a stand that are profitable to harvest.

Effect of the three types of property taxes on bare land values The soil expectation formula that we have employed in our tax discussion measures the value of bare land for timber production. Klemperer (1976) has noted that the three forms of tax discussed (ad valorem, productivity, and yield) do not have the same effect on the bare land value when all three raise the same dollar revenue over the rotation. Figure 13-8 shows the amount of property tax over the rotation under the three forms of the tax. The ad valorem tax starts low and then rises throughout the rotation as the value of the asset (land plus timber) increases. The productivity tax starts above the ad valorem tax, but stays constant over the rotation. Finally, the yield tax comes due entirely at rotation age when the timber is cut.

Since the timing of these taxes varies, the present value of the future

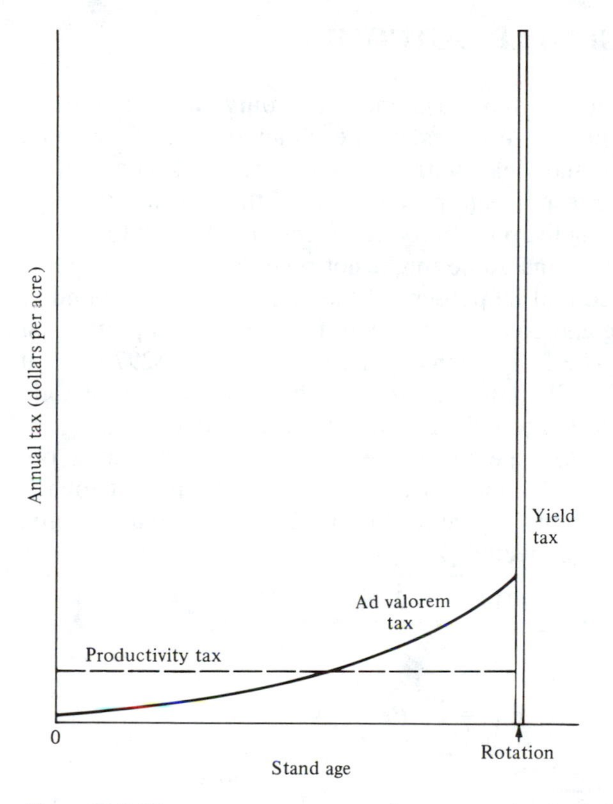

Figure 13-8 The tax per acre as a stand matures under three different forms of property tax. *(From Klemperer, 1976.)*

taxes under each system will differ for any interest rate greater than 0. The higher the proportion of tax dollars that come due toward rotation end, the lower will be their present value: the present value of future taxes will be least under the yield tax, greater with the ad valorem tax, and greatest for the productivity tax. Therefore given that taxes are capitalized into lower land value (as shown in our soil expectation formula under each tax), the maximum land value for bare forestland would be highest under the yield tax, next highest under the ad valorem tax, and lowest under the productivity tax.

If the tax authority wishes to encourage investment in timber production, the yield tax would seem the preferred property tax. While the taxing authority would be interested in the distribution of age classes across its tax area to see how different forms of the taxes would affect its flow of tax revenue over time, it probably is clear from this discussion why many people have advocated a yield tax as the major form of property tax on forestland.

CONSIDERING MULTIPLE OUTCOMES

Throughout this chapter, we have assumed that only one outcome is possible at each age or period. We presented each analysis as if we could predict the costs, prices, and yields with complete certainty. As an example, we assumed a single response to precommercial thinning in our even-aged analysis, not allowing the possibility that some stands might respond to precommercial thinning and some might not respond.

Suppose we believed that 80 percent of the stands would respond to precommercial thinning and 20 percent would not. Further, suppose that we calculate the SEV to be $314 when a stand responds and $297 when it does not. As discussed in Chap. 6, we can use these subjective probabilities (likelihoods) and returns to calculate an expected value.

As defined in Chap. 6, expected values are weighted averages obtained by multiplying the likelihood of each event by the payoff for the occurrence of that event, and summing the products across all events. Here the expected value (expected SEV) is

$$\text{SEV} = 0.8(314) + 0.2(297) = \$311$$

In general terms,

$$\text{EV} = \sum_{i=1}^{n} P_i r_i$$

where EV = expected value
P_i = the probability of event i
r_i = the payoff if event i occurs
n = the number of events

$$\sum_{i=1}^{n} P_i = 1.00$$

As we also discussed in Chap. 6, we may be faced with a sequence of probabilistic events to analyze: development of a stand is a sequence of probabilistic events over its life, such as occur in differential response to our cultural actions and the possibility of natural events such as insect attack.

A decision tree can be used to portray the sequence of probabilistic events. Numerous applications of decision trees have been made in forestry (Talerico et al., 1978, Hirsch et al., 1979, and Knapp et al., 1984). The approach can perhaps best be illustrated with an example taken from the work by Knapp et al. This example is quite similar to that shown in Fig. 6-2, with SEV now replacing "payoff."

Suppose we wish to find out how much the SEV for an acre of Douglas-fir site will be increased by being able to use herbicides for control of hardwood vegetation in the coast range of Oregon. The deci-

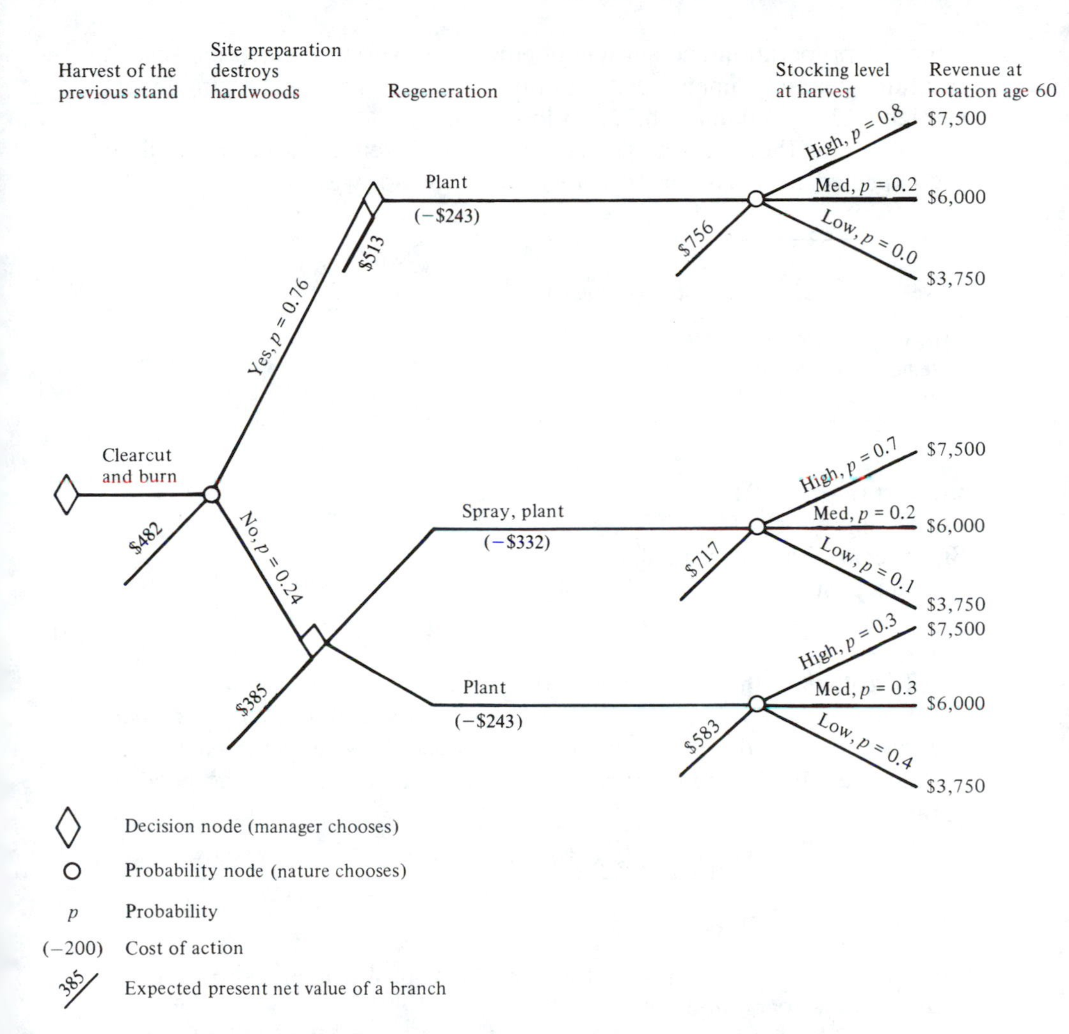

Figure 13-9 Decision tree showing calculation of expected values for stand regeneration under the "spray available" assumption.

sion tree when spraying is allowed is shown in Fig. 13-9. The first question is whether site preparation (burning after harvest) will adequately destroy the hardwoods that compete with Douglas-fir. Suppose that planting records yielded likelihoods of 0.76 for "yes" and 0.24 for "no." If "yes," the area is planted without further treatment. If "no," spraying to knock back the hardwoods might be done.

Each possible sequence of events leads to a distribution of the acre among three stocking levels of Douglas-fir at rotation age (60 years). As an example, the sequence of events that results in hardwoods being present after site preparation and spraying being done before planting causes, at rotation age, an estimated distribution of the acre among stocking levels of high = 0.7, medium = 0.2, and low = 0.1.

Assuming the following yields, costs, and revenues, we can calculate the expected SEV associated with whether or not we spray:

Stocking level	Yield at rotation age (age 60)
High	10,000
Medium	8,000
Low	5,000

Interest rate = 0.04
Price/cubic foot = \$0.75
Site preparation cut = \$120
Planting cost = \$100
Spraying = \$80

To calculate the SEV's under the different options, we work backwards through the decision tree, calculating expected values for each branch at each node. For example, the expected value of the harvest at rotation age for the branch through the tree that results from no need to spray is

$$\text{EV} = \frac{(\$0.75 \times 0.8 \times 10{,}000) + (\$0.75 \times 0.2 \times 8{,}000)}{(1 + 0.04)^{60} - 1}$$
$$= \$756$$

The expected values (SEV's) for the spray problem are shown on a slanted line at each node in Fig. 13-9.

The highest expected value at each node is used to calculate the expected value at the preceding node as the analysis moves backwards through the decision tree to the point of origin, i.e., the point at which an area has been clearcut and the manager faces the regeneration decision. As an example, the SEV from the different branches that result from the spray/no spray decision are 385 for spray and 340 for no spray. Since the SEV for the spray decision is higher, it appears on the diagram. When spraying is allowed, it should be done when hardwoods appear after site preparation to maximize SEV.

As Knapp et al. point out, "the expected value for one of the branches at that point distills all of the events, likelihoods, costs, payoffs, and subsequent decisions that the manager faces in making the initial decision that puts him or her on that branch."

The overall SEV when spraying is allowed is $482, as shown at the first node in Fig. 13-9. The analysis when spraying is not allowed is similar except that the spray/plant branch is not allowed and the overall SEV in this latter analysis is $470. Thus being able to spray when needed increases the SEV by $12.

In a world where little is certain, expected value techniques and other techniques that recognize the multiple outcome nature of forest management definitely have a place. The difficulties in obtaining the needed probability information should not be underestimated, but as more experience is gained in management, data for these probabilities are naturally produced. While the techniques have not yet been formally used very widely in stand decision models, that may change as we become more sophisticated.

QUESTIONS

13-1 You own one majestic yellow poplar tree that now contains 1,000 bd ft. Over the next 10 years, you expect it to increase by 500 bd ft. Assume that

(*a*) You wish to maximize your PNW over an infinite planning horizon using a guiding rate of return of 4 percent.

(*b*) You expect yellow poplar timber to be worth $200/M bd ft now and in the future.

(*c*) After you cut this tree, another yellow poplar tree will become established that will grow to 750 bd ft in 25 years and to 1,000 bd feet in 35 years.

Should you wait to cut your tree for at least 10 years or should you cut it now? First, make the calculations in terms of future values:

Tree value now	Tree value in 10 years	Tree value growth	Stock holding cost	Land holding cost	Total cost	Net gain from holding	Decision

Second, make the calculations in terms of present values:

	Cut $n = 0$	Leave $n = 10$
(1) SV_n		
(2) $\frac{SV_n}{(1+i)^n}$		
(3) MSEV		
(4) $\frac{MSEV}{(1+i)^n}$		
PNW [(2) + (4)]		

If the existing tree would not hinder establishment and growth of a young yellow poplar, would your decision change?

13-2 Suppose that you own two yellow poplar trees in a clump (*A* and *B*). They have the following growth and yield properties:

	Tree *A*	Tree *B*
Volume now	1,500	2,000
Volume in 5 years if leave both	1,900	2,300
Volume in 5 years if cut other	2,200	2,500

When both are cut, you plan to grow one yellow poplar in their place. Assuming the same prices and future generation yield as in Question 13-1, determine whether you should leave one or both trees for 5 more years or cut them now. Make the determination under both the future value and the present value approaches.

13-3 You are examining some land that you might buy to grow commercial timber. You figure that such an enterprise would have the following costs and returns per acre:

Planting cost $50
Precommercial thinning cost at age 10 $100
Commercial thinning return at age 30 $300
Final harvest return at age 50 $3,200

What is the maximum amount that you could pay for this land per acre assuming that

(*a*) You wish to maximize your present net worth?
(*b*) You have an infinite planning horizon?
(*c*) The rate of return on your next best investment is 5 percent?

13-4 Redo Question 13-3 under the assumption that you can also produce deer in addition to producing timber, according to the following yield table. Assume that each deer is worth $25.

Stand age	Output of deer per acre per year
0–10	2
11–50	1

What is the maximum amount that you could now pay for the acre?

13-5 You own 2 acres of timberland on which you would like to grow timber. You figure that such an enterprise would have the following costs and returns per acre:

Planting cost $100
Interest rate 4 percent
Revenue at final harvest:

Age	Net stumpage revenue
30	$ 900
40	1,500
50	1,700

Find the rotation age with the maximum soil expectation value. What is the maximum value? If you precommercially thinned, you believe that you could raise the net revenue at rotation age by 20 percent at a cost of $75 in year 10. Will precommercial thinning raise or lower your maximum soil expectation value? By how much?

13-6 Your land currently has two stands of timber on it, each occupying 1 acre, with the following inventory and growth characteristics:

Stand	Volume per acre now	Predicted volume per acre in 10 years	Net value/M now	Predicted net value/M in 10 years
A	3,000	5,000	$200	$220
B	5,000	6,000	250	250

Assume that future stands on your property will have the costs and returns of Question 13-5 and that you wish to maximize discounted financial return over an infinite planning horizon under a guiding rate of return of 4 percent.

Under these assumptions, should you keep these trees for 10 more years or cut them? First, make this calculation in terms of future values:

Stand	Value now	Value in 10 years	Value growth over 10 years	Stock holding cost	Land holding cost	Total cost	Net gain from holding	Decision

Then, make the calculation in terms of present value:

Stand *A*		Stand *B*	
Replace $n = 0$	Keep $n = 10$	Replace $n = 0$	Keep $n = 10$

13-7 How would the following changes in Question 13-6 tend to affect your decision to hold or replace the two stands in that problem?

(*a*) higher regeneration costs
(*b*) higher PCT costs
(*c*) higher value for future crops
(*d*) higher value in 10 years for the current stands
(*e*) an income tax of 25 percent

13-8 Suppose that we predict the following results for an acre of spruce in Colorado:

Age	Node	Initial or residual volume	Volume at age + 5	Residual volume at age + 5			
				All	1,500	1,000	0
				Harvest at age + 5 to achieve residual volume			
15		900	1,600	×	100	600	×
20		1,500	2,200	×			×
		1,000	1,600	×			×
25		1,500	2,000	×			
		1,000	1,550	×			
30		1,500	1,800	×	×	×	
		1,000	1,400	×	×	×	

(*a*) Fill in all harvest entries that are not ×ed out. Then draw a network with the residual volume at each age as the nodes, showing the harvest possibilities (and associated rewards) at each age. Finally, use forward recursive dynamic programming to solve the network for the combination of thinning and final harvest that maximizes the mean annual increment for a rotation of 30 years and for a rotation of 35 years. Which rotation gives the highest mean annual increment? What is the associated path? *Hint:* Maximize the volume harvested over each rotation using dynamic programming and then convert this maximum volume harvest over each rotation into a mean annual increment for comparison.

(*b*) Using the same data, redraw the network so that the rewards reflect an objective of maximizing present net worth over one rotation under an interest rate of 4 percent, a net stumpage price of $1.00/ft^3, and a regeneration cost of $100. Use forward recursive dynamic programming to search the network to find the combination of thinning and final harvest that maximizes present net worth for a rotation of 30 years and for a rotation of 35 years. Calculate the soil expectation value associated with each rotation. Which rotation gives the highest soil expectation value? What path does this rotation follow?

REFERENCES

Adams, D. M., 1976: A Note on the Interdependence of Stand Structure and Best Stocking in a Selection Forest, *Forest Sci.*, **22**:181–184.

——— and A. R. Ek, 1974: Optimizing the Management of Uneven-Aged Forest Stands, *Can. J. Forest Res.*, **4**:274–287.

Brodie, J. D., and C. Kao, 1979: Optimizing Thinning in Douglas-Fir with Three Descriptor Dynamic Programming to Account for Accelerated Diameter Growth, *Forest Sci.*, **24**:513–522.

Calish, S. R., D. Fight, and D. E. Teeguarden, 1978: How Do Nontimber Values Affect Douglas-Fir Rotations?, *J. Forestry*, **76**:217–221.

Clutter, J. R., J. C. Fortson, L. V. Pienaer, G. Brister, and R. L. Bailey, 1983: "Timber Management: A Quantitative Approach," Wiley, New York.

Curtis, R. O., G. W. Clendenon, and D. J. DeMars, 1981: "A New Stand Simulator for Coast Douglas-Fir: DFSIM User's Guide," U.S. Forest Service General Technical Report PNW-128.

Dreyfus, S. E., and A. M. Law, 1977: "The Art and Theory of Dynamic Programming," Academic Press, New York.

Duerr, W. A., 1960: "Fundamentals of Forestry Economics," McGraw-Hill, New York.

Duerr, W. A., and W. E. Bond, 1952: Optimum Stocking of a Selection Forest, *J. Forestry,* **50:**12–16.

Duerr, W. A., J. Fedkiw, and S. Guttenberg, 1956: "Financial Maturity: A Guide to Profitable Timber Growing," U.S. Department of Agriculture Technical Bulletin 1146.

Dykstra, D. P., 1984: "Mathematical Programming for Natural Resource Management," McGraw-Hill, New York.

Fairchild, F. R., 1935: "Forest Taxation in the United States," U.S. Department of Agriculture Miscellaneous Publication 218.

Haight, R., J. D. Brodie, and D. Adams, 1985: Optimizing the Sequence of Diameter Distributions and Selection Harvests for Uneven-Aged Stand Management, *Forest Sci.,* **31:**451–462.

Hirsch, S. N., G. F. Meyer, and D. L. Ratloff, 1979: "Choosing an Activity Fuel Treatment for Southwest Ponderosa Pine," U.S. Forest Service General Technical Report RM-67.

Klemperer, W. D., 1976: Impacts of Tax Alternatives on Forest Values and Investment, *Land Economics,* **52:**135–157.

———, 1981: "An Analysis of Selected Property Tax Exemptions for Timber," Report submitted to the Lincoln Institute of Land Policy, Cambridge, Mass.

Knapp, W., T. Turpin, and J. Beuter, 1984: Vegetation Control for Douglas-Fir Regeneration in the Suislaw National Forest: A Decision Analysis, *J. Forestry,* **82,**168–172.

Leuschner, W. A., 1984: "Introduction to Forest Resource Management," Wiley, New York.

McArdle, R. E., and W. H. Meyer, 1930: "The Yield of Douglas-Fir in the Pacific Northwest," U.S. Department of Agriculture Technical Bulletin 201.

Ritters, K., J. D. Brodie, and D. W. Hann, 1982: Dynamic Programming for Optimization of Timber Production and Grazing in Ponderosa Pine, *Forest Sci.,* **28:**517–526.

Sleavin, K. E., and K. N. Johnson, 1983: Searching the Response Surface of Stand Simulators under Different Objectives and Constraints . . . DFSIM as a Case Study, in "Proceedings of the National Silviculture Workshop on Economics of Silvicultural Investments," Timber Management, U.S. Forest Service, pp. 106–128.

Talerico, R. L., C. M. Newton, and H. T. Valentine, 1978: Pest-Control Decisions by Decision-Tree Analysis, *J. Forestry,* **76:**16–19.

Tarrant, R., B. Bormann, D. DeBell, and W. Atkinson, 1983: Managing Red Alder in the Douglas-Fir Region: Some Possibilities, *J. Forestry,* **81:**787–792.

CHAPTER

FOURTEEN

CLASSICAL APPROACHES TO FOREST REGULATION

The organization of a forest property to provide an even flow of timber products forms the heart of traditional forest management for timber production. To understand why the idea of even flow is so firmly embedded in forestry thinking and literature, we must trace the intellectual history of forestry in the United States back to its roots in Europe.

European foresters in the 1800s, especially German foresters, emphasized the notion of a forest organized to produce an even flow of timber as the way to deal with their particular situation. Their forests were largely publicly owned and had been continuously cut for hundreds of years. Wood product uses changed slowly. Self-sufficiency in timber production was a primary goal in each country's wood products policies. Stability was the overriding economic tenet of the day. All of these factors helped create a psychological climate in which organizing a forest for even flow became the guiding ideal in Europe's forest management.

Our early foresters—Fernow, Schenck, and Pinchot—were educated in Europe. Most went to German schools where "forest regulation" was at its zenith as an idea around which forestry was organized. Not surprisingly, this idea was transferred intact to the United States. Fear of a timber famine made this country's forestry ripe for embracing the European idea of a forest organized to produce an even flow of timber forever. Such a fear helped bring the forestry profession into existence, slowed the movement of public domain forestland into private ownership, and resulted in the creation of the national forests. Organizing a forest to produce a continuous flow of wood served as an intellectual rallying point for the new forestry profession just as the creation of the national forests was a political rallying point for laypersons concerned with overcutting and timber shortages.

The European source of professional education for early foresters in the United States combined with worry over timber famine created a climate in which the idea of forest regulation for continuous production could flourish. As stated in the first forest management book published in the United States (Roth, 1925), there are many reasons why it is desirable to obtain a regular harvest from the forest:

1. A yearly cut of approximately equal volume, size, quality, and value of timber provides a stable business planning base.
2. A current harvest, growth, and income are obtained from a forest growing stock no larger than necessary.
3. An approximate balance between yearly expenditures and receipts is obtained. Land and income taxes are levied annually, as are many other costs; financial management is facilitated by a reasonably current equation between income and outgo.
4. Safety from fire, insects, diseases, and other dangers is maintained because the forest is kept growing, vigorous, and usually well-distributed in size, age, and condition over the forest area.
5. There is maximum opportunity for correlation with other forestland uses—recreation, wildlife, watershed protection, and forage—on a stable planned basis.
6. Continuity of work load is ensured, thus providing regular employment.

Right or wrong, such a litany has been the driving force in forest management planning during much of this century.

A major purpose of this chapter is to explain the meaning of the "fully regulated forest"—a forest that produces an even flow of timber forever—for even-aged and uneven-aged stands. Given this idea, traditional approaches to determining the harvest for forests not yet regulated are described.

STRUCTURE OF A FULLY REGULATED FOREST

How is a forest organized for continued production? The best place to start our inquiry is at the end—to describe how the forest looks and behaves when the regulated structure is finally attained. Even though this idealized final structure is rarely attained, it serves to focus our thinking and provides a standard by which timber management progress has traditionally been measured.

The essential requirements of a fully regulated forest are that age and size classes be represented in such proportion and be consistently growing at such rates that an approximately equal annual or periodic yield of products of desired sizes and quality may be obtained in perpetuity. A progression of size and age classes must exist such that an approximately equal volume and size of harvestable trees are regularly available for cutting.

Even-Aged Management Systems

Even-aged management is keyed to the periodic birth and death of stands as determined by the rotation age. While rotation age is surely the dominant decision, many others are needed to characterize the structure, quality, and growth of stands on each acre.

Decisions needed in even-aged management

1. *Rotation length.* The interval between one regeneration harvest and the next regeneration harvest. The stand age at final harvest and the rotation interval between harvests are the same if new trees are successfully established the same year the existing stand is harvested. If establishment is delayed, age at harvest is less than the rotation interval.
2. *Commercial thinning.* The number and timing of entries and the amount and kind of trees removed at each intermediate entry prior to regeneration harvest.
3. *Species for regeneration.* Species and genetic stock selected to regenerate each stand type in the forest.
4. *Site preparation and regeneration method.* The combination of pre- and postharvest treatments scheduled to establish the desired species and control early growth.
5. *Other cultural treatment.* Precommercial thinning, release, and fertilization.

In the simplest case, a regulated even-aged forest is composed of a single species on a constant site under one management intensity with each age class forming one stand type. Let A be the number of acres in such a forest and R the selected rotation age. In this classical regulated forest, the acres in the forest are distributed such that A/R acres are in each age class from 1 to R years and there are R stand types in the forest. If all acres of each age class were contiguous, then this would be a forest containing R stands.

With this distribution, the acres in the oldest age class are final harvested and regenerated each year, becoming the acres in the youngest age class in the following year. All other age classes are treated as prescribed and annually increment one age class. With each stand type having equal acreage, the structure of the forest stays constant from year to year, the same number of acres are cut in each year, the same harvest volume is produced each year, and harvest equals growth.

Since each stand type goes through the same sequence of actions over its life as every other stand type, the sum of the actions that occur across the forest each year is also the sum of the actions that occur across each stand type over its life. Therefore, the harvest volume that is produced from the forest each year equals the harvest volume produced from each stand type over its life.

The equality between the annual harvest from the forest and the harvest over the life of each stand type can be used to formally define the structure of a regulated forest. We want to show that harvest equals growth in such a forest structure. The annual harvest from the forest is

$$H_t = (v_1a_1 + v_2a_2 + \cdots + v_{R-1}a_{R-1} + v_Ra_R)_t$$

where

H_t = total harvest in year t,
v_i = prescribed harvest per acre from a stand type that has attained age i in year t ($v_i \neq 0$ only for stand ages with a prescribed harvest entry),
a_i = area in stand type i in year t.

Stand types are defined by the age of the stand.

In a regulated forest by definition $a_i = A/R$ for all i and

$$a_1 = a_2 = \cdots = a_{R-1} = a_R$$

or

$$a_i = a \qquad \text{for all } i$$

Then in any year t there will always be the same number of acres in each stand type. We can write the regulated forest harvest as

$$H_t = (v_1a + v_2a + \cdots + v_{R-1}a + v_Ra)$$

$$H_t = a(v_1 + v_2 + \cdots + v_{R-1} + v_R)$$

and also that

$$H_t = H_{t+1} \qquad \text{for all } t$$

Mean annual increment of a stand is the average growth of a stand over its life and can be defined for a rotation age stand as

$$\text{MAI} = \frac{V}{R}$$

where

V = total volume harvested from an acre over its lifetime
$= v_1 + v_2 + \cdots + v_{R-1} + v_R$

Multiplying MAI by a/a, we obtain

$$\text{MAI} = \frac{aV}{aR}$$

Since $aV = H$ (forest harvest) and $aR = A$ (forest area),

$$\text{MAI} = \frac{H}{A}$$

and, rearranging,

$$H = \text{MAI} \times A$$

Growth in a regulated forest can be measured by the product of the forest acreage times the mean annual increment of an average stand over its life (MAI $\times$ A). Since H = MAI $\times$ A here, harvest equals growth in a regulated forest.

To better understand the rhythm of a regulated even-aged forest, consider a regulated forest and management system with the following specifications and management prescription:

Forest area	2,000 acres (A = 2,000)
Site quality	All one stand type, site II
Species	Slash pine, genetic stock xxx
Site preparation	Burn, chop, plant 2-0 stock at 10-ft by 10-ft spacing
Rotation	20 years, yield 20 cords (R = 20)
Thinning	One commercial entry at age 15, yield 5 cords

The distribution of acres by age and volume per acre in the regulated forest is shown in Fig. 14-1*a* just before harvest, with A/R = 2,000/20 = 100 acres being final harvested, planted, and site-prepared each year and another 100 acres being thinned. There are 100 acres in each age class from 1 to 20 years, accounting for all the acres in the forest.

As each acre produces 5 cords from thinnings and 20 cords from final harvest every 20 years, the mean annual increment (MAI) of each acre under this management system is (5 + 20)/20, or 1.25 cords per acre per year. The annual growth and harvest on the forest is MAI $\times$ A or 1.25 $\times$ 2,000 = 2,500 cords per year. Since this harvest can be perpetuated forever, it is also called the long-term sustained yield (LTSY) of this forest under the given management prescription. As a check, each year 100 acres are final harvested (producing 2,000 cords), and 100 acres are thinned (producing 500 cords), for a total harvest of 2,500 cords per year. Because growth equals harvest in a fully regulated forest, such a check can be made.

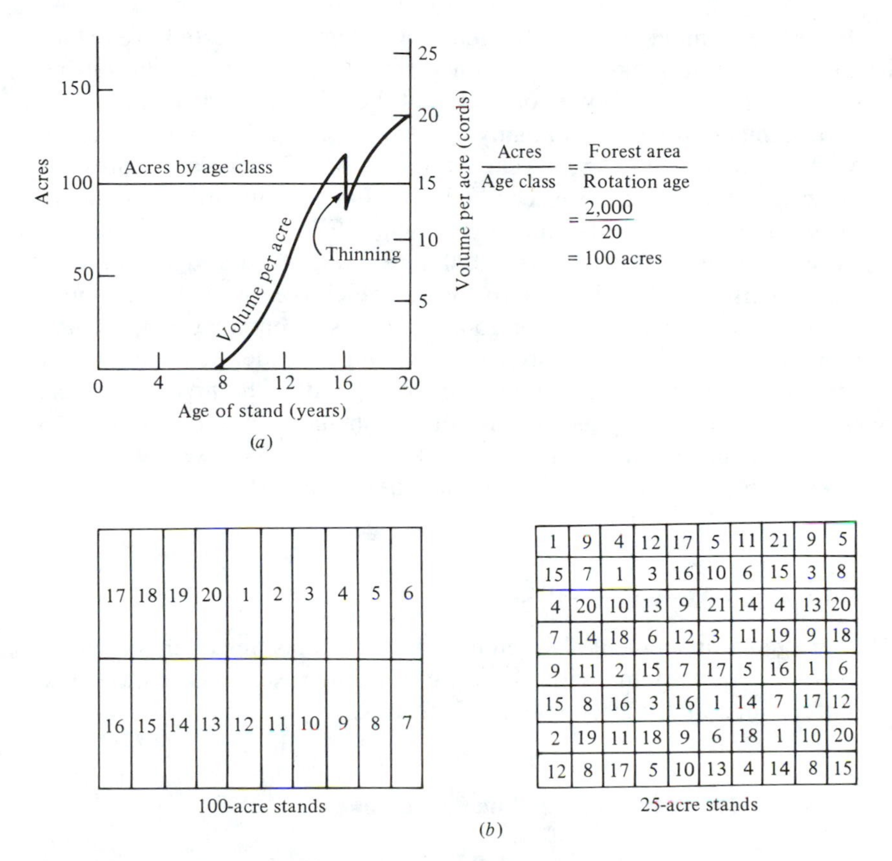

Figure 14-1 (*a*) Distribution of area by age and volume per acre of a fully regulated 2,000-acre forest under even-aged management on a 20-year rotation. (*b*) Stand age in two different layouts of cutting compartments of a fully regulated 2,000-acre forest on a 20-year rotation just before cutting, under even-aged management.

The important annual activities of this regulated forest, given a few assumptions about costs and prices, are

1. Final harvest	100 acres
2. Thinned	100 acres
3. Volume harvested	2,500 cords
4. Site prepared and planted	100 acres
5. Seedlings planted (435/acre at 10 ft by 10 ft)	43,500 seedlings
6. Timber sale revenue at $10/cord	$25,000
7. Regeneration cost at $150/acre to prepare and plant	$15,000
8. Fixed cost at $3/acre for foresters and protection	$6,000
9. Net annual income = $25,000 − $15,000 − $6,000	$4,000

Once the number of acres treated and volumes harvested are established, it is easy to make the additional input and output calculations as suggested in items 5 to 9. We could extend the list to include estimates of labor, logging equipment, and many other things. These annual activities are what an owner can expect year after year on this regulated forest.

This schedule of activities can be located in a virtual infinity of ways on the ground. Two such layouts are shown in Fig. 14-1*b*, the first shows the forest cut into 20 contiguous 100-acre stands and sequenced with military precision, while Fig. 14-1*b* shows age classes spread throughout the forest by creating 80 smaller 25-acre stands to enhance wildlife and aesthetic values. This second layout will produce the identical numbers of stand types, acres treated, and volumes harvested as the first. However, logging costs may be higher because of the smaller individual stand. As the stands become smaller and smaller, they eventually take the form of even-aged groups within an overall uneven-aged forest.

Uneven-Aged Management Systems

Uneven-aged management also requires a set of decisions to characterize its management system, with cutting cycle and reserve growing-stock level often seen as the predominant decisions.

Decisions needed in uneven-aged management

1. *Cutting cycle.* Number of years between harvest entry on each acre.
2. *Reserve growing stock level.* Residual volume or basal area per acre of the stand immediately after harvest.
3. *Stand structure.* Number of trees per acre by species and diameter that make up the reserve growing stock.
4. *Sustainability procedures.* Constraints established on harvesting and regeneration to ensure maintenance of the stand structure, and thus perpetuation of the harvest over all future cutting cycles.
5. *Other cultural treatment.* Release and fertilization.
6. *Species for regeneration.* Species and genetic stock selected for each stand type in the forest.
7. *Site preparation and regeneration method.* Combination of pre- and postharvest treatments scheduled to establish the desired species and control their early growth.

The first two decisions define the macrostructure of the regulated uneven-aged forest, while the next four affect the structure, quality, and growth performance of the stands on each acre. Sustainability procedures were added to this list of traditional variables simply to recognize that managerial decisions are required to perpetuate a forest and consider it fully regulated. Decisions 6 and 7 are needed if desired natural regeneration needs stimulation or when underplanting is needed to obtain adequate regeneration.

In the simplest case, a regulated uneven-aged forest is composed of a single species on a constant site under one management intensity. Let CC be the cutting cycle in years, RGS be the level of reserve growing stock, V_{CC} be the volume removed per acre at the end of the cutting cycle, and A be the forest acreage. A/CC acres will be entered for harvest annually in a regulated uneven-aged forest, removing all volume in excess of RGS per acre, which will produce the same harvest at each entry. The periodic annual increment (PAI) of an acre over a cutting cycle is V_{CC}/CC.* The acreage in the forest will be distributed such that A/CC acres will be in each "years since cutting" class between 1 and CC years. These "years since cutting" classes are the stand types for uneven-aged management. When harvested, a stand type will have grown for CC years since the previous harvest. Since all stand types have equal acreage, the structure of the forest stays constant from year to year, the same number of acres are cut in each year, the same harvest volume is produced each year, the PAI is the same for all acres, and harvest equals growth.

Since each stand type goes through the same sequence of actions over a cutting cycle as every other stand type, the actions that occur across the forest each year are also the actions that occur in each stand type over the cutting cycle. Therefore, the harvest volume that is produced from the forest each year equals the harvest volume produced from each stand type over the cutting cycle. Further, as can be demonstrated in a manner similar to that previously used for the even-aged regulated forest, the total annual harvest is equal to the product of the forest acreage times the periodic annual increment (PAI) of a stand type over the cutting cycle [$H = A(\text{PAI})$]. Also, the long-term sustained yield of the forest equals this harvest level. To better understand the rhythm of a regulated uneven-aged forest, consider a regulated uneven-aged forest with the following specifications and management prescription:

Forest area	2,000 acres (A = 2,000)
Site quality	All one stand type, site II
Species	Average genetic quality
Cutting cycle	10 years (CC = 10)
Reserve growing stock	16 M bd ft (RGS = 16)
10-year growth (V_{CC}) from RGS	6 M bd ft

* For the regulated uneven-aged forest PAI is the counterpart of MAI for even-aged forests.

The distribution of acreage in this forest by volume per acre and "years since cutting" at a point in time just before harvest is shown in Fig. 14-2*a*. There are 2,000/10, or 200 acres in each "year since cutting" class; the volume in each of these classes increases from 16.6 M bd ft on the acres that were cut the year before to 22 M bd ft on the acres that are just about to be cut.

The reserve growing stock level of 16 M bd ft is measured immediately after cutting and before any growth has occurred. Each year 200 acres would be selectively logged, removing an average of 22 minus 16, or 6 M bd ft per acre, for a total forest harvest of 1,200 M bd ft.

As each acre produces 6 M bd ft every 10 years, the periodic annual increment (PAI) of each acre under this management system is 6/10, or

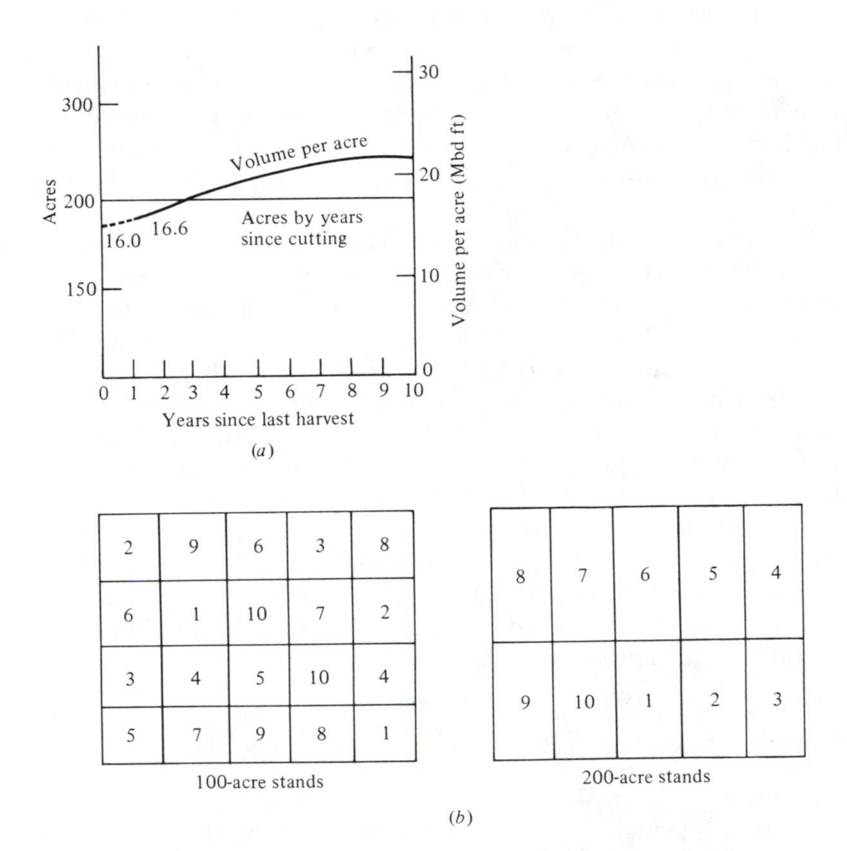

Figure 14-2 (*a*) Distribution of area by years since last harvest and volume per acre of a fully regulated 2,000-acre forest under uneven-aged management with a 10-year cutting cycle and a reserve growing stock of 16 M bd ft. (*b*) Years since last harvest for two different layouts of a fully regulated 2,000-acre forest on a 10-year cutting cycle just before cutting, under uneven-aged management.

0.6 M bd ft per acre per year. The long-term sustained yield level of the forest is the PAI × A, or 0.6 × 2,000 = 1,200 M bd ft per year. (As a check, each year 200 acres are harvested, producing 1,200 M bd ft per year. Again, because growth equals harvest in a fully regulated forest, such a check can be made.)

A schedule of activities for the regulated uneven-aged forest, similar to that for the regulated even-aged forest, could be generated. Probably, such a schedule would show more marking and sale administration and less site preparation and cultural activity.

Unlike the brute force simplicity of clearcut and plant, the number, size, and species of trees marked at each harvest entry become crucial to implementing and sustaining uneven-aged management and require considerable professional sophistication and experience to do well. Genetic control of the regenerated stand is limited since sprouting and natural seedlings usually are the main source of genetic material; typically, new genetic stock is not introduced by planting.

Harvesting compartments could be laid out in many ways on the ground. Two possible layouts are shown in Fig. 14-2*b*: on the right the forest is organized into 10 stands of 200 acres each, and on the left it is organized into 20 stands of 100 acres each. Both layouts contain 10 stand types.

STRATEGIES FOR FOREST REGULATION

Given the existing forest and a conception of the fully regulated forest that we would like to achieve, classical timber management scheduling addressed the questions of how many acres and how much volume to cut. Over the years of forest history, a great many methods for determining the cut were developed in various parts of the world, especially in Europe. Recknagel (1917) describes 18 methods; more have been devised since then. Many were never applied to any large extent in the woods because they were designed to meet more or less special situations. Even so, a broad survey of these methods will help us appreciate the approaches that past generations of foresters puzzled over and will also help us understand the nature of timber management scheduling problems. Most of these methods fall into one of two categories:

1. *Area control.* A regulated forest can be achieved in one rotation, but often at the sacrifice of achieving a stable harvest in the near future.
2. *Volume control.* A stable harvest can be achieved in the near future, but often at the sacrifice of not achieving a regulated forest after one rotation.

After presenting typical techniques for each regulation strategy, methods which combine area and volume control to exploit the advantages of both will be examined.

Area Control

The principle of area control is very simple: harvest and regenerate the same number of acres each year or period that would be harvested in a fully regulated forest. The resultant volume harvested is defined by the timber on the area scheduled for cutting each year.

Application to even-aged stands Area control is the easiest way to regulate an unmanaged forest and guarantee that the regulated structure is attained within one rotation. If the forest to be managed on the 60-year rotation is 2,500 acres in size, and if each year 2,500/60, or 41.7 acres are harvested and successfully regenerated, this forest will have the balanced age class distribution of the regulated forest immediately after the 60th harvest, regardless of the age, species, and volume structure of the initial forest.

Applied to a forest that is initially irregular in age class structure or site, area control can yield a fluctuating timber harvest volume and size while regulation is being achieved. Consider a loblolly pine forest that now contains 20 acres of 55-year-old timber and 10 acres of 15-year-old timber and that grows at a rate to produce the yields shown for site index 90 in Table 14-1. Application of strict area control under a 30-year rotation and a rule of taking the oldest timber first will result in the periodic harvest increasing and then decreasing (Table 14-2). In this example the forest is entered in the middle of each 10-year period to harvest one-third of the area and create a forest of three different age classes or stand types.

Table 14-1 Cubic measure per acre, including stump and tip but not bark for trees 3.6 in. and larger yield of even-aged loblolly pine

Age (years)	Site index (ft) 70	80	90	100	110	120
20	1,250	1,500	1,750	2,000	2,300	2,650
30	2,500	2,970	3,440	3,930	4,400	4,870
40	3,120	3,710	4,290	4,910	5,490	6,080
50	3,450	4,090	4,740	5,420	6,070	6,720
60	3,680	4,370	5,070	5,790	6,480	7,170
70	3,870	4,600	5,330	6,090	6,820	7,550
80	4,030	4,800	5,560	6,350	7,110	7,870

Source: Meyer (1942).

Table 14-2 Achieving a regulated forest through area control*

	Age (years)	Inventory: Acres	Inventory: Volume per acre (M ft³)	Inventory: Total volume (M ft³)	Cut: Acres	Cut: Volume (M ft³)	Periodic growth (M ft³)
Period 1							
Before cut	60	20	5.07	101.4	10	50.7	
	20	10	1.75	17.5			
				118.9			
After cut, before growth	60	10	5.07	50.7			2.6
	20	10	1.75	17.5			16.9
	0	10	0	0			0
				68.2			19.5
Period 2							
Before cut	70	10	5.33	53.3	10	53.3	
	30	10	3.44	34.4			
	10	10	0	0			
				87.7			
After cut, before growth	30	10	3.44	34.4			8.5
	10	10	0	0			17.5
	0	10	0	0			0
				34.4			26.0
Period 3							
Before cut	40	10	4.29	42.9	10	42.9	
	20	10	1.75	17.5			
	10	10	0	0			
				60.4			
After cut, before growth	20	10	1.75	17.5			16.9
	10	10	0	0			17.5
	0	10	0	0			0
				17.5			34.4
Period 4							
Before cut	30	10	3.44	34.4	10	34.4	
	20	10	1.75	17.5			
	10	10	0	0			
				51.9			
After cut, before growth	20	10	1.75	17.5			16.9
	10	10	0	0			17.5
	0	10	0	0			
				17.5			34.4

* Calculation of harvest, growth, and inventory from a 30-acre loblolly pine forest on site 90 to be managed on a 30-year rotation under area control. 10-year periods are used. Starting structure at year 5 (middle of first period) includes 20 acres at age 60 and 10 acres at age 20. Yields come from Table 14-1. In the middle of each period, cut A/R = 30 acres/3 periods = 10 acres/period, and then grow. Cut the oldest stands first.

After a full rotation (30 years, or three periods), the forest is completely regulated, with an equal number of acres in age classes 10, 20, 30 just before harvest and an equal number of acres in age classes 0, 10, and 20 just after harvest. In each period after regulation, as shown in period 4, harvest of the oldest age class (age 30) produces a volume of 34.4 M ft^3 and reduces the inventory from 51.9 to 17.5 M ft^3. Growth to the next period brings the inventory volume back to 51.9 M ft^3, then harvest of the oldest age class again produces 34.4 M ft^3, and so on forever.

A variety of twists on simple area control can help smooth volume flows during regulation. It is not necessary, for example, to cut the oldest timber each year. As far as the establishment of future age classes is concerned, it does not matter what age of timber is cut. In the case study on the Davy Crockett National Forest given below, a variety of age classes are cut, and some area objectives are not entirely achieved, in order to smooth volume flows and meet wildlife and recreation objectives.

Application to uneven-aged stands Area control can be used to guide the amount of acreage entered for harvesting each year in an uneven-aged forest. We saw earlier that once an uneven-aged forest is fully regulated, an equal area in one or more cutting compartments is entered each year of the cutting cycle, removing an equal volume per acre. For the unregulated forest, however, the number, size, and species of trees harvested depend on growth rates, the existing and desired species, and the diameter distribution of the stands making up the initial forest.

Applying area control as a way of bringing an unregulated uneven-aged forest into a regulated condition over one or more cutting cycles can cause irregular volume flows. Often it is not possible to achieve the desired diameter structure in a few cutting cycles; the needed residual trees may just not be there, or regeneration may not be as prompt or ample as needed. Thus area regulation will not guarantee a regulated forest over a set time period as it did for the even-aged system.

Application where different site qualities are present Seldom does a forest of any extent occur on land of the same site quality. Site variability complicates but does not upset the application of area control. If variations in site quality are more or less randomly distributed over the forest, as in a gently rolling area where bottom, slope, and ridge top sites are closely intermingled, an average site may be assumed and site differentials handled as a matter of on-the-ground adjustment. Where the forest includes lands of substantially different site qualities, recognition of these site differences is necessary in dividing the area up into stand cutting units for both even-aged and uneven-aged management systems.

The needed adjustments can be illustrated best by an example involving a regulated forest of varying sites but one common rotation length

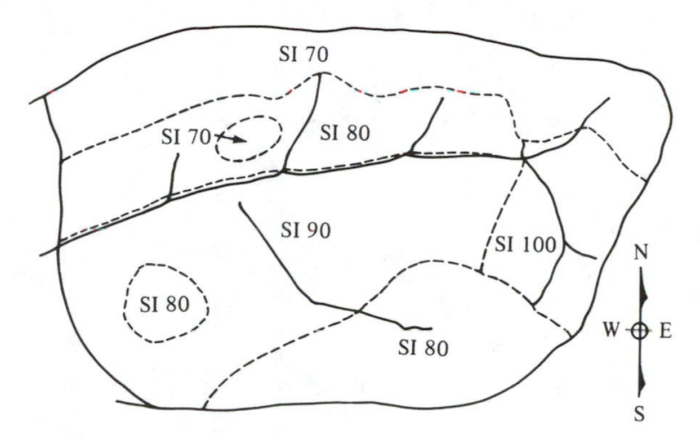

Figure 14-3 Site index map of a management area.

(Fig. 14-3). The area has been mapped by loblolly pine site index quality. As shown, the south-facing slopes along the north side of the drainage average low site quality. The site improves on the north slopes, and a cove of particularly high quality exists at the head of the drainage.

With even-aged management to regulate the forest by area control it would be necessary to cut, on the average, an area equal to the total area/rotation each year. The same would be true for the area of intermediate cuttings in each age group treated. The cutting areas would be of the same size from year to year only if all lands were of equal site quality. They would not be of the same size in this forest because of large differences in site quality over various portions of the area. The basic aim here is to harvest areas of *equal productivity* rather than of equal surface area. It would obviously be necessary to cut more acres of site index 70 land than of site index 100 land to get the same total yield. Including lands of more than one site quality in the same cutting area further complicates the picture.

A solution to the problem can be indicated for the forest shown here to illustrate a general approach. Assume: (1) the total forest area is 2,500 acres, (2) the forest is being managed on a 60-year rotation, (3) average site quality differentials are given by differences in cubic foot yields from Table 14-1.

Using these data, we can calculate the relative productivities of the different sites, the number of acres to be harvested each year under area control regulation, and the long-term sustained yield (LTSY) of the forest. All of these calculations are presented in Table 14-3. Note that all calculations are based only on the number of acres by site class of the forest and the rotation prescription used for management of the regulated forest. The inventory and the growth rate of the current forest do not enter the analysis.

Table 14-3 Productivity estimates and area control harvest acreage for a 2,500-acre forest with four site qualities on a 60-year rotation

Site index (1)	Area of forest (acres) (2)	Regenerated stand yield at age 60 (M ft³) (3)	Relative productivity to site 80 (4) = (3) ÷ 4.37	Acres of land equal in productivity to 1 acre of site 80 (5) = 1 ÷ (4)	Mean annual increment (MAI) (ft³) (6) = (3) ÷ 60 × 1,000	LTSY, the annual growth and yield from forest (ft³) (7) = (2) × (6)	Productivity equivalent forest of all site 80 (acres) (8) = (2) × (4)	Acres cut per year in each site for equal yields, rotation= 60 years (acres)* (9)	Years of cutting in stands of each site class (10) = (2) ÷ (9)
70	550	3.68	0.842	1.19	61.33	33,731	463.1	51.96	10.58
80	885	4.37	1.000	1.00	72.83	64,454	885.0	43.75	20.23
90	815	5.07	1.160	0.86	84.50	68,867	945.4	37.72	21.61
100	250	5.79	1.325	0.75	96.50	24,125	331.3	33.02	7.58
	2,500					191,177	2,625		60.0

* Acres cut in site class 80 = (site 80 equivalent forest area) ÷ (rotation) = 2,625/60 = 43.75.
Acres cut in sites 70, 90, 100 = (acres cut in site 80) ÷ (relative productivity). For site 70, acres to cut = 43.75 ÷ 0.842 = 52.0, or 43.75 × 1.19 = 52.0.

The relative productivity of each site is determined by choosing one of the sites to be the base or standard of comparison and relating all others to it. Any site can be chosen for this purpose; in the example, site 80 was chosen as the base site. The relative productivities of the other sites are determined by dividing their yields at the prescribed rotation by the yield of the base site (column 4). The number of acres needed in each site to produce the same yield as the base site (column 5) is the reciprocal of the relative productivity. The long-term sustained yield of the forest is calculated here by first finding the mean annual increment of each site (column 6), multiplying by the number of acres in each site, and summing over all forest acres to get the LTSY of 191,117 ft^3/yr (column 7).

The acreage to cut in any particular site under area control is calculated in two steps. First, the acreage in a hypothetical forest composed entirely of the base site that has a LTSY equivalent to that of the actual forest is found by weighting the actual acres in each site class by their relative productivities (column 8). This produces an equivalent site 80 forest of 2,625 acres. As a check, the growth of this forest would be $2{,}625 \times 72.83 = 191{,}177$ ft^3/yr, which is the potential growth of the actual forest. Under area regulation the equivalent forest would be cut in 60 years, harvesting 2,625/60, or 43.75 acres per year. Second, the number of acres to cut per year in the other sites in order to have a yield equivalent to that of site 80 (column 9) is determined by dividing the base site 80 acres cut by the relative productivities of the other sites. As a final check, the acreage in the actual forest should support exactly 60 years of cutting under area regulation. Therefore the years it takes to cut each site class at these acreage rates should sum to the rotation (column 10).

These calculations for site productivity differences have practical use to lay out cutting areas in the forest so that the total productivity of the area cut in a period from different parcels of different site quality is the same.

If this sample forest is handled on an uneven-aged basis, the problem is substantially the same. In a 10-year cutting cycle, approximately one-tenth of the total area would be cut over each year, with the cutting area size adjusted to create areas of approximately equal productivity.

In practice, site differences often cannot be determined with precision, and differences in density and treatment further complicate the picture. Nonetheless, the general fact that it takes more acres of poor land than of good land to yield the same volume in a regulated forest is important and demands recognition.

Application with different rotation lengths Up until now a single rotation length has been assumed for all cases. Usually rotation lengths will differ on a forest depending on the site and species group on a particular area or the uses planned for the area. Site influences growth rate. In cases where the desired rotation length keys on the culmination of mean annual increment, the chosen rotation will vary with the site. Species differ in growth pattern and products utilized, so they will often have different desired rotation lengths. As we discussed in Chap. 13, stands used primarily as visual backdrops, deer forage, or woodpecker habitat will often have their rotations adjusted to emphasize these features.

When a forest is to be regulated under area control with different rotation ages for different land groupings, the common approach is to regulate each land grouping separately. The procedure and results are like the single rotation and constant site case earlier, but now there are as many target age class distributions as there are rotations. The Davy Crockett case study below illustrates this multiple-rotation variation on the area control theme.

Evaluation of area control Area control's usefulness in achieving a fully regulated forest has made it a popular technique. Also, it is simple, direct, and has the virtue of portraying the harvest in terms of area to cut—something all regulation techniques must do eventually.

Traditionally its main disadvantage has been seen as a lack of control on volume harvested. And that is but the tip of the iceberg. For area control, by itself, speaks to *no* objectives save the desire to cut A/R acres per year. Rarely do ownership objectives collapse into such a simple policy. Area control, when used, is almost always combined with other procedures to reflect the diverse objectives of forestland owners.

Volume Control

In volume control, the essential decision involves how much volume to cut each year. The acres to cut are then chosen to satisfy this volume. Determination of the volume to cut is approached through the amount and distribution of the growing stock and its increment. Because these items are susceptible to mathematical treatment, foresters have concocted numerous formulas for determining the cut by volume control. While they are no longer much used, a good working knowledge of these methods is valuable both as a guide to historical practice and for the light they cast on understanding possible approaches to cut calculation.

Several formula approaches to volume control are presented. To illustrate their application, we use the same 2,500-acre forest as used for site quality adjustment, which has now been inventoried and had growth measurements made (Table 14-4). All four stand types are found to be evenly distributed over the different site qualities. The regulated forest data are a summary of our earlier information (Table 14-3).

The calculation to estimate the regulated forest inventory warrants additional explanation. An equivalent site 80 forest was found earlier to

Table 14-4 Current age, inventory, and growth of example 2,500-acre southern pine forest, site index 80

Stand type	Area (acres) (1)	Age (years) (2)	Volume per acre (ft^3) (3)	Inventory (M ft^3) (4) = [(1) × (3)] ÷ 1,000	Current per acre growth (ft^3) (5)	Total current growth (M ft^3) (6) = [(1) × (5)] ÷ 1,000
A Residual old growth	300	150	5,600	1,680	−50	−15
B Large sawtimber	500	70	4,200	2,100	10	5
C Old field poles	1,300	20	1,500	1,950	75	97
D New plantation	400	0	0	0	0	0
Totals	2,500			5,730		87
Regulated forest data, 60-year rotation						
Inventory	2,500 acres × 2.529 M ft^3 = 6,323 M ft^3					
Long-term sustained yield (LTSY)	191.2 M ft^3					
Mean annual increment (MAI)	76.47 ft^3/acre/yr					

contain 2,625 acres, which is 43.75 acres in each age class from 1 to 60 just prior to regeneration harvest of the oldest stand (Table 14-3). We can use the yield data for site 80 in Table 14-1 to estimate the inventory of the site 80 forest. Since we have yield data only at 10-year intervals, we will assume that all acres between two 10-year age classes are evenly distributed and have the average volume of these classes. For example, 1/6 of the forest is between 30 and 40 years old. We will assume these acres have an average volume of (3,710 + 2,970)/2, or 3,340 ft^3. A similar calculation for each of the six age classes is

Age class	Average volume (ft^3)	Portion of the forest	Weighted average per acre (ft^3)
0–10	0	1/6	0
11–20	750	1/6	125
21–30	2,235	1/6	372.5
31–40	3,340	1/6	556.67
41–50	3,900	1/6	650
51–60	4,230	1/6	705
		1.0	2,409.17

The average regulated inventory per acre in the forest is 2.409 M ft^3, and the total inventory on the 2,625-acre site 80 regulated forest is 2.409 × 2,625 = 6,323 M ft^3. Since the 2,625-acre site 80 forest produces a yield equivalent to that of the actual forest, their inventories also must be approximately the same at 6,323 M ft^3. For the actual forest, therefore, the average inventory per acre is 6,323/2,500 = 2.529 M ft^3.

Formula methods based on current growing stock and potential growth One elementary approach to volume control requires only an inventory of the existing forest and the potential growth of the managed forest of the future. Growth of the existing forest is not measured. Rather, this growth is assumed to be proportional to its inventory in the same ratio that growth is proportion to inventory in the regulated forest. Finally, it is further assumed that current growth is a good guide to the amount of timber to harvest.

This line of reasoning is called *Hundeshagen's formula* after the German forester who conceived it and reads:

$$\text{Cut} = Y_a = (Y_r/G_r) \times G_a$$

where Y_a = growth or harvestable yield in existing forest
Y_r = growth of regulated forest (long-term sustained yield)
G_r = growing stock in regulated forest
G_a = growing stock in existing forest

Y_r/G_r is the percentage rate of growth of the regulated forest, which can be extracted from the yield tables. To illustrate the formula we use the data from the 2,500-acre southern pine forest given in Table 14-4. Directly substituting into Hundeshagen's equation, we have

$$\text{Cut} = (191/6{,}323) \times 5{,}730 = 173.1 \text{ M ft}^3/\text{yr}$$

Since current growth on the forest is actually less than this cut ($87 < 173.1$), some of the cut ($173.1 - 87 = 86.1$ M ft^3 in the first year) must be obtained through reduction of the inventory. Over time we would expect actual growth to increase as second growth replaces the old growth and large sawtimber, but Hundeshagen's formula does not address the longer run pattern of growth and inventory associated with a cut of 173.1 M ft^3.

A method based on growing stock only As an approximation method, Hundeshagen's formula can be simplified and the necessity for a yield table eliminated. It has been observed that in an approximately fully regulated forest there is a fairly regular and often more or less linear increase in volume by age classes. This suggests the possibility that the growing stock can be represented by a right triangle, such as the triangle $VV'C$ in Fig. 14-4. If this is assumed, then the total growing stock G_r corresponds to the area of the triangle. The base is represented by a sample of R acres, so that the number of acres equals the number of years in the rotation. The altitude is the yield at rotation age Y_r, indicating the annual cut exclusive of thinnings. Applying the usual formula for the area

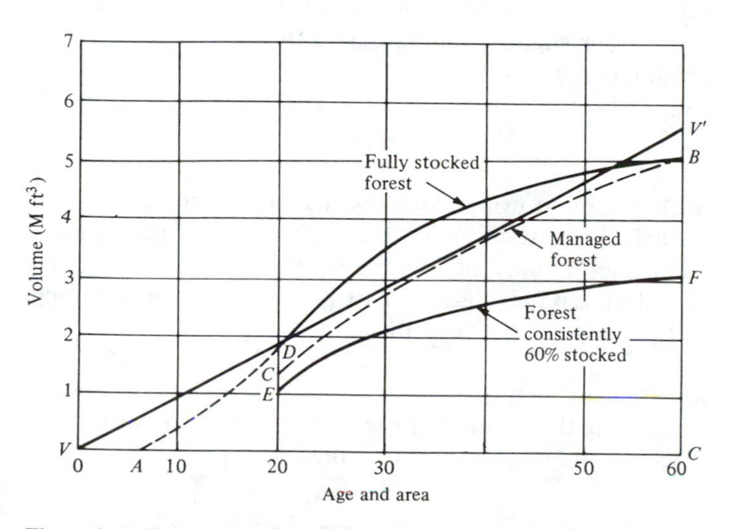

Figure 14-4 Growing stock structures for a regulated loblolly pine forest (site index 90).

of a right triangle,

$$G_r = \frac{R(Y_r)}{2}$$

This value for G_r can be substituted in Hundeshagen's formula, making it read

$$Y_a = \frac{Y_r(G_a)}{R(Y_r)/2}$$

which simplifies to

$$Y_a = \frac{2(G_a)}{R}$$

This substitution was made by von Mantel, and the formula commonly bears his name. It has been termed the method of "glorious simplicity." All that is needed to make an estimate of the annual yield is knowledge of the total growing stock and the rotation. The concept of a regulated forest as a right triangle is the simplest and most basic way of describing the nature of a managed and growing forest in contrast to a stand composed entirely of mature timber, which would be represented by a rectangle.

Using our data from Table 14-4, we can calculate the cut according to von Mantel as

$$Y_a = (2 \times 5{,}730)/60 = 191 \text{ M ft}^3$$

which is a bit higher than the proportionate Hundeshagen cut, but almost exactly equal to the LTSY level for the forest. This suggests that the cut is perhaps a bit high for our slightly understocked forest. Accuracy of the formula in application can be increased by subtracting the average number of years before measurable merchantable volume appears from the rotation.

Formula methods based on growing stock and its increment Because the current growing stock is rarely sufficient to establish an adequate volume control without considering current increment, the two measures have often been combined for cut calculation. This has been done in a number of ways, of which two are briefly described.

Austrian formula The so-called "Austrian formula" combines increment with a means of adjusting the volume of the growing stock either upward or downward. The formula in general terms reads

$$\text{Annual cut} = I + \frac{G_a - G_r}{a}$$

where I = annual increment (usually determined on the basis of net periodic annual increment)
G_a = present growing stock
G_r = desired growing stock, whether indicated by yield tables or some other empirical standard
a = adjustment period in years, which may be a full rotation

In application, the increment term is often modified to average present and expected increment,

$$I = \frac{I_p + I_e}{2}$$

where I_p = average increment of present growing stock
I_e = average expected increment of regulated forest growing stock

The easiest way to visualize the application of the formula is in terms of an uneven-aged forest. Stretch your imagination and assume that the total forest inventory and increment of Table 14-4 represent an uneven-aged forest,

$$I = \frac{87 + 191}{2} = 139 \text{ M ft}^3/\text{yr}$$

The present growing stock G_a is 5,730 M ft^3, and the desired average growing stock G_r is 6,323 M ft^3. An adjustment period a of 20 years is assumed, that is, it is estimated that the actual growing stock reasonably can be built up to the desired level in 20 years. Then

$$\text{Annual forest harvest} = 139 + \frac{5{,}730 - 6{,}323}{20}$$

$$= 139 + (-29.65) = 109.4 \text{ M ft}^3$$

Note that a part of the annual increment, 29.65 M ft^3 in this case, is retained on the ground to build up the reserve growing stock. If the situation is reversed, as in the case of an overdense natural stand, and the present average growing stock should be reduced, the formula will also bring this about, and the value of the $(G_a - G_r)/a$ portion of the formula becomes a plus. If, for example, G_a = 10,000 and G_r = 6,323 as before and other factors remain the same, then

$$\text{Annual cut per acre} = 139 + \frac{10{,}000 - 6{,}323}{20}$$

$$= 139 + 183.85 = 322.85 \text{ M ft}^3$$

The principle of gradually building up or reducing the average growing stock by cutting less or more than the current growth over some adjustment period has been widely applied. The Austrian formula gives a direct expression of this principle and is particularly applicable in uneven-aged forests.

Hanzlik formula The Hanzlik formula was developed to meet a common problem in the even-aged Douglas-fir stands of the Pacific Northwest: initiating management in unregulated forests that had contained mostly old growth. The formula reads:

$$\text{Annual cut} = \frac{V_m}{R} + I$$

where V_m = volume of merchantable timber above rotation age
R = rotation adopted for future stand in years
I = forest increment

As stated, the cut is made up of two parts: (1) that from timber above rotation age distributed over the rotation period, and (2) an estimate of increment. The first part of the cut could easily be calculated, because age and volume of the timber stands were normally known and a rotation had to be established as a planning base. Estimating the increment was more difficult. The knowledge of future net growth in presently unmanaged stands was generally scanty. Often normal yield tables adjusted for stocking and approach to normality were used, but the increment figures were at best a rough estimate.

Treating the residual old growth and the large sawtimber of stands A and B of Table 14-4 as over rotation age timber, V_m = 1,680 + 2,100, or 3,780 M ft^3. The total current increment is 87 and the rotation for future stands is 60 years. The cut according to the Hanzlik formula is

$$\text{Cut} = 3{,}780/60 + 87 = 150.0 \text{ M ft}^3$$

Comparing these four formula approaches to determine the cut gives widely varying answers and raises the question, "which is the best formula?"

Hundeshagen = 173.1 M ft^3
von Mantel = 191.0 M ft^3
Austrian = 109.4 M ft^3
Hanzlik = 150.0 M ft^3

An author of a 1950s federal timber management plan took an average of the four formulas, but that probably is not the answer. Mostly, this analysis shows how difficult it is to capture the complexities of a forest in a simple formula.

Area-volume check Volume control formulas, such as Hanzlik's, leave unanswered the question of whether the harvest they calculate actually can be maintained into the future. In forests dominated by even-aged stands, the proposed formula harvest can be checked by projecting it over a rotation. Such a procedure usually is called *area-volume check,* because the proposed volume is checked against the area available for cutting. To use the approach, we need:

1. All acres classified by age class
2. The ability to project growth of each age class until harvest
3. An established priority by which to harvest the stands such as "oldest first"
4. A rotation for future stands

A trial harvest, perhaps from Hanzlik's formula, is applied to the existing stands for one rotation. If these stands last less than one rotation under the harvest level, the trial harvest is lowered and the process begun again. If these stands last more than one rotation under the trial harvest, the trial harvest is raised and the check begins again. If the stands last just one rotation under the trial harvest, the harvest becomes the allowable cut.

To illustrate area-volume check, let us again employ the example loblolly pine forest of Table 14-2 that contains 20 acres of 60-year-old timber and 10 acres of 20-year-old timber that grows according to the yields shown in Table 14-1, and which we plan to manage on a 30-year rotation. Using Hanzlik's formula to set the initial trial harvest when we assume that growth averages 0.030 M ft^3 per acre per year, we obtain

$$\text{Annual cut} = \frac{5.07 \times 20}{30} + 0.03 \times 30 = \frac{101.4}{30} + 0.9 = 3.38 + 0.9$$

$$= 4.28 \cong 4.3 \text{ M ft}^3$$

where 5.07 is the volume per acre of the overmature (60-year-old) stands.

Simulating a harvest volume of 43 M ft^3 per 10-year period for three periods (Table 14-5*a*), we find that 4.4 acres of the original stands remain after harvest in the third period. Therefore we should raise the harvest volume.

Suppose that we try a harvest volume of 50 M ft^3 per period for three periods (Table 14-5*b*). Now we must cut 1.8 acres of age-20 stands in period 3, which is their second cutting. Therefore we should slightly lower the harvest.

Trying a harvest volume of 49 M ft^3 for three periods (Table 14-5*c*), we cut the existing stands over exactly 30 years. Therefore we have found

Table 14-5 Finding the even flow harvest level for three periods with area and volume check (all volumes in M ft³)*

(*a*) Iteration 1: Cut = 43 M ft³ per period

		Period 1				Period 2				Period 3				
		Inventory		Cut		Inventory		Cut		Inventory		Cut		
Age	Volume per acre	Acres	Total volume	Acres	Total volume	Acres	Total volume	Acres	Total volume	Acres	Total volume	Acres	Total volume	Area after harvest in period 3 (acres)
0														9.0
10						8.5	0			8.1	0			8.1
20	1.75	10	17.5							8.5	14.9			8.5
30	3.44					10	34.4							
40	4.29									10	42.9	5.6	24.1	4.4
50	4.74													
60	5.07	20	101.4	8.5	43									
70	5.33					11.5	61.3	8.1	43					
80	5.56									3.4	18.9	3.4	18.9	
Totals		30	118.9	8.5	43	30	95.7	8.1	43	30.0	76.7	9	43.0	30

(*b*) Iteration 2: Cut = 50 M ft^3 per period

		Period 1				Period 2				Period 3				
		Inventory		Cut		Inventory		Cut		Inventory		Cut		Area after harvest in period 3 (acres)
Age	Volume per acre	Acres	Total volume	Acres	Total volume	Acres	Total volume	Acres	Total volume	Acres	Total volume	Acres	Total volume	
0														12.5
10						9.9	0			9.4	0			9.4
20	1.75	10	17.5							9.9	17.3	1.8	3.2	8.1
30	3.44					10	34.4							
40	4.29									10	42.9	10	42.9	
50	4.74													
60	5.07	20	101.4	9.9	50									
70	5.33					10.1	53.8	9.4	50					
80	5.56									0.7	3.9	0.7	3.9	
Totals		30	118.9	9.9	50	30	88.2	9.4	50	30	64.1	12.5	50	30

Table 14-5 (Continued)

(*c*) Iteration 3: Cut = 49 M ft³ per period

		Period 1				Period 2				Period 3				
		Inventory		Cut		Inventory		Cut		Inventory		Cut		Area after harvest in period 3 (acres)
Age	Volume per acre	Acres	Total volume	Acres	Total volume	Acres	Total volume	Acres	Total volume	Acres	Total volume	Acres	Total volume	
0														11.1
10						9.7	0			9.2	0			9.2
20	1.75	10	17.5							9.7	16.9			9.7
30	3.44					10	34.4							
40	4.29									10	42.9	10	42.9	
50	4.74													
60	5.07	20	101.4	9.7	49									
70	5.33					10.3	54.9	9.2	49					
80	5.56									1.1	6.1	1.1	6.1	
Totals		30	118.9	9.7	49	30	89.3	9.2	49	30	65.9	11.1	49	30

* Calculation of harvest, growth, and inventory for a 30-acre loblolly pine forest on site 90 when area and volume check is used to set the even flow harvest level subject to cutting the existing stands over one rotation for future stands (30 years). 10-year periods are used. Starting structure at year 5 (middle of first period) includes 20 acres at age 60 and 10 acres at age 20. Yields come from Table 14-1. In the middle of each period cut and then grow to the next period. Cut the oldest stands first.

the harvest level that we seek: a harvest level such that we cut the forest once over a rotation.

Notice that the harvest level of 49 M ft^3 does not leave a regulated forest of 10 acres in each of the first three age classes after three periods. We have too many acres in age class 0 (11.1 acres) and too few acres in age class 10 (9.2 acres) and age class 20 (9.7 acres).

Also notice that this harvest level of 49 M ft^3 is higher than the long-term sustained yield of 34.4 M ft^3 (Table 14-2). Therefore this harvest level must eventually decline. Worry by people over how long a harvest level could be sustained caused them to wish to project the harvest level for a longer period than one rotation for future stands. If we wish to do so, however, we must choose new ending conditions to assess feasibility.

Previously we sought to find the harvest level that could be maintained for one rotation while cutting over the forest during that period. If, as an example, we wish to find the harvest level that we can maintain over four periods, we might use attainment of the growth or inventory volume of our regulated forest as the feasibility check after the fourth period.

The inventory of our regulated forest has the following characteristics (Table 14-2):

		Before harvest		After harvest	
Age class (years)	Volume/acre (M ft^3)	Acres	Total volume (M ft^3)	Acres	Total volume (M ft^3)
0				10	
10		10		10	
20	1.75	10	17.5	10	17.5
30	3.44	10	34.4		
			51.9		17.5

Using the volume after harvest on our regulated forest as the feasibility check, we should have 17.5 M ft^3 left after harvest in the last period of our simulation.

Suppose that we wish to find the volume that the forest can maintain over four periods. Starting with the sustainable harvest level that we found in our three-period simulation (49 M ft^3 per period), we find that we have almost no volume left on the forest after meeting the harvest of 49 M ft^3 in the fourth period (Table 14-6*a*). Lowering the harvest level to 39 M ft^3 per period, we find that we have more volume than we need left on the forest after harvest in the third period ($35.4 > 17.5$, Table 14-6*b*). Splitting the difference between 49 (first iteration) and 39 (second iteration), we next try a harvest level of 44 M ft^3 per period. That level leaves us with an

Table 14-6 Finding the even flow harvest level for four periods with area and volume check (all volumes in M ft³)*

(*a*) Iteration 1: Cut = 49 M ft³ per period

		Period 1				Period 2				Period 3				Period 4				
		Inventory		Cut		Inventory		Cut		Inventory		Cut		Inventory		Cut		Inventory volume after harvest
Age	Volume per acre	Acres	Total volume	Acres	Total volume	Acres	Total volume	Acres	Total volume	Acres	Total volume	Acres	Total volume	Acres	Total volume	Acres	Total volume	
0																		
10						9.7	0			9.2	0			11.1	0			
20	1.75	10	17.5							9.7	17.0			9.2	16.1	8.9	15.6	0.5
30	3.44					10	34.4							9.7	33.4	9.7	33.4	
40	4.29									10	42.9	10	42.9					
50	4.74																	
60	5.07	20	101.4	9.7	49													
70	5.33					10.3	54.9	9.2	49									
80	5.56									1.1	6.1	1.1	6.1					
Totals		30	118.9	9.7	49	30	89.3	9.2	49	30	66.0	11.1	49	30	49.5	18.6	49	0.5

(*b*) Iteration 2: Cut = 39 M ft^3 per period

		Period 1				Period 2				Period 3				Period 4				
		Inventory		Cut		Inventory		Cut		Inventory		Cut		Inventory		Cut		Inventory volume after harvest
Age	Volume per acre	Acres	Total volume	Acres	Total volume	Acres	Total volume	Acres	Total volume	Acres	Total volume	Acres	Total volume	Acres	Total volume	Acres	Total volume	
0																		
10						7.7	0			7.3	0			7.6	0			
20	1.75	10	17.5							7.7	13.5			7.3	12.8			12.8
30	3.44					10	34.4							7.7	26.5	1.1	3.9	22.6
40	4.29									10	42.9	2.6	11.2					
50	4.74													7.4	35.1	7.4	35.1	
60	5.07	20	101.4	7.7	39													
70	5.33					12.3	65.6	7.3	39									
80	5.56									5.0	27.8	5.0	27.8					
Totals		30	118.9	7.7	39	30	100.0	7.3	39	30	84.2	7.6	39	30	74.4	8.5	39	35.4

Table 14-6 (Continued)

(*c*) Iteration 3: Cut = 44 M ft³ per period

		Period 1				Period 2				Period 3				Period 4				
		Inventory		Cut		Inventory		Cut		Inventory		Cut		Inventory		Cut		Inventory volume after harvest
Age	Volume per acre	Acres	Total volume	Acres	Total volume	Acres	Total volume	Acres	Total volume	Acres	Total volume	Acres	Total volume	Acres	Total volume	Acres	Total volume	
0																		
10						8.7	0			8.3	0			9.4				
20	1.75	10	17.5							8.7	15.2			8.3	14.5			14.5
30	3.44					10	34.4							8.7	30.0	7.8	26.9	3.1
40	4.29									10	42.9	6.4	27.3					
50	4.74													3.6	17.1	3.6	17.1	
60	5.07	20	101.4	8.7	44													
70	5.33					11.3	60.2	8.3	44.0									
80	5.56									3.0	16.7	3.0	16.7					
Totals		30	118.9	8.7	44	30	94.6	8.3	44	30.0	74.8	9.4	44.0	30	61.6	11.4	44.0	17.6

* Calculation of harvest, growth, and inventory for a 30-acre loblolly pine forest on site 90 when area and volume check is used to set the even flow harvest level subject to leaving as much inventory after harvest in the fourth period as would be on the forest after it is regulated (17.5 M ft³). 10-year periods are used. Starting structure at year 5 (middle of first period) includes 20 acres at age 60 and 10 acres at age 20. Yields come from Table 14-1. In the middle of each period cut and then grow to the next period. Cut the oldest stands first.

inventory volume (17.6 M ft^3) very close to the 17.5 M ft^3 after harvest in the fourth period that we seek (Table 14-6*c*).

By increasing the number of 10-year planning periods from three to four, the sustainable harvest decreased from 49 to 44 M ft^3. As we continue to increase the number of planning periods, the sustainable harvest will continue to decrease until it approaches the long-term sustained yield of 34.4 M ft^3.

Evaluation of volume control The ability of volume control to provide a quick estimate of harvest level for an unmanaged forest and to assess whether such a harvest level can be maintained made it a popular way to calculate the cut. On the old-growth national forests of the west, volume control techniques found much use. Also, they were applied widely with uneven-aged management—the explicit adjustment of growing stock inherent in uneven-aged management proved a natural application.

Traditionally, volume control has been seen to have two major disadvantages. First, needed growth information on unmanaged stands often was hard to obtain. Second, a lack of formal control on area cut per year made it difficult to ensure that the forest was moving toward a regulated condition. In fact, volume control includes no objective beyond that of achieving some harvest volume over time. Rarely do owners have objectives that can be expressed so simply.

Area Control and Volume Control Combined

Area and volume control approaches necessarily are complementary—neither can provide a complete determination of the cut. Because they each deal with different facets of the problem, and because both have been popular, numerous attempts have been made to combine them. *Area and volume allotment* is the most general of the formally recognized combination of the two control methods. It provides no particular formulas and is not a specific procedure. Rather it is a framework in which the many facets involved in managing a forest property can be appraised and a decision reached. As nearly as it can be defined, area and volume allotment means setting a periodic allowable cut based on joint considerations of area, volume, and silvicultural conditions of specific stands, all evaluated in relation to the policies and desires of the owner. To the extent that it is a more or less definite method, it dates back to Cotta and Judeich. According to Schlich (1925), "the system was developed by degrees in Saxony from Cotta's time (about 1820) onward, and put in definite shape by Judeich."

General steps in the application of the procedure are:

1. Group all stands on the forest according to forest type, site, and age or size class.

2. Collect information on merchantable volume and silvicultural condition for each stand grouping.
3. Pick out for further analysis all stands that might be harvested or otherwise treated in the next few periods.
4. Collect and tabulate detailed information on these stands, including area, site, species mix, volume per acre, stand condition, and accessibility, plus any other information that might affect the harvest.
5. Make an area control calculation for the forest to estimate the annual regeneration harvest area. Calculate the average volume that would come off these acres, using the average volume per acre in mature stands. Perhaps also use a volume control approach to estimate volume that might be removed. Given these calculations, establish an overall guide to the annual acreage and volume that might be harvested.
6. Decide what stands to cut over the next few periods. This step gets to the heart of the matter. Roth (1925) described it as a sort of "wrestling match" in which silvicultural, financial, regulatory, and many other considerations are involved. Inevitably, a compromise results: not quite the needed acres for regulation are cut, not quite the volume called for by volume control is removed, not all the highest valued stands are taken, not all the ill and infected stands are treated.
7. On the basis of this negotiation, draw up a specific cutting budget for the first cutting period, usually 5 to 10 years, and draw up a tentative budget for the second period.
8. Cut the timber, compare the results to the expectations, and when the first cutting period ends, start the analysis all over again.

Area and volume allotment has many satisfying features:

1. Cutting is tied to the ground in terms of specific areas—something that must be done sooner or later under all approaches. Area and volume allotment forces this tie from the start.
2. The annual cut is guided by overall volume and area standards, but also by immediate silvicultural and financial necessities. Building the cut from the ground up, while allowing full interaction of all competing interests, ensures plan feasibility.
3. Major attention is given to stands most likely to receive immediate harvest. No matter what the future may bring, harvest in the near future comes from stands containing merchantable timber. Accurate data on their characteristics is essential for effective plan implementation. Information on other stands contributes largely through helping set an overall harvest level, and broad averages often suffice. Area and volume allotment explicitly recognizes that information and analysis needs for stands vary with the probability that action may be taken in the near future.

A Case Study in Combining Area and Volume Control: The 1980 Timber Management Plan of the Davy Crockett National Forest

During the 1970s all national forests in the south developed timber management plans using area control (U.S. Forest Service, 1980). These forests also had the objective of maintaining a harvest from period to period that did not decline.

One such forest is the Davy Crockett National Forest located in Houston and Trinity Counties of east Texas. The forest boundary encompasses 394,200 acres, of which 41 percent, or 161,500 acres, is national forestland. The remaining 232,700 acres largely contain private ownership, including homesites, farm and pasture land, and woodlands. Forest industries own some large blocks of timberlands intermingled with national forest land.

The forest has been divided into 120 compartments for management and planning. Timber stands within these compartments form the basic building blocks in the forest's information system.

Current vegetation The current forest cover is second-growth pine and hardwood species. Four major cover types are recognized, with loblolly pine the major cover type on commercial forestland (Table 14-7). Major hardwoods are white oak, red oak, and hickory in the upland hardwood type, and swamp chestnut oak, cherry bark oak, and sweet gum in the bottomland hardwood type. All four cover types contain some pine-hardwood combinations.

A majority of today's stands were established over a relatively short

Table 14-7 Acres of existing forest by cover type and age class on which timber production can regularly occur for the Davy Crockett National Forest

	Cover type			
Age class	Loblolly pine	Shortleaf pine	Upland hardwood	Bottomland hardwood
0–10	13,521	7,694	1,107	1,239
11–20	5,928	965	0	578
21–30	711	0	29	68
31–40	4,058	712	30	286
41–50	10,090	4,070	785	984
51–60	17,449	16,076	582	1,918
61–70	18,076	19,378	275	1,806
71–80	8,692	6,584	794	1,447
80+	3,689	3,130	77	649
Totals	82,214	58,609	3,859	9,424

period subsequent to heavy cutting during the early 1900s. As a result of this compacted time frame for establishment, over 65 percent of the stands have an average age of 50 to 80 years. Eleven age classes and the four cover types define 44 stand types (age × cover type) considered for planning.

Major forest uses Timber from the forest is harvested for lumber, plywood, pulp, and firewood. Slightly more than 2 million ft^3 of forest products were harvested annually in the late 1970s.

Dispersed recreation activities, such as camping, hiking, hunting, and fishing, predominate in the 166,600 visitor-days of recreation use that occur annually on the forest. On the moderately rolling terrain, visitor views are usually restricted to the foreground, with occasional glimpses beyond.

In the forest there are 450 species of wildlife, including gray and fox squirrels, white-tailed deer, bobwhite quail, and wild turkey; and 91 colonies of the rare and endangered red-cockaded woodpecker are known to exist on the forest. These colonies use 313 cavity trees and contain a total population of 250 birds.

The forest is heavily laced with streams, but the majority are not perennial and offer limited opportunity for fishing.

Timber management plan Even-aged management was chosen for all cover types, and a rotation length was developed for each. These rotations are 70 years for loblolly pine, 80 years for shortleaf pine, and 100 years for the hardwood cover types. They represent a compromise between shorter high-yielding rotations and longer amenity-oriented ones. For each cover type, its rotation indicates average final harvest age for future stands and number of years over which the current stands will be harvested. Under the area control framework used here, the cover type rotation is a key element in determining how many acres will be regeneration harvested per year. As the plan says, "the planned average to be cut and regenerated for a plan period is determined by dividing the total manageable acreage of a species group by the number of periods in the rotation of the group." Given the rotation age for a species group and the acres it covers, the acres that should be in each age class of the target forest can be easily determined (Table 14-8).

As an example, loblolly pine covers 82,200 acres with a proposed rotation of 70 years. Dividing the acreage by the rotation gives a number of 1,175 acres per year, or 11,750 acres per 10-year period, that should be harvested and that should be in each age class of the target forest (Table 14-8). According to this area control approach, 11,750 acres should be cut in the first period (the next 10 years). Actually, the plan calls for 10,088 acres to be harvested, with this harvest taking acres from all age classes above 30 years old (Table 14-9). According to the plan,

Table 14-8 Present and desired age class distribution for regulated forestland on the Davy Crockett National Forest (thousands of acres)

	Age class distribution											
Working group	0–10	11–20	21–30	31–40	41–50	51–60	61–70	71–80	81–90	91–100	Unclear	Total
Loblolly pine												
Present	13.5	5.9	0.7	4.1	10.1	17.4	18.1	8.7	3.4	0.3	—	82.2
Desired	11.8	11.8	11.8	11.7	11.7	11.7	11.7	—	—	—	—	82.2
Shortleaf pine												
Present	7.7	1.0	0	0.8	4.1	16.1	19.4	6.6	1.9	1.1	—	58.6
Desired	7.4	7.4	7.3	7.3	7.3	7.3	7.3	7.3	—	—	—	58.6
Upland hardwoods												
Present	1.1	0	—	—	0.8	0.6	0.3	1.0	0.0	0.1	—	3.9
Desired	0.3	0.4	0.4	0.4	0.4	0.4	0.4	0.4	0.4	0.4	—	3.9
Bottomland hardwoods												
Present	1.2	0.6	0.1	0.3	1.0	1.9	1.8	1.4	0.5	0.2	0.4	9.4
Desired	1.0	1.0	1.0	1.0	0.9	0.9	0.9	0.9	0.9	0.9	—	9.4
Totals												
Present	23.5	7.5	0.8	5.2	16.0	35.0	39.6	17.7	5.8	1.7	0.4	154.1
Desired	20.5	20.5	20.6	20.4	20.3	20.3	20.3	8.6	1.3	1.3	—	154.1

Source: U.S. Forest Service (1980).

Table 14-9 Davy Crockett National Forest, loblolly pine regulation progress table

	Age class distribution											
Period	0–10	11–20	21–30	31–40	41–50	51–60	61–70	71–80	81–90	91–100	80+	Unclear	Total
Desired	11,745	11,745	11,745	11,745	11,745	11,745	11,745	—	—	—	—	—	82,214
I. Actual	13,521	5,928	711	4,058	10,090	17,449	18,076	8,692	—	—	3,689	—	82,214
Regenerate	0	0	0	130	1,137	2,919	3,420	1,811	—	—	671	—	10,088
Balance	13,521	5,928	711	3,928	8,953	14,530	14,656	6,881	—	—	3,018	—	72,111
II. Actual	10,088	13,521	5,928	711	3,928	8,953	14,530	14,656	6,881	—	3,018	—	82,214
Regenerate	0	0	0	0	0	0	1,668	4,636	3,907	—	1,533	—	11,744
Balance	10,088	13,521	5,928	711	3,928	8,953	12,862	10,020	2,974	—	1,485	—	70,470
III. Actual	11,744	10,088	13,521	5,928	711	3,928	8,953	12,862	10,020	2,974	1,485	—	82,214
Regenerate	0	0	0	0	0	0	0	2,286	5,000	2,974	1,485	—	11,745
Balance	11,744	10,088	13,521	5,928	711	3,928	8,953	10,576	5,020	0	0	—	70,469
IV. Actual	11,745	11,744	10,088	13,521	5,928	711	3,928	8,953	10,576	5,020	—	—	82,214
Regenerate	0	0	0	0	0	0	0	1,725	5,000	5,020	—	—	11,745
Balance	11,745	11,744	10,088	13,521	5,928	711	3,928	7,228	5,576	0	—	—	70,469
V. Actual	11,745	11,745	11,744	10,088	13,521	5,928	711	3,928	7,228	5,579	—	—	82,214
Regenerate	0	0	0	0	0	0	0	2,102	4,067	5,576	—	—	11,745
Balance	11,745	11,745	11,744	10,088	13,521	5,928	711	1,826	3,161	0	—	—	70,469
VI. Actual	11,745	11,745	11,745	11,744	10,088	13,528	5,928	711	1,826	3,161	—	—	82,214
Regenerate	0	0	0	0	0	119	5,928	711	1,826	3,161	—	—	11,745
Balance	11,745	11,745	11,745	11,744	10,088	13,402	0	0	0	0	—	—	70,469
VII. Actual	11,745	11,745	11,745	11,745	11,744	10,088	13,402	—	—	—	—	—	82,214
Regenerate	0	0	0	0	0	0	11,745	—	—	—	—	—	11,745
Balance	11,745	11,745	11,745	11,745	11,744	10,088	1,657	—	—	—	—	—	70,469
VIII. Actual	11,745	11,745	11,745	11,745	11,745	11,744	10,088	1,657	—	—	—	—	82,214
Regenerate	0	0	0	0	0	0	10,088	1,657	—	—	—	—	11,745
Balance	11,745	11,745	11,745	11,745	11,745	11,744	0	0	—	—	—	—	70,469
IX. Actual	11,742	11,742	11,743	11,743	11,743	11,743	11,743	—	—	—	—	—	82,214

Source: U.S. Forest Service (1980).

> Selections in the loblolly pine working group were 1,657 acres less than desired. Sufficient acres qualified for regeneration to meet the desired goal. The pattern these stands took on the ground limited choices. They were frequently concentrated in contiguous areas. Many times portions of them had been regenerated in the previous 10 years. This made it difficult to select the acreage goal and still disperse regeneration areas in a desirable manner. The regeneration goal could have been met but only at considerable sacrifice in coordination with other resources.

This 10,088 acres is composed of stands identified within compartments. As an example, compartment 89 contains four stands of the loblolly pine species group scheduled for harvest during the first 5 years (Table 14-10). Linking the acres designated for harvest under area control to the candidate stands on the ground caused the compromise leading to proposal of fewer acres for harvest than called for by the regulatory framework. Also, the plan says,

> Both the upland and bottomland hardwood working group selections fell short of desired goals. The acreage in these working groups is relatively small and stands frequently serve critical wildlife habitat needs. Some were deferred because of close proximity to other regeneration areas. Others were held for short run mast needs pending development of hard mast supplies in other stands. Most acreage in the bottomland hardwood working group is concentrated on major creeks and along the Neches River. This tends to limit selections that still provide adequate dispersal of regeneration areas.

In total, the plan proposes regeneration harvesting 18,500 acres during the next 10 years, rather than the 20,400 acres called for under area

Table 14-10 Sample of stands selected for regeneration cut during 1980–1984 on the Davy Crockett National Forest considering multiple values

	Shortleaf pine		Loblolly pine		Upland hardwood		Bottomland hardwood	
Compartment	Stand number	Acres	Stand number	Acres	Stand number	Acres	Stand number	Acres
16	8	20			6	25	5	20
					9	40		
88	8	50	4	40				
	12	20	11	21				
89	21	50	1	50				
			2	80				
			13	120				
			16	35				
99			7	40			3	60
							8	80
							15	60

control. According to the regulation progress table for the loblolly pine species group, regeneration harvest acreage will equal that required for area control after the first 10-year period. Will that harvest pattern occur? Probably not. As they redo the plan every 5 to 10 years, U.S. Forest Service planners may decide that the area control framework is inadequate for their needs and use another approach. Even if they keep the approach described here, their multiple objectives combined with the spatial location of their timber stands often will require deviation from the regeneration harvest targets given by their regulatory framework.

Regeneration harvest will produce an average annual volume of 2.5 M ft^3 of sawtimber and roundwood per year. In addition, 1,500 acres of commercial thinnings per year will produce 0.5 million ft^3 annually, control density and species composition, and increase mast production on the remaining hardwoods. Over time, by spreading the harvest for each 10 years over a number of age classes (see Table 14-9 for the loblolly pine contribution), the forest projected a nondeclining yield of timber for the foreseeable future (Fig. 14-5). As volumes are built up across the forest, a substantial increase in harvest is projected, with 20 percent of the increase attributed to the introduction of genetically improved stock.

Regulation by Area and Volume Control: Some Example Problems

Regulation is easy enough to talk about, and the equations and cutting rules do not appear difficult to apply. While regulation is in one sense a

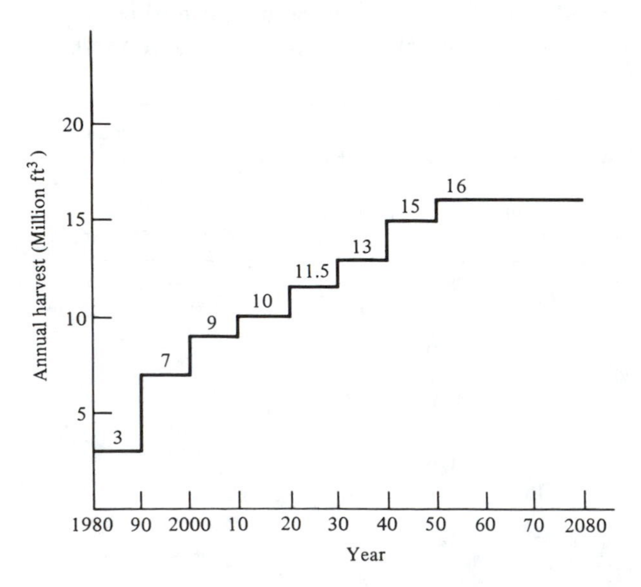

Figure 14-5 Projected annual harvest on the Davy Crockett National Forest.

"big picture" subject, it is also a subject full of incredible detail—the computations are an accountant's delight. The fine points come in deciding which stands in a real forest to cut each year, and in grasping the orderly but still complex dynamics of forest structure as a cutting and regulatory plan unfolds. To this end, numerical examples are presented for an even-aged forest under two different packages of regulatory rules followed by similar examples for an uneven-aged forest. Take the time to work through these examples as it helps nail down the strategies of forest regulation, to see the time dynamics of a forest, and to prepare for formulating these problems for analysis by mathematical programming.

An even-aged forest This forest is 40 acres in size and is best visualized as a set of five 8-acre compartments located next to each other, each compartment being all of one stand type and homogeneous in terms of age and volume. It might help to think of the compartments being laid out as a series of five adjacent stands along some country road. The inventory of this forest and the yield table appropriate to its managed stands are given in Table 14-11. In the interest of providing a compact example with annual

Table 14-11 Inventory, growth, and yield data for an even-aged forest

Initial inventory, Fall of year 1

Compartment	Age (years)	Volume per acre (M bd ft)	Area (acres)
A	5	40	8
B	0	0	8
C	0	0	8
D	1	5	8
E	2	6	8

Growth and yield of future stands

Age (years)	Volume per acre (M bd ft)	Age (years)	Volume per acre (M bd ft)
0	0	5	40
1	5	6	42
2	12	7	42
3	25	8	42
4	36	9	42

Forest data, 5-year rotation

Mean annual increment (MAI)	40/5 = 8 M bd ft/acre/yr
Long-term sustained yield (LTSY)	(MAI)(forest area) = 8 × 40 = 320 M bd ft
Acres cut for area control	A/R = 40/5 = 8 acres

harvests, the number of compartments was chosen to equal the rotation age, and the trees are assumed to grow extremely fast, reaching maturity in about 6 years. If it helps improve realism, you could think of these as multiyear periods rather than years. With the 5-year rotation, exactly one compartment has to be cut each year for the forest to be regulated within 5 years.

The five compartments are organized in Table 14-12 to track their changing age and volume over time and to present the schedule of forest activities, including the total harvest volume. Reading down the first five entries of the column for year 1 shows the forest age and volume structure in the first year. The numbers are identical to the inventory data presented in Table 14-11.

Our first example illustrates the application of unmodified area regulation harvest rules. Given these rules, the only choice in year 1 is to harvest compartment *A*, removing 40 units of volume per acre on each of the 8 acres for a total harvest of 320 M bd ft. After 1 year of growth, the age and volume structure of the forest is shown in the second column of the table. Compartment *A* was immediately reduced to a zero age stand

Table 14-12 Example 1. Area regulation of an even-aged forest

Harvest rules

1. Rotation age is 5 years.
2. Area regulation: cut and regenerate one 8-acre compartment each year.
3. Always harvest the compartment with the highest volume.
4. Cut, then grow.

	Year							
Compartment	1	2	3	4	5	6	7	8
	Age/volume per acre of compartments just prior to harvest							
A	5/40*	1/5	2/12	3/25	4/36	5/40*	1/5	2/12
B	0/0	1/5	2/12	3/25	4/36*	1/5	2/12	3/25
C	0/0	1/5	2/12	3/25*	1/5	2/12	3/25	4/36
D	1/5	2/12	3/25*	1/5	2/12	3/25	4/36	5/40*
E	2/6†	3/15*	1/5	2/12	3/25	4/36	5/40*	1/5
	Compartment cut/acres cut							
	A/8	*E*/8	*D*/8	*C*/8	*B*/8	*A*/8	*E*/8	*D*/8
	Compartment cut/volume cut per acre							
	A/40	*E*/15	*D*/25	*C*/25	*B*/36	*A*/40	*E*/40	*D*/40
	Total harvest							
	320	120	200	200	288	320	320	320

* Compartment regeneration harvested.

† These numbers are age/volume per acre.

by cutting in period 1, but after 1 year of growth, the stand is 1 year old with 5 M bd ft of volume, just like compartments *B* and *C*. Each of the first five rows tracks the history of what happens to each original compartment over time, while each column is a snapshot of the forest at the instant just before harvest in a particular year. Our harvest rules are easy to implement by scanning all compartments each year and picking the compartment with the highest volume as the one to harvest and regenerate. This procedure produces the indicated harvest sequence shown in the sixth row of the table and the falling and then rising pattern of total harvest shown in the eighth row. Compartment *E* started off only 6/12, or 50 percent stocked, hence the low estimate of volume in year 2 when this stand is cut. Compartments *A*, *B*, and *C* tie for the highest volume per acre in year 4, with a volume of 25 units per acre. Any of the three could have been chosen to harvest—we picked compartment *C*.

The forest is regulated after the harvest in period 5, with equal area in age classes 0, 1, 2, 3, and 4 years. The regulated structure is again revealed before cutting in year 6 with one compartment in each of age classes 1, 2, 3, 4, and 5. Regulation also is shown in harvest terms, with the forest harvest stabilizing at the long-term sustained yield level of 320 M bd ft in year 6.

The second example takes this same even-aged forest and adds the famous "nondeclining yield" requirement as a volume control restriction on the forest harvest. This nondeclining yield constraint here requires that the total harvest volume in any year cannot fall below the total harvest volume in the preceding year. The up and down total harvest volume of our previous example (row 8 of Table 14-12) is no longer acceptable. Now to solve the problem, we need to meter out the inventory of compartment *A* over several years to prevent a decline in forest harvest volume while waiting for the young stands to mature. After several unsuccessful tries with harvest levels that were too high, a first-year harvest volume of 200 M bd ft enabled a nondeclining yield of timber volume over the 8 years (Table 14-13). Stand *A* is initially harvested over 4 years in a sequence of 5, 2, 0.5, and 0.5 acres, as shown. Also, all of compartments *E*, *D*, *C*, and *B* are cut in years 2 through 5. The mixed age and volume structure of compartment *A* is again cut in year 6, bringing the forest to its regulated structure.

This result is not necessarily the optimal solution. It is simply the first solution we found that is feasible. In fact, this problem as stated does not give clear goals and criteria with which to judge optimality. If we added the goal of maximizing the first-period harvest, then after several trials we might decide that we are close to the optimal solution—but still we could not be sure. For practice, modify the initial inventory or the harvest rules to create different problems. It is also easy to create problems that have no solution.

Table 14-13 Example 2. Area regulation of an even-aged forest with a nondeclining yield volume control requirement

Harvest rules

1. Rotation age is 5 years.
2. Regulate by area control as soon as possible.
3. Volume harvested cannot fall below harvest in preceding period.
4. More or less than 8 acres can be cut during regulation.
5. Cut, then grow.

	Year							
Compartment	1	2	3	4	5	6	7	8
	Age/volume per acre of compartments just prior to harvest							
A_1	5/40*	1/5	2/12	3/25	4/36	5/40*	1/5	2/12
A_2		6/42*	1/5	2/12	3/25	4/36*		
A_3			7/42*	1/5	2/12	3/25*		
A_4				8/42*	1/5	2/12*		
B	0/0	1/5	2/12	3/25	4/36*	1/5	2/12	3/25
C	0/0	1/5	2/12	3/25*	1/5	2/12	3/25	4/36
D	1/5	2/12	3/25*	1/5	2/12	3/25	4/36	5/40*
E	2/6	3/15*	1/5	2/12	3/25	4/36	5/40*	1/5
	Compartment cut/acres cut							
	A/5	*A*/2, *E*/8	*A*/0.5, *D*/8	*A*/0.5, *C*/8	*B*/8	A_1/5, A_2/2, A_3/0.5, A_4/0.5	*E*/8	*D*/8
	Compartment cut/volume cut per acre							
	A/40	*A*/42, *E*/15	*A*/42, *D*/25	*A*/42, *C*/25	*B*/36	A_1/40, A_2/36, A_3/25, A_4/12	*E*/40	*D*/40
	Total forest harvest							
	200	204	221	221	288	290.5	320	320

* Compartment harvested.

An uneven-aged forest Application of area and volume control regulation approaches to uneven-aged forests follows patterns similar to those of even-aged forests, although the variables used—to classify the forest and predict response—change. Volume, stand structure, and time since harvest are the primary variables for identifying stands or predicting growth, rather than the stand age of even-aged forests. The data for this example are given in Table 14-14 and describe a 40-acre forest composed of four different 10-acre compartments. The compartments are homogeneous stand types, differing from one another in terms of stand structure and volume per acre. For this example, we assume that the stand structure of each type is sustainable and that annual volume growth can be predicted using only current volume as the independent variable.

Table 14-14 Inventory, growth, and yield data for an uneven-aged forest

Initial inventory, Fall of year 1

Compartment	Volume per acre (M bd ft)	Area (acres)
A	7	10
B	8	10
C	10	10
D	18	10

Growth of stands (M bd ft)

Volume of growing stock	Annual growth	Volume of growing stock	Annual growth
4	1.6	16	3.2
6	2.1	18	2.8
8	2.6	20	2.6
10	2.9	22	2.4
12	3.0	24	2.4
14	3.2	26	2.4

Cumulative growth and yield from a reserve growing stock of 8 M bd ft

Years of growth	Initial volume	Growth	Volume after 1 year
1	8	2.6	10.6
2	10.6	3.0	13.6
3	13.6	3.2	16.8
4	16.8	3.2	20.0

Forest data, 4-year cutting cycle

Reserve growing stock	8 M bd ft
Periodic annual increment (PAI) of regulated compartment	(20 − 8)/4 = 3 M bd ft/acre/yr
Long-term sustained yield (LTSY)	(PAI)(forest acres) = 3 × 40 = 120 M bd ft/yr
Acres entered for harvest each year = *A*/CC = 40/4 = 10	

A reserve growing stock (RGS) of 8 M bd ft and a cutting cycle of 4 years are used with the growth model to construct the specific yield table for the regulated forest (Table 14-14), with growth interpolated between the given values. Starting with the 8 M bd ft immediately after harvest, the growth model shows 2.6 M bd ft of growth for the first year, which when added will bring the stand volume to 10.6 M bd ft at the start of the second growth season. At the end of four growth periods the stand will have reached a volume of 20 M bd ft just prior to harvest in the fall of the fourth year. Then 12 M bd ft are harvested, returning it to the RGS level of 8 M bd ft. The 4-year growth of 12 M bd ft indicates a periodic annual increment (PAI) of 3 M bd ft per acre per year. This long-term sustained yield of the forest is (40 acres) × PAI = 120 M bd ft/yr.

The solution to the first uneven-aged example problem was found by directly following the problem statement rules as shown in Table 14-15. The first four rows track the volume history of each compartment and the last four give the harvest decisions and forest harvest. Stand *D* is cut first, followed in order by stands *C* and *B*. In year 4, stand *D* again has the highest volume and is cut. Stand *A*, which started out with less than the reserve growing-stock level, finally is cut in year 5. After harvest in year 5, the forest is regulated in structure and provides the long-term sustained yield level harvest of 120 M bd ft in year 6. If this example forest started with substantially less volume per acre in some of the compartments, the regulatory process would take longer and a lower forest harvest would be realized until all stands had volumes in excess of the reserve growing-stock level.

The second uneven-aged example adds volume control, requiring a minimum harvest of 80 M bd ft each year. Such a requirement could easily reflect a prior contract, the desire by a landowner to provide a minimum annual income, the need to supply a sawmill, or it could be the result of a volume control calculation. Anticipating that the strict requirements for

Table 14-15 Example 3. Area regulation of an uneven-aged forest

Harvest rules

1. One complete compartment must be cut each year.
2. All volume in excess of the RGS of 8 M bd ft is removed.
3. The compartment with the highest volume per acre is cut.

	Year						
Compartment	1	2	3	4	5	6	7
	Volume per acre just before harvest						
A	7	9.4	12.2	15.2	18.4*	10.6	13.6
B	8	10.6	13.6*	10.6	13.6	16.8	20.0*
C	10	12.9*	10.6	13.6	16.8	20.0*	10.6
D	18*	10.6	13.6	16.8*	10.6	13.6	16.8
	Compartment cut						
	D	*C*	*B*	*D*	*A*	*C*	*B*
	Acres cut						
	10	10	10	10	10	10	10
	Volume cut per acre						
	10	4.9	5.6	8.8	10.4	12	12
	Forest harvest						
	100	49	56	88	104	120	120

* Compartment harvested.

complete compartment harvest and removal down to 8 M bd ft would conflict with the minimum forest harvest requirement (can you see why?), the compartment requirements were relaxed and only a minimum harvest per acre was required. There are several ways to solve this problem; the one illustrated here was always to cut all 10 acres of each compartment but to vary the amount cut per acre and to cut more than one compartment per year for a few years. The forest harvest initially was set at the minimum of 80 M bd ft. Then different cutting plans were tried until one was found in which the cut never fell below 80 M bd ft, and area regulation was attained after the sixth-year harvest (Table 14-16).

To solve this problem, we examined the forest volume per acre structure in year 1, selected compartments to cut which would provide at least 80 M bd ft, used the growth model to grow the forest to the next year, and again looked for stands to cut. By cutting all 10 acres in a compartment whenever it was entered it was hoped that area regulation could be attained.

The details of the calculations are best seen by following the history

Table 14-16 Example 4. Area regulation of an uneven-aged forest with a minimum harvest level volume control requirement

Harvest rules

1. No cutting will reduce compartment volume below 8 M bd ft/acre.
2. At least 2 M bd ft/acre must be removed in each harvest for economic logging.
3. At least 80 M bd ft must be harvested each year.
4. The forest should attain area control regulation within 7 years.

	Year						
Compartment	1	2	3	4	5	6	7
	Volume per acre just before harvest						
A	7	9.4	12.2	15.2*	10.6	13.6	16.8
B	8	10.6	13.6*	10.6	13.6	16.8	20.0*
C	10	12.9*	12.5*	12.9*	13.8	16.9*	10.6
D	18*	12.9*	10.6	13.6	16.8*	10.6	13.6
	Compartment cut/acres cut						
	D/10	*D*/10, *C*/10	*C*/10, *B*/10	*A*/10, *C*/10	*D*/10	*C*/10	*B*/10
	Compartment cut/volume cut per acre (M bd ft)						
	D/8	*D*/4.9, *C*/3.2	*C*/2.5, *B*/5.6	*A*/7.2, *C*/2.0	*D*/8.8	*C*/8.9	*B*/12
	Forest harvest (M bd ft)						
	80	81	81	92	88	89	120

* Compartment harvested.

of an individual stand for a few years. Consider stand *C*, which has 10 M bd ft before harvest in the first year. After one year's growth (2.9 M bd ft from the growth model in Table 14-14), it reaches a volume of 12.9 M bd ft before harvest in year 2. In period 2, 3.2 M bd ft is cut, leaving a volume of 9.7 M bd ft. This residual volume grows 2.8 to the third-year level of 12.5 M bd ft, when it is again cut, removing 2.5 M bd ft and reducing it to 10.0 M bd ft. Stand *C* grows 2.9 M bd ft in year 4, reaching a volume of 12.9 M bd ft when it is again cut for a third time, removing 2 M bd ft and bringing it to 10.9 M bd ft. After two more years of 2.9 and 3.1 M bd ft growth, it accumulates to 16.9 M bd ft in year 6, when it is finally cut back to the reserve growing stock of 8 M bd ft. A similar but slightly less complicated history can be recreated for the other three compartments. As before, *this is just one feasible solution*. Others are possible and you might look for them by removing the constraint that one complete compartment must be cut each year.

To define and solve a problem with a specific objective, you could delete the minimum 80 M bd ft requirement and assume that the goal is to maximize the first-period harvest. Another interesting variation would be to discount the forest harvest volumes using an assumed price and interest rate, and try maximizing the present net worth of the forest. You will find that the computations rapidly become tedious with these additional considerations, as it will take several trials before you feel confident that you are close to a maximum, even with this simple four-stand forest.

Despite the simplifications needed to fit these problems on a textbook page, these examples show how growth models are used to move stands from one year to the next. They also show the complexity of harvest choice when volume control requirements are added to the simple area control strategy. It should be apparent that without running these calculations for several years into the future, you cannot tell whether the constraints are satisfied or if the forest is ever regulated. It may turn out that to satisfy volume control requirements the forest may settle on a stable cut cyclically varying age and volume structure rather than picture-perfect area regulation. But so what? Hopefully, studying these problems should also whet your interest in using computer-based techniques like mathematical programming to do all this accounting and actually find the optimum solution once a specific objective has been spelled out.

REFERENCES

Meyer, W. H., 1942: "Yield of Even-Aged Stands of Loblolly Pine in Northern Louisiana." Yale University School of Forestry Bulletin 51.

Recknagel, A. B., 1917: "Theory and Practice of Working Plans," 2d ed., John Wiley & Sons, Inc., New York.

Roth, Filibert, 1925: "Forest Regulation," 2d ed., George Wahr Publishing Company, Ann Arbor, Michigan.
Schlich, Sir William, 1925: "Manual of Forestry," 5th ed., vol. 3., Bradbury, Agnew and Co., Ltd., London.
U.S. Forest Service, 1980: "Final Environmental Statement for the Timber Management Plan, Davy Crockett National Forest."

QUESTIONS

14-1 A forest was recently inventoried and contained two even-aged stands. The owner wants it put under immediate *area* regulation with a rotation age of 40 years and a harvest every year. The inventory and yield data are as follows.

1. *Inventory*

Stand	Area (acres)	Age (years)	Volume per acre (M bd ft)
A	500	10	5
B	100	30	30

2. *Yield for this site*

Age (years)	Volume per acre (M bd ft)
0	0
10	5
20	15
30	30
40	50
50	70
60	80

(*a*) How many acres will be cut each year?
(*b*) How long will it take, at maximum, before the forest is regulated?
(*c*) What will be the MAI, LTSY, and annual harvest from the regulated forest?
(*d*) Assuming the oldest stands are cut first, what will be the harvest in the 5th, 10th, and 20th entries (after 5, 10, and 20 years of growth, respectively) of the regulation period?
(*e*) Sketch or show in a table the structure of this forest in terms of acres by age class after 10 years of growth and harvest.

14-2 A 900-acre forest made up of three existing stands is to be managed under an even-aged management system. The owner decides to use the area regulation technique and plans to cut through the existing stands, cutting the slowest growing stands first and the fastest growing stands last.

1. *Inventory of the existing stands*

Stand	Site quality	Area (acres)	Current age (years)	Current volume (units)	Average growth per acre per year (units)
A	I	300	10	5	+0.5
B	II	200	40	50	+1.0
C	II	400	70	65	−0.8

2. *Regenerated stand prescription.* An analysis of growth and economic data resulted in the following recommended prescription for all stands:

Year	Activity	Yield, site I	Yield, site II
0	Prepare and plant	—	—
20	Thinning	10	6
30	Regeneration harvest	38	24
	Total yield	48	30

3. *Economic data*

Stumpage price, all stands and entries $200/unit
Site preparation and plant cost $500/acre
Guiding interest rate 5 percent

(*a*) Calculate the relative productivity of each stand using stand *A* as the reference.

(*b*) Under strict area regulation, how many acres would be cut each year in each existing stand to make an equal productivity compartment when the forest is fully regulated?

(*c*) What is the LTSY level of balanced growth and harvest volume for this forest?

(*d*) Considering both today's inventory and the value of the bare land to produce timber using the regenerated prescriptions, what is the estimated market value today of an acre in each stand?

(*e*) Using the owner's cutting priorities and one rotation area regulation, what is the forest harvest, in units, in each of the following harvest entries (entry 5 is after five years of growth, etc.): (i) Entry 5. (ii) Entry 10. (iii) Entry 15. (iv) Entry 25. (v) Entry 40.

14-3 An uneven-aged stand that was harvested over four cutting cycles provided the following statistics (per acre):

Year	Volume before cut (bd ft)	Volume after cut (bd ft)	Periodic growth (bd ft)	Harvest (bd ft)
0	9,000	7,000		
10	10,500	8,000	3,500	2,500
20	12,500	9,000	4,500	3,500
30	13,000	8,000	4,000	5,000
40	12,500	8,500	4,500	4,000
			16,500	15,000

(*a*) Calculate the average net periodic annual increment (PAI) over the 40 years of forest growth.

(*b*) If this was a 100-acre forest, stabilized on a residual growing stock of 8,000 M bd ft after cutting and a 10-year cycle, what is your estimate of:

(i) The number of acres cut each year.
(ii) The volume cut per year per acre harvested.
(iii) The sustainable harvest from the forest.

14-4 Using area control, regulate an even-aged forest in one rotation. The following inventory, growth and yield data, and management prescription apply to this forest:

1. *Current inventory*

Age	Acres	Volume per acre (ft^3)
150	150,000	5,000
100	60,000	3,000

2. *Growth on current inventory,* 10 percent per decade.
3. *Management prescription for future stands,* an intermediate harvest at age 10 of 500 ft^3/acre and a regeneration harvest at a rotation age of 30.

4. *Yield of future stands*

Age	Volume per acre (ft^3)
0	0
10	1,500*
20	2,000†
30	3,000†
40	5,000†

*Volume before intermediate harvest.

†These volumes assume the intermediate harvest at age 10.

5. *No regeneration lag.* Regeneration starts as soon as stand is cut.

Use the following procedure:

1. Use 10-year periods.
2. Make the necessary harvest at the beginning of the period and then grow the stands to the next period.
3. For regeneration harvests, always cut the oldest stand first.

For the first three periods, report:

(*a*) Acres and volume cut by age class for both intermediate and final harvests and total acres and volume harvested.

(*b*) Growth by age class and total growth between periods.

(*c*) Inventory in each period, in terms of acres and volume per acre by age class, before harvest and after harvest but before growth. Also, report total inventory volume in each period.

Hint: The calculations of Table 14-2 serve as a guide.

14-5 You have been given the following data and assumptions.

1. *Data*

Current inventory

Stand	Age (years)	Area (acres)	Stocking (%)
1	15	300	100
2	25	400	100
3	45	500	100

Planning horizon: three periods
Period length: 10 years
Minimum harvest age: 10 years

Yield of fully stocked stands

Age (years)	Volume per acre (M ft^3)
0	0
10	0.50
20	1.50
30	2.50
40	3.00
50	3.25
60	3.50
70	3.75

2. Assumptions

All stands are cut at the midperiod age.

Cut acres in any of the three stands by clearcutting using an "oldest first" priority rule.

Maximize the even-flow volume harvested over the planning horizon subject to a constraint leaving the forest after harvest in the third period such that its growth between the third and fourth periods is at least as much as obtained from a forest regulated on a 30-year rotation. You determine this needed growth rate.

Do two iterations of the area and volume check approach using 1,000 as the trial harvest for the first iteration. Use the computation formats of Table 14-5 or 14-6. Would you raise or lower the harvest for the third iteration? Explain.

14-6 *Regulation of an uneven-aged forest*. You are the manager of a very rapidly growing uneven-aged forest that you want to regulate. You are given the following data:

1. *Initial inventory*

Compartment	Volume per acre (M bd ft)	Acres
A	6	100
B	10	100
C	14	100
D	24	100

2. *Growth of future stands* (M bd ft)

Volume of RGS	Annual growth
6	2.4
8	2.8
10	3.2
12	3.4
14	3.6
16	3.4
18	3.2
20	3.0
22	2.8
24	2.6
26	2.6
28+	2.4

3. *Growth and yield of 10 M bd ft reserve growing stock*

Years of growth	Initial volume (M bd ft)	Annual growth (M bd ft)	Volume after 1 year (M bd ft)
1	10	3.2	13.2
2	13.2	3.6	16.8
3	16.8	3.4	20.2
4	20.2	3.0	23.2

4. *Management prescription*

Cutting cycle — 4 years

Reserve growing stock — 10 M bd ft

PAI of regulated forest, $= \dfrac{23.2 - 10}{4}$ — 3.3 M bd ft/acre/yr

Long-term sustained yield, = (PAI)(forest acres) = 3.3(400) — 1,320 M bd ft/yr

Cut, then grow.

Using this information, regulate this uneven-aged forest under each of the following two sets of guidelines (use the computational format of Table 14–15):

(*a*) *Area control guidelines*

(i) One complete compartment should be cut each year.

(ii) No cutting shall reduce the RGS (residual growing stock) below 10 M bd ft/acre.
(iii) The compartment with the highest volume per acre must be cut.

(*b*) *Area control with minimum harvest level guidelines*

(i) No cutting shall reduce a compartment RGS below 10 M bd ft/acre.
(ii) Harvest may occur in more than one compartment in a year, but all of the compartment must be cut with the same volume per acre removed.
(iii) A minimum of 2 M bd ft/acre must be cut from any harvested acre.
(iv) At least 1,000 M bd ft in total must be harvested each year.
(v) The forest should be area regulated by year 8 (that is, the 100-acre compartment chosen to be harvested every year starting in year 9 must yield 1,320 M bd ft).
(vi) Harvest volume must be nondeclining from year to year.

14-7 You are a consultant for Mary's Forestry Consulting. Your client has a 60-acre forest to be regulated. You have divided the forest into the following six compartments:

Compartment	Age (years)	Volume per acre (M bd ft)	Area (acres)
A	0	0	10
B	50	48	10
C	0	0	10
D	10	4	10
E	20	12	10
F	20	12	10

The following growth and yield data apply to this timber and site quality:

Age (years)	Volume per acre (M bd ft)
0	0
10	4
20	12
30	24
40	40
50	48
60	54
70	58
80	60
90	60
100	60
110	60

Cut, then grow.

Problem (*a*) The owner wants area regulation with the same acreage harvested each decade. All harvests are at the beginning of each 10-year period. The owner also feels that the oldest compartment should be cut first. A rotation age of 60 years has been agreed upon. Use a table similar to Table 14-12 to show which compartment is cut in each decade and the harvest flow over time.

Problem (*b*) The owner of this piece of land has just read a paper on possible timber shortages in the near future and has therefore requested a nondeclining harvest flow. You may now cut any number of acres from any of the stands per decade and you need not cut the oldest stand first. Show how you will regulate this forest by area control within eight periods while attempting to achieve the goal of maximizing first-period harvest.

(*Hint:* For both problems (*a*) and (*b*) write down the set of harvest constraints and goals before you start.)

CHAPTER

FIFTEEN

FOREST MANAGEMENT SCHEDULING UNDER CONSTRAINTS

Analysis of forest management scheduling problems using linear programming as the solution technique has increased rapidly since the early 1970s. Various binary search simulation models have also been employed to identify harvest patterns that maximize harvest quantities over the planning period. A variety of useful and creatively titled computer programs have been developed to assist this effort—SIMAC, MAX MILLION, RAM, TREES, ECHO, MUSYC, and FORPLAN, to name but a few.

This chapter presents the more common decision variables and constraints used to portray management scheduling problems as a set of linear equations for solution by linear programming. The fundamental model I and model II options in problem structure, aggregate emphasis and multiple outcome formulations, and some binary search techniques are also presented. A solid mastery of the problem identification material presented in Chap. 6, the financial analysis procedures of Chaps. 7 and 8, and the traditional harvest scheduling techniques of Chap. 14 is assumed.

Much of this chapter is presented in tutorial fashion, using example problems that evolve with increasing complexity, each stage building on the preceding one. To really understand and enjoy working with this subject, you have to do some analysis of your own, actually set up some problems, and solve them. As most are linear programming problems, try to obtain access to a mainframe or personal computer that has a linear programming package. All of the problems in this chapter were solved using the LINDO (Schrage, 1981) system that allows problems to be entered as a set of free-format equations rather than the more unforgiving matrix approach common in large linear programming systems.

INTRODUCTION: THE JERRY WILCOX PROBLEM

To begin, consider the following timber harvesting problem presented as a word problem like those of Chap. 6. In order to identify the problem, we describe it both as a set of equations and as a detached coefficient matrix, solve the problem, and interpret the answer. The problem:

> Jerry Wilcox is a young farmer who recently purchased a 250-acre tract of timberland adjacent to his existing livestock ranch and commercial recreation operation. The tract is divided into three stands: *A* (50 acres of old growth with 52 M bd ft/acre), *B* (100 acres of cutover pole stands with 18 M bd ft/acre), and *C* (100 acres of young plantation with no merchantable volume). Jerry wants to plan the harvest over the next 30 years. Since he is building up the ranch, he wants to harvest as much of the timber as possible in the first 10 years to provide needed capital. Given this desire, he also wants to keep the land in timber production and to cut at least 2,500 M bd ft in each of the second and third 10-year periods to help send his children to college. To guarantee that something is left for the future, he decides that at least 3,000 M bd ft of merchantable inventory should be left on the ground after the harvest in period 3. All harvests in the first 30 years come from the original stands, although the regenerated stand volumes can count against the ending inventory requirement. As a final consideration, Jerry wants to pick out 60 acres from this timber tract to set aside for development as a park/campground to expand his recreation business. No timber harvesting is to be permitted in the park, and the park inventory cannot count as part of the required 3,000 M bd ft of ending inventory since it is not part of the producing forest.

The local consulting forester did some quick analysis with a locally calibrated, variable-density stand growth model and provided a table showing the average harvest volume per acre to expect from each stand in each of the three periods. He also estimated the inventory of regenerated timber at the end of the third period as it related to when the stand was cut.

Harvest and inventory volumes per acre (thousands of board feet)

Period original stand cut	Average volume harvested from existing stands			Ending inventory of regenerated stands	
	Stand *A*	Stand *B*	Stand *C*	Stands *A*, *B*	Stand *C*
1	51	20	10	35	60
2	48	23	30	10	20
3	45	30	60	0	5

We notice that the old-growth stand A is losing volume from mortality as it ages over the three periods, cutover stand B is slowly growing, and the youthful vigor and higher site quality of stand C are revealed in its rapid growth rate.

The first thing we will do with this problem is to define the decision variables. The only real choices involve deciding how much of each existing stand to cut in each of the three periods and how much of each to assign to park status. Accordingly the 12 decision variables for this problem are identified as follows:

Existing stand	Period cut 1	2	3	Leave uncut
A	A_1	A_2	A_3	A_U
B	B_1	B_2	B_3	B_U
C	C_1	C_2	C_3	C_U

For example, C_2 stands for the number of acres in stand C cut in period 2, while C_U is the number of acres of stand C permanently assigned to the uncut park status.* By combining the information from the yield table with these variables we can write equations to identify Mr. Wilcox's problem. His objective is to maximize the first-period harvest, or $51A_1 + 20B_1 + 10C_1$. The constraints are several.

Timber volume constraints are required on the harvest in periods 2 and 3 and on the ending inventory. The minimum harvest in period 2 requires that $48A_2 + 23B_2 + 30C_2$ be greater than 2,500, and for period 3 that $45A_3 + 30B_3 + 60C_3$ also be greater than 2,500. The ending inventory constraint requires that at least 3,000 M bd ft be left after harvest in the last period. The ending inventory is the sum of the residual volume on all acres of the forest not assigned to park status. The stands cut in period 1 will have more ending inventory than those cut in period 2 because the regenerated stands have an additional period to grow. A similar relationship exists between periods 2 and 3. The equation representing this ending inventory condition is $35A_1 + 35B_1 + 60C_1 + 10A_2 + 10B_2 + 20C_2 + 5C_3 \geq 3{,}000$, with all the volumes per acre coming from the regenerated stands.

Area control constraints are required to ensure that 60 acres be set aside for a park and that no stand have more acres assigned to its decision variables than it contains. The constraint that the area assigned to park status reach the minimum of 60 acres is $A_U + B_U + C_U = 60$. The unstated restriction that the number of acres assigned to decision variables for each

* Notice that we require that any acres assigned to timber production be cut during the three planning periods. There are no decision variables allowing the stands to be assigned to timber production but not cut during the three periods.

stand cannot exceed the total number of acres in the stand must be formalized. With three contiguous stands we need three equations, one for each stand:

$$A_1 + A_2 + A_3 + A_U = 50$$

$$B_1 + B_2 + B_3 + B_U = 100$$

$$C_1 + C_2 + C_3 + C_U = 100$$

Putting these constraints and the objective together as a summary problem statement, we get a set of equations ready for solution by linear programming:

1. Maximize $51A_1 + 20B_1 + 10C_1$

subject to

2. $48A_2 + 23B_2 + 30C_2 \geq 2{,}500$
3. $45A_3 + 30B_3 + 60C_3 \geq 2{,}500$
4. $35A_1 + 35B_1 + 60C_1 + 10A_2 + 10B_2 + 20C_2 + 5C_3 \geq 3{,}000$
5. $A_U + B_U + C_U = 60$
6. $A_1 + A_2 + A_3 + A_U = 50$
7. $B_1 + B_2 + B_3 + B_U = 100$
8. $C_1 + C_2 + C_3 + C_U = 100$

To make the problem statements more compact and facilitate effective analysis of the problem structure, the problems in the rest of this chapter are presented in matrix form. A detached coefficient matrix representing the Jerry Wilcox problem is presented in Table 15-1.

When this problem is solved as a linear programming problem, the results shown in Table 15-2 are obtained. Remembering that the objective is to maximize the first-period harvest, a solution with intuitive appeal emerges. High-volume but deteriorating stand *A* is all harvested in period 1 ($A_1 = 50$). Some of low-volume, slow-growing stand *B* is cut in periods 1 and 2, and the rest is assigned to satisfy the park requirement ($B_U = 60$). Stand *C*, which has low volume in period 1, but is rapidly growing, is saved to meet the minimum harvest requirements in periods 2 and 3 ($C_2 = 58.33$, $C_3 = 41.67$).

The shadow (dual) prices for the constraints show every constraint except the ending inventory to be binding. Relaxing the period 2 or 3 minimum harvest requirement by 1 M bd ft would allow the harvest in the first period to rise by a maximum of 0.87 or 0.43 M bd ft, respectively. Reducing the park requirement by 1 acre would allow one more acre of stand *B* to be harvested in period 1 for an increase of 20 M bd ft of harvest. The land constraints for stands *A* and *B* have shadow prices equal to their current volumes per acre: an additional acre of these stand types would be harvested in the first period.

Table 15-1 Matrix representation of the Jerry Wilcox harvest scheduling problem

	Columns												
	(1)	(2)	(3)	(4)	(5)	(6)	(7)	(8)	(9)	(10)	(11)	(12)	(13)
	Decision variables												
Rows	A_1	A_2	A_3	A_U	B_1	B_2	B_3	B_U	C_1	C_2	C_3	C_U	RHS
(1) Objective	51				20				10				Maximize
(2) Period 2 minimum harvest		48				23				30			≥2,500
(3) Period 3 minimum harvest			45				30				60		≥2,500
(4) Ending inventory	35	10			35	10			60	20	5		≥3,000
(5) Park acres				1				1				1	=60
(6) Acres stand *A*	1	1	1	1									=50
(7) Acres stand *B*					1	1	1	1					=100
(8) Acres stand *C*									1	1	1	1	=100

Table 15-2 Linear programming solution printout for the Jerry Wilcox harvest scheduling problem

(*a*) Problem input

```
MAX      51 A1 + 20 B1 + 10 C1
SUBJECT TO
        2)     48 A2 + 23 B2 + 30 C2 >=    2500
        3)     45 A3 + 30 B3 + 60 C3 >=    2500
        4)     35 A1 + 35 B1 + 60 C1 + 10 A2 + 10 B2 + 20 C2 + 5 C3 >=
               3000
        5)      AU +  BU +  CU =   60
        6)      A1 +  A2 +  A3 +  AU =    50
        7)      B1 +  B2 +  B3 +  BU =    100
        8)      C1 +  C2 +  C3 +  CU =    100
```

(*b*) Solution values for decision variables

```
        OBJECTIVE FUNCTION VALUE

 1)        2697.8260

VARIABLE        VALUE           REDUCED COST
       A1        50.0000              0.0
       B1         7.3913              0.0
       C1         0.0                16.0869
       A2         0.0                 9.2609
       B2        32.6087              0.0000
       C2        58.3333              0.0
       A3         0.0                31.4348
       B3         0.0                 6.9565
       C3        41.6667              0.0
       AU         0.0                31.0000
       BU        60.0000              0.0
       CU         0.0                 6.0869
```

(*c*) Shadow prices

```
ROW              SLACK            DUAL PRICES
        2)        0.0               -0.8695
        3)        0.0               -0.4348
        4)      709.7826             0.0
        5)        0.0              -20.0000
        6)        0.0               51.0000
        7)        0.0               20.0000
        8)        0.0               26.0869
```

Table 15-2 (Continued)

***(d)* Sensitivity analysis on solution**

RANGES IN WHICH THE BASIS IS UNCHANGED

COST COEFFICIENT RANGES

VARIABLE	CURRENT COEF	ALLOWABLE INCREASE	ALLOWABLE DECREASE
A1	51.0000	INFINITY	9.2608
B1	20.0000	4.4375	12.3333
C1	10.0000	16.0869	INFINITY
A2	0.0	9.2609	INFINITY
B2	0.0	4.6667	4.4375
C2	0.0	13.9130	6.0870
A3	0.0	31.4347	INFINITY
B3	0.0	6.9565	INFINITY
C3	0.0	26.0869	13.9130
AU	0.0	31.0000	INFINITY
BU	0.0	INFINITY	6.0870
CU	0.0	6.0870	INFINITY

RIGHTHAND SIDE RANGES

ROW	CURRENT RHS	ALLOWABLE INCREASE	ALLOWABLE DECREASE
2)	2500.0000	170.0000	750.0000
3)	2500.0000	340.0000	1500.0000
4)	3000.0000	709.7826	INFINITY
5)	60.0000	7.3913	60.0000
6)	50.0000	INFINITY	20.2795
7)	100.0000	INFINITY	7.3913
8)	100.0000	25.0000	5.6667

The reduced costs for the decision variables not in solution, such as C_1, measure how much their volume per acre at harvest would have to increase before that decision variable would be selected in the optimal solution. For example, if the volume coefficient of C_1 increased 16 M bd ft to reach a volume of 26 M bd ft, it would be included at some level in the new solution. Also, if one acre of C_1 were forced into solution ($C_1 \geq 1.0$) with the current 10 M bd ft per acre coefficient, then the reduced cost indicates that the objective function would decrease by 16 M bd ft. The range for C_1 shows it can increase 16 M bd ft per acre before the solution will change. For review, the more detailed discussion of solution information was presented in Chap. 6. With some study you should be able to explain the logic of many reduced-cost, shadow-price, and range values. It is not always obvious, however.

ELEMENTS OF FOREST MANAGEMENT SCHEDULING PROBLEMS

This quick plunge into the Jerry Wilcox problem gives a feel for the nature of harvest scheduling problems and briefly reviews equation and matrix problem representation. Linear programming is seen to give solutions, shadow prices, and sensitivity information. Now we consider forest management scheduling problems in general—the more typical approaches to defining decision variables and the most common objective functions and constraints found in harvest scheduling problems. We need something specific to talk about, so we will introduce our second and more complete tutorial example, the Daniel Pickett forest problem.

The Daniel Pickett Forest Problem

The Daniel Pickett forest could be anywhere, so the specifics of location, site quality, and species are not identified. Inventory, prescription choices, management policy, financial data, and some policy questions are described and provide the substance of the presentation that follows.

1. **Inventory**

Stand type	Acres	Site	Age	Volume per acre	Condition
A	1,000	Good	100	34	Well stocked, healthy
B	500	Poor	100	18.5	Diseased, cutover
C	1,000	Poor	10	5	Young, well stocked

2. **Stand management choices**
   Existing stands: clearcut and regenerate original stands (a regeneration harvest) within 40 years, or assign them to permanent park-reserve status.
   Future stands: manage the future stands after regeneration harvest of the existing stands by one of two prescriptions
      a. Prescription 1 (all sites): prepare site, plant, clearcut at age 30. Then regenerate again.
      b. Prescription 2 (good sites): prepare site, plant, release, commercially thin about 40 percent of volume at age 20, clearcut at age 40. Then regenerate again.

   Planning horizon: all costs and revenues for the next 70 years will be evaluated in the analysis.
3. **Primary management goal** Maximize the present net worth of the forest over the 70-year planning interval to the planning horizon.

4. **Management constraints**
   a. The harvest volume in any period should not vary more than 20 percent from the harvest volume in the preceding period.
   b. At least 200 acres must be set aside uncut in park-reserve status.
   c. At least 100 acres of stand type *A* must be reserved as uncut park to protect the habitat of an endangered owl species.
   d. The ending forest inventory, measured after harvest in the last period of the overall planning interval, must contain at least 500,000 ft^3 to ensure perpetuation of the forest harvest.
   e. No more than 700 acres can be harvested in each of the first three planning periods to give a good distribution of area by ages.
   f. At least 400 acres must be assigned to prescription 2 to demonstrate and evaluate the practice of intensive management on good sites.
5. **Financial data**
   a. Log prices at landing
   Healthy old growth: $4 per cubic foot
   Diseased old growth: $2 per cubic foot
   Young growth and regenerated: $2.50 per cubic foot
   b. Logging costs
   Old growth on good sites: $1 per cubic foot
   Regenerated and young on good sites: $0.75 per cubic foot
   Old growth on poor sites: $1.50 per cubic foot
   Regenerated and young on poor sites: $1.25 per cubic foot
   c. Site preparation, planting, and release costs
   Good sites: $500 per acre (brush control)
   Poor sites: $300 per acre (no brush control)
   d. Annual management cost for timber and park lands
   Good sites: $3 per acre per year
   Poor sites: $2 per acre per year
   e. Interest rates and inflation
   Interest rate: the real, noninflationary guiding rate of interest to use for present net worth calculations is 4 percent.
   Inflation: there will be no inflation and no real increases or decreases in the costs of management or the prices of logs from their present levels.
6. **Issues to evaluate**
   a. Evaluate the impact on this property of implementing a possible state-mandated harvest policy of nondeclining yield.
   b. Evaluate the opportunity cost of the 100-acre owl reserve requirement.
   c. Evaluate the impact of raising the minimum ending inventory from 500,000 ft^3 to 1,000,000 ft^3.

Formulating the problem Given this listing of goals, constraints, and issues, the inventory, the silvicultural choices, and economic data, how do we get started to develop a mathematical formulation of the problem? There are many possible ways to begin, but experience has shown that three critical decisions are typically left to the analyst and they are the first order of business. These are choosing (1) the kind and number of land types, stand types, stands, and management units to recognize in the model, (2) the number and length of planning periods used to portray the planning interval from the present to the planning horizon, and (3) the number of different cultural treatment choices for each existing and future stand over its lifetime and the extent to which the cultural treatment choice can change on a given acre or stand over the planning interval.

These three analytical decisions largely determine the number of decision variables and provide their definition. Once the decision variables are defined, then the objectives and constraints of the problem can be formulated as equations and the empirical data organized to provide appropriate coefficients for the variables in these equations.

There is no set or best way to choose land classes, time periods, or prescription options. This is where the art and ingenuity of problem formulation flourishes. What's best depends on your problem, the people, and the analytical resources available. A lot of creative but unpublished work goes on in the planning shops of industry and public agencies, and many times they work on problems similar to yours. Get to know and talk to other planners and analysts. Find out what they have done, what worked and what didn't, and borrow some of the better tricks and techniques. We can present only some of the possibilities as we now proceed to formulate the Daniel Pickett problem in traditional or Model I structure.

Land types, stands, stand types, and management units* Choosing the classes used to describe the land base is the first and perhaps the most important decision in organizing and setting up a harvest scheduling analysis. It determines whether or not location-specific control will be modeled, the sample design for inventory and measurements to take, how and which growth and economic data will be collected and tabulated, and the eventual size of the harvest scheduling problem. Much important discussion on this decision has already been presented in Chap. 2. Often computer capacity or budgets of time and money place limits on model size and restrict the number of land classes that can be recognized in the model. Then the land classes have to be selected with great care to high-

* The definition of these approaches to classifying land are given in Chap. 2, and it is important to be clear on their differences before proceeding with Chap. 15.

light or control for those impacts or outputs of greatest importance to the decision maker.

For example, if certain big-game winter range or potential wildernesss areas are of critical importance, then each area should be identified in separate land classes in the model. If identifying and scheduling activities on land that can be logged in wet or winter weather is critical, then the land should be evaluated for slope, drainage, and soil type affecting logging to define a suitable set of classes for analysis. If the existing timber inventory data and yield models will be used in the analysis and the available inventory data uses certain land types and timber conditions to classify the timberland, then these same classes will have to be recognized in the land classes used for modeling. Usually not all of the concerns that "must" be considered in a real-life forest planning problem can fit within the model size constraints, and some hard choices must be made by the analyst and decision maker.

The terms *analysis area* and *allocation and scheduling zone* are becoming standardized by the U.S. Forest Service in its models to describe the stand types and management units used to define decision variables. Companies often use management unit, land type, stand, tract, and other terms to describe and classify their land for planning. Because of the variability, be sure of the definition in use before starting a detailed discussion about a particular model or problem formulation.

Few real problems can be handled by a single approach to land classification, and a mixture of stands, stand types, and management units may be used and cross-referenced within the same model structure. Certain individual stands may be identified for initial harvest schedules, while generalized stand types may be used to model future harvests of the remaining timberlands without special regard for location.

Many public and some private owner problems need to use both the spatial definition of management units and the homogeneity of stand types to adequately evaluate options and issues. Management units such as a watershed, ownership parcel, or administrative planning unit might be identified for evaluating water, wildlife, and erosion impacts, for reporting, and to distribute workloads. Within these management units stand types might be used to estimate timber yields, forage production, and erosion rates.

Since the collection of many inventory, treatment response, and impact data requires knowing what land classes will be used, establishing these classes takes a lot of care and thought to avoid wasting time and money on unneeded information later on. It is kind of a chicken and egg problem, since the significance of different approaches to land classification is not clearly revealed until the management problem is modeled, some analysis is done, and preliminary results are available. But by this time it may be too late to revise the land classification strategy without great expense or completely starting over. There is no magic or easy answer; whatever you do will be a compromise. Think about it a lot and talk to others to see what they did.

For the Daniel Pickett forest problem the choice of land class is mandated by the inventory data and a model size constraint of fitting the matrix on one textbook page. The stand types *A*, *B*, and *C* will be identified as the three land classes in the model. Each stand type consists of several homogeneous but spatially separated stands, so the implied assumption of our model is that on-the-ground implementation of the plan can be allocated to individual stands without violating important constraints. If we recognized individual stands, we would have many more land classes in our model.

Keeping track of time For each harvest scheduling problem, a planning horizon must be specified and then the planning interval divided into a set of shorter periods in which the activities can occur. Depending on the needs of the decision maker, we may see planning horizons of 30 years or so for private industry in the South to 200 years for public lands in the West. Planning periods of 5 to 10 years are typical and equal-length periods are normally used. However it is not particularly difficult and is in many respects desirable to set up planning periods of different lengths. By making the first few periods 1 to 5 years long, good direction is given to implementation and budget preparation in the immediate future. The later periods can be longer (10 to 20 years) and still ensure general compliance with constraints. Since a replanning effort introducing new or modified goals and constraints is usually conducted every 5 to 10 years, it hardly seems worthwhile to fine-tune a model to identify annual activities 50 years hence. As with land classes, the number of planning periods directly affects model size, and this effect needs to be considered in model design.

For reasons of simplicity, we choose to divide the 70-year planning interval of the Daniel Pickett forest problem into seven equal 10-year planning periods. If we designate January 1985 as the beginning of the first planning period and assume that in each period the harvest age of each stand or stand type will be its age at the midpoint of the period, the time accounting for the problem is as shown in Table 15-3. The first four columns show how the planning interval is divided into periods and identified by planning and calendar time. Assuming that harvest occurs at the midpoint of the different periods, the age of the existing stands when they are cut is shown in the next three columns. In the last five columns we show the midperiod age of regenerated stands that were born immediately after cutting in each of the first five periods. For example, if original stand type *B*, now 100 years old, is cut in the middle of the third period, its age will be 125 years when cut. If the new regenerated stand created after this third-period harvest is again cut in the sixth period, it will have grown for

Table 15-3 Stand age, planning periods, and calendar time for the Daniel Pickett forest harvest scheduling problem of seven, 10-year planning periods starting January 1985

Calendar time	Planning time (years)	Planning period	Midperiod planning time (years)	Midperiod age of existing stands if uncut (years)			Midperiod age of future (regenerated) stands in years by period existing stand is cut				
				A	*B*	*C*	1	2	3	4	5
1985	0	1	5	105	105	15	0				
1995	10	2	15	115	115	25	10	0			
2005	20	3	25	125	125	35	20	10	0		
2015	30	4	35	135	135	45	30	20	10	0	
2025	40	5	45	145	145	55	40	30	20	10	0
2035	50	6	55	155	155	65	50	40	30	20	10
2045	60	7	65	165	165	75	60	50	40	30	20
2055	70										

three periods and will be 30 years old when cut. (We will assume that regeneration occurs immediately after regeneration harvest.)

Prescriptions In the context of formulating harvest scheduling models we consider a prescription to be the schedule of all development, cultural, and harvest activities that take place in each of the defined planning periods from the present to the planning horizon for a given stand, stand type, land type, or management unit.

For even-aged timber, these activities include the timber culture used such as site preparation, thinning, or regeneration harvest, and the schedule of this culture over the life of existing stands and future stands. For uneven-aged prescriptions, the schedule of culture and harvest is defined on the existing stand and continued over all planning periods.

This concept of a prescription thus includes both the kinds of activities and their timing. Moreover, it can be extended to include development such as roads and nontimber culture such as forage improvement, fuel reduction, habitat improvement, or erosion mitigation.

Because the timber management schedule prescription can cover the life of two or more tree stands growing on the same acre over the planning periods of the model, it can be thought of as being composed of two or more component silvicultural prescriptions for the stands sequentially growing on the acre. We also recognize that the unmodified term "prescription" is ambiguous and needs to be further defined as either a harvest schedule prescription or as the silvicultural prescription covering the life of one stand. In the examples of this chapter we will use the symbol X_{ij} to define a harvest schedule prescription (decision variable), where X = the stand or land type, i = the silvicultural prescription for the existing stand, and j = the silvicultural prescription for regenerated (future) stands to the planning horizon.

For the Daniel Pickett problem we will initially assume that once one of the two even-aged future stand silvicultural choices is implemented on an acre, the same choice will be used for all subsequent future stands on this acre. (In some later formulations we do not make this assumption.)

Existing and future stands Cultural treatment of existing and future stands is usually quite different. Existing stands are often simply left alone until they are entered for regeneration harvest. Future stands by contrast are often more intensively managed and may be given site preparation, planting, precommercial and commercial thinning, and various release treatments in addition to changing the genetic stock.

A prescription for a land class under even-aged management has two parts: (1) the activities on the existing stand and (2) the activities on the regenerated stand. Later we see that very important differences in model structure revolve around how we handle and keep track of the existing and regenerated prescriptions for a land parcel or land type. For uneven-aged management there is no regenerated stand and the prescription would have only one part, a schedule of activities for the existing stand over all planning periods to the planning horizon.

For the Daniel Pickett problem we will let each of the three existing stands be cut in any of the first four periods. After immediate regeneration, one or both of the two future stand silvicultural choices can be followed, depending on site quality.

Volume forecasts for prescriptions The amount of timber harvest and timber inventory in each period is of central concern, and estimates are needed for each land class and prescription combination. Available and appropriate growth and yield models are used to make these forecasts, as we discussed in Chaps. 2 to 5. The independent variables are the land or site characteristics, existing stand attributes, and the planned prescription for existing and regenerated stands.

For the Daniel Pickett forest, the stated prescription policy suggests we need volume projections on existing stands for the first four periods and a projection on regenerated stands for prescription 1 on both good and poor sites and for prescription 2 on good sites. These forecasts are displayed in Table 15-4. The inventory at the planning horizon is obtained by subtracting the cut volume from the before harvest inventory volume in the last planning period as shown in the last two columns of Table 15-4*b*.

Generation of decision variables The decision variables for a problem are found by enumerating all the possible unique permutations of land classes and the existing and regenerated prescriptions used for the problem. Recall that for the Jerry Wilcox problem we had only three possible harvest entry times for the original stands plus the park assignment to give a total of four prescriptions for each of three stand types, yielding 12 decision variables. For the current problem, stand type *A* can use two different regenerated stand prescriptions. Since the existing stand has to be cut within the first four periods, we have 4 (existing stand) × 2 (regenerated stand) choices, or eight unique timber management prescriptions plus the assignment to park-reserve status, yielding nine decision variables for stand type *A*. Stand types *B* and *C* both occur on poor sites. Hence prescription 2 is not an option, and both have only five decision variables. The set of decision variables for stand types *A*, *B*, and *C* along with the harvest volume and ending inventory and age projections are given in Table 15-5. The column labels, such as A_{11}, A_{12}, etc., are the names given to the decision variables.

Table 15-4 Midperiod yield of existing stands and yield of future stands for two prescriptions (hundred cubic feet)

(*a*) Existing stands

Planning period	Stand type *A*		Stand type *B*		Stand type *C*	
	Age	Volume	Age	Volume	Age	Volume
1	105	35	105	18	15	6
2	115	37	115	17.5	25	11
3	125	38.5	125	16.5	35	15
4	135	40	135	15	45	17.5

(*b*) Future stands

		Prescription 1				Prescription 2, good site		
		Good site		Poor site				
Periods of growth	Stand age	Volume before harvest	Volume cut	Volume before harvest	Volume cut	Volume before harvest	Volume cut	Volume after harvest
0	0	0	0	0	0	0	0	0
1	10	3	0	0	0	3	0	3
2	20	16	0	5	0	16	6*	10
3	30	26	26	12	12	19	0	19
4	40	—	—	—	—	26	26	0

* Yield from intermediate harvest. All other cuts are regeneration harvests.

We can now use these decision variables to write the goals and constraints of the Daniel Pickett forest problem as a set of linear equations and to illustrate traditional Model I formulation.

Objective Functions

Two primary goals often used in forest management problems are (1) to maximize the physical volume of sustainable timber harvest and (2) to maximize the present net worth of the forest. The first might occur when the forest is the principal supply of the log input for a processing plant. Maximizing net worth is more general and treats the forest as a separate enterprise or profit center. It also can consider explicitly income or benefits from nontimber outputs and services.

The goal of maximizing wood harvest is usually stated as maximizing the first-period timber harvest subject to a timber volume flow constraint. For the Daniel Pickett forest problem, maximizing the first-period harvest

Table 15-5 Decision variables, prescriptions, and midperiod volumes harvested for the Daniel Pickett problem (volumes in units of hundred cubic feet)

(*a*) Stand *A*

• Period existing stand is cut and regenerated	1	1	2	2	3	3	4	4	*U*
• Future stand prescription	1	2	1	2	1	2	1	2	
• Decision variable*	A_{11}	A_{12}	A_{21}	A_{22}	A_{31}	A_{32}	A_{41}	A_{42}	A_U
Period 1	35	35							
2			37	37					
3		6			38.5	38.5			
4	26			6			40	40	
5		26	26			6			
6				26	26			6	
7	26	6				26	26		
Ending inventory volume	0	10	16	3	3	0	0	19	
Ending age	0	20	20	10	10	0	0	30	

(*b*) Stands *B* and *C*

• Period existing stand is cut and regenerated	1	2	3	4	*U*	1	2	3	4	*U*
• Future stand prescription	1	1	1	1		1	1	1	1	
• Decision variable*	B_{11}	B_{21}	B_{31}	B_{41}	B_U	C_{11}	C_{21}	C_{31}	C_{41}	C_U
Period 1	18					6				
2		17.5					11			
3			16.5					15		
4	12			15		12			17.5	
5		12					12			
6			12					12		
7	12			12		12			12	
Ending inventory volume	0	5	0	0		0	5	0	0	
Ending age	0	20	10	0		0	20	10	0	

* Decision variable X_{ij}, where X = stand type, i = existing stand prescription, j = future stand prescription.

is written using the decision variables and volume per acre information in Table 15-5:

$$\text{Maximize } 35A_{11} + 35A_{12} + 18B_{11} + 6C_{11} \quad (15\text{-}1)$$

The goal of maximizing present net worth is much more comprehensive and evaluates the financial attractiveness of each decision variable. Since decision variable units are the allocation of 1 acre (or other unit area) of a given land class to a prescription, the discounted net return considering all benefits and costs to the planning horizon per acre associated with decision is the appropriate coefficient in a present net worth objective function. Using the financial analysis technique of Chap. 7, we need to identify the cash flow schedule for each decision variable and discount by the guiding rate of interest.

The Daniel Pickett forest problem considers timber as the only forest output to be valued in the objective function. You could argue the owls or the park have value; but we have not given prices for these outputs so they cannot be explicitly incorporated in the objective function. Instead they will be modeled as constraints and valued indirectly.

To illustrate calculation of present net worth coefficients, the details for decision variables A_{12} and B_{31} are worked out in Table 15-6. Variable A_{12} obtains revenue from harvests in periods 1, 3, 5, and 7 and has a regeneration cost in periods 1 and 5. We recognize the difference in log price and logging cost between old-growth and regenerated stands in calculating the stumpage price per cubic foot in column 4. After obtaining gross stumpage revenue in the harvest periods in column 5, regeneration costs are subtracted to get the net cash flow in column 7. Using the midperiod age and the interest rate of 4 percent, the discount factor $1/(1 + i)^t$ is used to get the present value of each period's activity in column 10. Finally the present values are summed, the capitalized values of the annual costs subtracted, and the present net worth coefficient for the A_{12} decision variable is estimated as $9,298. When rounded and expressed in hundreds of dollars to simplify the objective function, we use $93 (hundreds) per acre.

Decision variable B_{31} has harvests scheduled in periods 3 and 6, as shown in the last two rows of Table 15-6. Different costs, prices, and yields are used because of the character of the original stand and the poorer site quality, resulting in a present net worth of $297, which rounds in hundreds to $3 per acre for the coefficient of B_{31} in the objective function.

A similar calculation is made for every decision variable, and the resulting present net worth objective function is

$$\begin{aligned}\text{Maximize } & 95A_{11} + 93A_{12} + 65A_{21} + 66A_{22} + 45A_{31} + 46A_{32} \\ & + 32A_{41} + 30A_{42} - 0.75A_U + 8B_{11} + 5B_{21} + 3B_{31} \\ & + 2B_{41} - 0.5B_U + 7C_{11} + 8C_{21} + 7C_{31} + 5C_{41} - 0.5C_U \qquad (15\text{-}2)\end{aligned}$$

The negative coefficients for A_U, B_U, and C_U are the capitalized annual costs for managing these areas as parkland with no quantifiable benefits to compensate. To check your procedures and our arithmetic, it would be good practice to calculate a few of the other coefficients yourself. Other problems would have different cost and revenue items, but the general procedure is the same: establish the cash flow schedule for all quantifiable costs and benefits and discount to the present.

Table 15-6 Calculation of present net worth objective function coefficient for decision variable A_{12} and B_{31}

Period	Midperiod harvest volume per acre (ft^3) (1)	Log price ($\$/ft^3$) (2)	Logging cost ($\$/ft^3$) (3)	Stumpage price (4) = (2) − (3)	Stumpage revenue per acre (5) = (4) × (1)	Site preparation and planting cost (6)	Net revenue (7) = (5) − (6)	Midperiod age (8)	Present value factor at 4% (9)	Present value (10) = (7) × (9)
					Variable A_{12}					
1	3,500	4	1	$3	$10,500	$500	$10,000	5	0.82	$8,200
2	—	—	—	—	—	—	—	—	—	—
3	600	2.5	0.75	1.75	1,050		1,050	25	0.38	399
4	—	—	—	—	—	—	—	—	—	—
5	2,600	2.5	0.75	1.75	4,550	500	4,050	45	0.17	688
6	—	—	—	—	—	—	—	—	—	—
7	600	2.5	0.75	1.75	1,050		1,050	65	0.08	86
										$9,373
						Capitalized value of annual costs* = a/i = 3/0.04 = 75				− 75
								Present net worth		$9,298
								Rounded for model use		$9,300
					Variable B_{31}					
3	1,650	2	1.5	$0.5	$ 825	$300	$ 525	25	0.38	$ 199
6	1,200	2.5	1.25	1.25	1,500	300	1,200	55	0.12	144
										$ 343
						Capitalized value of annual costs* = a/i = 2/0.04 = 50				− 50
								Present net worth		$ 293
								Rounded for model use		$ 300

* The perpetual capitalization equation (7-11) is used as an easy first approximation. For finer resolution the terminating equation (7-8) could be used.

Constraints

While every problem is different, nearly all forest management scheduling problems have constraints on area available, volume and/or area control of the harvest, and some constraint to ensure an acceptable ending structure of the forest at the conclusion of the planning period. To create these constraints efficiently, the technique of accounting variables needs development.

Accounting variables Accounting variables are used to add up some quantity of interest, such as inventory, to facilitate the expression of constraints related to this quantity. Also the quantity can be reported in the linear programming solution, avoiding a lot of tedious hand calculations. The amount of harvest in each period from the whole forest is such a quantity of interest to forest owners. Volume control constraints restrict periodic harvests, and the amount of harvest is practical information about raw material supply. The harvest accounting row for the first period is no more than a restatement of the maximum first-period harvest objective function developed earlier. Define H_1 as the accounting variable that equals the first-period harvest. Using information from Table 15-5 again, we have

$$H_1 = 35A_{11} + 35A_{12} + 18B_{11} + 6C_{11}$$

This equation can be rearranged to put it in the standard form as

$$35A_{11} + 35A_{12} + 18B_{11} + 6C_{11} - H_1 = 0 \qquad (15\text{-}3)$$

When a solution is found for the decision variables, the value of our accounting variable H_1 must necessarily take on the value of the four terms on the left of this equation ($35A_{11} + 35A_{12} + 18B_{11} + 6C_{11}$) in order to balance the equation and meet the requirement that the total be zero. An accounting variable for the harvest in each of the seven periods can be defined and calculated in a similar fashion. For period 2 we define H_2 as the total harvest in the second period and calculate it by

$$37A_{21} + 37A_{22} + 17.5B_{21} + 11C_{21} - H_2 = 0 \qquad (15\text{-}4)$$

We can create accounting variables to add up anything of interest that can reasonably be expressed as a linear function of the decision variables. Suppose we wanted to know the number of seedlings that we have to order for regeneration in the first period and we know it takes 500 seedlings per acre on good sites and 700 on poor sites. Defining S_1 as the total number of seedlings needed in period 1, we could calculate the seedlings as

$$500A_{11} + 500A_{12} + 700B_{11} + 700C_{11} - S_1 = 0 \qquad (15\text{-}5)$$

We could proceed in a similar manner for sediment produced, logging labor required, or the cost of brush control 2 years later. Although from the modeling point of view accounting rows are effective and convenient, they carry a computational cost: for every accounting variable a new row and a new column is added to the matrix.

Volume control The most direct way to inject volume control in a problem formulation is to place an upper or lower limit on the absolute amount of harvest in a period. For example, assume the owner of the Daniel Pickett forest said that a minimum of 5,000 and a maximum of 12,000 units (a unit = 100 f^3) can be harvested in each period. This takes two equations, one each for the upper and the lower bound to put it into the problem. The two equations for the first period illustrate:

$$35A_{11} + 35A_{12} + 18B_{11} + 6C_{11} \geq 5{,}000$$

$$35A_{11} + 35A_{12} + 18B_{11} + 6C_{11} \leq 12{,}000$$

Here is an excellent place to make use of our accounting variables. When used with Eq. (15-3) the equations

$$H_1 \geq 5{,}000 \qquad H_1 \leq 12{,}000 \tag{15-6}$$

give us exactly the same constraint with fewer terms in the equations.

A second way that volume control constraints are stated is to define a relationship between the harvest in different periods. Most common is to say that the harvest in a period cannot vary more than a certain percentage from the harvest in the preceding period.

Suppose the specified volume control relationship allows a maximum variation of 20 percent between periods. Using our accounting variables, we can write $H_2 \geq 0.8H_1$, which states that the harvest in period 2 must be not less than 80 percent of the harvest in period 1, another way of saying the harvest can only drop 20 percent from period 1 to period 2. On the up side we need a second equation: $H_2 \leq 1.2H_1$, which states that the second period harvest must not be more than 120 percent of the first period harvest. Each sequential pair of periods need these two types of equations to impose the volume control constraint over the planning periods.

The general form of the equations limiting period-to-period variation is

$$\begin{aligned} H_2 &\geq (1 - a)H_1 \\ H_2 &\leq (1 + b)H_1 \end{aligned} \tag{15-7}$$

where a and b are the permitted proportional decrease and increase.

For a four-period problem where both a and b equal .20, we would need three pairs of equations:

$$H_2 \geq 0.8H_1 \qquad H_3 \geq 0.8H_2 \qquad H_4 \geq 0.8H_3$$
$$H_2 \leq 1.2H_1 \qquad H_3 \leq 1.2H_2 \qquad H_4 \leq 1.2H_3 \tag{15-8}$$

The nondeclining yield constraint used by the U.S. Forest Service says that the harvest cannot decrease from period to period. This means that $a = 0$ and b is unspecified, and we can write the requirement with only one equation per pair of periods as

$$H_2 \geq H_1 \qquad H_3 \geq H_2 \qquad H_4 \geq H_3 \tag{15-9}$$

The fully regulated forest is one with an equal cut in every period, which implies $a = 0$, $b = 0$, and these even flow constraints are

$$H_2 = H_1 \qquad H_3 = H_2 \qquad H_4 = H_3 \tag{15-10}$$

The sets of equations (15-6) to (15-10) are the usual ways to inject volume control into a problem. Volume control constraints often dominate a problem's solution or drive it infeasible, so take special care in using these restrictions when initiating analysis. One way to start the analysis is to set the allowable percent variation in Eq. (15-7) at 30 percent or so and then tighten down after a feasible solution is obtained to evaluate the cost of more restrictive volume control.

Area control Area control requires a minimum or maximum number of some classes of acres to be cut in different periods. Area control is used for many reasons much like the volume control constraints. In classic area regulation with forest area A and a rotation of R years, the total acreage cut in each period is set equal to $(A/R) \times$ (number of years per period). For area regulation, the 1,000 acres of stand type A in the Daniel Pickett forest will be split between a 30- and a 40-year rotation when regenerated, and the 1,500 acres of stand types B and C will all be managed on a 30-year rotation. These rotation lengths for regenerated stands indicate that between 1,000/40 and 1,000/30, or 25 to 33 acres of stand A can be cut per year (250 to 333 acres per 10-year period) to achieve area regulation in four periods. Similarly stand types B and C should have $1{,}500/30 \times 10$, or 500 acres cut each period. If the requirement for stand A is set at 280 acres per period (which implies most of the area will be assigned regenerated prescription number 1), the area control constraints for the first two periods to achieve this interpretation of area regulation are written

$$A_{11} + A_{12} = 280 \qquad B_{11} + C_{11} = 500$$
$$A_{21} + A_{22} = 280 \qquad B_{21} + C_{21} = 500 \tag{15-11}$$

Allocation constraints Area constraints for other purposes are specific to each problem. In the Daniel Pickett forest problem, for example, we are required to set aside at least 200 acres of the forest for park reserve and at least 100 acres of stand type A for the owls. The decision variables defined for such set asides are A_U, B_U, and C_U. Assuming owl acres qualify as park acres, the requirements can be written as

$$A_U \geq 100 \qquad A_U + B_U + C_U \geq 200$$

Ending forest structure Without some sort of additional considerations, the solution to the Daniel Pickett forest problem will call for the harvest in the last planning period of any merchantable inventory that makes a positive contribution to the objective function and whose cutting will not violate the problem constraints. Often this is an unwanted result and two approaches are commonly used to deal with it: (1) assigning a value to the ending inventory in the objective function, and (2) formulating constraints that require minimum or maximum levels of ending inventory.

The value of an acre of land existing after the last harvest can be calculated in much the same way as the land and stand were valued in Chap. 11. The value there was composed of two parts: (1) the present value of the existing timber when it is harvested at some planned future date, and (2) the discounted SEV value of the bare land assuming it will be used to produce a perpetual series of future crops. The calculated values for the ending inventory and SEV would be added to the PNW already calculated for the seven periods of the planning interval. For example, in Table 15-6, decision variable A_{12} has a PNW coefficient of \$9,300. If the ending inventory of 10 cubic units shown in Table 15-5*a* plus the SEV were calculated to have a present value of about \$300, then the PNW coefficient of A_{12} would be changed to \$9,300 + \$300, or \$9,600. For more discussion of this approach, see Clutter et al. (1983).

Valuing ending inventory is consistent with the owner's economic goals, but it does not ensure that merchantable growing stock will be left. Cutting all the merchantable growing stock in the last period under a nondeclining yield constraint, for example, may allow a higher harvest in all preceding periods. With discounting, the ending inventory is not worth much and its value could easily be more than offset by the gain in objective function value due to the higher harvest starting in the first period.

The economic efficiency goal of an owner may quite rationally accept full inventory liquidation; however, other goals of the owner or society may mandate that some merchantable-sized trees exist on parts of the forest after the planning horizon. In this case the direct approach of ending inventory constraints is needed. For the owner who wants the plan to be consistent with forest production after the planning horizon has been reached, inventory constraints are needed to plan for leaving some trees. There are many ways to do this. The approaches involve either constraints on the volume of timber left after the last period harvest, constraints on the amount of growth of timber left after the last period harvest, or constraints on the distribution of acreage by age or volume class of timber in the residual stand. The easiest approach sets a minimum total ending inventory requirement over all stands in the forest. The amount of the minimum requirement might logically be the average forest growing stock found in a regulated forest calculated using the procedures outlined in Chap. 14. Assuming that the ages of stands which meet an ending inventory constraint are reasonably well distributed, then something like a regulated forest would be left on the ground at the planning horizon.

In the fully regulated forest, stand types *B* and *C* will be under a 30-year rotation and stand type *A* will be split between a 30- and a 40-year rotation and the area will be distributed evenly by age class. Using the midperiod age for calculation, this means that after harvest the 1,500 acres of sites *B* and *C* will be distributed 500 acres at 0 years, 500 acres at 10 years, and 500 acres at 20 years. Similarly with stand *A*. If we assume that stand *A* has assigned 400 acres to prescription 2 and 600 acres to prescription 1, we can calculate the ending inventory as follows:

Stand type *A*

	Volume after harvest		Regulated acres by age class		Total volume (100 ft³)	
Age	R_{X1}	R_{X2}	R_{X1}	R_{X2}	R_{X1}	R_{X2}
0	0	0	200	100	0	0
10	3	3	200	100	600	300
20	16	10	200	100	3,200	1,000
30	—	19	—	100	—	1,900
			600	400	3,800	3,200

Stand types *B* and *C*

Age	Volume per acre after harvest R_{X1}	Regulated acres by age class R_{X1}	Total volume (100 ft³) R_{X1}
0	0	500	0
10	0	500	0
20	5	500	2,500
		1,500	2,500

The total inventory of the regulated forest is the sum of the stand-type ending inventories:

$$3{,}800 + 3{,}200 + 2{,}500 = 9{,}500$$

The owner might then reduce the 950,000-ft^3 inventory level to 500,000 ft^3 to be sure an initial feasible solution to the linear program could be found, with the idea of raising the requirement in subsequent problem runs.* The postharvest inventory per acre is taken from Table 15-4 and is shown for each decision variable in Table 15-5. Variable A_{22}, for example, was regenerated in period 6. Thus it is 10 years old in the middle of period 7 and has a volume of 3 cubic units. The ending inventory equation for all decision variables, measured in hundreds of cubic feet, is

$$10A_{12} + 16A_{21} + 3A_{22} + 3A_{31} + 19A_{42} + 5B_{21} + 5C_{21} \geq 5{,}000 \qquad (15\text{-}12)$$

Simply requiring a total ending inventory does not guarantee leaving a balanced, productive forest. For example, the solution could leave most acres cleared, while meeting the inventory requirement with a few acres of oldest regenerated age class. This would satisfy the constraint but not the intent.

To guarantee a desired structure, a more elaborate set of constraints is needed. These could be specified in terms of volume, but to illustrate another approach, we will set up constraints that require a balanced ending distribution of acres by age class.

To understand the construction of these constraints, look at the last row for each decision variable in Table 15-5, where the ending age after the seventh-period harvest is given. For example, variable A_{12} was last regenerated in period 5, so in period 7 it will be 20 years old. A regulated forest structure has approximately equal acreage in each age class. Assuming the whole forest uses a 30-year rotation, about 2,500/3, or 833 acres are in each of the three age groups of 0, 10, and 20 years after the harvest in each period. Prescription 2 on the better site runs to 40 years, and at least 400 acres must be in this prescription. If 400 acres are allocated to this prescription, this 400 acres when regulated will be distributed 100 acres to each of the age classes 0, 10, 20, and 30. For the forest as a whole this means 100 acres should be in the 30-year age class and the remaining 2,400 acres distributed 800 acres to age classes 0, 10, and 20.

If every forest acre is completely assigned to ending age classes, it is probable that the problem will become infeasible because of unforeseen interaction with other constraints. To relax the constraints a bit, a set of constraints that required 700, 700, 700, and 100 acres left in age classes 0, 10, 20, and 30, respectively, would ensure an acceptably well structured ending forest. For the Daniel Pickett forest these constraints are written using the information from Table 15-5:

* These total ending inventory calculations have also included the 200 acres constrained to remain uncut. If assumptions are made about which stand types and prescriptions they come from, then the regulated ending inventory can be reduced accordingly.

$$\begin{aligned} &\text{0-year age class:} \quad A_{11} + A_{32} + A_{41} + B_{11} + B_{41} + C_{11} + C_{41} \geq 700 \\ &\text{10-year age class:} \quad A_{22} + A_{31} + B_{31} + C_{31} \geq 700 \\ &\text{20-year age class:} \quad A_{12} + A_{21} + B_{21} + C_{21} \geq 700 \\ &\text{30-year age class:} \quad A_{42} \geq 100 \end{aligned} \tag{15-13}$$

This set of constraints may still not meet the owner's objectives because it forces all of the 30-year age class to come out of variable A_{42}, and there are twice as many variables producing a 0-year age class as the other age classes. This result occurs partly because of the requirement that all original stands be cut in the first four periods and because only two prescription options are being considered. Can you improve on these constraints? These ending requirements can be detailed by value, species groups, sites, watersheds, or any combination of land classes, depending on the need of the owner.

Area accounting If the forest acreage is fixed, these constraints need only say that the allocation of a given stand type to its decision variables must equal the initial acreage of that type. For stand B, this restriction is

$$B_{11} + B_{21} + B_{31} + B_{41} + B_U = 500 \tag{15-14}$$

An equality is used because we wish all acres to be assigned to some decision variable. Similar equations can be written for stands A and C.

The option might be open to purchase or otherwise add new land to the forest in the different stand types. If this were the case and the owner wanted to evaluate how much land to add, a decision variable for land acquisition could be defined. To illustrate, let B_P be a decision variable for the number of new acres of stand type B purchased. The sum of stand type B acreage allocations must now equal the sum of the original 500 acres plus the number of new acres purchased. A revised version of Eq. (15-14) that shows this is

$$B_{11} + B_{21} + B_{31} + B_{41} + B_U - B_P = 500 \tag{15-14a}$$

This technique of defining a variable to represent the addition of new inputs such as land, labor, equipment, or facilities is useful whenever the problem asks to simultaneously solve for the optimum input level and the allocation of these new and existing inputs to the production of different forest outputs. The cost and revenue impacts of the new input on the problem should of course be included in the relevant budget constraints (if any) in the objective function.

Relational constraints A relational constraint is used to relate one set of decision variables to another set. The volume control constraints were of this type, where we said that the harvest in a period was limited to some percentage of the harvest in the preceding period. As another illustration, suppose the owner of the Daniel Pickett forest wanted a ratio established between prescriptions 1 and 2 in each period, such that at least 1.5 acres of stand type A were assigned to prescription 2 for every acre of stand type A assigned to prescription 1. In words, we write this as

$$\text{(Acres in prescription 2)} \geq 1.5\text{(acres in prescription 1)} \quad (15\text{-}15)$$

Note carefully the location of the coefficient 1.5 in this statement; it is easy to get it backward. To check that the equation is the one we want, assign 100 acres to prescription 1 in Eq. (15-15). Solving, we see that a minimum of 150 acres would be assigned to prescription 2. Substituting the Daniel Pickett decision variables into Eq. (15-15), for the first period we have $A_{12} \geq 1.5A_{11}$. In standard form, this is written as $A_{12} - 1.5A_{11} \geq 0$. The set of equations for the first four periods would be

$$\begin{aligned} A_{12} - 1.5A_{11} &\geq 0 \\ A_{22} - 1.5A_{21} &\geq 0 \\ A_{32} - 1.5A_{31} &\geq 0 \\ A_{42} - 1.5A_{41} &\geq 0 \end{aligned} \quad (15\text{-}15a)$$

FORMULATION, SOLUTION, AND ANALYSIS OF THE DANIEL PICKETT FOREST PROBLEM

If we go back to the statement of the Daniel Pickett forest problem and use the problem formulation techniques developed above, a full statement of the initial problem in matrix form is developed and given in Table 15-7. The problem requires a matrix of 29 columns and 32 rows.

Accounting variables were defined for the harvest volume in each period, for the ending inventory, and for forest present net worth, as shown in the first nine rows of the matrix and columns 20 to 28. These accounting variables were defined because we knew from examining the problem description that constraints on harvests and inventory would be needed and that present net worth would be calculated for all problem formulations. Furthermore we wanted PNW, ending inventory, and the harvest volume in each period to be calculated and portrayed in the solution. The values and layout of the volume coefficients used in the harvest and inventory accounting rows are taken virtually unchanged from Table 15-5, where the decision variables were defined and associated with growth and yield forecasts.

With the decision variables and accounting variables defined, we can now proceed to the stated constraints and objectives of the Daniel Pickett forest.

> The harvest volume in any period should not vary more than 20 percent from the harvest in the preceding period.

This requires volume control constraints linking the harvest between successive periods. Since there are seven periods, we will need six pairs of upper and lower bound constraint equations of the form shown in Eq. (15-7). These 12 equations are rows 10 to 21 in the matrix of Table 15-7. Note that the equations are easily expressed by making use of the harvest accounting variables.

> At least 200 acres must be set aside uncut in park-reserve status.

This requirement is a simple area constraint on the decision variables, which sets aside land for parks. The equation of row 22 in Table 15-7 does this.

> At least 100 acres of stand type *A* must be reserved as park to protect the habitat of an endangered owl.

This requirement can be stated as an area constraint on A_U, the only decision variable that can meet this constraint, and is shown by the equation of row 23 in Table 15-7.

> The ending forest inventory measured after the harvest in the last period must be at least 500,000 ft^3 to ensure perpetuation of the forest harvest.

This requirement needs an equation that adds up the residual inventory after the period 7 harvest on all land not in park-reserve status. The volume and acres in park-reserve status cannot count in this constraint because the reserve status is permanent and the acres in it can never be cut. The needed inventory equation is already developed as accounting row 9, using the technique of Eq. (15-12), so the ending inventory requirement itself is the simple constraint $I \geq 5{,}000$, as shown in the equation of row 24 of Table 15-7.

> No more than 700 acres can be harvested in each of the first three planning periods to give a good distribution of area by ages.

This area control goal is directly expressed using equations that restrict the sum of all decision variables that harvest timber in each of the first three periods to not more than 700 acres. The three equations used are rows 25, 26, and 27 of Table 15-7. Note that the area harvest constraint in period 1 also imposes area constraint on period 4 and again on period 7 due to the linkage of original stand harvest to regenerated stand harvest.

Table 15-7

			Stand decision variables*													
			Stand A									Stand B				
			(1) A_{11}	(2) A_{12}	(3) A_{21}	(4) A_{22}	(5) A_{31}	(6) A_{32}	(7) A_{41}	(8) A_{42}	(9) A_U	(10) B_{11}	(11) B_{21}	(12) B_{31}	(13) B_{41}	(14) B_U
Accounting rows	(1)	H_1	35	35								18				
	(2)	H_2			37	37							17.5			
	(3)	H_3		6			38.5	38.5						16.5		
	(4)	H_4	26			6			40	40		12			15	
	(5)	H_5		26	26			6					12			
	(6)	H_6				26	26			6				12		
	(7)	H_7	26	6				26	26			12			12	
	(8)	I		10	16	3	3			19			5			
	(9)	PNW	95	93	65	66	45	46	32	30	−0.75	8	5	3	2	−0.5
Problem constraints																
	(10)	1–2	Lower bound (L) (periods)													
	(11)		Upper bound (U)													
Harvest volume control	(12)	2–3	L													
	(13)		U													
between periods	(14)	3–4	L													
	(15)		U													
	(16)	4–5	L													
	(17)		U													
	(18)	5–6	L													
	(19)		U													
	(20)	6–7	L													
	(21)		U													

This relationship is one to one for stand types B and C; an acre of B cut in period 1 is automatically cut again in periods 4 and 7. For stand A the link is not as rigid. Depending on the prescription chosen, an acre of A cut in period 1 could be recut again in either period 4 or 5.

At least 400 acres must be assigned to prescription 2 to demonstrate and evaluate the practice of intensive management on good sites.

The Daniel Pickett forest problem in Model I formulation

Stand C					Accounting variables period harvest accounting									
(15) C_{11}	(16) C_{21}	(17) C_{31}	(18) C_{41}	(19) C_U	(20) H_1	(21) H_2	(22) H_3	(23) H_4	(24) H_5	(25) H_6	(26) H_7	(27) I	(28) PNW	(29) RHS
6					−1									= 0
	11					−1								= 0
		15					−1							= 0
12			17.5					−1						= 0
	12								−1					= 0
		12								−1				= 0
12			12								−1			= 0
	5											−1		= 0
7	8	7	5	−0.5									−1	= 0
					−0.8	1								≥ 0
					−1.2	1								≤ 0
						−0.8	1							≥ 0
						−1.2	1							≤ 0
							−0.8	1						≥ 0
							−1.2	1						≤ 0
								−0.8	1					≥ 0
								−1.2	1					≤ 0
									−0.8	1				≥ 0
									−1.2	1				≤ 0
										−0.8	1			≥ 0
										−1.2	1			≤ 0

This restriction is an area constraint on all decision variables contributing acres to prescription 2, as shown by the equation of row 28 in Table 15-7.

Now that we have covered all the accounting constraint rows and the policy and multigoal constraints on outputs, we have only to add the input constraints that limit the acres cut or reserved to the total acreage in each stand type, as shown in rows 29, 30, and 31 of Table 15-7.

The final row of the problem matrix brings in the primary manage-

Table 15-7

			Stand decision variables*													
			Stand A									Stand B				
			(1) A_{11}	(2) A_{12}	(3) A_{21}	(4) A_{22}	(5) A_{31}	(6) A_{32}	(7) A_{41}	(8) A_{42}	(9) A_U	(10) B_{11}	(11) B_{21}	(12) B_{31}	(13) B_{42}	(14) B_U
Park acres	(22)										1					1
Owl acres	(23)										1					
Ending inventory	(24)															
Area	(25)	P_1	1	1								1				
control	(26)	P_2			1	1							1			
	(27)	P_3					1	1						1		
Prescription 2 acres	(28)			1		1		1		1						
Acres	(29)	A	1	1	1	1	1	1	1	1	1					
available	(30)	B										1	1	1	1	1
	(31)	C														
Objective	(32)															

* Decision variables A_{ij}, B_{ij}, C_{ij} designate the acres of existing stands A, B, C entered for first harvest in period i and using prescription j for regenerated stands. A_U, B_U, C_U are allocations of these stands to uncut park-reserve status.

† H_i = total harvest in year i, PNW = present net worth over 7 periods, I = ending inventory after harvest in period 7.

ment goal of the Daniel Pickett forest: maximizing the present net worth of the forest. It is a simple function of the accounting variable for present net worth, as shown by the equation of row 32 of Table 15-7.

Solution and Optimal Harvest Schedule

The basic specification of the Daniel Pickett forest problem as shown in Table 15-7 was solved by linear programming and showed that a maximum forest present net worth of $8,385,800 could be achieved while satisfying all constraints on the problem. The solution values for the decision variables are shown in Table 15-8 and provide the information needed for planning harvest and culture activities and for anticipating the volume of timber available to be sold in each period. The results are not unexpected. As much of stand A as possible should be cut in the first period to capture the high first-period present net worth values of the well-stocked old growth, with most of the acres being assigned to regener-

The Daniel Pickett forest problem in Model I formulation (Continued)

Stand C					Accounting variables period harvest accounting									
(15) C_{11}	(16) C_{21}	(17) C_{31}	(18) C_{41}	(19) C_U	(20) H_1	(21) H_2	(22) H_3	(23) H_4	(24) H_5	(25) H_6	(26) H_7	(27) I	(28) PNW	(29) RHS
				1										≥ 200
														≥ 100
												1		≥ 5,000
1														≤ 700
	1													≤ 700
		1												≤ 700
														≥ 400
														= 1,000
														= 500
1	1	1	1	1										= 1,000
													1	= Max

ation prescription 2 (A_{12}). The present value of stand type A drops \$3,000 per acre by waiting until period 2 for harvest, much more of a decline than for stand types B and C between periods 1 and 2 (see row 9 of Table 15-7). Thus as much of stand type A as possible should be taken in the first period to achieve the objective.

Harvest in stand type B is heavy in period 2, and stand type C is cut heavily in periods 3 and 4. Stand type B had the lowest present net worth and was logically chosen for the additional 100 acres of park-reserve needed over the 100 acres of stand type A required for owl habitat, which also counts as parkland. The ending inventory constraint is not binding; the ending inventory is 672,800 ft^3—providing an inventory surplus of 172,800 over the minimum requirement of 500,000 ft^3.

The harvest volume each period (the H_i columns) declines steadily over the first four periods, reflecting the higher present net worth payoff from harvesting the original stands as soon as possible. While this pattern might be disturbing to a mill manager who will be forced to look elsewhere

Table 15-8 Optimal solution to the Daniel Pickett forest problem

Decision variable*	Amount	Row	Row	Shadow price
A_{11}	232	1	H_1	0.8
A_{21}	—	2	H_2	0.14
A_{31}	—	3	H_3	1.0
A_{41}	—	4	H_4	0.09
		5	H_5	0.07
A_{12}	374	6	H_6	—
A_{22}	257	7	H_7	—
A_{32}	36	8	Inventory	—
A_{42}	—	9	PNW	1
A_U	100	10†	1–2 L	−0.98
		11	1–2 U	—
B_{11}	—	12	2–3 L	−1.05
B_{21}	400	13	2–3 U	—
B_{31}	—	14	3–4 L	—
B_{41}	—	15	3–4 U	—
B_U	100	16	4–5 L	—
		17	4–5 U	0.07
		18	5–6 L	—
C_{11}	—	19	5–6 U	—
C_{21}	43	20	6–7 L	—
C_{31}	664	21	6–7 U	—
C_{41}	293	22	Park acres	−5.02
C_U	—	23	Owl acres	−65.5
		24	Ending inventory	—
H_1	21,236	25	Area control 1	—
H_2	16,989	26	Area control 2	2
H_3	13,591	27	Area control 3	16
H_4	12,717	28	Prescription 2	—
H_5	15,261	29	Acres of *A*	70
H_6	14,657	30	Acres of *B*	4.5
H_7	12,742	31	Acres of *C*	6.6
I	6,728	32	Objective	—
PNW	83,858			

* Decision variables and rows as given in Table 15-7.

† Rows 10–21 are the lower and upper bound constraints on change in harvest volume between sequential periods.

for wood raw material, it is clearly the way to maximize the present net worth of the forest itself. The volume control constraint for the maximum 20 percent decline between periods 1 and 2 (row 10) and between periods 2 and 3 (row 12) had nonzero shadow prices, indicating that this constraint is limiting the volume of harvest in the early periods.

For on-the-ground implementation of this harvest schedule, the field foresters would follow the schedule of acres to cut in each stand type, as shown in Table 15-9, until the plan was updated. The number of acres to site prepare, plant, thin, and clearcut also can be directly associated with this regeneration harvest schedule and constitutes the guiding operational plan for the management of the forest until it is again planned.

Evaluation of Issues

The final step in our harvest scheduling analysis of the Daniel Pickett forest is to consider the impact on the solution of changing some key problem constraints to reflect possible external influences on management of the forest or to reflect internal policy changes. In each case, the problem solution to the initial problem defined and reviewed in Tables 15-7 to 15-9 is the basic or reference solution from which impacts are measured.

Management has requested an evaluation of three issues:

1. The imposition of a state-mandated harvest policy of nondeclining yield
2. The opportunity cost of the 100-acre owl reserve requirement
3. An internal policy of raising the ending inventory to 1,000,000 ft^3 for a more substantial residual forest

Each issue was modeled by altering the original problem of Table 15-7 and then finding a new optimal solution. The problem was then restored to the original and the next issue introduced. Nondeclining yield was modeled by deleting rows 10 through 21 of Table 15-7 and substituting six new

Table 15-9 Implementation schedule of acres for intermediate and regeneration harvest by stand type, prescription, and planning period on the Daniel Pickett forest

Stand type Prescription Type of harvest	*A* 1 Regeneration	*A* 2 Regeneration	*A* 2 Intermediate	*B* 1 Regeneration	*C* 1 Regeneration
Period 1	232	374		—	—
2	—	257		400	43
3	—	36	374	—	664
4	232	—	257	—	293
5	—	374	36	400	43
6	—	257		—	664
7	232	36	374	—	293

equations of the form $H_{t+1} \geq H_t$, where t is the time period, as discussed earlier [Eq. (15-9)]. The opportunity cost of owl habitat was estimated by changing the right-hand side value of row 23 from 100 acres to 0 acres. The increased ending inventory requirement was modeled by changing the right-hand side value of row 24 from 5,000 to 10,000.

The solutions to the original problem and the three new problems can be compared in Table 15-10. There are many differences between the solutions. Changing the volume control constraint from ±20 percent to

Table 15-10 Impact of three policies on the basic solution and harvest schedule for the Daniel Pickett forest

Decision variable	Basic problem: MAX PNW ± 20% I = 5,000 owls = 100	Policy I (NDY) MAX PNW NDY I = 5,000 owls = 100	Policy II (owl habitat): MAX PNW ± 20% I = 5,000 owls = 0	Policy III (ending inventory): MAX PNW ± 20% I = 10,000 owls = 100
A_{11}	232	—	238	218
A_{21}	—	—	—	88
A_{31}	—	6	—	7
A_{41}	—	61	—	—
A_{12}	374	412	395	311
A_{22}	257	314	311	135
A_{32}	36	106	56	—
A_{42}	—	—	—	141
A_U	100	100	0	100
B_{11}	—	—	—	6
B_{21}	400	—	300	335
B_{31}	—	—	—	—
B_{41}	—	400	—	59
B_U	100	100	200	100
C_{11}	—	—	—	166
C_{21}	43	256	89	141
C_{31}	664	508	644	692
C_{41}	293	235	267	—
C_U	—	—	—	—
H_1	21,236	14,436	22,163	19,597
H_2	16,989	14,436	17,730	15,677
H_3	13,591	14,436	14,184	12,542
H_4	12,717	14,436	12,728	15,050
H_5	15,261	14,436	15,274	16,085
H_6	14,657	14,436	15,809	12,868
H_7	12,742	14,436	13,221	10,294
I	6,728	6,368	6,827	10,000
PNW $(100)	83,858	73,666	90,409	77,649

strict nondeclining yield had the most pronounced impact on the optimum harvest schedule. The volume harvested in each period changed to an even flow schedule of 1,443,600 ft^3 per period, and the present net worth dropped 12 percent ($1,019,200). The harvest schedule itself made a few changes, the most noticeable being a shift from prescription 1 to prescription 2 for stand *A* and a two-period delay in the harvest of stand *B*. Shifting to the thinning with later harvest prescription made it easier to meet the demanding nondeclining harvest constraint.

Relaxing the owl constraint from 100 to 0 acres increased the present net worth 6 percent ($655,100). This increase could be interpreted as the opportunity cost of the owl constraint measured in terms of the reduced present net worth of commercial timber production. Early-period harvests rose moderately, and the major allocation change was to free up 100 acres of stand type *A* for timber production by obtaining all 200 acres of the park reserve from stand *B*.

Doubling the ending inventory requirement dropped the present net worth by 7.4 percent ($520,900) and caused some shifts in the harvest schedule. Decision variables A_{21} and A_{42} both had high ending inventory coefficients, and we expected that the acreage assigned to these prescriptions would show a net increase. It did.

ADVANCED PROBLEM FORMULATION TECHNIQUE

Many real-world harvest scheduling problems have been set up and solved to good effect with only the formulation technique and constraint types developed thus far from the Daniel Pickett forest problem. The complexity of analysis needed to handle the goals and constraints for some ownership situations, however, has stimulated the development of additional modeling techniques. Three important techniques introduced here are (1) Model II, (2) multiple outcomes, and (3) aggregate emphasis for group choices. Unfortunately the space available on one or two facing textbook pages limits exposition. Because of this, we show only the portions of the problem matrix that have important and less than self-evident changes. In all examples the data from the basic Daniel Pickett forest problem are used, and the problem matrix of Table 15-7 serves as our primary reference.

Models I and II

Johnson and Scheurman (1977) coined the terms Model I and Model II to label fundamentally different ways to define decision variables for a timber scheduling problem, the distinction being due to the way the regenerated stands are handled. We have been using Model I all along in this chapter. The regenerated stands are coupled directly to and identified by the existing stands to which they are associated. Model II detaches the regenerated stands from the existing stands and defines new decision variables for them.

Model I defines decision variables that follow the life history of an acre over all planning periods while Model II defines decision variables that follow the history of an acre over the life of a stand growing on the acre, from its birth through its death at final harvest. In Model II an acre may pass through several decision variables before reaching the planning horizon. To describe these two approaches to harvest scheduling, we first review the Model I formulation that we used for the Daniel Pickett problem, and then we show how the Model II formulation differs.

Consider only the decision variables A_{11} and A_{12} for stand type A, which regeneration harvest the existing stand in period 1. After the existing stand has been cut, the regenerated stands are managed by one of these two possible prescriptions. The relevant portion of the problem matrix (rows 1 to 7 and columns 1 and 2 in Table 15-7), which shows the yield and harvest schedule for A_{11} and A_{12}, is

Volume harvested by period and decision variable

	Decision variable	
Period	A_{11}	A_{12}
1	35	35
2		
3		6
4	26	
5		26
6		
7	26	6

In this Model I approach to the Daniel Pickett problem, each decision variable covers all the activities that could occur on the acre over the seven periods along with the associated yields and financial effects. Decision variable A_{11} shows a regeneration harvest of the existing stand in period 1 and then two complete 30-year rotations for the regenerated stand. Decision variable A_{12} shows a similar regeneration harvest in period 1 and one complete 40-year rotation and part of another for the future stand.

All of the activities after harvest of the existing stand in period 1 are associated with the future stands. To separate the future stands for Model

II, we define new decision variables just for future stands. Let R_{ijk} represent a future stand, where

i = period of stand birth
j = period of stand death by regeneration harvest
k = silvicultural prescription sequence followed

For all acres "born" in period 1 as a result of harvesting some of the existing stands, the variables describing what happens to these acres are R_{141} and R_{152}. The new stands born in period 1 are either regeneration harvested again in period 4 following the 30-year rotation of silvicultural prescription sequence 1 (R_{141}), or they are thinned in period 3 and regeneration harvested in period 5, following the second silvicultural prescription (R_{152}).

With each decision variable in Model II only tracking an acre over the birth to death life of a stand, we need equations to pass the acres from decision variables representing one generation to the decision variables representing the next generation. A relational constraint is needed:

$$\text{Acres of stands born in period } t = \text{acres of stands that die in period } t \qquad (15\text{-}16)$$

In terms of our new decision variables, the necessary equation is

$$R_{141} + R_{152} = A_{11} + A_{12}$$

which, rewritten in standard matrix form, appears as

$$-A_{11} - A_{12} + R_{141} + R_{152} = 0 \qquad (15\text{-}16a)$$

Since the future stand prescription is identified in the R_{ijk} variables, this identification does not need to be carried in the existing stand variables. Thus A_{11} and A_{12} collapse to the variable A_1, acres of stand type A harvested in period 1, and Eq. (15-16a) simplifies to

$$-A_1 + R_{141} + R_{152} = 0 \qquad (15\text{-}16b)$$

So far our model has an acre of stand type A that is cut in period 1 (A_{11} or A_{12}) and will be cut again in periods 4 or 5 (R_{141}, R_{152}). What happens to the acre next, after the regeneration harvest in period 4 or 5? More future stand decision variables are needed to represent the possible assignment of the acre over the periods of the planning interval—specifically, R_{471}, R_{4U2}, R_{5U1}, R_{5U2}, R_{7U1}, and R_{7U2}. These decision variables pick up a born-again acre in periods 4 or 5 and harvest it again in period 7 or leave it uncut but still identified by assignment to the different silvicultural prescription sequences. Figure 15-1 follows an acre of stand type A that is cut in period 1 through all possible sequences. We see that there are now five options, with path 1 being the same as our original Model I variable A_{11}

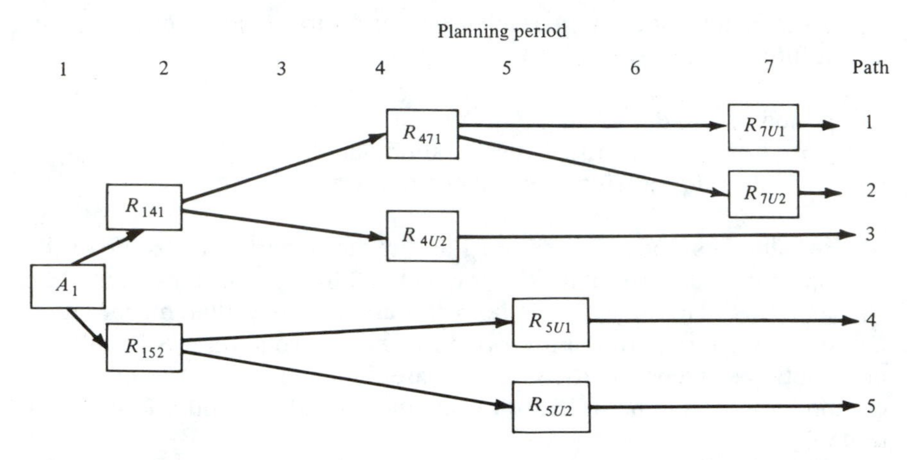

Figure 15-1 Possible paths by which an acre of stand type *A* that has been regeneration harvested in period 1 can pass through the Model II regenerated stand decision variables over the seven-period planning horizon.

and path 5 being the same as our Model I variable A_{12}. Notice that the other three paths allow an acre to switch between silvicultural prescriptions 1 and 2 during the planning interval. These additional paths could have been represented in Model I by defining three additional decision variables for stand type *A* in period 1.

A direct comparison of the Model I and Model II matrix representations for acres of stand type *A* cut in period 1 is shown in Table 15-11. The existing stand harvest components of decision variables A_{11} and A_{12} of Model I have been combined into a single variable A_1 in Model II, with eight new decision variables for the future stands added for Model II (R_{141} to R_{7U2}). If we want to compare Models I and II without allowing a change in regenerated prescriptions on an acre from rotation to rotation, delete columns R_{4U2}, R_{5U1}, and R_{7U2}. Study the volume coefficients, and you should be able to trace all the numbers from Model I into Model II. When Models I and II represent the same prescription path options, the solution to both problem formulations will give exactly the same objective function value and harvest schedule.

The acre accounting equations for future stands, also commonly called Model II transfer rows, are shown below the volume rows of the matrix. A row is given for every period in which there are acres born through regeneration harvest. The coefficient +1 in a row identifies the source or birth of the acres and the coefficient −1 identifies the disposition of these acres when they die through regeneration harvest. Every column except those that do not contain a birth and death has both a coefficient −1 and a coefficient +1, which corresponds to our requirement that all acres regeneration harvested in Model II have both birth and

Table 15-11 Model I and Model II decision variables, harvest, and acreage accounting equations for acres of stand type A regeneration harvested in period 1

	Model I decision variables					Model II decision variables										
Period	A_{11}	A_{12}	A_U	Accounting equations	Period	A_1*	A_U	R_{141}†	R_{152}	R_{471}	R_{4U2}	R_{5U1}	R_{5U2}	R_{7U1}	R_{7U2}	RHS
1	35	35		Harvest accounting rows	1	35										
2					2											
3		6			3				6							
4	26				4			26								
5		26			5				26							
6					6						6					
7	26	6			7					26			6			
				Future stand acreage (Model II transfer rows)	1	−1		1	1							= 0
					4			−1		1	1					= 0
					5				−1			1	1			= 0
					7					−1				1	1	= 0
	1	1	1	Existing stand type acreage		1	1									= 1,000

* Decision variable k_i is stand k regeneration harvested in period i.

† Decision variable R_{ijk} is a future stand born in year i, regeneration harvested in year j, and following prescription k.

death accounted for. The disposition of the area of existing stand type *A* is represented through the last equation, which controls the assignment of stand type *A* acres to its existing stand decision variables.

The ending structure of the forest in a Model II formulation is described by the acres and ending inventory in each of the decision variables that do not contain a regeneration harvest. The total solution acreage in these unharvested variables should exactly equal the total forest acreage.

Daniel Pickett example of Model II The Model II formulation of the full Daniel Pickett forest problem is shown in Table 15-12. Four important changes from the Model I version of the problem have been made to improve exposition and need to be recognized when making comparisons with the original problem as summarized in Table 15-7. These changes are made primarily to keep the problem compact enough to fit on a page; however, they allow a couple of new variations to be illustrated.

1. A third silvicultural prescription has been added, which enables the regenerated stand to grow unthinned for four periods with a harvest of 36 cubic units.
2. Acres can switch silvicultural prescriptions between successive regeneration harvests.
3. Regenerated decision variables that do not contain a regeneration harvest are combined over all silvicultural treatment sequences.
4. When regenerated, the poor-quality sites of original stand types *B* and *C* are permanently fertilized and raised to the good site quality of stand type *A* at a cost of $200 per acre. Without this assumption a complete second set of sixteen future stand variables for the poor permit sites would be needed.

The harvest accounting and acreage accounting rows develop in the same way as shown in Table 15-11, and again the repetitive pattern of coefficients in the matrix can be seen. The present net worth accounting row coefficients have changed in two ways. First, the net worth contributions of the regenerated stands are detached and shown under the regenerated stand decision variables. Recall that in Table 15-7 the present net worth coefficient of decision variable A_{11} was $95. The old decision variable A_{11} can be located in Table 15-12 by following the path A_1, R_{141}, R_{471}; when the present net worth coefficients for these three variables are added up, we get $82 + 10 + 3 = 95$.

Second, the present net worth coefficients of original stand decision variables for stand types *B* and *C* are reduced by subtracting the cost of fertilization. For example, the net revenue for the original stand component of variable B_3 (Table 15-6) is \$525. Subtracting the \$200 cost of fertilizing produces a new net revenue of \$325 at the midpoint of period 3. Discounted to the present, we have 325 × 0.38, or \$123.5. This is rounded (in hundreds) to \$1 in the model.

The ending inventory accounting equation has also changed. Ending inventory is now logically provided by the future stand decision variables that do not contain a final harvest. The coefficients themselves are the same as those used before in the Model I formulation.

Although the Model I and II formulations of the basic Daniel Pickett problem are not identical due to the four stated modifications for Model II, it is informative to compare the solutions. The equations of Table 15-12 were supplemented with additional columns and rows from Table 15-7. Specifically, columns 20 through 28 and rows 22 through 28 were added to complete the Model II formulation.

The solution to this Model II formulation is compared to the basic Model I solution in Table 15-13. We see that the objective function present net worth has increased 7.3 percent for Model II over Model I, to \$90,017, not a particularly dramatic change. The schedule of harvest entries in the three existing stand types is very similar. The most striking difference is the much higher harvest levels in periods 4 to 7. This is due to the fertilizer conversion of stand types *B* and *C* to high site quality and the much higher yields from the regenerated stands. The impact of discounting, as again displayed for these significantly higher future yields, did not appreciably change the present net worth.

Using Model II to consider multiple outcomes Until now we have assumed complete certainty in portraying stand management in forest management planning. Each management regime for a stand has been represented in terms of the stand being regenerated, given a series of cultural treatments and intermediate harvests as needed, and regeneration harvested at rotation age. Only one result (outcome) has been allowed at each age (Fig. 15-2*a*). The possibilities that some of the planting may fail or result in unintended vegetation (Fig. 15-2*b*), that a portion of the stand may be destroyed by fire (Fig. 15-2*c*), or that stand treatments, such as release, may not be successful on some acres (Fig. 15-2*d*) have not been explicitly recognized.

In recent forest planning efforts, foresters sometimes have tried to represent the expected values of these outcomes in the timber yield tables themselves. Each entry in the timber yield table would be reduced by a certain percentage to reflect the loss from fire, disease, and other factors that have important effects on timber yields. Such an approach may be useful for factors that reduce stand volume without destroying the stand, such as disease. Conversely, the approach may produce misleading

Table 15-12 Model II formulation of the Daniel Pickett forest problem

		Existing stands*														
		Stand *A*					Stand *B*					Stand *C*				
	Period born	−10	−10	−10	−10	−10	−10	−10	−10	−10	−10	−1	−1	−1	−1	−1
	Period died	1	2	3	4	*U*	1	2	3	4	*U*	1	2	3	4	*U*
	Prescription	*C*	*C*	*C*	*C*	*U*	*C*	*C*	*C*	*C*	*U*	*C*	*C*	*C*	*C*	*U*
	Decision variable	A_1	A_2	A_3	A_4	A_U	B_1	B_2	B_3	B_4	B_U	C_1	C_2	C_3	C_4	C_U
Accounting rows	(1) H_1	35					18					6				
	(2) H_2		37					17.5					11			
	(3) H_3			38.5					16.5					15		
	(4) H_4				40					15					17.5	
	(5) H_5															
	(6) H_6															
	(7) H_7															
	(8) *I*															
	(9) PNW	82	59	42	29	−0.75	5	3	1	1	−0.5	4	6	6	5	−0.5
Model II acre, transfer rows	(10) 1	−1					−1					−1				
	(11) 2		−1					−1					−1			
	(12) 3			−1					−1					−1		
	(13) 4				−1					−1					−1	
	(14) 5															
	(15) 6															
	(16) 7															
Acres available	(17) *A*	1	1	1	1	1										
	(18) *B*						1	1	1	1	1					
	(19) *C*											1	1	1	1	1

* Decision variable k_i represents existing stand *k* regeneration harvested in period *i*.

(Partial matrix)

Future stands on good sites*																
Period 1			Period 2			Period 3			Period 4			Period 5		Periods 6,7		
1	1	1	2	2	2	3	3	3	4	4	4	5	5	6	7	RHS
4	5	5	5	6	6	6	7	7	7							
1	2	3	1	2	3	1	2	3	1	2	3	1,3	2	All	All	
R_{141}	R_{152}	R_{153}	R_{251}	R_{262}	R_{263}	R_{361}	R_{372}	R_{373}	R_{471}	R_{4U2}	R_{4U3}	R_{5U1}	R_{5U2}	R_{6U}	R_{7U}	
	6															
26				6												
	26	36	26				6									
				26	36	26				6						
							26	36	26				6			
										19	26	16	10	3		
10	11	10	7	7	7	5	5	4	3	1			1			
1	1	1														= 0
			1	1	1											= 0
						1	1	1								= 0
−1									1	1	1					= 0
	−1	−1	−1									1	1			= 0
				−1	−1	−1								1		= 0
							−1	−1	−1						1	= 0
																= 1,000
																= 500
																= 1,000

* Decision variable R_{iik} represents a regenerated stand born in period i, regeneration harvested in period j or left as ending inventory (u), and following prescription k.

Table 15-13

Decision variable*	Model I base problem (from Table 15-8)	Model II problem of Table 15-12
A_1	606	600
A_2	257	268
A_3	36	32
A_4	—	—
A_U	100	100
B_1	—	—
B_2	400	400
B_3	—	—
B_4	—	—
B_U	100	100
C_1	—	100
C_2	43	32
C_3	664	668
C_4	293	200
C_U	—	—
R_{141}	—	272
R_{152}	—	428
R_{153}	—	—
R_{251}	—	88
R_{262}	—	612
R_{263}	—	—
R_{361}	—	88
R_{372}	—	612
R_{373}	—	—
R_{471}	—	88
R_{4U2}	—	384
R_{4U3}	—	—
R_{5U1}	—	—
R_{5U2}	—	516
R_{6U}	—	700
R_{7U}	—	700
H_1	21,236	21,594
H_2	16,989	17,275
H_3	13,591	13,820
H_4	12,717	14,239
H_5	15,261	17,087
H_6	14,657	20,505
H_7	12,472	21,295
I	6,728	14,557
PNW	83,858	90,017

* *A*, *B*, *C*, and *R* units are acres; *H* and *I* are cubic units; PNW is given in hundreds of dollars.

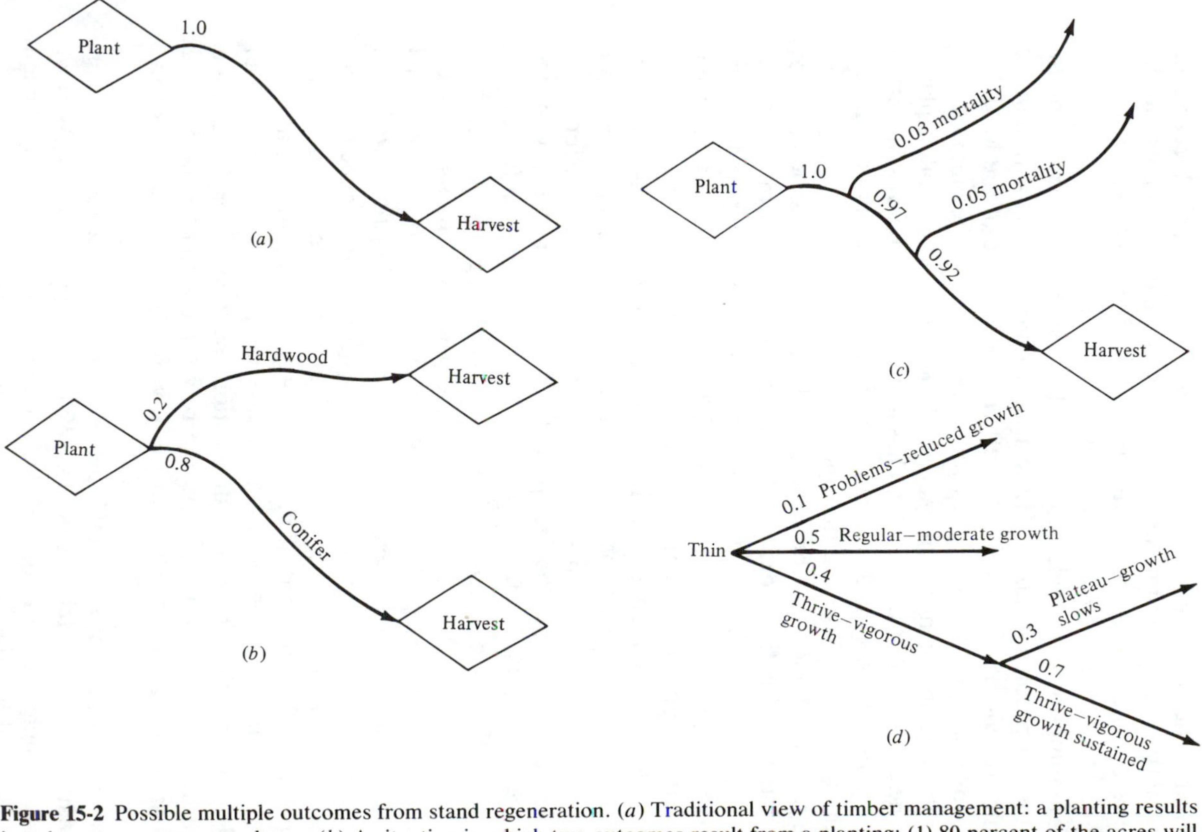

Figure 15-2 Possible multiple outcomes from stand regeneration. (*a*) Traditional view of timber management: a planting results in only one outcome at each age. (*b*) A situation in which two outcomes result from a planting: (1) 80 percent of the acres will produce conifer stands (conifer was the species planted), and (2) 20 percent will produce hardwood stands (in spite of the planting). (*c*) A situation in mortality occurs at two ages: (1) 3 percent of the stand is lost at one age, and (2) 5 percent of the remaining stand is lost at a later age. (*d*) A situation in which a commercial thinning results in multiple outcomes at two ages: three outcomes (problems, regular, and thrive) can occur at one age, and the thrive path will result in two outcomes (plateau and thrive) at a later age.

results when used for factors such as fire, which often destroy the stand entirely.

Consider the projected acreage–age class distribution. Recognizing fire mortality through reduction in stand yield can misrepresent the acreage–age class distribution over time as it does not recognize the cycling of acres caused by these fire losses.

The assumption of a single outcome in the Model II formulation of the Daniel Pickett problem (Table 15-12) can be seen in the disposition of the acres after regeneration harvest. A coefficient −1 in one of the Model II transfer rows signifies that the acres will be passed to the beginning of the next rotation when regeneration harvest of the current stand takes place. Taking the column for stand *A*, which has a regeneration harvest in period 3 (decision variable A_3), as an example, we have a disposition coefficient −1 in the transfer row for the third period.

Multiple outcomes can be represented in a harvest scheduling by breaking the disposition coefficients into components in the Model II transfer rows. Suppose that we wish to represent a 0.3 percent stand mortality rate per year (3 percent per decade) for all periods before regeneration harvest and we assume that acres immediately regenerate after such mortality. Again taking the column for stand *A*, which has a regeneration harvest in the third period, as an example, we would represent three disposition coefficients for this column in three different transfer rows: (1) a −0.03 in the row for the first period, (2) a −0.03 in the row for the second period, and (3) a −0.94 in the row for the third period.

Or take the case where reforestation after regeneration harvest results partially in the species we planted and partially in another species. To represent this other species in the Model II formulation of the Daniel Pickett problem, we would need to add a new set of transfer rows linked to a new set of future stand decision variables for the new species. Suppose that each planting results, on the average, in 90 percent of the desired species and 10 percent of undesired. Again taking this column for stand *A*, which has a regeneration harvest in the third period, as an example, we would represent two disposition coefficients for this column in the transfer rows: (1) a −0.90 in the row for the third period of the desired species, and (2) a −0.10 in the row for the third period of the undesired species.

Numerous other types of multiple outcomes can be represented in this variation of the Model II structure. While this approach is fairly new, it has been implemented in one of the versions of FORPLAN (Johnson et al., 1986) and has the potential for wide use in the future.

Comparative evaluation of Models I and II The rather heroic assumption about the effect of fertilizer on stands *B* and *C* allows us to illustrate an important result of Model II. Because all three original stands are now treated as good sites for regenerated stand yields, we see that all three original stands collapse their acres into the same regenerated stand decision variables. This means that it is no longer easy to tell from what stand a subsequent decision variable came. Where did an acre of, say, R_{141} come from? It could have been an acre of A, B, or C. The mathematical model does not tell us directly. Given the solution, we can trace the passage of an acre from rotation to rotation, but it requires more work than with Model I, where an acre is tracked as a single entity over the planning periods.

In addition, Model II requires more rows than Model I to represent the same problem. Model II formulations need future stand acreage accounting rows (transfer rows), while Model I formulations do not. Often many types of future stand accounting rows are needed to segregate species, site, and management emphasis. In these cases, Model II formulations have many more rows than Model I formulations. And linear programming solution time is especially sensitive to the number of rows.

So why use Model II if it has more difficulty keeping track of specific acres over time and has more rows to represent the same problem? Well, in some problems Model II provides a much more compact problem matrix with many fewer columns and only a moderate increase in row number. Admittedly this seems contrary to the model size results of Tables 15-7 and 15-12. Nevertheless, it is true and arises because all acres from similar stands harvested in a period collapse into a common pool of acres. If we had 500 existing stands of the same species, site quality, and management emphasis, a single set of regenerated variables could possibly serve all of them. Johnson (1977) compared model sizes for the two formulations and the results are striking (Table 15-14). As these results show, some Model I formulations have many more columns than Model II formulations of the same problem.

There are always overall size restrictions on the number of rows and columns that a given computer system and linear programming package can handle. A Model I formulation that adequately considers all goals and constraints and, at the same time, allows many silvicultural and other treatment options for the different land classes, may have too many columns. When the number of prescription options is reduced to make the problem fit computational limits, then either a feasible solution may not be found or the analysis may be open to criticism on the grounds that the decision variables used did not allow enough alternatives to be considered. In this situation a Model II formulation might be useful.

In summary, a Model I formulation requires fewer rows and provides more direct information on what happens to an acre from rotation to rotation, while a Model II formulation requires fewer columns. With no clearcut overall choice, the decision of which formulation to use depends

Table 15-14 Comparison of the numbers of rows and columns needed in Model I and Model II when each type-site has a beginning inventory of 15 age classes aged 6 to 20 periods, and the planning horizon and maximum time between regeneration harvests are both 20 periods

Number of stand types	Within a stand type, acres from beginning age classes combined when regeneration harvested at same time?	Minimum periods between regeneration harvests	Model I		Model II	
			Constraints (rows)	Activities (columns)	Constraints (rows)	Activities (columns)
1	yes	4	15	11,239	35	276
1	yes	6	15	2,904	35	245
1	yes	12	15	339	35	146
3	yes	4	45	33,717	75	828
3	yes	6	45	8,712	75	735
3	yes	12	45	1,017	75	438
1	no	4	15	11,239	315	2,460
1	no	6	15	2,904	315	1,995
1	no	12	15	339	315	939
3	no	4	45	33,717	915	7,380
3	no	6	45	8,712	915	5,985
3	no	12	45	1,017	915	2,817

on the problem being addressed and on the interests of the analyst and decision maker, which is why we discussed both model structures in this section.

Aggregate Emphasis for Group Choice

Implementing management plans means specific acres have to be assigned to specific prescriptions. If decision variables used in a linear programming harvest scheduling analysis are defined for stand types, then ground implementation gets little guidance from the analysis. Moreover, many important costs and impacts, which can vary substantially with site-specific implementation, often are not evaluated in the harvest scheduling model. For example, road development and logging costs are strongly affected by the topography, soils, and spatial distribution of stands within a logging unit, such as a small watershed. It matters which logging unit is developed first and how the development within the unit proceeds on the area over time. The amount of stream sedimentation or the effects on visual and wildlife habitat quality are likewise affected by the implementation sequence chosen.

One approach to incorporating these effects into the scheduling model is to divide the forest into logical management units for development or implementation purposes. Then these units are evaluated to decide on some implementable, on-the-ground allocation choices using maps, site visits, and other information. This preplanning may actually map the location of acres to be assigned to different decision variables and considered as a package by the scheduling model. The difference between such package choices for a management unit might be different road development strategies, early or late planning period entry into the management unit, or the allocation of the area to fundamentally different management uses, such as intensive timber production versus park or wilderness reserve.

Once the choices or "package deals" have been defined for the management units, they can be incorporated into the scheduling model, forcing the model to take an either-or decision, that is, assigning all of the management unit to one or the other of the available choices. The method of building these either-or choices into the model was developed by Crim (1980), and has been called "aggregate emphasis" technique. To illustrate the technique, we first look at a very simple choice problem and then introduce a more involved choice for the Daniel Pickett forest.

Consider a 300-acre parcel of stand type G. On-site examination of this parcel reveals that at least 200 acres of the parcel should be assigned either to intensive timber production or to extensive multiple use for combined timber, wildlife, and recreation production. Five decision variables are defined for stand G as follows:

G_{T1} = acres of G assigned to intensive timber production, prescription 1
G_{T2} = acres of G assigned to intensive timber production, prescription 2
$G_{\text{MU}3}$ = acres of G assigned to extensive multiple use, prescription 3
$G_{\text{MU}4}$ = acres of G assigned to extensive multiple use, prescription 4
G_5 = all other allocations of G

The requirement that at least 200 acres be assigned to either intensive timber or extensive multiple use means either

1. $$G_{T1} + G_{T2} \geq 200$$

or

2. $$G_{\text{MU}3} + G_{\text{MU}4} \geq 200$$

while satisfying the total acres available constraint,

$$G_{T1} + G_{T2} + G_{\text{MU}3} + G_{\text{MU}4} + G_5 = 300 \qquad (15\text{-}17)$$

To force the choice, we define new decision variables which represent these choices. Let

$$W_1 = 200 \text{ acres assigned to } G_{T1} \text{ and } G_{T2}$$

$$W_2 = 200 \text{ acres assigned to } G_{\text{MU}3} \text{ and } G_{\text{MU}4}$$

and require that the new variables take on values between 0 and 1 by the equation

$$W_1 + W_2 = 1 \qquad (15\text{-}18)$$

This last equation is the key to the aggregate emphasis formulation, as we want either W_1 or W_2 to be 1 and the other 0, effectively making the choice an all or nothing package deal.

We also must link the new aggregate decision variables to the regular scheduling decision variables and their associated acreage constraints. If W_1 is selected in the solution with a value of 1.0, for example, 200 acres of stand G are allocated to G_{T1} and G_{T2}. Since the acres have to add up, we use the equation

$$-200W_1 + G_{T1} + G_{T2} \geq 0 \qquad (15\text{-}19)$$

To insert the two aggregate choices W_1 and W_2 into a linear programming scheduling model requires adding at least three equations to the model in addition to the total acreage accounting constraint:

1. $-200W_1 + G_{T1} + G_{T2} \geq 0$ (15-19)
2. $-200W_2 + G_{\text{MU}3} + G_{\text{MU}4} \geq 0$ (15-19)
3. $W_1 + W_2 = 1$ (15-18)
4. $G_{T1} + G_{T2} + G_{\text{MU}3} + G_{\text{MU}4} + G_5 = 300$ (15-17)

The first two equations link the acreage allocations to the aggregate choice variables. The third equation is the aggregate choice accounting equation, and the fourth equation is the total acreage accounting equation.

Aggregate choices can be defined as mixtures of acres from different land types assigned to mixes of different prescriptions. Virtually any combination of decision variables can make up a group choice, with the caveat that each acre can be used only once in each aggregate choice.

Ideally we would like to require that all aggregate choice variables be integers and only take the values of 0 or 1. Then the aggregate choice accounting equation $W_1 + W_2 + \cdots + W_n = 1$ will work as intended, allowing only one variable to have a nonzero value of 1. This option is available in mathematical programming software that has mixed-integer capability. Unfortunately as of this writing (1985), analysts have to live with the potential for splits between the choice allocation variables and make appropriate adjustments when they occur. The mixed-integer formulation of forest planning problems generally takes too long to solve and is too expensive for the technique to be considered operational. Perhaps by the time you read this, more efficient mixed-integer software will be available.

Daniel Pickett example of aggregate choices Suppose the management of the Daniel Pickett forest decided to allocate a substantial portion of the forest to a public park. They felt that the earlier requirement of 200 acres was too small and did not recognize the physical arrangement of the forest. A map of the forest showing timber stand types, topography, and cultural features was prepared, and three logical management units were defined following the major ridges, dividing the square tract into natural watershed units (Fig. 15-3). The existing roads and timber stands are linked to past harvesting, and all of management unit I has been cut over at least once. All of the good-site, high-volume timberland of stand type *A* is found in the unroaded portions of units II and III. An evaluation for park and developed site potential showed that management units II and III were the only reasonable choices for park allocation. The present value costs of developing roads and facilities for the management units were established by on-the-ground layout and are shown in Table 15-15 along with the acreage distributions.

Using these new data along with the aggregate emphasis choice technique developed earlier, the reformulation of the Daniel Pickett problem

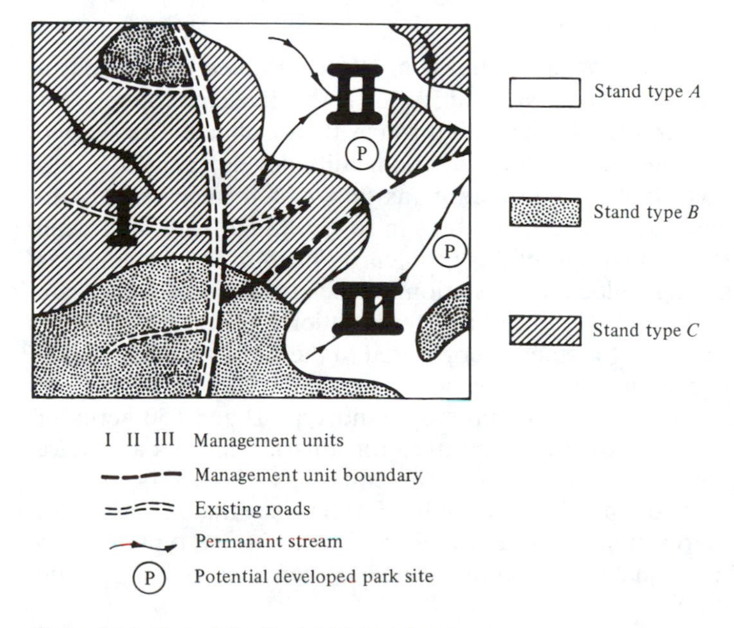

Figure 15-3 Map of the Daniel Pickett forest.

Table 15-15 Development costs and distribution of acreage by management units

(*a*) Acreage by stand type in each management unit

Management unit	Stand type *A*	*B*	*C*	Total
I	—	350	650	1,000
II	550	—	250	800
III	450	150	100	700
Total	1,000	500	1,000	2,500

(*b*) Present value of development costs

Management Unit:	I		II		III	
Use allocation:	Timber	Park	Park	Timber	Park	Timber
Road miles	0	NA	1.5	8.0	3	3.5
Cost per mile (dollars)	—	NA	6,667	10,000	5,000	5,000
Road cost (dollars)	—	—	10,000	80,000	15,000	17,500
Park site cost (dollars)	—	—	50,000	0	35,000	—
Total present cost (dollars)	0	0	60,000	80,000	50,000	17,500

to require that either management unit II or III be allocated to park status is shown in Table 15-16. The first modification is that five new columns for aggregate choice variables W_T and W_1 through W_4 are added. The second modification is the addition of the road and facility development cost into the present net worth objective function, as shown by the negative coefficients for the aggregate choice variables in row 9. (Such costs had been ignored in earlier formulations of this problem.) The third set of changes is the groups of acreage allocation equations associated with the aggregate choices in rows 10 to 15. Note that this formulation has also reduced the total number of choices for stand *A* compared to the earlier Model I and II formulations by eliminating regeneration prescription 2 as an option.

Management unit I has 350 acres of stand type *B* and 650 acres of stand type *C*. Since unit I cannot be used for a park, all of its acres are allocated to timber by variable W_T in the equations of rows 14 and 15. From the acre distribution given in Table 15-15, we see that management unit II has 550 acres of stand type *A* and 250 acres of stand type *C*. Two acreage allocation equations are thus needed to represent each aggregate choice for management unit II in this model. Choice variable W_1 allocates all of unit II to park, and the two allocation equations of rows 10 and 12 accordingly assign the acreage in unit II to prescriptions A_U and C_U. Aggregate choice W_2 allocates all of unit II to timber production and the two allocation equations of rows 13 and 15 assign the acreage in unit II to the timber production decision variables for stand types A and C. Similarly rows 10 to 15 set up the possible allocations for unit III. The aggregate choice equation for management unit I is in row 16, for unit II in row 17, and for unit III in row 18.

Although parks are required, they do not make a positive contribution to the present net worth. Therefore no solution would allocate any land to parks. This is resolved by the constraint equation in row 19, which requires either W_1 or W_3 to be in solution, ensuring that at least one of the aggregate choice options will be selected that allocates land to parks. Other constraints and variables of the Daniel Pickett problem are as before, as shown in rows 20 to 37.

Solution and discussion When the formulation of Table 15-16 was solved, the solution neatly allocated all of management unit III to park status. It also dropped the total 7-period harvest level to 78,024 ft^3, about 75 percent of the harvest provided by Model I basic problem as shown in Table 15-17. This is partly due to the reduced number of choices for stand A and partly due to more acres going to the park. Sometimes these aggregate choice problems will split the choice variables. For example, a solution where $W_1 = 0.37$ and $W_2 = 0.63$ would be entirely consistent with the constraint that $W_1 + W_2 = 1.0$. If such a "split" had happened, we might have been able to force an integer solution in the linear programming software by invoking a mixed integer programming option and specifying that W_1 and W_2 must be integers and have only the values 1 or 0. Fortunately, in most large problem applications of the aggregate emphasis technique, such as in the National Forest planning models of the 1980's, splits have been rare. Our Daniel Pickett problem is a good example of this, and we feel that the aggregate approach has considerable potential to formulate problems with important spatial constraints.

While we see some differences in the harvest schedule due to the reduced timber acreage—the solution pattern is robust, repeating the early entry into slow-growing stand type A and the late entries into young stand type C.

Many costs, such as roads, present serious analytical problems in forest planning since they are associated with management units and yet must be allocated on an average cost per acre basis in problem formulations that do not recognize management units. These costs are "joint" and apply to all acres in the unit. This presents the eternal dilemma of how

Table 15-16

Rows		Aggregate choice decision variables											
Management unit		I	II	II	III	III							
Emphasis		Timber	Park	Timber	Park	Timber	Model I Stand						
Variable		W_T	W_1	W_2	W_3	W_4	A_1	A_2	A_3	A_4	A_U	B_1	B_2
Harvest accounting	1						35					18	
	2							37					17.5
	3								38.5				
	4						26			40		12	
	5							26					12
	6								26				
	7						26			26		12	
Ending inventory	8							16	3				5
Present net worth	9	0	−600	−800	−500	−175	95	65	45	32	−.75	8	5
Area transfer for aggregate choices	10		−550		−450						1		
	11				−150								
	12		−250		−100								
	13			−550		−450	1	1	1	1			
	14	−350				−150						1	1
	15	−650		−250		−100							
Aggregate choices	16	1											
	17		1	1									
	18				1	1							
Park	19		1		1								

Aggregate choice formulation of the Daniel Pickett forest problem

decision variables								Accounting variables									
B_3	B_4	B_U	C_1	C_2	C_3	C_4	C_U	H_1	H_2	H_3	H_4	H_5	H_6	H_7	I	PNW	RHS
			6					−1									= 0
				11					−1								= 0
16.5					15					−1							= 0
	15		12			17.5					−1						= 0
				12								−1					= 0
12					12								−1				= 0
	12		12			12								−1			= 0
				5											−1		= 0
3	2	−.5	7	8	7	5	−.5									−1	= 0
																	= 0
		1															= 0
							1										= 0
																	= 0
1	1																= 0
			1	1	1	1											= 0
																	= 1
																	= 1
																	= 1
																	≥ 1

Table 15-16

Emphasis		Timber	Park	Timber	Park	Timber	Model I Stand						
Variable		W_T	W_1	W_2	W_3	W_4	A_1	A_2	A_3	A_4	A_U	B_1	B_2
Harvest flow constraints	20												
	21												
	22												
	23												
	24												
	25												
	26												
	27												
	28												
	29												
	30												
	31												
Owl	32										1		
Minimum ending inventory	33												
Area control	34						1					1	
	35							1					1
	36								1				
Objective	37	Objective: Maximize PNW											

Aggregate choice formulation of the Daniel Pickett forest problem (Continued)

decision variables								Accounting variables									
B_3	B_4	B_U	C_1	C_2	C_3	C_4	C_U	H_1	H_2	H_3	H_4	H_5	H_6	H_7	I	PNW	RHS
								−0.8	1								≥ 0
								−1.2	1								≤ 0
									−0.8	1							≥ 0
									−1.2	1							≤ 0
										−0.8	1						≥ 0
										−1.2	1						≤ 0
											−0.8	1					≥ 0
											−1.2	1					≤ 0
												−0.8	1				≥ 0
												−1.2	1				≤ 0
													−0.8	1			≥ 0
													−1.2	1			≤ 0
																	≥ 100
															1		≥ 5,000
			1														≤ 700
				1													≤ 700
1					1												≤ 700
																1	

Table 15-17 Comparison of harvest and activity between the optimal aggregate emphasis solution and the Model I base solution

Decision variable*	Model I base problem (from Table 15-8)	Aggregate emphasis problem
A_1	606	400
A_2	257	131
A_3	36	19
A_4	—	—
A_U	100	450
B_1	—	—
B_2	400	350
B_3	—	—
B_4	—	—
B_U	100	150
C_1	—	—
C_2	43	219
C_3	664	681
C_4	293	—
C_U	—	100
W_1	—	0
W_2	—	1.0
W_3	—	1.0
W_4	—	0
W_T	—	1.0
H_1	21,236	13,998
H_2	16,989	13,385
H_3	13,591	10,944
H_4	12,717	10,398
H_5	15,261	10,236
H_6	14,657	8,665
H_7	12,472	10,398
I	6,728	5,000
PNW	83,858	53,874
Total 7-period harvest	106,923	78,024
Reduction in total harvest		28,899
Percent reduction		27

*A, B, C, and W units are acres; H and I are cubic units; PNW is given in hundreds of dollars.

to allocate joint costs. If we know how these costs vary between management units, and recognize these differences through the use of aggregate choice variables, the joint costs can be recognized in the form they occur and the efficiency of the resulting solutions improved.

Land Allocation and Harvest Scheduling: Sequential or Simultaneous?

When forest management is concerned with producing two or more goods and services from the land base, both land allocation and harvest scheduling decisions are usually needed.

Land allocation involves deciding which stands or land units to give exclusive or restricted assignment for the production of different goods and services. An obvious example from the Daniel Pickett forest problem is deciding which acres go to park status and which acres go to intensive timber production. In a multiple-use management environment such allocation choices abound: coming up with a county zoning plan is the epitome of land allocation decision making. Even in a purely timber production environment, decisions are made by allocating land to the production of certain species, product sizes, or to reserve for bad weather harvesting. Allocation choices become increasingly difficult (and interesting) as the land in question becomes varied in its productive capability and when the desired outputs conflict and interact strongly on the same acre and between adjacent acres.

Harvest scheduling involves choosing which timber stands to harvest in each period and what silvicultural prescriptions to use on the existing and regenerated stands on the acres of forest that have been allocated to timber production. Most of this book is concerned with making harvest scheduling decisions, taking for granted that the land is allocated to timber production as a dominant-use goal.

If allocation is an important decision, and on public lands it is often the most critical decision, then the analytical process can choose to treat allocation and scheduling decisions either sequentially or simultaneously.

Sequential allocation and scheduling: two-stage analysis Here allocation decisions are made first by using a variety of information such as maps, air photos, site visits, and inventory data. Economic analysis of costs and relative value productivity can help the decision process, but it typically remains a visual, graphic analysis problem worked out in a pluralistic, political decision process. The end result is a map that designates the acceptable uses for each parcel of land in the property being planned. Then timber harvests can be scheduled on the subset of acres allocated to timber uses, while recognizing the subset of goals and constraints relating to timber production and harvesting. The areas of the forest allocated to nontimber uses can also be scheduled and developed for these other outputs.

To illustrate the two-stage procedure for allocation and scheduling, assume that a forest has two areas A and B, each capable of supporting timber production or recreation use through prescriptions T_i for timber and R_i for recreation. The sequential approach proceeds as follows:

Step 1. *Allocation*. Area A to timber and Area B to recreation.
Step 2. *Timber schedule*. Schedule harvests over the planning interval, considering only timber decision variables $A_{T1}, A_{T2}, \ldots, A_{Ti}$ on area A.
Step 3. *Recreation schedule*. Schedule recreation development and use over all planning periods, considering only recreation decision variables $B_{R1}, B_{R2}, \ldots, B_{Ri}$ on area B.

Simultaneous allocation and scheduling For joint or simultaneous analysis we want our mathematical decision model to help evaluate some or all of the allocation choices by considering the physical and economic interaction of the allocation and scheduling decisions. While still recognizing the political aspects of allocation decisions, such analysis also establishes an objective analytical framework. To parallel the little example given above, the simultaneous approach handles it all in one step:

Step 1. *Allocation and scheduling*. Schedule recreation and timber harvests over the planning interval in a model that explicitly and simultaneously considers decision variables $A_{T1}, A_{T2}, \ldots, A_{Ti}$, $A_{R1}, A_{R2}, \ldots, A_{Ri}$ for area A and decision variables $B_{T1}, B_{T2}, \ldots, B_{Ti}, B_{R1}, B_{R2}, \ldots, B_{Ri}$ for area B, given output targets for timber and recreation.

By defining recreation and timber decision variables for both areas, we let the model allocate areas to the two uses, while simultaneously considering the scheduling implications of these allocation decisions. Every example based on the Daniel Pickett forest problem used the simultaneous approach as we always considered the allocation to the uncut park-reserve status for all areas in our decision variable definitions. The aggregate emphasis technique was in fact developed to allow the model to guide the choice between realistic, implementable allocation and scheduling packages.

With the apparent superiority of simultaneous allocation and scheduling over sequential allocation and scheduling, why is it not always used when both allocation and scheduling are at issue? For three reasons, the sequential approach is often used. First, people and politicians easily understand and relate to the idea of allocating acres to particular uses such as a wilderness area, a grazing allotment, or a forest production area.

They know about buying and selling land and about zoning laws and plans. It makes sense to map and legislate allocations, stand on the acres, and say this is what will happen here. Second, by changing the allocation and recalculating the schedule in the sequential approach, the benefits and costs of different allocation-scheduling combinations can be determined. Even if the "best" overall combination is not found, this procedure does highlight the implications of different combinations and is sometimes easier to understand than the simultaneous approach. Third, the simultaneous approach can result in formulations so large that model size constraints demand that some aspects of the scheduling problem be consolidated. This happened on Table 15-16, when including the allocation variables necessitated the elimination of one of the scheduling prescriptions.

MATRIX GENERATORS, REPORT WRITERS, AND THE COMPUTER

Real harvest scheduling problems are large and virtually impossible to deal with in the manner of the illustrative problems of this book. Suppose a 50,000-acre forest is divided into 10 management units and evaluated with a model that considers 2 site qualities, 10 stand types by species and age condition, 10 planning periods, 1 existing stand prescription, and 2 regenerated stand prescription options. If all stand types and all prescription options are considered for each management unit, it would generate 10 × 2 × 10 × 10 × 1 × 2 for a minimum of 4,000 stand decision variables, not counting any accounting variables. If acres are identified by site, stand type, and management unit, this requires at least 2 × 10 × 10, or 200 area accounting rows. Assuming we can comfortably draw and see a 20 × 20 portion of a matrix on a sheet of letter-sized paper, the 4,000-column × 200-row problem matrix would take 2,000 sheets of paper. Taped on a wall, it would take a gymnasium-sized wall 142 ft long and about 9 ft high to lay it all out. And this is a small problem. You can imagine the tedium and the potential for errors in filling in this linear programming matrix by hand.

Fortunately the computer is here to help us, and this is what makes the detailed quantitative analysis of harvest scheduling problems possible. Most mainframe computers support a standard linear programming program which takes input in the form of the problem matrix and provides output in the form of solution values for the variables and the other information, as was shown in Table 15-2. MPSX (I.B.M.), FMPS (Univac), and MPSIII (Keytron Co. for IBM) are such programs. Surely effective programs with adequate capacity will also be available on microcomputers in the late 1980s or 1990s.

The real problem lies not in finding a linear programming package, but in easily and accurately creating the matrix that describes a particular harvest scheduling problem and writes it into the input format required by the linear programming software. We also need something to convert the long, boring, and confusing output from the linear program into some summary tables and graphs that tell us what the solution means over time. Programs called matrix generators and report writers are needed for this. The flowchart of a typical matrix generator–report writer system is shown in Fig. 15-4 and describes four stages in problem analysis. After the initial problem identification stage, which utilizes only your basic human brain to operate, collections of yield tables, economic tables, and other problem specifications provide input to a matrix generator program which in turn provides two outputs at stage 2—a matrix report and an input file to drive the linear program.

The matrix report is essentially an organized playback of data and the problem specifications provided by the user for checking against the origi-

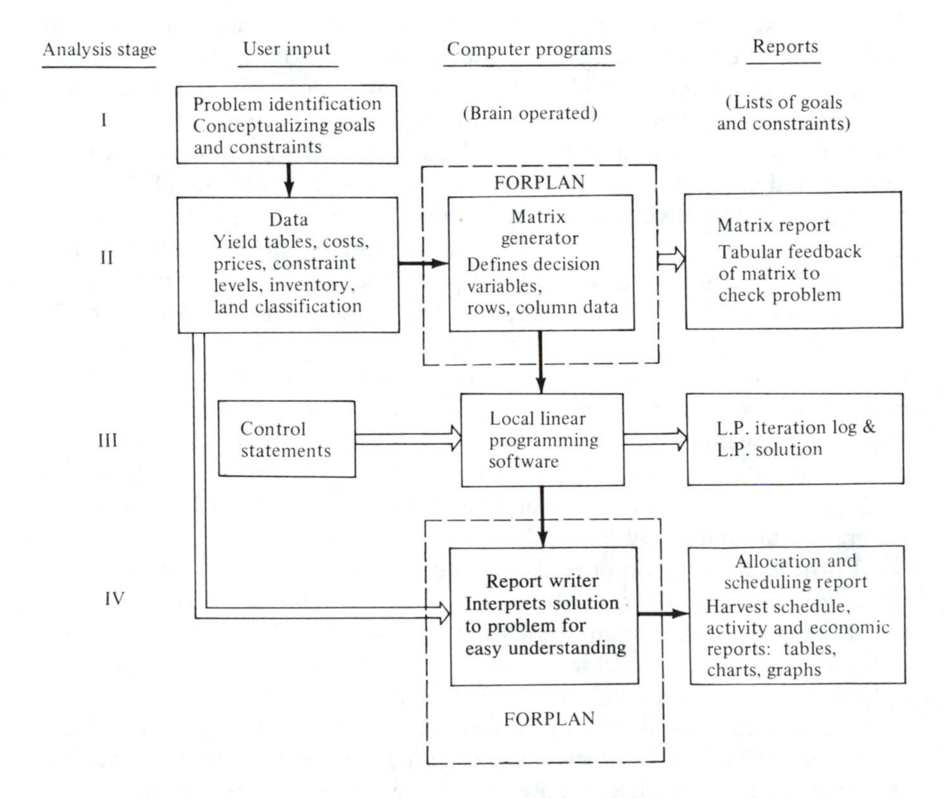

Figure 15-4 Flowchart for using the forest planning (FORPLAN) system for analysis of forest management scheduling problems. The FORPLAN computer programs consist of a matrix generator and a report writer as indicated.

nal problem conceptualization. It allows us to answer the question, "is this the problem you want the linear program to solve?" The problem identification and data input steps are approximated by considering the description of the Daniel Pickett forest problem at the beginning of this chapter along with the yield table (Table 15-4). The matrix generator defines the decision variables, makes the volume forecasts, calculates the present net worth, and establishes all the needed accounting variables and other constraint rows to produce the matrix of Table 15-7. If we like the problem statement shown by the matrix generator, then we can choose to pass the input file on to the linear program and obtain a solution.

The linear program also takes system control input and produces a solution log reporting progress in finding the solution. The log becomes particularly relevant when a feasible solution cannot be found and we need to figure out what is wrong. Given that an optimal solution is obtained, the report writer is invoked to summarize the output. The illustrative linear programming output of Table 15-2 represents the input to the report writer from the linear program, and the organized results of Tables 15-8, 15-9, and 15-10 represent some of the outputs that might be produced by a report writer.

There are many programs to do the matrix generation–report writing job. Most are customized and have never been published or documented. The programs may be separate, or they may be combined into a smoothly running, flexible package of programs that can run from data input to final reports in one pass, untouched by human hands. The sequentially developed series of programs used by the U.S. Forest Service is the best known and most elaborate of the matrix generation–report writing systems. Starting with Timber RAM (Navon, 1971) which handled only Model I problems, they progress through MUSYC (Johnson and Jones, 1979), which handled both Model I and Model II, FORPLAN I (Johnson, 1986), which also handles aggregate emphasis, and finally, FORPLAN II (Johnson et al., 1986), which adds a variety of options, including direct linkage of plans to budgets and a variety of aggregate emphasis options.

Increasingly complex problems have been defined and solved to plan the management of 1,000,000-acre forests for multiple outputs over 200-year planning horizons. Some of the recent problems successfully solved had matrix sizes of 500 to 3,000 rows and 5,000 to 100,000 columns.

MAX MILLION (Clutter et al., 1978) works directly with forest inventory data in a Model I formulation and is used extensively in the south. A variety of other systems such as CAPS, a Model II formulation of Crown Zellerbach, and a Model I system of Champion are employed as of 1985 by private industry.

Obtaining current and documented copies of these programs will always be difficult, and it is often no small task to install and support the programs on a computer system. Consultants, a few universities, and the public agencies will keep and revise the programs. Copies and needed training will be found through these institutions. The forestry profession has much to learn about the maintenance and transfer of this complex analysis technology.

Binary Search Approaches to Harvest Scheduling

Binary search is a simulation technique that uses a forest inventory data set and appropriate growth models to find the maximum even flow of volume or value that can be sustained over a finite planning interval subject to certain harvest flow and ending inventory constraints. Binary search is thus a counterpart to linear programming and provides an alternate way to solve some of the same harvest scheduling problems. The name *binary search* emerges because (1) there is only one decision variable per period, the level of harvest, and (2) there are only two choices in the problem, either increase or decrease that harvest. All other needed decisions, such as what prescription to apply to each stand type and the priorities for selecting stands for intermediate harvest by thinning and for regeneration harvest, are decided external to the simulation and given to the simulation algorithm as input.

The simulation process itself is simple, with the computational steps usually running as follows:

1. Set the amount of first-period harvest volume.
2. Select the stands to harvest from the inventory list to provide the needed harvest volume using priority rules. (For example, first thin all stands between 20 and 40 years to a residual volume of 12 cubic units and then obtain the remaining volume needed to meet the overall volume target by regeneration harvesting stands in the order of their age starting with the oldest.)
3. Grow all residual stands for one period, creating the forest structure by age and volume just before second-period harvest.
4. Repeat steps 2 and 3 for each subsequent period, checking at each iteration to see that the required harvest can be achieved. If not, return to step 1 and reduce the first-period harvest.
5. After the harvest in the final planning period, check the residual forest structure to see whether it meets ending requirements. If there is too much inventory, return to step 1 and increase the first-period harvest. If there is not enough ending inventory, return to step 1 and decrease the first-period harvest. Then repeat steps 2 through 5.
6. When the ending condition at step 5 converges to within a certain percentage of the desired ending condition, stop the process and report the latest first-period harvest level as the maximum sustainable harvest level.

Sounds familiar? It should. If you look back to the regulation problem examples of Chap. 14, you should see that many of the regulatory approaches used procedures similar to the first five steps above. (For example, see the area-volume check examples in Tables 14-5 and 14-6 and the second example for both the even-aged and the uneven-aged forests in Tables 14-13 and 14-16.)

The second area-volume check example demonstrates the use of the binary search approach to find a harvest level. In that example we sought the even-flow harvest that could be maintained for four periods while leaving as much inventory after harvest in the fourth period as would be found on the equivalent regulated forest (Table 14-6). First we tried a cut of 49,000 ft^3, which left too little inventory after harvest in the fourth period. Lowering the harvest to 39,000 ft^3, we tried again with the result that too much inventory was left. Splitting the difference between the two previous harvest levels, we tried a harvest of 44,000 ft^3, which left the right amount of inventory. So 44,000 ft^3 is the even flow harvest level that meets our conditions.

The search procedure used in the example is typical of binary search. As the search continues, the step size used in changing the initial harvest is reduced as the harvest level approaches one that can be sustained while just meeting the ending inventory conditions. Suppose that an initial harvest of 100 million ft^3 per decade is estimated along with an initial increment of 10 million ft^3 for a forest that actually can sustain a harvest of 123 million ft^3 per decade. Simulation of a harvest rate at that initial volume will show that the volume is sustainable, with surplus inventory left at the planning horizon. The initial harvest will be incremented to 110 million ft^3 and the simulation done again and to 120 million ft^3 with similar results. Simulation of 130 million ft^3 will exhaust the inventory at one of the planning periods or leave too little inventory at the planning horizon. Rather than decrease the harvest back to 120, the step size is halved and the next level tried is 125. Usually the step size is halved each time conditions cannot be met and the harvest increment must be reversed. In this case, the halving will result in trying 125, 122.5, 123.75, 123.125, ... The process continues until the step size drops below some predetermined amount, or the actual ending inventory comes within some amount of the desired ending inventory.

As a logical extension of area-volume check, a wide variety of binary search models were built during the 1960s and 1970s. Perhaps the most widely known binary search timber management scheduling model for public lands is SIMAC (Sassaman et al., 1972). This model was used for many years to set timber harvest and investment levels on land administered by the Bureau of Land Management in the west. In an upgraded form (in part to allow for the sequential even flow discussed below), it is still being used by the Bureau.

Sequential even flow Previously we have used binary search to find the even flow harvest sustainable over four periods, subject to leaving a certain amount of inventory volume after harvest in the fourth period (Table 14-6). Often users of binary search want to look at harvest policies other than strict even flow.

If the user wishes to allow an increasing harvest over time, such as when future stands will develop more volume than existing stands, or allow a decreasing harvest over time, such as when existing stands contain more volume than will develop in future stands, a variation of the simple even flow approach is used. This variation is labeled sequential even flow and can be viewed as a series of even flow simulations that overlap.

As before, sequential even flow seeks the maximum volume that can be harvested for a number of periods, while meeting specified ending conditions. That harvest level, however, is used only for one period. After taking the sustainable harvest for the period in question, the inventory is updated to the next period and the simulation again occurs. The harvest that can be sustained for a set number of periods into the future is thus calculated for each period in turn. The length of this sustainability check usually is shorter than the total number of periods in the planning interval. The shorter the length of the sustainability check, the steeper the possible change in harvest level between adjacent periods.

Chappelle (1966) first implemented this sequential even flow approach in a model called SORAC. Since then, a variety of models have incorporated this option, including the harvest scheduling model used by the Washington State Department of Natural Resources (Chambers and Pierson, 1973) and TREES (Johnson et al., 1975; Tedder et al., 1980). Chambers and his colleagues in Washington pioneered the use of the sequential approach for setting harvest levels on public lands, and much of the TREES structure developed from their work.

To illustrate the sequential even flow approach, let use again consider the binary search example from Chap. 14 that found an even flow harvest of 44,000 ft^3 for four periods subject to a constraint that 17,500 ft^3 be left as inventory after harvest in the fourth period (Table 15-18).

Since long-term sustained yield equals 34,400 ft^3, we would expect a drop in harvest to approximately this level after period 4. In fact, further analysis showed that harvesting a 44,000 ft^3 per period for four periods would require harvest at a level slightly below 34,400 ft^3 for a few periods thereafter (33,000 ft^3 per period) before moving permanently to the long-term sustained yield in period 7 (Fig. 15-5).

We can use sequential even flow to demonstrate a more gradual decline from the even flow level of 44,000 ft^3 for the first period to the long-term sustained yield of 34,400 ft^3. If we take the inventory that would

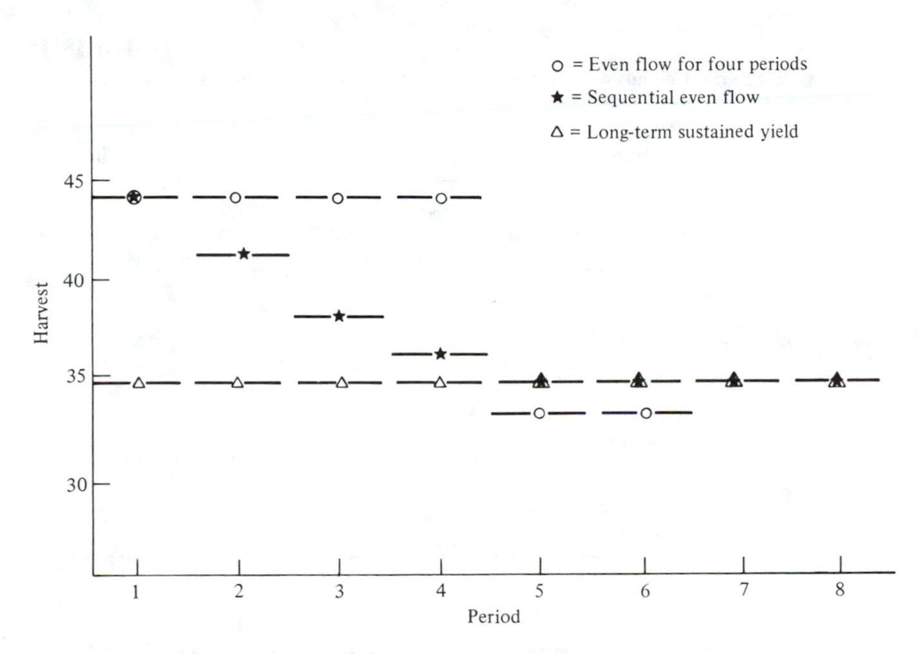

Figure 15-5 Comparison of even flow for four periods, sequential even flow, and even flow forever (long-term sustained yield).

result in period 2 from a harvest of 44,000 ft^3 in period 1 and search for the harvest level that could be sustained for four more periods subject to the same ending condition, we find that harvest level to be 41,000 ft^3 (Table 15-18). Continuing that procedure for the third and fourth periods, we develop a harvest pattern that stairsteps downward to the long-term sustained yield (Fig. 15-5).

Examining the production for six periods under both strategies, it is clear that the policy of even flow for four periods produces a greater total harvest volume than a sequential even flow policy. However, it also has a more abrupt adjustment to the long-term sustained yield level. We will discuss further these kinds of trade-offs in Chap. 16.

The economic harvest optimization model (ECHO) Some forest owners hold vast blocks of timberland from which they cut only a small percentage of the volume per year. The national forests, as an example, control most of the timber in many parts of the west, while offering less than 1 percent of the inventory volume per year. If any of these national forests were to greatly increase their stumpage offerings in some period, they might lower the stumpage price in the local market, that is, the forests face a downward-sloping demand for their stumpage.

Table 15-18

(*a*) Even flow harvest equals 44,000 ft^3 per period for 4 periods starting in period 1

		Period 1				Period 2			
		Inventory		Cut		Inventory		Cut	
Age	Volume per acre	Acres	Total volume	Acres	Total volume	Acres	Total volume	Acres	Total volume
0									
10						8.7			
20	1.75	10	17.5						
30	3.44					10	34.4		
40	4.29								
50	4.74								
60	5.07	20	101.4	8.7	44				
70	5.33					11.3	60.2	8.3	44.0
80	5.56								
Totals		30	118.9	8.7	44	30	94.6	8.3	44

(*b*) Even flow harvest equals 41,000 ft^3 per period for 4 periods starting in period 2, using

		Period 2				Period 3			
		Inventory		Cut		Inventory		Cut	
Age	Volume per acre	Acres	Total volume	Acres	Total volume	Acres	Total volume	Acres	Total volume
0									
10		8.7				7.7			
20	1.75					8.7	15.3		
30	3.44	10	34.4						
40	4.29					10	42.9	4.9	21.0
50	4.74								
60	5.07								
70	5.33	11.3	60.2	7.7	41				
80	5.56					3.6	20.0	3.6	20.0
Totals		30	94.6	7.7	41	30	78.2	8.5	41

To help determine the revenue-maximizing level of harvest when a stumpage owner faces a downward-sloping demand for his or her stumpage, Walker (1971, 1976) has suggested a variation of the binary search approach. In the binary search approach we find the level of harvest that can be maintained over time subject to certain ending conditions. In Walker's variation, called the *EC*onomic *H*arvest *O*ptimization model (ECHO), he finds the level of discounted marginal net revenue that can be

Example of sequential even flow using binary search

Period 3				Period 4				
Inventory		Cut		Inventory		Cut		Inventory volume after harvest
Acres	Total volume	Acres	Total volume	Acres	Total volume	Acres	Total volume	
8.3				9.4				
8.7	15.2			8.3	14.5			14.5
				8.7	30.0	7.8	26.9	3.1
10	42.9	6.4	27.3					
				3.6	17.1	3.6	17.1	
3.0	16.7	3.0	16.7					
30	74.8	9.4	44	30	61.6	11.4	44	17.6

the inventory that resulted from a harvest of 44,000 ft³ in period 1

Period 4				Period 5				
Inventory		Cut		Inventory		Cut		Inventory volume after harvest
Acres	Total volume	Acres	Total volume	Acres	Total volume	Acres	Total volume	
8.5				10.0				
7.7	13.5			8.5	14.9			14.9
8.7	29.9	4.9	16.8	7.7	26.5	7.1	24.7	1.8
				3.8	16.3	3.8	16.3	
5.1	24.2	5.1	24.2					
30	67.6	10.0	41	30	57.7	10.9	41	16.7

maintained over time, while using a planning horizon long enough such that the solution has stabilized at the economically optimum rotation age.

To understand Walker's technique, let us first look at the objective function of a timber management scheduling problem when the price received is a function of the quantity offered in a stumpage market. An objective function that seeks to maximize timber sale revenues over the planning interval can be written as

$$\text{Maximize} \sum_{t=1}^{N} P_t H_t$$

where t = year, $t = 1, 2, \ldots, N$
P_t = unit price of stumpage in year t
H_t = total units of volume harvested and sold in year t

When the guiding interest rate is nonzero, the objective is to maximize the discounted present value of these harvest revenues:

$$\text{Maximize} \sum_{t=1}^{N} \frac{P_t H_t}{(1 + i)^t} \tag{15-20}$$

where i is the guiding rate expressed as a decimal.

If the selling price declines as the harvest increases in a given year, we say this seller faces a downward-sloping demand curve. If the relationship of price to harvest is linear, we write the price or demand equation as

$$P_t = I_t - S_t H_t \tag{15-21}$$

where P_t = price per unit volume received in year t
I_t = the y-axis intercept of the equation in year t
S_t = the slope coefficient in year t
H_t = the total volume harvested in year t

Equation (15-21) says the demand relationship can shift from year to year as I_t and S_t take on different values in different years. For example, the explicit form of (15-21) might be

$P_{15} = 150 - 0.015H_{15}$ for year 15 and change to

$P_{20} = 175 - 0.011H_{20}$ in year 20

We can substitute the linear price equation into the discounted revenue equation (15-20) to get a nonlinear objective dependent on the coefficients of the demand relationship and the harvest level in the period

$$\text{Maximize} \sum_{t=1}^{N} \frac{(I_t - S_t H_t) H_t}{(1 + i)^t} \quad \text{or} \quad \sum_{t=1}^{N} \frac{I_t H_t - S_t H_t^2}{(1 + i)^t} \tag{15-22}$$

If we define harvesting cost per unit volume in year t as W_t, Eq. (15-22) becomes the total discounted net revenue equation

$$\text{Maximize} \sum_{t=1}^{N} \frac{I_t H_t - S_t H_t^2 - W_t H_t}{(1 + i)^t} \tag{15-23}$$

If we take the first derivative of (15-23) with respect to H_t, we obtain an expression for the marginal net discounted revenue per unit volume harvested in period t

$$\text{MNR}_t = \frac{I_t - 2S_tH_t - W_t}{(1 + i)^t} \tag{15-24}$$

In ECHO Walker finds the level of harvest in each period such that the marginal net discounted revenue, λ, is equal across all planning years

$$\frac{I_t - 2S_tH_t - W_t}{(1 + i)^t} = \frac{I_{t+1} - 2S_{t+1}H_{t+1} - W_{t+1}}{(1 + i)^{t+1}} \tag{15-25}$$

This relationship also implies that it is optimum to split the harvest of some stand between each set of consecutive years. This split stand is the last stand harvested in one year and the first stand to be harvested in the next year. A linkage equation across two years shows the essential relationship equating the marginal net discounted revenue per acre:

$$\frac{(I_t - 2S_tH_t - W_t)V_{kt}}{(1 + i)^t} = \frac{(I_{t+1} - 2S_{t+1}H_{t+1} - W_{t+1})V_{k,t+1}}{(1 + i)^{t+1}} \tag{15-26}$$

where V_{kt} is the per acre volume of stand k in year t.

Because for planning purposes we typically use periods longer than one year, the time index t needs to be converted to a period basis. For equal length periods, if j is the sequential period number from 1 to N and if p is the length of a period in years, then $t = jp$. Using this revision, the ECHO linkage of the marginal or split stand between period j and $j + 1$ is

$$\frac{(I_j - 2S_jH_j - W_j)V_{kj}}{(1 + i)^{jp}} = \frac{(I_{j+1} - 2S_{j+1}H_{j+1} - W_{j+1})V_{k,j+1}}{(1 + i)^{(j+1)p}} \tag{15-27}$$

To ease our exposition and discussion without loss of generality, assume that $W_j = 0$ for all j and then multiply both sides of Eq. (15-27) by the factor $(1 + i)^{jp}$ giving

$$(I_j - 2S_jH_j)V_{kj} = \frac{(I_{j+1} - 2S_{j+1}H_{j+1})V_{k,j+1}}{(1 + i)^p} \tag{15-28}$$

The relationship of (15-28) can be rearranged in a useful way to make the marginal net revenue in period $j + 1$ of stand k to be a function of marginal net revenue in period j, the volume growth rate of stand k, and the interest rate

$$(I_{j+1} - 2S_{j+1}H_{j+1}) = \frac{(I_j - 2S_jH_j)(V_{kj})(1 + i)^p}{(V_{k,j+1})} \tag{15-29}$$

Walker uses Eq. (15-29) to find the optimum rate of harvest for existing and future stands. To simplify our presentation here, we will only look at determining the optimum rate of harvest for existing stands, much like what was done in the introduction of area-volume check (Chap. 14). To find the rate of harvest that solves this equation in each period, Walker and others have used the following algorithm (Johnson and Scheurman, 1977):

1. Select an initial harvest H_1 for which $I_1 - S_1H_1 > 0$ and specify a value for N, j, p.
2. If $H_j < 0$, go back to the beginning and raise the harvest.
3. If $H_j >$ available inventory, go back to the beginning and lower the harvest.
4. Cut the stands according to some predetermined order. We then know what stand we are cutting when the harvest is filled and we can calculate $I_j - 2S_jH_j$.
5. Use Eq. (15-29) with everything known on the right-hand side. We can calculate $I_{j+1} - 2S_{j+1}H_{j+1}$ and then calculate the harvest required for the next period, H_{j+1}.
6. If $j < N$, increase j by 1, move all acres up one age class and go back to 2.
7. If $j = N$ and all acres have been cut over once, stop. Otherwise raise H_1 and go back to 2.

By varying the value of N, the impact of different liquidation periods on the discounted value of the harvest can be assessed.

To see how this algorithm works, let us use the example employed to introduce binary search in Chap. 14. In that problem, we had the following table.

Age	Acres	Volume per acre
0		
10		
20	10	1.75
30		3.44
40		4.29
50		4.75
60	20	5.07
70		5.33
80		5.56
90		

When gaps exist in the age classes, as shown here, the algorithm often encounters some computational difficulties whose resolution is beyond the scope of our discussion. To avoid such difficulties, we assume all 30 acres are 60 years old in the middle of the first period.

In the introduction to area-volume check in Chap. 14, we used area-volume check to find the rate of harvest that would cut the existing timber over three 10-year periods. Using area-volume check here over the same number of periods, we find the even flow rate of harvest to be approximately 53,000 ft^3 per period.

Period	Age	Volume per acre in existing stands (M ft^3)	Acres cut to achieve 53,000 ft^3/period
1	60	5.07	10.45
2	70	5.33	9.94
3	80	5.56	9.53
Total acres cut			29.92

We will use this harvest level as our guess of the initial harvest for the ECHO algorithm.

In our discussion let us assume that the firm we are studying faces a stumpage demand curve of the form $p = 1{,}000 - 5h_j$, where h_j is the harvest in period j measured in thousands of cubic feet, and p is the price received per thousand cubic feet harvested. Assume also that the curve does not shift between periods. In the terminology introduced previously, $I_j = 1{,}000$ and $S_j = 5$ for all periods.

With a harvest of 53,000 ft^3 the price in each period is $1{,}000 - 5(53) = 1{,}000 - 265 = \$735$ for a stumpage revenue in each period of $735 \times 53 = \$38{,}955$ and a total discounted return over all three periods for the even flow harvest of

$$\frac{38{,}955}{(1 + 0.04)^5} + \frac{38{,}955}{(1 + 0.04)^{15}} + \frac{38{,}955}{(1.04)^{25}} = 32{,}019 + 21{,}631 + 14{,}613$$

$$= \$68{,}263$$

Using this demand function, we can employ Walker's technique to find the optimum rate of harvest of the 30 acres of existing timber over three periods. Starting with 53,000 ft^3 (10.5 acres of 60-year-old timber) as our first guess as to the optimum harvest level for the initial period, we can calculate $I_1 - 2S_1H_1$ as $1{,}000 - 2(5)(53) = \$470$. With this marginal revenue per M ft^3 for the first period, we can use the linkage equation (15-29) to calculate the required marginal revenue for the second period:

$$I_2 - 2S_2H_2 = \frac{(I_1 - 2S_1H_1)V_{11}(1 + i)^{10}}{V_{12}}$$

or

$$1{,}000 - 2(5)H_2 = \frac{[1{,}000 - 2(5)(53)]5.07(1.04)^{10}}{5.33}$$

$$= 470\,\frac{5.07(1.48)}{5.33} = \$663$$

A marginal revenue of \$663 involves a harvest of $1{,}000 - 2(5)H_2 = \$663$ or $H_2 = (1{,}000 - 663)/10 = 33{,}700$ ft^3, which consumes 6.3 acres of the 70-year-old timber. Since the total acres to be harvested (10.5 + 6.3) are less than the total acreage of existing timber, we can move on to the third period. Using the second period's marginal revenue in the linkage equation, we can form the linkage between the second and third periods:

$$1{,}000 - 2(5)H_3 = \frac{663(5.33)(1.48)}{5.56} = \$940$$

A marginal revenue of \$940 involves a harvest of $H_3 = (1{,}000 - 940)/10 = 6{,}000$ ft^3, which consumes 1.1 acres of 80-year-old timber. Thus a harvest of 53,000 ft^3 in period 1 results in a total acreage harvest over the three periods of $10.5 + 6.3 + 1.1 = 17.9$ acres. Since this is less than 30 acres, we need to start over with a higher harvest.

Using an increment of 10,000 ft^3, we next try a first-period harvest of 63,000 ft^3 (Table 15-19). Since the total acreage cut is still less than 30 acres, we again raise the initial harvest by 10,000 ft^3 (Table 15-19). Now we run out of inventory in the third period. Therefore we halve the incre-

Table 15-19 Results from use of the ECHO linkage equation to determine the rate to harvest existing timber over 3 periods

Period	Volume per acre (M ft^3)	Marginal revenue (\$/M ft^3)	Harvest volume (M ft^3)	Harvest acreage
		Initial harvest = 53		
1	5.07	470	53	10.5
2	5.33	663	33.7	6.3
3	5.56	940	6.0	1.1
		Total acreage harvested		17.9
		Initial harvest = 63		
1	5.07	370	63	12.4
2	5.33	521	47.8	8.9
3	5.56	739	26.1	4.7
		Total acreage harvested		26.0
		Initial harvest = 73		
1	5.07	270	73	14.4
2	5.33	381	62	11.6
3	5.56	541	46	(8.27)*
		Total acreage harvested		(34.27)*
		Initial harvest = 68		
1	5.07	320	68.0	13.4
2	5.33	451	54.9	10.3
3	5.56	640	36.0	6.5
		Total acreage harvested		30.2

* Harvest exceeds inventory and cannot be met.

ment and lower the initial harvest by five to 68 (Table 15-19). Now we cut 30.2 acres over the three periods, which is approximately equal to the acreage we seek (30). We have found the pattern of harvest over three periods that will maximize the present net worth given the demand curve. That harvest, the associated stumpage revenue, and the present net worth are as follows:

Period	Harvest volume (M ft^3)	Price ($p = 1{,}000 - 5h$)	Stumpage revenue	Contribution to present net worth
1	68.0	660	$44,880	$36,889
2	54.9	726	39,857	22,131
3	36.0	820	29,520	11,074
				$70,094

In summary, harvesting the existing timber over three periods on an even flow basis has a present net worth of $68,263, while harvesting the timber over three periods under the ECHO approach has a present net worth of $70,094. Thus there is some cost associated with the even flow policy.

By trying out different liquidation periods for the existing timber, we can find the overall financially optimum rate of harvest for the existing timber. By including future stands, we can find both the financially optimum rate of harvest for the existing timber in light of the regenerated timber that will follow and the best rotation for these future stands.

All these calculations are done through the linkage equation. Since it ignores land-holding costs (see Chap. 13 for a discussion of land-holding costs) and requires that the stand priority for removal be known outside of its calculations, the results from the equation only approximate the financially optimum harvest and investment levels. Still, Walker's ECHO approach is a valued addition to our arsenal of tools for dealing with timber management scheduling.

Linear Programming versus Binary Search

Linear programming and binary search models can both be classified under the Model I–Model II dichotomy (Fig. 15-6).

Most linear programming approaches used Model I until Johnson and his colleagues started allowing a choice of model type. As seen from the Daniel Pickett example, the Model II structure is somewhat less intuitive and more complicated than the Model I structure, so it is not too surprising that most early modelers chose Model I.

On the other hand, binary search models built so far have generally

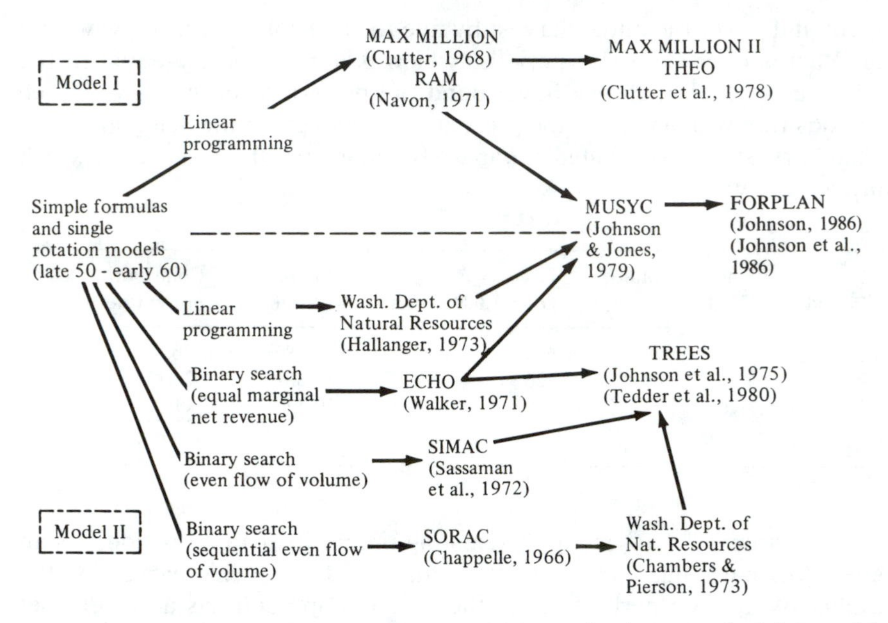

Figure 15-6 Chronological development of multiple-rotation harvest scheduling models. Only those models available to the public are shown.

been of the Model II variety. Probably this pattern of model development is caused by the Model I characteristic that does not allow acres whose stands have similar characteristics in the future, but which have different initial characteristics, to be combined when harvested in the same period. In the example (Table 15-18) the two existing stands differ in terms of age, but future stands on these acres have identical properties when regenerated in the same period. In Model I, acres from two existing stands regenerated in the same period must be carried separately through time.

We have seen that linear programming typically has hundreds of decision variables and can simultaneously schedule and allocate land to different uses and under different prescriptions while considering a variety of acreage control constraints and periodic constraints on inputs and outputs. Binary search has few decision variables and only makes the simple choice of the harvest amount inside the model. Many important decisions, such as land allocation and choice of prescription, have to be made externally, and the binary search solution cannot be considered optimum unless all the externally supplied decisions are also considered optimum. Johnson and Tedder (1983) summarize the relative advantages of the two methods as follows:

> Advantages of binary search over linear programming include the ability to (1) portray the inventory in greater detail, (2) shift acreage more simply in and out of the forest

inventory base, (3) produce analyses at lower cost, and (4) find feasible solutions more easily.

Advantages of linear programming over binary search include a better ability to (1) simultaneously consider alternative yield trajectories for the same acres, (2) portray unusual yield trajectories, (3) constrain portions of the inventory, and (4) find the optimum.

Both linear programming and binary search are viable harvest scheduling tools that are used extensively throughout the forest products industry and public agencies charged with managing forest resources. . . . Each has advantages for certain types of problems and neither is uniformly superior for all applications. Ignoring the potential for suboptimality with binary search, it is well suited for harvest scheduling problems that contain numerous inventory categories, standard yield functions, and no constraints on the harvest beyond an overall harvest flow. Linear programming, on the other hand, is well suited for problems that have relatively few inventory categories, unusual yield functions, and numerous constraints on the harvest.

ADVANCED SCHEDULING TECHNIQUES

Binary search is a heuristic (approximation) approach to solving harvest scheduling problems. It takes the shortcut of reducing the scheduling problem to a few decision variables and then exploiting the sequential nature of timber stand development to find the harvest level that meets certain constraints. As discussed in the previous section, this procedure has advantages and disadvantages in comparison to linear programming. Lower cost per run and the related ability to recognize the inventory in greater detail are major advantages, while the inability to consider alternative management intensities, consider constraints beyond harvest flow, and find the optimal stand priority for harvest are major disadvantages.

Recently, Hoganson and Rose (1984) have presented a heuristic that allows the consideration of alternative management intensities and finds the optimal stand priority for harvest, overcoming two major drawbacks of binary search, while retaining a low cost per run and the ability to consider the inventory in great detail. Given an objective of maximizing present net worth and a specified volume to harvest per period, Hoganson and Rose vary the price of stumpage in each period until they find a set of prices for the timber harvest such that the best time to harvest each stand to maximize its present net worth on an individual basis is also the best time to harvest the stands in aggregate to meet the overall harvest constraints.

This procedure is probably best seen through an example. Suppose that we had the following problem:

1. six 10-year periods
2. 0.0 discount rate
3. eight 100-acre stands with harvest volumes shown in Table 15-20*a*

Table 15-20 Harvest volumes for existing and future stands (cords/acre)

(*a*) Existing stands

	Period						
Stand	1	2	3	4	5	6	7
1	50	50	48	46	40	34	25
2	48	46	40	34	25	17	—
3	—	—	46	50	50	48	46
4	—	30	36	38	35	30	25
5	40	38	35	30	25	—	—
6	—	—	—	46	50	50	48
7	40	37	34	30	25	17	—
8	50	48	46	40	34	25	—

(*b*) Future stands

	Age						
	1	2	3	4	5	6	7
Yield table 1	—	—	50	50	48	46	42
Yield table 2	—	—	40	50	47	45	42

4. two future stand yield tables as shown in Table 15-20*b*
5. assignment of stands 1–4 to the first future stand yield table and assignment of stands 5–8 to the second future stand yield table
6. harvest target of 10,000 cords/period
7. net (marginal) revenue of $20/cord
8. an objective of maximizing present net worth

In the first iteration, $20/cord is used as the marginal revenue (price) in each period. Finding the best time to harvest the existing and future stands on an acre to maximize present net worth results in high harvests in periods 1 and 5, and low harvests in periods 2, 3, 4, and 6 (Table 15-21).

For the next iteration, a new marginal revenue must be developed for each period to force the harvest toward those periods that have too little volume harvested and away from those that have too much volume harvested. Hoganson and Rose present many different techniques that are used to make these adjustments. We will use an equation of the form:

$$M_{t,i+1} = M_{t-1,i+1} + (M_{ti} - M_{t-1,i}) + c_p \left(\frac{D_{ti} - D_{t-1,i}}{H_t}\right) M_{ti}$$

where M = estimate of marginal revenue
D = desired harvet volume − actual harvest volume
H = desired harvest volume
t = period number
i = iteration number
c_p = smoothing parameter (in this example −0.05)

Applied for six iterations (Table 15-22), this equation resulted in a price structure over time in each iteration that moves the harvested schedule toward the desired goal of 10,000 cords per period while minimizing the cost of achieving this constraint (Table 15-21). More iterations

Table 15-21 Results of three harvest scheduling iterations

		Period of harvest		Volume harvested in period					
Iteration	Stand	Existing	Regeneration	1	2	3	4	5	6
1	1	1	4	5,000			5,000		
	2	1	4	4,800			5,000		
	3	3	6			4,600			5,000
	4	3	6			3,600			5,000
	5	1	5	4,000				5,000	
	6	5	—					5,000	
	7	1	5	4,000				5,000	
	8	1	5	5,000				5,000	
	Total			22,800	0.0	8,200	10,000	20,000	10,000
4	1	2	5		5,000			5,000	
	2	1	4	4,800			5,000		
	3	3	6			4,600			5,000
	4	3	6			3,600			5,000
	5	1	5	4,000				5,000	
	6	5	—					5,000	
	7	1	5	4,000				5,000	
	8	1	5	5,000				5,000	
	Total			17,800	5,000	8,200	5,000	25,000	10,000
6	1	2	5		5,000			5,000	
	2	1	4	4,800			5,000		
	3	3	6			4,600			5,000
	4	3	6			3,600			5,000
	5	1	5	4,000				5,000	
	6	4	—				4,600		
	7	1	5	4,000				5,000	
	8	2	6		4,800				5,000
	Total			12,800	9,800	8,200	9,600	15,000	15,000

Table 15-22 Marginal revenue (price) per cord

Iteration (i)	Period (t) 1	2	3	4	5	6
1	20	20	20	20	20	20
2	20	21.37	20.88	20.77	20.17	20.77
3	20	19.96	20.33	20.73	20.13	18.24
4	20	21.33	21.20	21.49	20.28	18.94
5	20	22.15	21.81	22.31	19.89	19.39
6	20	22.49	22.18	22.25	20.40	19.04

would move the schedule further toward the desired level, and employing some of Hoganson and Rose's other procedures for adjusting the prices would help even more.

In their application, Hoganson and Rose employ their techniques in a more sophisticated way than that shown here. They combine the original revenue with the marginal revenue to guide selection at each iteration, put an ending value on any inventory left after harvest in the last period, and use dynamic programming to come up with the best regime for each stand at each iteration, given the prices for that iteration. But the notion of varying the prices until a set is found that meets the constraints is the same as shown here.

Hoganson and Rose's work clears away major criticisms aimed at heuristics like binary search. Now the best management intensity and harvest priority can be determined as part of the solution. The technique still retains the criticism that it has difficulties with constraints in addition to harvest amounts, but this pioneering work should breathe new life into the use of heuristics in harvest scheduling problems.

QUESTIONS

15-1 *Sara Shope Problem*. Sarah Shope, the owner of a 120-acre tract in Georgia, contacted a consulting forester and asked him to prepare a harvest schedule that would *maximize the total volume of harvested timber from her property over the next four decades* subject to certain constraints. A survey of the forest provided the data on the next page. (All land is of the same site quality.) For analysis purposes there are four planning periods of 10 years each. All harvest culture and treatment are at the midpoint of each decade.

A close discussion revealed that the following constraints were important to Sarah:

1. Harvest volume control: the harvest in a decade cannot deviate by more than 20 percent from the harvest in the preceding decade.
2. The inventory of merchantable standing timber after the fourth-decade harvest must be at least 1,600 M bd ft.

Stand	Condition	Area (acres)	Age (years)	Volume per acre (M bd ft)
A	Poorly stocked, diseased loblolly	40	35	8
B	Fully stocked, vigorous loblolly	80	55	32

3. After existing stands have been regeneration harvested, they are immediately planted and established. All regenerated stands are given the same prescription of plant, a thinning harvest at age 20, and a final harvest at age 30.
4. A 20-acre park in stand *B* is to be set aside uncut as a possible future homesite.

The consultant then evaluated the mid-decade yield of existing and future stands in M bd ft:

Existing stands			Future stands				
Decade cut	Stand *A*	Stand *B*	Decades of growth	Age when cut	Inventory before cut	Inventory after cut	Volume cut
1	10	40	0	0	0	0	0
2	13	50	1	10	0	0	0
3	15	58	2	20	12	7	5
4	16	65	3	30	30	0	30

With this information the consultant formulated the problem in a Model I structure as shown.

Questions

(*a*) The consultant provided the linear programming solution printout below to Sarah. Explain to her what each of the six numbered items means.

(*b*) Answer the following five questions using the solution as your reference.

1. Why does the dual price equal zero for the park constraint? Doesn't it seem reasonable that this should be a binding constraint?
2. If the ending inventory was increased from 1,600 to 2,000 M bd ft, what is the estimated opportunity cost in terms of total timber harvest?
3. How can the constraints on H_4 (rows 11 and 12) be satisfied when the decision variables A_4 and B_4 have a zero value?
4. Is this a Model I or a Model II formulation, and exactly how do you know for sure which it is?
5. Suppose new wildlife and visual concerns required that each stand be represented in the total harvest for the first period in proportion to how much of the total area is occupied by the stand with a maximum deviation of ±10 percent. Write constraint equations for the problem to ensure that this condition is satisfied.

Model I formulation for Question 15-1

Row		Activities										Harvest accounting					RHS
		A_1	A_2	A_3	A_4	A_U	B_1	B_2	B_3	B_4	B_U	H_1	H_2	H_3	H_4	I	
(1) Objective		45	18	15	16		75	55	58	65		*(1	1	1	1)		Maximize
(2) Harvest accounting	H_1	10					40					−1					= 0
(3)	H_2		13					50					−1				= 0
(4)	H_3	5		15			5		58					−1			= 0
(5)	H_4	30	5		16		30	5		65					−1		= 0
(6) Ending inventory	I_5	0	7	0	0	16	0	7	0	0	65					−1	= 0
(7) Harvest flow	LH_2											−0.8	1				≥ 0
(8) control	UH_2											−1.2	1				≤ 0
(9)	LH_3												−0.8	1			≥ 0
(10)	UH_3												−1.2	1			≤ 0
(11)	LH_4													−0.8	1		≥ 0
(12)	UH_4													−1.2	1		≤ 0
(13) Ending inventory	I															1	≥ 1,600
(14) Park											1						≥ 20
(15) Acres available	A	1	1	1	1	1											= 40
(16)	B						1	1	1	1	1						= 80

* Optional use of accounting variables to state the objective:

Maximize $H_1 + H_2 + H_3 + H_4$

(*c*) Recast the problem in a Model II structure. Define decision variables and present a complete detached coefficient matrix.

```
MAX        H1 + H2 + H3 + H4
SUBJECT TO
      2)  -  H1 + 10 A1 + 40 B1 =    0
      3)  -  H2 + 13 A2 + 50 B2 =    0
      4)  -  H3 + 5 A1 + 5 B1 + 13 A3 + 58 B3 =    0
      5)  -  H4 + 30 A1 + 30 B1 + 5 A2 + 5 B2 + 15 A4 + 65 B4 =    0
      6)     7 A2 + 7 B2 + 16 AU + 65 BU -  I =    0
      7)  - 0.8 H1 +  H2 >=   0
      8)  - 1.2 H1 +  H2 <=   0
      9)  - 0.8 H2 +  H3 >=   0
     10)  - 1.2 H2 +  H3 <=   0
     11)  - 0.8 H3 +  H4 >=   0
     12)  - 1.2 H3 +  H4 <=   0
     13)     I >=   1600
     14)     BU >=   20
     15)     A1 +  A2 +  A3 +  A4 +  AU =    40
     16)     B1 +  B2 +  B3 +  B4 +  BU =    80
END


   LP OPTIMUM FOUND  AT STEP    17

        OBJECTIVE FUNCTION VALUE

 1)        5286.89844

VARIABLE         VALUE          REDUCED COST
      H1        984.8921            0.0
      H2       1181.8704            0.0
      H3       1418.2444            0.0
      H4       1701.8918            0.0
      A1         37.0764            0.0
      B1         15.3532   1        0.0
      A2          2.9236            0.0
      B2         22.8773            0.0
      A3          0.0               9.2089
      B3         19.9327            0.0
      A4          0.0              24.8138
      B4          0.0              69.8175
      AU          0.0               7.5262   2
      BU         21.8368            0.0
       I       1600.0000            0.0
```

ROW	SLACK	DUAL PRICES
2)	0.0	−1.6698
3)	0.0	−1.3539
4)	0.0	−1.3178
5)	0.0	−1.1018
6)	0.0	−1.1759
7)	393.9565 **3**	0.0
8)	0.0	0.5581 **4**
9)	472.7478	0.0
10)	0.0	0.7600
11)	567.2974	0.0
12)	0.0	0.8982
13)	0.0	−1.1759 **5**
14)	1.8368	0.0
15)	0.0	26.3407
16)	0.0	76.4341

NO. ITERATIONS= 17

RANGES IN WHICH THE BASIS IS UNCHANGED

COST COEFFICIENT RANGES

VARIABLE	CURRENT COEF	ALLOWABLE INCREASE	ALLOWABLE DECREASE
H1	1.0000	0.9722	5.3680
H2	1.0000	2.3610	1.7943
H3	1.0000	32.8931	1.2384
H4	1.0000	INFINITY	0.9286
A1	0.0	56.6632	12.3827
B1	0.0	21.7843	110.2621
A2	0.0	12.3827	6.6437
B2	0.0	28.8486	31.1107
A3	0.0	9.2089	INFINITY
B3	0.0	235.5592	34.0196
A4	0.0	24.8138	INFINITY
B4	0.0	69.8175	INFINITY
AU	0.0	7.5262	INFINITY
BU	0.0	77.5968	49.4320
I	0.0	1.1759	INFINITY

RIGHTHAND SIDE RANGES

ROW	CURRENT RHS	ALLOWABLE INCREASE	ALLOWABLE DECREASE
2)	0.0	523.8242	54.0503
3)	0.0	646.0498	173.7155
4)	0.0	2598.2173	260.9387
5)	0.0	55.8723	435.3713
6)	0.0	2911.7947	121.2103
7)	0.0	393.9565	INFINITY
8)	0.0	64.8603	511.3967

```
  9)          0.0          472.7478       INFINITY
 10)          0.0           56.6725         382.3198
 11)          0.0          567.2974       INFINITY
 12)          0.0           55.8723         435.3713
 13)       1600.0000      2911.7947         121.2103   6
 14)         20.0000         1.8368       INFINITY
 15)         40.0000        20.3612           3.0051
 16)         80.0000         4.4989          44.7968
```

15-2 *Idaho Stud Problem.* The Idaho Stud Company of Coeur D'Alene recently purchased 10,000 acres of fir and larch stands which it planned to manage as a unit. The tracts purchased had three stand types, each with different site quality.

Stand	Condition	Area (acres)	Site	Current volume per acre (M bd ft)
A	Young plantation	2,000	I	4
B	Poorly stocked second growth	5,000	II	12
C	Poorly stocked second growth with old growth residuals	3,000	III	31

Corporate management wants a harvest plan for 5 decades that tells how many acres should be harvested in each stand and the total volume harvested in each decade. For planning purposes, all harvests are assumed to occur at the mid-decade times of 5, 15, 25, 35, and 45 years from now.

The forestry staff reviewed their stands, species, and the economic situation and selected three possible prescriptions for existing stands and two prescriptions for regenerated stands.

Existing stands

1. Clearcut and plant in periods 1, 2, or 3.
2. Commercially thin in period 1, clearcut and plant in period 3.
3. In period 1, permanently assign the land to uneven-aged management.

(Note that prescription 2 cannot be applied to stand *A* because of insufficient volume.

Regenerated stands. For stands treated under prescriptions 1 or 2 above, the subsequent regenerated stands can be assigned either of the following even-aged prescriptions:

1. Regeneration harvest at age 20.
2. Regeneration harvest at age 30.

These prescription options along with detailed plot data describing the average composition of the three stand types were evaluated in a yield simulator to produce yield forecasts for the three stands.

Yields (thousand board feet)

Existing stands	Years from now								
	5	10	15	20	25	30	35	40	45
	Even-aged prescriptions								
Unthinned stand *A*	6	10	14	22	29	37	44	48	
Thinned stand *B* (thin 4 M bd ft at year 5)	11	15	20	24	28	33	36	38	
Unthinned stand *B*	15	19	24	27	29	33	36	38	
Thinned stand *C* (thin 10 M bd ft at year 5)	24	27	30	34	36	38	39	40	
Unthinned stand *C*	34	36	38	39	40	40	41	41	
	Uneven-aged prescriptions								
Stand *A*									
Harvest	0		4		8		12		12
RGS	6		10		12		12		12
Stand *B*									
Harvest	5		8		9		10		10
RGS	10		13		13		14		14
Stand *C*									
Harvest	10		8		6		6		6
RGS	24		22		20		18		18

Regenerated stands	Years from birth				
	0	10	20	30	40
Site I (stand *A*)	0	5	10	27	40
Site II (stand *B*)	0	2	7	19	28
Site III (stand *C*)	0	0	3	9	18

Management constraints. At the end of 10 years, management will reconsider the plan. Until then, the following goals and constraints are to be used to guide the harvest plan over the next 5 decades.

1. The driving goal is to maximize harvest in the first period.
2. At least 20,000 M bd ft must be harvested in each period.
3. A minimum of 2,000 acres must be in uneven-aged management for stream buffers and protection of visually sensitive zones.
4. A maximum of 5,000 acres is permitted in uneven-aged management because company foresters do not believe it to be as productive as even-aged management.
5. Stand *C* is currently a good habitat for the rare and endangered Idaho bearcat, and by law some acreage must be left uncut for at least two periods while the researchers figure out what to do. Therefore a maximum of 1,000 acres can be clearcut harvested in each of the first two periods.
6. As an extra guarantee against overcutting at least 10,000 M bd ft of residual inventory *after* the fifth period harvest in 45 years is required.

Questions

(*a*) Make a list defining a set of Model I decision variables for this problem.

(*b*) Write a set of equations to present the objective function and constraints of this problem in a Model I formulation.

(*c*) If possible, solve the problem and evaluate the solution and harvest schedule for corporate management.

15-3 *Clearcut Restoration, Inc.* Clearcut Restoration, Inc., has assigned you to propose a management plan for 2,000 acres of cleared forestland outside of Stumpville, Calif. The company is interested in maximizing the present net worth, but it is also trying to improve its public image. All allocation and planting decisions are made immediately at the beginning of year 1, and all timber stands will be immediately regenerated at the time of harvest. The planning horizon is infinite.

Assume that this area has 1,200 acres of site I quality land and 800 acres of site II. You may plant either Plucky pine (PP) or Fantastic fir (FF) and may choose to thin or not to thin. Once an acre is allocated to a prescription, it will be perpetuated forever. The table below describes the yield in thousand board feet for each possible prescription and the number of years until each harvest.

Species	Planting cost per acre, year 0	Prescription	Site	Thin volume, year 30	Harvest volume	
					Year 40	Year 50
PP	$300	Thin, Yr 30 Cut, Yr 50	I	4		23
PP	300	Cut, Yr 40	I		20	
PP	200	Thin, Yr 30 Cut, Yr 50	II	3		17
PP	200	Cut, Yr 40	II		15	
FF	300	Thin, Yr 30 Cut, Yr 50	I	6		26
FF	300	Cut, Yr 40	I		22	
FF	200	Thin, Yr 30 Cut, Yr 50	II	2		14
FF	200	Cut, Yr 40	II		13	

Assume the net revenue received for Plucky pine is $300 per M bd ft and for Fantastic fir, $250 per M bd ft. Clearcut's guiding rate of return is 5 percent.

The land in question is under a number of constraints. The California Water Board requires you to leave a stream buffer strip in the site I land to protect water quality. This takes 75 acres. The Sierra Club is offering a lump sum of $650 per acre for scenic easements (payable now, year 0) to grow timber and not cut it.* In addition it will pay $400 per acre (payable now) for a wildlife easement to let land grow to brush for wildlife diversity. Clearcut Restoration, Inc., wants at least 400 total acres but not more than 600 acres dedicated to these easement programs. To show balance in their environmental support, for every acre of

* This does not include the 75 acres for the stream buffer strip.

brush there must be at least 3 acres preserved for scenic quality. In addition, the company would like you to grow at least 1 acre of fir for every 4 acres of pine on site II to help promote wildlife diversity in that area.

(*a*) Formulate this problem in Model I structure with a present net worth objective function. Show it as a detached coefficient matrix.

(*b*) If you are in a position to solve the problem, modify the constraint set appropriately to evaluate the opportunity costs of the scenic easement program.

15-4 *Model I versus Model II.* A forest contained two existing stands, both growing on the same site land.

Prescriptions for existing stands

1. Clearcut and plant period 1.
2. Clearcut and plant period 2.

Prescriptions for regenerated stands

1. Clearcut and plant after every 2 periods of growth.
2. Clearcut and plant after every 4 periods of growth.

Stand	Acres
A	100
B	100

Planning horizon, 5 periods.

(*a*) Make a list of the decision variables needed to analyze this problem in a model I structure. Define fully in terms of the period when the stand is cut and the regenerated prescription used.

(*b*) Write all the constraint equations needed to keep full accounting of the acres available to harvest for all decision variables in the model I structure.

(*c*) Make a list of the decision variables needed to analyze this problem in a model II structure. Define fully in terms of the period when the stand is cut and the period a stand is regenerated.

(*d*) Write all the constraint equations needed to keep full accounting of the acres available to harvest for all decision variables in the model II structure over the planning horizon.

15-5 You are planning the management of a 900-acre portion of a forest which has the following characteristics:

Area distribution

		Management zone		
Stand type	Current stand age	*A* North of ridge crest	*B* South of ridge crest	*C* Crater of volcano
I	35 years	50	250	
II	55 years	300	100	200
		350	350	200

Location

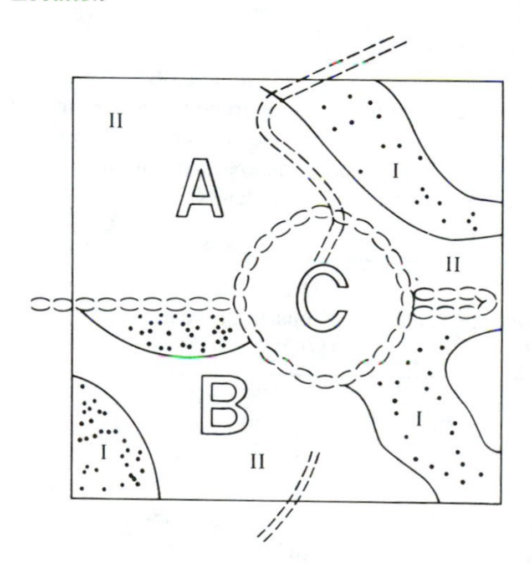

MANAGEMENT ZONE

A – North of ridge crest

B – South of ridge crest

C – Crater of volcano

Ridge crest

Proposed roads

I Stand type I

II Stand type II

Your problem is whether to manage each zone for timber or allocate it to wilderness. If zone *C* is allocated to timber, then zone *A* must also be allocated to timber because of the design of the road system since access to zone *C* is through zone *A*. Completed road costs are as follows (present net worth):

Zone *A* segment, $33,000
Zone *B* segment, $25,000
Zone *C* segment, $7,000

Based on a survey by the local hiking club, you estimate the amount and the present net worth of recreation for the three management zones under timber and wilderness allocations to be as follows:

	Allocated to timber			Allocated to wilderness		
	A	*B*	*C*	*A*	*B*	*C*
Thousand visitor-days per year	2	3	1	5	7	3
Present net worth (thousands)	$8	$12	$5	$35	$49	$25

On those areas allocated to timber production, you also wish to plan a timber management schedule for three 10-year periods. You will harvest in the middle of each period using clearcutting. The timber in both stand classes is mature enough to be harvested in the first, second, or third periods. Regeneration occurs in the same year as harvest. The minimum rotation age for future stands is 20 years, but longer rotations are also possible.

Yield and value information for both existing and future stands is:

Stand age (years)	Yield per acre (thousand ft³)	Stumpage value ($/ft³)	Timber sale Preparation cost ($/ft³)
10	0.5	0.2	$0.1
20	1.0	0.4	0.1
30	1.5	0.6	0.1
40	1.8	0.8	0.1
50	2.0	1.0	0.05
60	2.2	1.0	0.05
70	2.3	1.0	0.05
80	2.4	1.0	0.05

The guiding rate of interest is 4 percent. There are no reforestation costs.

(*a*) Given that the entire forest is allocated to timber production, set up in a detached coefficient matrix the model I structure that will allow you to maximize the present net worth of all revenues over three 10-year periods.

Constraints

1. Acreage accounting constraints
2. A constraint that the harvest cannot decline more than 10 percent from period to period
3. A constraint that not more than $8,000 can be spent on timber sale preparation in the first period
4. A constraint that not more than 200 acres can be clearcut in the first period

Define all decision variables and identify the constraint rows in your matrix.

(*b*) Now modify the problem by assuming that you must decide whether to manage each zone for timber production or wilderness. Set up a formulation that considers the timber versus wilderness decision for each zone as an aggregate choice, and the timber scheduling decisions on a stand-type basis using a Model I structure.

Your formulation should maximize the present net worth for three 10-year periods subject to:

1. Aggregate choice and acreage accounting constraints
2. A constraint that at least one zone must be allocated to wilderness.
3. A constraint that zone *C* cannot be allocated to timber production unless zone *A* is allocated to timber production.
4. Constraints to ensure that a minimum of 300 acres of timber at least 50 years old will exist on the forest after harvest in each period for any possible combination of timber and wilderness allocations.
5. Constraints to ensure that not more than 20 percent of the acres allocated to timber production across the forest can be harvested in the first period.

REFERENCES

Chambers, C. J., and R. N. Pierson (Eds.), 1973: "Sustainable Harvest Analysis, 1971 and 1972," Washington State Department of Natural Resources Harvest Regulation Report 5.

Chappelle, D. E., 1966: A Computer Program for Scheduling Allowable Cut Using either Area or Volume Regulation during Sequential Planning Periods, U.S. Forest Service Research Paper PNW-33.

Clutter, J. L., et al., 1968: "MAX MILLION—A Computerized Forest Management Planning System," School of Forest Resources, University of Georgia, Athens.

———, J. C. Fortson, and L. V. Pienaar, 1978: "MAX MILLION II—A Computerized Forest Management Planning System, User's Manual," School of Forest Resources, University of Georgia, Athens.

Crim, S., 1980: "Separate vs. Combined Resource Allocation and Scheduling: A Case Study of Two National Forests," Ph.D. Dissertation, Department of Forestry, Colorado State University, Fort Collins.

Hallanger, W., 1973: The Linear Programming Structure of the Problem, in C. J. Chambers and R. N. Pierson (Eds.), "Sustainable Harvest Analysis, 1971 and 1972," Washington State Department of Natural Resources Harvest Regulation Report 5, pp. 67–71.

———, H. L. Scheurman, and J. Beuter, 1975: Oregon Timber Resource Model—A Brief Introduction, in J. Meadows, B. Bare, K. Ware, and C. Row (Eds.), "Systems Analysis and Forest Resource Management," Society of American Forestry, Washington, D.C., pp. 148–163.

Hoganson, H. H., and D. W. Rose, 1984: A Simulation Approach for Optimal Timber Management Scheduling, *Forest Sci.*, **30**:220–238.

Hrubes, R. J., and D. I. Navon, 1976: "Application of Linear Programming to Downward Sloping Demand Problems in Timber Production," U.S. Forest Service Research Note PSW-135.

Johnson, K. N., 1977: A Comment on "Techniques for Prescribing Optimal Timber Harvest and Investment under Different Objectives," *Forest Sci.*, **23**:444–445.

———, 1986: "FORPLAN (Version I)—An Overview," Land Management Planning, U.S. Forest Service, Fort Collins, Colo. (mimeographed).

———, and D. B. Jones, 1979: "A User's Guide to Multiple Use-Sustained Yield Resource Scheduling Calculation (MUSYC)," Timber Management, U.S. Forest Service, Fort Collins, Colo. (mimeographed).

———, and H. L. Scheurman, 1977: "Techniques for Prescribing Optimal Timber Harvest and Investment under Different Objectives—Discussion and Synthesis," Forest Science Monograph 18.

———, and P. L. Tedder, 1983: Linear Programming vs. Binary Search in Periodic Harvest Level Calculation, *Forest Sci.*, **29**:569–582.

———, H. L. Scheurman, and J. Beuter, 1975: Oregon Timber Resource Model—A Brief Introduction, in J. Meadows, B. Bare, K. Ware, and C. Row (Eds.), "Systems Analysis and Forest Resource Management," Society of American Forestry, Washington, D.C., pp. 148–163.

———, T. Stuart and S. Crim, 1986: "FORPLAN (Version II)—An Overview," Land Management Planning, U.S. Forest Service, Fort Collins, Colo. (mimeographed).

Navon, D. I., 1971: "Timber RAM . . . A Long-Range Planning Method for Commercial Timber Lands under Multiple-Use Management," U.S. Forest Service Research Paper PSW-70.

Sassaman, R. W., E. Hold, and K. Bergsvick, 1972: "User's Manual for a Computer Program for Simulating Intensively Managed Allowable Cut," U.S. Forest Service General Technical Report PNW-1.

Schrage, L., 1981: "Linear Programming Models with LINDO," Scientific Press, Palo Alto, Calif.

Tedder, P. L., J. S. Schmidt, and J. Gourley, 1980: "TREES: Timber Resource Economic Estimation System, Volume I—A User's Guide for Forest Management and Harvest Scheduling," Forest Research Laboratory Research Bulletin 31a, Oregon State University, Corvallis.

Walker, J. L., 1971: "An Economic Model for Optimizing the Rate of Timber Harvesting," Ph.D. Thesis, University of Washington, Seattle, *Diss. Abstr.* **32**(5):2276A (microfilm no. 71-28489).

———, 1976: ECHO: Solution Technique for a Nonlinear Economic Harvest Optimization Model, in J. Meadows, B. Bare, K. Ware, and C. Row (Eds.), "Systems Analysis and Forest Resource Management," Society of American Forestry, Washington, D.C.

CHAPTER

SIXTEEN

ANALYSIS OF HARVEST SCHEDULES

The schedule of planned future treatments and harvests from a forest is what owners and their constituencies see and get excited about. To the owner it conveys meaning about output schedules, cash revenues, other benefits, workloads, and costs. To the constituents the schedule conveys how their welfare and interests will be affected.

For many private owners the implication of different goals and constraints, separately and in combination, is not known until the problem is actually modeled and analyzed and the resultant output schedules are produced. In this case the harvest schedules are used primarily to explore and evaluate the importance of different goals and constraints and arrive at a negotiated decision as to how the owner's problem is finally specified. For example, it is easy, initially, to say, "let harvest vary by no more than 10 percent." But when it turns out that the reduction in the present net worth due to a 10 percent limit is three times that of a 20 percent limit, an owner may get a lot more flexible.

For public forests, the different goal and constraint sets used to define planning problem alternatives often reflect the viewpoint and desires of their different constituencies, be they environmentalists, local government, forest industry, or budget-cutting federal legislators. The different output schedules from the solutions provide the comparative information used for the administrative and political resolution of conflicting desires.

We have already produced some timber harvest schedules; let us look at them again. Figure 16-1*a* shows the four harvest schedules we generated by solving the regulation examples in Chap. 14, and Fig. 16-1*b* shows

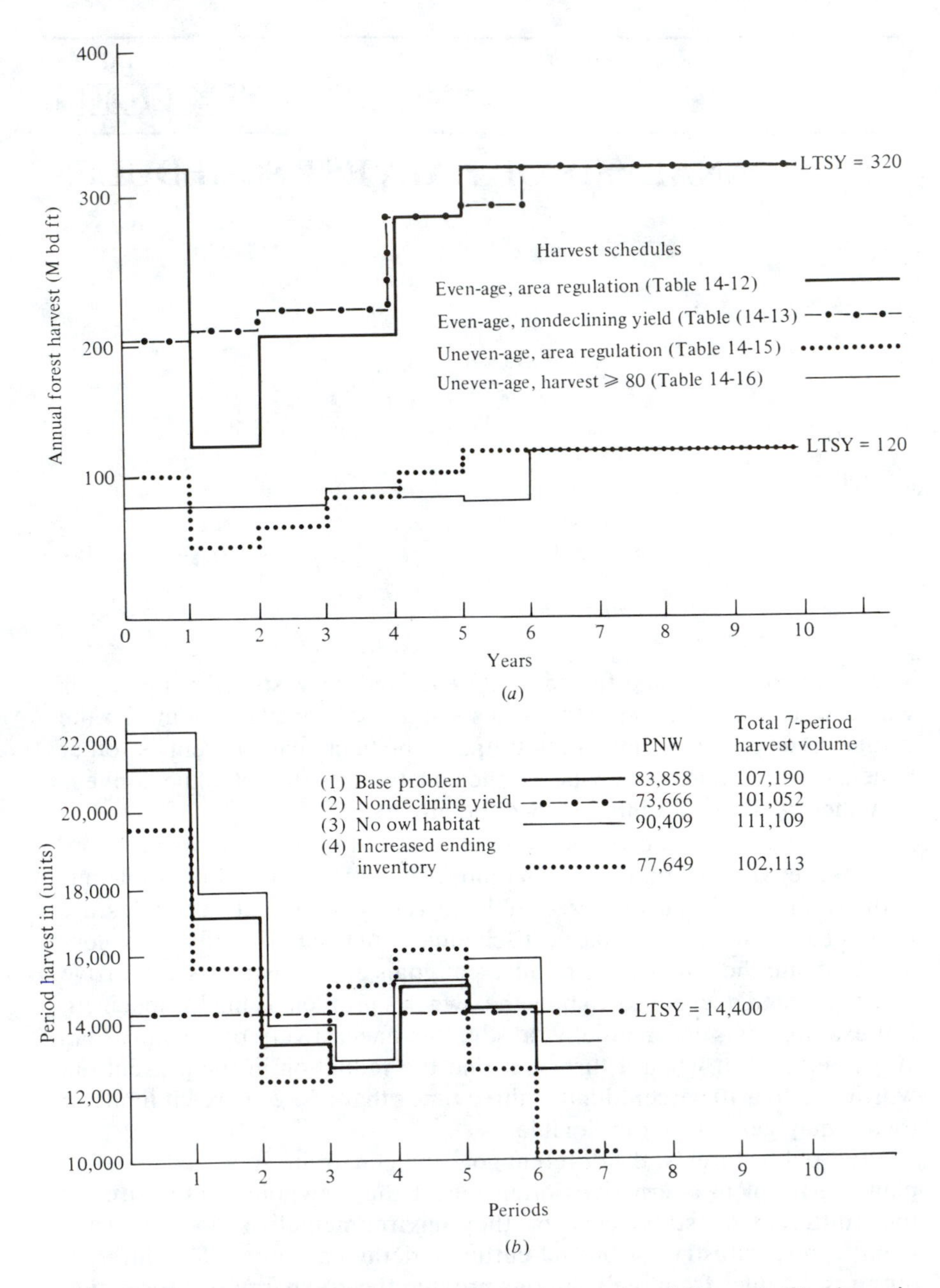

Figure 16-1 (*a*) Harvest schedules from regulation examples of Chap. 14. (*b*) Alternative harvest schedules for the Daniel Pickett forest problem (Table 15-10).

the four alternative schedules produced by our more elaborate analysis of the Daniel Pickett forest.

Both the even-aged and the uneven-aged forests in Fig. 16-1*a* were fully regulated after the harvest in year 6, as shown by the subsequent even harvest at the long-term sustained yield level. In each case the owner has a choice: (1) to take a heavy harvest in the first year and let the harvest decline substantially in the next 3 to 5 years, or (2) to restrain the first-year harvest and have the cut gradually build to the long-term sustained yield level with no significant declines. If the present net worth of the early harvest strategy is higher, the choice comes down to the owner's trade-off of present net worth versus harvest stability and need for midperiod volume.

The schedules for the Daniel Pickett forest show a variety of choices. The Daniel Pickett problem has the economic objective of maximizing the present net worth, so that it is not surprising that the less constrained solutions chose high harvest levels in the early periods. Since the analysis was restricted to seven periods, we can only speculate about the exact harvest schedule after the planning period. We expect the future harvests under the 20 percent variation restriction to continue cycling up and down around the long-term sustained yield level of 14,400 M bd ft, as shown by schedules 1, 3, and 4. The solution under the nondeclining yield constraints immediately went to the long-term sustained yield harvest level, and it is a good bet that it would stay there if we extended the planning horizon. If this set of solutions represented the options for a public forest, then the choice would require negotiation and weighing the importance of interests that favor timber production, present net worth, owls, harvest stability, and a classically regulated forest. Within limits, more of each is feasible; what is the compromise mix?

The Daniel Pickett forest schedules also display the tantalizing question that occurs at least once to every forest owner and forest economics student: "Why fool around with future timber harvests when we can cut all the timber right now, put the money in the bank, and live off the interest?" To be sure, this often looks like the best financial deal on paper, and the forest industry often took such an approach in the early decades of this century. Immediate liquidation is not implemented much anymore, except on small ownerships, for three reasons. First, putting a large amount of timber on a local market may cause the stumpage price to decline, causing the income to be smaller than expected. Second, it is not easy to find enough loggers and sawmills to liquidate a large forest. Third, many owners are providing a steady source of logs to certain mills—their own or those with which they have a long-term agreement. Such a roller-coaster harvest schedule may be so disruptive to these mills, and the communities they support, as not to seem a worthwhile option to the landowners.

Short-term liquidations often wind up by selling the land and timber together anyway, in which case the trees are still growing on the land and the timber management scheduling problem is simply passed on to a new owner. Nevertheless, the question about whether to liquidate the existing timber recognizes the dominant effect that volume control constraints can have on problem solutions and challenges forest owners to understand the implications of these constraints.

A HARVEST SCHEDULING PRIMER*

To help you understand the implications of different harvest scheduling formulations, we will work through a series of examples covering policies and constraints often used in public and private forestry. First we project the harvest schedule under a "base harvest policy" similar to that of the U.S. Forest Service:

1. A harvest schedule that exhibits nondeclining yield (NDY) at or below long-term sustained yield (LTSY) capacity
2. A regeneration harvest age at or beyond culmination (maximum) of mean annual increment (CMAI)
3. A planning horizon equal to twice the CMAI rotation length

Then we project the harvest pattern under two often suggested modifications to the base policy: (1) releasing the constraint that the harvest must achieve nondeclining yield at or below the long-term sustained yield capacity (here called "departure from nondeclining yield"), and (2) releasing the constraint that stands cannot be regeneration harvested until culmination of mean annual increment.

Finally, we examine the impact of management intensification on the harvest schedule under the base policy.

In all cases we assume that our objective causes us to harvest the maximum amount of timber available in each period under the constraints and that scheduling restrictions from other outputs (such as water and wildlife) do not limit harvest on the land allocated to timber production.

Two Types of Harvest Schedules

For a particular land allocation and set of yields for existing and future stands, a harvest schedule under base harvest policy usually will reveal either a surplus or a deficit of timber in existing stands. A schedule indi-

* Most of what follows is adapted from Johnson (1981).

cates a surplus of timber in existing stands when lands allocated to timber production can provide more harvest volume from existing stands (including growth) than needed to maintain the harvest at the long-term sustained yield until the stands created when the existing stands are cut become available for harvest. We will call this condition "a surplus of existing timber." A schedule indicates a deficit of timber in existing stands when lands allocated to timber production cannot provide enough harvest volume from existing stands (including growth) to maintain the harvest at the long-term sustained yield until the stands created when the existing stands are cut become available for harvest. We will call this condition "a shortage of existing timber."

For each combination of management intensity and constraints, we will create a harvest schedule when there is a surplus of existing timber and when there is a shortage of existing timber. While the existing timber in the examples given below is all old growth, the results follow equally well when the existing timber is a combination of old and young stands. Classification into a surplus or shortage condition depends on the relationship between production from existing stands and production from future stands, not on the age of the existing timber.

Usually the harvest scheduling analyses shown below are done with one of the computer-based timber management scheduling models, such as FORPLAN, to represent the objectives and constraints and to keep track of the timber stands and their growth rates by species, site, age, and management intensity. To enable these calculations to be easily made here, we have recognized only one species, site, and management intensity in each analysis. Also, we have assumed that the existing timber (the old growth) does not produce any net growth. We recognize that more richness in detail of the problem is needed in actual application, but this would not, in general, change the conclusions drawn.

The Harvest Schedule under the Base Harvest Policy

With a surplus of existing timber Assume that your forest has the following attributes:

1. *Inventory*. 200 acres of old growth with a volume of 20 M bd ft per acre and showing no net growth.
2. *Growth of future stands*. All stands culminate at age 100 with a volume per acre equal to 15 M bd ft.

$$\begin{aligned}\text{LTSY} &= \frac{\text{total growth over life of future stand}}{\text{rotation age for future stand}} \times \text{acres involved}\\ &= \frac{15 \text{ M bd ft/acre}}{100 \text{ years}} \times 200 \text{ acres}\\ &= 30 \text{ M bd ft/year}\end{aligned}$$

3. *Harvest method*. Clearcutting with no intermediate harvest.

To assess the maximum harvest under the base harvest policy, it is useful to calculate the harvest that would occur under area control. With the forest area A equal to 200 acres and a rotation for future stands R of 100 years, $A \div R = 200$ acres $\div$ 100 years = 2 acres/year. Cutting the existing stands at 2 acres/year would result in an annual harvest of 2 acres $\times$ 20 M bd ft/acre = 40 M bd ft/year for 100 years. With a long-term sustained yield of 30 M bd ft and the base policy requirement that the harvest stay at or below the long-term sustained yield, an annual harvest of 40 units is not acceptable. Therefore under the base harvest policy, the maximum harvest will equal the long-term sustained yield (30 M bd ft/year). As a result, the old growth will be converted in 133 years [(200 acres $\times$ 20 M bd ft/acre) $\div$ (30 M bd ft harvest/year) = 133 years] (Fig. 16-2*a*). This schedule shows a "surplus of existing timber" because harvest begins at the long-term sustained yield, and it takes more than one rotation to cut the existing timber under the base harvest policy.

With a shortage of existing timber Suppose that you have the same forest, except that now you have an old-growth volume equal to 8 M bd ft per

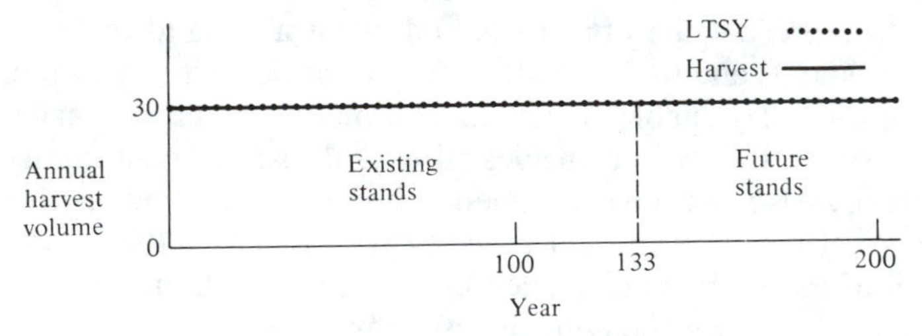

7.5 percent of forest area and volume is cut in the first decade

(*a*)

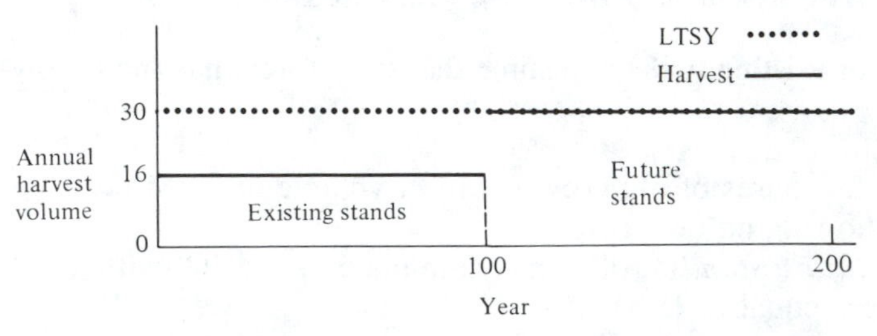

10 percent of forest area and volume is cut in the first decade

(*b*)

Figure 16-2 Maximum harvest under base harvest policy. (*a*) With a surplus of existing timber. (*b*) With a shortage of existing timber.

acre instead of 20 M bd ft. Again, let us start our analysis with a calculation of the harvest that could occur under area control. By cutting $A \div R = 200 \div 100$ acres/year for 100 years, we could harvest a volume of 8 M bd ft/acre × 2 acres/year = 16 M bd ft/year for these 100 years. Then the harvest would jump to the long-term sustained yield of 30 (Fig. 16-2*b*). This harvest pattern meets the requirements of the base harvest policy and gives the maximum harvest under that policy. Because harvest must begin below the long-term sustained yield under the base harvest policy, this schedule shows a "shortage of existing timber."

Looking again at the schedules produced in Chaps. 14 and 15, it appears that the regulation examples from Chap. 14 (Fig. 16-1*a*) would fall in the "shortage of existing timber" category and the Daniel Pickett forest (Fig. 16-1*b*) in the "surplus of existing timber" category.

Departure from Nondeclining Yield

In recent years "departures" have been a rallying cry for Western forest industry in their dealings with the public land agencies. By departures we here mean releasing the constraints embedded in the first part of the base harvest policy presented above: that the harvest schedule must exhibit nondeclining yield at or below the long-term sustained yield. While departures have mainly been viewed as a way to increase the harvest, they also can be used to decrease the land needed to support the harvest provided under the base harvest policy. Therefore departures should be of interest not only to groups that want a higher harvest (such as the forest industry), but also to groups that want more land reserved from timber production (such as the preservation groups).

Departure to increase the harvest With a surplus of existing timber, the most obvious departure to examine is that of area control. With area control the existing timber can be harvested at an annual rate of 40 M bd ft/year [(200 acres ÷ 100 years) × 20 M bd ft/acre = 40 M bd ft/year] for 100 years. Starting in year 101, the harvest will then drop to the long-term sustained yield of 30 M bd ft/year and continue at that rate forever (Fig. 16-3*a*). The increase of 10 M bd ft/year for the first 100 years under this harvest policy represents "surplus" volume that will be lost under the base harvest policy, a total of 1,000 M bd ft (10 M bd ft/year × 100 years = 1,000 M bd ft). This departure will produce 1,000 M bd ft of volume more than the base harvest policy because (1) acres will be more rapidly converted from stands that are not growing (old growth) to stands that are growing (young growth), and (2) the young growth will be cut at the age of maximum average growth as opposed to being cut at an age above maximum average growth. (Under the departure, harvest of young growth will begin at age 100 in year 101; under the base harvest policy,

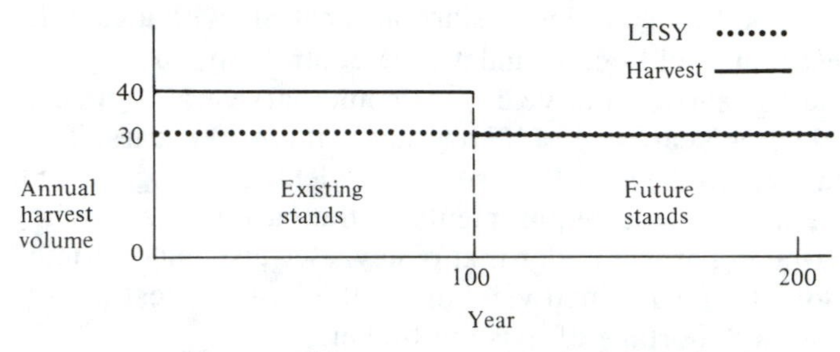

10 percent of forest area and volume is cut in the first decade

(*a*)

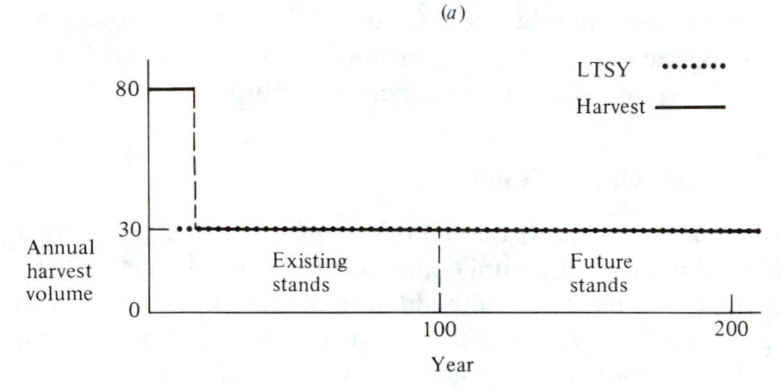

20 percent of forest area and volume is cut in the first decade

(*b*)

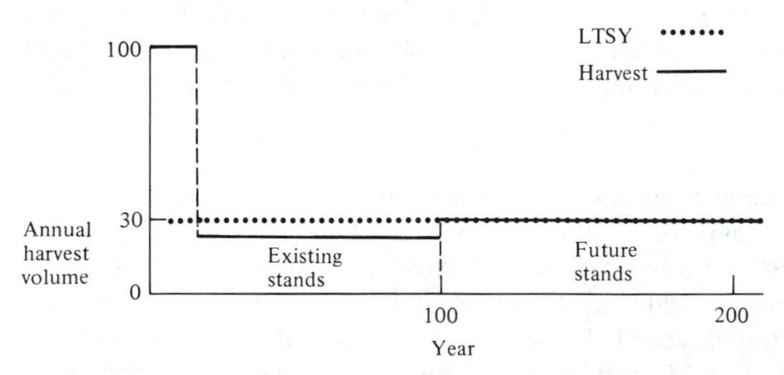

25 percent of forest area and volume is cut in the first decade

(*c*)

Figure 16-3 Departures that increase the harvest when there is a surplus of existing timber. (*a*) The surplus is harvested over 100 years without ever dropping below the long-term sustained yield. (*b*) The surplus is harvested over 20 years without ever dropping below the long-term sustained yield. (*c*) The surplus plus an additional volume is harvested over 20 years and then the harvest must drop below the long-term sustained yield.

harvest of young growth will begin at age 133 in year 134 and then gradually decline to age 100.)

Does this departure still allow any production to be lost? In answering this question remember that the base harvest policy will produce 30 M bd ft/year × 200 years = 6,000 M bd ft over 200 years and 6,030 M bd ft over 201 years. This departure will produce (40 M bd ft/year × 100 years) + (30 M bd ft/year × 100 years) = 7,000 M bd ft over 200 years and 7,030 M bd ft over 201 years. Maximum production over 201 years will involve cutting all 200 acres in the first year and then cutting them all again at year 101 and year 201 for a total production over 201 years of (20 M bd ft/acre × 200 acres) + (15 M bd ft/acre × 200 acres) + (15 M bd ft/acre × 200 acres) = 4,000 + 3,000 + 3,000 = 10,000 M bd ft. Thus the departure of area control still has a volume opportunity cost of 2,970 M bd ft over 201 years as compared to the volume-maximizing cutting policy.

As another departure, we might want to harvest the surplus over 20 years. In this instance the extra harvest will be 50 M bd ft/year (1,000 M bd ft ÷ 20 years = 50 M bd ft/year), which when added to the base harvest of 30 M bd ft/year gives a total harvest of 80 M bd ft/year for the 20 years (Fig. 16-3*b*). Since acres are released to young growth faster than under area control for the first 20 years (4 acres per year instead of 2 acres per year), more timber will be available at rotation age in years 101 to 120 than needed to harvest at long-term sustained yield. A slight surplus will then exist, which could be harvested through another departure at that time.

We might increase the harvest still higher, such as to 100 M bd ft/year for 20 years, but a decline below long-term sustained yield equal to the extra increase above 80 M bd ft/year will follow. An additional 400 M bd ft over the surplus are removed in the first 20 years [(20 years × 100 M bd ft) − (20 years × 30 M bd ft) = 1,400 M bd ft in excess of long-term sustained yield or 400 M bd ft in excess of surplus]. To compensate, we here take a decline of 5 M bd ft/year below the long-term sustained yield for the remaining 80 years until the young growth becomes available (Fig. 16-3*c*).

With a shortage of existing timber condition defined above, inventory is 8 M bd ft per acre and any increase in harvest above the starting level given by the base policy must be offset by a decrease below that starting level later (Fig. 16-4).* Suppose that we wish to harvest at long-term sustained yield (at 30 M bd ft/year) for the first 20 years instead of the 16 M bd ft possible under the base harvest policy. We will have to reduce the harvest by a like total [(30 − 16 M bd ft/year) × 20 years = 280 M bd ft] over the next 80 years. If we take the reduction equally, we can harvest

* When the timber deteriorates when held, as through bug attack, this statement is not correct. In this latter case, volume not cut immediately is lost and a departure from the nondeclining yield might enable a capture of volume that would otherwise be lost, without any reduction in the future harvest that would otherwise occur.

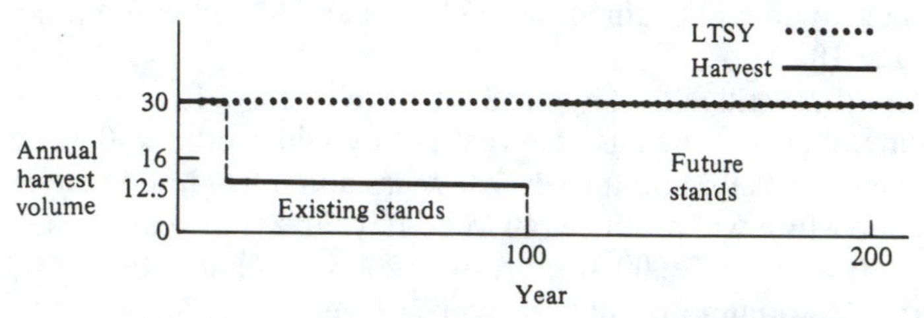

Figure 16-4 A departure that increases the harvest when there is a shortage of existing timber, with the resulting decline in the harvest level below that of the base harvest policy.

12.5 M bd ft/year (16 M bd ft/year − 280 M bd ft ÷ 80 years = 16 − 3.5 = 12.5 M bd ft/year) for the remaining 80 years until the young growth matures.

Releasing acres to young growth faster than under area control for the first 20 years (30 M bd ft/year ÷ 8 M bd ft/acre = 3.75 acres/year instead of 16 M bd ft/year ÷ 8 M bd ft/acre = 2 acres/year) will provide more timber at rotation age in years 101 to 120 than needed to harvest at long-term sustained yield. A slight surplus will then exist which, again, could be harvested through a departure at that time.

Departure to reduce the land base needed to maintain the harvest With a surplus of existing timber, the maximum harvest under the base harvest policy equals 30 M bd ft per year. Assume now that the land allocated to timber production is reduced to 75 percent of the original acreage (from 200 to 150 acres). By allowing the harvest to rise above the long-term sustained yield for the first 100 years, the harvest can still be maintained at 30 M bd ft/year for these years [(150 acres ÷ 100 years) × 20 M bd ft/acre = 30 M bd ft/year]. It will then drop to the new lower long-term sustained yield of 22.5 M bd ft/year [(150 acres ÷ 100 years) × 15 M bd ft/acre = 22.5 M bd ft/year] (Fig. 16-5).

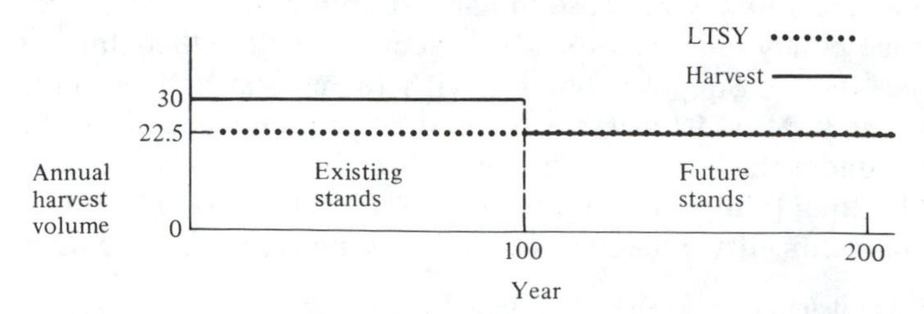

Figure 16-5 A departure, when there is a surplus of existing timber, that maintains the harvest level of the base harvest policy for 100 years on a smaller land base.

The old-growth surplus helps achieve the harvest level of the base harvest policy under the original land base when that land base is reduced by 25 percent. After the old growth has been used up, the harvest drops to the 25 percent lower long-term sustained yield.

With a shortage of existing timber, a departure on the reduced land base also can maintain the harvest level of the base harvest policy under the original land base, but the timber volume available will allow this departure for a shorter period of time than with the surplus condition and at higher cost in terms of the reduction in harvest that must follow.

Rotations Short of Culmination

Under the base harvest policy, regeneration harvest must not occur until the stands have reached culmination of mean annual increment. This rule, which uses a biological criterion instead of an economic criterion to set minimum rotation length, has often come under fire from the forest industry. While the industry hopes that the use of shorter rotations will allow an increase in harvest in the near future over that of the base harvest policy, it is also true that these shorter rotations could be used to maintain the same harvest as the base policy with less land. Both uses of shorter rotations are shown below.

When the base harvest policy is applied to a situation where there is a surplus of existing timber, it takes more than one rotation to harvest this existing timber (Fig. 16-2*a*). Rotations for future stands generally are beyond culmination of mean annual increment, and shortening the minimum harvest age below CMAI will have no effect. Therefore the discussion below relates only to situations with a shortage of existing timber.

Shorter rotations to increase the harvest Suppose that we use a rotation of 80 years for future stands, instead of the 100 years required under the base policy, and that the volume at 80 years equals 10 M bd ft per acre. Cutting the existing stands over 80 years instead of 100 years will give an annual harvest of 20 M bd ft/year [(200 acres ÷ 80 years) × 8 M bd ft/acre = 20 M bd ft/year] with 2.5 acres being harvested each year (Fig. 16-6*a*). Harvest after year 80 will climb to the long-term sustained yield of 25 M bd ft/year [(200 acres ÷ 80 years) × 10 M bd ft/acre = 25 M bd ft/year]—a 17 percent lower long-term sustained yield than that of a 100-year rotation.

Shorter rotations to reduce the land base needed to maintain the harvest Under the base harvest policy, the maximum harvest equals 16 M bd ft per year for 100 years and 30 M bd ft per year thereafter. Suppose that we allocate 20 percent of the timberland to wilderness. The starting nondeclining yield under the base harvest policy will then fall from 16 to 12.8 M

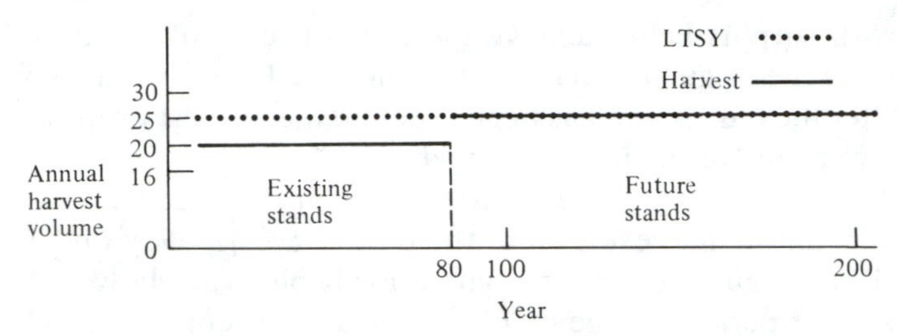

12.5 percent of forest area and volume is cut in the first decade

(*a*)

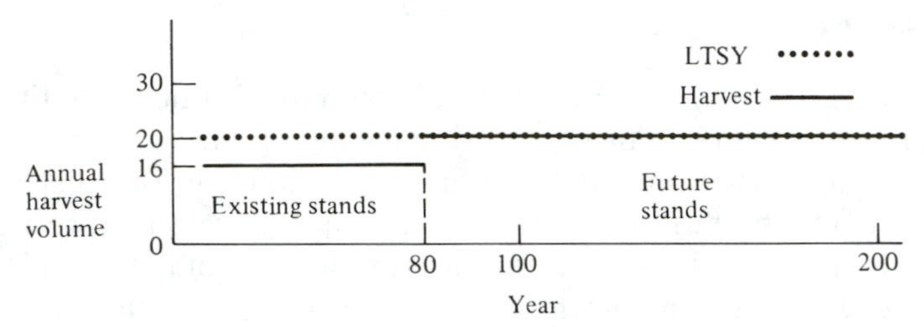

12.5 percent of forest area and volume is cut in the first decade

(*b*)

Figure 16-6 Using rotations short of culmination in future stands when there is a shortage of existing timber. (*a*) To increase the harvest over that of the base harvest policy. (*b*) To maintain the harvest of the base harvest policy on a smaller land base.

bd ft/year [(160 acres ÷ 100 years) × 8 M bd ft/acre = 12.8 M bd ft/acre] and the long-term sustained yield will fall from 30 to 24 M bd ft/year [(160 acres ÷ 100 years) × 15 M bd ft/acre = 24 M bd ft/year].

Suppose that in addition we now allow the harvest of future stands at age 80 rather than at age 100 on the reduced land base. This will enable the starting nondeclining yield level to return to 16 M bd ft/year [(160 acres ÷ 80 years) × 8 M bd ft/acre = 16 M bd ft/year] with a harvest of 2 acres/year [(16 M bd ft/year ÷ 8 M bd ft/acre) = 2 acres/year]. After 80 years the harvest will climb to the long-term sustained yield of 20 M bd ft/year (2 acres/year × 10 M bd ft/acre = 20 M bd ft/year) (Fig. 16-6*b*).

We have attained the starting nondeclining yield level of the base harvest policy with 20 percent fewer acres by allowing for a minimum rotation for future stands at 80 years instead of 100 years. The long-term sustained yield will fall by 33 percent, but the 16 M bd ft nondeclining yield harvest level under the base harvest policy can be maintained.

Management Intensification

Management intensification includes any cultural treatments such as genetic improvement, improved regeneration measures, brush release, precommercial thinning, commercial thinning, and fertilization that increase the yield or make it occur sooner, beyond those already included in the harvest schedule being examined. Public agencies traditionally have reacted to requests for a higher harvest with the idea that if management can be intensified, a higher harvest will be possible. Management intensification can also be used, however, to reduce the land base needed to maintain the harvest found under the base harvest policy. Both uses of management intensification are explored below, with the emphasis on intensifying management in the managed stands of the future.

Management intensification to increase the harvest Suppose that we intensified management of the forest with a surplus of existing timber in a way that increased the future stand yield at age 100 from 15 to 18 M bd ft/acre. Now the long-term sustained yield will equal 36 M bd ft/year [(200 acres ÷ 100 years) × 18 M bd ft/acre = 36 M bd ft/year] rather than 30 M bd ft/year [(200 acres ÷ 100 years) × 15 M bd ft/acre = 30 M bd ft/year] as discussed earlier. Under the base harvest policy, the maximum harvest will increase from 30 to 36 M bd ft/year, and the existing timber will last for 111 years [(200 acres ÷ 36 M bd ft/year) × 20 M bd ft/acre = 111 years] instead of 133 years, as occurred previously (Fig. 16-7*a*). Such an immediate increase in harvest under the base harvest policy made possible through a commitment to more investment in future stands often is called the allowable cut effect (ACE).

Assume that the existing stands are harvested over 100 years rather than 111 years under this higher management intensity through a departure. They will now be cut at a rate of 40 M bd ft/year [(200 acres ÷ 100 years) × 20 M bd ft/acre = 40 M bd ft/year]. In year 101 the harvest will then drop to 36 M bd ft/year and continue at that rate forever. Now the volume lost under the base harvest policy, as compared to a departure that cuts existing stands over 100 years, equals (approximately) 400 M bd ft (4 M bd ft/year × 100 years = 400 M bd ft) (Fig. 16-7*b*).

By increasing the management intensity, we increase the LTSY, enabling an increase in harvest under the base harvest policy and a reduction in the amount of "surplus" volume lost through that policy. If management intensification increases the predicted future stand yield to 20 M bd ft/acre, a volume equal to that in existing stands, the surplus will disappear altogether as the periodic harvest under the base policy rises to 40 M bd ft/year. If the predicted future stand yield rises above 20 M bd ft/acre, the harvest pattern will switch from that of a surplus of existing timber (harvest stands at LTSY, and it takes longer than one rotation to cut existing stands at that rate) to that of a shortage of existing timber (nondeclining yield level starts below LTSY).

With the shortage of existing timber condition shown in our original example (Fig. 16-2*b*), increasing future stand yield at age 100 from 15 to 18

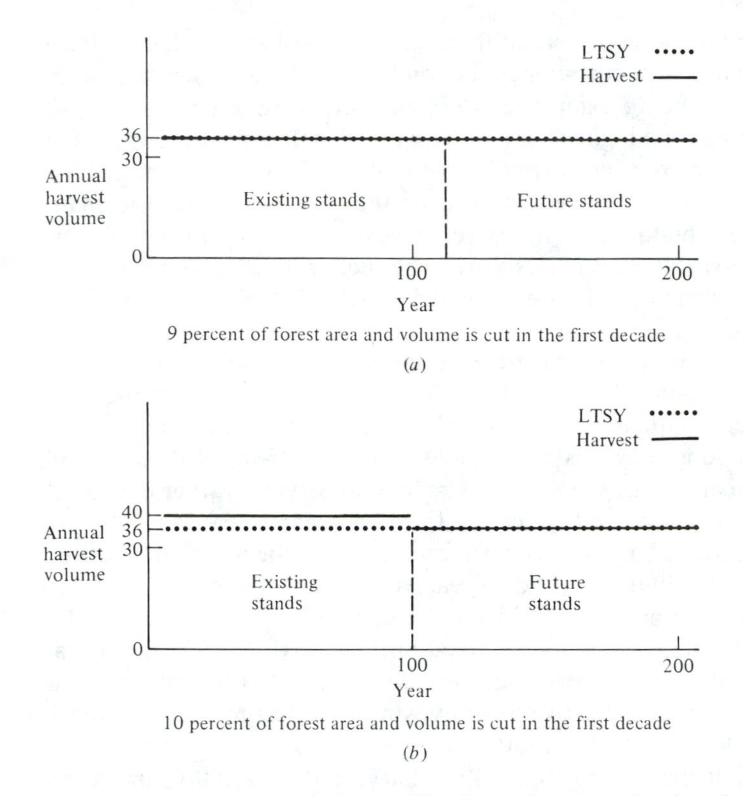

Figure 16-7 Using management intensification in future stands when there is a surplus of existing timber. (*a*) To increase the harvest under the base harvest policy. (*b*) To reduce the surplus available through a departure.

M bd ft/year with minimum harvest age still set at 100, as done previously to increase the harvest under the surplus condition, will not increase the starting nondeclining yield. Rather, increases in management intensity that make available more young-growth volume at earlier ages, such as through thinnings, or through cultural treatments lowering the CMAI age, will enable an increase in the starting nondeclining yield level. Suppose that one of our management intensities for managed stands culminates at 80 years old with a volume of 10 M bd ft per acre (a situation similar to that shown in Fig. 16-6*a*, but now with the stand culminating at age 80). Instead of cutting the old growth over 100 years at a harvest of 16 M bd ft per year, the existing stands will only need to last 80 years, enabling the starting nondeclining yield to jump from 16 to 20 M bd ft per year. The long-term sustained yield will decline to 25 M bd ft/year [(200 acres ÷ 80 years) × 10 M bd ft/acre = 25 M bd ft/year].

To look at another example of management intensification with a shortage of existing timber, suppose that one of our management intensities for future stands culminates at age 100 with a volume of 15 M bd ft per acre as before, but now that volume is removed by an intermediate harvest of 1.6 M bd ft per acre at ages 60, 70, 80, and 90 and a final harvest of 8.6 M bd ft per acre at 100. Starting at year 61, thinnings from stands planted as the old growth was harvested can replace some of the old-growth volume being held until these later years, allowing the old-growth volume to be cut sooner. Under the above yield schedule, harvest under the base harvest policy can start at 20 M bd ft per year (Fig. 16-8) instead of the 16 M bd ft per year of the original management intensity (Fig. 16-2*b*).

Management intensification to reduce the land base needed to maintain the harvest With a surplus of existing timber, management intensification can enable a higher long-term sustained yield and thus a higher first-period harvest. Rather than to increase the harvest and long term sustained yield 20 percent (30 to 36 M bd ft/year) under the base harvest policy through management intensification, which increases the yield at age 100 from 15 to 18 M bd ft per acre, we could reduce the land base to 30 ÷ 36, or 83 percent (200 acres × 0.83 = 166.7 acres) of its original size and maintain the harvest and long-term sustained yield at 30 M bd ft per year.

With a shortage of existing timber, management intensification that enables shorter rotations or intermediate harvests in managed stands will permit a higher starting nondeclining yield level under the base harvest policy. Rather than increasing the harvest 25 percent (16 to 20 M bd ft/year) through management intensification as discussed above for a shortage of existing timber, we could reduce the land to 16/20, or 80 percent of its original size (200 × 0.8 = 160 acres) and maintain the starting nondeclining yield level of 16 units per year found previously under the base

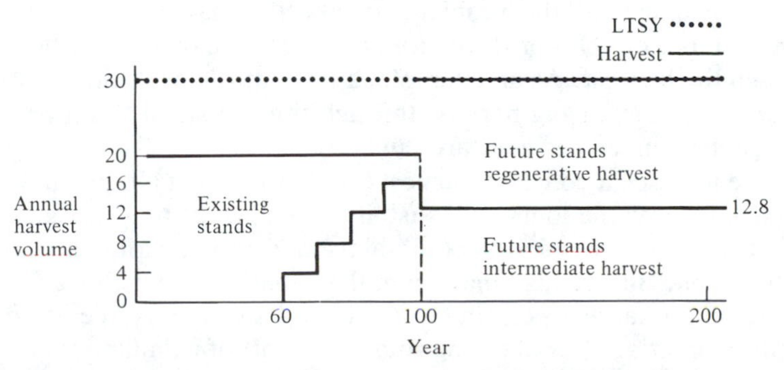

12.5 percent of forest area and volume is cut in the first decade

Figure 16-8 Using management intensification (commercial thinning) in future stands to increase the harvest under the base harvest policy when there is a shortage of existing timber.

harvest policy. The harvest would eventually climb to a long-term sustained yield of 24, which is 20 percent lower than that under the original land base.

Management intensification in existing stands With the existing stands in our example being wholly comprised of old growth, it is not surprising that we have concentrated here on understanding the volume flow effects of management intensification in future stands. Except for fertilization, and perhaps salvage, little management intensification is undertaken in old growth. Often, though, the existing timber is composed of a mixture of old and young growth. How would intensification in this young growth, or increased fertilization and salvage in the old growth, affect our results?

With a surplus of existing timber, management intensification in existing stands will not affect the results under the base policy because there is too much existing timber already. It will, however, increase the volume available through a departure.

With a shortage of existing timber, management intensification in existing stands will provide more volume when it is needed in much the same way as commercial thinning in future stands (Fig. 16-8). Therefore the intensification will increase the volume available in the near future.

Additional Harvest Scheduling Patterns

Two additional harvest scheduling patterns are commonly seen in timber management planning (Fig. 16-9). First, we may see the harvest stairstep up to the long-term sustained yield during harvest of the existing timber or harvest of the future stands, rather than to jump up in one step to the long-term sustained yield once all the existing timber has been cut (Fig. 16-9*a*). This result can occur for a number of reasons. It may not best meet the objectives specified for the problem to take the maximum harvest permitted in each period. Or constraints other than those on harvest flow may hold down the harvest. As an example, area control constraints may be used to direct an increasing harvest through time as stands with progressively higher volumes per acre are cut.

Second, we may see a constant harvest for all, or almost all, planning periods at a level below the long-term sustained yield (Fig. 16-9*b*). Again it may be that the objectives or the constraints cause the maximum harvest not to be taken. But we also may be in the situation in which we are taking the maximum harvest permitted under the base policy. We often see this result in practice after allowing rotations short of culmination, on a forest that has a shortage of existing timber, to the degree that the minimum rotation age is no longer a binding constraint on the harvest of the existing timber. Toward the end of planning periods, however, longer rotations closer to culmination begin to be used for one reason or another,

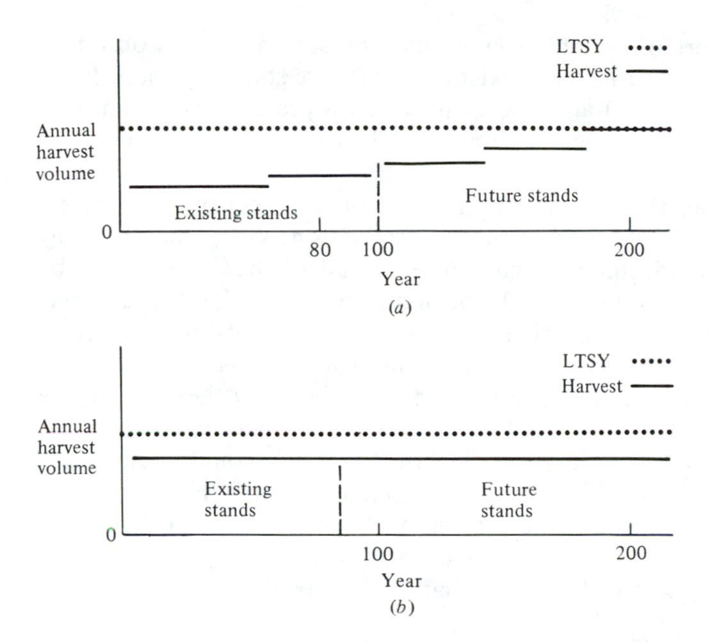

Figure 16-9 Additional types of harvest schedules. (*a*) A harvest schedule that stairsteps up to long-term sustained yield. (*b*) A harvest schedule that is constant for all planning periods at a level below the long-term sustained yield.

which keeps the long-term sustained yield up. In the kind of management intensification pattern shown in Fig. 16-9*b*, further management intensification to increase the harvest must simultaneously provide more volume and provide it sooner.

Simultaneous Allocation and Scheduling

Until now, we have set the amount of land that would be allocated to timber production before determining the harvest schedule that would result under different conditions. In much recent analysis, especially on public lands, determination of how much land to allocate to timber production is made simultaneously with determination of the harvest schedule. This more complicated analysis is difficult to do in our simple example, but we can point out a few of the commonly found results.

To explain this simultaneous allocation and scheduling twist, let us use the same example forest with which we have been working, but with the following refinements:

1. We specifically wish to maximize the present net worth over time from actions on the forest.

2. Three-quarters of the old growth has a present net worth on a per-acre basis (considering both existing and future stands) greater than 0 (supermarginal land) and one quarter has a present net worth on a per-acre basis less than 0 (submarginal land).

Without a harvest flow constraint, we would expect only the supermarginal land to be allocated to timber production. With a harvest flow constraint like nondeclining yield, however, all of the forest might be allocated to timber production. The benefits from the higher harvest level due to including the submarginal land could easily outweigh the costs from having eventually to harvest that land. This is especially true since the supermarginal land could be harvested in the early periods and the submarginal land harvested in the later periods. Thus the "benefits" could be obtained immediately, while the "costs" would occur much later. With discounting, this time span between obtaining the benefits and incurring the costs could make inclusion of the submarginal land look very good indeed—even though such a result would mean that some time in the future we would be faced with harvesting stands that, on their own merits, lose money!

Placing constraints like nondeclining yield on a forest planning problem can, understandably, affect the allocation when the allocation is determined simultaneously with the harvest schedule. In our problem, a nondeclining yield constraint could cause more land to be allocated to timber production, and perhaps away from other uses such as wilderness. Preservation groups often push for a nondeclining yield constraint because they feel that it will hold down the near-term harvest. With the linkage that the nondeclining yield constraint sets up between the harvest level now and the land allocated to timber production, however, these groups may wish to examine more closely their advocacy of nondeclining yield.

THE ALLOWABLE CUT EFFECT

The allowable cut effect (ACE) is an old idea recently quantified in detail through harvest scheduling models. It can generally be defined as the increase in early harvest from an existing forest which results from changed assumptions about the productivity of the future forest. The word allowable comes from the U.S. Forest Service's allowable cut analysis which calculates the maximum amount of timber that can be offered for sale each year.

We have already seen examples of the allowable cut effect in our discussion on the use of management intensification to increase the harvest or decrease the land needed to maintain the harvest. When we assumed that the future stands would have a higher and/or earlier yield, we saw that current harvest could be increased while still respecting the base harvest policy. This is not an unreasonable or uncommon possibility to consider. Every time research discovers productivity-enhancing technology, it allows current activity to change in anticipation. When any of us uses the argument that our expectations of future yields or income have risen as the rationalization for increasing current consumption, we are invoking a variation of the allowable cut effect. When the student runs out and buys a new car in anticipation of the job expected on graduation, this is ACE. The only difference is that credit allows us to consume now when we do not have any capital (the old-growth equivalent). Buy now, pay later!

The conditions for the timber allowable cut effect were defined in Schweitzer et al. (1972) as follows:

1. The allowable cut is calculated.
2. The allowable cut is actually sold and cut.
3. A volume control harvest policy is in force.
4. Ample merchantable timber is available in existing stands to support increased harvests.

Many forest owners fit the first three conditions (although to differing degrees), and many can meet the fourth requirement. If decisions about the amount of timber harvest are made in this context, then there are several ways the amount of harvest might be increased by changing the assumptions about the productivity of the future forest. Three of these ways are discussed next: (1) management intensification, (2) improved utilization, and (3) forest combination.

Management Intensification

We have already worked out the mechanics of how a commitment to management intensification in future stands can allow an increase in current harvest under a harvest flow constraint. It is worth noting that many of these cases involve only changed assumptions about treatments or activities that will take place in the future and for which it is further assumed that money will be available as needed. For instance, precommercial thinning of regenerated stands is critical to productivity in regenerated lodgepole pine and naturally regenerated mixed conifers. Most of it, however, will not take place until many years after the plan has been implemented. Other activities, such as site preparation and reforestation of brushfields or understocked stands, may occur in the near future. All can result in a projection of increased yield from future stands.

Such ability to raise the timber harvest immediately in anticipation of higher future yields was met with great enthusiasm by public agencies and some private firms. Now a way existed to have both nondeclining yield and a high harvest level. And investments in intensive management, something dear to the heart of most foresters, became easier to justify.

To see why the case for intensive management is easier to make with the allowable cut effect, let us look again at management intensification where there is a surplus of existing timber (Fig. 16-7*a*). In that case, management intensification raised the volume at culmination from 15 to 18 M bd ft per acre and allowed the cut under the base harvest policy to increase from 30 to 36 M bd ft per year. Assume that our objective is to maximize the present net worth at a 4 percent interest rate under the base harvest policy, that we need to invest an extra $50 per acre at regeneration to produce an extra 3 M bd ft at rotation age 100, and that timber is worth $250 per thousand board feet. Assuming that we will make this increased investment each time we regenerate the acre forever, we can compare the traditional per-acre analysis using the soil expectation approach with that using the allowable cut effect.

Traditional per-acre analysis If we invest $50 at year 0 to produce an additional 3 M bd ft worth $250 each at age 100, the soil expectation value of the investment is

$$\frac{3(250) - 50(1.04)^{100}}{(1.04)^{100} - 1} = \frac{750 - 2525}{49.5} = \frac{-1775}{49.5} = \$-36$$

Conclusion. The investment does not pay since it will make the present net worth decline by $36 per acre.

Allowable cut effect analysis If we make this investment on a surplus forest, however, we can immediately raise the harvest under the base harvest policy by 3 M bd ft $\div$ 100 year rotation = 0.03 M bd ft per year, or 0.03 $\times$ $250 = $7.50 per year forever.

The present value of an increased $7.50 per year forever at a 4 percent interest rate is 7.50 $\div$ 0.04 = $187.50. The present value of $50 invested every 100 years forever is $[50(1.04)^{100}] \div [(1.04)^{100} - 1] = 2{,}525 \div 49.5 =$ $51, giving a net return of $187.5 − $51 = $136.5.

Conclusion. The investment does pay since it will make the present net worth increase by $136.5 per acre.

By switching from our traditional per-acre analysis to one that includes the allowable cut effect, we have changed a poor investment (PNW = $−36.0) into a good one (PNW = $136.5). No wonder foresters were happy to discover the allowable cut effect; they could now justify on economic grounds the investments that they wanted to make.

Economists were not so happy (Teeguarden, 1973; Walker, 1977). They pointed out that the investment analysis that includes the allowable cut effect replaced the actual return from the investment ($750 every 100 years) with the return from relaxing the harvest flow constraint ($7.50 every year forever) and that what the analysis really showed was how costly (and foolish) this constraint was. True enough, through a departure from the harvest flow constraint in the base policy, it is possible to harvest more timber in the near future without the investment (Fig. 16-3), and it is probable that this departure would have a higher present net worth than the schedule that includes the intensive management (Fig. 16-7*a*). Now it is also true that the departures all show a decline of one form or another in the future. But is the ''cost'' of the decline greater than the ''cost'' of the harvest flow constraint? Is there value to projecting a stable harvest flow in a dynamic economy? We cannot settle that argument here, but it does illustrate that justifying management intensification through the allowable cut effect can be trickier than it first seems.

In addition, these calculations bring up another issue. The soil expectation calculation done for the traditional per-acre analysis is a continuation of the stand decision tools covered in Chap. 13. We used that analysis to find present net worth–maximizing prescriptions when considering an acre in isolation. Now we are considering an acre when its outputs can affect actions on other acres through the harvest flow constraint. And we find that what would reduce the present net worth of a forest when each of its acres is considered alone could increase the present net worth of the forest when the acres are bound together by constraints. By ameliorating the effect of the constraints on attainment of the forestwide objective (present net worth), the proposed investment is changed from seeming to be a bad idea to seeming to be a good one.

Often you will find that the ''best'' prescription for an individual acre will not appear as the best prescription when constraints bind the actions of the forest together—whether these constraints are on maximum harvest, maximum budget, or minimum owls. Making sure that prescriptions that are ''best'' when considering forestwide effects are included in the analysis is an art. It requires much study of the harvest scheduling solution and the associated shadow prices and reduced costs—especially since the ''best'' prescription for an acre can change as each constraint is added or removed or its constraining value is changed. It also means that soil expectation calculations have but limited value to decide what prescriptions to include in a forest-wide analysis that has a present net worth objective. This uncertainty about what prescription will be best suggests a strategy of providing the scheduling model a broad menu of feasible prescriptions to choose from for each stand type and management unit.

Improved Utilization

In the mid-1970s many public agencies and private firms in the west switched from calculating their allowable cuts in board feet to calculating them in cubic feet, with a resulting increase in the board-foot cut under harvest flow constraints. The number of board feet per cubic foot increases with tree size. In addition, cubic-foot measures may recognize volume and growth in smaller trees more than do board-foot measures. The combined effect, when harvest is calculated in cubic feet and the average tree size will decrease over time, is a higher current harvest in board-foot measure. Therefore by calculating volumes in cubic feet rather than board feet, the apparent extent of the forest resource is increased.

Table 16-1 gives a simple example of the effect on an even-flow allowable cut of switching from board feet to cubic feet. This table (lines 2 and 3) shows that more cubic volume will have to be cut from the smaller trees of the future to recover the same volume in board feet as produced now. It also shows (lines 2, 4, and 5) that the allowable cut to obtain an even flow in cubic feet (183) results in the board-foot cut for the first period increasing from 1,000 to 1,098 and the board-foot cut in the future decreasing from 1,000 to 915.

Fight and Schweitzer (1974) demonstrate the allowable cut effect of changing utilization standards on the Columbia master unit in western Oregon administered by the Bureau of Land Management. They calcu-

Table 16-1 Example of the effect on the even flow harvest level due to changing the standard for measuring timber volumes from board feet to cubic feet

Item	Harvest now (average tree is large)	Harvest in the future (average tree is small)
1. Average number of board feet estimated for each cubic foot of tree volume	6	5
2. Assumed perpetual allowable harvest level under even flow constraint when board feet is used	1,000	1,000
3. Number of cubic feet that would be harvested (line 2 ÷ line 1)	167	200
4. Perpetual allowable harvest level under even flow constraint when cubic feet is used; approximated by (167 + 200) ÷ 2	183	183
5. Equivalent harvests in board feet (line 4 × line 1)	1,098	915

Source: Fight and Schweitzer (1974).

lated an even flow allowable cut using the international $\frac{1}{8}$-in. log rule and then using cubic feet, measuring the same trees to the same top limit. The results were then expressed in both board feet and cubic feet and graphed for the 40-decade projection period (Fig. 16-10).

The board-foot harvest on the Columbia master unit could be immediately increased by slightly more than 9 percent if an even flow allowable cut were calculated in cubic feet instead of board feet (Fig. 16-10*a*). Over time, the board feet in the cubic-foot even flow allowable cut would gradually decline until reaching an ultimate level about 9 percent below the board-foot even flow level. This decline results because a shrinking proportion of the wood can be recovered as lumber as the average diameter decreases. Further, under the cubic-foot cut, stands ultimately are har-

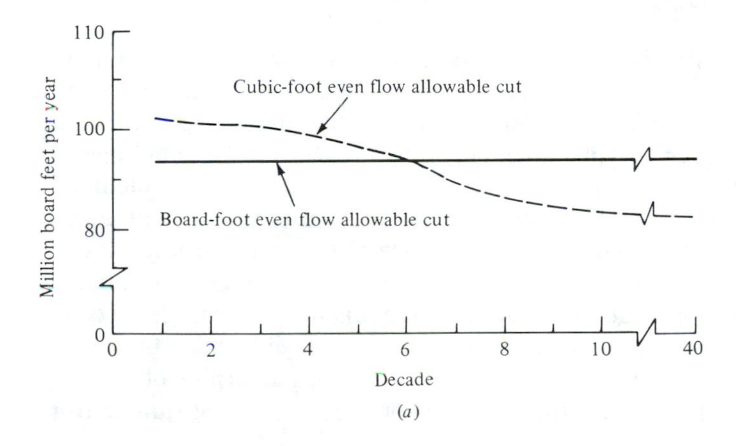

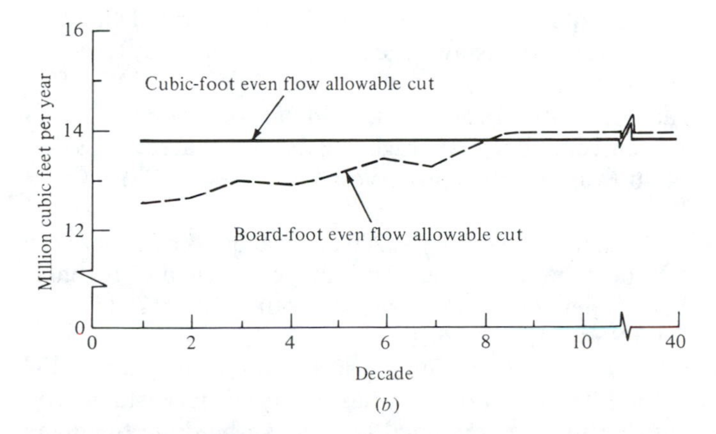

Figure 16-10 Allowable cut on the Columbia master unit of the Bureau of Land Management in western Oregon. (*a*) Board-foot and cubic-foot allowable cuts when both are expressed in board feet. (*b*) Board-foot and cubic-foot allowable cuts when both are expressed in cubic feet. *(From Fight and Schweitzer, 1974.)*

vested near culmination of mean annual cubic-foot volume growth, which produces less board feet than the maximum board-foot growth.

On the other hand, measuring both even flow allowable cuts in cubic feet shows an increasing cubic-foot component of the board-foot harvest over time (Fig. 16-10*b*).

More generally, utilization standards that count the highest proportion of wood grown in small trees relative to large trees will lead to the highest levels of allowable cut when the average diameter of the harvest will decrease over time. Such changes in the utilization standard reflect a view that ways have been found or will be found to use more and more of the small trees of the future.

Forest Combination

Another way to increase the allowable cut is to combine two or more forests to create one larger forest for purposes of calculating the allowable cut. This can be done administratively by actually purchasing more land, or simply on paper for planning purposes. However the combination takes place, the effect is almost always that the allowable cut calculated for the combined properties is higher than the sum of the allowable cuts for the forests considered separately. Since the owners do not have to change any of the on-ground activities or invest in intensified management, this impact is about as close to a free lunch as we can get. To see how this result occurs, consider two forests managed under the previously mentioned base harvest policy where one has a surplus of existing timber (forest *A*) and the other has a shortage of existing timber (forest *B*).

Forest A. 200 acres; rotation is 50 years; yield of regenerated stands is 20 M bd ft per acre; current stands average 30 M bd ft per acre; LTSY = $(200 \div 50) \times 20 = 80$ M bd ft; current harvest = LTSY = 80 M bd ft.

Forest B. 100 acres; rotation is 50 years; yield of regenerated stands is 20 M bd ft per acre; current stands average 5 M bd ft per acre; LTSY = $(100 \div 50) \times 20 = 40$ M bd ft; current harvest = 2 acres × 5 M bd ft = 10 M bd ft.

Forests A + B. 300 acres, inventory is 200 acres at 30 M bd ft per acre and 100 acres at 5 M bd ft per acre; rotation is 50 years; yield of regenerated stands is 20 M bd ft per acre; LTSY = $(300 \div 50) \times 20 = 120$ M bd ft; current harvest = LTSY = 120 M bd ft.

After combining forests, the total long-term sustained yield is now the upper limit on the total harvest. Forest *A* has plenty of harvestable volume, and all the cutting for the combined forests can be done for many years in its 30 M bd ft per acre stands. The total harvest from the two forests is increased from 90 to 120 M bd ft, a 33 percent boost.

Such an effect on allowable cut occurred when the Shelton cooperative sustained yield unit, created in 1946, brought under joint management 111,000 acres of the Olympic National Forest and 226,000 acres of industrial forestland owned by Simpson Timber Company in northwest Washington (Hrubes, 1976). The federal land was predominantly old growth, and the Simpson land was predominantly cutover. Simpson would have faced certain shutdown of its mills in the area without the additional stumpage provided by the agreement.

The joint allowable cut was eventually set at 135 million bd ft, while the annual allowable cut levels on the national forest and Simpson lands, if calculated independently, would have been 60 and 30 million bd ft, respectively. Therefore the creation of the Shelton unit generated an allowable cut effect of 50 percent.

On the other hand, combining forests with similar age structures often will result in insignificant allowable cut effects. Combining three national forests in eastern Oregon would result in a total increase in the first-decade harvest of only 4 percent (Table 16-2).

This latter example displays one of the potential drawbacks of combining forests to calculate the allowable cut: the gains in harvest come at the expense of harvest stability in the components. Here two of the forests (Malheur and Umatilla) in the combined calculation show a significant increase in first-decade harvest over the individual calculation, while the third (Wallowa-Whitman) shows a significant decrease. Then, in a few decades, this pattern will reverse. On these spatially separate forests such changes may prove disruptive to U.S. Forest Service personnel and the dependent communities.

The Shelton unit allowable cut also shows swings in the source of the harvest over time from the two components. By and large, the harvest was to come from U.S. Forest Service land until the young growth matured on private industry land. The intermingled ownership pattern of the two ownerships, however, meant that little spatial dislocation would be associated with these changes in source of the harvest.

Table 16-2 First-decade allowable cut for three national forests in eastern Oregon before and after they are combined

	Individual calculations		Combined calculation	
National forest	Million ft^3	Million bd ft	Million ft^3	Million bd ft
Malheur	342.38	2,248.65	479.13	3,156.18
Wallowa-Whitman	367.11	2,169.97	158.43	830.88
Umatilla	313.97	1,843.26	396.63	2,397.18
Total	1,023.46	6,261.88	1,034.19	6,384.24

Source: Hrubes (1976).

Legitimacy of the Allowable Cut Effect

The allowable cut effect owes its existence to the volume flow constraints on the harvest schedule. If these constraints restrict the harvest in the near future, we would expect that some methods of increasing this early harvest using the allowable cut effect will be suggested. Before the methods are accepted though, a number of questions need to be asked. Does the decision maker understand the cost of the harvest flow constraints? Is there a way to meet the objectives represented by the constraints other than through the allowable cut linkage?

Perhaps the constraints on harvest flow occur because the property supplies timber to the company sawmill. If wood can be purchased on the open market for less than the cost of any needed investments to increase forest productivity, such an option should be seriously considered.

If the decision maker understands the costs of these constraints, and alternatives to meet them other than through the allowable cut effect have been analyzed and rejected, he or she must decide whether to use the allowable cut effect to justify an increase in harvest.

To help make this decision, let us group the actions producing the allowable cut effect into two classes:

1. *Class I.* Actions such as reforestation which are implemented in the current year or are funded in approved short-term budgets
2. *Class II.* Actions such as precommercial thinning that are planned or assumed to take place in the future

Class I activities are at least sure things; they have already taken place or are paid for, and we know that their effect on yield is a reality. Including the allowable cut effect of these activities into planning could hardly be questioned unless the actions are a policy (like area combinations) which could be easily undone by future policymakers.

Class II activities, which are the majority, are innately suspect. They can range from serious good-faith proposals to mere speculation or wishful thinking. Implementation is just starting, and little or no money is in hand to guarantee that they will be implemented. Counting on class II activities to take place is an act of faith, and responsible planners should at least look hard at the likelihood that the owner will carry through.

For public agencies, the issue surrounding class II activities is pretty much a question of how much funding is implied for future implementation and how likely is it to be forthcoming from appropriations or retained timber sale revenue. The whims of politics can be guessed at, particularly regarding legislation preventing or requiring certain activities, such as herbicides or buffer strips along streams, but this is always difficult.

With private companies and individual owners, their financial prospects, past track record of intensive management, and availability of future investment capital from internal and outside sources would be the key to confidence in the reality of class II activities.

In the end, all you can do is to look back at past performance, contemplate current leadership and trends, and make your own forecasts. It is a fine line that separates responsible forest planning from simply playing the "harvest scheduling game" with a hidden agenda of maximizing (or minimizing) early-period harvest.

TYPE *A* AND TYPE *B* PROBLEMS REVISITED

In Chaps. 6 and 8 we developed the type *A* and type *B* classification of processes for developing alternative problem solutions. The type *A* situations occur when a single unique specification of goals and constraints can be made, the feasible region defined, and the optimum solution found. Type *B* situations occur, by contrast, when the owner cannot initiate problem solving with a unique set of goals and constraints, and several different sets are defined to generate the different alternatives. The optimum solution to each alternative set of goals and constraints can be found, but the real planning question or issue centers on which of the many alternative problem specifications is the correct or best one.

As we move through harvest scheduling with the Daniel Pickett problem and consider the possible variations in forest output schedules obtained by changing problem assumptions, it is clear that most forest management problems are approached as type *B* problem situations. From initial specification, mathematical programming is used to find the optimum solution. Then assumptions and constraints are changed and the problem is again solved.

If you are heading a type *B* analysis (or evaluating the work of another analyst who is), be aware that this is what you are doing. Caution should be used in attaching meaning to differences in objective function values and other information when comparing the solutions to different alternatives.

Many important and valued forest outputs, such as wildlife or harvest stability, are often modeled by constraint and not explicitly valued in the objective function because acceptable prices cannot be derived or agreed upon. In this case a timber-valued objective function can at best show opportunity cost information for changing levels of the output modeled by constraint as measured by timber output foregone. Moreover, if a whole package of constraints is changed between alternatives, then the opportunity cost usually cannot be assigned to any one constraint.

Because (or if) the objective function for a public forest planning problem does not include the benefits and costs of all forest outputs, it is likewise incorrect to use it directly to compare different solutions or alternatives. If benefit of wildlife is not considered in the objective function, for example, it is misleading and conceptually incorrect to look at calculated present net worth values from the problem solution and say that the low wildlife alternative is more economically efficient. In terms of total net benefit, a higher wildlife alternative may provide greater total net benefit; it is just that subjective judgments are needed to make the valuation.

Private corporations face the same issue when comparing solutions to problems when the objective function only reports the present net worth of timber harvested while the alternatives vary by different levels of volume control constraints. Here the strategic values of future wood supplies are not explicitly priced and must be subjectively estimated.

In practice, the models for timber management scheduling optimization discussed in this book are a set of tools for exploring the implications of different assumptions and policies for the management of a forest resource and mapping out these options for the owners. The good analyst will be a diligent and comprehensive explorer, make sure everyone knows what assumptions are going into the model, and leave the needed subjective value judgments to the owners.

QUESTIONS

16-1 An inventory of a 200-acre forest and an analysis of prescription possibilities for regenerated stands provided the yield and economic information tabled below.

Questions (*a*) to (*c*) assume stands *A* and *B* use uneven-aged prescriptions.

(*a*) If the combined forest of existing stands *A* and *B* were managed under uneven-aged prescription 1, how many acres would be cut each year?

(*b*) Given the available yield data and assuming the highest volume stands are entered first under area control, what is the best estimate of forest harvest under uneven-aged prescription 1 from stands *A* and *B* for the years 1, 4, 6, 11? (In year 1, the stands have grown 1 year, etc.)

(*c*) Is the same uneven-aged prescription economically optimal for both stands *A* and *B*?

Questions (*d*) to (*k*) assume that all stands use even-aged prescriptions.

(*d*) For a goal of maximum physical productivity, which of the three even-aged prescriptions is best?

(*e*) By the goal of economic efficiency, which of the three even-aged prescriptions is best?

(*f*) What is the long-term sustained yield of the forest under prescription 3?

(*g*) Two possible strategies are to manage stands (1) all under prescription 2 or (2) all under prescription 3. For each strategy calculate the harvest volumes to be received in the years 1, 5, 15, 25, 35, 45, 65. Use area regulation and harvest the oldest stands first. (Existing stands have grown 1 year in year 1, 5 years in year 5, etc.)

(*h*) Do the harvest schedules in your answer to question (*g*) appear to be *deficit* or *surplus* with respect to a base harvest schedule of nondeclining yield and harvest ≤ LTSY? Explain.

Forest data, existing stands

	Stand type	Area (acres)	Age (years)	Volume per acre (M bd ft)	Growth per acre per year (M bd ft)	Character
10	A*	50	150	100	−1.0	Old growth
10	B*	50	30	30	+0.4	Young, natural
20	C*	100	10	0	See yield tables	New, planted

* All existing stands have the identical site quality.

Harvest yield data for future stand prescriptions (thousand board feet)

Prescription options	Years after planting (even aged)							
	0	10	20	30	40	50	60	
Even-aged								
1. Clearcut, plant, regeneration harvest at 40	0	0	0	0	50	0	0	
2. Clearcut, plant, regeneration harvest at 50	0	0	0	0	0	70	0	
3. Clearcut, plant, thin at 30, regeneration harvest at 50	0	0	0	15	0	50	0	
	Time in years from now							
	0	10	20	30	40	50	60	
Uneven-aged								
1. Cutting cycle 10 years, reserve growing stock 30 M bd ft								
Applied to stand *A*	70*	6	8	10	12	12	12	. . .
Applied to stand *B*	0*	8	12	12	12	12	12	. . .
2. Cutting cycle 20 years, reserve growing stock 20 M bd ft								
Applied to stand *A*	80*		12		18		20	. . .
Applied to stand *B*	10*		18		20		20	. . .

Economics

1. Costs
   a. Even-aged—regeneration $300/acre; annual $5/acre.
   b. Uneven-aged—annual $10/acre.
2. *Stumpage* (no inflation)
   a. Even-aged—$120/M bd ft.
   b. Uneven-aged—$180/M bd ft.
3. Guiding rate of interest, 4 percent.

* The harvest yield of the first uneven-aged entry at time 0 is equal to the existing inventory minus the reserve growing stock called for in the prescription. If entry is delayed past the first year, the volume of the existing inventory will change according to its growth rate. The sequence of subsequent harvest yields will remain the same, only delayed by the same amount the first entry is delayed.

(*i*) Assume the entire forest is regenerated by even-aged prescription 1 under a volume control harvest policy of (1) $H_t \leq 0.8$ LTSY and (2) $H_{t+1} \geq H_t$. What is the annual harvest in each of the first 5 years if the oldest stands are cut first?

(*j*) Suppose a new supergenetic clone is discovered that increases the *regenerated* yield of even-aged prescription 3 from 50 to 70 M bd ft at age 50. What is the allowable cut effect (ACE) of this clone discovery to the forest under the harvest policies of question (*i*)?

(*k*) Suppose the owner could buy 100 acres of nearby cutover and newly planted forest. It could be combined with the existing forest and managed under the policies of question (*i*) without using the improved genetic stock.

(1) What is the allowable cut effect of this proposed purchase?

(2) If the new forest area cost $600 per acre, should the owner buy it?

16-2 Assume that your forest has the following characteristics:

1. 400 acres of existing stands with a volume of 10 units per acre and showing no net growth
2. Growth of future stands: all stands culminate at age 50 with a volume of 20 units per acre
3. Harvest method: clearcutting with no intermediate harvest
4. Planning horizon: 100 years

Determine the maximum harvest.

Calculate and diagram the maximum harvest in the following situations.

(*a*) Under the base harvest policy.

(*b*) Under area control.

(*c*) Under a policy of cutting the maximum amount of timber in the first 10 years subject to the harvest never dropping below the starting nondeclining yield level.

(*d*) Under a policy of nondeclining yield at or below the long-term sustaining yield with rotations shortened to 40 years. At 40 years years the stand will yield 15 units per acre. Assume that you want to harvest the long-term sustained yield upon reaching of the planning horizon, which means that you should not recut 20 percent of the acres cut in each of the first 40 years to allow some stands to grow beyond 40 years of age.

16-3 Assume that your forest has the following characteristics:

1. 400 acres of existing stands with a volume of 30 units per acre and showing no net growth
2. Growth of future stands: all stands culminate at age 50 with a volume of 20 units per acre
3. Harvest method: clearcutting with no intermediate harvest
4. Planning horizon: 100 years

(*a*) Calculate and diagram the maximum harvest in the following situations.

(1) Under the base harvest policy.

(2) Under area control.

(3) Under a policy of cutting the maximum amount of timber in the first 10 years subject to the harvest after that never dropping below the long-term sustained yield.

(*b*) How much harvest is lost over 101 years by adhering to the base harvest policy?

(*c*) Do (1), (2), and (3) of question (*a*) when all stands culminate at age 50 with a volume of 30 units per acre.

REFERENCES

Fight, R. D., and D. L. Schweitzer, 1974: What If We Calculate the Allowable Cut in Cubic Feet? *J. Forestry*, **72:**87–89.

Hrubes, R. J., 1976: "National Forest Working Circles: A Question of Size and Ownership Composition," U.S. Forest Service General Technical Report PSW-16.

Johnson, K. N., 1981: "A Harvest Scheduling Primer," (mimeographed).

Schweitzer, D. L., R. W. Sassaman, and C. H. Schallau, 1972: Allowable Cut Effect: Some Physical and Economic Implications, *J. Forestry*, **70:**415–418.

Teeguarden, D. E., 1973: The Allowable Cut Effect: A Comment, *J. Forestry*, **71:**224–226.

Walker, J. F., 1977: Economic Efficiency and the National Forest Management Act of 1976, *J. Forestry*, **75:**715–718.

ANSWERS

CHAPTER 2

2-1 (*a*) The *stand* is a real, contiguous parcel of land that is all similar or homogeneous with respect to the physical, biological, and developmental attributes of land used to define a particular *stand type*. In management planning, the *stand* is the entity which is actually located and treated or harvested. The *stand type* is an abstraction, a classification concept used to build forecasting or predicting models—such as a yield model—to estimate the behavior or productivity of the land under different treatments and to generalize results to all stands of the same stand type.

(*b*) It would be a good classification if the response to treatment or disturbance by outputs that are important to the owner, such as timber or sediment, can be consistently and accurately predicted for each land class.

2-2 *Stand types*

$$(18 \text{ stand types}) \times (6 \text{ districts}) \times (4 \text{ prescriptions}) \times (5 \text{ initiation times}) = 2160 \text{ decision variables}$$

Nontimber types

$$(10 \text{ land types}) \times (6 \text{ districts}) \times (6 \text{ prescriptions}) = 360 \text{ decision variables}$$

$$\text{Total} = 2160 + 360 = 2520 \text{ decision variables}$$

2-3 C = categorical, I = in-place

(*a*) C: Timber yields are mostly a function of stand types.

(*b*) C, I: Categorical for sediment rates per unit area, in-place for proximity of harvest areas to streams to estimate sediment transport.

(*c*) C, I: Categorical to determine soil moisture, compaction, climate, etc. In-place to relate to existing road system and timber.

(*d*) C: Can be handled mostly on a stand-type basis since squirrels are not migratory.

(*e*) C, I: Bears move to different habitats with the seasons—proximity and spatial arrangements are important. Categorical to estimate habitat quality at each site.

(*f*) I: Depends mainly on the location of the road system and the adjacent logsheds.

(*g*) C: Like timber—forage production is estimated by acres of each stand type multiplied by the forage production per acre.

(*h*) C, I: Stand conditions and location are both important—stand types determine the size and amount of material that can be thinned; the relationship of stands to roads partly determines logging costs.

You can disagree on any of these first-cut answers depending on what you assume about the context of the issue or problem—much more detailed and thoughtful answers about specific situations are needed and encouraged.

CHAPTER 3

3-1 This is *net* growth since mortality is not added into the equation. Mortality is real growth that occurs between V_2 and V_1, and is lost because the trees that died are not counted in V_2.

3-2

Inventory results

Tree number	First inventory	Second inventory	Survivor growth	Mortality	Harvest cut	Ingrowth	Net growth
1	21	41	20				20
2	42	53	11				11
3	27	38	11				11
4	—	19				19	19
5	97	—		97			−97
6	86	—			86		
7	—	24				24	24
Total	273	175	42	97	86	43	−12
Growth estimates	(V_1)	(V_2)		(M)	(C)	(I)	
(1) Gross increment including ingrowth $V_2 + M + C - V_1$	−273	+175		+97	+86		= 85
(2) Gross increment of initial volume $V_2 + M + C - I - V_1$	−273	+175		+97	+86	−43	= 42
(3) Net increment including ingrowth $V_2 + C - V_1$	−273	+175			+86		= −12
(4) Net increment of initial volume $V_2 + C - I - V_1$	−273	+175			+86	−43	= −55
(5) Net increase $V_2 - V_1$	−273	+175					= −98

3-3

Stand age	Periodic annual net growth	Net yield	Mean annual net growth	Periodic annual mortality	Periodic annual gross growth	Gross yield	Mean annual gross growth
20		600	(30)			700	(35)
	45			20	65		
30		(1050)	35			(1350)	45
	(55)			25	(80)		
40		1600	40			2150	53.75
	15.25			(22.25)	(37.5)		
50		1752	35			2525	50.5

3-4

Year	Growing stock after harvest	Net periodic annual growth	Periodic volume harvested	Periodic mortality	Gross periodic annual growth
0	(2500)				
		(200)		200	(220)
10	4500		(0)		
		(250)		(300)	280
20	(1500)		5500		
		(300)		(100)	310
30	4500		(0)		
		(275)		(100)	285
40	5250		(2000)		

3-5 Mean annual increment is the *average* rate of growth over the life of a stand. For the average to *increase,* increments of growth must be at a rate higher than the current average. This means current or periodic increment (PAI) must be greater than (MAI) if MAI is rising. Conversely, for the average to decrease, the increments of growth must be at a rate less than average.

It follows that at the limit, when MAI is neither rising nor falling, periodic growth increments will be the same as the average (MAI = PAI). At that age MAI is at a maximum since it has ceased to rise and begins to fall. Mathematical proofs are also available.

3-6 On the better sites the trees *can* develop larger crowns and root systems. Hence fewer but much faster growing dominant trees occur.

3-7 (*a*) $\log N_i = \log k - aD_i \log e$

$$\log N_i = 1.9 - 0.07D_i$$

$$\log k = 1.9$$

$$k = 79.43$$

$$-a \log e = -0.07$$

$$a = \frac{0.07}{\log e} = \frac{0.07}{0.43429} = 0.16118$$

and, using Eq. (3-2),

$$N_i = 79.43 \exp(-0.16118D_i)$$

(*b*) $q = \exp(2a)$ [Eq. (3-3)]

$$= \exp[2\ (0.16118)]$$

$$= 1.38$$

3-8 *Step 1*. Use Eq. (3-6) to determine the number of trees in the largest (14 in.) size class in the residual stand (N_{max}).

DBH	b_i†	q	$(D_{max} - D_i)/w$	Number of trees‡	Basal area per acre
2	0.02	1.4	6	7.53	0.151
4	0.08	1.4	5	5.38	0.430
6	0.19	1.4	4	3.84	0.730
8	0.34	1.4	3	2.74	0.932
10	0.54	1.4	2	1.96	1.058
12	0.78	1.4	1	1.4	1.092
14	1.07	1.4	0	1	1.070
					5.463

† b_i = basal area per tree
‡ trees = $q^{(D_{max} - D_i)/w}$

$$N_{max} = \frac{200}{5.463} = 36.6$$

Step 2. Given N_{max} use Eq. (3-4) to determine the number of stems needed in each class for a basal area of 200.

DBH	Number of stems	b_i	Basal area per acre
2	275.7	0.02	5.5
4	196.9	0.08	15.8
6	140.5	0.19	26.7
8	100.3	0.34	34.1
10	71.8	0.54	38.8
12	51.3	0.78	40.0
14	36.6	1.07	39.2
			200.1

CHAPTER 4

4-1

Plot	Tree	Tree site (feet)	Average site (feet)
A	1	126	
	2	108	
	3	130	121.3
B	1	110	
	2	126	
	3	135	123.6
C	1	80	
	2	90	
	3	94	88

If all three plots represent the stand, the average site for the stand is 110.9 feet. We might investigate to determine the reasons plot C has a much lower site index rating.

4-2 The current stand at 3,500 ft^3 is fully stocked relative to the yields of Fig. 4-3. The growth of site 140 between 60 and 100 years is 15,200 − 10,000 = 5,200. The growth of site 100 between 60 and 100 years is 7,400 − 3,500 = 3,900.

(*a*) The *increase* in expected yield at age 100 is 5,200 − 3,900 = 1,300.

(*b*) The MAI at age 100 with fertilization is

$$\frac{3{,}500 + 5{,}200}{100} = \frac{8{,}700}{100} = 87$$

4-3 $n = 5$ trees, $a = \frac{1}{25}$ acre

D_i	D_i^2	BA
7	49	.2673
9	81	.4418
13	169	.9218
17	289	1.5763
20	400	2.1817
66	988	5.3889

(*a*) Average diameter = 66/5 = 13.2 in., quadratic mean diameter $\bar{d}_q = \sqrt{988/5} = \sqrt{197.6} = 14.06$ in.

(*b*) Stand basal area = 5.3889 × 25 = 134.72 ft².

(*c*)
$$\text{SDI} = N\left(\frac{\bar{d}_q}{10}\right)^{1.605} = 125\left(\frac{14.06}{10}\right)^{1.605} = 215.98.$$

(*d*) Relative density RD = $\text{BA}/\sqrt{\bar{d}_q} = 134.72/\sqrt{14.06} = 35.93$.

(*e*) CW = 5 + 1.3 DBH, $a = 0.04$ acres

Tree DBH	CW	$\left(\frac{\text{CW}}{2}\right)^2$	MCA/acre
7	14.1	49.70	3,903
9	16.7	69.72	5,476
13	21.9	119.90	9,417
17	27.1	183.60	14,420
20	31.0	240.25	18,869
			52,085

CCF = 52,085/43,560 × 100 = 119.6. The CCF says 119 percent of the space is occupied.

4-4 (*a*) 700 TPA = 43,560/700 = 62.23 ft²/tree; square spacing = $\sqrt{62.23} = 7.89$ ft.

(*b*) BA of a 12-in. tree = $0.005454D^2 = 0.005454(12)^2 = 0.7854$ ft².

$$\text{TPA} = 160/0.7854 = 203.7.$$

$$\text{Spacing} = \sqrt{\frac{43{,}560}{203.7}} = \sqrt{213.8} = 14.6 \text{ ft.}$$

4-5 Using Eq. (4-11)

$$C = \frac{15.4}{\sqrt{\text{BA}}} = \frac{15.4}{\sqrt{160}} = \frac{15.4}{12.65} = 1.22$$

"Diameter times" rule is spacing between trees = 1.22 D_i.

CHAPTER 5

5-1 Stand volume = 5,000 ft^3; stocking = 5,000/7,050 = 0.709; yield table growth 50 to 80 years = 11,350 − 7,050 = 4,300

(*a*) Proportionate growth

$$V_{80} = 5{,}000 + 0.709(4{,}300) = 8{,}048.7$$

(*b*) Captures all growth potential

$$V_{80} = 5{,}000 + 4{,}300 = 9{,}300$$

(*c*) Stocking at age 80 = 0.709 + 3(0.05) = 0.859

$$V_{80} = 0.859(11{,}350) = 9{,}750$$

Note the "rule of thumb" approach of (*c*) implies more growth (4,750) than the normal stand (4,300).

5-2

$$\ln V = 4 - \frac{100}{S} - \frac{12}{A} + 0.9 \ln \text{BA} \qquad (1)$$

$$\ln \text{BA}_2 = \frac{A_1}{A_2} \ln \text{BA}_1 + 5\left(1 - \frac{A_1}{A_2}\right) \qquad (2)$$

Substitute BA_2 from Eq. (2) into Eq. (1).

$$\ln V_2 = 4 - \frac{100}{S} - \frac{12}{A_2} + 0.9\left[\frac{A_1}{A_2} \ln \text{BA} + 5\left(1 - \frac{A_1}{A_2}\right)\right]$$

$$= 4 - \frac{100}{S} - \frac{12}{A_2} + 0.9\frac{A_1}{A_2} \ln \text{BA}_1 + 4.5 - 4.5\frac{A_1}{A_2}$$

$$= 8.5 - \frac{100}{S} - \frac{12}{A_2} + \frac{A_1}{A_2}(0.9 \ln \text{BA}_1 - 4.5)$$

5-3 $A_1 = 35$, $A_2 = 50$, $\text{BA}_1 = 130$, $S = 70$.

(*a*)
$$\log \text{CV}_2 = 1.52918 + 0.002875(70) - 6.15851\left(\frac{1}{50}\right) + 2.291143\left(1 - \frac{35}{50}\right) + 0.93112 \log 130\left(\frac{35}{50}\right)$$

$$= 1.52918 + 0.20125 - 0.12317 + 0.68734 + 1.37783$$

$$= 4.1187$$

$$\text{CV}_2 = 4{,}704$$

(*b*)
$$\log \text{CV}_1 = 1.52918 + 0.002875(70) - 6.15851\left(\frac{1}{35}\right) + 0.93112 \log B$$

$$= 1.52918 + 0.20125 - 0.17596 + 1.96833$$

$$= 3.5228$$

$$\text{CV}_1 = 3{,}333$$

$$\text{PAI} = \frac{\text{CV}_2 - \text{CV}_1}{15} = \frac{4{,}704 - 3{,}333}{15} = 91.4$$

5-4

Residual basal area	Product	Annual net growth	Price/ft³	Annual value growth	Total value growth
60	Saw	33.29	\$0.90	\$29.96	
	Poles	18.63	0.20	3.73	\$33.69
80	Saw	38.68	0.90	34.81	
	Poles	12.20	0.20	2.44	37.25
100	Saw	51.94	0.90	46.75	
	Poles	4.63	0.20	0.93	47.68

The basal area of 100 ft² has, substantially, the highest value growth at \$47.68/acre per year. This is due to the larger sawtimber component at the higher stocking levels, but the higher stocking also has a higher capital carrying cost to be considered before making a decision (Chap. 13).

5-5

DBH class midpoint	Starting number of trees	BA	Ingrowth	Upgrowth	Mortality	Ending number of trees
6	200	39.26	35.05	95.35	16.74	122.96
8	140	48.87		66.74	11.72	156.89
10	100	54.54		47.67	8.37	110.70
12						46.67
	440	142.67	35.05	209.76	36.83	437.22

5-6

		4-in. DBH Classes				
DBH class midpoint (inches)	Starting number of trees	Periodic DBH growth (inches)	M (%)	0 Class	1 Class	2 Class
2	20	3	75	0.25	0.75	0
6	12	4	100	0	1.0	0
10	6	3	75	0.25	0.75	0
14	2	1	25	0.75	0.25	0
18+	—	—	—			

DBH midpoint	Number of trees moving 0 Class	1 Class	Ending number of trees
2	5	15	5
6	0	12	15
10	1.5	4.5	13.5
14	1.5	0.5	6.0
18+	—	—	0.5

5-7 Critique:
Site The equation states BAG = $-0.2S$
This means if site increases in quality the growth will decrease. This is *not* biologically defendable.
Age BAG = $-0.07A + 0.0001A^2$. Solving for different values of A yields

A	Δ BAG
30	−2.01
60	−3.84
80	−4.96

This says the stand rate of growth diminishes as the stand gets older. This is biologically defendable.
Basal Area Density BAG = $0.04\text{BA} + 0.001\text{BA}^2$
This says growth is a continuously increasing function of BA. There is no peak growth at some finite density. *Not* biologically defendable. Calculating over the range 50–400 we see this.

BA	Δ BAG
50	4.5
100	14
200	48
400	176

5-8 This could be a very long list and discussion. In addition to questions of the model's accuracy and biological defendability as considered in question 5-7, some starting points to consider are the following

1. *Species* — Is the model for the same species genetic stock or species mix as the subject stand?
2. *Geographic similarity* — Was the data base for the model collected from similar soils, climates, latitude, and longitude?
3. *Past stand history* — Did the subject stand have a similar history of culture, diseases, site preparation, etc. as the stands of the yield model?
4. *Alternatives* — Is there an alternative growth model that better meets the criteria?

5. *Stand future*	Will the subject stand be managed the same as the stands used in the yield model? i.e., a "normal" model isn't too helpful for stands that will be heavily thinned.
6. *Merchantability and utilization standards*	Are the same standards for the subject used in the yield model or are conversions needed?
7. *Available data for the subject stand*	Inventory? Tree lists? Increment? The data variables currently existing or financially available for the subject stand must be the same as required to enter the yield model.

Given that the match between yield models and subject stands typically shows deficiencies on some or all of these counts, the question resolves to which is the best of the available models. In truth we do not have much quantitative information to evaluate the aggregate significance of these problems, so the model user must decide which issue is the most important to minimize.

CHAPTER 6

6-1 *Decision variables*

X_1 = number of beer and bean feeds, X_3 = number of trees sold,
X_2 = number of cords delivered, X_4 = miles of truck use rented.

Equations

Goal: Maximize $150X_1 + 100X_2 + 5X_3 - 0.5X_4$
Subject to the constraints:

Capital budget	$75X_1 + 30X_2 + 2X_3 + 0.5X_4 \leq 1{,}500$
Bean cooks	$5X_1 \leq 20$
Skilled labor	$5X_2 + 0.2X_3 \leq 100$
Grunt labor	$10X_2 + X_3 \leq 500$
Truck miles	$50X_2 + 2X_3 - X_4 \leq 500$
Contract feeds	$X_1 \geq 2$
Contract trees	$X_3 \geq 150$

Detached Coefficient Matrix

Row	Column (1) X_1	(2) X_2	(3) X_3	(4) X_4	Sign	(5) RHS
(1) Goal	150	100	5	−0.5		
(2) Budget	75	30	2	+0.5	≤	1,500
(3) Cooks	5				≤	20
(4) Skilled labor		5	0.2		≤	100
(5) Grunts		10	1.0		≤	500
(6) Miles		50	2	−1	≤	500
(7) Contract feeds	1				≥	2
(8) Contract trees			1		≥	150

6-2

a. Timber $300X_1 + 50X_2 + 800X_3 \geq 50{,}000$
b. Grazing $0.4X_4 + 0.1X_5 \geq 100$
c. Budget $2X_1 + 0.5X_2 + 10X_3 + 2X_4 + 0.1X_5 + 5X_6 + 0.5X_7 + 0.1X_8 + 20X_9 \leq 5{,}000$
d. Goal Maximize

$$58X_1 + 9.5X_2 + 150X_3 + 0X_4 + 0.4X_5 - 0.5X_6 - 0.4X_7 + 0X_8 + 55X_9$$

e. Balanced allocations $X_1 + X_2 + X_3 \geq 200$
$X_4 + X_5 + X_6 \geq 200$
$X_7 + X_8 + X_9 \geq 200$
f. Dollar output (acres to timber & range) $\geq$ 10 (acres recreation)
$(X_1 + X_2 + X_3 + X_4 + X_5 + X_6) \geq 10(X_7 + X_8 + X_9)$
or $X_1 + X_2 + X_3 + X_4 + X_5 + X_6 - 10X_7 - 10X_8 - 10X_9 \geq 0$
g. Grazing country (AUM) $\geq$ 20 (M bd ft)
$(0.4X_4 + 0.1X_5 + 0.9X_6) \geq 20(0.3X_1 + 0.05X_2 + 0.8X_3)$
$0.4X_4 + 0.1X_5 + 0.9X_6 \geq 6X_1 + X_2 + 16X_3$
or $-6X_1 - X_2 - 16X_3 + 0.4X_4 + 0.1X_5 + 0.9X_6 \geq 0$
h. Feasibility No, the problem statements (a) to (g) are full of inconsistencies. Two of them are:
(1) Equation (a) requires 50 M bd ft of timber. This, in turn, requires 1,000 AUM because of equation (g). If *all* land were in grazing we would obtain only 382 AUM!

$$700(0.4) + 210(0.1) + 90(0.9) = 382 \text{ AUM}$$

(2) Equation (e) requires 200 acres of recreation allocation. Equation (f) would then require 2,000 acres in timber and grazing. There are only 1,000 acres in the management unit!

Can you find more?

6-3 *Coefficients and Decision Variables.* First calculate the round trip cost from each site to each mill. For example, from site 2 to mill C we have:

$$20 \text{ miles} \times 2 = 40 \text{ miles round trip}$$

$$40 \text{ miles} \times \$1.50/\text{mile} = \$60 \text{ per trip}$$

Similar calculations give the table:

Round trip cost/load					Shipment decision variables			
	Mill					Mill		
Logging site	A	B	C	Site capacity	Logging site	A	B	C
1	24	45	150	20 loads	1	X_{11}	X_{12}	X_{13}
2	30	51	60	30 loads	2	X_{21}	X_{22}	X_{23}
3	90	75	45	25 loads	3	X_{31}	X_{32}	X_{33}
Mill requirement (loads)	25	25	25					

The equation formulation
Minimize

$$24X_{11} + 45X_{12} + 150X_{13} + 30X_{21} + 51X_{22} + 60X_{23} + 90X_{31} + 75X_{32} + 45X_{33}$$

Subject to

$$X_{11} + X_{12} + X_{13} \le 20$$

$$X_{21} + X_{22} + X_{23} \le 30$$

$$X_{31} + X_{32} + X_{33} \le 25$$

$$X_{11} + X_{21} + X_{31} \ge 25$$

$$X_{12} + X_{22} + X_{32} \ge 25$$

$$X_{13} + X_{23} + X_{33} \ge 25$$

Linear Programming Solution
Objective function value is 3,030.

Variable	Value	Reduced cost
X_{11}	20	0
X_{12}	0	0
X_{13}	0	111
X_{21}	5	0
X_{22}	25	0
X_{23}	0	15
X_{31}	0	60
X_{32}	0	24
X_{33}	25	0

Row	Slack	Dual prices
2	0.0	6
3	0.0	0
4	0.0	0
5	0.0	−30
6	0.0	−51
7	0.0	−45

6-4

(*a*) Decision variables

X_1 =acres allocated to even-aged management

X_2 =acres allocated to uneven-aged management

Objective
maximize $140X_1 + 100X_2$
subject to the constraints:

(1) Land	$X_1 + X_2 \le 640$
(2) Stream buffers	$X_2 \ge 150$
(3) Even/uneven	$-0.5X_1 + X_2 \ge 0$
(4) Maximum even-aged	$X_1 \le 600$
(5) Suitability for even-aged	$X_1 \ge 200$

(*b*) Graph see Fig. P6-4.

(*c*) *Binding constraints*
The optimal solution occurs at point C where the land constraint and the ratio constraint are binding.

(*d*) New prices change the slope of the objective function and the optimal solution as shown on the graph. The new solution at point B has more uneven-aged acres and greater revenue. The new binding constraints are land and suitability.

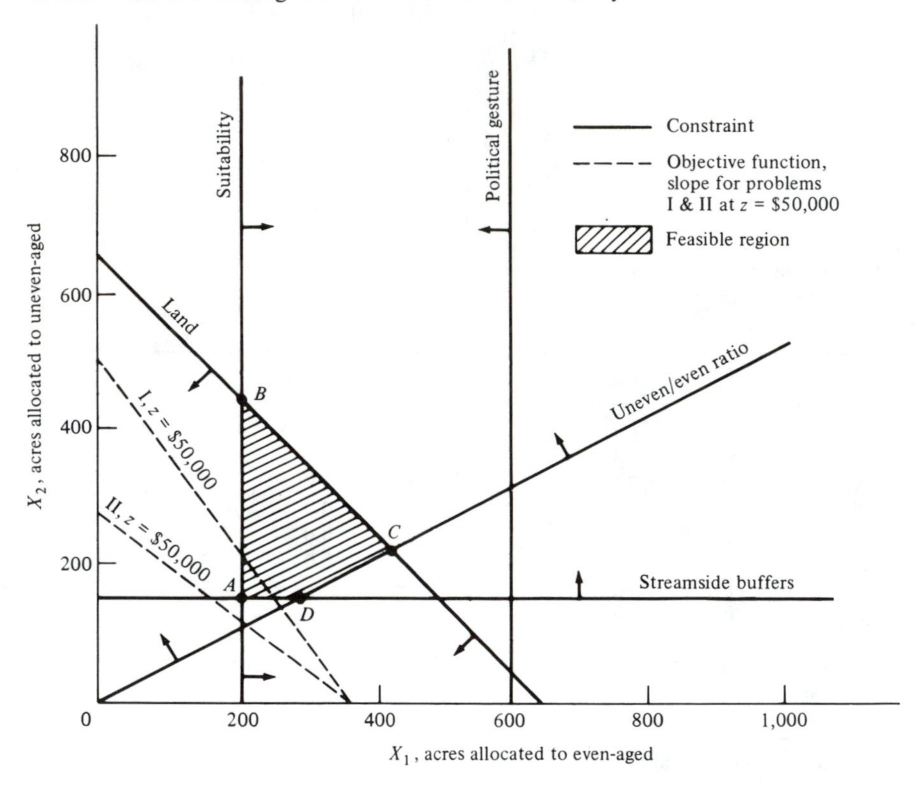

OPTIMUM SOLUTIONS

Problem	Net revenue/acre Even	Net revenue/acre Uneven	Acre allocations Even	Acre allocations Uneven	Objective value	Point
I	$140	$100	426.67	213.33	81,066	C
II	$140	$175	200	440	105,000	B

Figure P6-4 Graphical solution to the Quincy lumber problem

6-5

Row 1 An increase in 1 acre of available land will increase the value of the enterprise by a maximum of $20 in present net worth.

Row 2 The maximum harvest constraint appears not binding and a small change is not likely to affect the value of the solution. Look further to see if the slack for this row is nonzero to be sure.

Row 3 Increasing the turkey habitat requirement by 1 acre will decrease the value of the enterprise by a maximum of $300 in present net worth.

Row 4 Raising the minimum required wilderness visitor days by one visitor day will decrease present net worth by a maximum of $0.035.

6-6

(*a*) Cut 60 acres of stand A in period 1 and 40 acres in period 2. Cut 12 acres of stand B in period 2 and 188 acres in period 3.

(*b*) 4584 M bd ft.

(*c*) Although all acres are cut, the stands cut in periods 1 and 2 will likely regenerate by natural or deliberate means. If so, we should have some residual inventory by the end of the third period. We would need yield tables for regenerated stands to quantify ending inventory.

(*d*) The reduced cost for variable X_{21} (stand B in period 1) is 2.88. This means the yield per acre would have to increase 2.88 M bd ft from 7 to 9.88 M bd ft before any of stand B would be cut in period 1.

(*e*) The current dual or opportunity cost is 1.16 M bd ft of total harvest per M bd ft in the first period. The range is -100 to $+400$ for this row so the constraint could be dropped 100 to 500 for a gain of 100×1.16 or 116 M bd ft of total harvest. Beyond this change the basis will also change and we do not know what the new shadow price will be.

(*f*) From the ranges for row 2, stand A could increase 10 acres or decrease 40 acres before the basis would change. For stand A, areas between 60 and 110 acres would leave the basis is unchanged.

6-7 (*a*) Net revenue matrix (net revenue = sale revenue − road cost)

Action	State of nature s_1	s_2	Row maximum	Row minimum	Expected value
a_1	100,000	−50,000	100,000*	−50,000	25,000
a_2	60,000	−10,000	60,000	−10,000	25,000*
a_3	20,000	20,000	20,000	20,000*	20,000
Equal probability	0.5	0.5			

(*b*) The best decision for the optimist is action a_1 with a payoff under s_1 of $100,000. The best decision for the pessimist is to choose action a_3, which produces an assured positive return of $20,000 under either s_1 or s_2. The rationalist who believes states s_1 and s_2 are equally likely, chooses either action a_1 or action a_2, both with the highest expected value payoff of $25,000.

(*c*) With the new probability information on the states ($s_1 = 0.7$, $s_2 = 0.3$), the expected values are recalculated using these probabilities:

a_1 $\quad 0.7(100{,}000) + 0.3(-50{,}000) = 55{,}000$

a_2 $\quad 0.7\,(60{,}000) + 0.3(-10{,}000) = 39{,}000$

a_3 $\quad 0.7\,(20{,}000) + 0.3(20{,}000) = 20{,}000$

and we see that action a_1 now has the highest expected value and would be preferred by the rationalist. The optimist and pessimist would probably stick to their original choices.

6-8 *The Decision Tree*

Let s_1 = dry, prices rising
s_2 = dry, prices falling
s_3 = wet, prices rising
s_4 = wet, prices falling

Manager's choice	Nature's choice		Probabilities	Outcomes
D_1: a_1 Pine	4,200	s_1	0.16	3,000
		s_2	0.24	5,000
		s_3	0.24	3,000
		s_4	0.36	5,000
a_2 Super pine	3,400	s_1	0.16	10,000
		s_2	0.24	−1,000
		s_3	0.24	10,000
		s_4	0.36	−1,000
a_3 Fast oak	3,000	s_1	0.16	6,000
		s_2	0.24	6,000
		s_3	0.24	1,000
		s_4	0.36	1,000

Expected value of each action

$$EV(a_1) = 0.16(3{,}000) + 0.24(5{,}000) + 0.24(3{,}000) + 0.36(5{,}000) = 4{,}200^*$$

$$EV(a_2) = 0.16(10{,}000) + 0.24(-1{,}000) + 0.24(10{,}000) + 0.36(-1{,}000) = 3{,}400$$

$$EV(a_3) = 0.16(6{,}000) + 0.24(6{,}000) + 0.24(1{,}000) + 0.36(1{,}000) = 3{,}000$$

Plant with regular pine if you think these are good estimates of outcomes and their probabilities.

CHAPTER 7

7-1 *Earning rate problem*
Using the formula $(1 + i)^n = V_n/V_0$, with V_0 = \$250 and n = 100 years, solve for i

$$\text{at \$100 price, } V_n = \$\ 4{,}000, \quad \frac{V_n}{V_0} = 16, \quad i = 0.0281$$

$$\text{at \$200 price, } V_n = 8{,}000, \quad \frac{V_n}{V_0} = 32, \quad i = 0.0353$$

$$\text{at \$300 price, } V_n = 12{,}000, \quad \frac{V_n}{V_0} = 48, \quad i = 0.0395$$

$$\text{at \$400 price, } V_n = 16{,}000, \quad \frac{V_n}{V_0} = 64, \quad i = 0.0425$$

7-2 (*a*) Present value of harvest. Use the formula $V_0 = V_n/(1 + i)^n$, with $V_n = \$5{,}000$, $n = 60$ years.

at $i = 0.03$, $V_0 = \$848.67$

at $i = 0.06$, $V_0 = \$151.57$

at $i = 0.09$, $V_0 = \$\ 28.40$

(*b*) The present net worth is present value of harvest less the $80 planting cost

at $i = 0.03$, $848.67 − 80 = $768.67

at $i = 0.06$, 151.57 − 80 = $ 71.57

at $i = 0.09$, 28.40 − 80 = −$ 51.60

7-3 Calculate the present net worth and the SEV value for each rotation age and interest rate.

			Present net worth of one rotation		SEV	
Year	Volume	Harvest value	@ 4%	@ 8%	@ 4%	@ 8%
30	2	400	23.3	−60.2*	33.7	−66.9*
50	8	1,600	125.1	−65.9	145.6*	−67.3
70	15	3,000	92.7	−86.3	99.0	−86.7
90	40	8,000	134.5*	−92.1	138.5	−92.2
110	70	14,000	87.3	−97.1	88.4	−97.1
130	90	18,000	9.9	−99.2	9.9	−99.2

(*a*) The rotation maximizing present net worth is 90 years at 4 percent and 30 years at 8 percent.

(*b*) The rotation maximizing SEV is 50 years at 4 percent and 30 years at 8 percent.

(*c*) At an 8 percent guiding rate present net values are negative for either criterion and investing in this ponderosa pine plantation would not be a good idea. At the 4 percent guiding rate we see SEV recommending a shorter rotation because it is also considering the returns from subsequent future rotations. We would need to know more about the investor's decision context to say which was the better criterion or rotation recommendation. The dip in present values at age 70 is due to a slight dip or wave in the yield curve.

7-4 (*a*) 30-year rotation
Increased cost = \$50
Increased value yield at 30 years = 0.3(2) = 0.6 M bd ft worth \$120
Present value of \$120 @ 4% = \$36.99, @ 8% = \$11.93
Present net value @ 4% = \$36.99 − \$50 = −\$13.01. Do not buy seed.
Present net value @ 8% = \$11.93 − \$50 = −\$38.07. Do not buy seed.
(*b*) 50-year rotation
Increased cost = \$50
Increased value yield at 50 years = 0.3(8) = 2.4 M bd ft worth \$480
Present value of \$480 @ 4% = \$67.54, @ 8% = \$10.23
Present net value @ 4% = \$67.54 − \$50 = \$17.54. Buy seed.
Present net value @ 8% = \$10.23 − \$50 = −\$39.77. Do not buy seed.
(*c*) 90-year rotation
Increased cost = \$50
Increased value yield at 90 years = 0.3(40) = 12 M bd ft worth \$2,400
Present value of \$2,400 @ 4% = \$70.34, @ 8% = \$2.35
Present net value @ 4% = \$70.34 − \$50 = \$20.34. Buy seed.
Present net value @ 8% = \$2.35 − \$50 = −\$47.64. Do not buy seed.

7-5 Value of the future yield increase = V_n

$$\begin{array}{lr} 5 \text{ M bd ft} \times \$100 = & 500 \\ 10 \text{ cords} \times \$10 \quad = & 100 \\ \hline V_n = & \$600 \end{array}$$

(*a*) Present value and earning rate

$$(1 + i)^n = \frac{V_n}{V_0} = \frac{\$600}{\$75} = 8.0$$

at $n = 40$ years, $(1 + i)^n = 8.0$ for $i = 0.0534$ so the earning rate is 5.34%.
(*b*) The present net worth

$$V_0 = \frac{V_n}{(1 + i)^n} = \frac{\$600}{(1.10)^{40}} = \frac{\$600}{45.26} = \$13.26$$

Present net worth = \$13.26 − \$75 = −\$61.74

(*c*) No. I would not recommend PCT as the earning rate is less than the alternative rate and the present value of the revenue (\$13.26) is considerably less than the present value of the costs (\$75).

(*d*) The maximum you could spend is simply the present value of the future return at the given interest rate. In this case, \$13.26.

Check
$$(1 + i)^n = \frac{600}{13.26} = 45.26$$

which, for $n = 40$, has $i = 0.10$.

7-6 Schedule of cash revenues

	Year	Payment	Year	Payment
Dec	1986	15,000	2001	15,000
	1989	15,000	2004	15,000
	1992	15,000	2007	15,000
	1995	15,000	2010	15,000
	1998	15,000		

This problem is solved using Eqs. (7-4) and (7-7) for periodic series payments. (It could also be directly solved by discounting or compounding each item.)

(*a*) Using Eq. (7-7), the estate value in December 1985 for $i = 0.10$ is

$$V_{86} = 15{,}000 + 15{,}000 \frac{1.10^{24} - 1}{(1.10^3 - 1)(1.10^{24})} = \$55{,}716$$

$$V_{85} = V_{86}/1.10 = \$50{,}651$$

(*b*) The future value of this revenue series in December 2010 is calculated using Eq. (7-4) and $i = 0.08$

$$V_{2010} = 15{,}000(1.08)^{24} + 15{,}000 \frac{1.08^{24} - 1}{1.08^3 - 1} = \$403{,}604.4$$

Yes, there will be enough in the trust fund to send the sons to Harvard.

7-7 This is a present value of a series payment problem. The implied value of the wildlife benefits per acre per year are at least the timber yield reduction less the increase in forage sales of $10 − $1.50 = $8.50. So, the wildlife value must be at least $8.50/acre/year or $21,250 for the analysis area/year.

The present value equation (7-8)

$$V_0 = a \frac{(1 + i)^n - 1}{i(1 + i)^n} = 21{,}250 \frac{(1 + 0.07)^{50} - 1}{0.07(1 + 0.07)^{50}}$$

$$= \$21{,}250 \frac{29.457 - 1}{0.07(29.457)}$$

$$= \$21{,}250(13.80)$$

$$= \$293{,}250$$

is the present value of needed wildlife benefits to compensate for timber income losses.

This answer is the opportunity cost of wildlife enhancement in terms of timber revenues forgone. It most decidedly does not tell us anything about the *value* of the increased wildlife.

7-8 $V_{20} = 3{,}000$, inflation rate, $k = 0.05$ guiding rate, $r = 0.07$.

$$\text{Present value } (V_0) \text{ (today's dollars)} = \frac{V_{20}}{[(1 + r)(1 + k)]^{20}}$$

$$= \frac{3{,}000}{(1 + 0.07 + 0.05 + 0.0035)^{20}}$$

$$= \frac{3{,}000}{(1.1235)^{20}} = \$292.18$$

7-9 General inflation rate = 5%, pulpwood inflation rate = 3%, sawtimber inflation rate = 9%, guiding rate = 4%. *Today's prices:* pulpwood = \$15/cord, stumpage \$50/M bd ft. Yields: at age 20, pulpwood, 20 cords worth \$300 at today's prices; age 30, pulpwood, 10 cords worth \$150, sawtimber, 20 M bd ft worth \$3,000 at today's prices.

Present value of pulpwood harvest. Use Eq. (7-12) with $r = 0.04$, $k = 0.05$, and $p = 0.03$.

$$\text{20-year harvest, } V_0 = \frac{300(1.03)^{20}}{(1.04)^{20}(1.05)^{20}} = \frac{300(1.806)}{5.8137} = \$93.19$$

$$\text{30-year harvest, } V_0 = \frac{150(1.03)^{30}}{(1.04)^{30}(1.05)^{30}} = \frac{150(2.427)}{14.018} = \$25.97$$

Present value of sawtimber harvest. Use Eq. (7-12) with $r = 0.04$, $k = 0.05$, and $p = 0.09$.

$$\text{30-year harvest, } V_0 = \frac{3{,}000(1.09)^{30}}{(1.04)^{30}(1.05)^{30}} = \frac{3{,}000(13.267)}{14.018} = \$2{,}839.32$$

$$\text{Total present value} = 2{,}839.32 + 25.97 + 93.19 = \$2{,}958.48$$

7-10 (*a*) *PNW prescription 1*

$$\frac{400}{(1 + 0.05)^{30}} - 100 = 92.55 - 100 = -\$7.45$$

PNW prescription 2

$$\frac{800}{(1 + 0.05)^{30}} - \frac{50}{(1 + 0.05)^{10}} - 100 = 185.10 - 30.69 - 100 = \$54.41$$

PNW prescription 3

$$\frac{1{,}200}{(1.05)^{30}} - \frac{30}{(1.05)^{20}} - \frac{80}{(1.05)^{10}} - 100 = 277.65 - 11.30 - 49.11 - 100 = \$117.24$$

(*b*) Treating the minimum prescription as the "without" treatment
PNW thinning age 10 only

$$54.41 - (-7.45) = \$61.86$$

PNW thinning and fertilizing

$$117.24 - (-7.45) = \$124.69$$

(*c*) *SEV values of the three prescriptions*

$$\text{SEV} = \frac{a}{(1 + i)^t - 1}, \qquad t = 30, \quad i = 5\%$$

Prescription #1

	a	SEV
3%: $a = 400 - 100(1.03)^{30} =$	157.27	110.19
5%: $a = 400 - 100(1.05)^{30} =$	−32.19	−9.69
7%: $a = 400 - 100(1.07)^{30} =$	−361.22	−54.64

Prescription #2

	a	SEV
3%: $a = 800 - 100(1.03)^{30} - 50(1.03)^{20} =$	466.98	327.19
5%: $a = 800 - 100(1.05)^{30} - 50(1.05)^{20} =$	235.15	70.79
7%: $a = 800 - 100(1.07)^{30} - 50(1.07)^{20} =$	−154.70	−23.40

Prescription #3

	a	SEV
3%: $a = 1{,}200 - 100(1.03)^{30} - 80(1.03)^{20} - 30(1.03)^{10} =$	772.47	541.22
5%: $a = 1{,}200 - 100(1.05)^{30} - 80(1.05)^{20} - 30(1.05)^{10} =$	506.68	152.52
7%: $a = 1{,}200 - 100(1.07)^{30} - 80(1.07)^{20} - 30(1.07)^{10} =$	70.21	10.62

7-11 (*a*) Schedule of cash revenues in current dollars

	Year						
Option	0	10	20	30	40	50	...
1	1,250	—	2,500	—	2,500	—	...
2	0	3,500	—	2,500	—	2,500	...

The SEV(U) equation (7-18) can be modified to deal with the initial adjustment harvest.

Option 1

$$\text{PNW} = 1{,}250 + \frac{2{,}500}{(1.06)^{20} - 1} = \$2{,}382.75$$

Option 2

$$\text{PNW} = \left[3{,}500 + \frac{2{,}500}{(1.06)^{20} - 1}\right] \frac{1}{(1.06)^{10}}$$

$$= (3{,}500 + 1{,}132.76)\,\frac{1}{1.7908} = 2{,}586.98$$

In this case it would pay to delay the harvest for 10 years.

(*b*) The PNW calculation for part (*a*) must be modified to allow for price increases. Combining it with Eq. (7-12) and letting $(1 + h) = (1 + p)/(1 + k)$

$$\text{PNW} = \left(R_L + \frac{a}{[(1 + r)/(1 + h)]^w - 1}\right) \frac{1}{[(1 + r)/(1 + h)]^L}$$

where R_L = revenue from initial harvest
a = periodic revenue from regulated harvest
L = years before first harvest
w = regulated harvest period in years
p = rate of price inflation of item evaluated
k = average rate of inflation
r = guiding rate in real terms
h = real rate of price increase (decrease)

Option 1

$$R_L = 1{,}250, \quad a = 2{,}500, \quad L = 0, \quad w = 20$$

$$p = 0.07, \quad k = 0.04, \quad r = 0.05, \quad h = 1.02885$$

$$\text{PNW} = \left(1{,}250 + \frac{2{,}500}{(1.05/1.02885)^{20} - 1}\right)\frac{(1.02885)^0}{(1.05)^0} = \$6{,}228$$

Option 2

$$R_L = 3{,}500, \quad a = 2{,}500, \quad L = 10, \quad w = 20$$

$$p = 0.07, \quad k = 0.04, \quad r = 0.05, \quad h = 1.02885$$

$$\text{PNW} = \left(3{,}500 + \frac{2{,}500}{(1.05/1.02885)^{20} - 1}\right)\frac{(1.02885)^{10}}{(1.05)^{10}} = \$6{,}933$$

CHAPTER 8

8-1 (*a*)

Project	Cost (V_0)	Revenue (V_n)	n	IRR (%)
1	85	1,100	30	8.9
2	70	1,000	20	14.2
3	150	5,000	40	9.2
4	60	400	10	20.9

(*b*) Present net worth per acre, normalized for length and reinvestment by extending all projects by reinvesting all revenues at 8 percent until year 40

Project	Cost (V_0)	V_n	Value at 40 years (V_{40}) $V_n(1.08)^{40-n}$	$\frac{V_{40}}{(1.10)^{40}}$	Normalized present net worth (PNW)
1	85	1,100	2,374.8	52.47	−32.53*
2	70	1,000	4,660.95	102.98	32.98
3	150	5,000	5,000	110.47	−39.53
4	60	400	4,025	88.93	28.93

* For example, PNW project 1 = 52.47 − 85 = −32.53.

Project #2 has the highest normalized PNW.

(*c*) Realizable rate of return (RRR) [Eq. (7-19)]

$$(1 + \text{RRR})^n = \frac{\sum_{t=1}^{n}[R_t(1 + r)^{n-t}]}{\sum_{t=0}^{n}[C_t/(1 + r)^t]}$$

Project	Present value of costs	V_{40}	RRR (%)
1	85	2,374.8	8.68
2	70	4,660.95	11.06
3	150	5,000.0	9.16
4	60	4,025	11.08

(*d*) Projects 2 and 4 could be used for investment since their RRR is greater than the guiding rate of 10 percent

Project 2, 100 acres × \$70 per acre = \$7,000
Project 4, 200 acres × \$60 per acre = \$12,000
Total budget = \$19,000

8-2 (*a*) Technical coefficients (a_{i2}) for agriculture

Sector	Agriculture purchases (x_{i2})	x_{i2}/z_2 agriculture a_{i2}
(1) Forest products	70	70/400 = 0.175
(2) Agriculture	100	100/400 = 0.25
(3) Other manufacturing and services	50	50/400 = 0.125
(4) Households	130	130/400 = 0.325
(5) Imports	50	50/400 = 0.125
	z_2 = 400	400/400 = 1.0

The a_{i2} column of technical coefficients we calculated for agriculture could be thought of as its current production function since they describe the proportionate mix of inputs used per dollar of sales.

(*b*) The rank is based first on the dollars of exports generated by the \$100 increase in total sales. Then this export increase is weighted by the export multiplier for each sector to get the total impact on the internal, regional economy.

Sector	Exports per $100 of total sales	Export impact multiplier	Impact of a $100 increase in total sales on the internal sectors*	Rank
(1) Forest products	58.33	2.79	162.74	1
(2) Agriculture	50	3.19	159.50	2
(3) Other manufacturing and services	25	2.55	63.75	3
(4) Households	24.29	1.80	43.72	4

* For example, 162.74 = 58.33 × 2.79

Industrial sectors with proportionately large exports and fairly high type II multipliers are important to a regional economy. When such industries also have a high level of total sales, we call them an economic base industry as they are the foundation of the regional economy.

(*c*) The total impact of a $100 increase in agricultural exports is equal to the sector multiplier × increase in exports or 3.19 × 100 = $319. It is distributed to the internal sectors as follows

Sector	Export impact multiplier	Sales increase
(1) Forest products	0.42	42
(2) Agriculture	1.46	146
(3) Other manufacturing and services	0.53	53
(4) Households	0.78	78
	3.19	total $319

8-3

1. Type *A* process, second economic efficiency situation. The goal is to achieve the target 100 AUM at least cost.
2. Probably a type *B* process, first economic efficiency situation. Being public there are probably several different goals and thus benefits associated with the candidate campgrounds, for example: maximum total RVD consumed, maximum cash receipts, or maximum benefit index. They could calculate *B*/*C* ratios for these different criteria and choose an efficient set for restoration under each ranking criterion. The big argument will be choosing which goal to optimize—a type *B* issue.
3. Type *A* process—third economic efficiency situation. The timber staff will likely evaluate and rank all options by PNW or RRR with a single goal: economic efficiency. The results of this assessment will likely be used in higher level problem solving of a type *B* process where different goals (PNW, growth, market share, bond rating) and different constraints (capital budget, borrowing, regional marketing strategy) are involved.
4. Type *A* process, second economic efficiency situation. This could be simply a question of finding the sequence of available candidate stands that minimizes timber sale prepara-

tion, administration, and road costs. If environmental impacts were an issue, then it could easily become a type *B* process.

5. Type *B* process, third economic efficiency situation. This is the classicial multigoal, multiconstraint situation with complex alternatives leading to the type *B* process. If budgets are unconstrained, then it is the third efficiency situation.
6. Type *B* process, first efficiency situation. This private owner likely has many goals and the different prescriptions will likely give differential emphasis to them, for example: steady income, capital revenue, hunting, visual or noise abatement value. If the owner could say flat out that "I want to maximize SEV, or, sustainable income," then it could become a type *A* process.
7. Type *A* process, second or third efficiency situation. The goal in this case is likely either to minimize the cost of supplying a set quota of seedlings, or to maximize the PNW of the nursery if it operates as an independent profit center. In this latter case the budget would not be given and the design would want to consider alternatives that included seedling sales to outside purchasers.
8. Type *A* process, second efficiency situation. Apparently a straight cost minimizing problem with well-defined goals and resource constraints. The only complication would be when the diet also affects the flesh color, vigor, flavor, or other attributes of the fish and thus their value to the fisherman. In this case the problem could expand to a type *B* process if the different goals and constraints of the fisherman clientele are considered.
9. Type *A* process, second situation. Like the hatchery, if the dimensions of the sale are known and it is given that it will be logged, then it is a cost minimization problem. If other goals are important, like keeping company loggers busy, establishing a good image with the environmental community, or encouraging an independent helicopter logging firm to stay in business, then this could also become a type *B* process.
10. Type *A* process, first efficiency situation. The farmer has given resources of fields, time, and operating funds and would do an income or PNW maximization analysis. Or she could complicate this problem with other goals like producing farm wildlife habitat or the need to make a balloon mortgage payment on July 1 and shift to a type B process.
11. Type *A* process, second situation. Finally, an almost pure cost minimization problem. Unless fresh is better, or the owner likes machinery, or enjoys watching potatoes being shredded, or . . .
12. Type *B* process, third situation. These two specialists will approach the problem with different goal sets and explicit and implicit constraints. They will likely negotiate a set of options reflecting different goal and constraint assumptions and then bring in others to subjectively choose.

CHAPTER 12

12-1 (*a*) Terms-of-trade table, exchange rates: units for units.

Forest Output	Units	Sawlogs	Water	Forage	Recreation	Sport fishing	Commercial fishing	Dollars
Sawlogs	M bd ft	1.0	0.01	0.0036	0.0057	0.0083	0.02	0.01
Water	Acre · feet	100	1.0	0.363	0.571	0.833	2.0	1.0
Forage	AUM	275	2.75	1.0	1.571	2.291	5.5	2.75
Recreation	RVD	175	1.75	0.636	1.0	1.458	3.5	1.75
Sport fishing	RVD	120	1.2	0.436	0.686	1.0	2.4	1.20
Commercial fishing	Pounds	50	0.5	0.182	0.286	0.417	1.0	0.5
Dollars	1 Dollar	100	1	0.363	0.571	0.833	2.0	1.0

(*b*) If sawtimber is exchanged for $100 per M bd ft then the dollar price exchange rates are shown in the last row, imputing prices of $1 per acre · foot to water $0.57/RVD to recreation and $0.83/RVD to sport fishing; *given the initial barter exchange rates.*

12-2 Recreation Demand

(*a*) Per capita at-home demand (zone data were plotted). You may get slightly different estimates depending on your plot.

Source of observation	Total visit cost (gate fee + travel cost)	Number of visits per year
(estimate)	0	16
(estimate)	5	12
Zone 1	10	9
Zone 2	15	7
(estimate)	20	6
Zone 3	25	5
(estimate)	30	4.3
(estimate)	35	3.6
Zone 4	40	3
Zone 5	45	2
(estimate)	50	1
(estimate)	55	0

(*b*) *At-gate demand for persons in zone 3,* Travel cost = $20.

Gate fee	Total cost	Visits per year
0	20	6
5	25	5
10	30	4.3
15	35	3.6
20	40	3
25	45	2
30	50	1
35	55	0
40	60	0

Total annual consumption at different gate fees: at-gate aggregate demand

		Gate Fee ($)									
Zone	Population	0	5	10	15	20	25	30	35	40	45
1	200	2,400	1,800	1,400	1,200	1,000	860	720	600	400	200
2	20,000	180,000	140,000	120,000	100,000	86,000	72,000	60,000	40,000	20,000	—
3	10,000	60,000	50,000	43,000	36,000	30,000	20,000	10,000	—	—	—
4	100,000	360,000	300,000	200,000	100,000	—	—	—	—	—	—
5	500,000	1,500,000	1,000,000	500,000	—	—	—	—	—	—	—
Total visits		2,102,400	1,491,800	864,400	237,200	117,000	92,860	70,720	40,600	20,400	200

(*c*) Aggregate demand, all zones.

Per capita visitation rates per zone at different gate fees

	Travel	Gate fee ($)									
Zone	cost	0	5	10	15	20	25	30	35	40	45
1	5	12	9	7	6	5	4.3	3.6	3	2	1
2	10	9	7	6	5	4.3	3.6	3	2	1	0
3	20	6	5	4.3	3.6	3	2	1	0	—	—
4	35	3.6	3	2	1	0	—	—	—	—	—
5	40	3	2	1	0	—	—	—	—	—	—

(*d*) *Consumer Surplus* Using the aggregate demand curve from (*c*) (in the format of Table 12-4)

Gate fee	Visits	Incremental visits	Gate receipts	Incremental willingness to pay	Total willingness to pay
			(Thousands of Dollars)		
0	2,102,400	—	0	—	19,832
		610,600		1,527	
5	1,491,800		7,459		18,305
		627,400		4,705	
10	864,400		8,644		13,600
		627,200		7,840	
15	237,200		3,558		5,760
		120,200		2,014	
20	117,000		2,340		3,746
		24,140		543	
25	92,860		2,322		3,203
		22,140		609	
30	70,720		2,122		2,594
		30,120		979	
35	40,600		1,421		1,615
		20,200		757	
40	20,400		816		858
		20,200		858	
45	200		9		

At a gate fee of $15

$$\text{Total willingness to pay} = \$5{,}760{,}000$$

$$\text{Average willingness to pay} = \frac{5{,}760{,}000}{237{,}200} = \$24.28$$

$$\text{Consumer surplus} = 5{,}760{,}000 - 3{,}558{,}000 = \$2{,}202{,}000$$

CHAPTER 13

13-1 Future value approach:

Tree value now	Tree value in 10 years	Value growth	Stock holding cost	Land holding cost	Total cost	Net gain from holding	Decision
200	300	100	96.05	43.24	139.29	−39.29	Cut now

$$\text{Stock holding cost} = 200(1.04)^{10} - 200 = 96.05$$

$$\text{SEV}_{25} = \frac{150}{(1.04)^{25} - 1} = 90.04$$

$$\text{SEV}_{35} = \frac{200}{(1.04)^{35} - 1} = 67.89$$

$$\text{Land holding cost} = 90.04(1.04)^{10} - 90.04 = 43.24$$

Present value approach:

		Cut $n = 0$	Leave $n = 10$
(1)	SV_n	200	300
(2)	$\frac{\text{SV}_n}{(1 + i)^n}$	200	202.67
(3)	MSEV	90.04	90.04
(4)	$\frac{\text{MSEV}}{(1 + i)^n}$	90.04	60.83
	PNW[(2) + (4)]	290.04	263.50

Decision: cut now.

If the existing tree does not hinder establishment and growth of a young yellow poplar, the decision is reversed: leave the tree because value growth (100) is greater than stock holding cost (96.05) and the land holding cost is zero.

13-2 Future value approach:

		Leave both		Leave tree *B*	Leave tree *A*
		Tree *A*	Tree *B*		
(1)	Value now	300.00	400.00	400	300
(2)	Value in 5 years	380.00	460.00	500	440
(3)	Value growth	80.00	60.00	100	140
		140.00			
(4)	Stock holding cost	64.98	86.64	86.64	64.98
		151.62			
(5)	Land holding cost	19.50		19.50	19.50
(6)	Total cost	171.12		106.14	84.48
(7)	Net gain	−31.12		−6.14	55.52

Leave tree *A*, cut tree *B*.

Present value approach:

		Leave both		Leave *B*		Leave *A*		Leave none	
		Tree *A* $n = 5$	Tree *B* $n = 5$	Tree *A* $n = 0$	Tree *B* $n = 5$	Tree *A* $n = 5$	Tree *B* $n = 0$	Tree *A* $n = 0$	Tree *B* $n = 0$
(1)	SV_n	380.00	460.00	300.00	500.00	440.00	400.00	300.00	400.00
(2)	$\frac{SV_n}{(1+i)^n}$	312.34	378.10	300.00	410.98	361.66	400.00	300.00	400.00
		690.44		710.98		761.66		700.00	
(3)	$\frac{MSEV}{(1+i)^n}$	74.01		74.01		74.01		90.04	
PNW		764.45		784.99		835.67		790.04	

Leave tree *A*, cut tree *B*.

13-3

$$SEV = \frac{NI_t}{(1+i)^t - 1}, \qquad i = 0.05, \quad t = 50$$

$$\begin{aligned} NI_t &= R_{50} + T_{30}(1+p)^{20} - C_R(1+p)^{50} - C_{PCT}(1+p)^{40} \\ &= \$3,200 + \$300(1.05)^{20} - 50(1.05)^{50} - \$100(1.05)^{40} \\ &= \$3,200 + \$795.99 - \$573.37 - \$704.0 \\ &= \$2,718.62 \end{aligned}$$

$$SEV = \frac{\$2,718.62}{(1.05)^{50} - 1} = \frac{\$2,718.62}{10.47} = \$259.72/\text{acre}, \quad \text{the maximum you should pay}$$

13-4

$$\text{SEV} = \text{SEV}_{\text{Timber}} + \text{SEV}_{\text{Deer}}$$

$$= 259.72 + \frac{50\left(\dfrac{(1.05)^{10} - 1}{0.05}\right)(1.05)^{40} + 25\left(\dfrac{(1.05)^{40} - 1}{0.05}\right)}{(1.05)^{50} - 1}$$

$$= 259.72 + 711.45$$

$$= 971.17/\text{acre}$$

13-5

Without precommercial thinning

Age	Gross stumpage revenue	Net revenue at rotation	SEV
30	900	$900 - 100(1.04)^{30} = 575.66$	$575.66/[(1.04)^{30} - 1] = 256.60$
40	1,500	$1{,}500 - 100(1.04)^{40} = 1{,}019.9$	$1{,}019.9/[(1.04)^{40} - 1] = 268.32$*
50	1,700	$1{,}700 - 100(1.04)^{50} = 989.33$	$989.33/[(1.04)^{50} - 1] = 162.01$

With precommercial thinning

Age	Gross stumpage revenue	Net revenue	SEV
30	$0.20(900) + 900 = 1{,}080$	$1{,}080 - 100(1.04)^{30} - 75(1.04)^{20} = 591.33$	263.59
40	$0.20(1{,}500) + 1{,}500 = 1{,}800$	$1{,}800 - 100(1.04)^{40} - 75(1.04)^{30} = 1{,}076.65$	283.25*
50	$0.20(1{,}700) + 1{,}700 = 2{,}040$	$2{,}040 - 100(1.04)^{50} - 75(1.04)^{40} = 969.25$	158.72

SEV Comparisons (with − without PCT)

Age	No PCT (A)	W/PCT (B)	DIFF (B − A)
30	256.6	263.59	6.99
40	268.32	283.25	*14.93 ∴ best regime → PCT and R40
50	162.01	158.72	−3.29

* Best regime among those being compared.

13-6

Future value approach

Stand	Value now	Value in 10 years	Value growth over 10 years	Stock holding cost	Land holding cost	Total cost	Net gain from holding	Decision
A	600	1,100	500	288.15	136.03	424.18	75.82	Leave
B	1,250	1,500	250	600.31	136.03	736.34	−486.34	Cut

Stand	Stock holding cost	Land holding cost
A	$600(1.04)^{10} - 600 = 288.15$	(a) * SEV = MSEV = 283.25
B	$1{,}250(1.04)^{10} - 1{,}250 = 600.31$	(b) $283.25(1.04)^{10} - 283.25 = 136.03$

Present value approach

		Stand *A*		Stand *B*	
		Replace	Keep	Replace	Keep
(1)	SV	600	1,100	1,250	1,500
(2)	$SV/(1.04)^{10}$	600	743.12	1,250	1,013.35
(3)	MSEV	283.25	283.25	283.25	283.25
(4)	$MSEV/(1.04)^{12}$	283.25	191.35	283.25	191.35
PNW[(2) + (4)]		883.25	934.47*	1,533.25*	1,204.70
			keep	cut	

13-7 Effects on decision.

(*a*) Higher regeneration costs. The lower SEV implies lower land holding cost, which encourages *holding*.

(*b*) Higher PCT costs. The lower SEV implies lower land holding costs, which encourages *holding*.

(*c*) Higher value for future crops. The higher SEV implies higher land holding cost and lower net gain from holding, which encourages *replacement*.

(*d*) Higher value in 10 years for current stands. The higher capital value growth implies higher net gain from holding, which encourages holding.

(*e*) A 25 percent income tax. The lower SEV implies lower land holding cost, which encourages holding.

13-8

Age	Node	Initial or residual volume	Volume at age +5	Residual volume at age +5: All	1,500	1,000	0
				Harvest to achieve residual volume at age +5			
15		900	1,600	×	100	600	×
20		1,500	2,200	×	700	1,200	×
		1,000	1,600	×	100	600	×
25		1,500	2,000	×	500	1,000	2,000
		1,000	1,550	×	50	550	1,550
30		1,500	1,800	×	×	×	1,800
		1,000	1,400	×	×	×	1,400

(*a*)

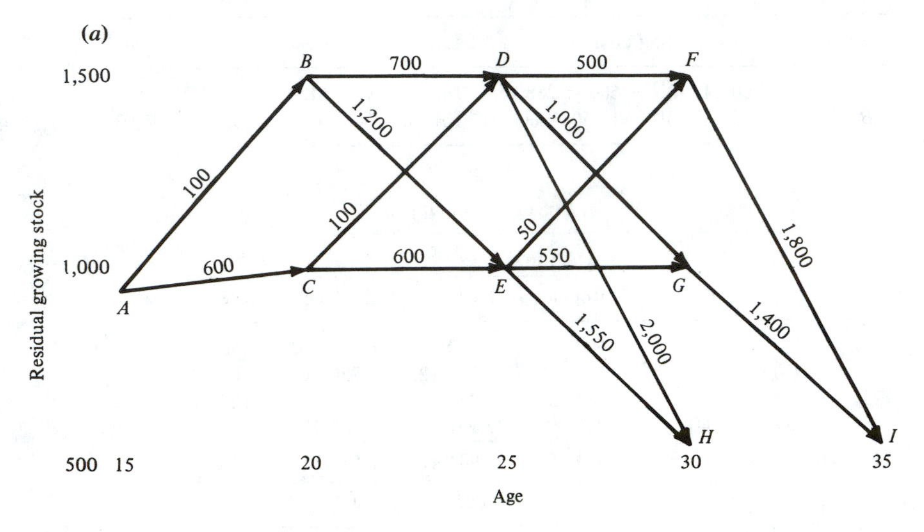

Age 15

$$R_A = 0$$

Age 20

$$R_B = R_A + r_{AB} = 0 + 100 = 100, \qquad P_B = A$$

$$R_C = R_A + r_{AC} = 0 + 600 = 600 \qquad P_C = A$$

Age 25

$$R_D = \max \begin{Bmatrix} R_B + r_{BD} \\ R_C + r_{CD} \end{Bmatrix} = \max \begin{Bmatrix} 100 + 700 \\ 600 + 100 \end{Bmatrix} = \max \begin{Bmatrix} 800 \\ 700 \end{Bmatrix} = 800, \quad P_D = B$$

$$R_E = \max \begin{Bmatrix} R_B + r_{BE} \\ R_C + r_{CE} \end{Bmatrix} = \max \begin{Bmatrix} 100 + 1{,}200 \\ 600 + 600 \end{Bmatrix} = \max \begin{Bmatrix} 1{,}300 \\ 1{,}200 \end{Bmatrix} = 1{,}300, \quad P_E = B$$

Age 30

$$R_F = \max \begin{Bmatrix} R_D + r_{DF} \\ R_E + r_{EF} \end{Bmatrix} = \max \begin{Bmatrix} 800 + 500 \\ 1{,}300 + 50 \end{Bmatrix} = \max \begin{Bmatrix} 1{,}300 \\ 1{,}350 \end{Bmatrix} = 1{,}350, \quad P_F = E$$

$$R_G = \max \begin{Bmatrix} R_D + r_{DG} \\ R_E + r_{EG} \end{Bmatrix} = \max \begin{Bmatrix} 800 + 1{,}000 \\ 1{,}300 + 550 \end{Bmatrix} = \max \begin{Bmatrix} 1{,}800 \\ 1{,}850 \end{Bmatrix} = 1{,}850, \quad P_G = E$$

$$R_H = \max \begin{Bmatrix} R_D + r_{DH} \\ R_E + r_{EH} \end{Bmatrix} = \max \begin{Bmatrix} 800 + 2{,}000 \\ 1{,}300 + 1{,}550 \end{Bmatrix} = \max \begin{Bmatrix} 2{,}800 \\ 2{,}850 \end{Bmatrix} = 2{,}850, \quad P_H = E$$

Age 35

$$R_I = \max \begin{Bmatrix} R_F + r_{FI} \\ R_G + r_{GI} \end{Bmatrix} = \max \begin{Bmatrix} 1{,}350 + 1{,}800 \\ 1{,}850 + 1{,}400 \end{Bmatrix} = \max \begin{Bmatrix} 3{,}150 \\ 3{,}250 \end{Bmatrix} = 3{,}250, \quad P_I = G$$

$$30 \text{ yr. rotation MAI} = \frac{2{,}850}{30} = 95 \leftarrow \text{rotation that maximizes mean annual increment}$$

$$35 \text{ yr. rotation MAI} = \frac{3{,}250}{35} = 92.9$$

Path $A \rightarrow B \rightarrow E \rightarrow H$ with a 30 yr. rotation is the maximum path.

(*b*)

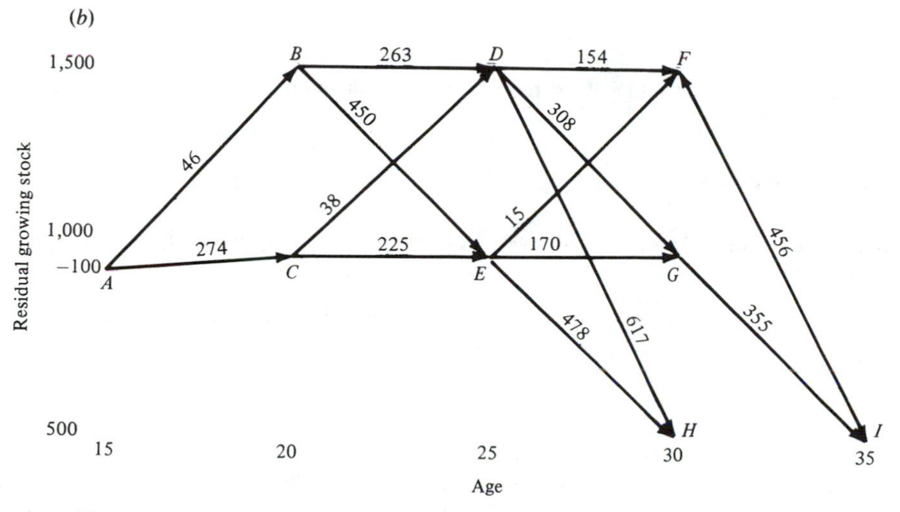

Age 15

$R_A = -100$ planting cost

Age 20

$$R_B = R_A + r_{AB} = -100 + 46 = -54, \qquad P_B = A$$

$$R_C = R_A + r_{AC} = -100 + 274 = 174, \qquad P_C = A$$

Age 25

$$R_D = \max \begin{Bmatrix} R_B + r_{BD} \\ R_C + r_{CD} \end{Bmatrix} = \max \begin{Bmatrix} -54 + 263 \\ 174 + 38 \end{Bmatrix} = \max \begin{Bmatrix} 209 \\ 212 \end{Bmatrix} = 212, \qquad P_D = C$$

$$R_E = \max \begin{Bmatrix} R_B + r_{BE} \\ R_C + r_{CE} \end{Bmatrix} = \max \begin{Bmatrix} -54 + 450 \\ 174 + 225 \end{Bmatrix} = \max \begin{Bmatrix} 396 \\ 399 \end{Bmatrix} = 399, \qquad P_E = C$$

Age 30

$$R_F = \max \begin{Bmatrix} R_D + r_{DF} \\ R_E + r_{EF} \end{Bmatrix} = \max \begin{Bmatrix} 212 + 154 \\ 399 + 15 \end{Bmatrix} = \max \begin{Bmatrix} 366 \\ 414 \end{Bmatrix} = 414, \qquad P_F = E$$

$$R_G = \max \begin{Bmatrix} R_D + r_{DG} \\ R_E + r_{EG} \end{Bmatrix} = \max \begin{Bmatrix} 212 + 308 \\ 399 + 170 \end{Bmatrix} = \max \begin{Bmatrix} 520 \\ 569 \end{Bmatrix} = 569, \qquad P_G = E$$

$$R_H = \max \begin{Bmatrix} R_D + r_{DH} \\ R_E + r_{EH} \end{Bmatrix} = \max \begin{Bmatrix} 212 + 617 \\ 399 + 478 \end{Bmatrix} = \max \begin{Bmatrix} 829 \\ 877 \end{Bmatrix} = 877, \qquad P_H = E$$

$$\text{SEV}_{30} = \frac{877(1.04)^{30}}{(1.04)^{30} - 1} = 1268 \leftarrow \text{MSEV}$$

Age 35

$$R_I = \max \begin{Bmatrix} R_F + r_{FI} \\ R_G + r_{GI} \end{Bmatrix} = \max \begin{Bmatrix} 414 + 456 \\ 569 + 355 \end{Bmatrix} = \max \begin{Bmatrix} 870 \\ 924 \end{Bmatrix} = 924, \qquad P_I = G$$

$$\text{SEV}_{35} = \frac{924(1.04)^{35}}{(1.04)^{35} - 1} = 1{,}238$$

Path $A \rightarrow C \rightarrow E \rightarrow H$ with a 30 yr. rotation is the maximum path.

CHAPTER 14

14-1 (*a*) Rotation = 40 years; area = 600 acres; acres cut by area regulation = A/R = 600/40 = 15.

(*b*) 40 years, one rotation, at a maximum.

(*c*) Regulated forest MAI per acre = 50/40 = 1.25 M bd ft/acre/year; forest LTSY = 600 × 1.25 = 750 M bd ft/year; forest harvest = 15 × 50 = 750 M bd ft/year.

(*d*)

Year	Stand cut	Age when cut	Volume per acre	Acres cut	Harvest volume
5	*B*	35	40*	15	600 M bd ft
10	*A*	20	15	15	225 M bd ft
20	*A*	30	30	15	450 M bd ft

* Interpolated from yield table.

Years to cut: stand $A = \frac{500}{15} = 33.3$, stand $B = \frac{100}{15} = 6.7$.

(*e*) Structure after 10 years of growth and harvest.

Age class	Acres
1	15
2	15
3	15
4	15
5	15
6	15
7	15
8	15
9	15
10	15
⋮	
20	450
	600

14-2 (*a,b*) Relative productivity

Stand	Site	Acres	Total yield at 30 years	Relative productivity	Acres if all site I	Acres cut/year	Years to cut
A	I	300	48	1.000	300	22.5*	13.33†
B	II	200	30	0.625	125	36.0‡	5.55
C	II	400	30	0.625	250	36.0	11.11
					675		29.99

* Stand $A = \frac{675}{30} = 22.5$

† Years to cut = 300/22.5 = 13.33

‡ Stand B = 22.5/0.625

(*c*) $\text{LTSY} = 300 \times \frac{48}{30} + 600 \times \frac{30}{30} = 1{,}080$

(*d*) Bare Land Values

$$\textit{Stand A}\ \text{SEV} = \frac{(38 \times 200) + (10 \times 200)(1.05)^{10} - 500(1.05)^{30}}{(1.05)^{30} - 1} = \frac{7{,}600 + 3{,}258 - 2{,}161}{3.32} = \$2{,}619$$

$$\textit{Stand B, C}\ \text{SEV} = \frac{(24 \times 200) + (6 \times 200)(1.05)^{10} - 500(1.05)^{30}}{(1.05)^{30} - 1} = \frac{4{,}800 + 1{,}955 - 2{,}161}{3.32} = \$1{,}384$$

Inventory values			Market values		
Stand	Volume per acre	Value per M bd ft	Inventory value per acre	Land value (SEV)	Total values
A	5	200	$1,000	$2,619	$3,619
B	50	200	$10,000	$1,384	$11,384
C	65	200	$13,000	$1,384	$14,384

(*e*) *Harvest Schedule*

Year	Stand cut	Age when cut	Volume per acre	Acres harvested	Volume harvested
5	*C*	75	61	36	2,196
10	*C*	80	57	36	2,052
15	*A*	25	12.5	22.5	281
25	Regen *B*	65	75	36	2,700
	thin *C*	20	6	36	216
					2,916
40	Regen *C*	30	24	36	864
	thin *A*	20	10	22.5	225
					1,089

14-3 (*a*) Average PAI = 16,500/40 = 412.5 M bd ft/acre/year.
(*b*) (i) Area/cutting cycle = 100/10 = 10 acres
(ii) The 10-year periodic growth per acre is estimated to be approximately 4,500 bd ft. Note that both times the RGS of 8,000 grew 4,500 in 10 years. The periodic harvest equals this growth.
(iii) The LTSY of the whole forest is about 450 × 100 = 45,000 bd ft/year.

14-4

	Age	Inventory (M) Acres	Inventory ($M\ ft^3$) Vol/Acre	Inventory Total Vol	Cut M Acres	Cut Volume	Growth
				(MM)		(MM)	(MM)
Period 1							
Before cut	150	150	5	750	70	350	
	100	60	3	180			
				930			
After cut, before growth	150	80	5	400			40
	100	60	3	180			18
	0	70	0	0			105
				580	70	350	163

	Age	Inventory (M) Acres	Inventory (M ft³) Vol/Acre	Inventory Total Vol	Cut M Acres	Cut Volume	Growth
Period 2							
Before cut	160	80	5.5	440	70	385	
	110	60	3.3	198			
	10	70	1.5	105	70	35	
				743			
After cut, before growth	160	10	5.5	55			5.5
	110	60	3.3	198			19.8
	10	70	1.0	70			70.0
	0	70	0	0			105.0
				323	140	420	200.3
Period 3							
Before cut	170	10	6.05	60.5	10	60.5	
	120	60	3.63	217.8	60	217.8	
	20	70	2.0	140			
	10	70	1.5	105	70	35	
After cut, before growth				523.3			
	20	70	2.0	140			70
	10	70	1.0	70			70
	0	70	0	0			105
				210	140	313.3	245

Starting in the fourth period, growth = harvest = 245 M M ft³.

14-5 Regulated forest LTSY = (2.5/30) × 1,200 = 100 M ft^3 per year or 1,000 M ft^3 per 10-year period. This is the growth target between the third and fourth periods.

Iteration 1 (Cut = 1,000). All volumes in M ft^3.

		Period 1				Period 2				Period 3				Forest after harvest		
		Inventory		Cut		Inventory		Cut		Inventory		Cut				
Age	Volume/ acre	Acres	Total volume	Acres	Total volume	Acres	Total volume	Acres	Total volume	Acres	Total volume	Acres	Total volume	Acres	Total volume	10-year periodic growth
		Cutting oldest first														
0														309.1	0	154.6
10	0.50					307.7	153.9			301.3	150.7			301.3	150.7	301.3
20	1.50	300	450							307.7	461.6			307.7	461.6	307.7
30	2.50	400	1,000			300	750									
40	3.00					400	1,200	109	326.9	300	900	18.1	54.2	281.9	845.7	70.5
50	3.25	500	1,625	307.7	1,000					291	945.8	291	945.8			
60	3.50					192.3	673.1	192.3	673.1							
70	3.75															
		1,200	3,075	307.7	1,000	1,200	2,777	301.3	1,000	1,200	2,458.1	309.1	1,000	1,200	1,458	834.1*

* Growth of 834.1 is less than the target 1,000. Also, 281.9 acres of original stands are still left after the third period harvest. Therefore raise the harvest. Since 281.9/1,200 is about 25 percent of the forest area, raise the harvest 25 percent to 1,250 for the next iteration.

Iteration 2 (Cut = 1,250). All volumes in M ft^3.

		Period 1				Period 2				Period 3						
		Inventory		Cut		Inventory		Cut		Inventory		Cut		Forest after harvest		
Age	Volume/ acre	Acres	Total volume	Acres	Total volume	Acres	Total volume	Acres	Total volume	Acres	Total volume	Acres	Total volume	Acres	Total volume	Growth
0	0.50													406.8	0	203.4
10	0.50					384.6	192.3			397.4	198.7			397.4	198.7	397.4
20	1.50	300	450							384.6	576.9			384.6	576.9	384.6
30	2.50	400	1,000			300	750									
40	3.00					400	1,200	282	846.1	300	900	288.8	866.5	11.2	33.6	2.8
50	3.25	500	1,625	384.6	1,250					118	383.5	118	383.5			
60	3.50					115.4	403.9	115.4	403.9							
70	3.75															
		1,200	3,075	384.6	1,250	1,200	2,546.2	397.4	1,250	1,200	2,059.1	406.8	1,250	1,200	809.2	988.2*

* Growth of 988.2 is less than 1,000. Also, 11.2 acres of original stands are still left after the third period. Therefore raise the harvest about 11.2/1,200 or 1 percent.

14-6 (*a*) Uneven-aged regulation under area control. Tabled values are unit volume per acre before/after harvest.

Compartment	Year 1	2	3	4	5	6	7	8
A	6	8.4	11.2	14.5	18.1	21.3/10*	13.2	16.8
B	10	13.2	16.8/10*	13.2	16.8	20.2	23.2/10*	13.2
C	14	17.6/10*	13.2	16.8	20.2/10*	13.2	16.8	20.2
D	24/10*	13.2	16.8	20.2/10*	13.2	16.8	20.2	23.2/10*
Compartment cut / Volume cut per acre	*D*/14	*C*/7.6	*B*/6.8	*D*/10.2	*C*/10.2	*A*/11.3	*B*/13.2	*D*/13.2
Total harvest								
Cut per acre × 100 acres	1,400	760	680	1,020	1,020	1,130	1,320	1,320
8-year total harvest = 8,650								

* Harvested compartment.

14-6 (*b*) Uneven-aged regulation under area control with minimum harvest levels. Tabled values are unit volume per acre before/after harvest.

Compartment	Year 1	2	3	4	5	6	7[2]	8
Harvest schedule[1]								
A	6	8.4	11.2	14.5	18.1	21.3/19.3*	22.4/10*	13.2
B	10	13.2	16.8/10*	13.2	16.8/10*	13.2	16.8	20.2
C	14	17.6/10*	13.2	16.8/10*	13.2	16.8	20.2	23.2/10*
D	24/14*	17.6/15.2*	18.7/15.5*	19.0/15.8*	19.2/15.8*	19.2/10*	13.2	16.8
Compartment cut / Volume cut per acre	*D*/10	*C*/7.6 *D*/2.4	*B*/6.8 *D*/3.2	*C*/6.8 *D*/3.2	*B*/6.8 *D*/3.4	*A*/2.0 *D*/9.2	*A*/12.4	*C*/13.2
Total harvest (100 acres per unit)	1,000	1,000	1,000	1,000	1,020	1,120	1,240	1,320
8-year total harvest = 8,700								

[1] Solutions with higher total harvests are possible. This is just a starting point. This problem is useful for group competition to see who can obtain the highest first year or 8-year total harvest.

[2] This forest has a regulated structure after the seventh year harvest.

* Compartment harvested.

14-7 Even-aged regulation

(*a*) Immediate area regulation; cutting oldest compartment first. Tabled values are compartment age/volume per acre before harvest.

Goal: Regulate in 6 periods

Constraints:

1. Cut 10 acres each year, all in the same compartment.
2. Cut *oldest* stand first.
3. Rotation = 60 years.

Compartment	Period 1	2	3	4	5	6[1]	7	8
A	0	10/4	20/12	30/24	40/40*	10/4	20/12	30/24
B	50/48*	10/4	20/12	30/24	40/40	50/48*	10/4	20/12
C	0	10/4	20/12	30/24	40/40	50/48	60/54*	10/4
D	10/4	20/12	30/24	40/40*	10/4	20/12	30/24	40/40
E	20/12	30/24	40/40*	10/4	20/12	30/24	40/40	50/48
F	20/12	30/24*	10/4	20/12	30/24	40/40	50/48	60/54*
Harvest schedule								
Compartment cut / Volume cut per acre	*B*/48	*F*/24	*E*/40	*D*/40	*A*/40	*B*/48	*C*/54	*F*/54
Total harvest	480	240	400	400	400	480	540	540

* Compartment harvested.

[1] Forest has a fully regulated structure after the sixth period harvest.

(*b*) Area regulation with volume objectives. Tabled values are compartment age/volume per acre before harvest.

Goal: Maximize first period harvest.

Constraints:

1. The forest must have a regulated structure after the eighth period harvest, based on a 60-year rotation.
2. Harvest volume must be nondeclining from year to year.

14-7 (*b*) Area regulation – nondeclining yield. Tabled value = age/volume (acres).

Compartment	1	2	3	4	5	6	7	8
A_1 A_2	0/0	10/4	20/12	30/24	40/40	50/48	10/4 (2.4) 60/54 (7.6)	10/4
B_1 B_2 B_3 B_4 B_5	50/48*	10/4 (6.7) 60/54* (3.3)	20/12 (6.7) 10/4 (2.0) 70/58* (1.3)	30/24(6.7) 20/12 (2.0) 10/4 (0.6) 80/60 (0.7)	40/40 (6.7) 30/24 (2.0) 20/12 (0.6) 10/4 (0.2) 90/60 (0.5)	10/4	20/12	30/24
C	0/0	10/4	20/12	30/24*	10/4	20/12	30/24	40/40
D	10/4	20/12	30/24*	10/4	20/12	30/24	40/40	50/48
E_1 E_2 E_3 E_4	20/12	30/24	40/40*	10/4 (3) 50/48 (7)	20/12 (3) 10/4 (3) 60/54 (4)	30/24 (3) 20/12 (3) 10/4 (0.8) 70/58 (3.2)	10/4	20/12
F	20/12	30/24*	10/4	20/12	30/24	40/40	50/48	60/54*

Compartment	1	2	3	4	5	6	7	8
Harvest schedule								
(Compartment cut) / (acres cut) / (Volume cut per acre) *Example:* *B*/8/48 means: 8 acres of compartment *B* cut at 48 M bd ft per acre	$B/6.7/48$	$B_2/2.0/54$ $F/10/24$	$B_3/0.6/58$ $D/10/24$ $E/3/40$	$B_4/0.2/60$ $C/10/24$ $E_2/3/48$	$B_1/6.7/40$ $B_2/2.0/24$ $B_3/0.6/12$ $B_4/0.2/4$ $B_5/0.5/60$ $E_3/0.8/54$	$E_1/3/24$ $E_2/3/12$ $E_3/0.8/4$ $E_4/3.2/58$ $A/2.4/48$	$A_1/2.4/4$ $A_2/7.6/54$	$F/10/54$
Volume removed by compartment	$B/321.6$	$B/108$ $F/240$	$B/34.8$ $D/240$ $E/120$	$B/12$ $C/240$ $E/144$	$B/354$ $E/43.2$	$E/296.8$ $A/115.2$	$A/420$	$F/540$
Total harvest	321.6	348	394.8	396	397.2	412	420	540

The first period harvest of 321.6 is fairly low. You should be able to raise it and still satisfy the problem constraints.

CHAPTER 15

15-1 (*a*) Six items in the linear programming solution.

1. 15.35 acres of stand type B will be regeneration harvested in period 1.
2. None of stand type A is currently allocated to uncut status. If it had a volume coefficient in the objective function of over 7.53 units per acre, some of A_U would enter solution. Since such a price or value would not represent harvested timber (A_U means uncut), it would have to be Sarah's value for the uncut inventory in terms of the objective function (cubic feet of harvest). Not a very clear meaning.
3. Row 7 is the lower bound harvest flow constraint between periods 2 and 1. This means 394 units of volume in excess of this minimum have been harvested in period 2. By the zero slack in row 8, we see that harvesting is, in fact, at the upper limit.
4. Row 8 is the upper limit to harvest between periods 2 and 1 and we see it has a zero slack and a nonzero shadow price. This means it is a binding constraint on the solution. If the upper limit were raised by 1 unit, the total harvest would have a maximum increase of 0.558 units of volume.
5. Row 13 is the minimum ending inventory requirement of 1,600 units. If this were raised 1 unit to 1,601, the total harvest would decline by a maximum of 1.18 units.
6. The ending inventory row, currently with a RHS value of 1,600 could increase 2,911 units to 4,511 or decrease 121 units to 1,479 without changing the basis of the problem solution. The shadow prices for the inventory constraint would remain constant within this range at 1.18 units.

(*b*) Short answer

1. The park constraint is row 14 which requires $B_U \geq 20$. In this particular solution $B_U =$ 21.83 acres in order to satisfy the ending inventory constraint so the park constraint is more than satisfied and thus nonbinding.
2. We saw earlier that the ending inventory could increase 2,911 units without changing the basis. Therefore an increase of 400 units would not change the basis and the shadow price is a constant for such a change. The estimated opportunity cost is

$$400 \times 1.18 = 472 \text{ units of reduced timber harvest}$$

3. All existing stands are cut by period 3, and the harvests from stands A and B that were regenerated in periods 1 and 2 meet the period 4 harvest needs (see row 5).
4. This is a Model I because the prescriptions and yields for regenerated stands are directly under the yields for the existing stands in the decision variable columns for existing stands.
5. Stand A = 40 acres, stand B = 80 acres. So we want 2 acres harvested in stand B for every acre harvested in stand A, or $B = 2A$. To keep the balance with upper and lower 10 percent limits:

$$(1) \quad B_1 \geq 0.9(2A_1), \quad \text{or,} \quad B_1 \geq 1.8A_1$$

$$(2) \quad B_1 \leq 1.1(2A_1), \quad \text{or,} \quad B_1 \leq 2.2A_1$$

(*c*) Model II structure will separate the regenerated prescription from the existing stand prescription and present them as separate decision variables:

Let E_i be acres of existing stand E harvested in period i, where $i = 1, 2, 3, 4, U$ (U = uncut)

Let R_{ijk} be acres of future stands,
where i = period of birth, $i = 1, 2, 3, 4$
j = period of death, $j = 4, U$ (U = uncut)
k = silvicultural prescription, $k = 1$, thin 5 units at age 20 and regeneration harvest 30 units at age 30. $k = 2$ assigns the regenerated stand to uncut status.

The model formulation is shown in Table P15.1. All stands born in periods 3 and 4 have their decision variables combined and represented by R_{3UU} and R_{4UU} since there is no harvest or inventory.

15-2 (*a*) *Decision variables*
X_{ik} where X = letter name of existing stand, i = period existing stand is regeneration harvested, k = prescription followed for existing and regenerated stands.

$X = A, B, C$
$i = 1, 2, 3$
$k = 1, 2, 3, 4, U$

Table of *k*

Rotation of regenerated stand	Existing stand prescription		
	1	2	U
20	1	3	U
30	2	4	

Table P15.1 Model II formulation

Row			Activities																					
			Original stands										Regenerated stands						Harvest accounting					
			A_1	A_2	A_3	A_4	A_U	B_1	B_2	B_3	B_4	B_U	R_{141}	R^*_{1U2}	R_{241}	R^*_{2U2}	R_{3U2}	R_{4U2}	H_1	H_2	H_3	H_4	I	RHS
(1)	Objective																		1	1	1	1		Maximize
(2)	Harvest	H_1	10					40											−1					= 0
(3)	&	H_2		13					50											−1				= 0
(4)	inventory	H_3			15					58			5								−1			= 0
(5)	account	H_4				16					65		30		5							−1		= 0
(6)		I					16					65		30	7	12	0	0					−1	= 0
(7)	Model	T_1	−1					−1					1	1										= 0
(8)	II area	T_2		−1					−1						1	1								= 0
(9)	transfer	T_3			−1					−1							1							= 0
(10)	rows	T_4				−1					−1		−1					1						= 0
12	Harvest	LH_2																	−0.8	1				≥ 0
13	flow	UH_2																	−1.2	1				≤ 0
14	control	LH_3																		−0.8	1			≥ 0
15		UH_3																		−1.2	1			≤ 0
16		LH_4																			−0.8	1		≥ 0
17		UH_4																			−1.2	1		≤ 0
18	Ending inventory	I																					1	≥ 1,600
19	Park											1												≥ 20
20	Acres	A	1	1	1	1	1																	= 40
21	available	B						1	1	1	1	1												= 80

* R_{1U2} and R_{2U2} represent a prescription not in Model I. It allows acres regenerated in periods 1 and 2 to remain uncut.

Possible decision variables for Question 15-2(*a*)

Stand prescription	X k	A 1	A 2	A 3	A 4	A U	B 1	B 2	B 3	B 4	B U	C 1	C 2	C 3	C 4	C U
Harvest age	(i)															
	1	A_{11}	A_{12}	—	—	A_{1U}	B_{11}	B_{12}	—	—	B_{1U}	C_{11}	C_{12}	—	—	C_{1U}
	2	A_{21}	A_{22}	—	—	—	B_{21}	B_{22}	—	—	—	C_{21}	C_{22}	—	—	—
	3	A_{31}	A_{32}	—	—	—	B_{31}	B_{32}	B_{33}	B_{34}	—	C_{31}	C_{32}	C_{33}	C_{34}	—

(*b*)

```
MAX       H1
SUBJECT TO
       2)  -  H1 + 6 A11 + 6 A12 + 15 B11 + 15 B12 + 4 B33 + 4 B34 + 5 B1U
         + 34 C11 + 34 C12 + 10 C33 + 10 C34 + 10 C1U =     0
       3)     8 B1U + 8 C1U + 14 A21 + 14 A22 + 4 A1U + 24 B21 + 24 B22
         + 38 C21 + 38 C22 -  H2 =     0
       4)    10 A11 + 7 B11 + 28 B33 + 28 B34 + 9 B1U + 3 C11 + 36 C33
         + 36 C34 + 6 C1U + 8 A1U + 29 A31 + 29 A32 + 29 B31 + 29 B32
         + 40 C31 + 40 C32 -  H3 =     0
       5)    27 A12 + 19 B12 + 10 B1U + 9 C12 + 6 C1U + 10 A21 + 12 A1U
         + 7 B21 + 3 C21 -  H4 =     0
       6)    10 A11 + 7 B11 + 7 B33 + 10 B1U + 3 C33 + 6 C1U + 27 A22 + 12 A1U
         + 19 B22 + 9 C22 + 10 A31 + 7 B31 + 3 C31 -  H5 =     0
       7)    5 A12 + 2 B12 + 7 B34 + 14 B1U + 3 C34 + 18 C1U + 5 A21 + 12 A1U
         + 2 B21 + 10 A32 + 7 B32 + 3 C32 -  I =     0
       8)     H1 >=   20000
       9)     H2 >=   20000
      10)     H3 >=   20000
      11)     H4 >=   20000
      12)     H5 >=   20000
      13)     B1U +  C1U +  A1U >=   2000
      14)     B1U +  C1U +  A1U <=   5000
      15)     C11 +  C12 <=   1000
      16)     C21 +  C22 <=   1000
      17)     I >=   10000
      18)     A11 +  A12 +  A21 +  A22 +  A1U +  A31 +  A32 <=   2000
      19)     B11 +  B12 +  B33 +  B34 +  B1U +  B21 +  B22 +  B31 +  B32
        <=    5000
      20)     C11 +  C12 +  C33 +  C34 +  C1U +  C21 +  C22 +  C31 +  C32
        <=    3000
```

(*c*)

VARIABLE	VALUE	REDUCED COST
H1	139285.700000	.000000
A11	500.000000	.000000
A12	1214.286000	.000000
B11	.000000	.000000
B12	5000.000000	.000000
B33	.000000	11.000000
B34	.000000	11.000000
B1U	.000000	3.714285
C11	1000.000000	.000000
C12	.000000	.000000
C33	.000000	6.285715
C34	.000000	6.285715
C1U	2000.000000	.000000
A21	.000000	.000000
A22	285.714300	.000000
A1U	.000000	1.428571
B21	.000000	4.714286
B22	.000000	4.714286
C21	.000000	.000000
C22	.000000	.000000
H2	20000.000000	.000000
A31	.000000	6.000000
A32	.000000	6.000000
B31	.000000	15.000000
B32	.000000	15.000000
C31	.000000	16.285720
C32	.000000	16.285720
H3	20000.000000	.000000
H4	139785.700000	.000000
H5	24714.290000	.000000
I	52071.430000	.000000

ROW	SLACK OR SURPLUS	DUAL PRICES
2)	.000000	−1.000000
3)	.000000	−.428571
4)	.000000	.000000
5)	.000000	.000000
6)	.000000	.000000
7)	.000000	.000000
8)	119285.700000	.000000
9)	.000000	−.428571
10)	.000000	.000000
11)	119785.700000	.000000
12)	4714.286000	.000000
13)	.000000	−2.857143
14)	3000.000000	.000000
15)	.000000	17.714280
16)	1000.000000	.000000
17)	42071.430000	.000000
18)	.000000	6.000000
19)	.000000	15.000000
20)	.000000	16.285720

15-3 (*a*)

Plantation decision variables and SEV values for prescriptions

Decision variable	Prescription	Site	Rev_{30}	Rev_{40}	Rev_{50}	C_0	SEV @ 5%
X_1	PP $T_{30}R_{50}$	I	$1,200		$6,900	$300	$634.57*
X_2	PP R_{40}	I		$6,000		300	643.90
X_3	PP $T_{30}R_{50}$	II	900		5,100	200	496.00
X_4	PP R_{40}	II		4,500		200	511.90
X_5	FF $T_{30}R_{50}$	I	1,500		6,500	300	672.40
X_6	FF R_{40}	I		5,500		300	560.93
X_7	FF $T_{30}R_{50}$	II	500		3,500	200	241.93
X_8	FF R_{40}	II		3,250		200	304.97

* For example: $\text{SEV } X_1 = [6{,}900 + 1{,}200(1.05)^{20} - 300(1.05)^{50}]\dfrac{1}{(1.05)^{50} - 1} = \$634.57.$

Row	Pine plantations				Fir plantations				Easements				RHS
	X_1	X_2	X_3	X_4	X_5	X_6	X_7	X_8	X_9	X_{10}	X_{11}	X_{12}	
Objective	635	644	496	512	672	561	242	305	350*	400	450†	400	
Acres site I	1	1			1	1			1	1			≤1125‡
Acres site II			1	1			1	1			1	1	≤800
Maximum easements									1	1	1	1	≤600
Minimum easements									1	1	1	1	≥400
Fir/pine balance site II			−0.25	−0.25			1	1					≥0
Brush/tree easement									1	−3	1	−3	≥0

where

X_1–X_8 = plantation decision variables

X_9 = uncut pine scenic easement site I

X_{10} = brush easement site I

X_{11} = uncut pine scenic easement site II

X_{12} = brush easement site II

* present value = 650 − 300 = $350

† present value = 650 − 200 = $450

‡ = the 75-acre buffer strip was subtracted from the 1,200 acres of site I

(*b*) Clearcut restoration inc. problem optimal solution

```
MAX      635 X1 + 644 X2 + 496 X3 + 512 X4 + 672 X5 + 561 X6 + 242 X7
         + 305 X8 + 350 X9 + 400 X10 + 450 X11 + 400 X12
SUBJECT TO
      2)      X1 +  X2 +  X5 +  X6 +  X9 +  X10 <=   1125
      3)      X3 +  X4 +  X7 +  X8 +  X11 +  X12 <=   800
      4)      X9 +  X10 +  X11 +  X12 <=   600
      5)      X9 +  X10 +  X11 +  X12 >=   400
      6)   - 0.25 X3 - 0.25 X4 +  X7 +  X8 >=   0
      7)      X9 - 3 X10 +  X11 - 3 X12 >=   0
END


    LP OPTIMUM FOUND  AT STEP      5

         OBJECTIVE FUNCTION VALUE

 1)          1124240.0000

VARIABLE        VALUE           REDUCED COST
       X1          0.0              37.0000
       X2          0.0              28.0000
       X3          0.0              16.0000
       X4        320.0000            0.0
       X5       1125.0000            0.0
       X6          0.0             111.0000
       X7          0.0              63.0000
       X8         80.0000            0.0
       X9          0.0             301.4000
      X10          0.0             251.4000
      X11        400.0000            0.0
      X12          0.0              50.0000


ROW             SLACK            DUAL PRICES
      2)           0.0             672.0000
      3)           0.0             470.6000
      4)         200.0000            0.0
      5)           0.0             -20.6000
      6)           0.0            -165.6000
      7)         400.0000            0.0

NO. ITERATIONS=    5
```

15-4 (*a*) Decision variables model I

Variable	Stand	Period harvested	Regenerated prescription
A_{11}:	A	1	1
A_{12}:	A	1	2
A_{21}:	A	2	1
A_{22}:	A	2	2
B_{11}:	B	1	1
B_{12}:	B	1	2
B_{21}:	B	2	1
B_{22}:	B	2	2

(*b*) Acre accounting equations model I

$$A_{11} + A_{12} + A_{21} + A_{22} \leq \text{acres of } A$$

$$B_{11} + B_{12} + B_{21} + B_{22} \leq \text{acres of } B$$

(*c*) Decision variables model II

Existing stands			Regenerated stands			
Variable	Stand	Period cut	Variable	Period born	Period cut*	Prescription
A_1:	A	1	R_{131}	1	3	1
A_2:	A	2	R_{152}	1	5	2
B_1:	B	1	R_{241}	2	4	1
B_2:	B	2	R_{2U2}	2	U	2
			R_{351}	3	5	1
			R_{3U2}	3	U	2
			R_{4U1}	4	U	1
			R_{4U2}	4	U	2
			R_{5U1}	5	U	1
			R_{5U2}	5	U	2

* U = left as ending inventory.

(*d*) Area accounting and transfer rows 1 model II

$$A_1 + A_2 \leq \text{acres of } A$$

$$B_1 + B_2 \leq \text{acres of } B$$

$$-A_1 - B_1 + R_{131} + R_{152} = 0$$

$$-A_2 - B_2 + R_{241} + R_{2U2} = 0$$

$$-R_{131} + R_{351} + R_{3U2} = 0$$

$$-R_{241} + R_{4U1} + R_{4U2} = 0$$

$$-R_{351} - R_{152} + R_{5U1} + R_{5U2} = 0$$

15-5 Data and decision variables for questions *a* and *b*.

Variable	Period cut or allocated	Regenerated prescription	Timber yields MCF P_1	P_2	P_3	Present values
X_{11}	1	1	1.8			\$ 1,036
X_{12}	1	2	1.8		1.0	1,149[1]
X_{13}	2	1, 2		2.0		1,055[2]
X_{14}	3	1, 2			2.2	784
X_{1U}	1	uncut				
X_{21}	1	1	2.2			1,718
X_{22}	1	2	2.2		1.0	1,831
X_{23}	2	1, 2		2.3		1,213
X_{24}	3	1, 2			2.4	855
X_{2U}	1	uncut				
A_T	Zone *A* allocated to timber					−\$25,000[3]
A_W	Zone *A* allocated to wilderness					+ 35,000[4]
B_T	Zone *B* allocated to timber					− 13,000
B_W	Zone *B* allocated to wilderness					+ 49,000
C_T	Zone *C* allocated to timber					− 2,000
C_W	Zone *C* allocated to wilderness					+ 25,000

Example PNW calculations i = 4 percent

[1] $X_{12}\ \dfrac{(1.8)(700)}{1.217} + \dfrac{1.0(300)}{2.67} = 1{,}149$

[2] $X_{13}\ \dfrac{(2.0)950}{1.801} = 1{,}055$

[3] $A_T = -\text{road cost} + \text{recreation revenue}$
$= -33{,}000 + 8{,}000 = -25{,}000.$

[4] $A_W = \text{recreation revenue} = 35{,}000$

15-5 (*a*) Detached coefficient matrix where all area is allocated to timber production using aggregate choice variables

Row		A_T	B_T	C_T	X_{11}	X_{12}	X_{13}	X_{14}	X_{1U}	X_{21}	X_{22}	X_{23}	X_{24}	X_{2U}	H_1	H_2	H_3	RHS
(1) Objective:	Max.	−25,000	−13,000	−2,000	1,036	1,149	1,055	784		1,718	1,831	1,213	855					
(2) Harvest accounting	H_1				1.8	1.8				2.2	2.2				−1			= 0
(3)	H_2						2.0					2.3				−1		= 0
(4)	H_3					1.0		2.2			1.0		2.4				−1	= 0
(5) Area accounting	Stand I	−50	−250		1	1	1	1	1									= 0
(6)	Stand II	−300	−100	−200						1	1	1	1	1				= 0
(7) Maximum cost in period one					180	180				110	110							≤ 8,000
(8) Maximum area regenerated in period 1					1	1				1	1							≤ 200
(9) Harvest flow control															−.9	1		≥ 0
(10)															−1.1	1		≤ 0
(11)																−.9	1	≥ 0
(12)																−1.1	1	≤ 0

(b) Detached coefficient matrix showing zone allocation choices between timber and wilderness

Row		Timber and wilderness zone allocation variables						Stand I timber variables					Stand II timber variables					
		A_T	A_W	B_T	B_W	C_T	C_W	X_{11}	X_{12}	X_{13}	X_{14}	X_{1U}	X_{21}	X_{22}	X_{23}	X_{24}	X_{2U}	RHS
(1) Objective	(Max)	−25,000	+35,000	−13,000	+49,000	−2,000	+25,000	1,036	1,149	1,055	784		1,718	1,831	1,213	855		
(2) Area accounting	Stand I	−50		−250				1	1	1	1	1						= 0
(3)	Stand II	−300		−100		−200							1	1	1	1	1	= 0
(4) Aggregate	A	1	1															= 1
(5) Choices	B			1	1													= 1
(6)	C					1	1											= 1
(7) Required wilderness			1		1		1											≥ 1
(8) Link C_T to A_T		1				−1												≥ 0
(9) Required area of	P_1		300		100		200								1	1	1	≥ 300
(10) 50 year old +	P_2		350		350		200				1	1				1	1	≥ 300
(11) timber	P_3		350		350		200					1					1	≥ 300
(12) Maximum timber area cut period 1		−70		−70		−40		1	1				1	1				≤ 0

CHAPTER 16

16-1 (*a*)

$$\text{area cut per year} = \frac{\text{management area}}{\text{cutting cycle}}$$

$$= \frac{100 \text{ acres (stands } A \text{ \& } B)}{10 \text{ years}}$$

$$= 10 \text{ acres}$$

(*b*)

Year	Stand cut	Acres	Harvest* volume/acre	Entry	Total volume cut
1	*A*	10	69	first	690
4	*A*	10	66	first	660
6	*B*	10	2.4†	first	24
11	*A*	10	6	second	60

* Stand *A* loses 1 M bd ft of volume for every year of delayed entry. At year 4, the cut is Vol − RGS or (100 − 4) − 30 = 66. Stand *B* by contract adds 0.4 M bd ft per acre per year.

† You might argue that this is too little a volume to economically harvest and this should be 0.0.

(*c*) Estimate the present net worth of each prescription for each stand. Because of variable yields a combination of calculations is needed.

	Stand *A*, prescription 1			Stand *B*, prescription 1		
Year	Volume	Value	Present value	Volume	Value	Present value
0	70	$12,600	12,600	0	—	
10	6	1,080	730	8	1,440	973
20	8	1,440	658	12	2,160	3,039†
30	10	1,800	555			
40	12	2,160	1,387*			
			15,930			$4,012

$$^* \text{ SEV}_{40} = \left[2{,}160 + \frac{2{,}160}{(1.04)^{10} - 1}\right] \frac{1}{(1.04)^{40}} = \$1{,}387$$

$$^\dagger \text{ SEV}_{20} = \left[2{,}160 + \frac{2{,}160}{(1.04)^{10} - 1}\right] \frac{1}{(1.04)^{20}} = \$3{,}039$$

Year	Stand *A*, prescription 2			Stand *B*, prescription 2		
	Volume	Value	Present value	Volume	Value	Present value
0	80	$14,400	14,400	10	1,800	1,800
20	12	2,160	986	18	3,240	1,479
40	18	3,240	675	20	3,600	1,380*
60	20	3,600	630*			
			$16,691			$4,659

$$\text{PV of annual costs} = \frac{10}{0.04} = \$250$$

* Also SEV values.

Since the same $250 of capitalized annual costs would be deducted from all the revenue totals, the ranking of prescriptions is unchanged by ignoring costs. Prescription #2 gives a higher present net worth for both stands—primarily because of the higher yields in year 0.

(*d*) Maximum physical productivity means maximum MAI

$$\text{Prescription 1} = \frac{50}{40} = 1.25$$

$$\text{Prescription 2} = \frac{70}{50} = 1.40 \text{ * maximum}$$

$$\text{Prescription 3} = \frac{65}{50} = 1.30$$

(*e*) Calculate the SEV of each prescription

Prescription 1

$$\text{SEV} = \frac{(50 \times 120) - 300(1.04)^{40} - 5\left[\frac{(1.04)^{40} - 1}{0.04}\right]}{(1.04)^{40} - 1}$$

$$= \frac{6{,}000 - 1{,}440 - 475}{3.80}$$

$$= \frac{4{,}085}{3.80} = \$1{,}075$$

Prescription 2

$$\text{SEV} = \frac{(70 \times 120) - 300(1.04)^{50} - 5\left[\frac{(1.04)^{50} - 1}{0.04}\right]}{(1.04)^{50} - 1}$$

$$= \frac{8{,}400 - 2{,}132 - 763}{6.107}$$

$$= \frac{5{,}505}{6.107} = 901$$

Prescription 3

$$\text{SEV} = \frac{(50 \times 120) + (15 \times 120)(1.04)^{20} - 300(1.04)^{50} - 5\left[\frac{(1.04)^{50} - 1}{0.04}\right]}{(1.04)^{50} - 1}$$

$$= \frac{6{,}000 + 3{,}944 - 2{,}132 - 763}{6.106}$$

$$= \frac{7{,}049}{6.107} = \$1{,}154$$

so prescription 3 has the highest SEV value.

(*f*) Forest = 200 acres, MAI prescription 3 = 1.30, LTSY = 200 × 1.30 = 260 M bd ft/year.

(*g*) *Harvest schedule*

Prescription 3, rotation = 50 years

regenerated yields = 50 M bd ft age 50 (regeneration harvest)
15 M bd ft age 30 (thinning)

$$\text{acres cut per year} = \frac{A}{R} = \frac{200}{50} = 4 \text{ acres}$$

Prescription 2, rotation = 50 years

regenerated yields = 70 M bd ft at 50 years

$$\text{acres cut} = \frac{A}{R} = \frac{200}{50} = 4 \text{ acres}$$

Regeneration harvest priority and cutting years, both prescriptions

Stand	Acres	Cutting order*	Regulation years to cut (R = 50)
A	50	1	12.5
B	50	2	12.5
C	100	3	25
			50 years

* Based on existing stand age.

Harvest schedules

Year	Stand cut	Prescription 3 Acres cut	Volume acre	Harvest volume	Prescription 2 Acres cut	Volume acre	Harvest volume
1	*A*	4	100 + (−1)(1) = 99	396	4	99	396
5	*A*	4	100 + (−1)(5) = 95	380	4	95	380
15	*B*	4	30 + 0.4(15) = 36	144	4	36	144
25	*B*	4	30 + 0.4(25) = 40	160	4	40	160
35	*C*	4	45*	120	4	66*	140
(thinning)	*A*	4	15	60 / 180			
45	*C*	4	54*	180	4	73*	264
(thinning)	*B*	4	15	60 / 240			
65	*B*	4	50	200	4	70	280
(thinning)	*C*	4	15	60 / 260			

* Estimated from regenerated yield tables; i.e., if a harvest of 50 at age 50 is possible, we guessed that a 45-year-old stand would have 45 units of volume.

The harvest schedules are quite similar since the rotation is the same and stand *C* is already planted. Prescription 3 had more entries due to the thinning.

(*h*) For prescription 3:

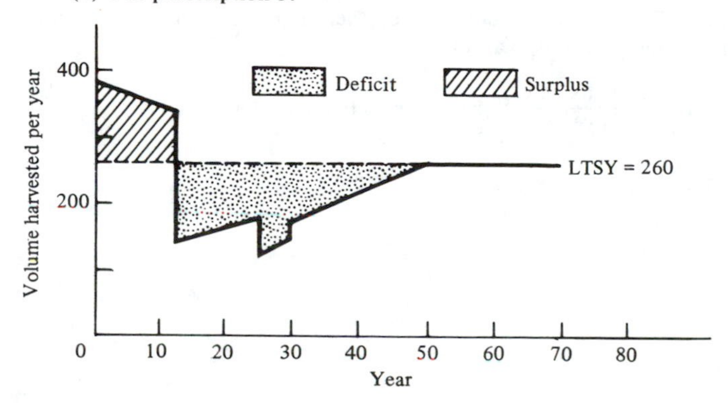

This schedule suggests the early cutting (years 1–12.5) provides a harvest that exceeds the LTSY by a little less than the deficit of harvest under LTSY between age 12.5 and 50. To close to call without more analysis—a slight deficit is the best guess.

(*i*) Prescription 1 LTSY = 200 × 1.25 = 250

$$\begin{aligned} H_t &\leq 0.8\text{LTSY}, \qquad t = 1, \ldots, \infty \\ &\leq 0.8(250) \\ &\leq 200 \end{aligned}$$

Since unconstrained area regulation would have us cutting 5 acres of existing stand *A*, which averages 97 M bd ft over the first 5 years. This is 5 × 97 = 485 M bd ft more or less. These constraints will require the harvest to be held at 200 M bd ft in each of the first 5 years, a significant decline.

(*j*) The genetic discovery will increase yield to 70 and the LTSY from 250 to (70/40) × 200, or 350.

$$\begin{aligned} H_t &\leq 0.8\text{LTSY} \\ &\leq 0.8(350) \\ &\leq 280 \end{aligned}$$

The harvest can increase 280 − 200 = 80 M bd ft/year. This is the amount of the allowable cut effect (ACE). By inspection, the nondeclining yield constraint can barely be satisfied at an annual harvest of 260 M bd ft. We need more analysis to be sure how much of the 80 M bd ft increase in cut can be captured from current inventory.

(*k*) The 100 acres would increase the forest LTSY by 100 × 1.25 or 125 M bd ft. For about 5 years stand *A* could be heavily cut to increase early harvest by this amount under harvest. However, the nondeclining yield constraint will be violated after year 5 when stand *A* is cut out—so only part of the increased LTSY can actually be captured for an allowable cut effect.

If we estimate the cut can increase from 200 to 250 or 50 M bd ft per year, this is worth 0.5 × 120 or $60/acre/year of increased revenue. This is a/i = 60/0.04 or $1,500 of SEV for the ACE effect. Plus we calculated earlier that such acres under prescription 1 were worth $1,075. Sounds like a good buy! It's a reasonably legitimate ACE since the land is already planted.

16-2 (*a*) Base harvest policy (old growth deficit since there is not enough old growth to harvest at LTSY until young growth is available).

$$\text{LTSY} = \frac{20 \text{ units/acre}}{50 \text{ years}} \times 400 \text{ acres} = 160 \text{ units/1 year}$$

Old growth will be harvested in 50 years at the rate of (400 acres/50 years) × 10 units/acre = 80/year.

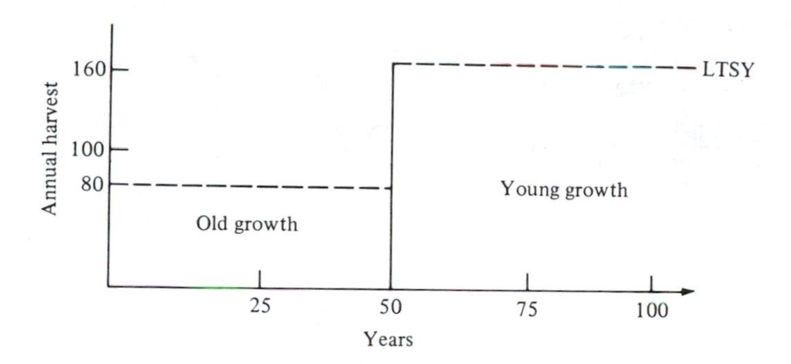

(*b*) *Area Control:* The old growth will be harvested at 80 units/year for 50 years. Thereafter (beginning age 51), the young growth will be harvested at 160 units/year.

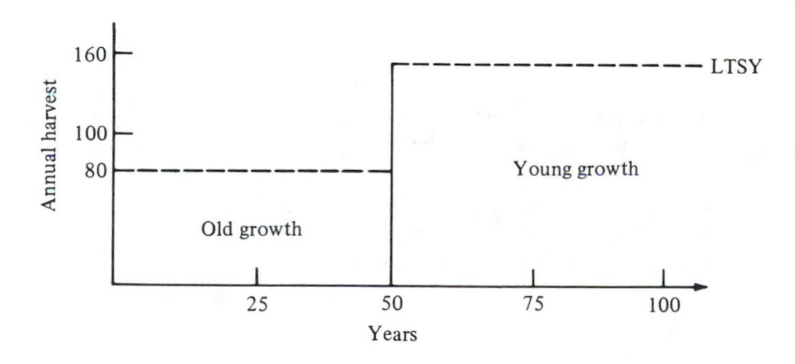

(*c*) Since there is no surplus old growth, harvest level will be the same as for *a* and *b*.

(*d*) By shortening the rotation to 40 years from 50 years, the old growth will be harvested at 100 units/year.

$$\frac{400 \text{ acres}}{40 \text{ years}} \times 10 \text{ units/acre} = 100 \text{ units/year}$$

Thereafter, the young growth will be harvested at:

$$0.8 \times \frac{400 \text{ acres}}{40 \text{ years}} \times 15 \text{ units/acre} = 120 \text{ units/year}$$

120
100
Annual harvest
Old growth
Young growth
25
50
75
100

16-3 (a.1) Base harvest policy (old growth surplus since more than enough old growth is available to harvest at LTSY until young growth is available).

$$\text{LTSY} = \frac{20 \text{ units/acre}}{50 \text{ years}} \times 400 \text{ acres} = 160 \text{ units/year}$$

Under the base harvest policy the old growth will be harvested in 75 years.

$$\frac{400 \text{ acres}}{160 \text{ units/year}} \times 30 \text{ units/acre} = 75 \text{ years.}$$

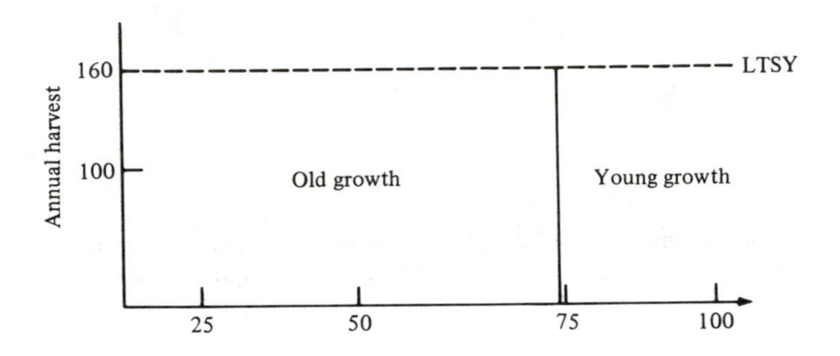

(a.2) *Area Control:* The old growth will be harvested at 240 units/year for 50 years. Thereafter (starting year 51), the young growth will be harvested at 160 units/year.

$$\frac{400 \text{ acres}}{50 \text{ years}} \times 30 \text{ units/acre} = 240 \text{ units/year}$$

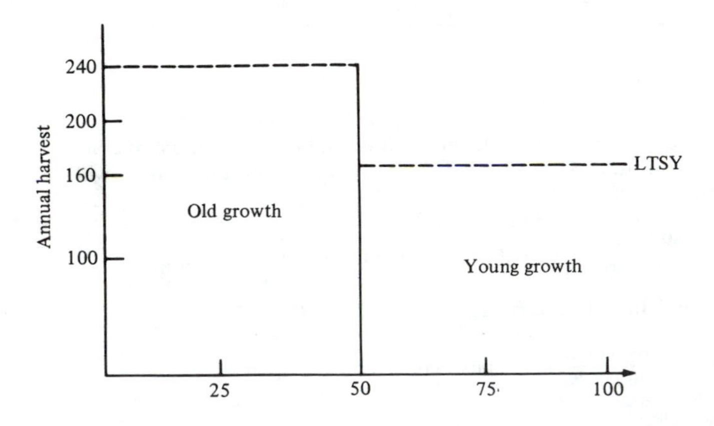

(a.3) In order to harvest the surplus over 10 years, the extra harvest will be 400 units/year,

$$\frac{(400 \times 30) - (160 \times 50)}{10 \text{ years}} = 400 \text{ units/year}$$

which, when added to the base harvest of 160 units/year gives an annual harvest of 560 units/year for the first 10 years.

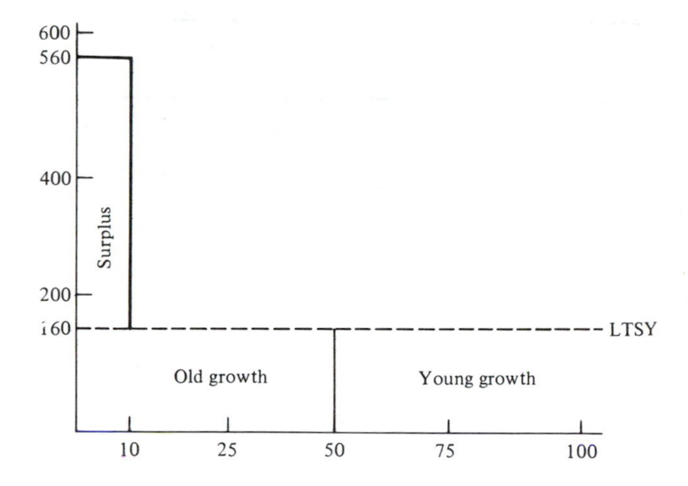

(*b*) Base harvest policy will produce over 101 years:

$$160 \times 101 \text{ years} = 16{,}160 \text{ units}$$

$$\text{Maximum production} = \text{cut all acres in year 1 and again in years 51 and 101.}$$

$$= 400 \times 30 + 400 \times 20 + 400 \times 20 = 28{,}000.$$

$$\text{Harvest loss} = 28{,}000 - 16{,}160 = 11{,}840.$$

(*c*) Base Harvest Policy

$$\text{LTSY} = \frac{30 \text{ units/year}}{50 \text{ years}} \times 400 \text{ acres} = 240 \text{ units/year}$$

The old growth will be harvested in

$$\frac{400 \text{ acres}}{240 \text{ units/year}} \times 30 \text{ units/acre} = 50 \text{ years}$$

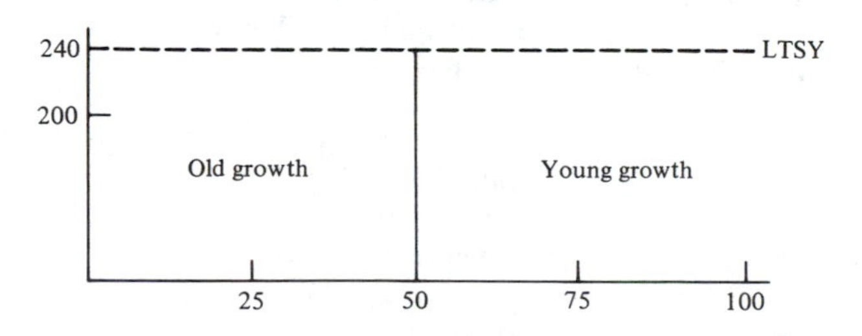

The old growth will be harvested at 240 units/year for 50 years. Thereafter (beginning year 51), the young growth will be harvested at the same rate since the yield for both is the same (see graph above).

Since there is no surplus old growth, no increased harvest will occur during the first decade under area control (see graph above). Over 101 years, base policy produces $240 \times 101 = 24{,}240$. The maximum production equals the cut at year 1, 51, 101 for $400 \times 30 + 400 \times 30 + 400 \times 30 = 36{,}000$. Loss $= 36{,}000 - 24{,}240 = 11{,}760$.

INDEX